Volcanic Gas Reservoir Characterization

Volcanic Gas Reservoir Characterization

Qiquan Ran

Yongjun Wang

Yuanhui Sun

Lin Yan

Min Tong

ELSEVIER

AMSTERDAM • BOSTON • HEIDELBERG • LONDON
NEW YORK • OXFORD • PARIS • SAN DIEGO
SAN FRANCISCO • SINGAPORE • SYDNEY • TOKYO
Gulf Professional Publishing is an imprint of Elsevier

Gulf Professional Publishing is an imprint of Elsevier
225 Wyman Street, Waltham, MA 02451, USA
The Boulevard, Langford Lane, Kidlington, Oxford, OX5 1GB, UK

First edition **2014**

Library of Congress Cataloging-in-Publication Data
A catalog record for this book is available from the Library of Congress

British Library Cataloguing in Publication Data
A catalogue record for this book is available from the British Library

For information on all Gulf Professional publications
visit our web site at store.elsevier.com

Printed and bound in USA
14 15 16 17 18 10 9 8 7 6 5 4 3 2 1

ISBN: 978-0-12-417131-2

Contents

Foreword

Natural gas is a kind of efficient, economic, and clean energy. With the rapid development of our economies, the demand for natural gas is climbing year by year. With the development and advancement of exploration technology, the world's largest volcanic gas reservoirs were discovered in China. As an important part of natural gas resources in China, the discovered resources in volcanic gas reservoirs reach up to several trillion cubic meters, the proved gas in place is up to several hundred billion cubic meters, and the production keeps growing annually. The exploration and development of natural gas in China is in a period of vigorous development. As a new domain, the rational development of volcanic gas reservoirs plays a more and more important role in the natural gas industry.

Compared with conventional sandstone or carbonate gas reservoirs, the geological conditions of volcanic gas reservoirs are much more complicated, and their exploitation is difficult. Volcanic reservoirs, formed after multistage volcanic eruption, are composed of multilevel, superimposed, complex internal structural units with various rock types; their lithology and lithofacies are complex and change laterally. It is difficult to identify and predict effective reservoirs because of the complex mechanism of reservoir genesis; the distribution of pores, cavities, and fractures; and the great change in reservoir thickness and physical properties. The complexity of geological conditions results in diverse gas-water distributions, complex porous-flow mechanisms and development characteristics, big differences in productivity, and great difficulty in rational development. Meanwhile, there are few worldwide development examples of volcanic gas reservoirs, and this implies a low degree of research, no successful experience, and a lack of mature theoretical and technological bases. It is a world-class puzzle. In order to realize the rational development of volcanic gas reservoirs, the top priority is to understand volcanic gas reservoirs. Volcanic gas reservoir characterization is the key to solve the problem. Therefore, it is imperative to do more research on volcanic gas reservoir characterization, as this is an important riddle to solve.

In recent years, the authors participated in scientific research and production practice for volcanic gas reservoir development in the Songliao basin and the Junggar basin, such as the research that focused on basic theories, key technologies, development plan design, and production performance tracking. Particularly in volcanic gas reservoir characterization, through the application of numerous data from field outcrops, laboratory experiments, and field production, we innovatively developed characterization methods and related technologies with independent intellectual property rights. Accordingly, we obtained much more knowledge of volcanic reservoirs, guided the rational development successfully, and accumulated abundant practical experiences.

Because hydrocarbon reserves and production of volcanic reservoirs are far less than those of sedimentary reservoirs, little attention was paid to the study of these reservoirs. At present, the research thoughts and methods for sedimentary reservoirs are still applied to solve the problems of volcanic gas reservoirs, but these thoughts and methods cannot meet the actual needs of volcanic gas reservoir development. In order to fill this technology gap, the authors summarized the latest research achievements, incorporated relevant research results, and finally composed this book for reference. This book, a product of theoretical research and production practice, systematically introduces the research thoughts, methods, and technologies of volcanic gas reservoir characterization. It covers the authors' research achievements related to volcanic gas reservoir development, including the reservoir characterization technology, development plan design, and field application examples in volcanic gas reservoirs in Daqing, Jilin, and Xinjiang.

This book is composed of 12 chapters. Chapter 1 is the Introduction, which describes the importance, technical concept, and research ideas of volcanic gas reservoir characterization. Chapters 2, 10, and 11 introduce the concept and research ideas of internal structure in volcanic gas reservoirs by systematically describing the identification and characterization of volcanic formation, volcanic edifice, volcanic massif, accumulation and permeation units, and microporous structures. Chapter 3 introduces the concepts, research ideas, identification, division, and correlation methods of volcanic stratigraphic sequences. Chapters 4, 5, 6, 8, and 9 introduce the identification and prediction technology of lithofacies, lithologies, fractures, gas-bearing layers, and water zones as well as effective reservoirs. Chapter 7 introduces the parameter interpretation technology for fractured volcanic gas reservoirs, and Chapter 12 introduces the geological model building technology of volcanic gas reservoirs.

Chapter 1 is written by Qiquan Ran, Lin Yan, and Yuanhui Sun. Chapter 2 is written by Qiquan Ran, Yuanhui Sun, and Yongjun Wang. Chapter 3 is written by Yuanhui Sun and Qiquan Ran. Chapter 4 is written by Yuanhui Sun and Qiquan Ran. Chapter 5 is written by Yongjun Wang and Lin Yan. Chapter 6 is written by Yongjun Wang and Qiquan Ran. Chapter 7 is written by Yongjun Wang and Yuanhui Sun. Chapter 8 is written by Yongjun Wang, Yuanhui Sun, and Min Tong. Chapter 9 is written by Qiquan Ran, Yongjun Wang, and Min Tong. Chapter 10 is written by Qiquan Ran and Yongjun Wang. Chapter 11 is written by Lin Yan and Yongjun Wang. Finally, Chapter 12 is written by Lin Yan and Yuanhui Sun.

Thanks for the care and help from the leaders and experts of PetroChina Research Institute of Petroleum Exploration and Development. And thanks for the support and help from the leaders and experts of the Natural Gas Department and Science & Technology Management Department of PetroChina Exploration and Production Company, and the Science & Technology Management Department of CNPC, Daqing Oilfield Company, Jilin Oilfield Company, and Xinjiang Oilfield Company. Dr. Lei Zhengdong, Dr. Zhu Liyao, and Dr. Chen Fuli and colleagues, the members of the volcanic rock project team, have done much work for this project. Professor Zhong Dakang, Professor Ouyang Yonglin, Professor Wang Pujun, and Dr. Zhang Jing and colleaguesalso contributed to this project. Professor Qiu Yinan, a famous development geologist in China, and Professor Guo Shangping, a CAS academician, gave us much help for the composition of this book. We express our sincere appreciation to them herein! The book cited or extracted some contents from the papers and literature of other researchers, so thanks to those researchers too!

At present, volcanic gas reservoir development is still in the early stage. Because of complex geological conditions and new research challenges, many problems concerning volcanic gas reservoir characterization should be solved through further research. Suggestions for improvement will be gratefully received.

The authors
November 2010 in Beijing

Preface I

Volcanic rocks have been known as hydrocarbon reservoir rocks for nearly a century. However, petroleum geologists paid little attention to these rocks in the past because far less hydrocarbon reserve was found and little production was obtained from them than from sedimentary reservoirs. Particularly, the research completed on petroleum development geology and reservoir geology is insufficient in regard to volcanic reservoirs, and only a small amount of research has been published worldwide.

Recently, some large volcanic gas fields, such as the Xushen gas field in the Songliao basin and the Kelameili gas field in the Junggar basin, have been discovered and developed in China, and they provide an important platform that Chinese petroleum geologists can use to make some progress on volcanic reservoir geology. A research team composed of young doctors was set up in the Research Institute of Petroleum Exploration and Development (RIPED), PetroChina. These young scientists worked in the gas fields with local geologists for many years; ultimately, they worked out a set of integrated volcanic gas reservoir characterization techniques that technically ensure the rational development of volcanic gas fields. They also contributed some new theories of volcanic gas reservoir geology. This book embodies their academic achievements in this area.

Reservoir characterization technique is the technical manifestation of the application of reservoir geology theory in oil and gas field development. As far as I know, this is the first book to systematically introduce the volcanic gas reservoir characterization in China and in other countries. It expounds the technical content and research ideas of volcanic gas reservoir characterization. In this book, the internal structure of volcanic reservoir is creatively classified for the first time. Based on petrology, modern well logging, and seismic reservoir prediction, the techniques of identification, correlation, tracking, and lateral prediction are developed for various reservoir hierarchies, including volcanic formation, edifice, massif, lithology facies, and accumulation-permeation units. This book also introduces the well logging interpretation techniques for the identification of volcanic lithology, facies, fractures, gas layers and aquifers, effective reservoirs, and the correspondent reservoir parameters. Finally, the book explores the 3D geological model building technology of volcanic gas reservoirs.

I believe the publication of this book will be of great value to the characterization and exploitation of volcanic gas reservoirs. Certainly, this book will catch the attention of petroleum geologists and oil/gas reservoir engineers from all over the world. As a veteran in the area of oil and gas field development, I am willing to promote the publication of this book and eagerly want to see this book coming into the world in the near future.

Qiu Yinan
October 27, 2010

Preface II

The world largest volcanic gas reservoirs were discovered in the Songliao basin and the Junggar basin in China. As a new domain, the exploitation of volcanic gas reservoirs is becoming more and more important in the natural gas industry. Compared with conventional gas reservoirs, the geological background of these reservoirs is much more complicated. The characterization and understanding of these reservoirs are the keys and restrict conditions to their rational development and puzzled the geologists all over the world.

At present, there are fewer examples, experiences, and theoretical and technical bases for the development of volcanic gas reservoirs. Considering the special geological features and development difficulties of these reservoirs, the authors carried out elaborate scientific research with a multidisciplinary group, and presented the special technical thought, method, and technology for volcanic gas reservoir characterization, which establishes a fundamental basis for the research on volcanic reservoir geology and development. The authors described the internal structure characteristics of volcanic rocks, which are quite different from those of clastic rocks; revealed the grade and relationship of volcanic stratigraphic sequence; developed the identification and prediction methods of effective volcanic reservoirs and interpretation technology of reservoir parameters; and built the volcanic rock geological model constrained by multimedia and complicated internal structure. These innovative achievements have been successfully applied in production and have effectively guided the development of volcanic gas reservoirs.

Based on the summary of scientific research and technical development achievements, along with the collection of valuable practical experiences and important understandings, the authors completed this book, *Volcanic Gas Reservoir Characterization*. This book covers five aspects: the identification and characterization of the internal structure of volcanic rock, the classification and correlation of volcanic rock stratigraphic sequencing, effective volcanic reservoir identification and prediction, the parameter interpretation of multimedia volcanic reservoirs, and volcanic reservoir geological model building.

This book features plentiful contents, extensive methodology, strong practicality, and high innovation. The publication of this book will play an important role in the rational development of volcanic gas reservoirs and will serve as an important reference for the development of similar complicated gas reservoirs.

Guo Shangping
CAS Academician
November 2010

Capsule summary

This book exhibits systematically the theoretical study and field application of volcanic gas reservoir characterization, including research ideas, methods, and techniques of volcanic gas reservoir characterization, as well as applications and understanding of the technology in practice.

This book will be a useful reference for graduate and advanced undergraduate students in petroleum geology and oil and gas reservoir engineering. It will also be of great value to researchers in oil and gas field development.

CHAPTER

Introduction

1

The world's largest volcanic gas reservoirs have been discovered in China. Their rational development can potentially mitigate the energy supply and demand imbalance and accelerate the growth of the natural gas industry in China. Meanwhile, this development is of great significance for the advancement of natural gas production technology in general and for the development of similar gas reservoirs.

Due to their complex geological settings, the development of volcanic gas reservoirs has received little study up to now, with only limited experiences accumulated in this area. It is difficult for us, therefore, to put volcanic gas reservoirs into production with current technologies. To develop volcanic gas reservoirs effectively, a comprehensive knowledge base needs to be built up about this type of reservoir first, with reservoir characterization being the focal point. However, the complex petrogenic processes and internal architecture of volcanic gas reservoir rocks imply that the volcanic gas reservoir characterization is different from the characterization of conventional gas reservoirs in terms of concepts, methods, and techniques. Based on the production practice of large-scale volcanic gas reservoirs in the Daqing, Jilin, and Xinjiang regions, we propose in this book new concepts, methods, and techniques based on numerous experiments, modeling, and other scientific investigation for the characterization of volcanic gas reservoirs. Meanwhile, this book also summarizes the latest research results, and attempts to standardize volcanic gas reservoir characterization procedures and techniques, which have been applied with various degrees of success in the production of volcanic gas reservoirs and have accelerated the development of exploitation technology.

1.1 CURRENT STATUS AND SIGNIFICANCE OF VOLCANIC GAS RESERVOIR DEVELOPMENT

1.1.1 Resources of volcanic gas reservoirs and development status in the world

Since the first discovery of volcanic hydrocarbon reservoirs in the San Joaquin Basin in California, United States, in 1887, a large number of volcanic hydrocarbon reservoirs have been found worldwide, such as the hydrocarbon-bearing basins in Japan, the United States, Venezuela, Cuba, the former Soviet Union, and China [1–3]. By the end of 2003, a total of 169 volcanic hydrocarbon reservoirs had been discovered, with 65 oil and gas shows and 102 oil seepages, and the proven hydrocarbon resources amount to more than 1.5 billion tons of oil equivalent [4].

Volcanic gas reservoirs are distributed mainly in China, Japan, the United States, Brazil, Namibia, Congo, and Indonesia. Most of the discovered volcanic gas reservoirs, however, have not been put into real development. Only the Higashi-Kashiwazaki Volcanic gas field (1968) and the Southern Nagaoka Volcanic gas field (1978) in Japan have produced for a sufficiently long duration with good results.

Nonetheless, the geological research on volcanic gas reservoirs has been sporadic, and investigation on development technology remains at a preliminary level. The available research results could only meet the initial production demand [5,6].

1.1.2 Resources and development status of volcanic gas reservoirs in China

It has been over half a century since the first discovery of volcanic oil reservoirs in the northwestern margin of Junggar basin in 1957, although large-scale discoveries in the Songliao and Junggar basins were not made until 2004, with estimated natural gas resources up to several trillion cubic meters and 3P(proved, probable, and possible) reserves approaching 1 trillion cubic meters. Hitherto these are the largest volcanic gas reservoirs known in the world.

The exploration of volcanic gas reservoirs in China experienced a challenging period of probing, innovation, and development. In the absence of related experiences as well as theoretical and technical guidance, we had to start from scratch our theoretical and technological research as well as and production trials in the Xushen gas field of Daqing, the Changling gas field in Jilin, and the Kelameili gas field in Xinjiang Province. In the process, we solved various problems in volcanic gas reservoir development and in a short period established several billion cubic meters of annual natural gas deliverability.

1.1.3 Significance of volcanic gas reservoir development

As an important part of natural gas resources in China, the rational development of volcanic gas reservoirs is of great significance to the domestic economy, especially in the following aspects:

1. *Mitigate energy supply and demand imbalance and optimize energy infrastructure in China.* Energy is one of the fundamental resources for national economic growth. With the rapid development of economy, the energy demand in China is increasing swiftly. In 2009, annual crude oil import in China exceeded 0.2 billion tons for the first time, accounting for about 52% of domestic consumption and exceeding the currently acceptable international warning line of 50%. Thus, rational development of volcanic gas reservoirs will relieve the pressure of rapidly growing oil and gas demand and mitigate the energy supply and demand imbalance in China. Natural gas is also a type of clean and efficient energy and is of great significance for optimizing national energy infrastructure and promoting the rapid development of a "low-carbon" economy.
2. *Increase gas resources and promote the development of natural gas industry in China.* The exploration and development of natural gas in China is progressing rapidly, and the golden time for natural gas development is expected in the near future. We have the largest volcanic gas reservoirs in the world, with a huge exploration and development potential. As an important part of natural gas resources in China, the rational development of volcanic gas reservoirs is of great significance for the development of the natural gas industry.
3. *Provide energy support for revitalizing economy in northeastern China and implement the major development program in western China.* In the Xushen and Changling gas fields, billions of cubic meters of annual gas deliverability has been established to supply natural gas to Changchun and Harbin through the Northeast Pipeline Network, which provides important energy support for industrial revitalization and economic development in Northeast China. The Kelameili gas field

will supply more than 1 billion cubic meters of natural gas annually to cities in western China, such as Urumqi, which plays an important role in promoting national harmony and maintaining social and economic stability.

1.2 GEOLOGICAL FEATURES OF VOLCANIC GAS RESERVOIRS AND CHALLENGES IN RESERVOIR CHARACTERIZATION

1.2.1 Geological features of volcanic gas reservoirs

Due to their fundamentally different petrogenic condition, internal architecture, lithology and lithofacies, pore-space types, and fracture features, volcanic rocks are very different from sedimentary rocks in terms of their reservoir distribution and gas-water relationships.

1. Special petrogenesis and complex internal architecture

Volcanic rocks are the products of eruption and deposition, and they are controlled by multiple factors including eruption mode, eruption energy, magma properties, eruption amount, eruption frequency, paleotopography, and subsequent alteration. Volcanic rocks have multilevel internal architectures and can be divided into various units, from large to small scales, as volcanic buildup, volcanic edifice, volcanic massif, volcanic lithofacies, and volcanic lithology. Each architectural unit has various shapes, different scales, and complex superposition relationships (e.g., cross-cutting and stacking patterns). As a result, volcanic reservoir rocks are characterized by unique petrogenesis and complex internal architecture compared to sedimentary reservoir rocks.

2. Various reservoir rocks and complex lithology and lithofacies

The components of volcanic rocks are complex and contain almost all the chemical elements and cation oxides in the earth's crust. As a result of volcanic eruption and diagenetic processes, there are diverse volcanic rock structures and constructions with various rock types and complex lithology. In terms of the volcanic eruption mode, eruption energy, and rock assemblage features, six volcanic lithofacies are recognized: eruptive facies, effusive facies, extrusive facies, volcanic conduit facies, subvolcanic lithofacies, and volcanic sedimentary facies. In terms of the occurrence and petrographic characteristics, these can be subdivided into a dozen subfacies. As is typical of an event deposition, volcanic lithofacies can change rapidly in rock types, shapes, superposition relationships, and scales.

3. Development of reservoir pores, cavities and fractures, various pore-fracture assemblages, complex reservoir accumulation, and permeation patterns

The volcanic reservoir space comprises mainly vesicles, intergranular pores, dissolution pores, and various fractures. Due to the complex shapes of pores and fractures and highly variable pore sizes, volcanic reservoirs are multimedium reservoirs with developed pores, cavities, and fractures. Pores are major reservoir spaces in volcanic reservoirs, and throats and fractures are flow passages. Various pores and fractures lead to different pore-fracture assemblages. Meanwhile, different petrogenesis, shape, size, and the degree of matrix-pore growth lead to various throat types with complex shapes and large variations in throat radius. Therefore, the reservoir accumulation and permeation pattern in a volcanic reservoir is expected to be complex.

4. *Vertical and lateral continuity and strong heterogeneity of pores*

The complex shape, superposition relationships, and spatial distribution of each architectural unit in volcanic rocks lead to scattered distribution of reservoirs, poor vertical and lateral continuity, and limited horizontal distribution, controlled by internal architecture. In addition, volcanic reservoirs have strong interlayer heterogeneity, which, together with the great changeability of physical properties in effective reservoirs, results in strong intrastratal heterogeneity.

5. *Gas-Water distribution controlled by structure, internal architecture, multi-medium porosity, and complex gas-water relationships*

The gas-water distribution in volcanic gas reservoirs is controlled by structure, internal architecture, and multiple porous media, with a broad gas-over-water spatial relationship. However, the gas-water contacts for different volcanic edifices and different volcanic massifs are not the same. As multiple porous media reservoirs, the gas saturation thresholds in different volcanic matrices and fractures are different, also leading to different gas-water distribution patterns, resulting in complex gas-water relationships in volcanic gas reservoirs.

1.2.2 Challenges in volcanic gas reservoir characterization

The unique geological conditions and complex geological features require new concepts, approaches, and methods for volcanic gas reservoir characterization, which is drastically different from conventional sedimentary reservoir rocks. In the absence of existing theoretical guidance and technical references, our research on volcanic gas reservoir characterization started from scratch and met numerous challenges, especially in the following aspects (Table 1.1).

1. *Difficulties in identification and characterization of complex internal architecture*

Volcanic rock bodies are composed of complex multilevel internal architectural units, which have different sizes and geological features from one other, and the difficulties in their identification and characterization are also different. Large-scale architectural units with a simple geometry and superposition relationships are highly different from surrounding rocks and can be easily identified and characterized with readily detectable geological and geophysical logging and seismic characteristics. On the contrary, small-scale architectural units with complex geometric and superposition relationships are only slightly different from surrounding rocks and thus difficult to identify and characterize on the basis of geological and geophysical logging and seismic characteristics. Moreover, volcanic edifices with complex petrogenesis and various types are more difficult to identify and characterize because their diagnostic features are usually destroyed by external factors such as tectonic activities.

2. *Challenges in characterization of microporous structures due to variability of reservoir space and complex pore-cavity-fracture assemblage*

There are many varieties of volcanic reservoir space, each characterized by unique shapes and volumes, degree of development, and complexity of pore-cavity-fracture assemblage, which make it difficult to characterize the capacity of reservoir space accurately. Compared to sedimentary rocks, the pore throats in volcanic rocks have more complex origins, more variable shapes, a wider range of sizes, and more complex pore-throat assemblages. This provides challenges to categorize the shape, size, and permeability of throats in volcanic rocks. The complexity of pores, cavities, fractures, and throats leads to complex reservoir accumulation and fluid-flow patterns in volcanic gas reservoirs, and it is

Table 1.1 Geological Features and Characterization Challenges for Volcanic Gas Reservoirs

Research Contents	Geological Feature		Difficulties in Volcanic Gas Reservoir Characterization	
	Sedimentary Rock	*Volcanic Rock*	*Characterization Content*	*Challenges*
Internal structure	Internal architectural features of "mud-in-sand" or "sand-in-mud"	Multi-level internal architectures composed of volcanic formation—volcanic edifice—volcanic massif—volcanic lithofacies—volcanic lithologies.	Shape, scale, superposition relationship and spatial distribution of volcanic formation, volcanic edifice, volcanic massif, volcanic lithofacies, and accumulation-permeation unit	From large- to small-scaled internal architecture, more complex shape and superposition relationships, smaller difference from surrounding rocks, and higher difficulty in identification and characterization
Microporous structure	1. Dominated by intergranular pores and dissolved pores 2. Mainly pore-shrinking type throat 3. Simple pore-fracture assemblage 4. Simple reservoir accumulation and fluid-flow pattern	1. Vesicles, intergranular pores, dissolved pores and fractures developed, with multimedium characteristics 2. Various throat types 3. Various pore-fracture assemblages 4. Complex reservoir accumulation and permeation pattern	1. Reservoir spaces, genesis type, shape, size of throats, accumulation and permeation ability 2. Reservoir accumulation and fluid-flow pattern, accumulation and permeability, and distribution 3. Reservoir microscopic producibility	1. Difficulty in characterization of accumulation capacity due to various reservoir spaces and complex pore-cavity-fracture assemblage relationships 2. Difficulty in characterization of fluid-flow capacity due to complex throat genesis and throat combination 3. Difficulty in evaluation of reservoir production capacity due to multiple pore-fracture assemblages and complex reservoir accumulation and fluid-flow pattern
Effective reservoir	1. Simple lithology and less effective reservoir types	1. Complex lithology and diversified effective reservoir types	1. Distribution of favorable lithology or lithofacies zones	1. Volcanic rocks with the same components have similar seismic

Continued

Table 1.1 Geological Features and Characterization Challenges for Volcanic Gas Reservoirs—cont'd

Research Contents	**Geological Feature**		**Difficulties in Volcanic Gas Reservoir Characterization**	
	Sedimentary Rock	***Volcanic Rock***	***Characterization Content***	***Challenges***
	2. Fewer fracture types 3. Relatively simple gas-water relationships 4. Weak heterogeneity 5. Controlled by sand body and physical properties, with laminar distribution	2. Diversified fracture types 3. Complex gas-water relationship 4. Strong heterogeneity 5. Controlled by complex internal structure, with irregular distribution	2. Distribution of fracture zones 3. Gas-water distribution 4. Classification and identification of effective reservoirs 5. Spatial distribution of effective reservoirs	response, and these rocks are difficult to predict from seismic data 2. Fractures are difficult to predict due to rapid changing in lithology, complex internal architecture, and boundary influence of lithology and internal architecture 3. Difficulty in identification of low-resistivity gas layers 4. Complex internal structure and irregular distribution of effective reservoirs make it difficult to realize body-controlled seismic inversion
Reservoir parameter	1. Limited change in rock matrix parameters 2. Simple reservoir accumulation and permeation pattern 3. Relatively simple conduction mechanism 4. Mainly simple structural fractures	1. Great change in rock matrix parameters 2. Complex reservoir accumulation and permeation pattern 3. Complex conduction mechanism 4. Various complex fractures	1. Interpretation of matrix porosity, permeability; and gas saturation 2. Interpretation of fracture width, density, and length, as well as porosity, permeability, and gas saturation	1. Complex reservoir spaces, throats, and accumulation and permeation pattern, leading to complex permeability characteristics make it difficult to interpret permeability 2. Complex rock conduction mechanism; fluid distribution and accumulation state in various pores make it difficult to interpret fluid saturation

Research Contents	Geological Feature		Difficulties in Volcanic Gas Reservoir Characterization	
	Sedimentary Rock	*Volcanic Rock*	*Characterization Content*	*Challenges*
Geological model building	1. Reservoir distribution controlled by sedimentary facies 2. Fluid distribution controlled by sand body and reservoir physical properties	1. Reservoir distribution controlled by multilevel internal architecture 2. Fluid distribution controlled by structure, internal architecture, and multimedium	1. Multilevel structural model 2. Multilevel reservoir framework mode; 3. Attribute model of multimedium reservoirs under framework model constraint 4. Gas-water distribution model in matrix and fractures under constraints of structure, internal architecture, and reservoir attributes	1. Difficulty in building structural and reservoir framework models because of complex internal architecture and difficulty in occurrence control and surface trend control 2. Difficulty in building attribute model because multimedium reservoir distribution is controlled by internal architecture 3. Difficulty in building gas-water distribution model because gas-water distribution is controlled by structure, internal architecture, and reservoir attributes

difficult to evaluate the characteristics pertinent to accumulation, permeability, and productivity of the reservoir patterns, respectively. Furthermore, volcanic reservoirs have multiple pore-fracture assemblages and extremely strong heterogeneity, making it difficult to evaluate accurately the effectiveness and producibility of volcanic reservoirs.

3. Challenges in identification and prediction of effective volcanic gas reservoirs due to diverse reservoir types and strong heterogeneity

Volcanic reservoirs have complex components and highly variable rock textures and structures that are difficult to distinguish using geophysical logging data. Volcanic reservoirs have various fracture types with different shapes, scales, and growth characteristics. It is difficult to identify blast fractures, contraction fractures, and dissolution fractures, respectively. There are various effective reservoir types, which are interbedded with various low-resistivity gas layers and high-resistivity aquifers, thus rendering effective reservoirs difficult to identify.

Volcanic rocks have a wide range of lithologies, but seismic responses for volcanic rocks with similar composition are nearly identical. This makes it difficult to predict volcanic lithology distribution using seismic data. In volcanic reservoirs, blast fractures, contraction fractures, weathered fractures, and sutured fractures with a small size and limited extension are also difficult to detect through seismic data. Moreover, the interference as a result of rapid changes in lithological boundaries and complex internal architectural boundaries seriously hinders fracture prediction using seismic data. For volcanic rocks with complex internal architecture, it is difficult to carry out reservoir parameter inversion under the control of these structures. Moreover, strong heterogeneity, nonregular distribution, thickness fluctuation, and low seismic resolution contribute to the challenges in predicting effective thin reservoirs, making it more difficult to achieve classified prediction in effective reservoirs.

4. ***Difficulties in interpreting parameters of multimedium volcanic gas reservoirs***
Parameters of volcanic rock matrix vary greatly with various pore types, different pore sizes, and pore distribution, resulting in different geophysical logging responses. The lithology and reservoir space characteristics exert a strong influence on porosity interpretation. Various volcanic reservoir spaces have different shapes, sizes, distribution patterns, and connection modes. Similarly, various throat types are reflected by their different shapes, sizes, sinuosities, and pore-throat assemblages with complex pore structures. Volcanic reservoir accumulation and fluid-flow patterns are highly variable, leading to different fluid-flow characteristics and large differences in permeability. This, in combination with the interference of fractures, presents difficulties for interpreting volcanic reservoir permeability. Also, the highly variable electric conduction modes and combination types, together with complex conduction paths, give the resistivity logs complicated response characteristics. In various pore types, fluid distribution and accumulation state are complex, causing complicated nuclear magnetic logging responses and greatly hampering saturation interpretation.

5. ***Difficulties in building 3D geological models for multimedium volcanic gas reservoirs with complex internal architecture***
Multilevel internal architectures, together with complex distribution and superposition relationships of individual structures, present difficulties for building an architectural reservoir framework model with occurrence and trend surface control. Volcanic reservoirs have a complex petrogenesis, and the reservoir distribution is controlled by internal architecture, with multiple media such as pores, cavities, and fractures. It is thus difficult to characterize reservoir parameters quantitatively and build attribute models. Volcanic gas reservoirs also have complex gas-water relationships, and the gas-water distribution pattern is controlled by structures, internal architecture, and reservoir characteristics. This contributes to the challenges of building gas-water distribution models.

1.3 SIGNIFICANCE OF VOLCANIC GAS RESERVOIR CHARACTERIZATION AND ITS TECHNICAL CONCEPT

1.3.1 Significance of volcanic gas reservoir characterization

The abundant volcanic natural gas resources with large reserves in China impart great significance to rational development of volcanic gas reservoirs. The key to volcanic gas reservoir development lies in its geological understanding, which depends on reservoir characterization. By taking into consideration the geological features, research difficulties, and production practice, volcanic gas reservoir characterization techniques have been developed recently through innovative research concepts and multidiscipline approaches, and they have contributed effective solutions to many key technological problems. For example, the characterization of internal architecture and microporous structures, identification and prediction of effective volcanic reservoirs, dual-medium parameter interpretation, and internal architecture helped constrain geological modeling of multimedium volcanic reservoirs. Based on reservoir characterization, the shape, scale, and superposition relationships of internal architectural units at each scale of volcanic reservoirs are clarified; the scale and relationships of volcanic stratigraphic sequences are revealed; the spatial distribution of effective volcanic reservoirs are predicted; and the three-dimensional (3D) geological model is built with the constraints of multimedium and complex internal architecture. Reservoir characterization provides the basis and guidance for drilling site optimization, horizontal well trajectory design, well-controlled dynamic reserves evaluation, and optimization of technology development strategies to establish a foundation for increasing drilling success rate, well production, and overall development level of gas fields.

1.3.2 Technical concept of volcanic gas reservoir characterization

1. Definition of conventional reservoir characterization

At the International Seminar for Reservoir Characterization first held in 1985, the reservoir characterization was defined as "a method to determine reservoir properties, identify geological information and spatial change quantitatively." The ultimate goal is to understand the reservoir heterogeneity accurately and to improve reservoir management. Conventional reservoir characterization includes stratigraphic correlation, structural features and distribution, type and distribution of sedimentary facies, reservoir-caprock features and distribution, reservoir types and fluid distribution, hydrocarbon reserves calculation, hydrocarbon reservoir temperature and pressure system, and driving characteristics [7]. Present reservoir characterization is an integrated multidisciplinary approach, including seismology, geophysical logging, geology, computer science, outcrop study, modern sedimentology, high-density well pattern data study, production performance research, qualitative and quantitative methods of geostatistics, and neural network [8]. Wu believed that reservoir characterization was to study subsurface heterogeneous reservoirs quantitatively based on multidiscipline data. He divided reservoir characterization into four aspects as characteristics identification, pattern recognition, vertical interpretation, and lateral prediction [7].

2. Technical concept of volcanic gas reservoir characterization

Conventional reservoir characterization involves mainly stratigraphic correlation, structural interpretation, sedimentary facies analysis, reservoir evaluation, and gas reservoir type study. Due to their unique petrogenesis and complex geological settings, volcanic gas reservoir characterization is significantly different from conventional reservoir characterization in terms of the research subjects and technical concept. Volcanic gas reservoir characterization mainly aims to solve the following problems.

1) Dissection of volcanic internal architecture

Complex internal architecture is a fundamental geological feature of volcanic gas reservoirs and an important factor controlling spatial distribution of these reservoirs. Dissection of the volcanic internal architecture makes it possible to clearly determine the relationships of volcanic internal architectural units at each level and to characterize the shape, scale, superposition relationships, and spatial distribution of such internal architectural units. These steps enable the construction of a framework model to constrain volcanic reservoir seismic inversion and geological modeling and to establish a foundation for volcanic stratigraphic sequence division and correlation.

2) Volcanic stratigraphic sequence division and correlation

Stratigraphic sequence is the basis for the division of development layer series. On the basis of the constraint of internal architecture and the principle of "hierarchical control and stepwise correlation," the volcanic stratigraphic sequence division and correlation are carried out, and the relationships among volcanic stratigraphic sequence, internal architecture, and development layer series are distilled to provide the basis for a logical division of development layer series.

3) Characterization of volcanic reservoir microporous structures

Reservoir microporous structures refer to the shape, size, distribution, and connectivity of pores, throats, and fractures. Characterization of volcanic reservoir microporous structures can reveal effectively the shape, scale, accumulation, and permeability of reservoir space and conduits in volcanic reservoirs. Investigations at this stage aim to characterize the accumulation and permeation capacity, as well as distribution features of various reservoir accumulation and permeation patterns, to evaluate reservoir effectiveness and fluid mobility quantitatively, and hence to provide a basis for evaluating reserve producibility and calibrating ultimate recovery efficiency.

4) Identification and prediction of effective volcanic reservoirs

In comparison with sedimentary reservoirs, the characteristics of effective volcanic reservoirs are much more complicated. Lithological and lithofacies analyses can help predict the spatial distribution of volcanic lithofacies and point to favorable lithological and lithofacies zones. Fracture identification and prediction can clarify the fracture growth level, fracture effectiveness, and fracture trend in volcanic rocks, predict the distribution of fracture zone, and guide the deployment of high-yielding wells. Prediction of effective reservoirs also helps determine the spatial distribution of various effective reservoirs in volcanic rocks and to optimize drilling locations and horizontal well trajectory.

5) Identification of gas layers and aquifers in volcanic gas reservoirs

Identification of gas layers and aquifers is a major step for determining gas and water distribution, which provides a basis for identification of the gas-water system in volcanic gas reservoirs and determination of gas reservoir types and gas-bearing area.

6) Interpretation of volcanic reservoir parameters

The interpretation of volcanic reservoir parameters aims to clarify the accumulation and fluid-flow capacity and gas-bearing characteristics of volcanic gas reservoirs, thus establishing a framework for the quantitative characterization of reservoir properties, evaluation of in situ gas reserves, and geological modeling of gas reservoirs.

7) 3D geological modeling of volcanic gas reservoirs

Three-dimensional geological modeling of complex internal architecture and dual-medium characteristics of volcanic rocks is the integrated representation derived from volcanic gas reservoir characterization. It is applied mainly to evaluation of in situ gas, to guide the planning of well location and horizontal well trajectory, and to provide models for numerical simulation.

1.3.3 Techniques of volcanic gas reservoir characterization

For different geological features and research subjects, the approaches for volcanic gas reservoir characterization are different from those used for conventional reservoir characterization. In view of the challenges in volcanic gas reservoir characterization, as discussed earlier, we follow a step-by-step approach to characterize volcanic gas reservoirs, progressing from qualitative to quantitative and from macroscopic and microscopic levels, using geological, geophysical logging, seismic, and dynamic data to carry out the study in the following steps: internal structure, stratigraphic sequence, microporous structure, reservoir identification, fluid identification, parameter interpretation, and finally construction of geological model (Figure 1.1).

1. Technique for dissecting volcanic internal architecture

Based on the division of multilevel internal architectures and the established identification criteria for each architectural unit, the internal architectures are described in the order of single well, cross-section profile, and two-dimensional and three-dimensional rendering, using geological, geophysical logging, and seismic data. The shape, scale, and superposition relationships of the architectural units are

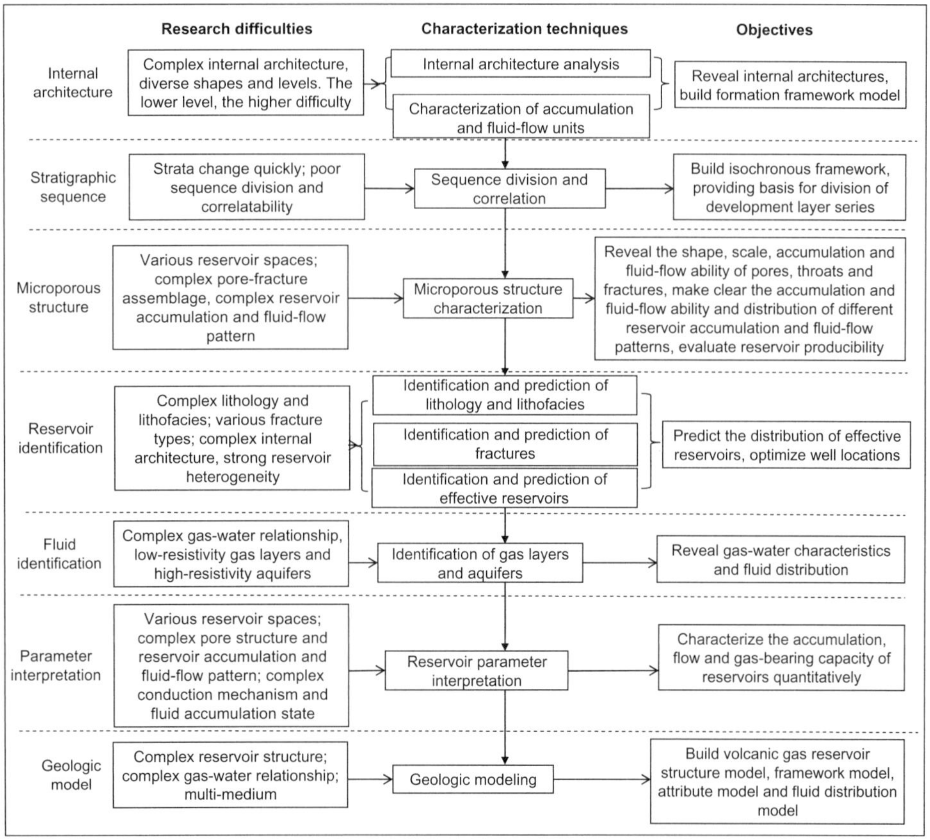

FIGURE 1.1

Research approaches and technical procedures of volcanic reservoir characterization.

characterized quantitatively, and the distribution of each such unit is revealed, which will provide a basis for the division of development layer series, reservoir seismic inversion, accumulation-permeation unit characterization, and construction of geological model.

2. Techniques for volcanic stratigraphic sequence division and correlation

The identification of volcanic stratigraphic sequence is based on the determination of stratigraphic sequence markers with level-control principle; the correlation of each sequence can be determined by streamlining the relationships among stratigraphic sequence, internal architecture, and gas reservoir layer series. Within the architectural constraint, volcanic stratigraphic sequences can be divided and correlated, the spatial distribution of volcanic stratigraphic sequences delineated, and the volcanic stratigraphic framework ultimately established.

3. Technique for volcanic reservoir microporous structure characterization

Through integration of core logs, thin section analysis, experimental study, analyses of reservoir space type, throat features, reservoir accumulation and fluid-flow patterns, and microscopic producibility, the microporous structural characteristics in volcanic reservoirs are defined, and the shape, scale, assemblage features, and distribution are characterized quantitatively, which in turn provides a basis for the producible reserve evaluation of volcanic gas reservoirs.

4. Techniques for effective reservoir identification and prediction

The construction of a geophysical logging identification model of volcanic rock components, rock texture/structure, and rock type is based on image analysis, TAS (total alkaliversus silica) plot, and dual-variable cross-plot analysis. Volcanic lithology may be identified by ECS (Eliemental Capture Spectroscopy) logging, acoustic and electrical image logging, and conventional geophysical logging. On the basis of fracture study and logging response analysis, acoustic and electrical image logging and conventional geophysical logging may be used to identify volcanic rock fractures, calculate fracture parameters, evaluate fracture growth degree and effectiveness, and determine fracture occurrence. Combined with various static and dynamic data, the identification markers for various effective reservoirs are determined and used to identify effective reservoirs in volcanic rock qualitatively. The establishment of cutoff standard and classified evaluation standard of effective reservoirs will lead to the quantitative identification of effective volcanic reservoirs.

Volcanic lithologies can be predicted by seismic reflection profile analysis, waveform classification, and frequency-divided inversion, using the results from single well identification, calibration of well-seismic assemblage, and analysis of seismic response characteristics. The fracture distribution in volcanic reservoirs can be predicted by poststack seismic attribute analysis, prestack azimuth treatment, and fracture parameter inversion. The reservoirs can be predicted level by level using seismic inversion under the constraint of internal architecture. The distribution of effective lithology and lithofacies, fractures, and effective reservoir space can be ultimately defined.

5. Techniques for identification of gas layers and aquifers in volcanic gas reservoirs

On the basis of mud logs indication for gas layers, aquifers and dry layers, formation tests, and geophysical logging response characteristics, the gas layers and aquifers in volcanic rocks are identified qualitatively and quantitatively in combination with gas-bearing data from wells.

6. Technique for interpreting reservoir parameter logs

By combining core experiments with nuclear magnetic logging, acoustic and electrical image logging, and conventional logging data, the porosity, permeability, and gas saturation in volcanic rock matrix and fractures can be interpreted through integration of theoretical and statistical models to define the physical properties and gas-bearing features of volcanic gas reservoirs.

7. Technique for 3D geological modeling of volcanic gas reservoirs

Core data, geophysical logging, mud logs, formation test, and production performance data are used as constraints for well point calculation. Internal architectural features, such as volcanic eruption cycles, volcanic edifices, volcanic massif, volcanic lithofacies, and accumulation-permeation units, are used as constraints for interwell calculation. In addition, seismic attribute volumes and inverted parameter volumes are used as co-constraints. The volcanic gas reservoir structure model, reservoir framework model, reservoir attribute model, and fluid distribution model are constructed using Kriging interpolation, sequential indicator simulation, and stochastic simulation. In this process, geological modeling technique is also developed for dual-medium volcanic gas reservoirs with complex internal architectures.

References

[1] Hunter BE, Davis DK. Distribution of volcanic sediments in the Golf coastal province-significance to petroleum geology. Trans Golf Coast Assoc Geol Soc 1979;29(1):147–55.

[2] Hawlander HM. Diagenesis and reservoir potential of volcanogenic sandstone-cretaceous of Surat basin. Aust Sediment Geol 1990;66(3/4):181–95.

[3] Homvc JF. Hydrocarbon exploration potential within intraplate shear-related depocenters: Deseado and San Julian basins. AAPG Bull 2001;85(10):1795–816.

[4] Petford N, Mccaffrey KJW. Hydrocarbons in crystalline rocks. London: Geological Society of London; 2003.

[5] He Y, Wu Y, Wu N. Study on volcanic hydrocarbon reservoirs. Pet Geol Oilfield Dev Daqing 1998; 18(14):6–14.

[6] Zou C, Zhao W, Jia C, et al. Formation and distribution of volcanic hydrocarbon reservoirs in Chinese sedimentary basins. Petrol Explor Develop 2008;35(3):257–71.

[7] Wu S. Reservoir characterization and model building. Beijing: Petroleum Industry Press; 2010, p. 10.

[8] Lv X, Li J. Hydrocarbon reservoir characterization technique. Beijing: Petroleum Industry Press; 2005.

CHAPTER

Internal Architecture of Volcanic Gas Reservoirs

2

Complex internal architecture is the primary geological feature of volcanic gas reservoirs. In the framework of its definition and classification, the identification markers of architectural units at various scales should be established to predict their spatial distribution. Furthermore, the shape, size, and superposition relationships (e.g., cross-cutting orders and stacking patterns) of various architectural units can be quantitatively characterized based on data acquired from the outcrops, seismic survey, and dense well grid. This will facilitate a clear delineation of the distribution patterns of the architectural units, which are the basis for the classification of development layer series, reservoir seismic inversion, characterization of accumulation-permeation units, and construction of geological models.

2.1 CONCEPT AND LEVELS OF VOLCANIC ROCK INTERNAL ARCHITECTURE

2.1.1 Concept of volcanic rock internal architecture

Internal architectures reflect the relationships among geological units in space and time [1]. In their study of the Zhujiang deep-water submarine fan system in the South China Sea, Wu and colleagues defined internal architecture as the relationship among the geological characteristics (such as 3D geometry, lithology, and physical properties) of sedimentary bodies and their individual internal layers and units, and their lateral and vertical spatial distributions as well as the occurrence sequences [2]. Liu defined internal architecture of the buried hill oil reservoirs in Bohai Gulf as the distribution patterns of internal stratigraphy, lithology, and faults [3].

Much work has been done on volcanic rock internal architecture [4–11]. However, there has been no clear definition for volcanic rock internal architecture in Chinese or international literature. In this book, volcanic rock internal architecture is defined for the first time, based on gas reservoir development and the compositional features of volcanic rock, as *the genetic relationship, occurrence sequence, superposition relationships, as well as the geometry and size of architectural units at various levels from volcanic formation to volcanic edifice, volcanic massif, volcanic lithofacies, and volcanic lithology*.

2.1.2 Levels of volcanic rock internal architecture

Volcanic rock internal architecture is a multilevel feature. Based on the petrogenetic and compositional characters of volcanic rocks, with reference to existing literature, outcrop survey, and field practice, volcanic rock internal architecture can be divided into five categories, including, from large to small, volcanic formation, volcanic edifice, volcanic massif, volcanic lithofacies, and volcanic lithology.

1. Volcanic formation

The term *formation*, first introduced into geological literature by Fyuksel in the mid-18th century, refers to the association of regional rock strata with similar composition and spatial position [12]. Meng proposed that the concept of formation was to be applied to the study of the origin and distribution of the paragenetic rock association in the stratified lithospheric crust [13]. In 1981, a volcanic formation was defined as a regular association of several volcanic rocks with similar origin and common characteristics [14]. Yuan and colleagues regarded a volcanic formation (or volcanic rock association, volcanic rock series) as a set of volcanic rock associations formed on the crustal surface of a given tectonic unit at a given evolving stage [15]. Sun and Peng considered a magmatic formation a regular and natural association of rocks closely associated in origin with close paragenetic relationships in a given tectonic environment and at a given geological evolution stage [16].

In this book, a volcanic formation refers to *a set of rock assemblages formed through superposition and connection of volcanic rocks that are closely associated in petrogenesis and syngenetics spatially under a given tectonic setting at a given geological stage.* The volcanic rocks of the Yingcheng Formation in the Songliao basin (Figure 2.1) and the Carboniferous volcanic rocks in the Junggar basin are typical volcanic formations.

2. Volcanic edifice

From various perspectives, a volcanic edifice has been given different definitions and interpretations [17–22], including (1) aggregate of the products derived directly from volcanic activity near the magmatic conduits; (2) residues of all volcanic and subvolcanic rocks that fill the volcanic conduits in or associated with the same eruption stage, and the residues of the rocks near the volcanic conduits; (3) an

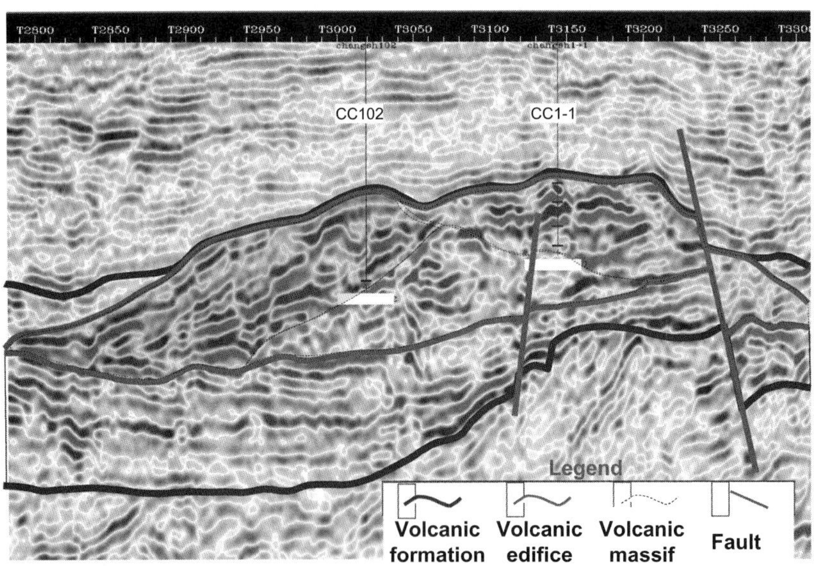

FIGURE 2.1

Seismic response characteristics of volcanic rock internal architectures at various levels.

assemblage composed of various products derived from the volcanic activities within a given time unit; (4) various extrusive-intrusive products near volcanic conduits formed in the same eruption period, as well as the overall shape of the products in a specific spatial and architectural relationship that reflects the eruption mechanism; and (5) the combination of various volcanic products, which, formed in a given geological time from the same eruption source, accumulated around the eruption point with a given shape and paragenetic association relationship.

In this book, a volcanic edifice is regarded as a geological architectural unit secondary to a volcanic formation (see Figure 2.1). A volcanic edifice is characterized by (1) eruption from the same magma source or same magma chamber; (2) mutually associated petrogenesis, with magma migrating in an identical main conduit; eruption and effusion being the main volcanic process, with some shallow intrusion; (3) the rock body being continuous in time sequence, formed within a specific time interval; and (4) spatially related paragenesis, comprising rocks in the volcanic craters, conduits, eruption surfaces, and other associated volcanic deposits in the same assemblage. The volcanic crater is a pathway for magma eruption, and the volcanic conduit is a pathway for magma migration from the magma chamber to the surface. Eruptive rocks are the major components of a volcanic edifice, which is formed by the accumulation of lava and volcaniclastic rocks derived from the eruption.

3. Volcanic massif

Currently, there are various definitions of *volcanic massif*, from a monolithological body in a narrow sense to a volcanic formation in a broad sense. [23–30] Wang considered that volcanic massif was formed through multiple eruptions in most cases, and the ancient weathered interbeds were usually formed between successive volcanic rock layers because of the long hiatuses between eruptions [31]. An alternative definition is that the internal architecture of a volcanic massif is dependent on the spatiotemporal association of volcanic lithofacies and genetic facies, and the deformation depth of magmatic products should also be taken into consideration [32].

In reference to the previous definitions and considering the applicability in gas reservoir development, a volcanic massif is regarded in this study as a geological architectural unit between a volcanic edifice and volcanic lithofacies in scale (see Figure 2.1) and is defined as a succession of genetically associated volcanic rock bodies, including the rock association of explosive facies–effusive facies–volcanic sedimentary facies with strong to weak eruption energy, or a succession of genetically associated volcanic rock bodies of effusive facies–explosive facies–effusive facies–volcanic sedimentary facies with a weak-strong-weak eruption energy sequence. Usually there are weathered crusts, loose layers, or sedimentary interbeds of variable thicknesses in the top and bottom parts of a volcanic massif.

4. Volcanic facies

The term *volcanic facies* was introduced in the Russian geological literature in early years [33], and it was divided into primary eruptive facies, subvolcanic facies, and volcanic conduit facies according to the occurrence and geometry of the volcanic massif. Li and associates regarded volcanic facies as "a rock body formed under certain natural conditions and thermal conditions (which are reflected in the architectural features of the rock body)." [18] Wang and coworkers defined volcanic facies as "the overall spatial distribution and the occurrence of volcanic products as well as their appearance and characteristics in the outcrop." [22] Qiu and colleagues considered volcanic facies "a general term for the characteristics of the products derived from volcanic activity under a given environment." [17]

In this book, volcanic facies is defined as the spatiotemporal distribution and geological characteristics of the same type of volcanic activities and their products. A volcanic facies is a geological architectural unit subsidiary to a volcanic massif or as a component of a volcanic massif. Lithofacies analysis can reveal the spatial distribution of volcanic rocks and the genetic relationships among various lithological associations and such analysis is an important part in the study of volcanic petrogenesis and physical properties [20].

5. Volcanic lithology

A volcanic lithology is a geological architectural unit subsidiary to volcanic facies and is a basic architectural unit recognized on the basis of unique rock type, composition, texture, and architecture of volcanic rocks. Variability in these characters is much more pronounced in volcaniclastic rocks than in volcanic lavas. As a result, the classification of volcanic lava lithologies is usually based on rock flow units that are formed through condensation and consolidation of a single magma eruption [33], and adjacent rock flow units are separated by a hiatus. Differences in cooling conditions cause variations in rock texture, architecture, and related physical properties in different parts of a rock flow unit [34], resulting in various types of volcanic lava, such as the upper vesicular rhyolite, middle tight rhyolite, and lower rhyolite with deformed rhyotaxitic architecture.

2.2 CHALLENGES AND TECHNICAL CONCEPTS FOR DISSECTING VOLCANIC ROCK INTERNAL ARCHITECTURE

2.2.1 Challenges in dissecting volcanic rock internal architecture

Because the composition, geometry, superposition relationships, and size of volcanic internal architectural units are different at various levels, the difficulties in identification, dissection, and characterization vary accordingly.

A high-level internal architectural unit contrasts sharply with the surrounding rocks, with distinct geological features and wireline log signatures. Because of its large size, clear seismic response, and well-defined boundaries characterized by the strong traceable reflections of discontinuities, a high-level architectural unit is usually easy to identify. In comparison, a low-level internal architectural unit does not show much difference from the surrounding rocks, with ambiguous geological features and wireline log signature. Due to its small scale and ambiguous seismic response features as well as poor boundary traceability, a low-level architectural unit is difficult to identify. With descending architectural hierarchy and scale from volcanic formation to volcanic lithology, the geological features, wireline log signatures, and seismic responses become increasingly more ambiguous and the boundary traceability more reduced, thus making architectural identification and dissection increasingly more difficult.

Similarly, the characteristics, geometry, and superposition relationships of a high-level architectural unit are relatively simple, which, in addition to its large size, greatly facilitates its identification, dissection, and quantitative characterization. A low-level architectural unit, however, is small in size and has more subtle and intricate properties, thus providing a tough challenge for its identification, dissection, and quantitative characterization. With descending hierarchy and scale from volcanic formation to volcanic lithology, the difficulty for quantitative characterization increases accordingly.

2.2.2 Technical concepts for dissecting volcanic rock internal architecture

As is shown in Figure 2.2, dissection of volcanic rock internal architecture and characterization of the rock body geometry, scale, superposition relationships, and distribution patterns are carried out in steps and phases. The proposed procedures take into consideration the volcanic architectural features and dissection difficulties, and the goals are achieved by integrating data acquired from a gravity survey, a magnetic survey, outcrop studies, core and geophysical log data, a seismic survey, and high-density well grid data, guided by the technical concept of proceeding from large to small scales, and from single well to cross section, then to 2D and 3D rendering.

1. Establishment of identification indicators

Based on geological, geophysical, and architectural features of volcanic rocks, the volcanic rock characteristics reflected in gravity, magnetic, and seismic surveys as well as geophysical logging can be revealed clearly to establish lithological and geophysical logging indicators as well as seismic facies patterns for identification of architecture units at various levels, thus providing a basis for well identification and spatial prediction.

2. Identification of volcanic rock internal architecture

Based on the identification indicators of volcanic rock internal architecture, and following the technical procedure of *point-line-plane-volume*, the distribution characteristics of architectural units at various levels can be effectively revealed by means of single well and profile identification, followed by planar and spatial prediction.

1) Single-Well identification

Lithological and geophysical logging indicators, rock composition, texture and architecture, facies sequence association, and the value, shape, and smoothness of geophysical wireline logs are used together to identify the architectural units of various levels in single wells, to delineate the unit boundaries, and to determine the vertical distribution characteristics of various units, all of which lays the foundation for profile identification.

2) Profile identification

On the basis of integrated well-seismic calibration, taking into account the seismic facies patterns of various architectural units and their indications, the architectural units at various levels are identified in sections, the unit boundaries are tracked, and a clear picture of their sectional distribution characteristics is obtained with the help of an attribute analysis of seismic reflective sections such as waveform, amplitude, frequency, and phase. The results from this step lead to planar identification.

3) Planar prediction

The sensitivity analysis of seismic attributes is conducted for architectural units of various levels to optimize the sensitive attributes or attribute association. Attributes are extracted under the constraint of geological and well-log characteristics, and the planar distributions of various architectural units are delineated first at a preliminary level. Thereafter, through further combined analysis and repeated verification with geological features, the planar distribution of the architectural units of various levels is predicted at a comprehensive level, which provides the basis for spatial prediction.

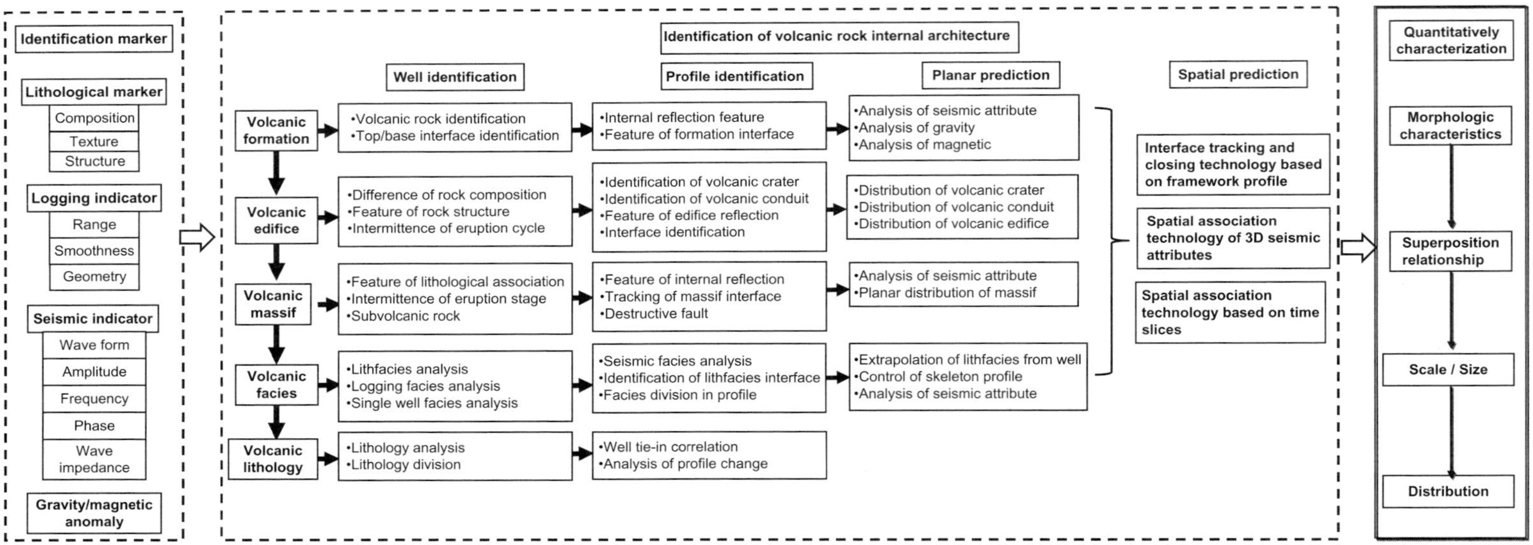

FIGURE 2.2

Flowchart of dissection techniques for volcanic rock internal architecture.

4) Spatial prediction

On the basis of well identification, profile identification, and planar prediction, three methods are used to predict the spatial distribution of architectural units at various levels: (1) establishing a skeletal profile network covering the entire area, tracking architectural interfaces based on profile identification of architectural units; (2) extracting 3D data volumes of various sensitive attributes using appropriate extraction techniques, and combing them according to the spatial distribution of architectural units derived from attribute analysis and optimization; and (3) analyzing the time series and formation slices of seismic data using spatial association technology to complete the prediction.

3. Characterization of internal architecture

Results from architectural unit identification at various levels, in combination with field outcrop survey, seismic data interpretation, and dissection of dense well gird data, are used to quantitatively characterize the geometry, superposition relationships, and the scale of architectural units; to reveal the regularity of their spatial distribution; and to lay a foundation for reservoir prediction, gas reservoir type evaluation, and geological model construction.

Volcanic lithology and facies will be discussed in Chapters 4 and 5. This chapter will focus on the identification and characterization of three high-level units, the volcanic formation, volcanic edifice, and volcanic massif.

2.3 IDENTIFICATION AND CHARACTERIZATION OF VOLCANIC ROCK FORMATION

2.3.1 Identification of volcanic rock formation

In terms of its geological features, a volcanic rock formation can be distilled to establish identification indicators, using all the data obtained from cores, geophysical logs, seismic survey, gravity survey, and magnetic survey. These indicators are then used to identify a volcanic formation from single well to a profile, and then a plane and 3D space; to reveal the spatial distribution characteristics of the volcanic formation; and to provide a basis for the characterization of the geometry and scale of the formation.

1. Identification indicators

A volcanic rock formation is composed of volcanic rocks generated in multiple eruptive cycles or at multiple stages, whereas the surrounding rocks are sedimentary rocks. Through analysis of fully cored wells and response characteristics of wireline logs, seismic survey, gravity survey, and magnetic survey, the indicators of lithology, well logs, as well as seismic, gravity, and magnetic surveys can be established for the identification of volcanic formations, surrounding rocks, and interfaces (Table 2.1).

1) Lithology indicators

A volcanic rock formation is composed of volcanic rocks and subvolcanic rocks with various compositions (from acidic to basic), textures (welded, brecciated, tuffaceous and porphyritic, etc.), and architectures (vesicular, rhyolitic, etc.), and interbedded with some volcanic sedimentary rocks (Figure 2.3 and Table 2.1).

The surrounding rocks are composed of regionally distributed normal sedimentary rocks, typically with horizontal, parallel, wavy beddings (see Figure 2.3 and Table 2.1).

Table 2.1 Indicators for Identification of Volcanic Formation

Identification Objects	Lithology	Wireline Logs			Seismic Survey					Gravity Survey	Magnetic Survey
		Value Range	Main Curve Shape	Smoothness	Waveform	Amplitude	Frequency	Phase	Impedance		
Volcanic formation	Volcanic rocks	High resistivity High density High interval transit time	Box shape, bell-funnel shape	Smooth-dentoid	Chaotic	Strong	Low	Discontinuous	High	High	High
Surrounding rocks	Sedimentary rocks	Low resistivity Low density High interval transit time	Various shapes	Dentoid -saw tooth like	Laminar	Weak	High	Continuous	Low	Low	Low
Interfaces	Volcanic sedimentary rocks Sedimentary volcanic rocks	Varies between that of volcanic rock and sedimentary rock	Digitate	Dentoid	Strong reflection of unconformity interface					Value varies between that of volcanic rock and sedimentary rock	

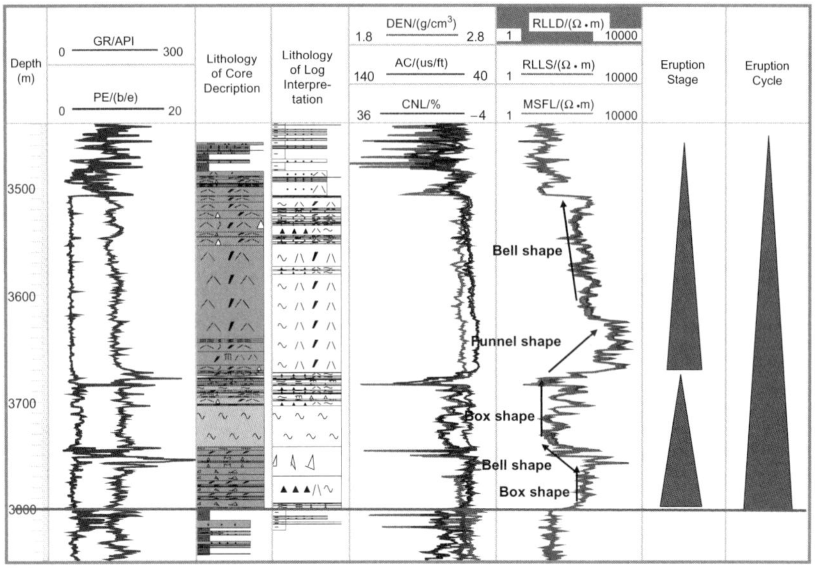

FIGURE 2.3

Lithological and log characteristics of volcanic formation (Well XX1-2).

Normally, various sedimentary volcanic rocks and volcanic sedimentary rocks are developed close to the interfaces of a volcanic formation, with petrological features of both volcanic rocks and sedimentary rocks.

2) Wireline log indicators

Volcanic rocks tend to have a great thickness, thus exhibiting typical features of *high resistivity, high density, low interval transit time, and low neutron,* and their curves are smooth-dendroid, mostly in box shape or bell-funnel shape (see Figure 2.3 and Table 2.1).

Sedimentary strata are usually much thinner, and their log signatures are characterized by *low resistivity, low density, high interval transit time, and high neutron,* with dentoid-serrated logging curves of various shapes (see Figure 2.3 and Table 2.1).

Sedimentary volcanic rocks or volcanic sedimentary rocks are relatively thin bedded, and their petrological and logging characteristics fall between those of volcanic and sedimentary rocks, with logging curves that are mostly in a dentoid or digitate shape (see Figure 2.3 and Table 2.1).

3) Seismic response characteristics

Volcanic rocks are characterized by great thickness, high rock density, and high acoustic velocity. In seismic responses, a volcanic formation exhibits an overall strong amplitude, low frequency, and poor continuity, usually with chaotic reflection, and a wave impedance higher than 9000 g/cm^3• m/s, even up to 16,000 (g/ cm^3) • (m/s) (Figure 2.4 and Table 2.1).

In contrast, the surrounding sedimentary rocks are typically thin bedded and have a low rock density and relatively low acoustic velocity. Their seismic responses exhibit a weak amplitude, high frequency,

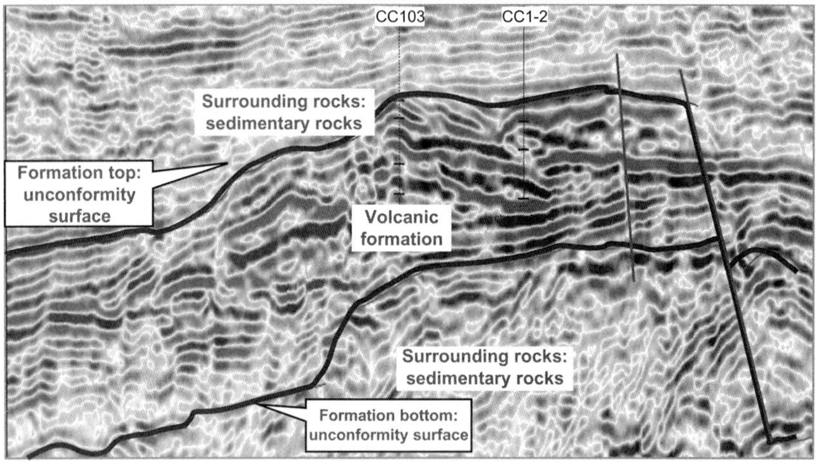

FIGURE 2.4

Seismic response characteristics of volcanic formation.

and fairly good continuity, normally with laminar reflection, and wave impedance between 5000 and 12,000 (g/ cm^3) • (m/s), which increases with depth, even up to 13,000 (g/ cm^3) • (m/s) (see Figure 2.4 and Table 2.1).

Sedimentary volcanic rocks and volcanic sedimentary rocks have response characteristics intermediate between volcanic and sedimentary rocks, and the interface between a volcanic formation and surrounding rocks mostly exhibit strong reflection features of a discontinuity.

4) Gravity and magnetic characteristics

Volcanic rocks typically have a high density and high content of ferromagnesium minerals. Thus, the gravity and magnetic forces of a volcanic formation are higher than those of sedimentary rocks, exhibiting a positive anomaly in the gravity field and an abnormally high value in the magnetic field (see Table 2.1).

2. Well identification

The lithological and wireline log indicators, as discussed earlier, are used to identify internal volcanic rocks, internal sedimentary rocks, and the external surrounding rocks, using integrated information obtained from coring, mud logging, and wireline logging to delineate the top/bottom interfaces of volcanic formation and to reveal the features of its vertical distribution. This forms the basis for profile identification and planar prediction.

1) Identification of volcanic rocks and preliminary determination of top/bottom contacts of volcanic formation

With petrological features and logging responses as guidelines, the information derived from coring, mud logs, and well logs (Elemental Capture Spectroscopy (ECS) logs, acoustic-electric imaging logs, and conventional logs) are utilized together to distinguish among volcanic, sedimentary volcanic, volcanic sedimentary, and sedimentary rocks, and to determine the top/bottom contacts or interfaces of a volcanic formation at a preliminary level.

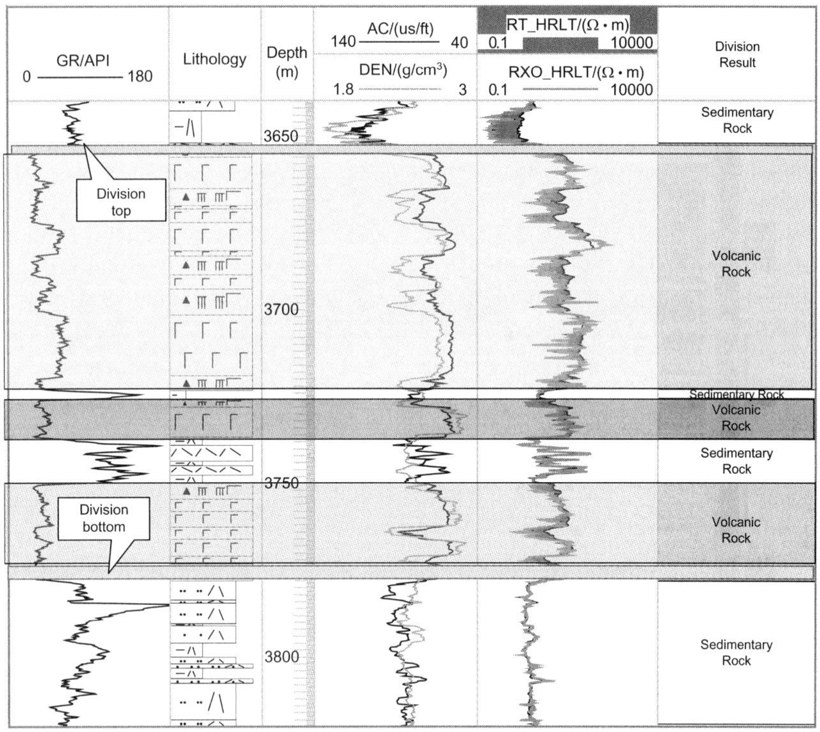

FIGURE 2.5

Well identification map of volcanic formation (Well DD171).

Figure 2.5 is a comprehensive map of Well DD171, in which basalt, basaltic breccia lava, basaltic welded breccia, sedimentary tuff, and tuffaceous sandstone were penetrated in the well. Based on an analysis of mud logging and well logging data, three sets of volcanic rocks are recognized, including basalt and basaltic welded breccia lava assemblage in the interval of 3653 to 3723.3 m, basalt in the interval of 3726.05 to 3736.6 m, and basalt and basaltic breccia lava assemblage in the interval of 3750.05 to 3778.05 m. A preliminary determination is also achieved about the top and bottom contact (interfaces) of the volcanic formation, possibly at 3653 m and 3778.05 m, respectively, with a maximum thickness of 125.05 m for the volcanic rock formation.

2) Identification of internal sedimentary rock in volcanic formation and external surrounding rock to confirm the formational top and base interfaces

In contrast to the external surrounding rocks, the internal sedimentary rocks in a volcanic formation were formed during the intermittence of the eruption cycle or eruption stage, characterized by the *high content of volcanic material, small thickness, and narrow planar distribution*. These sedimentary rocks are mainly sedimentary volcanic rocks or volcanic sedimentary rocks, and their geophysical logs exhibit a higher resistivity, density, and lower neutron values, and shorter interval transit time than those of normal sedimentary rocks.

In combination with the analysis of geological time and the distribution of volcanic edifices as well as the correlation with adjacent wells, the top and bottom contacts of the volcanic formation are further determined. In Well DD171 (see Figure 2.5), there are two suites of sedimentary tuffs developed among three suites of volcanic rocks, which are interbedded with several exotic clasts, and their thicknesses are about 2.75 m and 13.45 m, respectively. Their resistivity and density values in wireline logs are notably higher than those of the upper and lower thick sedimentary rocks, and their distribution is limited within the volcanic edifice. These sedimentary tuffs, therefore, are interpreted as deposits accumulated during the intermittent period of a volcanic formation. In this well, the strata above 3653 m and below 3778.05 m are formed mainly by non-equigranular arkoses, which are normal sedimentary rocks. Hence, the analyses demonstrated the presence of a basic volcanic formation containing mainly basalt in this well, with the top and bottom contacts of the volcanic formation at 3653 m and 3778.05 m, respectively.

3. Profile identification

On the basis of single well identification, a volcanic formation can be identified in seismic profiles by downhole seismic calibration and seismic reflection analysis, and the formation top and bottom boundaries are determined.

1) Identification of volcanic formation

The seismic profiles of volcanic rocks have such reflection features as strong amplitude, low frequency, and poor continuity with a chaotic pattern. Based on downhole seismic calibration, the distribution scope of volcanic formation in seismic profile can be determined preliminarily based on well constraints and characteristic analysis of the geometry and continuity of seismic events.

To illustrate the analytical process, the volcanic rocks of Yingcheng Formation in the CC gas field is taken as an example (see Figure 2.4) to show the determination of its distribution in the seismic profile using the reflection features of strong amplitude, moderate-low frequency, and poor continuity. In Figure 2.4, Well CC103 is located at a local subhigh, which may have developed near a volcanic conduit, and Well CC1-2 is located at the junction of two volcanic massifs.

2) Discrimination of top and bottom interfaces of volcanic formation

The top and bottom interfaces of a volcanic formation typically have a strong reflection of a discontinuity surface. Based on well constraints, these interfaces can be identified and discriminated in seismic profiles. In Figure 2.4, the top interface is calibrated, identified, and delineated in Wells CC103 and CC1-2; the bottom interface is identified by means of seismic waveform feature and the contact relationship.

4. Planar prediction

The 2D planar distribution of volcanic formations can be predicted at a comprehensive level by analyzing gravity, magnetic, and seismic attributes because volcanic formations typically have a high density, high seismic wave velocity, high gravity, and strong magnetic force.

1) Analysis of seismic attributes

Seismic attributes refer to the geometric, kinematic, kinetic, or statistic features of the seismic waves, which are derived from mathematical transformation of prestack or post-stack seismic data, and these attributes are the comprehensive reflection of underground geological architecture, lithology, fluid, and

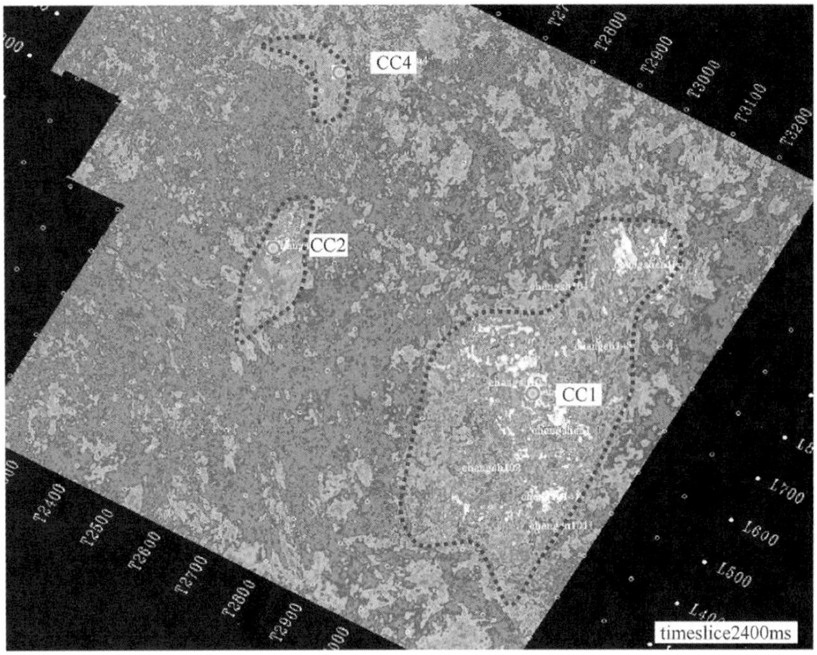

FIGURE 2.6

RMS amplitude attribute slice of volcanic formation.

so on. The basic seismic attributes include the features of waveform, amplitude, frequency, and phase of seismic waves, which can evolve into many derived attributes or associated attributes.

Because the values of seismic wave attributes vary with changing stratigraphic parameters, the abnormal value of these parameters can be determined through the analysis of seismic attributes. Compared with their surrounding rocks, volcanic rocks have such features as strong amplitude, poor continuity, and moderate to low frequency. For this reason, they exhibit a high value of root mean square (RMS) amplitude (Figure 2.6) and a chaotic and weak coherence of coherent attribute. Using these attributes, the distribution scopes of a volcanic formation at different depths can be displayed through time slicing and formation slicing, and thus the planar distribution can be predicted.

2) Magnetic analysis

A volcanic formation typically has a high magnetic anomaly. Through deduction and interpretation of the magnetic anomalies observed, the boundaries and scope of the volcanic formation can be delineated roughly to predict the spatial distribution and variation of the volcanic formation. Figure 2.7 is a map realized from magnetic survey from the deep part of the southern Songliao basin, showing a basically consistent magnetic anomaly and the spatial extent of the volcanic formation.

3) Gravity analysis

Taking into account the high density and high gravity characteristics of a volcanic formation, the boundaries and distribution of a volcanic formation in plan view can be revealed at a preliminary level to predict its regional distribution and variation by means of gravimetric measurement and analysis.

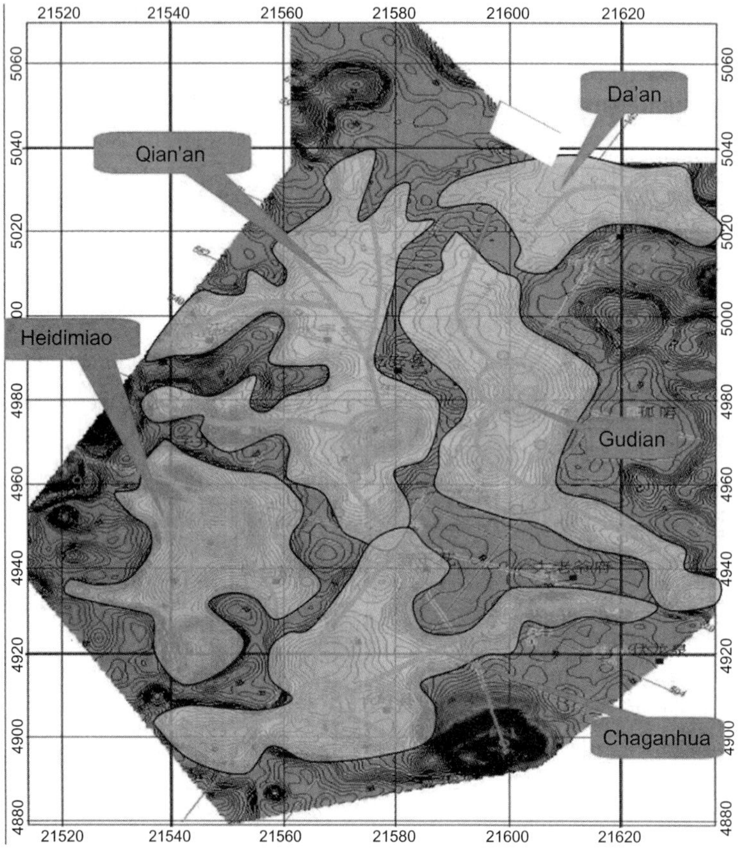

FIGURE 2.7

Magnetic anomaly of volcanic formation.

Figure 2.8 is a map of gravity survey derived from the deep part of the southern Songliao basin, showing the superposition relationships of areas with gravity anomaly and magnetic anomaly, which is basically consistent with the regional distribution pattern of the volcanic formation.

5. Spatial prediction

Single well identification, profile identification, and 2D planar prediction, as discussed earlier, can be used as the basis to further predict the 3D spatial distribution of a volcanic formation, using the two methods that follow.

1) Tracking and closing technique of volcanic formational top and bottom interfaces

Based on the identification of interfaces in wells, the formational contacts or interfaces can be tracked between wells in seismic profiles with the constraint of planar distribution. The skeletal profile network, which forms an effective control over the study area, is set up to track and close

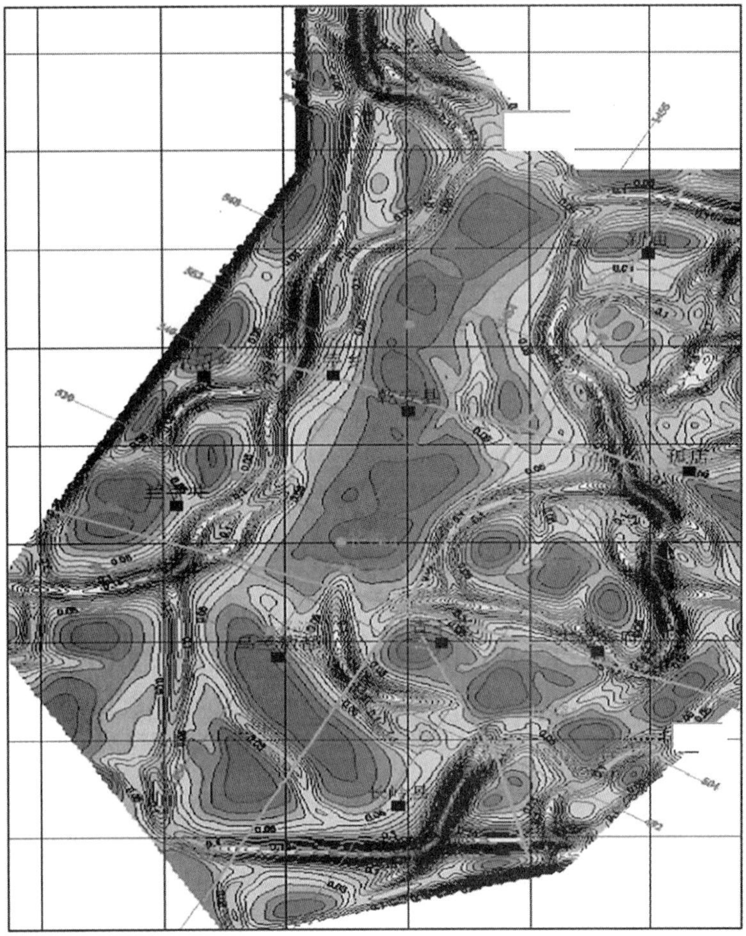

FIGURE 2.8

Gravity anomaly of volcanic formation.

the volcanic formation top and bottom interfaces and then to predict the 3D spatial distribution of the volcanic formation.

Figure 2.9 is a spatial distribution map of the volcanic formational top and bottom interfaces in the CC gas field. Based on this map, the geometry and scale of the spatial distribution of volcanic formation can be depicted effectively.

2) Spatial combination technique of 3D attributes volume

This technique includes four steps: (1) seismic attribute discrimination criteria of volcanic rocks are determined through sensitivity analysis; (2) various sensitive attributes are extracted in 3D space using the extraction technique of seismic attributes; (3) the discrimination criteria are used to extract and

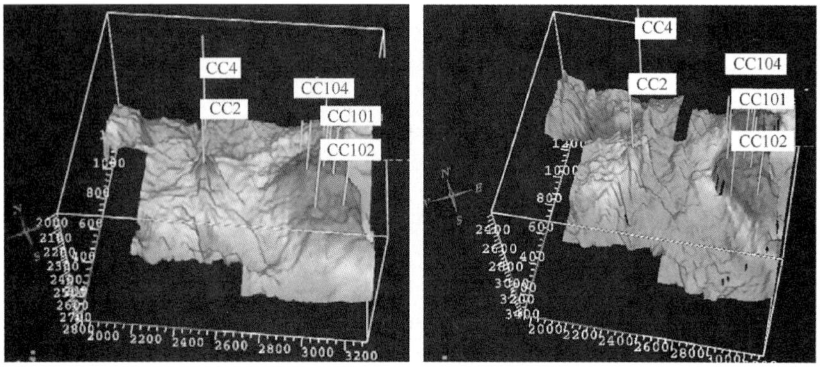

(a) (b)

FIGURE 2.9

Spatial distribution map of the volcanic formation (CC gas field). (a) Spatial distribution map of volcanic formational top. (b) Spatial distribution map of volcanic formational bottom.

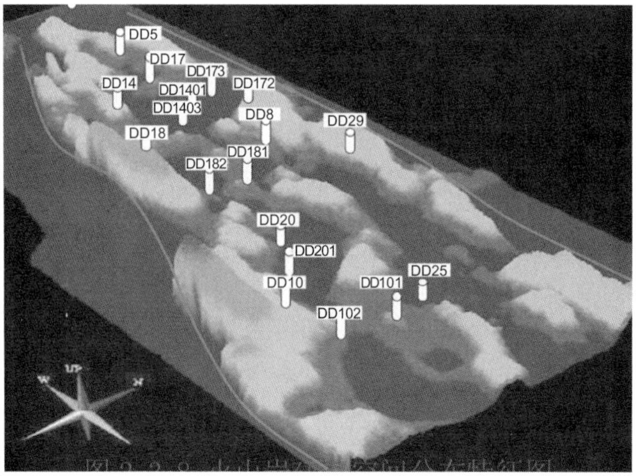

FIGURE 2.10

Spatial distribution map of the volcanic formation (DD gas field).

integrate spatial abnormal points that reflect the sensitive attributes of volcanic formation according to their spatial positions; and (4) the predicted results of multiple attributes are superposed and optimized to reveal the spatial distribution of the volcanic formation.

Figure 2.10 is the spatial distribution of volcanic formation predicted by combining high amplitude abnormal points of 3D amplitude attribute volume in the DD gas field. This demonstrates that the predicted result corresponds reasonably well to the volcanic rock distribution revealed by well data, and it reflects adequately the 3D geometry and scale of the volcanic formation.

2.3.2 Characterization of volcanic formation

A volcanic formation is a very large-scaled architectural unit, which cannot be revealed fully by the conventional outcrop survey and dissection of a dense well grid. In general, geological investigation and seismic prediction are used to characterize the shape, superposition relationships (cross-cutting or stacking patterns), and distribution of a volcanic formation.

1. Shape characterization of volcanic formation

For broadly distributed volcanic formations, usually seismic data are collected only in some areas. Therefore, geological investigation is combined with seismic prediction to characterize the shape of a volcanic formation. Results from these investigations help establish the connection between the shape of the volcanic formation with a specific eruption mode, magmatic properties, and paleotopographical features.

1) Shape of volcanic formation formed by centered eruption

A volcanic formation derived from a single centrally eruptive volcanic edifice typically manifests as an uplifted mound, dome, or lens typically, being circular or elliptical in plan view and as a bead string along the fault zone. A volcanic formation formed by eruption of acidic magma is mainly shaped as a steep-sloped mound and as small circular or elliptic bodies in plan view; a volcanic formation formed by eruption of basic magma is usually shaped as a flat shield and as a large circular or elliptical body in plan view.

A volcanic formation formed by multiply combined central volcanic edifices commonly manifests as an irregular mound with increased area, forming strips, ad bands or irregular elliptical patterns in plan view (Figure 2.11).

2) Shape of volcanic formation formed by fissured eruption

Most of the volcanic formations produced by fissured eruption are irregularly laminar. The formations formed by effusion of acidic magma are thick, usually in the shape of a steep wedge, showing a strip-like geometry with a small areal extent in plan view, whereas those formed by effusion of basic magma are mainly flat laminar, in the shapes a bands or strips, with a large areal extent in plan view.

3) Paleotopographical impact

The magma is in liquid state when erupted and flows to lower areas while it is condensed and consolidated continuously. A volcanic formation formed in a flat place has a shape with good symmetry and less variation in thickness, whereas that formed in a lower area will show an asymmetrical shape, large slope; and small thickness. Therefore, the paleotopographical relief would have a great impact on the shape of volcanic formation and made its geometry complex.

2. Size characterization of volcanic formation

The size of a volcanic formation is associated with the volcanic eruption energy and duration, with great variation in thickness and area. In the XX gas field, the volcanic formation has an areal extent up to several 10,000 square kilometers and a thickness of 300 to 1500 m, with an average thickness of

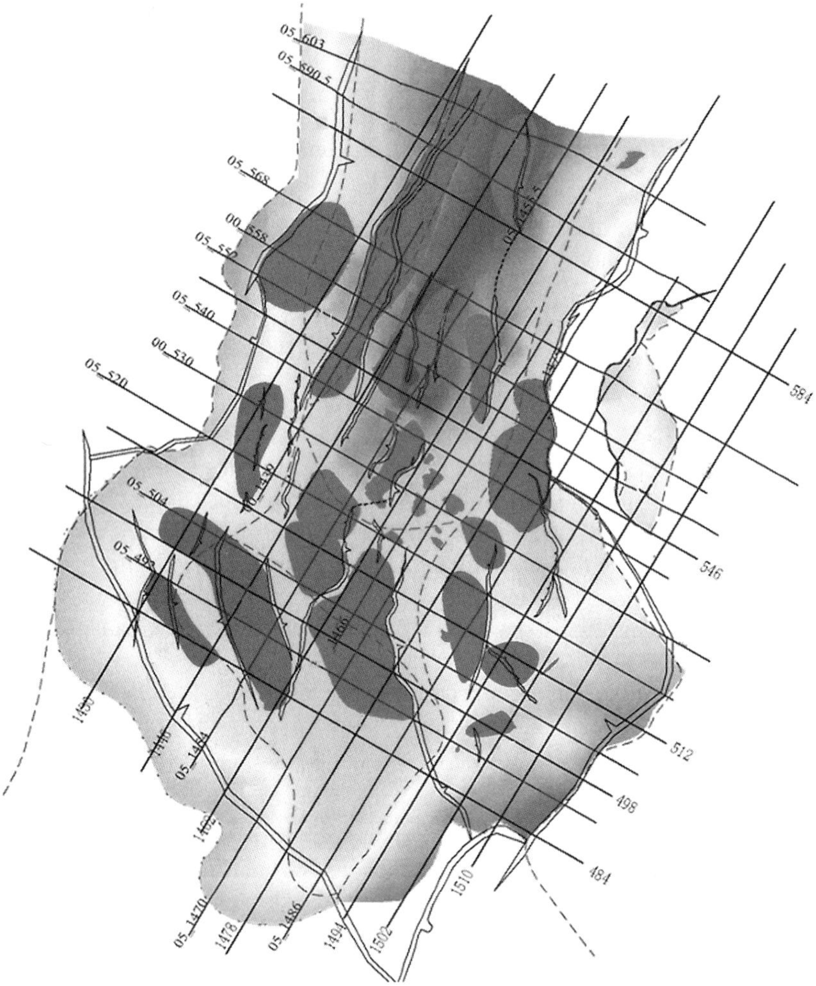

FIGURE 2.11

Shape of volcanic formation in the Changling faulted depression.

800 m. In the CC gas field, the volcanic formation covers an area of 11,000 km^2, with a thickness ranging between 250 and 1800 m and an average of 900 m. The late Jurassic volcanic formation in Jibei has a large size, covering several 10,000 square kilometers and its thickness varies greatly from 144 m to 6263 m. In the West Hill of Beijing District, the late Jurassic volcanic formation has a limited distribution and the thickness of volcanic rocks ranging from several hundreds meters to more than 4000 meters [35].

These examples show that the sizes of volcanic formations vary significantly, with thickness ranging from more than 100 meters to 6000 meters and areal coverage from several square meters to several 10,000 square kilometers.

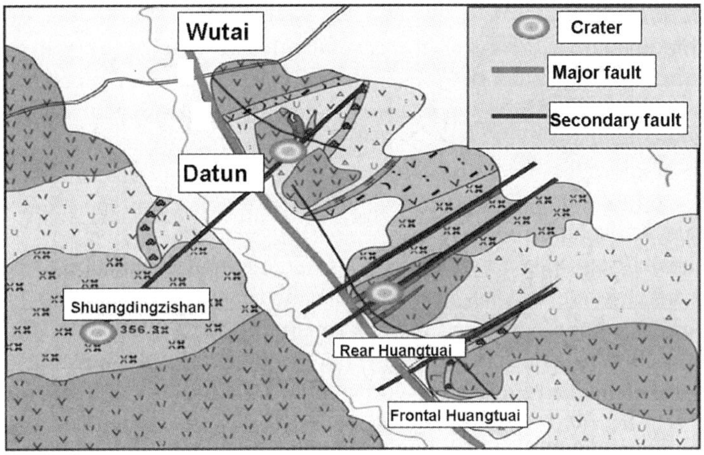

FIGURE 2.12

Relationships between volcanic formation and faults (Jiutai, Jilin Province, China).

3. Distribution of volcanic formation

Vertically, the distribution of a volcanic formation is controlled by basin tectonic evolution, which is developed primarily during the faulting and depression episodes of the basin. For instance, the volcanic formations in the XX and CC gas fields were developed in association with faulting and depression of this rift basin.

Horizontally, a volcanic formation is distributed mainly near major faults, particularly in the junction of two groups of faults, with a spreading direction basically consistent with the extending direction of the major faults (Figure 2.12).

2.4 IDENTIFICATION AND CHARACTERIZATION OF A VOLCANIC EDIFICE

A volcanic edifice is a basic architectural subunit of volcanic formation, and it determines the distribution, type, and size of volcanic gas reservoirs. On the basis of volcanic edifice classification, the identification indicators for volcanic craters, conduits, and periclines should be established first. Next, by following the approach from point to line, and then to 2D plane and 3D space, a volcanic edifice can be identified step by step at various levels from a single well to a profile, then to a 2D plane and 3D space, to reveal its spatial distribution and to characterized its superposition relationship and size.

2.4.1 Volcanic edifice classification

In terms of petrogenesis, volcanic edifices are divided into three categories: centered type, fissured type, and combined type [36]. They can be subdivided into various subtypes based on their various shapes and architectures.

1. Centered volcanic edifice

A centered volcanic edifice is formed by the accumulation of magma erupted from one or several columnar conduits around the crater, manifesting mostly as equiaxial or subequiaxial circles in plan view. It can be subdivided into various types such as conic, shield, laminar, and domed edifices according to its external morphology.

1) Conic volcanic edifice

As the name implies, a conic volcanic edifice has conic external morphology, composed mainly of volcanic clastic rocks derived from high-energy volcanic eruptions. This type of volcanic edifice is circular in plan view with good symmetry, a high primary dip angle, and a great thickness in the cone margin, but with a small diameter. In terms of the number of craters or conduits, it can be subdivided into monoconic and multiconic types:

1. *Monoconic volcanic edifice.* A monoconic volcanic edifice has a single crater, with one volcanic conduit in a dotted, symmetrical distribution, with its strata inclined inward. The slope angle of the cone margin decreases from top to bottom (Figure 2.13). A typical example of a monoconic volcanic edifice is found in the Well Block DD-10.
2. *Multiconic volcanic edifice.* A multiconic volcanic edifice has multiple craters and volcanic conduits, which are mostly distributed in a bead-string pattern along a major fault with a linear symmetry and with many depressions developed between cones (Figure 2.14). A good example of a multiconic volcanic edifice is observed in the Well Block DD-14.

2) Shield volcanic edifice

A shield volcanic edifice is formed by low-energy multiple effusions (occasionally weak explosion), characterized by the slow and gentle accumulation of basic magma in the form of a shield (Figure 2.15). The marginal rock layers of a shield volcanic edifice dip gently (normally only several degrees) with a small thickness and large distribution diameter.

FIGURE 2.13

Monoconic volcanic edifice.

FIGURE 2.14

Multiconic volcanic edifice.

FIGURE 2.15

Shield volcanic edifice.

3) Domed volcanic edifice

A domed volcanic edifice is formed by slow extrusion of high-viscosity, acidic, neutral-acidic, or basic magma (Figure 2.16). It is steep in the margin, flat at the center, with bell-, mound-, or tabular-shaped rock bodies developed on its top.

2. Fissured volcanic edifice

A fissured volcanic edifice is usually distributed in a strip along a fault or fissure. Its crater and volcanic conduit are not clearly defined. Its lithology is simple, consisting mainly of basic basalt and neutral andesite with some dacite or rhyolite. In general, extensive lava plateau, plain, and platform are formed by effusion of magma along the fissured conduits. Horizontally, a fissured volcanic edifice is almost equiaxial with wide and broad top and gentle flanks, and the rock layers dip outward with a gentle

FIGURE 2.16

Domed volcanic edifice.

FIGURE 2.17

Fissured volcanic edifice.

curvature (Figure 2.17). The volcanic edifice in the Well Block DD-17 in Xinjiang is a typical example of a fissured volcanic edifice.

3. Combined volcanic edifice

Normally, a combined volcanic edifice (Figure 2.18) is formed under three circumstances: (1) the volcano remains unchanged but has migrating major volcanic conduits and craters, and the accumulated volcanic massifs shift with these conduits and craters and overlap on top of one other or extend in a given direction; (2) subsidiary volcanic massifs such as parasitic cone and volcanic domes occur above the parasitic or lateral volcanic conduits; and (3) many large-scale lava domes occur within a caldera as a result of late volcanic activity.

2.4.2 Identification of a volcanic edifice

A volcanic edifice consists of a volcanic crater, conduit, and pericline [18,37]. The volcanic crater and conduit are important markers for identification of a volcanic edifice. Thus, it is necessary to establish lithological, logging, and seismic indicators to identify volcanic craters, conduits, and periclines; to

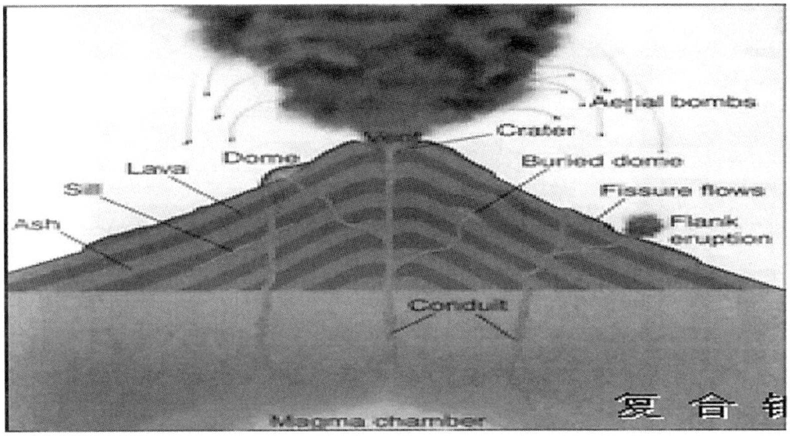

FIGURE 2.18

Schematic map of a combined volcanic edifice.

identify the volcanic edifices and track the edifice interfaces with the information obtained from coring, logging, and seismic survey; and hence to reveal the spatial distribution of a volcanic edifice, providing a basis for a quantitative characterization of its size, shape, and superposition relationships.

1. *Identification indicators*

The internal characteristics and external boundaries of volcanic craters, conduits, and periclines are of fundamental importance for volcanic edifice identification. The lithological, wireline log, and seismic indicators for identifying volcanic craters, conduits, and periclines can be established through analysis of lithological characteristics, well logging, and near-well seismic responses (Table 2.2).

1) Lithological indicators

Volcanic craters. It is a circular depression formed around the bocca with the accumulation of volcanic ejecta. Magma tends to retract at the end of an eruption, leading to volcano collapse and resulting in complex internal lithology in a crater. The major lithological indicators include (Figure 2.19a) (1) volcanic bombs, volcanic breccia, and tuff distributed in a circle surrounding the crater, interbedded with the contemporaneously lagged breccia or epigenetic clasts, with conglomerate texture, breccia texture, tuffaceous texture, and bedding; and (2) occurrence of a mushroom- or cloud-shaped extrusive dome or massif with a massive architecture.

Volcanic conduits. Located below the crater, it acts as the pathway for magma migration and is filled by the retracted magma finally. The lithological indicators include (1) volcanic lava, welded breccia, and subvolcanic rock in occurrence of a stock, vein, and upright column in the volcanic neck subfacies, with xenoliths found occasionally. The rock textures include an upright rhyolitic texture (Figure 2.19b), a large columnar joint (Figure 2.19c), a porphyritic texture, and a condensation margin texture, among others; and (2) cryptoexplosive breccia distributed like a broken branch and an irregular vein in the cryptoexplosive breccia subfacies (Figure 2.19d), with cryptoexplosive breccia texture, autoclastic porphyritic texture, and cataclastic texture.

Table 2.2 Identification Indicators of a Volcanic Edifice

Identification Objects		Typical Lithology			Characteristics of Well Logging Response			Features of a Seismic Response			
		Rock Types	Geometry of Rock Assemblages	Architectures/ Textures	Range	Major Behavior	Smoothness	Waveform	Amplitude	Frequency	Continuity
Volcanic crater		Circularly distributed volcanic clastic rocks intercalated with epigenetic sedimentary clastic rocks	Subparallel intercalation	Agglomerate texture, volcanic breccia texture, tuffaceous texture, bedding	High and low interbedded	Finger shaped	Dentoid-serrate	Laminar locally, arclike sag or graben-like pulldown	Strong	High	Good
		Extrusive dome or massifs	In shape of mushroom, cloud	Massive architecture, rhyolitic architecture	High resistivity, high density, low transit time	Box shaped	Smooth-microserrated				
Volcanic conduit	Volcanic neck	Lava, welded breccia, subvolcanic rocks, xenolith	Subupright columnar	Columnar joint, upright rhyolitic architecture, porphyritic texture	Medium/high resistivity, medium/high density, medium/low transit time	In shape of box, bell + funnel composite	Smooth-microserrated	In shape of umbrella, funnel, column and line	Medium-weak	High	Poor
	Crypto-explosive breccia	Crypto-explosive breccia	In shape of broken branch, irregular vein	Crypto-explosive breccia architecture, autoclastic porphyritic texture, cataclastic texture	Medium/low resistivity, medium/low density, medium/high transit time	Box + funnel composite shape	Dentoid-serrate				
Pericline	Proximal crater zone	Agglomerate, breccia, welded breccia, vesicular lava	In shape of wedges, lens and blocks	Volcanic agglomerate architecture, breccia architecture, vesicular architecture	Medium resistivity, density and transit time	In shape of bell, box	Smooth-microserrated	Chaotic, convergent upward	Weak-medium	Low	Poor
	Distal crater zone	Tuff, micro-vesicular lava intercalated with volcanic sedimentary rocks	Laminar	Tuffaceous texture, rhyolitic architecture, bedding	Medium/low resistivity, medium/low density, medium/high transit time	In shape of box, finger	Microserrated-dentoid	Pseudolaminar	Medium-strong	High	Fairly good
External envelope		Lithological interfaces, surfaces of occurrence variation and unconformity surfaces			Traditional Gamma Ray (GR), Density (DEN), resistivity abrupt change, Formation micro image (FMI) occurrence change			Strong reflection of unconformity surface			

FIGURE 2.19

Typical lithological indicators of volcanic craters and conduits. (a) Tuff interbedded with epigenetic clast. (b) Upright rhyolitic architecture. (c) Columnar joint. (d) Cryptoexplosive breccia.

Periclines. It is composed of volcanic rocks distributed around the crater and volcanic sedimentary rocks within the volcanic edifice. Volcanic conglomerate, volcanic breccia, or vesicular lava are predominant in the proximal crater zone, mostly as wedges, lenses, or blocks, whereas volcanic tuff and microvesicular lava are typical of the distal crater zone and appear in a laminar form. Farther away from the crater, the grain size of volcanic clasts decreases gradually, the vesicles in lava become smaller, and the dip angle of the rhyolitic bodies decreases gradually.

The envelope of a volcanic edifice is marked mainly by discontinuity surfaces due to variations in lithology or stratigraphic succession.

2) Characteristics of a well logging response

The characteristics of a well logging response in the volcanic crater, conduit, pericline, and external envelope of a volcanic edifice are important bases for the determination of volcanic edifice types.

Logging response characteristics of a volcanic crater. The lithology within a volcanic crater is complex and variable. The wireline logs of volcanic rocks are typified by high resistivity, high density, and low interval transit time, in contrast to the low resistivity, low density, and high interval transit time for sedimentary rocks. If volcanic rocks are interbedded with epigenetic clastic rocks in a crater, the logging curves will show a dentoid, serrated, or digitate pattern (Figure 2.20). In a crater composed of extrusive dome or volcanic massifs, the curves will be in the shape of a flat, microserrated box.

Logging response characteristics of a volcanic conduit. Lava, welded breccia, subvolcanic rocks, together with minor xenoliths are developed mainly in the volcanic neck, with relatively consistent compositions and textures. The logging curves primarily have features of moderate-high resistivity, moderate-high density, and moderate-low interval transit time, their patterns being mostly that of a flat, microserrated box and a combined bell and funnel (Figure 2.21a).

The cryptoexplosive breccia tract is occupied by the specific cryptoexplosive breccia. If the interstitial space of breccia is filled by cognate magma, the logging curves will show moderate-high resistivity, moderate-high density, and moderate-low interval transit time and will be shaped as a

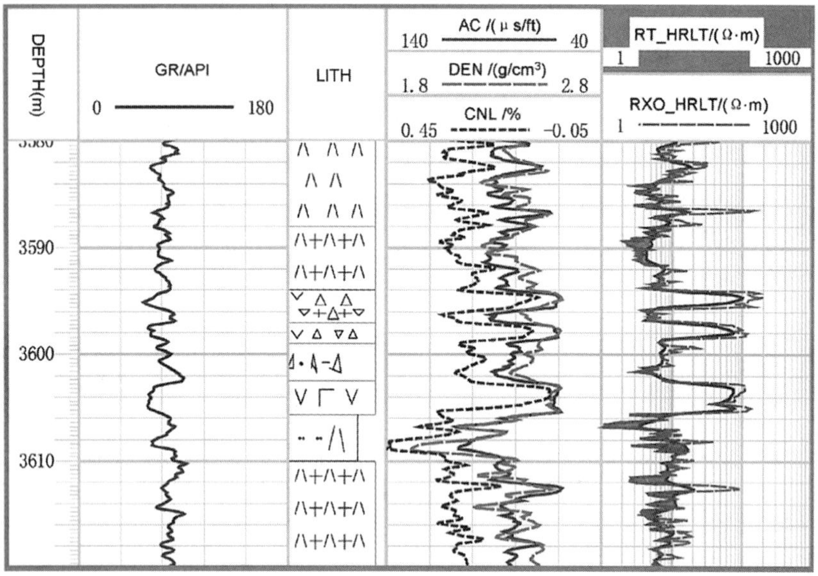

FIGURE 2.20

Logging response characteristics of a volcanic crater (Well DD-8).

microserrated box. In breccias that are not filled or filled with low-resistivity surrounding rocks, the curves will have features of moderate-low resistivity, moderate-low density, and moderate-high interval transit time and will have dentoid-serrate combined box-funnel shapes (Figure 2.21b).

Logging response characteristics of a pericline. In the proximal zone of a volcanic crater, massive volcanic rocks are predominant, characterized by consistent shapes of logging curves: flat, microserrated bell and box, with moderate amplitude (Figure 2.22, left).

In the distal zone of the crater, thin beds of tuff and microvesicular lava are intercalated with volcanic sedimentary rock, showing distinct interbedding. The shape of logging curves varies greatly, from microserrated-serrated box and serrated-digitate, reflecting in general a moderate-low resistivity, moderate-low density, and moderate-high interval transit time (Figure 2.22, right).

Envelope of a volcanic edifice. The envelope of a volcanic edifice is its interface or contact with the surrounding sedimentary rocks or with another volcanic edifice. The logging indicators for identification are as follows.

Interface of lithological variation. An abrupt contact between sedimentary rocks and volcanic rocks corresponds to an abrupt change from low resistivity, low density, and high interval transit time to high resistivity, high density, and low interval transit time in logging curves (Figure 2.22, blue line).

An abrupt change in composition at the contact between adjacent volcanic rock bodies, such as an abrupt contact between acidic volcanic rock and basic volcanic rock, will be reflected in logging curves as an abrupt transition from high gamma, low density, and low neutron to low gamma, high density, and low neutron (Figure 2.23).

An abrupt interface between an intrusive rock and an extrusive rock, or between regional lava and volcanic conglomerate or volcanic breccia, normally corresponds to some changes in curve shape and smoothness of conventional logging curves. For volcanic breccia, the curves are strongly serrated,

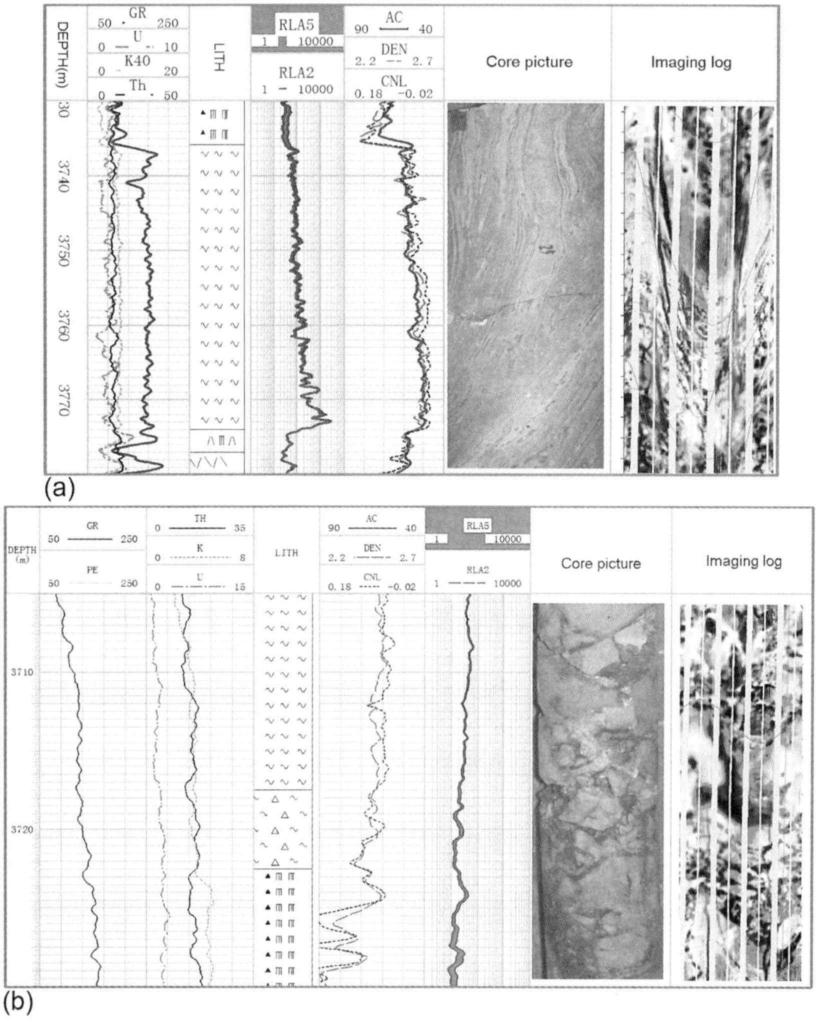

FIGURE 2.21

Characteristics of cores and logging curves in volcanic conduits. (a) Characteristics of cores and logging curves of volcanic neck subfacies (Well CC 1-1). (b) Characteristics of cores and logging curves of cryptoexplosive breccia subfacies (Well CC 103).

and the interfaces cannot be identified precisely without the acoustic-electric imaging log and the variation analysis of rock texture and architecture.

Interface of a rock occurrence variation. The bedding of sedimentary rocks and the rhyolitic architecture in the pericline are normally consistent with the strike of the pericline, and the variation of their trends reflects the change of pericline strike. It is difficult to identify through conventional logging. An acoustic-electric imaging log is often needed to detect the occurrence in geological structures.

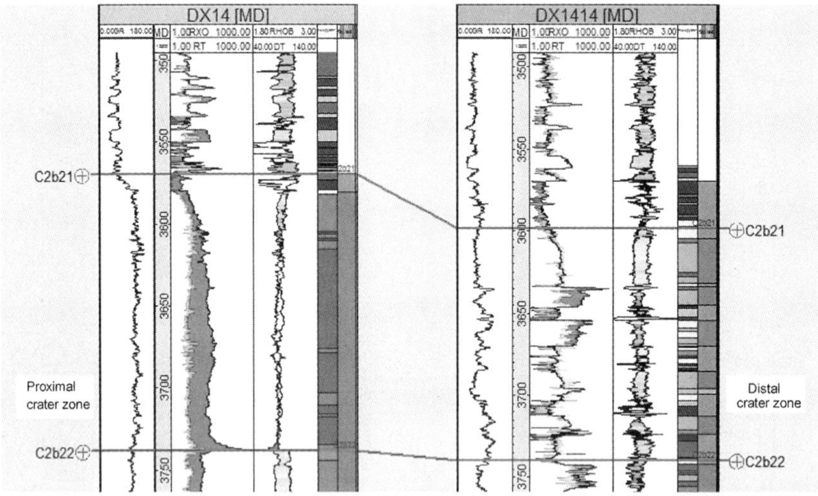

FIGURE 2.22

Logging response characteristics of proximal and distal crater strata (Well DD14 and DD1414).

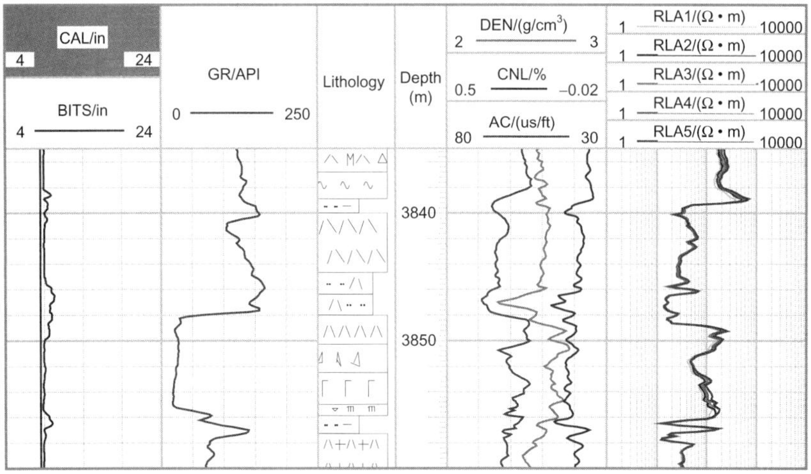

FIGURE 2.23

Abrupt interface in volcanic rock compositions of various volcanic edifices (Well CC2).

3) Seismic response characteristics of a volcanic edifice

Seismic response characteristics of a volcanic crater. The outer margin of a volcanic crater formed by centered eruption is marked by an uplifted rim due to in situ accumulation of volcaniclastic mass while the inner part is a depression due to collapse. In seismic profiles, therefore, the seismic reflection of the centered crater is primarily an arclike depression or graben-like sag that is locally laminar in shape, with a strong amplitude, high frequency, and a continuous pattern (Figure 2.24a). Horizontally, it appears nearly circular, with poor coherence or abrupt phase variation bands (Figure 2.24b). If the crater is completely filled with volcanic mass, the depression feature will disappear, becoming an inward inclined or flattened event.

Lava accumulation along both flanks and retraction of magma in center will result in an M-shaped reflection in the volcanic crater of fissured eruption, and extrusive perlite is developed on the top, with the central axis aligning with the crater (Figure 2.25). However, the characteristics of seismic response are not clear due to the narrow width of the fissure.

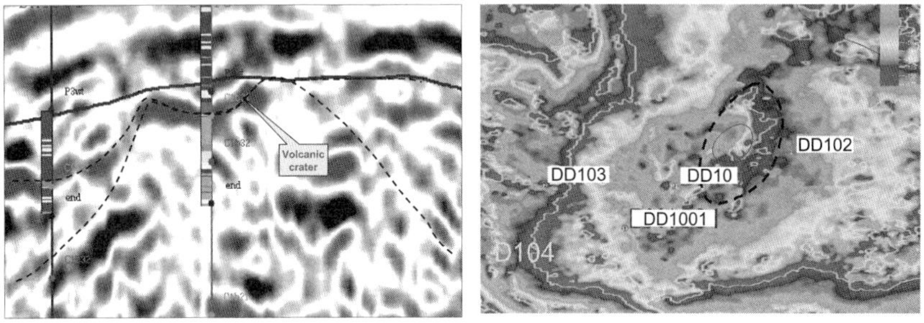

(a) (b)

FIGURE 2.24

Seismic response characteristics of a centered crater. (a) Characteristics of a seismic reflection profile. (b) Characteristics of the instantaneous phase.

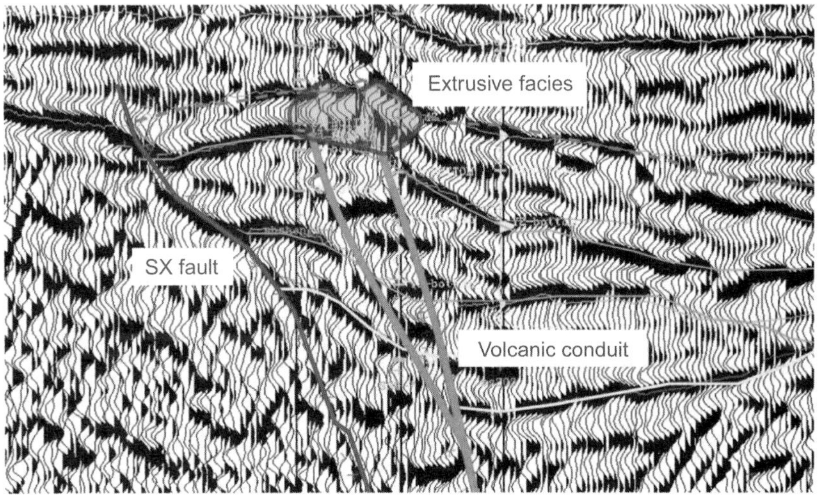

FIGURE 2.25

Characteristics of seismic response in a fissured volcanic crater and conduit.

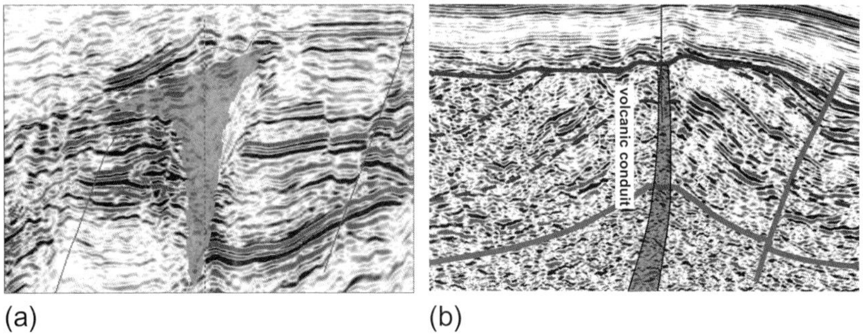

FIGURE 2.26

Seismic response characteristics of a volcanic conduit. (a) Umbrella shape. (b) Column and line shape.

Seismic response characteristics of a volcanic conduit. Due to its contrast with the surrounding strata, a volcanic conduit exhibits a nearly upright or divergent upward reflection pattern in a seismic profile, with its internal part featuring a moderate-weak amplitude, high frequency, and poor continuity. The seismic reflection patterns of volcanic conduits formed by centered eruption manifest mostly in the form of an upward diverging umbrella (Figure 2.26a) or nearly straight-up column (Figure 2.26b). A short conduit will be expressed as a funnel that is wide in the upper part and narrow in the lower part (Figure 2.26a). The volcanic conduits formed by centered eruption with a small diameter will appear as high-angle lines in the seismic profile, associated with a deep major fault.

Seismic response characteristics of periclines. The seismic reflection of centered volcanic edifices is mostly in the shape of mound, shield, and dome, with a radial symmetry and some variations in different portions, such as chaotic seismic reflection, upward convergence, moderate-weak amplitude, low frequency, and poor continuity in the proximal crater zone; quasi-laminar reflection, outward extension, medium-strong amplitude, medium-high frequency, and fairly good continuity in the distal crater zone; with gradual transition outward to parallel or subparallel architectures (Figure 2.27a, b, and c).

Most fissured volcanic edifices exhibit wedged and laminar reflections with strong event amplitude, low frequency, and good continuity; their symmetry is related to paleotopography, and asymmetrical wedges are well developed.

The contacts of a pericline with surrounding rocks are usually marked by angular unconformities, and seismic reflection of the external envelope is usually a strong discontinuity surface, locally continuous and traceable (see Figure 2.27).

2. Identification of a volcanic edifice in single wells

Within the boundary constraints of a volcanic formation, its rock composition, texture, architecture, and logging response characteristics can be analyzed, based on lithological and logging indicators, to identify a volcanic edifice in a single and to delineate its boundary.

1) Preliminary delineation of the top and bottom interfaces of a volcanic edifice based on lithological contacts

In volcanic rocks, the variation in Si, Fe, and Al contents reflects different magmatic sources, whereas Th, Ca, K, Na, and U contents indicate different diagenetic environments. Within the formation, two types of lithological interfaces are related to an edifice—that is, a contact between volcanic

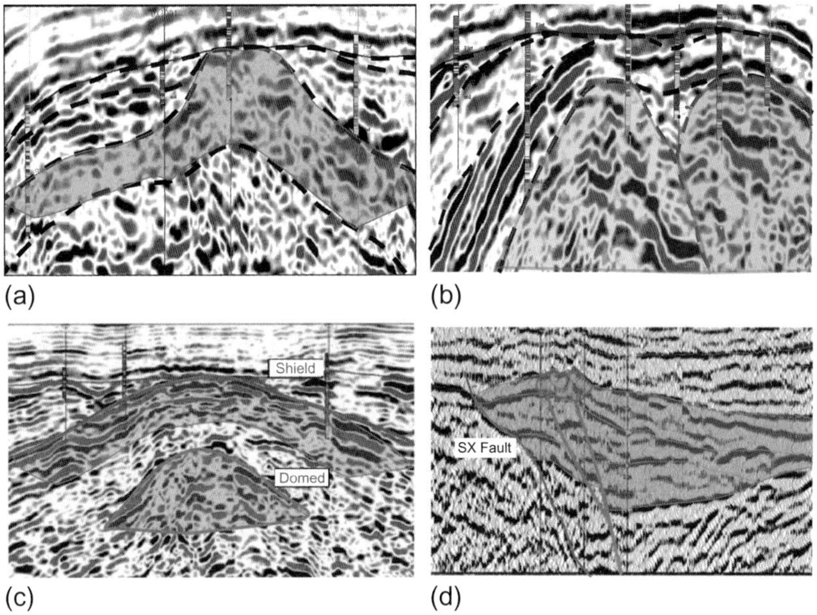

FIGURE 2.27

Seismic response characteristics of periclines. (a) Moundy reflection. (b) Composite moundy reflection. (c) Domed and shield reflection. (d) Wedged reflection.

sedimentary rock and volcanic rock—and interface marked by compositional change between in volcanic rock bodies. If the compositions of the volcanic rock units above and below an interface differ, the volcanic rocks belong to different edifices. If the compositions of the volcanic rocks above and below a volcanic sedimentary rock unit are the same, the rock composition, thickness, and spatial extent of the volcanic sedimentary rocks need to be investigated further to determine whether it can be regarded as an interface.

Figure 2.28 is an identification example of a volcanic edifice in Well CC1-1. Here, the top interface of volcanic formation lies at 3668 m, which is also the top surface of the volcanic edifice encountered in this well. The lithology in the interval of 3668 to 3714 m is rhyolitic ignimbrite, the acidity of which is consistent with that of the lower rhyolite. However, taking 3714 m as the interface, the gamma spectrometry log indicates that the K content in the upper and lower volcanic intervals differs by 50% and Th content differs by twice as much. The alkalinities and diagenetic environments are different up and down this interface, suggesting that these are two different volcanic edifices. From this profile identification (trace line on far right), the upper volcanic rock is identified as a distal crater zone of yc_2^5 volcanic edifice, and the lower volcanic rock is identified as a proximal crater zone of yc_2^3 volcanic edifice.

2) Diagnostic rock texture and rock occurrence as basis for identification of a volcanic edifice

Volcanic rock texture and architecture are two important indicators for identification of a volcanic edifice. In the proximal crater zone, cryptoexplosive breccia and upright rhyolitic architecture indicate a volcanic conduit. In the distal crater zone, the varying occurrence trend of the upper and lower rock layers is an interface indicator between two volcanic edifices.

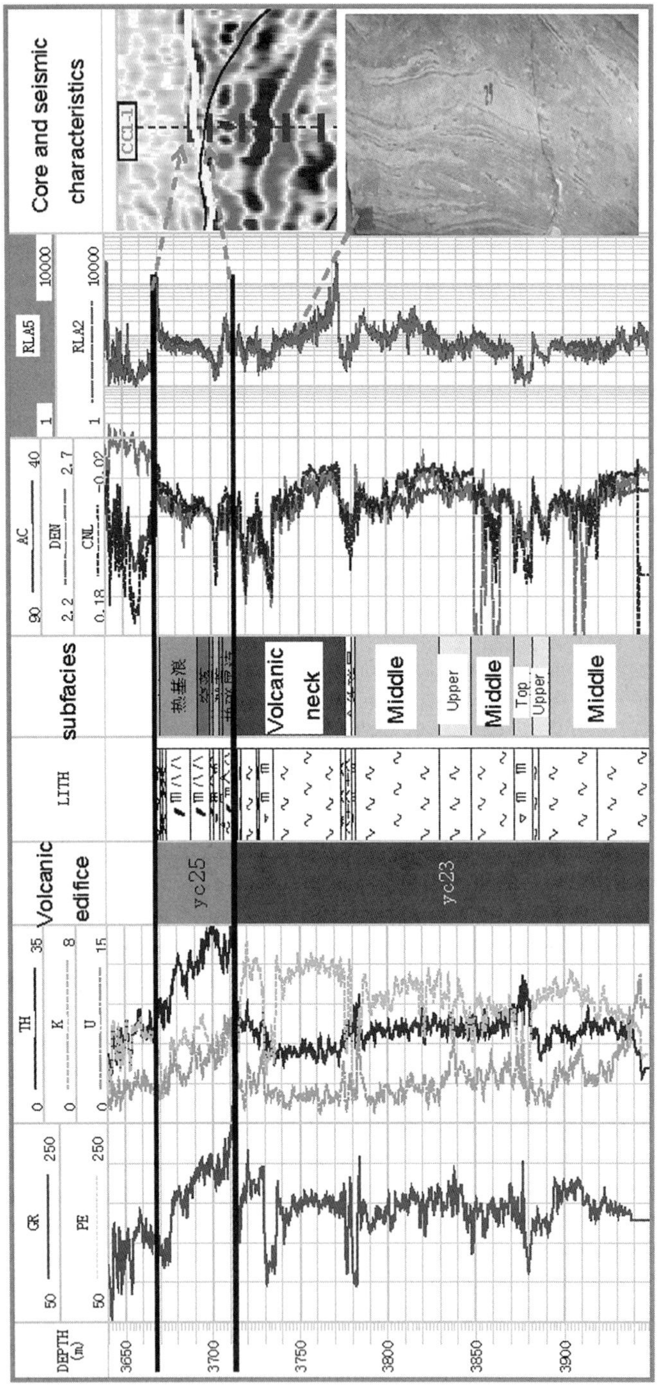

FIGURE 2.28

Well identification of a volcanic edifice (Well CC1-1).

In Well CC1-1, the proximal crater zone of yc_2^3 volcanic edifice was encountered, and the nearly upright rhyolitic architecture was developed in the cored interval 3727.5 to 3736.45 m of the lower rhyolite as well as in the Full-bore Micro Scanner Imager (FMI) imaging map (Figure 2.28, trace line on far right). It is thus determined tentatively that a volcanic conduit was encountered in this well.

3) Integration of various information using conventional logging to identify volcanic craters and conduits

Volcanic craters, conduits, and periclines can be identified based on well location and identification of rock composition and texture/architecture using conventional logging response patterns. As Figure 2.28 shows, a proximal crater zone in a pericline of yc_2^5 volcanic edifice was encountered in the interval of 3668 to 3714 m in Well CC1-1, the volcanic neck of yc_2^3 volcanic edifice was encountered in the interval of 3714 to 3777 m, and the distal crater zone was encountered below 3777 m.

3. Profile identification of a volcanic edifice

On the basis of single well identification and well-seismic integration, volcanic craters, conduits, and periclines are identified in the seismic response profiles according to the seismic facies pattern of the volcanic edifice by an analysis of seismic reflection characteristics.

1) Identification of volcanic craters and conduits and determination of volcanic edifice indicators

Volcanic craters and conduits have unique and significant seismic response characteristics that can be used to identify the volcanic edifice. Therefore, the identification of volcanic craters and conduits can be a key step for improving the success rate of volcanic edifice identification.

Figure 2.29 shows an example for identification of the volcanic edifice in the Well Block XX 23. Rhyolite, volcanic breccia, ignimbrite, and tuffaceous sandstone are present in Well XX23, and they are interbedded with each other vertically. Their logging curves exhibit microserrated–serrate box and serrated-digitate shapes, which are typical of a volcanic crater. In seismic profiles, local highs in the Well Block XX 23 are demonstrated in this well, and the near-wellbore seismic responses show an upward-radiated columnar reflection with a weak amplitude, high frequency, and discontinuous event, which are characteristic responses of a volcanic conduit. This points to the presence of a volcanic conduit in the Well Block XX 23, which serves as an indicator for identifying a volcanic edifice.

2) Identification of periclines, tracking of structural boundaries, and profile recognition of a volcanic edifice

With the "source-control concept" of a volcanic edifice as a guide and the volcanic crater and conduit as central reference points, periclines can be identified and their morphology delineated in seismic reflection profiles based on the seismic facies patterns of various periclines. Furthermore, the structural boundaries are tracked based on the discontinuity interfaces between periclines and the surrounding rocks or other volcanic edifices to reveal the profile of a volcanic edifice. In Figure 2.29, taking the columnar conduit as a reference center point, volcanic edifice XX23 is revealed in overall morphology as an umbrella-like reflection in seismic profiles.

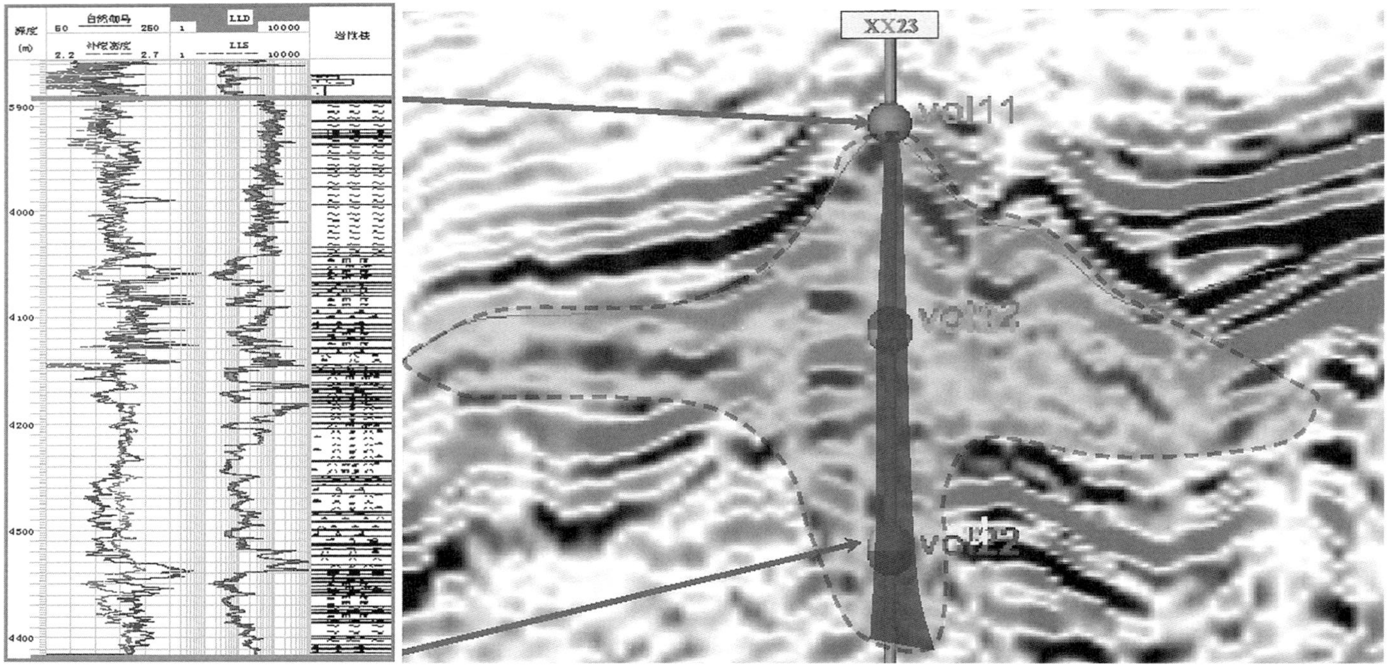

FIGURE 2.29

Identification of a volcanic edifice through well-seismic integration.

4. Planar prediction of a volcanic edifice

There are two methods used primarily for planar prediction of a volcanic edifice:

1) Method 1: Seismic attribute analysis

A volcanic conduit has an upward-radiated reflection shape, whereas a pericline features increasing amplitude, higher frequency, and better continuity with increased distance from the center. As a whole, a volcanic edifice has the attributes of poor coherence, weak amplitude, and high dip angle (Figure 2.30). Therefore, the 2D distribution of a volcanic edifice can be predicted by means of

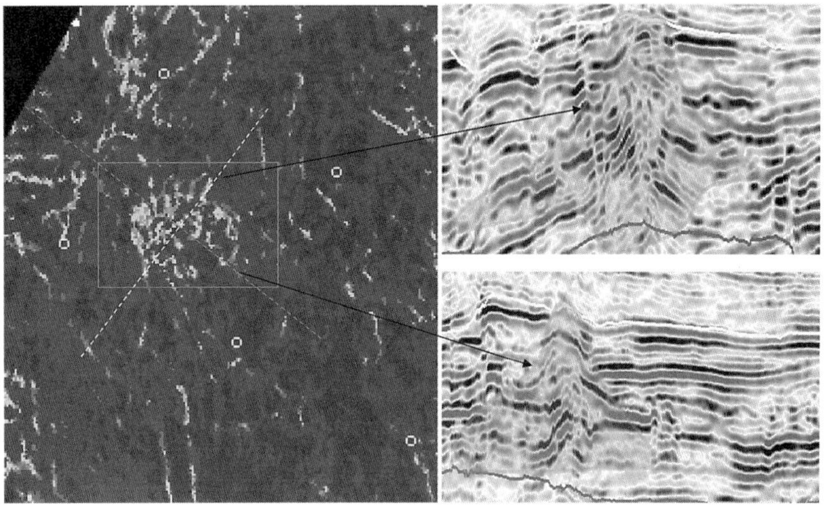

FIGURE 2.30

Dip angle attribute characteristic of a volcanic edifice.

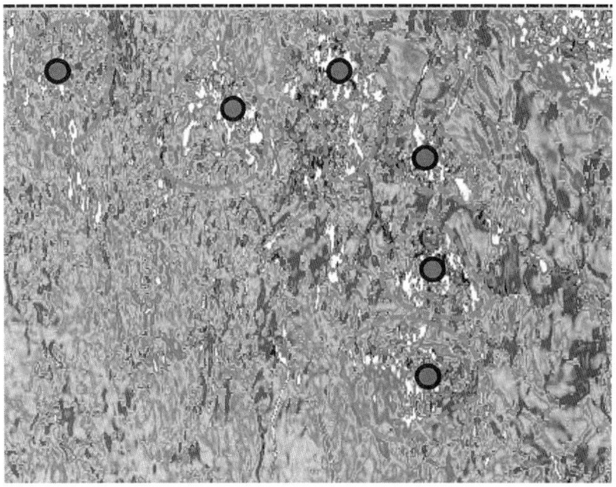

FIGURE 2.31

Prediction of the planar distribution of a volcanic edifice by dip angle attribute (XX gas field).

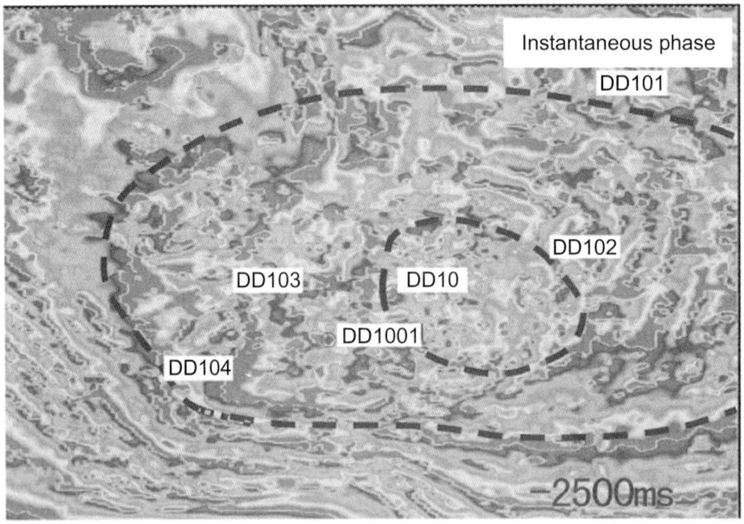

FIGURE 2.32

Prediction of the planar distribution of a volcanic edifice by instantaneous phase slice (DD gas field).

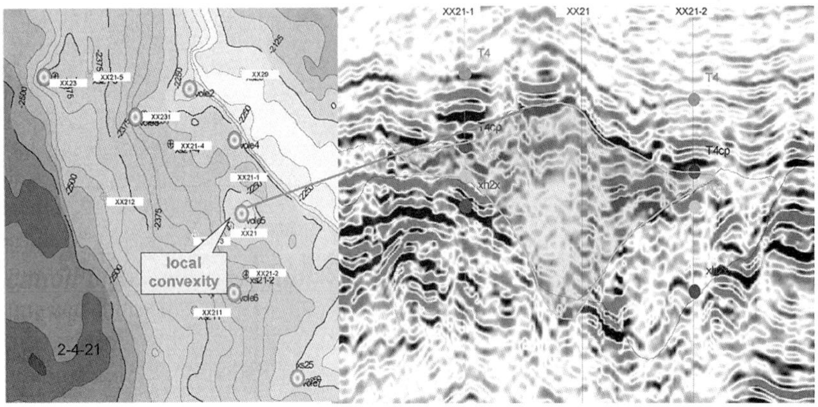

FIGURE 2.33

Microarchitecture response characteristics of a volcanic edifice top.

horizon-oriented extraction of sensitive seismic attributes (Figure 2.31) or extraction of time series of seismic attributes (Figure 2.32).

2) Method 2: Microarchitecture analysis

A volcanic edifice has an uplift feature against a local architecture background, displaying a positive microarchitecture. On the basis of this feature, together with interpretation of the top architectural surface, the area with a positive microarchitecture is analyzed to achieve a preliminary determination of the planar position of a volcanic edifice or crater.

Figure 2.33 is an architectural topographic map of volcanic rocks in the Well Block XX 21, in which the locally uplifted positive microarchitectures correspond quite well to the volcanic craters. According to the architectural configuration, seven local highs are delineated, including three minor anticlines and two minor nose architectures with distinct features that indicate the crater positions; and one minor anticline and one minor noselike architecture with clear features are the potential crater's positions.

5. Spatial prediction of a volcanic edifice

On the basis of single well and profile identifications and planar prediction, two methods are used for the integral prediction of spatial distribution of a volcanic edifice.

1) Method 1: Interface tracking and closing technique based on identification of a skeleton profile

Following the successful identification of a volcanic edifice in single wells and profiles and the establishment of a skeleton profile network, the interface of a volcanic edifice is tracked and closed to realize a 3D spatial identification of the volcanic edifice. Figure 2.34 is a 3D spatial distribution map of the volcanic edifice in the Well XX 21 block and the results of an analysis that lays the foundation for characterization of the shape, size, and superposition relationships of the volcanic edifice.

2) Method 2: Spatial association technology based on time series of seismic data

A volcanic edifice is an assemblage of various integral parts around the center of a volcanic conduit, which is continuous in time and distributes regularly in space [19]. However, a time slice reflects the instantaneous characteristic of seismic reflection. Therefore, on the basis of indicators for seismic facies identification of a volcanic edifice, the time series of seismic data is extracted continuously at a given time interval, and the spatial distribution of the volcanic edifice can be effectively predicted by slice identification and spatial association.

Figure 2.35 shows the seismic time series at 2300 ms (top), 2330 ms, 2360 ms and 2390 ms (base) in the Well Block DD 10. The figure shows that (1) all volcanic edifices have a relatively regular subelliptical shape; (2) downward from the top, the area of a volcanic edifice expands gradually toward the northeast; and (3) the volcanic rock in the Well Block DD 10 is a monoconic volcanic edifice prolonged extended to the northeast.

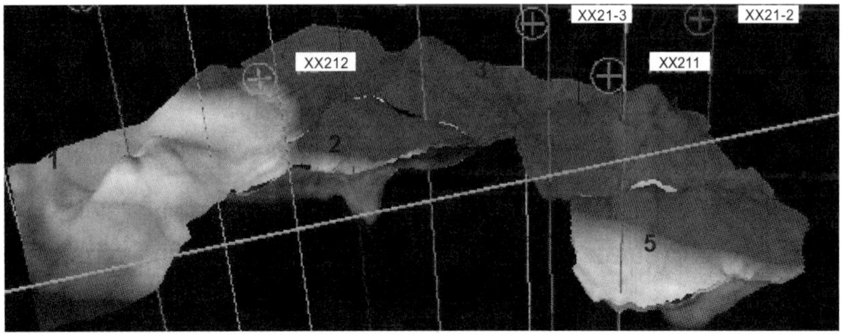

FIGURE 2.34

Spatial prediction of the volcanic edifice (XX gas field).

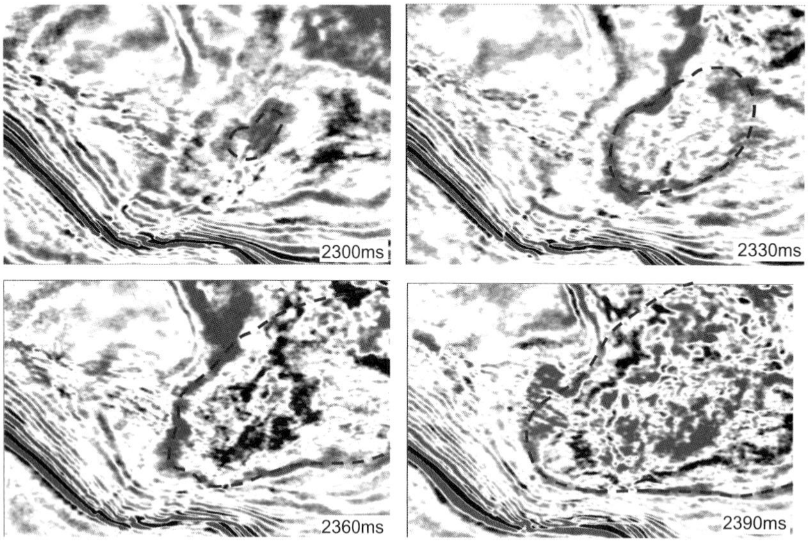

FIGURE 2.35

Prediction of a volcanic edifice spatial distribution by the spatial association technique of time slices (DD gas field).

2.4.3 Characterization of volcanic edifices

On the basis of volcanic edifice identification, using outcrop data and seismic information, the size, shape, and superposition relationships of a volcanic edifice are further characterized to reveal the distribution characteristics and patterns.

1. Shape characterization of volcanic edifices

Field outcrop study is an important step for intuitive and precise delineation of a volcanic edifice exposed on the surface. Seismic data explore the deep subsurface and can be interpreted by various methods; therefore, a seismic survey is the primary method for investigating subsurface volcanic edifices. In this book, 15 outcrops of the Yingcheng formation volcanic edifices in the Guanmashan and Fangniugou areas of Jiutai, Jilin Province, and several subsurface volcanic edifices in five well blocks of XX gas field were investigated comprehensively and the morphological features of volcanic edifices are revealed (Table 2.3).

1) Monoconic volcanic edifices

Monocones are the predominant morphology of centered volcanic edifices. About 66.6% of the volcanic edifices in outcrops are monoconic, and the typical example is the Guanmashan volcanic edifice. Approximately 50% of the subsurface volcanic edifices are monoconic, such as the XX 12 volcanic edifice. This type of volcanic edifice has a single crater, with a symmetrical shape, inwardly inclined rock layers, and a decreasing slope downward (see Table 2.3, Figures 2.13 and 2.27a). For the edifices in outcrops, the cone has slope angles ranging from 8.8 to 33.7 degrees, and for the subsurface edifices identified from seismic data, the angles are 8.4 to 29.7 degrees.

Table 2.3 Morphological Parameters of a Volcanic Edifice

Shape	Edifice Type	Morphological Feature	Slope (degrees)		Percentage (%)	
			Outcrop Survey	Seismic Identification	Outcrop Survey	Seismic Identification
Monoconic	Centered	Monocrater with symmetric shape, rock inclined inward and slope gets gentle downward	8.8-33.7	8.4-29.7	66.6	50
Multiconic	Centered, combined	The ejectas from different craters overlapped with poor symmetry and complex morphology	10.7-30	6.4-22.6	6.7	8.3
Shield	Centered, fissured	Monoconic crater with broad top and gentle flanks, small thickness but large distribution diameter	5.3-21.8	1.5-10	13.3	25
Domed	Centered	No crater at top, in shape of dune with steep slope or short and steep lava flow	26-80	24-70	6.7	8.3
Laminar	Fissured	Crater not obvious, broad top and gentle flanks, rock inclined outward, flat and cambering	4.3-18	3.6-9.7	6.7	8.3

2) Multiconic volcanic edifices

Multicones are significant features for centered and combined volcanic edifices. About 6.7% of the outcrop volcanic edifices are multiconic, which is typified by the Jianshanzi volcanic edifice; about 8.3% of the subsurface volcanic edifices are multiconic, with the typical example being the XX20 volcanic edifice. This type of volcanic edifice is usually formed by mutual superposition of ejecta from multiple craters, with poor symmetry and complex shape (see Table 2.3, Figures 2.14 and 2.27b). The cone slope angles range from 10.7 to 30 degrees in outcrop edifices and 6.4 to 22.6 degrees in subsurface edifices based on seismic data.

3) Shield volcanic edifices

Shield shape is a major shape of the centered and fissured volcanic edifices. This category comprises about 13.3% of the outcrop volcanic edifices (e.g., the Bakeshu volcanic edifice) and about 25% of the subsurface ones (e.g., the XXS3 and S4 volcanic edifices). A shield volcanic edifice has only one crater, with a broad top and gentle flanks, small thickness, a wide distribution diameter, and good symmetry (see Table 2.3, Figs. 2.15 and 2.27c). In outcrop shield edifices, the cone has slope angles ranging from 5.3 to 21.8 degrees, and those in the subsurface range from 1.5 to 10 degrees.

4) Domed volcanic edifices

Domed shapes are common for centered volcanic edifices. About 6.7% of the outcrop volcanic edifices are domed, as characterized by the Santai volcanic edifice; about 8.3% of the subsurface volcanic edifices are domed, such as the XX5 volcanic edifice. This type of volcanic edifice has no crater on the top and has a steep slope or short and steep lava flow (see Table 2.3, Figs. 2.15 and 2.27c). The cone slope angles range from 26 to 80 degrees in outcrop edifices and 24 to 70 degrees for those in the subsurface identified from seismic data.

5) Laminar volcanic edifices

The laminar shape is typical of fissured volcanic edifices, which comprise about 6.7% of the outcrop volcanic edifices (such as the Kaoshantun volcanic edifice) and about 8.3% of the subsurface ones (XX20 volcanic edifice). This type of volcanic edifice has no obvious crater on top, a broad top and gentle flanks, and rock layers inclined outward (see Table 2.3, Figs. 2.15 and 2.27c) and gently curved. Cone slope angles are 4.3 to 18 degrees for outcrop edifices and 3.6 to 7.7 degrees for subsurface ones.

2. Size characterization of volcanic edifices

1) Size of outcrop volcanic edifices

1. *Monoconic volcanic edifices.* These edifices are mostly moderate to small in size and have a long axis (2.8 to 8.9 km long, averaging 4.8 km) or a short axis (1.1 to 3.1 km, averaging 2.1 km). The height is 91 to 203 m (averaging 144 m) and the area is 6 to 13 km^2 (averaging 8.7 km^2) (Figure 2.36 and Table 2.4).
2. *Multiconic volcanic edifices.* This type of edifice has a greater size than the monoconic ones. In Jianshanzi, the outcrop multiconic volcanic edifice has a long axis of 7.5 km, a short axis as 5 km, an exposed height of 257 m (two to three times that of the monoconic volcanic edifice), and an area of 37.1 km^2 (over four times that of the monoconic edifice) (see Figure 2.36 and Table 2.4).
3. *Shield volcanic edifices.* Their long axis ranges from 3.3 to 8.8 km (with an average of 6.2 km) and their short axis ranges from 2.1 to 4.5 km (with an average of 3.6 km); the height ranges from 73 to 120 m (with an average of 90 m) and the areal extent is 8 to 20 km^2 (with an average of 18 km^2). Shield volcanic edifices are composed mainly of acidic volcanic rocks, and their sizes are relatively small compared to those of neutral-basic volcanic edifices. Compared to conic volcanic edifices, however, shield volcanic edifices have the characteristics of a large distribution area and small thickness (see Figure 2.36 and Table 2.4).
4. *Domed volcanic edifices.* These edifices have a small distribution area with great thickness. In Santai, an outcrop domelike volcanic edifice has a long axis of 3 km, a short axis of 1.8 km, a height of 134 m, and an area of 6 km^2, with an overall size that is less than that of a monoconic volcanic edifice.
5. *Laminar volcanic edifices.* The characteristics of this type of edifice are similar to those of the shield type but differ only slightly in the ratio of long axis to short axis. In Kaoshantun, the outcrop laminar volcanic edifice has a long axis of 8 km, a short axis 1.6 km, a height of 80 m, and covers an area of 13 km^2.

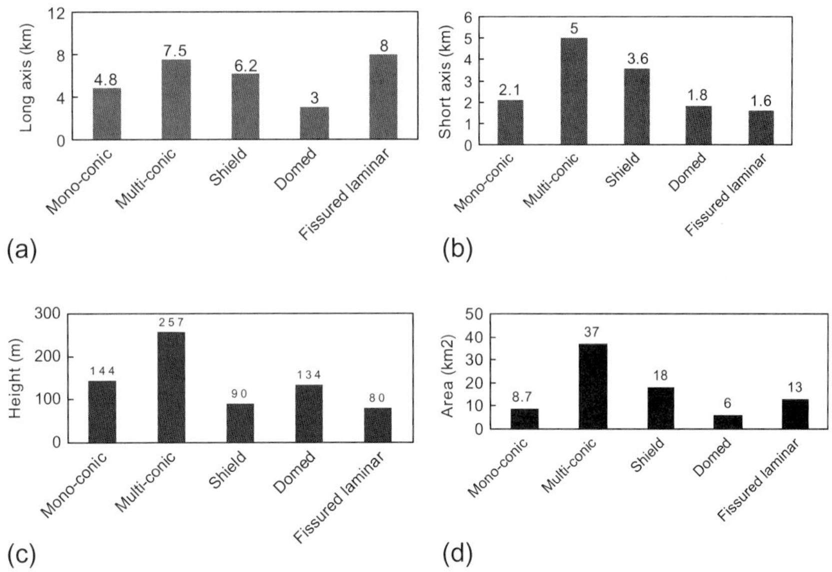

FIGURE 2.36

Parameters of outcropped palaeovolcanic edifice size. (a) Long axis. (b) Short axis. (c) Height. (d) Area.

Table 2.4 Parameters of Outcropped Palaeo-Volcanic Edifice Size

Shapes of Outcropped Palaeovolcanic Edifices	Statistic Quantity	Long Axis, km		Short Axis, km		Height, m		Area, km^2	
		Range	Average	Range	Average	Range	Average	Range	Average
Monoconic	9	2.8-8	4.8	1.1-3.1	2.1	91-203	144	6-13	8.7
Multiconic	1		7.5		5		257		37
Shield	3	3.3-8.8	6.2	2.1-4.5	3.6	73-120	90	8-20	18
Domed	1		3		1.8		134		6
Laminar	1		8		1.6		80		13

2) Size of subsurface volcanic edifices

Subsurface volcanic edifices are identified and their sizes are characterized using seismic data:

1. *Monoconic volcanic edifices.* Their long axis ranges from 6 to 10.6 km (with an average of 8.5 km); their short axis ranges from 3.9 to 5.1 km (average, 4.5 km); height ranges from 300 to 550 m (average, 420 m); and area ranges from 23.3 to 49.2 km^2 (average 34.4 km^2) (Figure 2.37 and Table 2.5). Their sizes are notably greater than those of the outcrop monoconic volcanic edifices, which may be related to outcrop completeness.
2. *Multiconic volcanic edifices.* Their long axis is 14 km, short axis is 6 km, height is 450 m, and area is 68.9 km^2. Their sizes are twice as large as those of the monoconic volcanic edifices (see Figure 2.37 and Table 2.5).

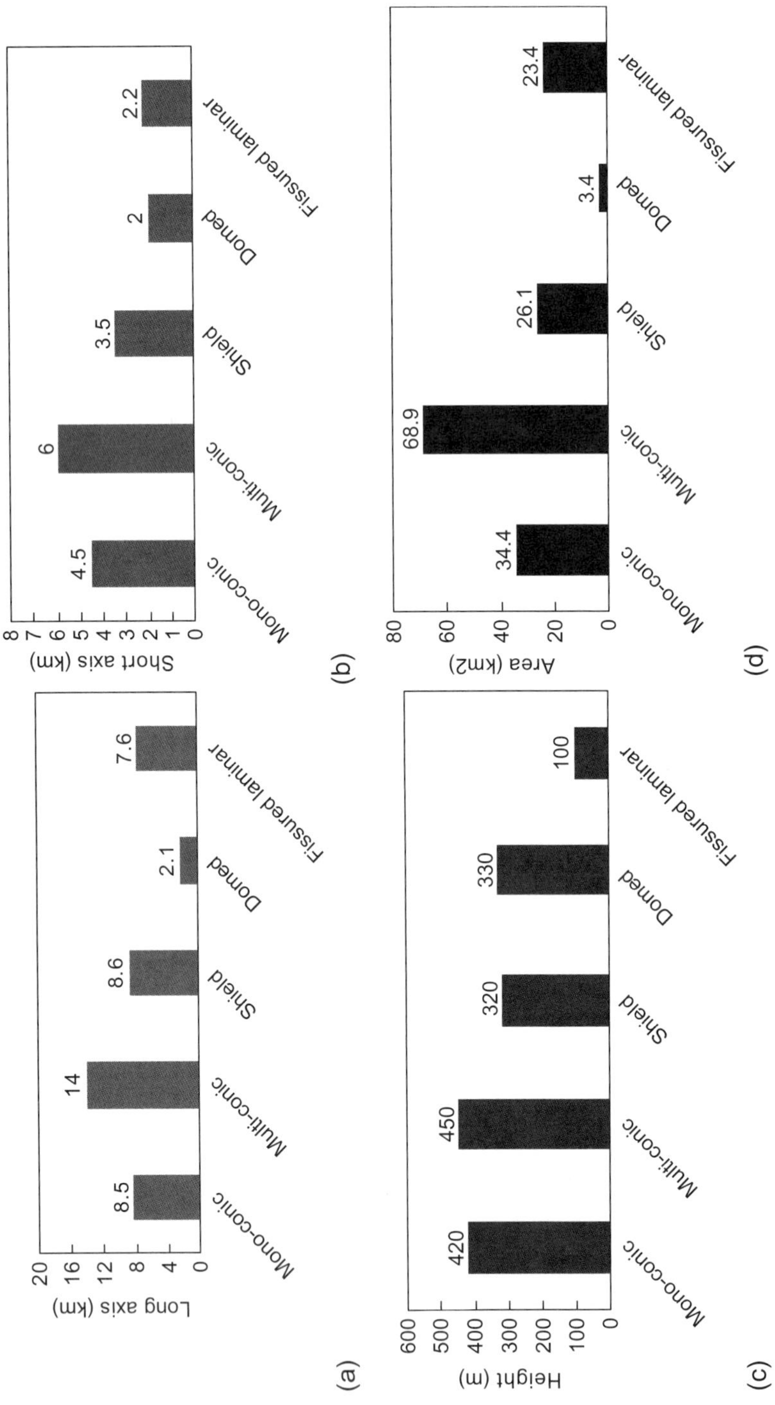

FIGURE 2.37

Size parameters of outcrop palaeovolcanic edifices. (a) Long axis. (b) Short axis. (c) Height. (d) Area.

Table 2.5 Size Parameters of Subsurface Palaeovolcanic Edifices

Shape of Subsurface Palaeovolcanic Edifices	Statistic Numbers	Long Axis, km		Short Axis, km		Height, m		Area, km^2	
		Range	Average	Range	Average	Range	Average	Range	Average
Monoconic	6	6-10.6	8.5	3.9-5.1	4.5	300-550	420	23.3-49.2	34.4
Multiconic	1		14		6		450		68.9
Shield	3	6.6-12.6	8.6	2.2-4.2	3.5	200-400	320	13.7-43.4	26.1
Domed	1		2.1		2		330		3.4
Laminar	1		7.6		2.2		100		23.4

3. *Shield volcanic edifices.* Their long axis has a range of 6.6 to 12.6 km (with an average of 8.6 km); their short axis ranges from 2.2 to 4.2 km (average, 3.5 km); height ranges from 200 to 400 m (average, 320 m); and area ranges from 13.7 to 43.4 km^2 (average, 26.1 km^2) (see Figure 2.37 and Table 2.5).
4. *Domed volcanic edifices.* Their long axis is 2.1 km long, short axis is 2 km, height is 330 m, and area is 3.4 km^2. These parameters are smaller than those of monoconic edifices (see Figure 2.37 and Table 2.5).
5. *Laminar volcanic edifices.* Their long axis averages 7.6 km, short axis averages 2.2 km, height averages 100 m, and an area coverage averages 23.4 km^2 (see Figure 2.37 and Table 2.5).

3) Characteristics of volcanic edifice size

In summary, acidic volcanic edifices have the following basic features:

1. The sizes of subsurface volcanic edifices are generally larger than those of volcanic edifices in outcrops, a characteristic that is related to outcrop completeness.
2. The multiconic volcanic edifice has the largest size, whereas the domed one has the smallest size.
3. Both shield and laminar volcanic edifices have a large areal coverage and small height, in contrast to domed volcanic edifices; conic edifices are intermediate in size between the other two types.
4. Compared with shield volcanic edifices, laminar edifices have a higher ratio of long axis to short axis.

3. Characterization of superposition relationships of volcanic edifices

The superposition relationships (cross-cutting or stacking patterns) of volcanic edifices can be classified into three categories based on the timing of their genesis and spatial arrangements.

1) Vertical superposition or stacking

Vertical superposition refers to the stacking relationship of an epigenetic volcanic edifice over an eogenetic volcanic edifice (Figure 2.38), which can result in the following cases:

1. After a long hiatus following the formation and burial of a centered or fissured volcanic edifice, an identical crater erupts again and a new volcanic edifice forms without damaging the eogenetic edifice.

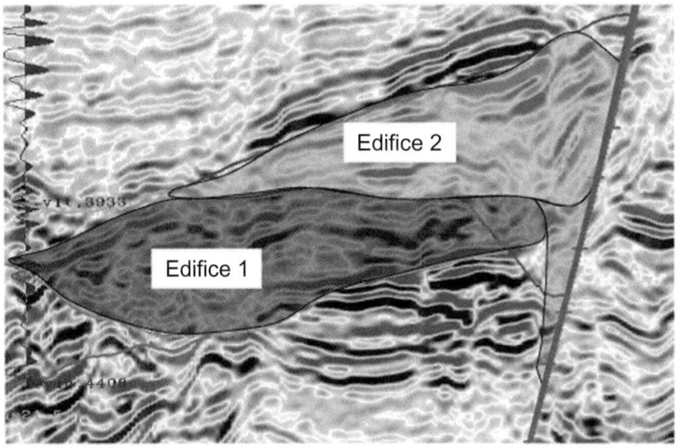

FIGURE 2.38

Vertical superposition.

2. For two adjacent craters with different heights, the lower crater erupts first and a volcanic edifice forms; this is followed by the eruption and formation of a higher edifice. As a result, the epigenetic edifice superposes partially over the eogenetic volcanic edifice.

The vertical superposition relationship is simple to identify and delineate, as is represented by the main superposition relationships in the example from the Songliao basin.

2) Lateral stacking

When two adjacent volcanic craters with slightly different heights erupt sequentially, the resultant volcanic edifices are normally in a relationship of lateral superposition or stacking, with the epigenetic one superposed over the eogenetic one (Figure 2.39).

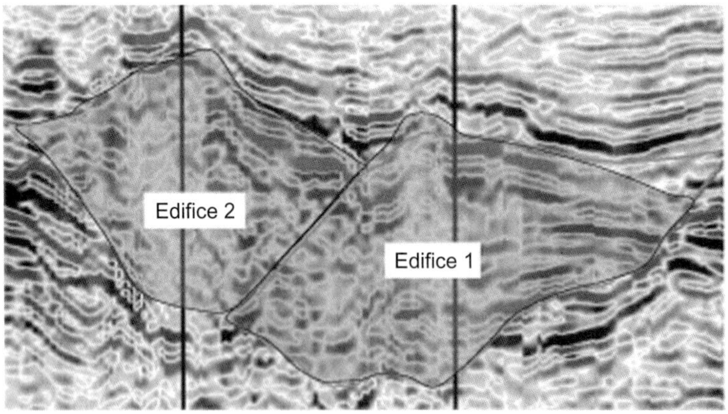

FIGURE 2.39

Lateral superposition.

3) Cross superposition

When two or more adjacent volcanic craters erupt simultaneously, a cross superposition relationship may be found in multiple volcanic edifices. This type of superposition relationship is complex and thus difficult to identify and delineate, particularly when two volcanic edifices have similar rock compositions.

2.5 IDENTIFICATION AND CHARACTERIZATION OF A VOLCANIC MASSIF

A volcanic massif is a basic geological unit controlling the accumulation-permeation unit and gas/water systems, which is of significance for determining the heterogeneity within a gas reservoir and predicting effective reservoirs. On the basis of genetic classification, volcanic massifs are identified in single wells, profiles, and planes by integration of geological, logging, and seismic data to reveal the spatial distribution of volcanic massifs and to characterize the shape, size, and superposition relationships of various volcanic massifs.

2.5.1 Classification of volcanic massifs

Volcanic massifs are classified into four categories based on the nature of their genesis and geological features (Figure 2.40):

Category 1: Lithological separation during dormancy period. Within a volcanic edifice formed from multistage eruptions, volcanic rock denudation or epigenetic clastic deposition often occurs during volcanic dormancy, leading to volcanic rock bodies separated by erosional surface or volcanic sedimentary rocks to form multiple volcanic massifs (Figure 2.40a).

Category 2: Alternation of strong-weak eruption energy. In the process of extensive and continuous eruption, the eruption energy may alternate from strong to weak several times, forming alternating surfaces of strong-weak or weak–strong energy changes, which separate the volcanic edifice into multiple volcanic massifs (Figure 2.40b).

Category 3: Isolated massif. An isolated massif refers to a shallow and ultra-shallow intrusion with identical volcanic rock in area, stage, and source. The volcanic rocks, subvolcanic rocks, and intrusive rocks formed in the same volcanic activity are usually under the control of the same volcanic edifice, which are the parts of this volcanic edifice (Figure 2.40c).

Category 4: Fault barrier. An integral volcanic edifice or massif is separated by faults into many different volcanic massifs as a result of tectonic destruction (Figure 2.40d).

2.5.2 Identification of volcanic massifs

Based on the classification scheme described, volcanic massifs are identified in single wells, profiles, 2D planes, and 3D space by establishing indicators for their identification to reveal their distribution characteristics and to form the basis for their quantitative characterization.

1. Identification indicators

The geological, logging, and seismic indicators for identifying internal lithological assemblages and external boundaries of volcanic massif are established based on their geological characteristics.

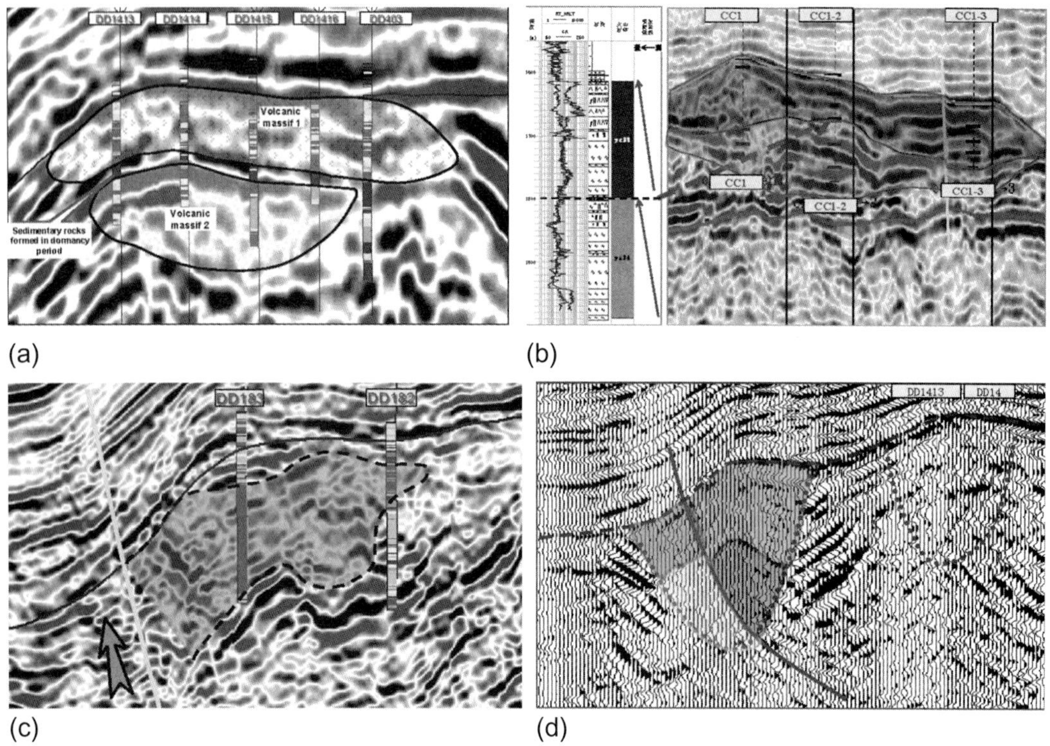

FIGURE 2.40

Genetic categories of volcanic massifs. (a) Category 1: Lithological separation during dormancy period. (b) Category 2: Alternation of strong-weak eruption energy. (c) Category 3: Isolated massif. (d) Category 4: Fault barrier.

1) Indicators for lithological assemblage

Lithological separation formed during dormancy period

- Internal lithological assemblage: A suite of volcanic rocks formed in an eruption stage. Massif boundary: Weathered crust or sedimentary rocks.

Separation of facies sequence alternation

- Internal lithological assemblage: For lithological assemblages with strong to weak alternation of eruption energy, the indicators (from bottom to top) include volcanic conglomerate, volcanic breccia, tuff, and lava, and vice versa for weak to strong alternation.
- Massif boundary: Boundaries marking energy alternations act as indicators, including strong-weak assemblage → strong-weak assemblage; strong-weak assemblage → weak-strong assemblage; weak-strong assemblage → strong-weak assemblage; and weak-strong assemblage → weak-strong assemblage.

Independent massif

- Internal lithological assemblage: Subvolcanic rocks or extrusive rocks.

- Massif boundary: Subvolcanic rocks (extrusive rocks), sedimentary rocks; eruptive rocks, subvolcanic rocks (extrusive rocks).

Fault barrier

- Internal lithological assemblage: Various volcanic rocks or subvolcanic rocks.
- Massif boundary: Fault and fractured zone.

2) Logging indicators

As mentioned earlier, there are four categories of lithological assemblages of volcanic massif/ ***Volcanic rocks (or subvolcanic rocks)→ Weathered Crust or Volcanic Sedimentary Rocks.*** The weathered crust or sedimentary rocks, generated during dormancy, divide the volcanic rock into two volcanic massifs. In Well DD1415 (e.g., see Figure 2.41a), the upper volcanic massif shows high

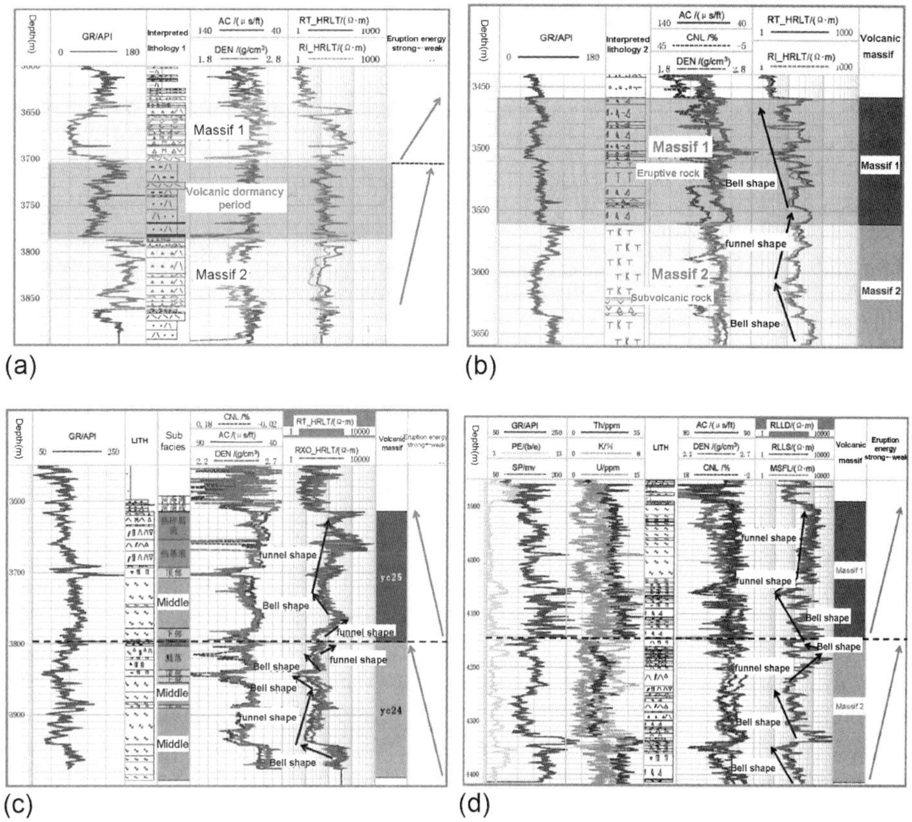

FIGURE 2.41

Logging response characteristics of volcanic massifs. (a) Volcanic rock to volcanic sedimentary rock (Well DD1415). (b) Eruptive rock to subvolcanic rock (Well DD182). (c) Weak/strong assemblage to weak/strong assemblage conversion (Well CC1-2). (d) Strong/weak assemblage to strong/weak assemblage conversion (Well XX23).

resistivity, high density, and low interval transit time on logging curves, whereas the lower volcanic massif shows low resistivity, low density, and high interval transit time. The curve pattern changes from microserrated bell to serrate-digitate or bell shapes from bottom to top, indicating the commencement of a new eruption stage.

Eruptive rocks→subvolcanic rocks (extrusive rocks). On logging curves, they exhibit some minor changes in radiography, resistivity, density, and interval transit time. For example, in Well DD182, an increase in gamma value and a slight increase in resistivity and density values can be observed (Figure 2.41b). The specific changes are associated with eruptive rock types, a diagenetic environment, and the physical properties of subvolcanic rocks.

Eruption energy alternation. The volcanic eruption energy is positively correlated to the degrees of volcaniclastic rock development and the grain size of volcanic clasts, with an alternation surface showing a bell or funnel shape on logging curves. For example, the alternation between weak-strong assemblage and weak-strong assemblage in Well CC1-2 exhibits an interface between two funnel-shaped curves (Figure 2.41c); the alternation between strong-weak assemblage and strong-weak assemblage in Well XX23 exhibits an interface in two bell-shaped curves (Figure 2.41d). The stronger the explosion energy, the more severe the dentoid shape of the curve.

Fault barrier. On logging curves, a fault will manifest as formation thinning (even absent) or thickening (even repeated). In continuously developed sedimentary layers, whether a faulting point is encountered in a well can be determined through the correlation with adjacent wells. However, because volcanic layers change rapidly, it is difficult to determine the faulting point in wells through thickness variation. Yet if a fault or fractured zone is encountered by wells, the intervals with low resistivity, low density, and high interval transit time can be observed on conventional logging curves, which are characterized normally by a serrated shape. In electric imaging, an autoclastic breccia texture can be found.

3) Seismic indicators

Internal reflection of massifs. In terms of seismic reflections, a volcanic massif of effusive facies has a laminar feature of weak amplitude, low frequency, and fairly good continuity (Figure 2.41b); the reflections of explosive facies have a chaotic feature of weak amplitude, low frequency, and discontinuity (Figure 2.41a), and extrusive and subvolcanic facies have a chaotic feature of medium-weak amplitude, medium frequency, and poor continuity (Figure 2.41c).

Boundary reflection characteristics. In seismic profiles, the interface of volcanic rocks (or subvolcanic rocks)→weathered crust (volcanic sedimentary rocks) shows a strong reflection of discontinuity surface with continuous events and good traceability (Figure 2.41a).

The interface of extrusive rocks→subvolcanic rocks and the interface of strong-weak energy alternation in extrusive rocks exhibit a medium-weak reflection of discontinuity surface with discontinuous events and poor traceability (Figure 2.41b).

In seismic profiles, fault is represented by event termination, strain-slip break, twist, merge, split, quantity change, and polarity reversal (Fig 2.41d).

2. Well identification

Constrained to the top and bottom interfaces of volcanic edifices, volcanic massifs can be identified and the interfaces delineated by integrating coring, mud logging, and various logging data, as well as using lithological and logging indicators as guiding tools.

1) Identifying weathered crust and volcanic sedimentary rocks to determine intermittent stages of eruption

From lithological identification and logging response analysis, weathered crust and volcanic sedimentary rocks are identified within a volcanic edifice to determine the intermittent stages of eruption, providing a basis for massif identification.

Figure 2.42 is a composite logging map of Well DD103 in the DD gas field. In this well, syenite porphyry was encountered for the first time at 3034 m; in a succession downward, two suites of volcanic sedimentary rocks were penetrated in the interval from 3034 m to 3200 m, including a suite of tuffaceous conglomerate in 3055 to 3080 m and a suite of sedimentary volcanic breccia in 3098 to 3131 m, with a resistivity value as low as 10 Ω•m, a transit time as high as 80 μs/ft, a compensated neutron value as high as 26%, and a natural gamma value as low as 45 to 60 American Petroleum Institute unit (API) on logging curves. The tuffaceous sandy conglomerate interval was cored, and these two intervals have proven to be typical volcanic sedimentary rocks. Thus, the subvolcanic rocks are divided into three subvolcanic massifs.

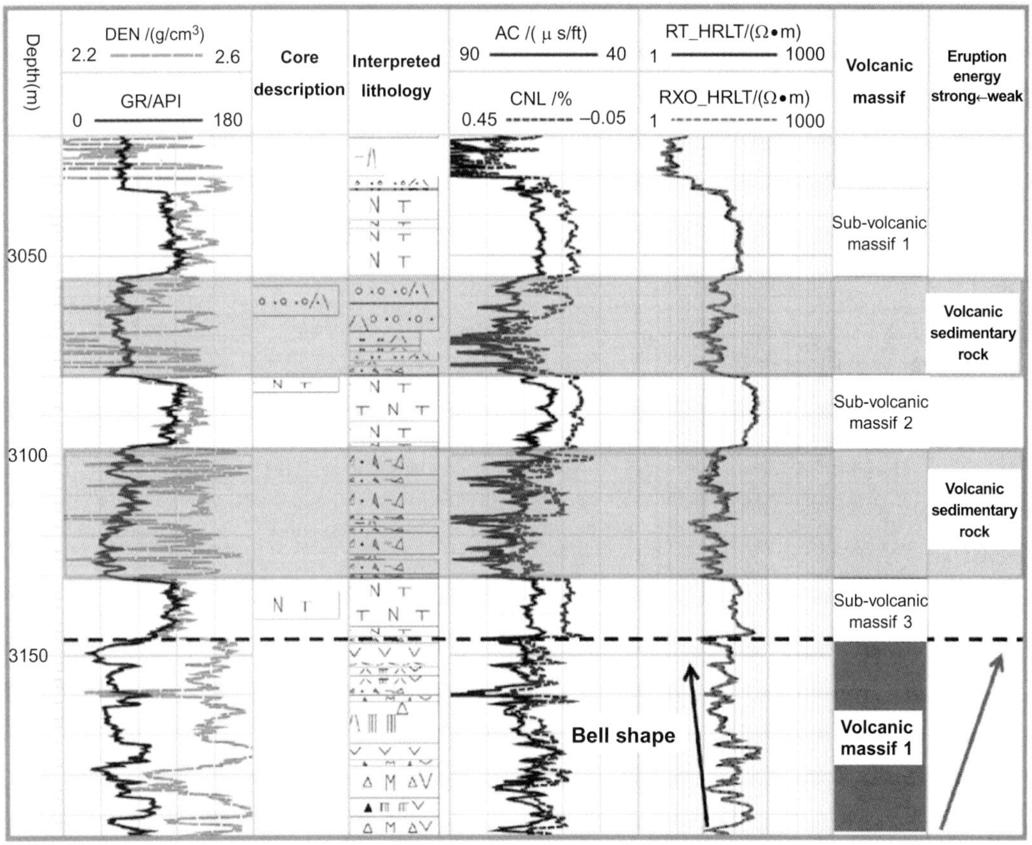

FIGURE 2.42

Well identification for volcanic massifs (Well DD103).

2) Identifying extrusive rocks and subvolcanic rocks to determine distribution boundaries of subvolcanic massifs

Compared with extrusive rocks, subvolcanic rocks are characterized by high resistivity, low neutron, low density, and low interval transit time in logging curves, and by a consistent microserrated box shape (see Figure 2.42). With regard to the volcanic rock distribution pattern, therefore, the rock bodies are identified based on the minute difference between extrusive rock and subvolcanic rock on logging curves to further determine the distribution boundaries of subvolcanic massifs.

3) Classifying lithological assemblage to establish a conversion mode of eruption energy

From the results of lithological identification, lithological assemblages are classified on the basis of eruption energy, and the alternations of eruption energy are established.

In Well DD103 (see Figure 2.42), the lower volcanic rocks are the typical lithological assemblage of andesitic volcanic breccia–breccia tuff–andesite, with gradually weakening eruption energy and logging curves in shapes of flattened bells. This interval of lithological assemblage constitutes the first volcanic massif in this well.

In Well CC1-2 (Figure 2.41c), two lithological assemblages were developed. In the lower part, it is a lithological suite of Rhyolit-rhyolitic volcanic breccia, with energy alternating from weak to strong. In the upper part, it is a rhyolit-rhyolitic ignimbrite suite, with energy alternating also from weak to strong. The logging curve is shaped as a microserrated funnel, with the alternation surfaces marked by the top and bottom interfaces of the funnel.

In Well XX23 (Figure 2.41d), two lithological assemblages were also developed. In the lower part, it is a suite of rhyolitic volcanic breccia–rhyolitic crystal tuff, with energy alternating from strong to weak and logging curves in a composite bell-funnel shape. The upper part is a suite of rhyolitic volcanic breccia–rhyolite, with energy alternating also from strong to weak and the logging curve shaped also as a composite form of bell-funnel; the alternation surfaces coincide with the top and bottom interfaces of two sections of mini-bell-shaped curve.

4) Delineation of the interfaces of volcanic massifs to determine their vertical distribution

On the basis of the preceding analyses, the interfaces of volcanic massifs are further delineated to determine the range and vertical distribution of volcanic massifs. In Well DD103, four volcanic massifs are identified, from top to bottom, including a subvolcanic massif in 3034 to 3055 m, a subvolcanic massif in 3080 to 3098 m, a subvolcanic massif in 3131 to 3145 m, and a volcanic massif in 3145 to 3195 m.

5) Profile identification

On the basis of boundary constraints of volcanic edifices, seismic facies patterns, and well identification of various types of volcanic massifs, volcanic massifs can be identified in seismic profiles and massif boundaries are tracked through well-seismic integrated calibration and analysis of seismic response to reveal the distribution of volcanic massifs in cross-sectional profiles.

Figure 2.43 is a case study of volcanic massif identification based on seismic profiles of Well DD18 in the DD gas field. In Well DD1824, three volcanic massifs are identified in total, including the upper trachyte of effusive facies, the thin-bedded tuff in the middle, and the lower syenite porphyry. In the adjacent well DD182, four volcanic massifs are identified, including, in descending order, andesitic volcanic breccia, syenite porphyry, andesitic ignimbrite, and basaltic ignimbrite. Through well

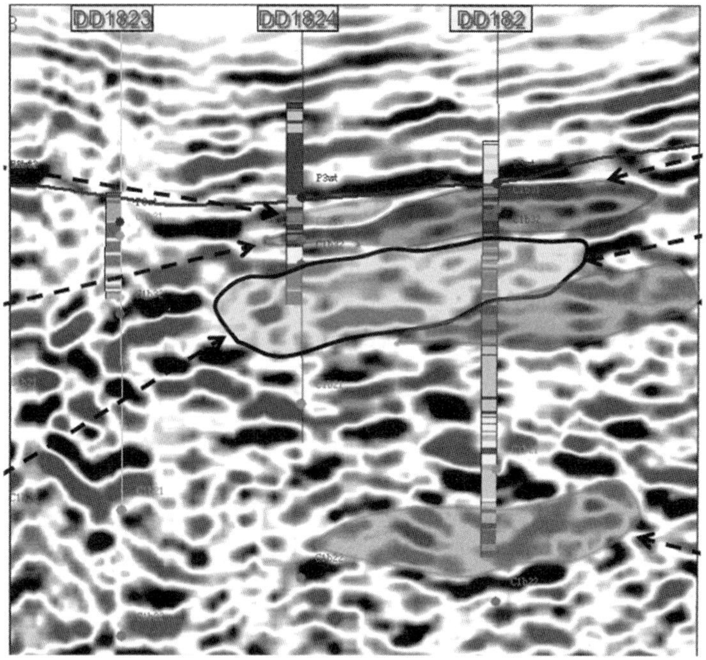

FIGURE 2.43

Profile identification of volcanic massifs.

calibration and combined seismic response of near-well trace, five volcanic massifs are identified in total, with the syenite porphyry massif in the middle being the largest, which separates the original volcanic massif into two volcanic massifs.

4. Planar prediction

Volcanic massifs are identified and their planar distribution is predicted through seismic response analysis, such as sensitivity analysis and the extraction of seismic attributes.

1) Analysis of sensitive attributes in volcanic massifs

Various types of volcanic massifs have different sensitive seismic attributes. For example, the homogeneous dome-shaped volcanic massif in the Well Block DD18 is most clearly displayed by its phase attribute but is reflected by amplitude and coherence attributes. In the Well Block DD17, the volcanic massifs are mainly of the shield type, which is a clearly revealed amplitude, but poorly shown in phase attributes. It is critical, therefore, to analyze and optimize the sensitivity of different seismic attributes, taking into account the geological features of volcanic massifs.

2) Planar prediction

Sensitive seismic attributes are extracted from the target horizon, and the volcanic massif distribution in plan view is predicted using the response characteristics of seismic attributes. For other attributes, strong coherence and low dip angle are generally typical of a volcanic massif, whereas weak coherence and high dip angle are diagnostic for the external boundary. As a result, the change of attribute values reflects the configuration of the volcanic massif. For an instantaneous phase attribute, the phase characteristic is consistent within a volcanic massif, whereas an abrupt phase change is presented in the external boundary (Figure 2.44). In terms of waveform classification, an abrupt change in waveform marks the external configuration of a volcanic massif (Figure 2.45). A strong amplitude value is exhibited internally in a volcanic massif.

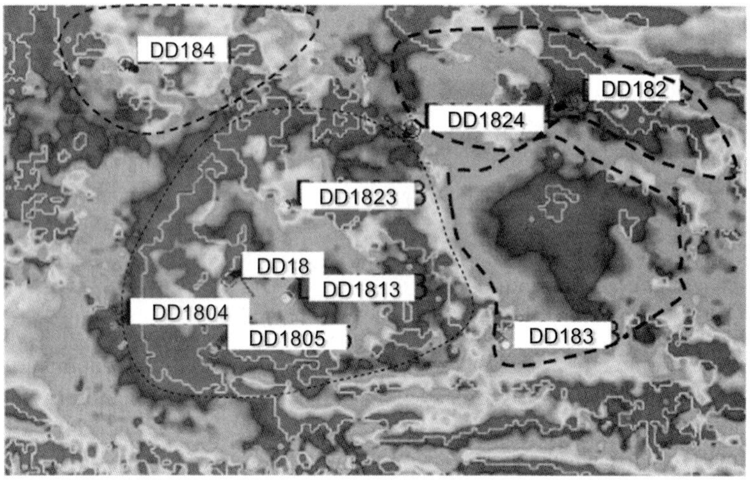

FIGURE 2.44

Prediction of subvolcanic massif distribution by phase attribute.

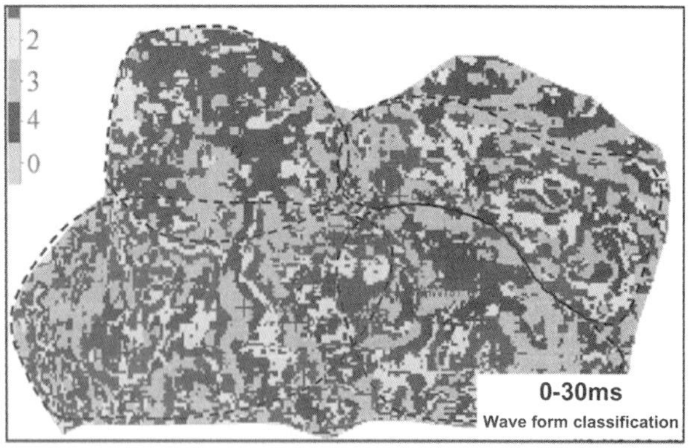

FIGURE 2.45

Prediction of subvolcanic massif distribution by attribute of waveform classification.

3) Comprehensive multiattribute analysis

The prediction of a volcanic massif using seismic attributes may rely on multiple interpretations. The accuracy of prediction, therefore, can be enhanced by integrating the predictions based on various attributes. The methods shown in Figures 2.44 and 2.45 are consistent with each other, showing that the classification of phase and waveform can effectively help predict the planar distribution of volcanic massifs, leading to better results through data integration.

5. Spatial prediction

On the basis of well and profile identification and planar prediction, the skeletal profile network is set up and the distribution of volcanic massif in 3D space is effectively predicted through the tracking and closing technique of the volcanic massif interface. Figure 2.46 is the spatial distribution map of a subvolcanic massif in the Well Block DD18 predicted using this technique, which reveals the spatial extent and superposition relationships of the subvolcanic massifs, providing information on the massif architecture for reservoir prediction and geological model building.

2.5.3 Characterization of volcanic massifs

Based on the results of volcanic massif identification, the outcrop and seismic data can be used further to quantitatively characterize the size, shape, and superposition relationships of volcanic massifs and reveal the distribution patterns of various types of volcanic massifs.

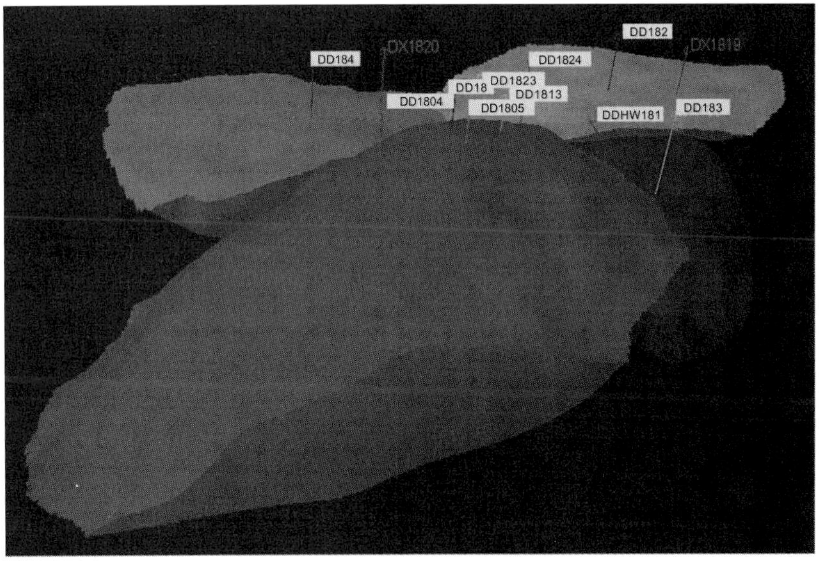

FIGURE 2.46

Spatial distribution prediction of a volcanic massif.

1. Shape characterization of volcanic massifs

A volcanic massif is an architectural subunit of a volcanic edifice. Within the edifice framework, the shapes of various volcanic massifs can be characterized based on profile identification as well as the planar and spatial prediction of volcanic massifs.

1) Volcanic massifs formed by centered eruption

Centered volcanic edifices present mostly in shape of cone, shield, and dome.

Volcanic massifs in a volcaniclastic cone are formed by in situ accumulation of a volcanic mass, which is mainly distributed near the crater. In seismic profiles, they exhibit mostly mound, umbrella, and lensoid shapes (Figure 2.47a and Figure 2.48) and appear in plan view as subcircular, elliptical, or irregular patterns, with a small length-to-width ratio.

Within a lava shield, volcanic massifs are generally located in the slope-break zone or nearby sags, mostly exhibiting lamellar and wedge shapes in seismic profiles (Figure 2.47c) and a circular or elliptical pattern in plan view. The domed volcanic massifs appear as columnar and massif reflection shapes in seismic profiles and exhibit a minimum-sized, subcircular pattern in plan view.

Subvolcanic massifs are generally located close to the volcanic conduit, characterized by columnar, wedge, and lensoid shapes. They may manifest as a strip, a band, or a lens in seismic profiles and appear subcircular or striplike in plan view.

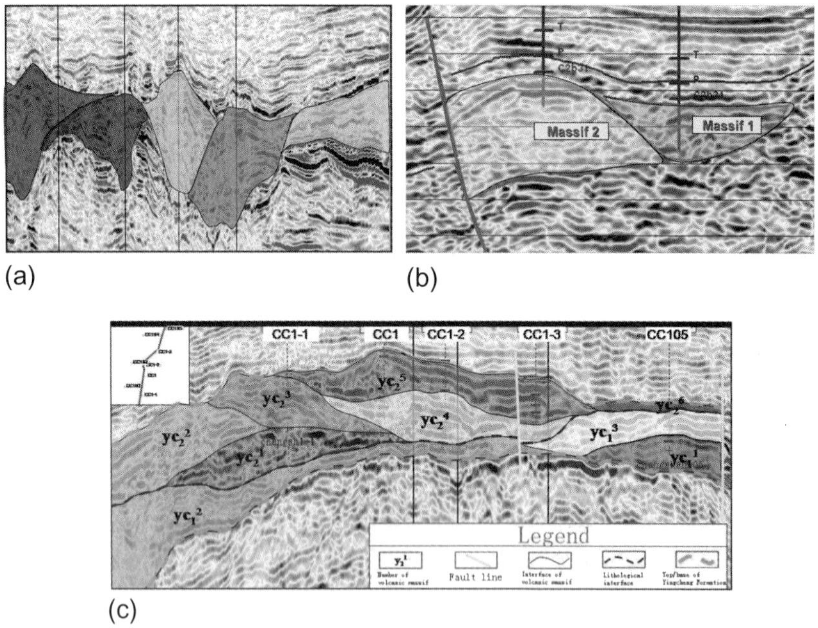

FIGURE 2.47

Sectional shape of volcanic massifs. (a) Volcanic massifs formed by centered eruption (volcanic clast). (b) Volcanic massifs formed by fissured eruption. (c) Volcanic massifs formed by centered eruption (volcanic lava).

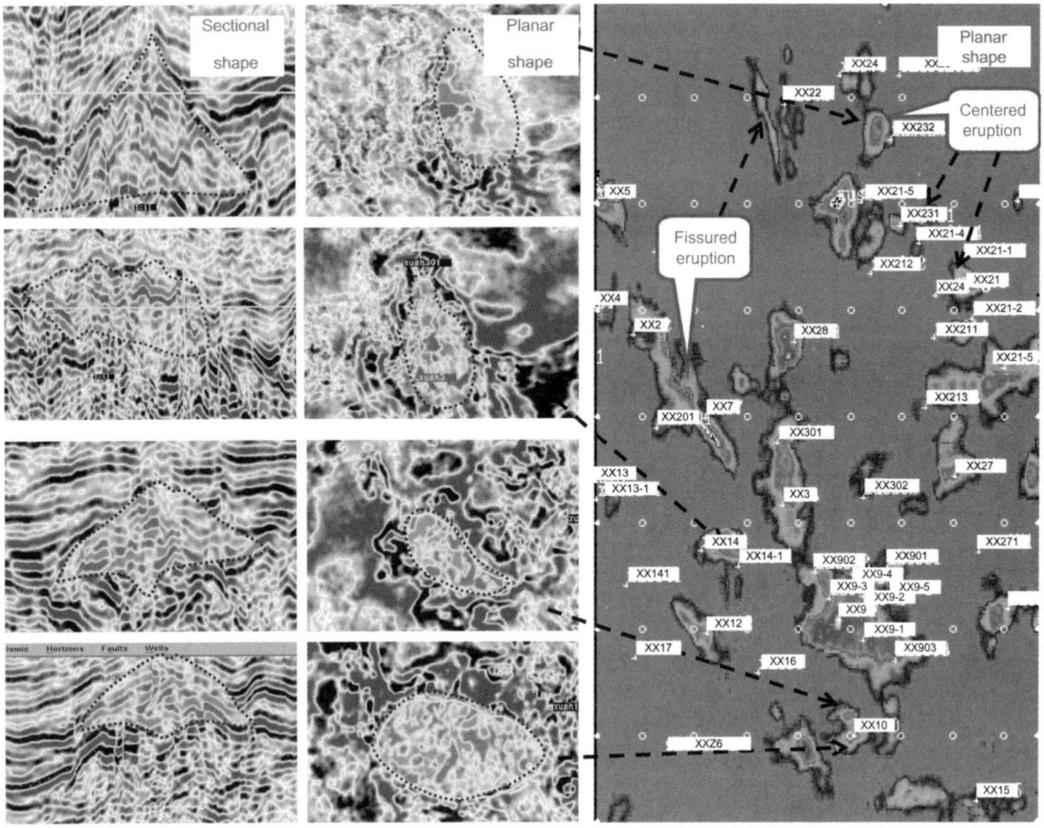

FIGURE 2.48

Planar shape of volcanic massifs.

2) Volcanic massifs formed by fissured eruption

Volcanic massifs formed by fissured eruption usually appear as a wedge, basin, or lamella in seismic profiles (Figure 2.47b) and as a line, a strip, a band, or an irregular pattern in plan view, with a high length-to-width ratio (see Figure 2.48).

2. Characterization of superposition relationships of volcanic massifs

In terms of their relative spatial positions, the superposition relationships (cross-cutting or stacking patterns) of volcanic massifs can be divided into vertical, lateral, inset, and isolated categories. The common interface relationships between adjacent massifs may be extensive contact, striplike contact, or point contact.

1) Vertical superposition (stacking)

Within a volcanic edifice, if the volcanic conduit and crater do not migrate after the eogenetic massif has formed, the epigenetic magma erupts again along the same conduit and crater and a new volcanic massif will form above the eogenetic body, resulting in a vertical stacking relationship between the two

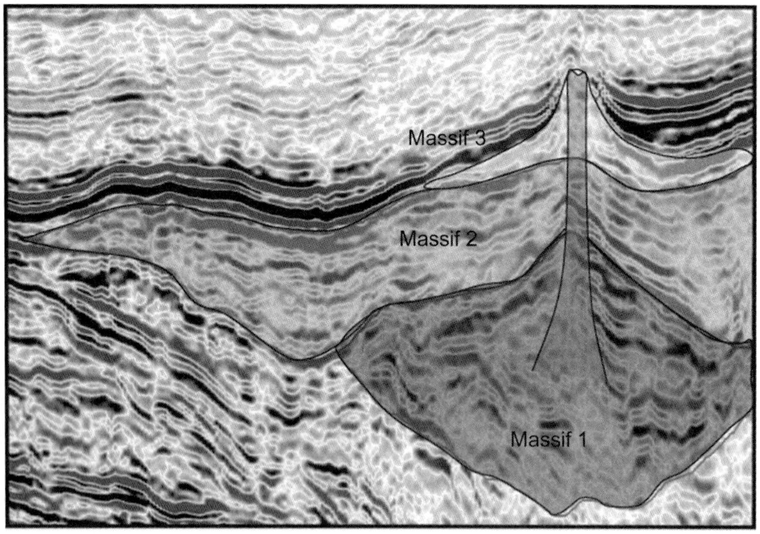

FIGURE 2.49

Vertical superposition mode.

massifs (Figure 2.49). If no epigenetic clast is deposited on the surface of the eogenetic massif, the two volcanic massifs formed in different stages will have extensive contact. In the event that some epigenetic clast is deposited on the surface of the eogenetic massif or the eogenetic massif is covered partially by the surrounding rocks, a striplike contact or point contact will form between the two massifs.

In the two adjacent volcanic edifices, the lower one formed first, and magma from the crater in the upper location erupted and effused directly over the lower edifice, thus the volcanic massifs in these two edifices formed a vertical superposition (see Figure 2.38). Similarly, according to the coverage of the epigenetic clast or surrounding rocks over the eogenetic volcanic massifs, the contact mode includes three types: extensive contact, striplike contact, and point contact.

2) Lateral superposition/stacking

Lateral superposition is an important stacking pattern of volcanic massifs. Lateral superposition can be found in the following four cases: (1) within a volcanic edifice, the volcanic conduit and crater migrate laterally after the eogenetic massif has formed, and the epigenetic massif superimposes laterally over the eogenetic one; (2) when two adjacent craters erupt sequentially, the volcanic rock bodies stack laterally on each other; (3) when a volcanic massif is cut by fault, the newly formed volcanic massifs superpose laterally on the former one; and (4) there is lateral stacking between subvolcanic or parasitic volcanic massif and the major volcanic massif.

Lateral stacking usually occurs between volcanic massifs in the flanks of a volcanic edifice, and the two adjacent massifs may have extensive contact (see Figure 2.47), a striplike contact (Figure 2.50a), or a point contact (Figure 2.50b).

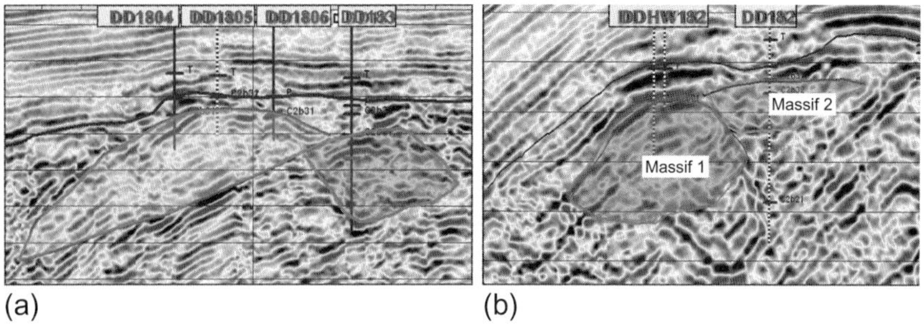

FIGURE 2.50

Lateral contact modes of volcanic massifs. (a) Striplike contact. (b) Point contact.

3) Inset superposition

This is a complex superposition relationship or stacking pattern that occurs in the following two cases: (1) when adjacent craters erupt alternatively at different stages, an inset superposition relationship will usually form in the flanks of volcanic massifs (see Figure 2.47); (2) when a volcanic massif is cut by a subvolcanic massif into two parts, so that one massif will be inserted into another (see Figure 2.43). Inset superpositions usually show a striplike contact or point contact, rarely an extensive contact.

4) Isolated superposition

A small volcanic edifice formed from an isolated, single eruption is a volcanic massif by itself. Alternatively, when the volcanic massifs are surrounded and separated by sedimentary rocks or volcanic sedimentary rocks, these volcanic massifs became isolated (see Figure 2.48).

3. Size characterization of volcanic massif

The sizes of various volcanic massifs can be determined through characterization of their geometric parameters, such as the long axis, short axis, height, and area.

1) Overall size

Among the 26 volcanic massifs identified in this study, their long axes range from 2.3 to 10.6 km (average, 5.5 km), short axes range from 0.8 to 7.7 km (average, 3 km), height ranges from 125 to 1149 m (average, 423 m), and area ranges from 2.1 to 60.5 km^2 (average 15.7 km^2). Different volcanic massifs vary significantly in size, and different types of volcanic massifs also have various sizes (Table 2.6).

2) Volcanic massif size of acidic explosive facies

The volcanic massifs of acidic explosive facies have long axes ranging from 2.9 to 6.5 km (average, 4.1 km), short axes ranging from 0.8 to 3.7 km (average, 2.2 km), height ranging from 130 to 560 m (average, 290 m), and area ranging from 2.1 to 18.7 km^2 (average, 8.4 km^2). Their sizes are relatively small in general (Table 2.6 and Figure 2.51).

3) Volcanic massif size of acidic effusive facies

Their long axes range from of 3.6 to 10.6 km (average of 7.3 km), short axes range from 1.7 to 7.7 km (average, 4 km), height ranges from 158 to 1149 m (average, 557 m), and area ranges from 6.5 to 60.5 km^2 (average 26 km^2). This type of massif has the largest size (see Table 2.6 and Figure 2.51).

Table 2.6 Parameters of Volcanic Massif Size

Rock Composition	Massif Type	Massif quantity	Long Axis, km		Short Axis, km		Height, m		Area, km^2	
			Range	Avg.	Range	Avg.	Range	Avg.	Range	Avg.
Acidic	Explosive	9	2.9-6.5	4.1	0.8-3.7	2.2	130-560	290	2.1-18.7	8.4
	Effusive	11	3.6-10.6	7.3	1.7-7.7	4	158-1149	557	6.5-60.5	26
Neutral	Subvolcanic	4	2.3-4.6	3.1	1.8-2.3	2.1	251-747	490	3.3-8.4	5
Basic	Effusive	2	6.2-6.4	6.3	2.2-2.3	2.3	125-178	152	12-12.6	12.3

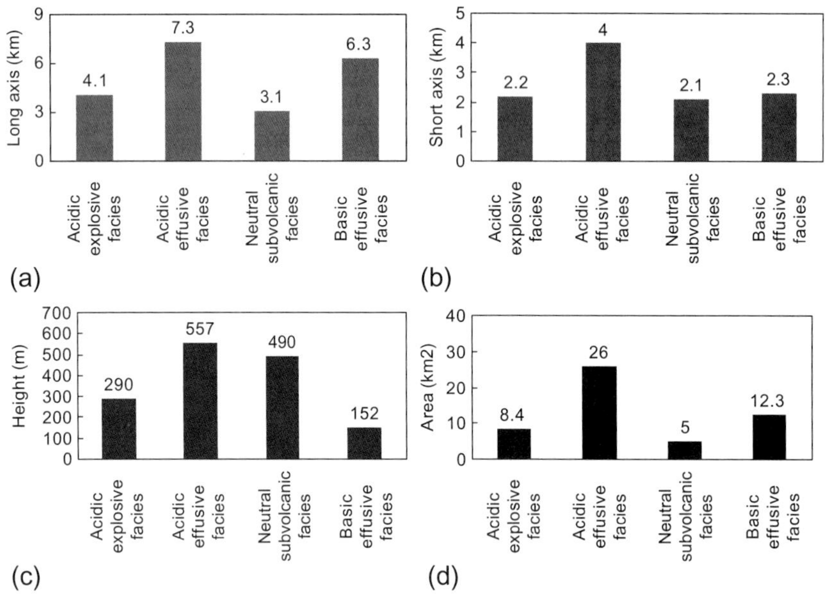

FIGURE 2.51

Size parameters of various types of volcanic massifs. (a) Long axis. (b) Short axis. (c) Height. (d) Area.

4) Size of neutral subvolcanic massif

Their long axes range from 2.3 to 4.6 km (average, 3.1 km), short axes range from 1.8 to 2.3 km (average, 2.1 km), height ranges from 251to 747 m (average, 490 m), and area ranges from 3.3 to 8.4 km^2 (average, of 5 km^2). Their sizes are among the smallest (see Table 2.6 and Figure 2.51).

5) Volcanic massif size of basic effusive facies

Their long axes range from 6.2 to 6.4 km (average, 6.3 km), short axes range from 2.2 to 2.3 km (average, 2.3 km), height ranges from 125 to 178 m (average, 152 m), and area ranges from 12 to 12.6 km^2 (average, 12.3 km^2). Their sizes are relatively large in general (see Table 2.6 and Figure 2.51).

This discussion demonstrates that the sizes of volcanic massifs in the study area have the following characteristics: (1) the size of volcanic massif is smaller than or equal to that of the host volcanic edifice; (2) the distribution area of volcanic massifs in effusive facies have the largest size, followed by explosive massifs, and the subvolcanic massifs are the smallest; and (3) the volcanic massifs in effusive facies and subvolcanic massifs have the greatest thickness, followed by volcanic massif in explosive facies.

2.6 TECHNICAL APPLICATIONS

The dissection technology of internal architecture of volcanic rocks can be applied to reveal the characteristics of internal architecture at various levels of volcanic gas reservoirs and to clarify the spatial distribution characteristics of volcanic formations, edifices, massifs, lithofacies, and lithology, which play the following important roles for the effective development of volcanic gas reservoirs.

2.6.1 Providing guidance for a logical division of development layer series

Because of the usually rapid changes in volcanic strata, the conventional geological correlation methodology cannot deal with the complex changes of geological characteristics. Under the constraint of internal architecture, the relationships between each level of architectural units and stratigraphic sequences (eruption cycles, stages, rhythms, and condensation units) can be streamlined to guide the logical division of development layer series in volcanic gas reservoirs.

2.6.2 Effectively predicting reservoir distribution to construct geological models

Due to the generally strong heterogeneity of volcanic reservoirs, their prediction by means of seismic inversion, which is a method suitable for conventional sedimentary reservoirs, becomes error prone for volcanic reservoirs, and the control of internal architecture over reservoir distribution and gas/water distribution cannot be highlighted by conventional modeling. On the basis of framework constraints, therefore, "volume controlled" reservoir inversion and "volume-facies-controlled" geological modeling can be achieved through the further dissection of internal architecture to effectively predict the pay-zone distribution in volcanic gas reservoirs and to build geological models more effectively.

2.6.3 Providing a basis for an appropriate analysis of gas-water relationship

A volcanic gas reservoir is generally under regional architectural control, with gas in the upper part and water below. The distribution of a gas reservoir is controlled by the spatial arrangements of volcanic edifices, and various volcanic edifices have different gas-water contacts. Generally, multiple gas-water systems are developed within a gas reservoir and the distribution of each gas-water system is constrained by the spatial distribution pattern of volcanic edifices. Therefore, the dissection of internal architecture provides a basis for an appropriate analysis of gas-water relationships in volcanic gas reservoirs.

References

[1] Stow DAV, Johansson MJ. Deep-water massive sands: nature, origin and hydrocarbon implication. Mar Petrol Geol 2000;17(2):145–74.
[2] Wu C, Peng D, Pang X, et al. Study on the microscopic internal architecture of Zhujiang deep-water fan system in South China Sea. Sediment Geol Tethyan Geol 2007;27(3):28.
[3] Liu Y. Study on architectures and reservoirs in early evaluation of oil and gas reservoirs. Information Summary of Sino-Foreign Cooperation of Bohai Petroleum Company Vol II (Geological Exploration Album) (internal research report in Chinese) 2007;27(3):184.
[4] Suo X, Li F. Carbonifeous geological architectures and volcanic rock distribution of malang depression in Santanghu basin. Xinjiang Petrol Geol 2007;30(4):463–6.
[5] Zheng H, Li T, et al. Discussion on characteristics and origin of Cenozoic volcanic rock area in the Northern Qingzang Plateau. Geoscience 2010;24(1):131–9.
[6] Tang H, Pang Y, Bian W, et al. Quantitative analysis of Cretaceous Yingcheng formation volcanic edifice in Songliao basin. Acta Petrol Sin 2008;29(6):841–5.
[7] Pang Y, Bi X, Shao R, et al. Early development characteristics and controlling factors of volcanic gas reservoirs. Acta Petrol Sin 2009;30(6):882–6.
[8] Jiao G, Luo X, Yin C, et al. Forming conditions and exploration direction of deep gas reservoir in Songliao basin. Nat Gas Ind 2009;9:28–31.
[9] Zheng R, Hu C, Dong X. Analysis of buried hill internal architecture and hydrocarbon accumulation conditions in Liaoxi Depression. Lithological Oil and Gas Reservoirs 2009;21(4):10–8.
[10] Shi W, Kong M, Song Z. Interior of Sheling formation mixed-source fan delta of Liangjia area in Yitong basin. Geosciences 2008;33(3):365–70.
[11] Ren Y, Lv J, Li Y. Detailed dissection of Bogu-1 buried-hill internal architecture. Mar Geol Lett 2007;23 (3):19–23.
[12] Meng X, Ge M. New concepts and principle analysis of sedimentary formation. Regional Geology of China 1985;5:210.
[13] Meng X. Analysis of sedimentary formation and its parageneous mines. Beijing: Geology Press; 1979, p. 1–2.
[14] Geology Dictionary Office of Ministry of Geology. Geology dictionary—geochemistry of minerals and rocks volume. Beijing: Geology Press; 1981, 122.
[15] Yuan J, Zhu S, Zhai Y. Ore mineralogy. Beijing: Geology Press; 1985, p. 156–57.
[16] Sun N, Peng Y. Igneous petrology. Beijing: Geology Press; 1985, p. 266–75.
[17] Qiu J, Tao K, Zhao J, et al. Volcanic rocks. Beijing: Geology Press; 1996, p. 23–28.
[18] Li S, Wang T. Volcanic rocks. Beijing: Geology Press; 1981, p. 219–220.
[19] Hou Q, Zhao Z, Wang L. Volcanic gas reservoirs. Beijing: Science Press; 2009, p. 1, 5.
[20] Wang P, Feng Z, et al. Volcanic rocks in basins-lithology, lithofacies, reservoir, gas reservoir and exploration. Beijing: Science Press; 2008, p. 64, 65, 71.
[21] (Spain) AnguitaVolcanic geology of Canary Islands in Spain. Translated by Wu H, Men G, Liu W, et al, Beijing: Petroleum Industry Press; 2009, p. 6–14.
[22] Wang D, Zou X. Volcanic petrology. Beijing: Science Press; 1982, p. 3, 10–18.
[23] Zeng F, Luo X, Xiao H, et al. Seismic integrated identification and prediction of deep volcanic massifs in Songliao basin. Prog Explor Geophys 2009;32(4):270–4.
[24] Sun D. Prediction of deep volcanic massifs in the Northern Songliao basin by use of the third generation coherence technology. Inner Mongolia Petrochemicals 2008;18:145–7.
[25] Li Y, Song Z, Li Q, et al. Study on identification of volcanic massif seismic characteristics and application. Minerals and Rocks 2008;3:105–10.
[26] Chen R, Wang P, Liu W, et al. Control of Xujiaweizi fault terrace zone over volcanic massifs and sedimentary facies belt. Oil and Natural Gas Geology 2003;24(2):126–9.

[27] Cai X, Tan S. Igneous rock distribution and hydrocarbon accumulation rules of Changlin faulted depression in the Southern Songliao basin. Petrol Geophys 2002;41(3):363–6.
[28] Tang Z, Qiao W. Evaluation methodology of lithological traps in volcanic massifs, Northern Songliao basin. Petroleum Geology and Oilfield Development in Daqing 2001;20(4):9–11.
[29] Mong Q, Men G. Prediction method and its application of deep volcanic massifs and lithofacies in Songliao basin. Petroleum Geology and Oilfield Development in Daqing 2001;20(3):21–4.
[30] Yu J. Lithofacies interpretation and reservoir study of cryptovolcanic massifs. Petrol Explor Develop 1995;22 (3):24–9.
[31] Wang X. Analysis of geomechanics in rock mass engineering of dam foundation. Beijing: Science Press; 1985, p. 4.
[32] The information research institute of the former ministry of geology & mineral resources. Mining Mineralogy and Mineralogical Mapping. Fujian: Fujian Science & Technology Press; 1986, p. 215.
[33] Zhu Y. Excellent transactions of Liaohe Oilfield exploration & development research institute. Beijing: Petroleum Industry Press; 2002, p. 34.
[34] Wang S, Li X, Shi Q, et al. Principle and methodology of logging evaluation in complex reservoirs. Beijing: Petroleum Industry Press; 2006, p. 138.
[35] Lu F, Wu Q. Composition, architecture and interaction between lithosphere and stratigraphy of lower lithosphere in typical areas of East China. Wuhan: China University of Geosciences Press; 2005, p. 70, 71.
[36] Regional geology & mining geology department. Manual of regional geological investigation methodology in volcanic rock areas. Beijing: Geology Press; 1987, p. 116–142.
[37] Zhang Y, He S, Zhou X, et al. Identification of seismic reflection characteristics of volcanic edifices in deep gas field of Southern Xingcheng. Acta Geosci Sin 2008;29(5):578.

CHAPTER

Volcanic Rock Sequence Division and Stratigraphic Correlation

3

Volcanic rocks and strata have complex structures. In this chapter, the definition and division of volcanic rock sequences are discussed first, and their identification markers are established for stratigraphic sequence classification, which is carried out using a *level-by-level control* approach. Furthermore, the relationships between stratigraphic sequences and internal architectures (or gas reservoirs) are distilled to determine correlation relationships at each level of sequences. The delineation and correlation of volcanic rock sequences are guided by the principle of level-by-level control and constrained by internal architectures. This leads to a better understanding of the spatial distribution of volcanic rock sequences and helps establish a foundation for an appropriate division of development layer series.

3.1 CONCEPT AND LEVELS OF VOLCANIC ROCK SEQUENCE

3.1.1 Concept of volcanic rock sequence

The basic concept of sequence was initially proposed in the late 18th century, with the strata of a sequence being bounded by top and bottom unconformities. In 1949, Sloss and colleagues formally used the term *tectonic sequence* for the first time and regarded sequence as a stratigraphic succession bounded by major regional unconformities. In 1988, van Wagoner and coworkers defined sequence stratigraphy as "the study of rock relationships within a chronostratigraphic framework of repetitive, genetically related strata bounded by surfaces of erosion or nondeposition, or their correlative conformities [1,2]." The early concept of sequence stratigraphy was derived from studies on marine deposits at passive continental margins, and hence it cannot be used to guide volcanic rock sequence classification. Shanley and colleagues pointed out that nonmarine sequence stratigraphic analysis can follow the general concept and methods of marine sequence analysis, but direct application is not advisable [3].

Up to now, there has been a lack of clear definition for volcanic rock sequence. Different scholars appear to have different concepts about volcanic rock sequence division. Taking into consideration the regularity of volcanic activity and especially volcanic rock formation, the former Regional Geology and Mineral Department of China classified volcanic rock sequences into four levels (or orders)—rock series, cycle, rhythm, and pulses—a scheme used during the conventional geological mapping projects [4]. On the basis of their spatiotemporal superposition relationships, the former Chinese Geological Survey Bureau divided the deposits from volcanic eruptions into four levels of cycles, following the methods and scheme of sedimentary sequence stratigraphy: cycle I (or first-order cycle) refers to an entire spatiotemporal process of basin evolution; cycle II (second-order cycle) stands for a specific stage magmatic evolution; cycle III (third-order cycle) represents a dormant episode of volcanic activity; cycle IV (fourth-order cycle) reflects eruption pulses of volcanic activity, equivalent to eruptive rhythm [5]. Xie divided nonmarine volcanic rock sequences into four levels (or orders) as cycle, formation, lithofacies, and stratum, and considered that a cycle corresponds to a volcanic edifice, a formation to an evolutionary stage of volcanic edifice, lithofacies to a specific volcanic eruption type, and a stratum being the smallest stratigraphic unit of volcanic deposits [6]. Huang and associates adopted a three-level scheme—cycle, period, and lithofacies—and used these terms in their study on the volcanic rocks of the Yingcheng Formation in the Songliao basin. In this scheme, a cycle corresponds to a volcanic edifice, a period to a lithofacies assemblage, and a lithofacies to a specific mode of volcanic rock formation, comprising one or more volcanic rock strata [7].

In light of the research results discussed previously [8–14], especially those pertinent to the volcanic rock sequences in the Songliao and Junggar basins, a volcanic rock sequence is defined in this study as various levels (or orders) of stratigraphic units with internal petrogenetic links and corresponding spatiotemporal relationships, bounded by regional unconformity surfaces on the top and at the bottom.

3.1.2 Levels of volcanic rock sequences

According to this definition, volcanic rock strata can be divided into four levels (analogous to orders of conventional sedimentary sequences), including, from large to small scale, volcanic eruption cycle, volcanic eruption period, volcanic eruption rhythm, and volcanic cooling unit.

1. *Volcanic eruption cycle*

A volcanic eruption cycle refers to *all cogenetic volcanic eruption products with given architectural patterns formed at different stages of a single volcanic activity in the volcanic formation.*

A volcanic eruption cycle is often related to a tectonic cycle. One or multiple volcanic edifices can be developed within one volcanic eruption cycle. Tectonic and eruption hiatuses usually occur between two successive cycles, expressed as a period of dormancy in volcanic eruption, resulting in the development of sedimentary interbed or weathered crust of various thicknesses. Thus, there usually exists an unconformity or erosional surface. A volcanic formation may comprise multiple eruption cycles, among which one eruption cycle can last tens of thousand years, with a lateral distribution range of tens to hundreds of kilometers and a volcanic rock succession reaching hundreds to thousands of meters. A volcanic eruption cycle usually begins with a near-crater facies association and forms a contact with the underlying strata through various unconformity surfaces; the cycle ends with a facies association relatively distal from the crater, which also forms a contact with the overlying strata through various unconformity surfaces, marked by sedimentary interbeds.

2. *Volcanic eruption period*

A volcanic eruption period refers to all volcanic products derived from one relatively continuous eruption during the formation of a volcanic edifice.

A volcanic eruption period is related to eruption energy change during the formation of a volcanic edifice. Two eruption periods are separated by an eruption dormancy marked by small discontinuity surfaces, such as a weathering crust, an unconsolidated layer, or sedimentary interbeds. One volcanic eruption period can last several thousand years, with lateral distribution up to several kilometers to tens of kilometers, and a volcanic rock succession reaching tens to hundreds of meters in thickness. The internal architecture representing a volcanic period typically shows continuous or quasicontinuous variation in the phase sequence, which corresponds to the intensity of volcanic activity. In the case of a weak-strong-weak eruption energy mode, for example, the corresponding phase sequence is typically effusive facies–explosive facies–effusive facies (extrusive facies/volcanic conduit facies); for a strong-weak eruption energy mode, the phase sequence is explosive facies–effusive facies (extrusive facies/volcanic conduit facies).

3. Volcanic eruption rhythm

A volcanic eruption rhythm refers to a volcanic rock association formed in a single eruption within an eruption period.

A volcanic eruption rhythm is related to the periodical variation in volcanic edifice eruption. The difference between two rhythms is represented by the variation in volcanic rock types, rock texture, and volcaniclastic grains. The duration of a volcanic eruption rhythm is relatively short, with a lateral distribution ranging from hundreds to thousands of meters, and a rock succession reaching several meters to tens of meters in thickness. According to the volcanic lava phase sequence association, the volcanic eruption rhythm can be divided into three types in terms of its internal architecture.

1) Volcanic lava rhythm

A single intensive effusion can produce a volcanic effusive facies comprising a lower, middle, upper, and top subfacies, whereas a multiple intensive effusive facies reflects the internal variation of effusive volcanic activity. On the basis of subfacies, therefore, a volcanic lava rhythm can be divided in terms of its phenocryst mineral composition, rock texture and structure variation of volcanic lava.

2) Volcaniclastic rock rhythm

A single intensive explosion can produce volcaniclastic rocks with different grain sizes and, therefore, a volcaniclastic rock rhythm always starts from volcanic conglomerate produced by the strongest energy and ends with dust- or ash-grade tuff produced by the weakest energy.

3) Interbedded rhythm of volcanic lava and clastic rock

A single intensive effusion can produce volcaniclastic rock and volcanic lava one after another, hence an interbedded rhythm of volcanic lava and clastic rock starts with volcaniclastic rocks reflecting strong energy and ends with volcanic lava from weak energy, thus forming a succession of volcaniclastic rock–volcanic clastic lava–volcanic lava in from bottom to top.

4. Cooling units

A cooling unit refers to the volcanic rock formed in the same eruption period but composed of multiple eruptive pulses, when the time interval between two pulses is sufficiently long to cool the volcanic rock surface of the earlier pulse to surface temperature to produce a cold-hot interface between two successive cooling units, with a cooling unit bounded in between. An unconsolidated layer can usually be found between two neighboring cooling units, with a soft top structure visible at the upper boundary and a hard bottom structure at the lower boundary. The lateral extent of a cooling unit can reach tens to hundreds of meters, with a thickness from several meters to tens of meters.

3.2 CHALLENGES AND TECHNICAL SOLUTIONS

3.2.1 Challenges

Volcanic rocks have special petrogenetic conditions. The accumulation mode and diagenetic processes are different from those of sedimentary rocks. This makes it difficult to apply the "stratified" stratigraphic correlation method for sedimentary rocks to volcanic rocks with complex stratigraphic structures and variations. For volcanic rock sequences, the difficulties or challenges in their identification, division, and correlation are different at various levels.

For high-level sequences (the volcanic formation and eruption cycle), the internal geological features, logging characteristics, and seismic responses are distinct, and the sequence boundary is marked by consistent sedimentary rocks, weathered crust, and volcanic ash, or by abrupt change in lithology. As a result, this type of sequence can be easily identified. For low-level sequences (eruption period, eruption rhythm, and cooling units), the internal geological features, logging characteristics, and seismic responses tend to be vague, and the sequence boundary is represented mainly by a thin volcanic sedimentary rock unit, small-scale weathered crust, transitional interface of lithological association, and cold-hot interfaces that change quickly with poor continuity, which make their identification difficult.

The high-level stratigraphic sequences are generally developed on a regional scale, with two successive sequences separated by a long volcanic dormancy and marked by clear time difference and spatial continuity; the sequence boundary is distinct and can be easily identified and correlated regionally. The low-level stratigraphic sequences are commonly confined to a local scale, with complex internal structures, separated by a short dormancy period; the sequence boundary is inconspicuous and difficult to identify and trace, tasks that are usually done through single-well identification and extrapolated using seismic responses. The internal sequences of different volcanic edifices are usually different, which poses a touch challenge to their regional stratigraphic correlation. The lower the sequence level, the greater the challenge for its division and correlation.

3.2.2 Technical solutions

In view of the difficulties and challenges in volcanic rock sequence division and correlation, the architectural constraints of volcanic rocks and sequence markers of different levels need to be used to identify, classify, and correlate volcanic rock sequences, following the principle of *level-by-level control and correlation* (Figure 3.1).

1. Establishing volcanic rock sequence identification markers

The geological characteristics of volcanic rock sequence, as well as the petrological properties, logging, and seismic response within the sequence and at the boundary, can be analyzed and distilled to generate geological markers, logging markers, and seismic facies patterns to identify a volcanic rock sequence at different levels and to establish a foundation for volcanic rock sequence identification.

2. Volcanic rock sequence identification

Based on the lithological and logging markers and the principle of level-by-level control, the core data and logging curves are applied together to identify the eruption cycle, eruption period, eruption rhythm, and cooling unit in wells at each level. Next, according to the seismic facies pattern, well-seismic

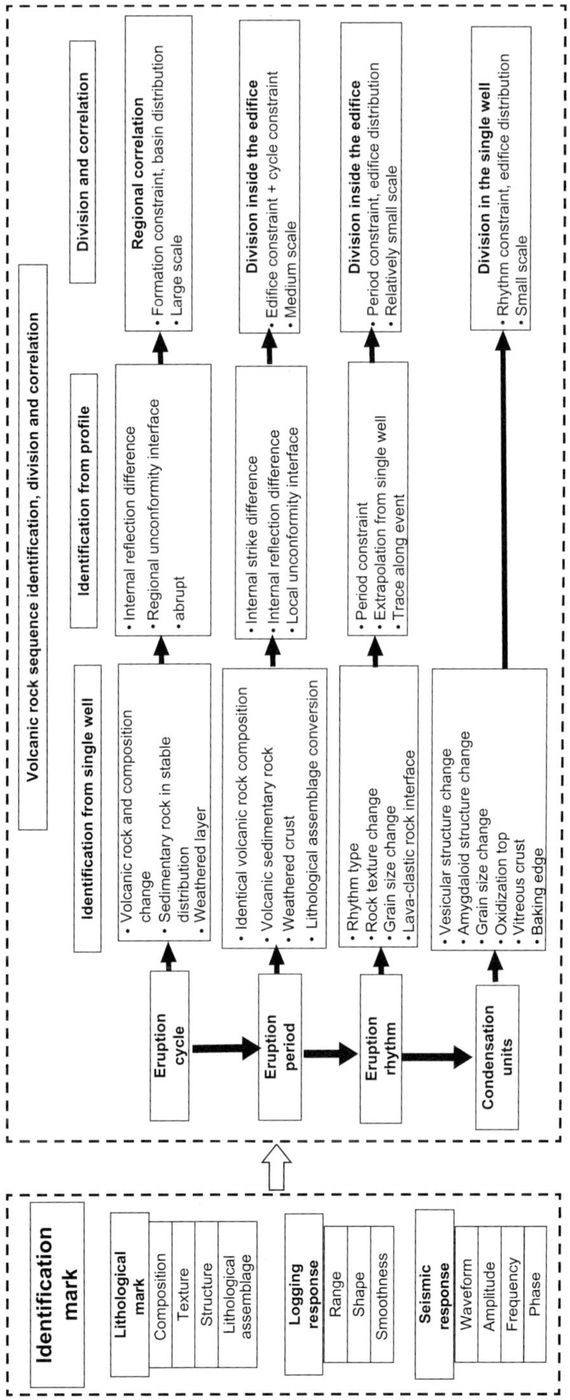

FIGURE 3.1

Technical flowchart for volcanic rock sequence division and correlation.

combination and seismic response characteristics, and volcanic architectural constraints, the volcanic rock sequence is identified and the sequence boundary is traced on the seismic profile.

3. Volcanic rock sequence division and stratigraphic correlation

The relationships among a volcanic rock sequence, its internal architecture, and the development layer series can be established based on the geological characteristics of a volcanic gas reservoir. Then the constraint mechanism, scale, and distribution range at each level of the sequences are correlated and analyzed. In terms of the scale of their correlation, volcanic rock sequences are classified into three types:

Type 1: A sequence that can be correlated regionally is generally an eruption cycle.
Type 2: A sequence that can be locally delineated under the constraint of a volcanic edifice may be recognized as an eruption period or eruption rhythm.
Type 3: A sequence that can be identified only in single wells is usually a cooling unit.

On this basis, the sequence identification results can be applied to every well, and seismic reflections can be used as the constraints for interwell extrapolation; hence, the volcanic rock sequence can be divided and correlated following the principle of level-by-level control and correlation.

3.3 VOLCANIC ROCK SEQUENCE IDENTIFICATION

Identification markers can be used in conjunction with systematic integration of such single-well data as cores, mud logs, and wireline logs to identify sequences in single wells, with sequence boundaries defined using seismic profiles, thus establishing a basis for sequence division and correlation.

3.3.1 Identification markers

Data from cores, wireline logging, mud logging, and seismic survey are employed together to establish the petrological, geophysical, and seismic identification markers for various levels of volcanic rock sequences (Table 3.1).

1. Identification markers of volcanic eruption cycle

The internal lithological differences within a cycle and cycle interfaces are the bases for identifying a volcanic eruption cycle.

1) Geological markers

A volcanic eruption cycle is the product of the same volcanic activity with a similar rock composition and continuous spatiotemporal distribution. Different cycles are separated by regional volcanic dormancy and a depositional hiatus.

Lithology within the cycle. The volcanic activity within a single eruption cycle is basically continuous but can be separated into many stages. Within one eruption cycle, for example, the eruption period is represented by volcanic rocks, and the dormancy period is represented by volcanic sedimentary rocks.

Table 3.1 Identification Marks for Volcanic Rock Sequence

Identification object			Petrological characteristics	Control factors	Logging Response Characteristics					Seismic Reflection Characteristics			
					Range			Typical Shape	Smoothness	Waveform	Amplitude	Frequency	Phase
					Resistivity	Density	Acoustic Wave						
Eruption cycle		Internal feature	Volcanic rock with similar composition	Volcanic formation	High	High	Low	Box shape	Micro-serrated~smooth	Chaotic	Strong	Low	Discontinuous
			Volcanic sedimentary rocks		Medium-low	Medium-low	Medium-high	Finger shape, Funnel shape	Serrated	Laminar	Medium-weak	Relatively high	Good continuity
		Cycle interface	Sedimentary rocks		Low	Low	High	Diverse shape	Serrated	Stratified	Weak	High	Continuous
			Weathered crust (large scale)		Low	Low	High	Box shape, Funnel shape	Micro-serrated	Chaotic	Medium-strong	Low	Discontinuous
			Stable volcanic ash layer		Low	Medium-low	Medium-high	Box shape, Finger shape	Micro-serrated~serrated	Laminar	Strong	Medium	Discontinuous
			Abrupt boundary of volcanic rock composition		Abrupt gamma ray (GR), density (DEN), and neutron change					Weak reflection of local unconformity			
Eruption period		Internal feature	Volcanic rock with identical composition	Volcanic edifice	High	High	Low	Box shape and bell shape	Serrated-micro-serrated	Chaotic	Strong	Low	Discontinuous
		Period interface	Volcanic sedimentary rocks		Medium-low	Medium-low	Medium-high	Finger shape and funnel shape	Serrated	Laminar	Medium-weak	Relatively high	Good continuity
			Weathering crust		Medium-low	Medium-low	Medium-high	Finger shape	Serrated	Medium-strong reflection of local unconformity			
			Alternation surface of lithological assemblage		Indistinct					Indistinct			
Eruption rhythm	Volcanic lava rhythm	Internal feature	Rhythmic lava assemblage	Volcanic massif	High	High	Low	Bell shape-funnel shape	Micro-serrated-smooth	Laminar	Medium-strong	Low	Poor continuity
		Rhythm boundary	Rock structure variation plane		Small change					Indistinct			
	Volcaniclastic rock rhythm	Internal feature	Rhythmic volcaniclastic lithological assemblage		High	High	Low	Box shape, Bell shape, Funnel shape	Serrated-micro-serrated	Chaotic	Medium-weak	Low	Discontinuous
		Rhythm boundary	Grain size variation plane		Small change			Curve shape change		Indistinct			
	Interbedded rhythm	Internal feature	Interbedded lava and volcaniclastic rock		High	High	Low	Clock shape, funnel shape, finger shape	Serrated-micro-serrated	Stratified	Strong	Medium-High	Discontinuous
		Rhythm boundary	Lava-volcaniclastic rock interface		Small change			Curve shape change		Indistinct			
Cooling unit		Internal feature	Volcanic rock	Volcanic rock facies	High	High	Low	Finger shape	Micro-serrated-serrated				
		Unit boundary	Vesicular structure		Medium-low	Medium-low	Medium-high	Bell shape, Funnel shape	Micro-serrated				
			Amygdaloid structure		High	High	Low	Finger shape, Box shape	Micro-serrated	Difficult to differentiate			
			Oxidization top, vitreous crust and baked edge		Indistinct								

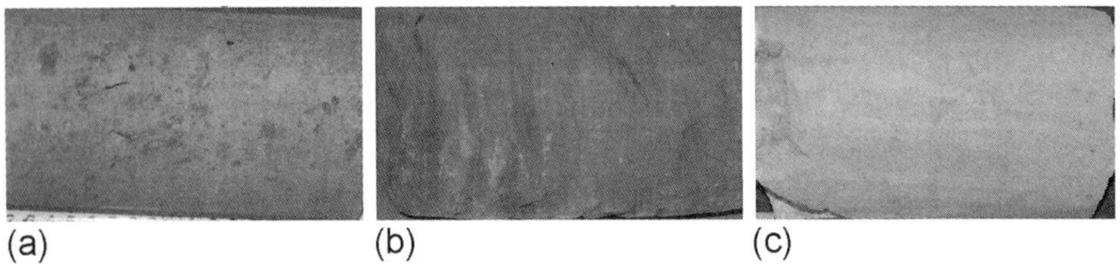

FIGURE 3.2

Lithological characteristics of a volcanic eruption cycle (Well DD1414). (a) Volcanic sedimentary rock. (b) Mudstone. (c) Tuff.

1. *Volcanic rocks in the eruption period.* The volcanic rock composition is mostly consistent, but there are multiple rhythmic lithology assemblages, which mainly reflect the variation patterns of eruption energy.
2. *Volcanic sedimentary rocks in the dormancy period.* There are two major volcanic sedimentary rock types between successive edifices and within one edifice, characterized by a high content of volcanic material, small thickness, and small areal distribution (Figure 3.2a).

Volcanic eruption cycle interface characteristics. There are four types of volcanic eruption boundaries, and different boundaries show different lithological characteristics and contacts.

1. *Sedimentary rock interface.* During a volcanic dormancy, various sediments accumulate, forming a regional sedimentary succession, characterized by high volcanic material content at the base and on the top, but low content in the middle (Figure 3.2b).
2. *Weathering crust interface.* The volcanic cap of an early eruption cycle is always weathered and eroded during the dormancy period, forming a weathering crust, which differs from normal volcanic rocks in its composition, color, and petrophysical properties. For example, the reactive components (e.g., FeO, K_2O, and Na_2O) tend to be leached, leading to a relative enrichment of inert components (e.g. TiO_2 and MgO).
3. *Volcanic ash interface.* A volcanic eruption starts with a high amount of volcanic ash and ends with exotic volcanic ash during dormancy. Therefore, the cogenetic volcanic ash base and exotic volcanic ash top act as the basal and top markers of one eruption cycle, respectively.
4. *Interface of abrupt change in volcanic rock composition.* The magmatic components in different volcanic activity periods may be different; thus, the interface of an abrupt change (such as change from acid to basic) can act as the marker of an interface between two cycles.

2) Logging response characteristics

The logging characteristics within a cycle are as follows:

1. *Volcanic rock in the eruption period.* The logging characteristics of volcanic rocks include high resistivity, high density and low interval transit time; the logging curves are typified by smooth-dentoid and box shapes (Figure 3.3).
2. *Volcanic sedimentary rock in dormancy period.* The rock features are intermediate between volcanic and sedimentary rocks; the logging characteristics include medium-low resistivity, medium-low density, and medium-high interval transit time; the logging curves are mostly serrated, digitate, or funnel in shape, but borehole enlargement is reflected by sharp local peaks (Figure 3.3b).

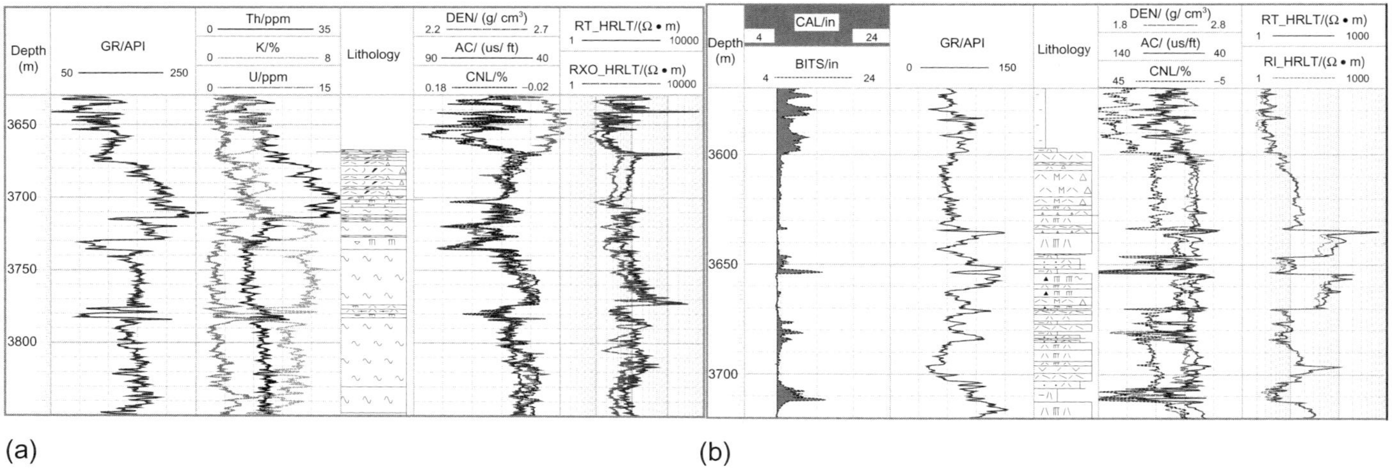

FIGURE 3.3

Logging response characteristics of volcanic eruption cycle. (a) Weathering crust (Well CC1-1). (b) Cycle interface and internal sedimentary rock.

The logging characteristics of a cycle interface are as follows:

1. *Sedimentary rock interface*. The logging characteristics of sedimentary rocks formed during the dormancy period include low resistivity, low density, and high interval transit time, reflecting the typically thick rock body, and enlarged borehole diameter. The logging curves are mostly dentoid-serrated box and bell in shape (see Figure 3.3b).
2. *Weathering crust interface*. After weathering, erosion, and leaching of reactive components, enrichment of inert components, and radioactive material absorption due to fluid movement, the weathered layer will show a decrease in K content, increase in Th and U contents and radioactivity, decline in resistivity and density, rise in interval transit time and neutron responses, and accentuated dentoid curve. The radioactivity curves are mostly bell shaped, and resistivity and petrophysical property curves are mostly funnel like (Figure 3.3a).
3. *Volcanic ash layer boundary*. Clay-grade volcanic tuff is characterized by very fine grain sizes and high interstitial and intramineral water content; the logging curve responses include low resistivity, low density, high neutron, and high interval transit time, characterized mostly by a smooth-microserrated box shape.
4. *Interface of abrupt change in volcanic composition*. The volcanic rock composition variation is reflected by its radioactivity, density, and neutron responses. With a change in acid-to-basic lithology, the radioactivity decreases, whereas the density and neutron increase.

3) Seismic reflection Characteristics

Seismic reflection characteristics within the cycle

1. *Volcanic rock formation*. The general characteristics include internal reflection of strong amplitude, low frequency, discontinuous event, and chaotic external reflection, with reduced amplitude for weathered crust. Different zones show different reflection characteristics: a volcanic edifice near the crater shows high seismic frequencies and poor event continuity, whereas a volcanic edifice away from the crater shows low reflection frequencies, good event continuity, and parallel or subparallel texture. The volcanic rock seismic reflection for different compositions is also different: an acid cycle shows a chaotic reflection waveform, low frequencies, and poor continuity, whereas a basic cycle shows high frequencies, good continuity, and a parallel, subparallel texture (Figure 3.4).
2. *Sedimentary rock formation*. Major characteristics include sublaminar reflections with weak amplitude, high frequencies, and a continuous event. The lower sedimentary rock unit, however,

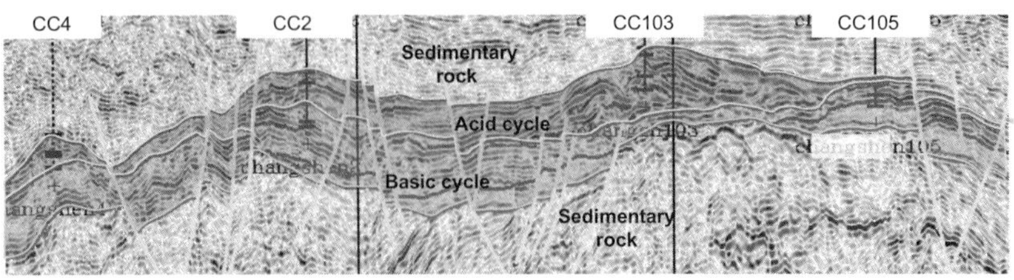

FIGURE 3.4

Seismic reflection characteristics of a volcanic eruption cycle (CC gas field).

tends to show weak amplitude, high frequencies, and poor continuity due to the shielding effect of volcanic rocks (see Figure 3.4).

Seismic reflection characteristics of cycle interface. The interfaces between a weathered crust, sedimentary rocks, and volcanic rocks usually show a strong reflection of angular unconformity; the interface of volcanic composition change and the interface between fine tuff and other volcanic rocks mostly show unconformity but are characterized by poor reflection energy, continuity, and traceability (see Figure 3.4).

2. Identification markers of volcanic eruption period

1) Geological markers

A volcanic eruption period comprises volcanic rocks of a single, relatively continuous eruption within a volcanic edifice, and the lithological assemblages at different eruption periods and their sedimentary characteristics for the dormancy period are important identification markers.

Lithological characteristics within period. A volcanic eruption period differs from a volcanic eruption cycle in its relatively continuous volcanic activity. Within the period, volcanic rocks have a largely identical composition. In terms of magmatic property and eruption energy, there may be many types of lithological assemblages, such as a volcanic lava assemblage, an interbedded assemblage of volcaniclastic rock and volcanic lava, and a volcaniclastic rock assemblage. A period usually ends with quiet effusive lava or an early deposition of clay-grade tuff.

Period boundary characteristics. Volcanic dormancy refers to a break between successive periods of eruption within a volcanic edifice. Therefore, the rock bodies formed in different dormancy periods are different in lithology, shape, scale, and distribution.

Volcanic sedimentary rock interface. During an eruption dormancy between two periods, a volcanic edifice will receive clastic sediments derived from its own erosion or from surrounding rocks. These sedimentary rocks can be distinguished from those formed during an eruption dormancy between two cycles in having a higher content of volcanic material, with smaller thickness and limited distribution.

Weathering crust interface. During a volcanic dormancy, volcanic rocks formed earlier will be exposed to weathering and erosion to form a concave or convex erosional surface. Intense chemical weathering usually results in a characteristic lateritic weathering crust. In contrast to the weathering layer in a cycle-level dormancy, the exposure time during a period-level dormancy is short, leading to the formation of a small-scale weathering crust within the volcanic edifice, with a limited areal distribution (Figure 3.5).

Transitional interface of lithological assemblage. When the eruption dormancy is short or the weathering is not intense, the volcanic sedimentary rocks and weathering crust will be underdeveloped, and two successive eruption periods will be separated by a transitional interface marked by a diagnostic lithological assemblage. Depending on the eruption energy variation, the transitional interface (from bottom to top) includes the volcanic tuff → volcanic breccia transitional interface, volcaniclastic rock → volcanic lava transitional interface, volcanic lava → volcanic clast transitional interface, and volcanic lava → volcanic lava transitional interface, and so on.

2) Logging response characteristics

1. *Logging characteristics within period.* The logging characteristics of volcanic rocks in an eruption period are typically high resistivity, high density, and low interval transit time; the

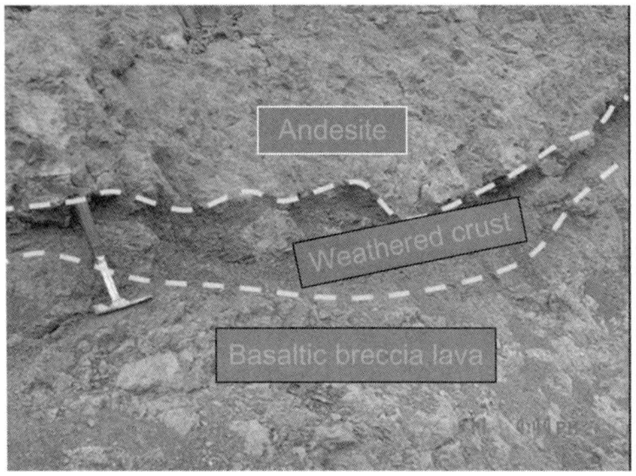

FIGURE 3.5

Weathering crust in the volcanic eruption period.

logging curves are mostly microserrated ~ serrated box and bell shaped. The particular shapes are related to the lithological assemblages and their petrophysical properties.

2. *Logging characteristics of period interface:*
   a. *Volcanic sedimentary interface.* Logging characteristics of volcanic rocks representing a dormancy include medium-low resistivity, medium-low density, and medium-high interval transit time; the curve is mostly serrated-digitate and funnel like, and partially bayonet like due to the generally small thickness.
   b. *Weathering crust interface.* Logging characteristics of volcanic rocks in the dormancy include medium-low resistivity, medium-low density, and medium-high interval transit time. Due to its small thickness, the curve tends to be mostly serrated-digitate or bayonet like.
   c. *Transitional interface of lithological assemblage.* Near the transitional interface of lithology assemblage, the logging response is related to the transition type. For example, for a transition of fine tuff → volcanic breccia, the logging curves show increased resistivity and density and decreased neutron and interval transit time, and the curves are serrated; for a transition of volcaniclastic rock → volcanic lava, the logging curve changes from a serrated bell shape to a microserrated box shape.

3) Seismic response characteristics

1. *Seismic characteristics within period.* Volcanic rocks generally have seismic reflection characteristics of strong amplitude, low frequency, and poor continuity; and the waveform is chaotic.
2. *Seismic reflection characteristics of period interface:*
   a. *Volcanic sedimentary rock interface.* Seismic reflection characteristics of a thick succession of volcanic sedimentary rocks are typically of medium-weak amplitude, relatively high frequency, and relatively good continuity, and the waveform is nearly laminar; the seismic reflection characteristics of a thin volcanic rock unit are of medium-strong amplitude and good continuity along the unconformity.

b. *Weathering crust interface*. The seismic reflection characteristics show discontinuity and medium-strong reflection along the local unconformity.
c. *Transitional interface of lithology assemblage*. Due to its identical volcanic composition and similar physical properties, a transitional interface of lithology assemblage is indistinct in seismic profile and difficult to identify and trace.

3. Identification markers of volcanic eruption rhythm

Layer and rhythm are both unstable on a regional scale and can be generally correlated within a gas reservoir [1]. The identification markers of volcanic eruption rhythm include the following features.

1) Geological markers

1. *Lava rhythm*. For the rhythmic volcanic lava assemblage, the degree of crystallization is, from bottom to top, in the order of low-high-low, the degree of vesicle development in the order of relatively high-low-high, and the vesicle size in the order of medium-small-large.
2. *Volcaniclastic rock rhythm*. For the rhythmic volcaniclastic lithology assemblage, the clastic size changes, from bottom to top, from large to small, and the physical property changes from good to poor.
3. *Interbedded rhythm of volcanic lava and volcaniclastic rock*. This rhythm type is represented mainly by alternations of volcanic lava, volcanic clastic lava, and volcaniclastic rock, reflecting variations in eruption energy.

2) Logging response characteristics

Lava rhythm. In response to the geological characteristics of a lava rhythm, the logging curves usually comprise three segments. The lower segment is mostly characterized by medium-low resistivity, medium-low density, and medium-high interval transit time, with the curve having mostly a serrated funnel shape, but this segment is usually absent; the middle segment is characterized by high resistivity, high density, and low interval transit time, with the curve having mostly a microserrated boxlike shape; and the upper segment is characterized by low resistivity, low density, and high interval transit time, with the curve being mostly in the shape of a serrated bell (Figure 3.6).

Volcaniclastic rock rhythm. The logging curves generally have a serrated funnel and box shape, and the reservoir physical property shows a spiraling decrease from bottom to top. The logging characteristics of the lower volcanic breccia include low resistivity, low density, and high interval transit time, and the curve is mostly in the shape of a serrated box; the logging characteristics of the middle tuff breccia or breccia tuff include medium-high resistivity, medium-high density, and medium-low interval transit time, with finer-grained tuff showing lower resistivity and density values, but greater interval transit time. The logging curves are microserrated box and funnel in shape (Figure 3.7).

Interbedded rhythm of volcanic lava and volcaniclastic rock. The logging curves are in various shapes. The lower volcaniclastic rock unit has such logging characteristics as medium-low resistivity, medium-low density, and medium-high interval transit time, and the curves are mostly denticulate ~ serrated bell and funnel in shape; the middle clastic lava has logging characteristics of low resistivity, low density, and high interval transit time, with the curves being mainly microserrated to serrated box in shape; the upper volcanic lava unit has the logging characteristics of medium-high resistivity, medium-high density, and medium-high interval transit time, and the curves are mainly

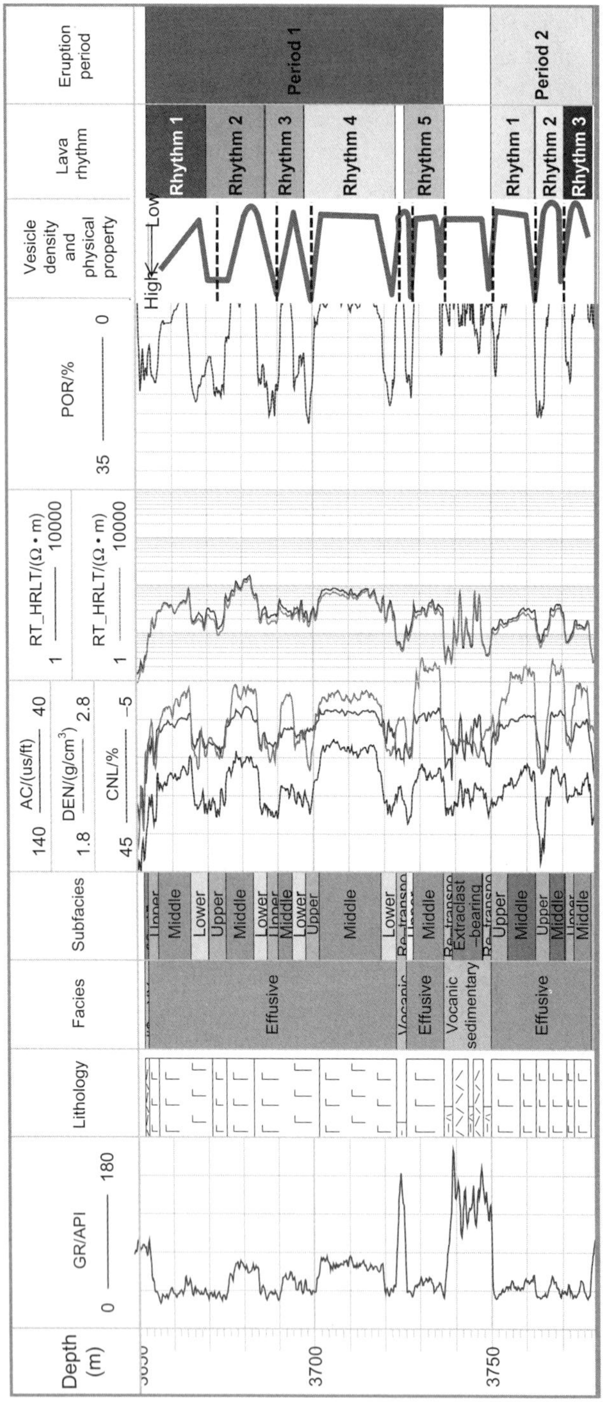

FIGURE 3.6

Logging response characteristics of volcanic lava rhythm (Well DD171).

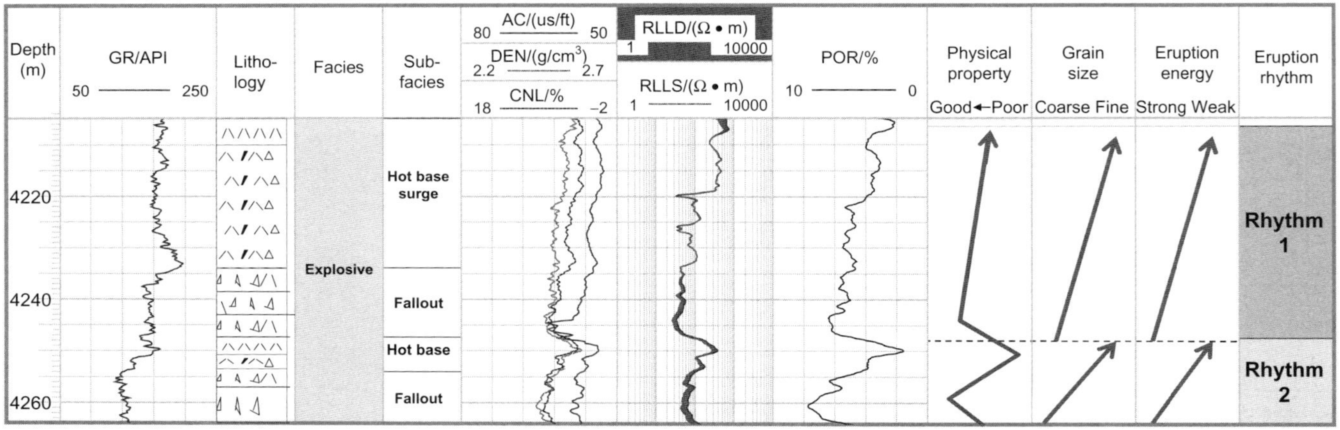

FIGURE 3.7
Logging characteristics of volcaniclastic rock rhythm (Well XX23).

smooth to microserrated bell and box in shape; the top vesicular lava has logging characteristics of low resistivity, low density, and high interval transit time, with the curves being mainly microserrated-digitate and bell shaped (Figure 3.8).

3) Seismic markers

The volcanic lava rhythm mainly mirrors the sublaminar reflection characteristics of medium-strong amplitude, low frequency, and good continuity; the seismic reflection energy for volcaniclastic rock rhythm is relatively weak, having mostly chaotic reflection characteristics of medium-weak amplitude, low frequency, and poor continuity; the stratification of interbedded rhythm of volcanic lava and volcaniclastic rock is relatively good, mostly corresponding to the laminar reflection characteristics of strong amplitude, medium-high frequency, and good continuity.

4. Identification markers of cooling unit

1) Geological markers

When the time interval between two pulses of eruption is sufficiently long, there will be a cold-hot interface between two successive cooling units. A cooling unit is the basic component within volcanic rocks, marking a hiatus between two cooling units. The geological markers include the following:

1. *Soft top and hard bottom structures.* Due to different cooling conditions at the top, in the middle, and at the bottom of a cooling unit, soft top structures can be found in the upper interface and a hard bottom structure in the basal interface.
2. *Vesicles and amygdaloids.* When lava erupts, a sudden pressure drop results in degassing of abundant volatile components and usually the formation of numerous vesicles near the lava top in a globular and sometimes irregularly cloudy shape. Such volatile contents are relatively low at the base, and a small amount of gas can escape only when the lower rock is baked. Thus, compared to the top, the basal vesicles are less common and have a different shape, usually tubular, curved tubular, or elliptical. The cooling in the middle part is slow, with crystallization being the main diagenetic feature, and the vesicles are underdeveloped.
3. *Oxidization top.* After a volcanic eruption, the volcanic rock surface is in contact with air to facilitate strong oxidization, converting Fe^{2+} into Fe^{3+} and producing a brownish red oxidization top. However, the volcanic rock bottom is not exposed and remains in a reducing environment; hence, it is usually green. This color feature of "red top and green bottom" can serve as a marker of stratification.
4. *Vitreous crust and molten slag crust.* The surface lava flow produces a thin vitreous crust on its top due to fast cooling, and this thin crust is generally dark, greenish black for basic lava and mauve to brown for neutral-acid lava. The crust is usually several millimeters to several centimeters thick. The top of some neutral-acid lava contains a large amount of volatiles, which escapes quickly to produce a coarse vitreous crust on the top.
5. *Baked edge.* Hot magma in the eruption bakes the underlying rocks that cooled early to produce a thin baked edge marked by color fading.

2) Logging response characteristics

Because of its consistent composition, the rock texture and structure of a cooling unit will change only slightly at the top and bottom, which leads to small logging variations for the unit. The change in vesicles and amygdales causes regular changes in lava physical properties, resistivity, density,

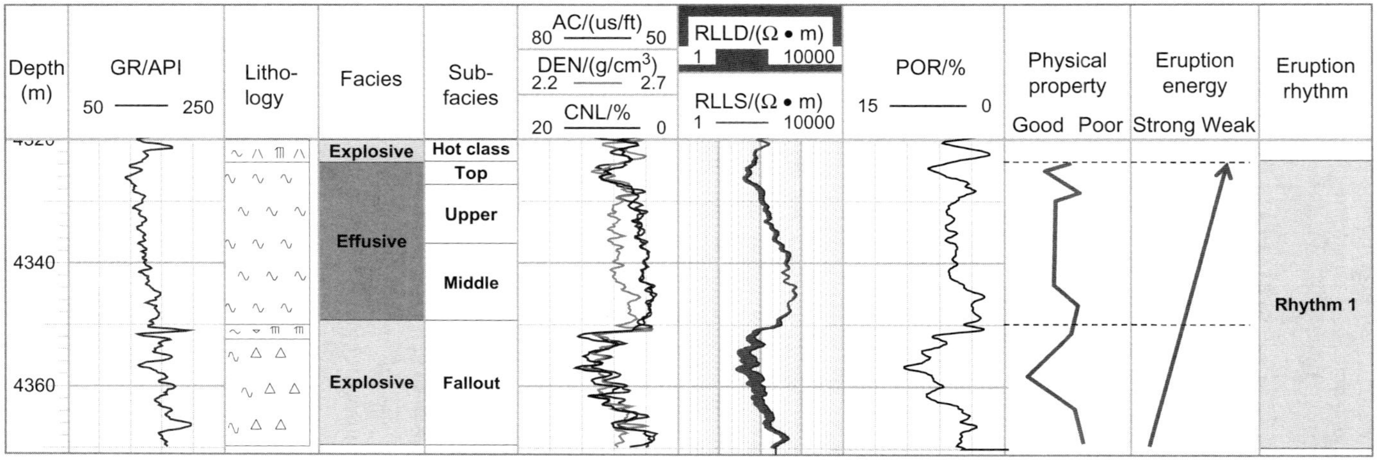

FIGURE 3.8

Logging characteristics of the interbedded rhythm of volcanic lava and volcaniclastic rock (Well XX301).

and interval transit time. In combination with rock texture and structure identification, the logging data can be used to identify a cooling unit.

3.3.2 Single-well identification

According to the geological markers and logging characteristics, the level-by-level control method can be used to identify the volcanic rock eruption cycle, period, rhythm, and cooling unit in single wells, as is demonstrated here using Well XX14 as an example.

1. Determining the environmental characteristics of volcanic eruption in single wells

Volcanic rocks of the Yingcheng Formation in the XX gas field constitute a suite of acid volcanic deposits, characterized by gray to off-white rhyolite, breccia lava, welded breccia, welded tuff, crystalline tuff, volcanic breccia interbedded with thin black mudstone, and silty mudstone, and the volcanic formation is divided into three eruption cycles according to regional eruption breaks. Well XX14 lies in the southwest of the volcanic formation and reveals the third, the second, and the top part of the first cycles.

2. Identifying lithology and rock composition, eruption dormancy, and eruption cycle

Regionally distributed sedimentary rocks and weathered layer, continuously distributed fine tuff, abrupt interface of volcanic rock composition, and extrusive facies are key indicators for recognizing eruption dormancy and differentiating volcanic cycles.

In Well XX14, there are mainly acid volcanic rocks with nearly identical composition. However, separated by a volcanic sedimentary rock unit in the 3949-3987 m interval, the Th content in the upper volcanic rocks is almost twice as high as in the lower part; and K content is relatively low, showing different alkalinity in the upper and lower volcanic rocks; meanwhile, this well encounters volcanic conduit facies and extrusive facies in the basal 30 m. According to the sedimentary rocks, rock compositions, and extrusive facies markers, therefore, three eruption cycles can be recognized, with the boundaries at 3948 m and 4125 m from top to bottom, respectively (see the fourth column on the right in Figure 3.9).

3. Finding marker of eruption dormancy within volcanic edifice and identifying eruption period

In a volcanic edifice, the direct criteria for determining eruption dormancy and identifying eruption period include the volcanic sedimentary rocks, weathering crust, and transitional interface of lithological assemblage. In Well XX14, the volcanic edifice is a centered type composed mainly of effusive facies with some explosive facies. Here, there are three eruption cycles with no volcanic sedimentary rocks or typical weathering crust inside; hence, the lithological assemblage conversion is employed to identify the eruption period.

In Well XX14, a volcaniclastic rock ~ rhyolite assemblage and a volcanic breccia ~ tuff assemblage occur from top to bottom. In cycle 1, the volcanic conduit facies and extrusive facies belong to different eruption periods, with logging curves being microserrated bell in shape. In cycle 2, there is a volcanic breccia ~ breccia tuff ~ tuff assemblage and volcanic breccia ~ breccia lava ~ rhyolite ~ volcanic

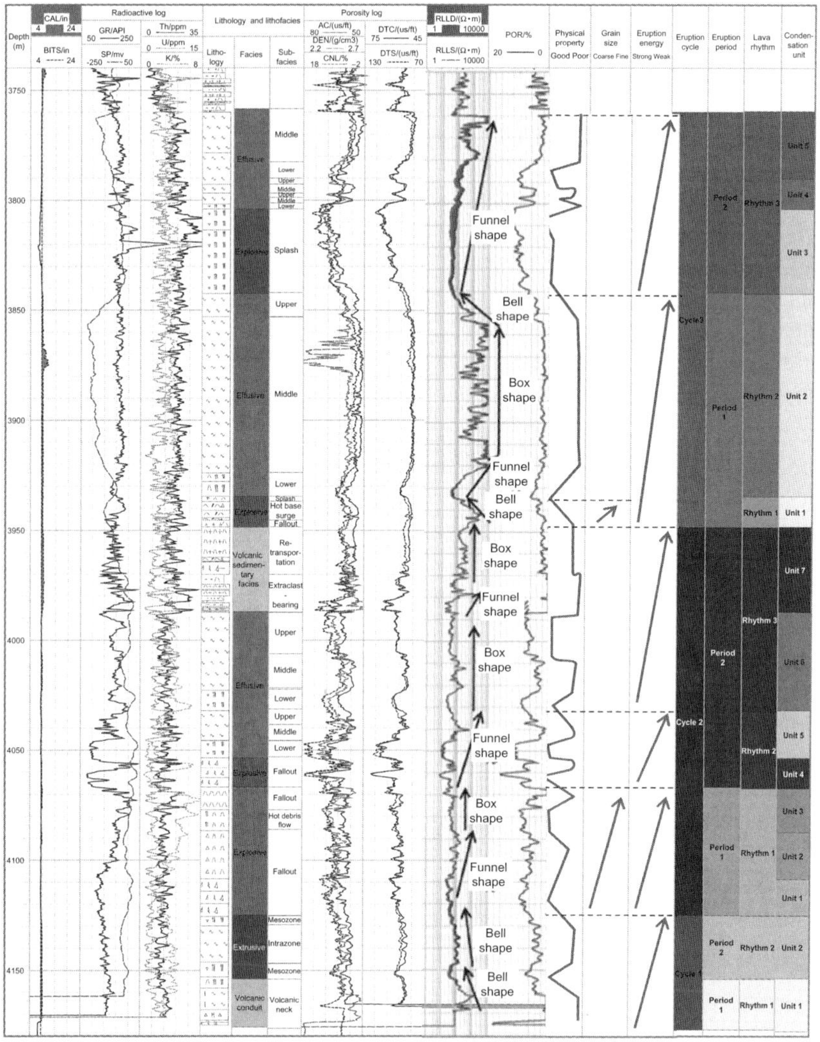

FIGURE 3.9

Identification of volcanic eruption cycle in a single well (Well XX14).

sedimentary assemblage, with logging curves being mainly microserrated box and funnel in shape, and the boundary is at 4066.8 m. In cycle 3, there is a crystalline tuff ~ tuff lava ~ rhyolite assemblage, with logging curves having bell ~ funnel ~ box ~ bell shapes; there is another breccia lava ~ rhyolite assemblage, with logging curves being funnel shaped. Thus, the lithological assemblage and logging curve characteristics indicate the presence of six eruption periods in three volcanic rock cycles in Well XX14, with each cycle containing two eruption periods (see the third column on the right in Figure 3.9).

4. Identifying volcanic eruption rhythm based on petrophysical properties, grain size variation, and lithological assemblages

On the basis of contact relationships between volcanic lava and volcaniclastic rock, as well as characteristics of each rhythm, various volcanic eruption rhythms can be identified, and the volcanic eruption rhythm distribution in single wells can be revealed.

In the noncored intervals, three types of volcanic rock rhythm can be effectively identified according to the variation rule of volcanic lava petrophysical properties from medium-small-large, volcaniclastic rock grain size from large to small, and lithological assemblage of interbedded volcanic lava and volcaniclastic rock from bottom to top. In Well XX14, eight eruption rhythms are developed in six eruption periods (see the second column on the right in Figure 3.9). Two interbedded rhythms in cycle 1 are consistent with the period scale. Period 1 in cycle 2 comprises one volcaniclastic rock rhythm, with logging curves being serrated box-funnel in shape; and period 2 consists of two interbedded rhythms, with logging curves being two clusters of serrated box-funnel combination. Period 1 in cycle 3 contains one volcaniclastic rhythm, with logging curves being microserrated bell in shape with small thickness. It also contains one interbedded rhythm, with logging curves showing a combined funnel-box-bell shape; and period 2 develops one interbedded rhythm, with logging curves being smooth-microserrated funnel in shape.

5. Recognizing cooling boundary and identifying cooling unit of volcanic rock

Soft top and hard bottom structures, oxidization top, vitreous crust, molten slag crust, baked edge, vesicle, and amygdaloid are the typical markers for distinguishing cooling units. Because the logging characteristics of most markers are indistinct, the abrupt contact of rock type, regular variation of vesicle and physical properties, as well as logging curves can be used to identify cooling units.

In Well XX14, there are 12 abrupt lithological interfaces reflecting eruption energy change, among which two interfaces reflect the regular change in petrophysical property. A comprehensive analysis under the constraint of eruption rhythm framework, therefore, recognized 14 cooling units in this well (see the first column on the right in Figure 3.9), among which there are two units in cycle 1, with the same scale as that of a rhythm and period; seven units in cycle 2, including two volcanic breccia units, one breccia tuff unit, one fine tuff unit, two rhyolite units, and one volcanic sedimentary rock unit; and five units in cycle 3, including one crystal tuff unit, three rhyolite units, and one breccia lava unit.

3.3.3 Profile identification

1. Identification of volcanic eruption cycle

Volcanic eruption cycles can be identified according to their seismic response mode, in conjunction with regional geology and seismic reflection analysis.

1) Sedimentary rock as cycle separator

The alternating occurrence of sedimentary and volcanic rocks reflects the episodic nature of volcanic activity. For example, three volcanic eruption cycles were found in the volcanic rocks of the DD gas field, which reflect three coupled sets of volcanic and sedimentary rocks. The bulk of these cycles is represented by volcanic rocks in their lower part. The upper sedimentary rock unit constitutes the barriers in these cycles. On the seismic profile, the volcanic eruption cycle shows alternating strong-weak

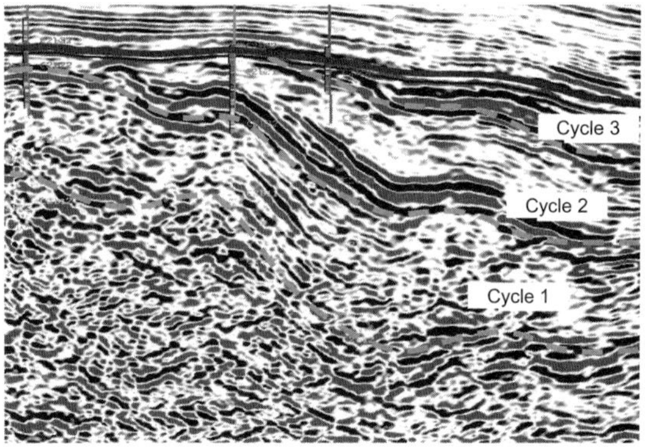

FIGURE 3.10

Seismic reflection characteristics of the sedimentary rock cycle interface.

seismic reflection characteristics throughout the area, and the volcanic-sedimentary rock interface shows a strong reflection of regional discontinuity on seismic profiles (Figure 3.10).

2) Weathering crust as cycle separator

Uplift and erosion of the early cycle result in a prominent regional weathered layer, which is the interface of paraconformity or angular unconformity. In the XX gas field, for example, there are three eruption cycles with predominantly acid volcanic rocks, and the cycles are in contact with each other through paraconformity; the reflection characteristics inside the cycles are similar in seismic profiles, and the cycles are separated by a discontinuous and traceable unconformity (Figure 3.11).

3) Magma evolution cycle

This type of cycle reflects the periodical evolution of magma compositions. For example, the volcanic rocks in CC gas field comprise two eruption cycles, with an discontinuous unconformity between the lower basic cycle and the upper acid cycle. In seismic profiles, there are different seismic responses

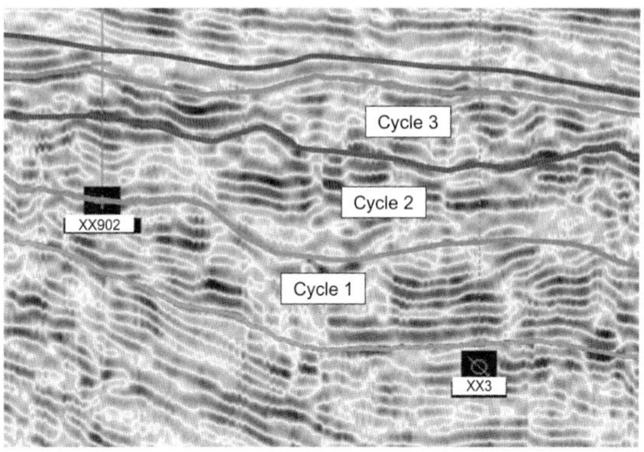

FIGURE 3.11

Unconformity contact of the cycle interface.

inside each volcanic eruption cycle, characterized mainly by pseudolaminar reflections with medium-high frequency, medium-weak amplitude and good continuity in the lower part of each cycle, and chaotic reflections with strong amplitude, low frequency, and discontinuity in the upper part of each cycle (see Figure 3.4).

2. *Identification of volcanic eruption period*

For different volcanic edifices, the eruption periods are different, and their identification should be carried out within the framework of volcanic edifice. In different eruption periods, the lithological assemblage, rock texture, structure, and strike are different. There are diagnostic characteristics of volcanic sedimentary rocks for volcanic dormancy, local weathering crust, and transitional interface between successive eruption periods, forming locally unconformable contacts between volcanic periods. The seismic reflections, therefore, are typically chaotic; they are of weak amplitude, low frequency, and discontinuity inside each period, and the boundaries between successive periods are discontinuity surfaces with strong reflection characteristics. An eruption period inside a volcanic edifice can be identified based on the strong reflections in the bottom and top parts and weak reflections in the middle.

A volcanic edifice, composed mainly of basic volcanic rocks and formed by fissured–centered eruption, was found in the Yingcheng Formation in Well Block CC2 (Figure 3.12). This edifice contains three suites of volcanic rocks of 12-m to 30-m thickness and two suites of volcanic sedimentary rocks of 10-m to 13-m thickness, which suggests three eruptions separated by two eruption hiatuses. Hence, this edifice can be divided into three volcanic eruption periods. In the seismic profile, the internal part of this edifice shows laminar reflections with relatively strong amplitude, medium frequency, and good continuity. The integrated well-seismic calibration shows that the reflections of the basalt formation in stages 1 and 3 have a relatively weak amplitude and good continuity; and the reflections of basaltic clastic rocks in stage 2 have a relatively strong amplitude and poor continuity; and the period interface shows strong reflections of the local paraconformity under the constraint of the volcanic edifice (see Figure 3.12).

3. *Identification of volcanic eruption rhythm*

The largely homogeneous composition of a volcanic eruption rhythm leads to indistinct seismic reflection characteristics for the rhythm interface. With the constraint of the volcanic edifice and eruption period, therefore, the eruption rhythm can be identified by tracing the event, in combination with integrated well-seismic calibration, as is shown in the following case study on volcanic edifice YC_2^1 in the CC gas field (Figure 3.13).

1) Identification from single well

Volcanic edifice YC_2^1 encountered in Well CC102 consists of a lower rhyolite unit about 150 m thick, a middle volcanic breccia unit about 70 m thick, and an upper rhyolite unit about 120 m thick. In terms of the volcanic sedimentary rock characteristics typical of volcanic dormancy, two periods are recognized (see the right part in Figure 3.13).

Constrained by eruption period, volcanic rock in the lower subfacies of the early effusive facies is not recognized but generally shows a complete vesicle sequence (underdeveloped → developed) and a petrophysical property sequence (poor → good), typical of a volcanic lava rhythm. Volcanic rocks of the late effusive facies also have a complete vesicle and petrophysical property sequence, although the

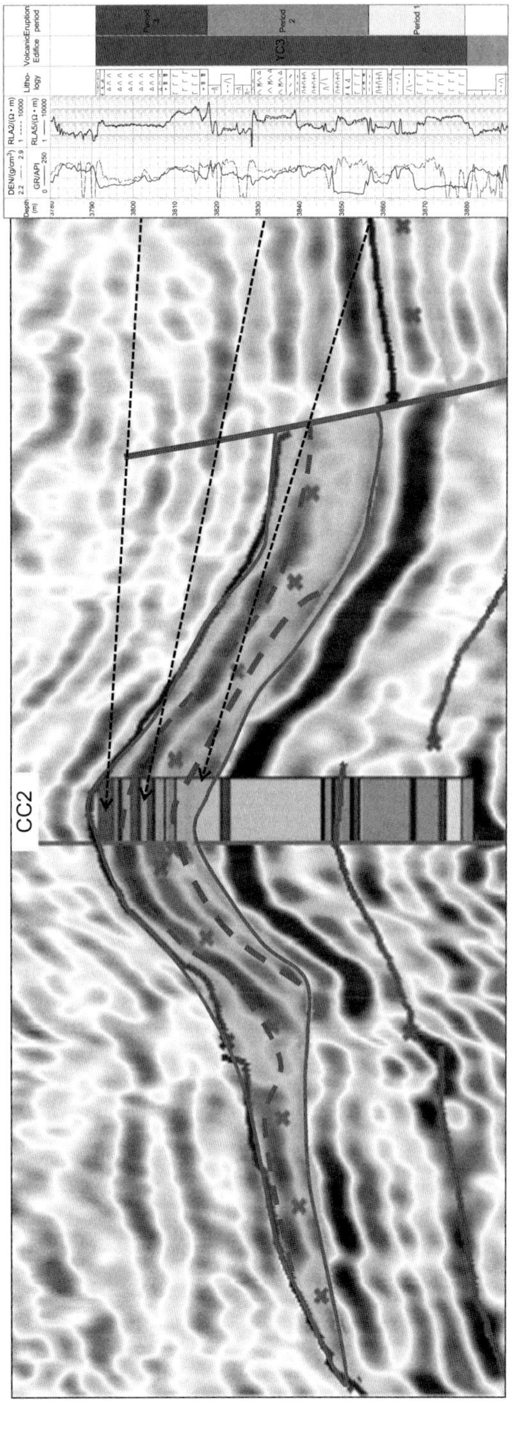

FIGURE 3.12

Volcanic eruption period identified in the seismic profile.

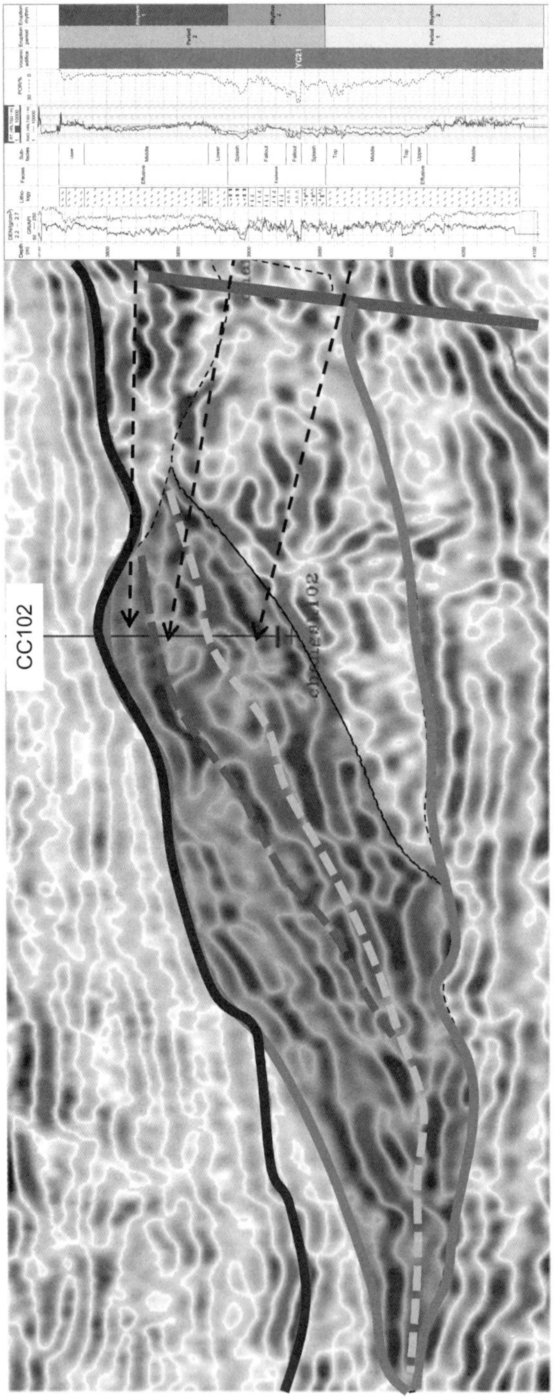

FIGURE 3.13
Volcanic eruption rhythm identification in the seismic profile.

late explosive facies does not have a complete grain size variation sequence but has a complete petrophysical property variation sequence. Thus, one volcanic lava rhythm and one volcaniclastic rock rhythm are recognized, and the volcanic edifice in this well contains two volcanic eruption periods and three eruption rhythms (see Figure 3.13).

2) Identification from seismic profile

Volcanic eruption periods can be identified within a volcanic edifice on the basis of well-seismic calibration, differences in internal reflection characteristics, and a strong reflection of local unconformity, With this constraint, the identification of an eruption rhythm can begin from a single-well seismic profile according to its pseudolaminar reflection of the volcanic lava rhythm, chaotic reflection of volcaniclastic rock rhythm, and paraconformity of the rhythm interface.

In Figure 3.13, the yellow line represents the boundary between two eruption periods inside the YC_2^1 volcanic edifice, and the red line represents the boundary between two eruption rhythms within the second eruption period. The upper and lower volcanic lava rhythms show laminar reflections with strong amplitude, low frequency, and good continuity; the middle volcaniclastic rock rhythm shows chaotic reflections with medium-strong amplitude, low frequency, and discontinuity; the upper and lower interfaces show reflections of a local unconformity, but there are event crossings in interface tracing, making it difficult to identify the volcanic eruption rhythm interface based on the seismic profile.

The cooling unit is small and distributed in a limited area; due to its low resolution, the seismic data are rarely used for its identification and division.

3.4 VOLCANIC ROCK SEQUENCE DIVISION AND STRATIGRAPHIC CORRELATION

Through volcanic rock sequence identification, correlation, and analysis, the volcanic rock sequence and internal architecture, as well as their relationships to gas reservoirs, can be established, and the stratigraphic sequences can be correlated at different levels. With the constraint of internal architecture, the volcanic rock sequences can be classified and the formation correlated level by level.

3.4.1 Corresponding relationships among volcanic rock sequence, internal architecture, and gas-bearing zone

The corresponding relationships among the volcanic gas-bearing zone, stratigraphic sequence, and internal architecture of volcanic rocks can be established based on an analysis of their internal architecture and stratigraphic sequences (Table 3.2).

1. Gas-bearing formation corresponding to the volcanic formation

A gas-bearing formation comprises continuous deposits within a level 2 depositional cycle, characterized by similar depositional origin and rock types, consistent gas and water features, and a cap layer of thick mudstone. The top and bottom boundaries of a gas-bearing formation are consistent with its time interface, and one gas-bearing formation comprises several gas zones.

Table 3.2 Corresponding Relationship of Volcanic Gas-Being Zone to Stratigraphic Sequence and Internal Structure

Levels	Internal Structure	Stratigraphic Sequence	Development Zone
1	Volcanic formation	Volcanic formation	Gas-bearing formation
2	Volcanic edifice	Eruption cycle	Gas zone
3	Volcanic massif	Eruption period	Gas subzone
4	Volcanic rock facies	Eruption rhythm	Single layer
5	Volcanic lithology	Condensation units	Accumulation-permeation unit

A volcanic formation is a volcanic rock assemblage with a closely related petrogenesis, continuous spatiotemporal occurrence, similar rock type, and consistent gas and water characteristics; its top and bottom contacts with surrounding rocks are marked by regional angular unconformities, consistent with the stratigraphic age interface, and one volcanic formation comprises multiple volcanic eruption cycles or volcanic edifices.

Within the volcanic gas reservoir range, the volcanic formation corresponds to the gas-bearing formations.

2. Gas zone corresponding to the volcanic eruption cycle and edifice

In level 3 depositional cycles, the gas-bearing layers with similar depositional environments, distributions, rock properties, petrophysical properties, and natural gas properties are classified into one *gas zone* but separated by relatively thick and stable interbeds. The *gas zone interface* is consistent with the sedimentary cycle interface, and one gas *zone* comprises several gas *subzones*.

A volcanic eruption cycle is a volcanic rock assemblage erupted in the same period from the same magma source in one or multiple edifices. The petrogenic environment, distribution, rock property, petrophysical property, and fluid property are relatively similar. The top and bottom layers are stable sedimentary rocks or a weathered crust. One eruption cycle comprises several volcanic edifices and internal eruption periods.

Within a volcanic gas reservoir, the volcanic eruption cycle and edifice correspond to the gas zone.

3. Gas subzone corresponding to the volcanic eruption period and volcanic massif

The enriched gas layers within a gas zone are grouped into one *gas subzone*, consistent with the horizons in a level 4 depositional cycle. The gas subzones are separated by stable barriers, and one *gas subzone* includes many *gas layers*.

A volcanic eruption period is represented by a volcanic rock assemblage formed in the relatively concentrated explosions within a volcanic edifice, its bottom and top are separated by relatively stable sedimentary rocks or weathered crust, and there are several layers composed of volcanic eruption rhythms inside the period.

In a volcanic gas reservoir, the volcanic eruption period and volcanic massif correspond to gas subzone.

4. Single layer corresponding to volcanic eruption rhythm and volcanic facies

A continuous gas interval separated by impermeable beds can be grouped into one single-layer [15]. A single layer corresponds to a horizon in a level 5 depositional cycle. In practice, the classification criteria for single layers should be the same within one development district, but they can be different in different development districts in a gas field [16]. Generally speaking, the independent layers with half the area of the development district can be grouped into single layers, and one single layer can contain several accumulation-permeation units.

Volcanic eruption rhythms are the volcanic rock assemblages formed by one concentrated eruption of a volcanic period. They have the lithological characteristics of a facies sequence and contain several *cooling units*.

In a volcanic gas reservoir, volcanic eruption rhythms and volcanic rock facies correspond to single layers.

5. Accumulation-permeation unit corresponding to cooling unit and volcanic lithology

A *cooling unit* is developed within the lithofacies equivalent to the volcanic lithology; and the cooling unit and volcanic lithology correspond to the accumulation-permeation unit.

3.4.2 Correlation of volcanic rock sequence

On the basis of the corresponding relation described previously and according to the geological features, constraints, scale, and distribution range of the volcanic rock sequence, the correlation at each sequence level is systematically determined (Table 3.3).

Table 3.3 Volcanic Rock Sequences

Stratigraphic Sequence	Constraint Mechanism	Distribution Range	Scale		Correlation
			Thickness	Extension	
Eruption cycle	Volcanic formation	Inside the basin	Hundreds to thousands of meters	Tens to hundreds of kilometers	Regional correlation
Eruption period	Volcanic edifice	Inside the edifice	Tens to hundreds of meters	Several to tens of kilometers	Division inside the edifice
Eruption rhythm	Eruption period		Several meters to tens of meters	Hundreds to thousands of meters	
Cooling unit	Eruption rhythm	In the massif	Several meters to tens of meters	Tens to hundreds of meters	Extrapolation from single well

1. Volcanic eruption cycle

A volcanic eruption cycle represents the product of volcanic activities at different stages within a volcanic formation, separated by long-term volcanic dormancy with distinct time difference and spatiotemporal continuity. It serves as a geological basis for dividing the gas-being zones of a volcanic gas reservoir. Such a cycle can be hundreds to thousands of meters thick and extend for tens to hundreds of kilometers, with a large areal range, or even covering an entire basin. Meanwhile, there are stable sedimentary rocks or a weathered crust between successive cycles. Therefore, volcanic eruption cycles can be correlated regionally within the range of a volcanic formation.

2. Volcanic eruption period

A volcanic eruption period consists of the product of one continuous eruption within a volcanic edifice, which is separated by short-term volcanic dormancy and characterized by spatiotemporal continuity (i.e., without recognizable time difference within the rock unit). It forms the geological basis for dividing the gas subzones of a volcanic gas reservoir. A period may be represented by rocks from tens to thousands of meters thick and extend laterally from several meters to tens of kilometers, with a moderate distribution range within the volcanic edifice. Therefore, the stratigraphic division and correlation of volcanic eruption periods can be carried out inside the range of a volcanic edifice.

3. Volcanic eruption rhythm

A volcanic eruption rhythm is the product of one concentrated eruption inside the eruption period and is separated by the regular interface of lithology assemblages. It serves as a geological basis for dividing the gas layers of a volcanic gas reservoir. A rhythm can be several meters to tens of meters thick and extend for hundreds to thousands of meters, with a small distribution area within the volcanic rocks of a given eruption. Therefore, stratigraphic division and correlation of volcanic eruption rhythms can be carried out within the framework of an eruption period.

4. Cooling unit

A cooling unit is the product of one eruptive pulse within an eruption rhythm, separated by a cold-hot interface of the volcanic rock. A cooling unit can be several meters to tens of meters thick and extend for tens to hundreds of meters, and the distribution range is small and changes quickly. The division and extrapolation in the drill intervals and single wells are usually used to predict the spatial distribution of a cooling unit.

3.4.3 Volcanic rock sequence division and stratigraphic correlation

The stratigraphic division and correlation complement each other and are inseparable [16]. As is shown in the case study of Well Block XX21, the division and correlation of the volcanic eruption cycle, period, and rhythm are carried out step by step based on sequence identification, following the principle of level-by-level control and correlation.

1. Regional correlation of volcanic eruption cycle

Regional correlation of an eruption cycle is carried out using its geological, logging, and seismic identification markers, following the identification procedure from single-well to interwell correlation and then profile tracing.

1) Interwell correlation of eruption cycle

With the constraint of regional geological framework and cycle identification in single wells, the interwell correlation of a volcanic eruption cycle can be conducted in conjunction with analyses of seismic response characteristics and unconformity contact.

In Well Block XX21, there are three eruption cycles in the volcanic rocks of the Yingcheng Formation. These include the second cycle encountered in Wells XX21-2, XX21, XX21-1, XX21-4 and XX231; the third cycle in Well XX23; and the bulk of the third cycle and the second cycle below 4316 m in Well XX21-5. The first cycle is not encountered by any wells. Thus, the interwell correlation of the volcanic eruption cycle can be established (Figure 3.14).

2) Regional correlation of volcanic eruption cycle based on profile delineation

Built on the interwell correlation pattern, the eruption cycle can be identified and traced in seismic profiles to determine its distribution and contact relationships on the basis of integrated well-seismic calibration and the formation occurrence and seismic reflection within the cycle.

As shown in Figure 3.16, three eruption cycles in Well Block XX21 all show an incline trend to the northwest: the first cycle is composed mainly of volcanic and volcanic sedimentary rocks with laminar reflections of strong amplitude, medium-low frequency, and good continuity in the seismic profile; the second cycle is composed of the volcanic rocks of mixed but mainly effusive facies, and the seismic profile shows reflections of weak amplitude, low frequency, and poor continuity and parallel to subparallel waveform; and the third cycle is composed of interbedded effusive facies and explosive volcanic facies, and the seismic profile shows reflections of strong amplitude, low frequency, and poor continuity and overlies the second cycle in lateral contact.

3) Volcanic edifice as framework constraint of volcanic eruption period

This can be demonstrated using the second cycle as an example. Wells XX231 and XX21 penetrated a large amount of autoclastic lava breccia, which shows almost perpendicular columnar reflections in the seismic profile. In combination with the seismic response of peripheral structures, at least two volcanic edifices were identified as parts of the second cycle in Well Block XX21 (Figure 3.15), thus providing a basis for volcanic period and rhythm division.

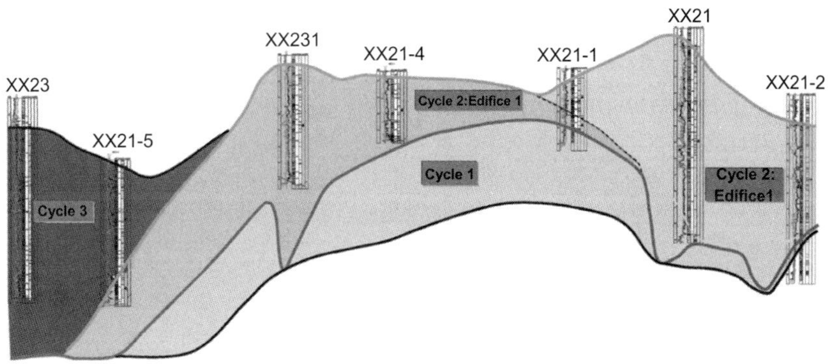

FIGURE 3.14

Well-tie correlation of the volcanic eruption cycle in Well Block XX21.

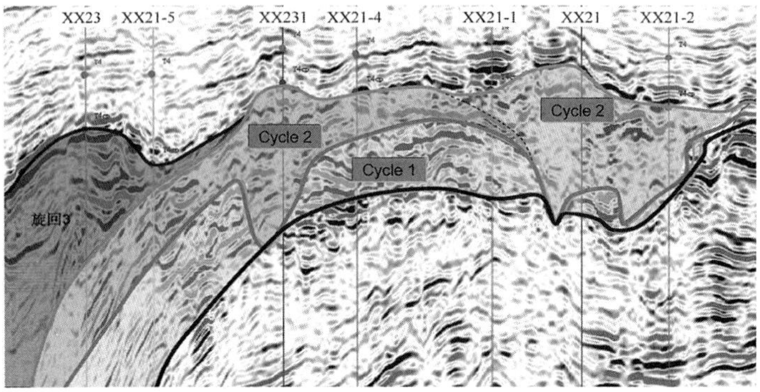

FIGURE 3.15

Seismic profile of the volcanic eruption cycle in Well Block XX21.

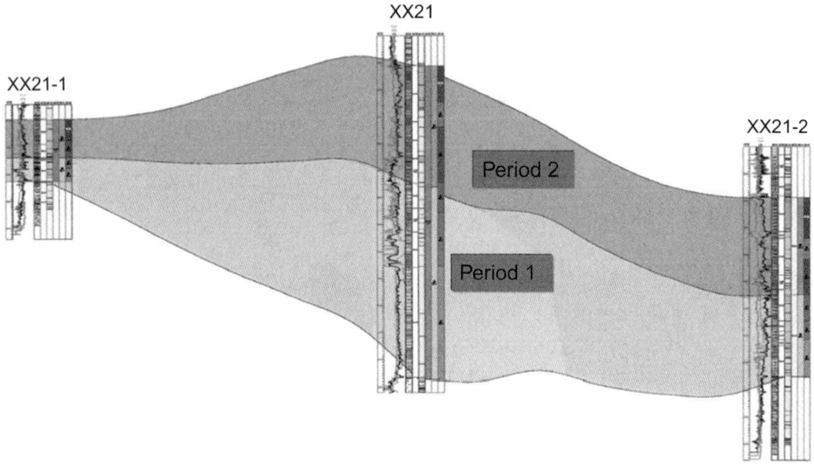

FIGURE 3.16

Interwell correlation of the volcanic eruption period in Well Block XX21.

2. Eruption period division and correlation constrained by volcanic edifice

A volcanic eruption period reflects the spatiotemporal relationship of the eruption products in different periods separated by dormancy inside the edifice. Under the constraint of volcanic edifice, therefore, the combined interwell pattern of an eruption period can be revealed through single-well identification and interwell correlation. This is followed by well-seismic calibration and volcanic eruption period division in the profile. The volcanic edifice encountered in the second cycle in Well Block XX21 is selected for the division.

1) Single-well identification and preliminary division of eruption period

On the basis of the logging characteristics of single wells, the rock identification can be used to recognize volcanic sedimentary rock, weathered crust, and the transitional interface of lithological assemblage to divide the volcanic eruption period in single wells.

Among the three wells that encountered XX21 volcanic edifice, Well XX21 penetrated a volcanic conduit represented mainly by autoclastic breccia lava. Well XX21-2 encountered mainly welded volcanic breccia and rhyolite, without apparent volcanic sedimentary rock; but the main massifs include two lithological assemblages: volcanic breccia~rhyolite~tuff lava and volcanic breccia~breccia lava~rhyolite, corresponding to two eruption periods, respectively. In Well XX21-1, there is a thin suite of volcanic sedimentary rock (3819 to 3823 m) in the second cycle. Thus, the volcanic rocks in this well are divided into two periods, corresponding to the first and second periods in Well XX21-1, respectively.

2) Interwell correlation to determine interwell combination pattern of eruption period

On the basis of single-well identification, volcanic edifice distribution pattern, and variations in lithology and lithofacies, the interwell correlation of eruption period can be conducted to determine the interwell combination pattern of an eruption period.

The XX21 volcanic edifice is formed by a centered eruption, the volcanic conduit lies to the north of Well XX21, and the eruption mode includes alternating explosion and effusion. According to the circular distribution geometry of centered eruption, the interwell combination pattern of the eruption period is established (see Figure 3.16) to form the basis for profile division.

3) Division of volcanic eruption period based on well-seismic combination

Through single-well division and integrated well-seismic calibration, different reflectors are identified in seismic profiles, the local unconformity is traced, and the eruption period within the volcanic edifice is systematically divided.

There are two different seismic reflectors inside the XX21 volcanic edifice (Figure 3.17); the upper part shows laminar reflections with strong amplitude, low frequency, and good continuity; the lower part has chaotic reflections with weak amplitude, high frequency, and discontinuity; and the interface shows strong reflections of a local unconformity. The seismic reflection characteristics are consistent

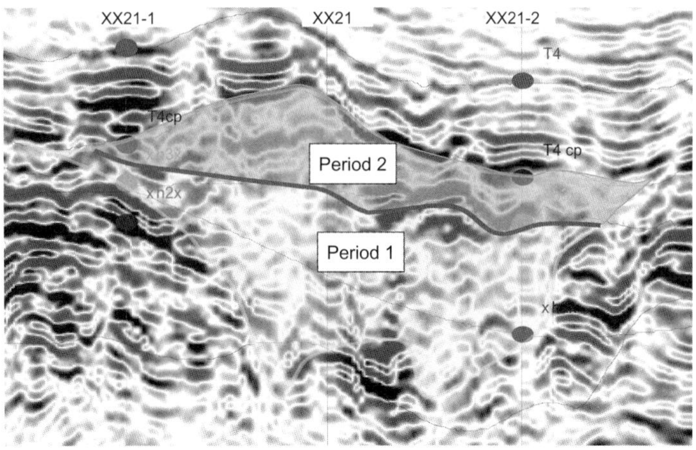

FIGURE 3.17

Eruption period division profile under the constraint of the volcanic edifice.

with the single-well calibration, and a comprehensive analysis confirms the presence of two eruption periods inside the XX21 volcanic edifice.

3. Rhythm division constrained by volcanic edifice and eruption period

A volcanic eruption rhythm reflects the spatiotemporal relationships between the concentrated eruption products within a volcanic edifice. Due to the minor differences in terms of reflection characteristics and indistinct interfaces inside the volcanic rock, the division of volcanic eruption rhythms can be extrapolated from single-well identification under the constraint of the volcanic edifice and the eruption period.

1) Single-well identification

The division of volcanic eruption rhythms relies on their geological and logging characteristics, as well as the single-well lithological assemblages and logging curve patterns.

Figure 3.18 is an example of eruption period division in Well XX21-2 that encounters the same volcanic edifice in Wells XX21 and XX21-1. Through integrated well-seismic analysis, the volcanic edifice is divided into two eruption periods and six eruption rhythms. The lithology and logging curve patterns indicate that Well XX21-2 encountered four eruption rhythms, from the second to the fifth rhythm. Rhythm 2 is characterized by interbedded volcanic breccia and rhyolite, with logging curves being serrated bell in shape; rhythm 3 consists of interbeds of welded tuff, tuff lava, and rhyolite, with logging curves being microserrated box and bell in shape; rhythm 4 comprises interbedded volcanic breccia and volcanic lava, with some volcanic conduit subfacies, and its logging curves have a serrated bell-funnel shape; rhythm 5 is typified by welded breccia, with logging curves being smooth bell-funnel in shape with good petrophysical properties in the top and bottom parts, but poor petrophysical properties in the middle. The single-well identification lays a foundation for profile division.

2) Profile division

On the basis of single-well identification, seismic reflection variation, the constraint of eruption period, and interactive analysis of interwell correlation and seismic profile tracing, the interwell combination pattern of a volcanic eruption rhythm can be systematically delineated, and the profile division is realized to determine its spatial distribution.

Figures 3.19 and 3.20 are the maps of interwell correlations and seismic profiles for the division of eruption rhythms within the volcanic edifice in Well Block XX21. These maps show two eruption periods inside the volcanic edifice, with three rhythms in each of the periods.

Rhythm 1 comprises two volcanic conduits and the surrounding rhyolite in Well XX21. The conduits are characterized by reflections of weak amplitude, low frequency, and discontinuity and an umbrella-shaped configuration. The rhyolite shows laminar reflections with strong amplitude, low frequency, and good continuity, with a high formation dip angle.

Rhythm 2 is composed of volcanic breccia and rhyolite, with its seismic profile showing chaotic reflections of medium-strong amplitude, high frequency, and poor continuity due to the volcanic conduit. This rhythm unit is encountered in Wells XX21, XX21-2, and XX21-1.

Rhythm 3 is represented by interbedded rhyolite and welded tuff; the seismic profile shows chaotic reflections of medium-strong amplitude, low frequency, and poor continuity, but weak amplitude and high frequency in the volcanic conduit. This rhythm unit is encountered in Wells XX21 and XX21-2.

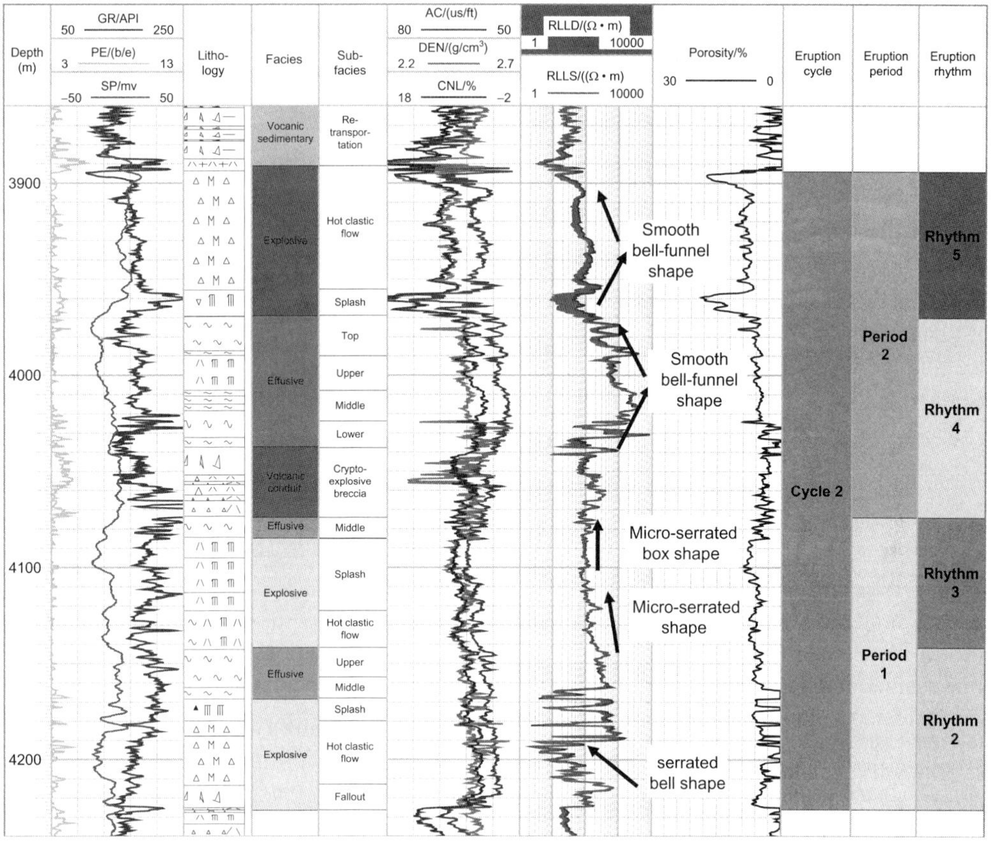

FIGURE 3.18

Eruption rhythm division in a single well (Well XX21-2).

Rhythm 4 comprises interbedded volcanic breccia and volcanic lava. Due to the presence of volcanic conduit (to the north of Well XX21), the seismic profile shows laminar reflections with strong amplitude, low frequency, and good continuity, but weak amplitude and high frequency in the volcanic conduit. This rhythm unit occurs in Wells XX21 and XX21-2.

Rhythm 5 consists of welded breccia and rhyolite, with laminar seismic reflections of strong amplitude, low frequency, and good continuity. This rhythm unit is encountered in Wells XX21, XX21-1, and XX21-2.

Rhythm 6 is represented by interbeds of welded breccia and breccia lava in Wells XX21 and XX21-1; its seismic reflections are chaotic and of medium amplitude, high frequency, and poor continuity.

The six volcanic eruption rhythms discussed here have contact with each other through paraconformity that can be traced discontinuously in the seismic profile. In the process of eruption rhythm division based on seismic profiles, interwell analysis and seismic profile analysis constrain and supplement each other, and continuous modification and optimization can contribute to appropriate profile division.

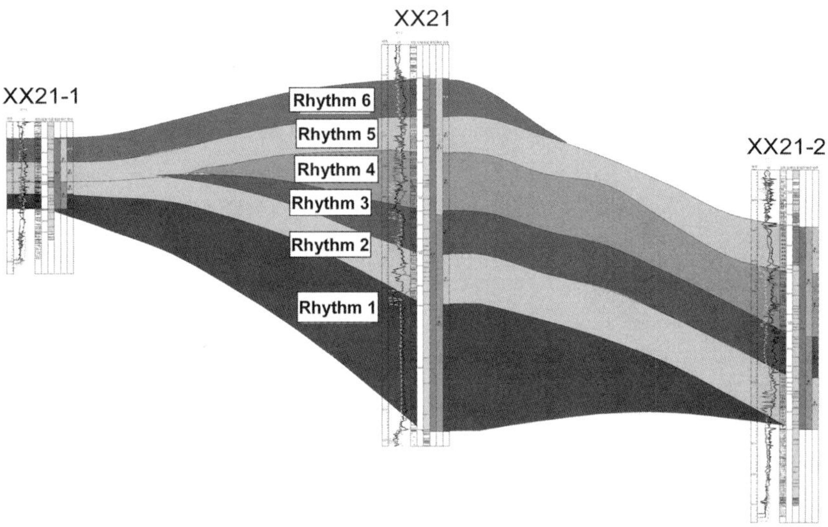

FIGURE 3.19

Eruption rhythm on a well-tie correlation profile.

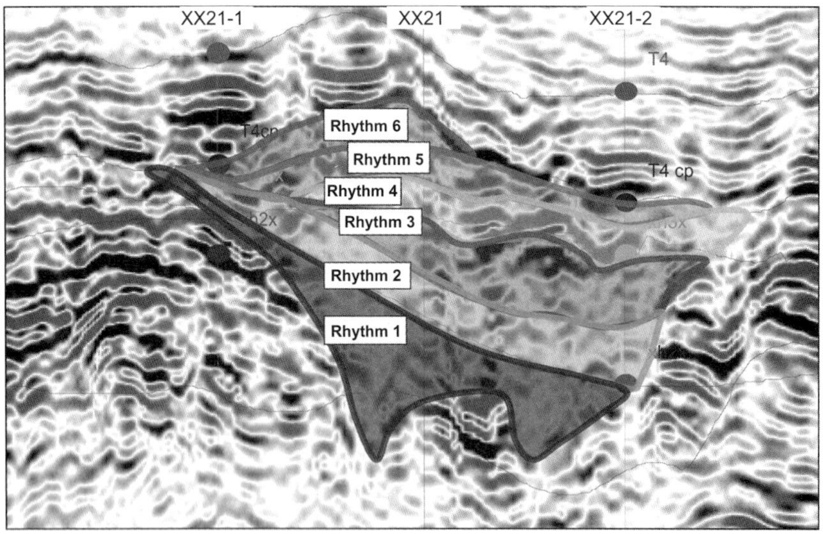

FIGURE 3.20

Eruption rhythm division in a seismic profile constrained by eruption period.

Volcanic rock sequence division and correlation form the basis for dividing the development zones, and complex geology and internal architectures are major impeding factors in the technical development of volcanic rock sequence division and correlation. On the basis of a geological analysis of volcanic rocks and architectural characterization, this book develops the techniques of volcanic rock sequence division and correlation, and lays a foundation for analyzing gas and water relationships and optimizing the division of development zones.

References

[1] Yue W, Bi Z, Jiao S, et al. Sequence stratigraphy in the Mesozoic Terrestrial basin in Jiangsu and Anhui along Yangtze River. Beijing: Geology Press; 2002, p. 38.
[2] Van Wagoner JC, Psomentier HW, Mitchum RM, et al. An overview of the fundamentals of sequence stratigraphy and key definitions. The Society of Economic Paleontologists and Mineralogists Spec Pub 1988;42:39–45.
[3] Shanley KW, McCabe PJ. Perspective on the sequence stratigraphy of continental strata. AAPG Bull 1994;78(4):544–68.
[4] Regional and Mineral Geology Department. Guideline and attached figures for regional geological survey in the volcanic rock area. Beijing: Geology Press; 1987, p. 168–73.
[5] China Geological Survey. Regional geological survey in China's provinces at the end of the 20th century. Beijing: Geology Press; 2003, p. 68.
[6] Xie J, Tao K. Mesozoic volcanic geology and volcanic-intrusive complexes in the southeastern continent of China. Beijing: Geology Press; 1996, p. 7.
[7] Hou Q, Zhao Z, Wang L. Volcanic gas reservoir. Beijing: Science Press; 2009, p. 1–3.
[8] Liu B. Discussion on the correlation methods of datum-level cycle and sedimentary cycle sequence. Acta Sedimentol Sin 2002;20(1):112–7.
[9] Li J, Deng W. Field work procedure for regional geologic survey. Beijing: Geology Press; 1980, p. 138, 139.
[10] Zhao C, Meng W, Jin C, et al. Volcanic rock and hydrocarbon in Liaohe basin. Beijing: Petroleum Industry Press; 1999, p. 48, 49.
[11] Wang P, Wu H. Volcanic facies of the SongLiao basin: sequence, model and the quantitative relationship with porosity and permeability of the volcanic reservoir. Journal of Jilin University (Earth Science Edition) 2006;36(5):805–12.
[12] Cheng R, Wang P. Sequence stratigraphy with fills of volcanic rocks in Xujiaweizi Faulted Depression of SongLiao basin, Northeast China. Journal of Jilin University (Earth Science Edition) 2006;36(5):469–74.
[13] Wang P, Liu W. Integrated characterization of lithology and lithofacies of volcanic reservoir in Shengping-Xingcheng gas field (Internal research report in chinese); 2006.
[14] Zhang C. Hierarchy analysis in reservoir research. Oil Gas Geol 1992;13(3):344–50.
[15] Liu J. Geological basis for oil and gas field development. Beijing: Petroleum Industry Press; 2006, p. 328.
[16] Qiu Y, Xue S. Hydrocarbon reservoir evaluation technology. Beijing: Petroleum Industry Press; 1997, p. 248, 249.

CHAPTER

Identification and Prediction of Volcanic Facies

4

Volcanic facies is of great importance for reservoir prediction because it controls the type, degree of development, and distribution characteristics of pores and fractures in volcanic rocks; affects the type and intensity of later diagenesis; and determines the distribution of favorable facies belts [1]. The identification criteria and models for different lithofacies are established on the basis of volcanic facies classification. Following the technical approaches of proceeding from single wells to a profile, then to a 2D plane and finally to 3D space, volcanic facies can be identified and their distribution predicted, including quantitative characterizations of their size, shape, and superposition relationships, as well as their distribution patterns. This will help establish a basis for predicting favorable reservoirs and building 3D geological models that reflects the characteristics of volcanic facies.

4.1 CLASSIFICATION OF VOLCANIC FACIES

4.1.1 Classification of lithofacies

Facies are the overall geological characteristics that reflect the genesis of a rock body [2]. Volcanic facies are the spatial distributions of volcanic products, the mode of occurrence, and the appearance and characteristics of these products, controlled by factors such as the modes of volcanism, eruption types, magma composition, geological structures, and paleogeographical settings [3].

M. Usov, Yu. Kuznetsov, and V.I. Pavlinov pioneered a detailed classification of volcanic facies. In 1973, Kropachev and Starostni proposed a new classification scheme, which constituted the foundation for classifying volcanic facies [4]. In 1978, Kepugef introduced the term *volcanic facies* into the geology literature and classified volcanic rocks into three facies based on their petrogenic conditions and rock body morphology, including primary eruptive facies, subvolcanic facies and volcanic conduit facies [5], a classification scheme adopted also by Li and Wang [6]. Fisher and Schmincke classified volcaniclastic rocks into four facies: volcaniclastic flow facies, volcaniclastic rock facies, eruptive alluvial facies, and ash flow facies [7]. In a more elaborative classification proposed by Tao and Qiu, 11 volcanic facies were recognized: eruptive-effusive facies, fallout facies, volcaniclastic flow facies, surge facies, lahar facies, collapse facies, extrusive facies, volcanic crater-neck facies, subvolcanic facies, cryptoexplosive breccia facies, and eruptive sedimentary facies [8]. On the basis of various modes of volcanic material transportation, Jin and colleagues classified volcanic rocks into four facies: explosive facies, eruptive-collapse and eruptive-effusive facies, extrusive and subvolcanic facies, and eruptive-sedimentary facies [9]. Xie and coworkers recognized 13 volcanic lithofacies: eruptive-effusive facies, explosive fallout facies, volcaniclastic flow facies, explosive-effusive facies, base surge facies, volcanic debris flow facies, eruptive sedimentary facies, volcanic neck facies, extrusive facies, subvolcanic facies, cryptoexplosive breccia facies, intrusive facies, and volcanic lake facies [10].

Liu and colleagues classified volcaniclastic rocks into four groups: eruption fallout deposits, volcaniclastic flow deposits, lahar deposits, and volcanic base-surge deposits [11]. Liu and colleagues classified the volcanic rocks in the Dabie Mountains into four facies: explosive facies, eruptive-effusive facies, eruptive-sedimentary facies, and subvolcanic facies [12]. Wang and associates classified the volcanic rocks of the Yingcheng Formation in Xujiaweizi into five volcanic facies: volcanic conduit facies, explosive facies, eruptive-effusive facies, volcanic-sedimentary facies, and extrusive facies [13].

A systematic and standardized volcanic facies classification scheme should take into account the following aspects: (1) volcanic eruption pattern; (2) volcanic eruption environment; (3) surface settings for the accumulation of volcanic products; (4) mechanisms of volcanic explosion and transportation and accumulation of volcaniclastic debris; (5) mechanisms of magmatic intrusion at a given depth; and (6) the specific location of the volcanic edifice [14]. By integrating the results of previous work with consideration of the geological features of volcanic rocks in the Songliao and Junggar basins, this book recognizes six volcanic facies: explosive facies, effusive facies, extrusive facies, volcanic conduit facies, subvolcanic facies, and volcanic sedimentary facies, among which a total of 19 subfacies are also identified (Table 4.1).

4.1.2 Lithofacies models

Lithofacies models are the generalization of lithofacies assemblages, with the conceptualized and simplified models to demonstrate the interrelationships between various volcanic facies [15]. The distribution of volcanic lithofacies is obviously controlled by the settings of the eruption "sources" and different levels of eruption energy. Different lithofacies have different geological and geographical settings and hence different assemblage patterns.

1. Lithofacies pattern of acidic volcanic rocks in Songliao basin

Due to the complexity and peculiarities of various volcanoes and the impact of later transformations, the volcanic facies have diverse distribution patterns.

In the acidic volcanic rocks of the Yingcheng Formation in the Songliao basin, for example, the volcanic edifices are mainly of three types: volcaniclastics, lava, and combined lava-volcaniclastics. Based on their types and distribution characteristics within the volcanic edifices, the overall distribution pattern of the lithofacies pattern is established (Figure 4.1) [16] as follows: The horizontal facies

Table 4.1 Classification of Volcanic Lithofacies

Facies	Subfacies
Explosive facies	Splash subfacies, hot debris flow subfacies, hot base surge subfacies, and fallout subfacies
Effusive facies	Top subfacies, upper subfacies, middle subfacies, and lower subfacies
Extrusive facies	Intrazone subfacies, mesozone subfacies, and outer zone subfacies
Volcanic conduit facies	Volcanic neck subfacies and cryptoexplosive breccia subfacies
Subvolcanic facies	Intrazone subfacies, mesozone subfacies, and outer zone subfacies
Volcanic sedimentary facies	Xenoclast-bearing subfacies, re-transported subfacies, and coal-bearing tuff subfacies

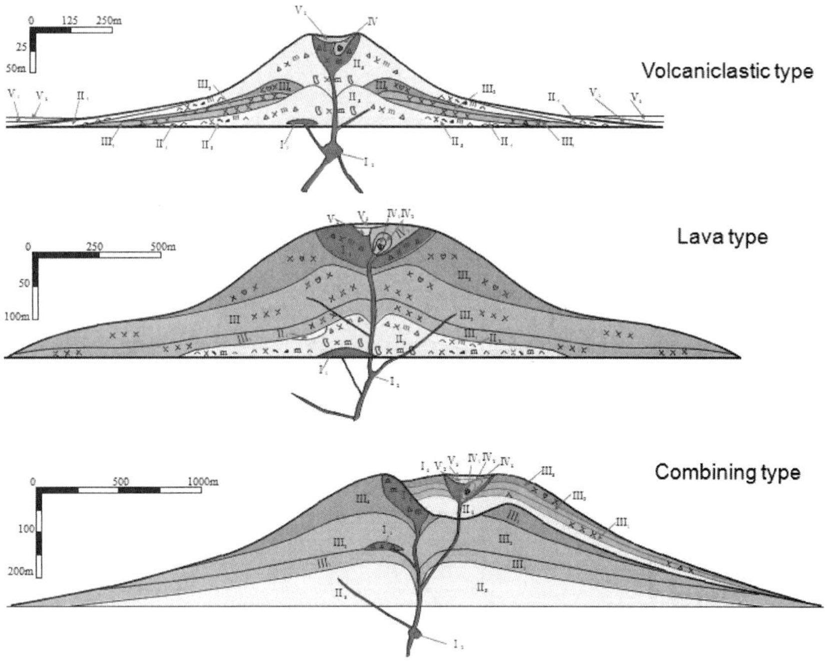

FIGURE 4.1

Lithofacies distribution of major volcanic edifice in Songliao basin (after Xu Zhengshun).I—volcanic conduit facies; I_1—volcanic neck subfacies; I_2—subvolcanic subfacies; I_3—cryptoexplosive breccia subfacies; II—explosive facies; II_1—fallout subfacies; II_2—hot base surge subfacies; II_3—hot debris flow subfacies; III—eruptive-effusive facies; III_1—lower subfacies; III_2—middle subfacies; III_3—upper subfacies; IV—extrusive facies; IV_1—intrazone subfacies; IV_2—mesozone subfacies; IV_3—outer zone subfacies; V—volcanic sedimentary facies; V_3—coal-bearing tuff subfacies; V_2—re-transported subfacies; V_1—xenoclast-bearing subfacies.

succession is characterized, from proximal to distal locations relative to the crater, by volcanic conduit facies (extrusive facies and subvolcanic facies) → explosive facies → effusive facies → volcanic sedimentary facies; whereas the vertical facies succession is represented, in ascending order, by subvolcanic facies → volcanic conduit facies → explosive facies → effusive facies → extrusive facies → volcanic sedimentary facies. In general, the complexity of lithofacies types decreases with increasing distances from the crater.

2. Vertical distribution characteristics

In general, three vertical distribution patterns or trends can be recognized in the volcanic rocks: weakening-upward sequence, strengthening-upward sequence, and alternating sequence.

1) Weakening-upward sequence

The eruption energy weakens gradually, with explosive facies developed first, followed by effusive facies (Figure 4.2). The explosive facies comprises, in ascending order, a succession of fallout subfacies, hot base surge subfacies, hot debris flow subfacies, and splash subfacies. The effusive facies

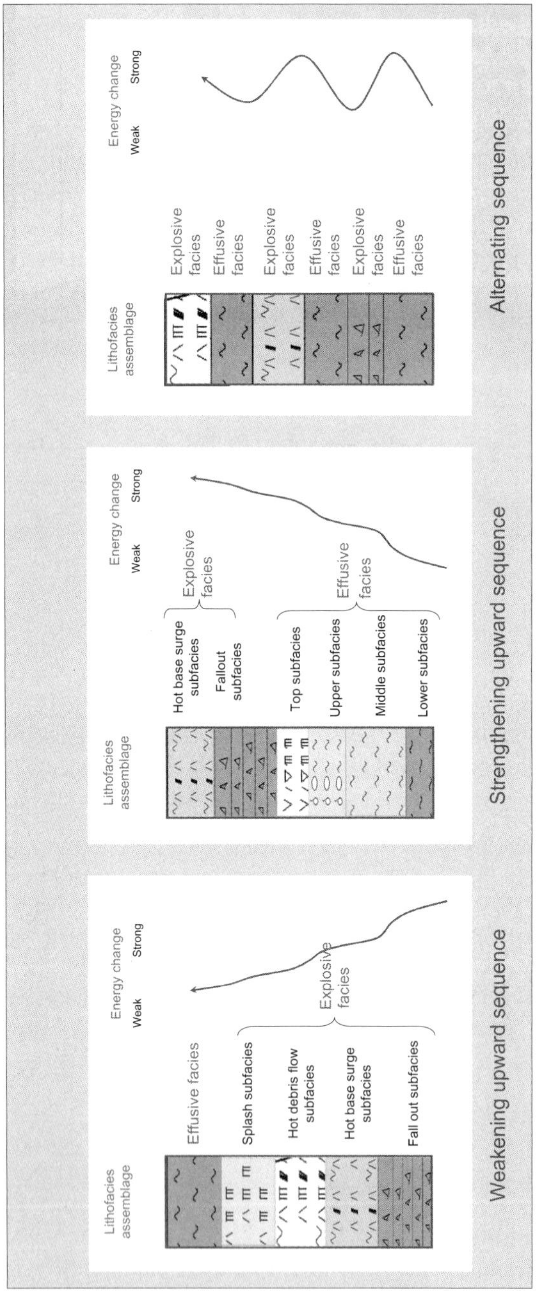

FIGURE 4.2

Vertical stacking pattern of volcanic facies.

consists of, in ascending order, a succession of the lower subfacies, middle subfacies, upper subfacies, and top subfacies. This lithofacies model is typical of acidic volcanic rocks.

2) Strengthening-upward sequence

The eruption energy increases gradually, with effusive facies developed first, followed by explosive facies (see Figure 4.2). In ascending order, the effusive facies is comprised of a succession of lower subfacies, middle subfacies, upper subfacies, and top subfacies, whereas the explosive facies comprises a fallout subfacies and hot base surge subfacies. This lithofacies pattern is characteristic of neutral-basic volcanic rocks.

3) Alternating sequence

The eruption energy alternates between strong and weak, indicating unstable eruption energy, resulting in explosive facies alternating with effusive facies (see Figure 4.2).

3. Horizontal distribution characteristics

With increasing distances from the crater, volcanic facies can be divided laterally into three lithofacies belts: crater facies, proximalcrater facies, and distal crater facies:

1) Crater facies

This facies is characterized by alternations of explosive facies, effusive facies, extrusive facies, volcanic conduit facies, and subvolcanic facies, with rapid changes in various lithofacies and complex stacking patterns or superposition relationships (Figure 4.3).

2) Proximal crater facies

This facies comprises mainly effusive facies and explosive facies, with a large thickness, relatively stable distribution, good continuity, and a primary paralaminar vertical superposition relationship (see Figure 4.3).

3) Distal crater facies

This facies consists of explosive facies, effusive facies, or volcanic sedimentary facies, with simple lithology, small thickness, stable distribution, good continuity, and predominantly laminar distribution (see Figure 4.3).

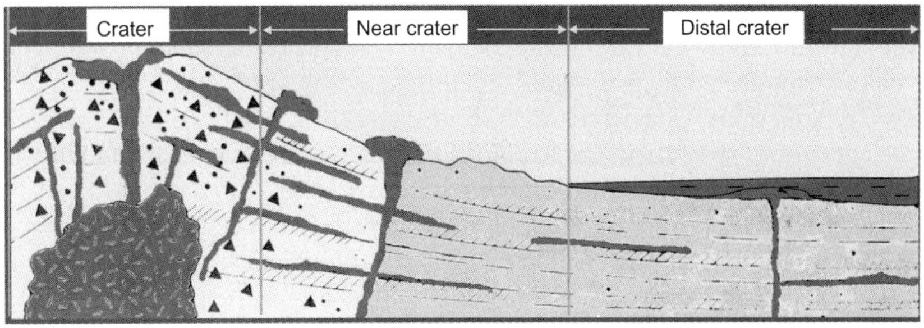

FIGURE 4.3

Lateral distribution pattern of volcanic facies.

4.2 RESEARCH CHALLENGES AND TECHNICAL SOLUTIONS

4.2.1 Research challenges

The formation of volcanic facies is controlled by multiple factors such as magma properties, eruption modes, transportation mechanisms, paleogeomorphology, and diagenetic processes [17–19], which lead to challenges for their identification and prediction because of petrogenetic complexity and various lithofacies with rapid facies change. The challenges can be summarized as follows:

1. There are many types of volcanic facies with complex mechanisms of petrogenesis, which translates to multiple schemes of facies identification and classification. For cored intervals, each lithofacies is identified based mainly on its lithological, textural, and structural markers. However, there is no one-to-one corresponding relationship between volcanic lithofacies and volcanic lithology, texture, and structure. Systematic analysis is needed by taking into consideration the location of cored intervals in the volcanic edifice. For noncored intervals, volcanic facies are identified based mainly on geophysical logging data, but logging data are affected by many factors such as borehole environments, logging instruments, and geological changes, which can be interpreted in multiple ways. It is thus difficult to identify and classify volcanic facies accurately.
2. Challenges in the prediction of volcanic lithofacies are due to rapid changes in volcanic facies and subtle differences in seismic reflections between different lithofacies, without clearly traceable interfaces. Compared with sedimentary rocks, the formation of volcanic rocks is characterized by rapid accumulation. Volcanic rocks inside the same volcanic edifice are generally syngenetic, resulting in largely homogeneous lithological compositions, similar density and acoustic velocity between different lithofacies, and hence minimal differences in seismic reflection characteristics between different lithofacies. Moreover, rapid changes of volcanic lithology and thickness are reflected also by the presence some thin-layer lithofacies, which are usually difficult to recognize due to the lower vertical resolution of seismic data, leading to a low accuracy of lithofacies prediction. In addition, the superposition relationships between volcanic facies are complicated, often lacking consistent, identifiable, and continuously traceable interfaces between adjacent lithofacies, posing further challenges for identifying and delineating volcanic facies.
3. Challenges in characterization of volcanic facies as a result of the complex shapes, scales, and superposition relationships and various types of facies sequence combination. Volcanic facies have multiple types, complicated shapes, significant differences in scales, and complex facies sequence combination relationships. Characterization of volcanic facies using outcrop data is restricted by the completeness of exposure and the workload of outcrop research; characterization of volcanic facies based on seismic data is influenced by the impact of seismic data on the resolution of lithofacies and subfacies; characterization of volcanic facies with dense well grid data is restricted by the location of dense well grids and the acquisition of coring, logging, and performance data. Furthermore, volcanic lithofacies change rapidly with poor facies sequence integrity and complex assemblage relationships. This makes it challenging to characterize the geometric parameters of volcanic facies.

4.2.2 Technical solutions

Volcanic lithofacies in single wells and cross sections can be identified by taking into consideration the characteristics and prediction difficulties of volcanic facies by using the classification scheme of rock types and lithology identification, volcanic facies markers, and analysis core, wireline log, and seismic data. The planar (2D) and spatial (3D) distribution patterns of lithofacies are predicted by establishing the corresponding relationships between lithofacies and seismic facies. Volcanic lithofacies are quantitatively characterized by systematically analyzing outcrop, seismic, and dense-grid well data to reveal the shape, scale, and superposition relationships of volcanic facies (Figure 4.4). The main concepts and approaches are as follows:

1. Lithological classification and identification markers are distilled to define the typical textures, structures, and log curve shapes of volcanic facies and to identify the locations of volcanic edifices and lithofacies in single wells.
2. Through seismic calibration based on single-well lithofacies, the seismic facies pattern of volcanic lithofacies is established; volcanic lithofacies are identified according to seismic reflection characteristics in skeletal profiles, based on well point constraints and seismic facies patterns.
3. The planar distribution of volcanic lithofacies is predicted based on the identification of single-well facies and profile facies, under the guidance of the lithofacies pattern, and through well point extrapolation, skeleton profile control, and seismic attribute analysis.

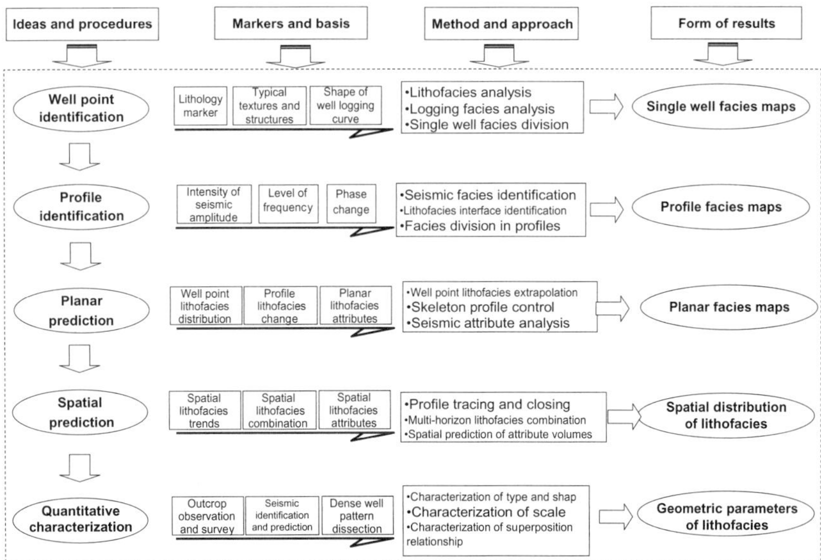

FIGURE 4.4

Technical procedures for identification and characterization of volcanic lithofacies.

4. A skeletal profile network is established for the entire study area, and the spatial distribution of lithofacies is revealed through lithofacies tracing, multihorizon lithofacies linkage, and extraction of lithology-sensitive seismic attribute volumes.
5. The shape, scale, and superposition relationships of volcanic facies are characterized by systematically integrating outcrop, seismic, and dense-grid well data.

4.3 SINGLE-WELL IDENTIFICATION AND CLASSIFICATION

4.3.1 Identification markers

As the foundation for single-well facies recognition and division, the identification markers for volcanic lithofacies are established on the basis of petrogenetic analysis, incorporating lithological, textural, and structural characteristics, and log responses of different volcanic lithofacies.

1. Geological markers

1) Explosive facies

Magma contains a large number of volatiles and exerts tremendous pressure on surrounding rocks. Its explosion within surrounding rocks results in the accumulation of volcaniclastic materials of various sizes. The explosive facies may be formed in any volcanic phases, but it is mostly found in the lower part of acidic eruption cycles and in the upper part of basic eruption cycles. The typical lithology of explosive facies consists mainly of volcanic agglomerate, volcanic breccia, tuff, and welded breccia (Figure 4.5), with corresponding lithological textures. Typical structures include graded beddings and ballistic trajectory drop-pebble structure. The explosive facies comprises four subfacies, including, in ascending order, the fallout, hot base surge, volcaniclastic flow, and splash subfacies. Different subfacies have different petrogenetic, lithological, textural, and structural characteristics (Table 4.2).

2) Effusive facies

This facies forms during the middle period of a volcanic eruption cycle, from magmatic melt containing crystallized material and syngenetic breccia driven by the propulsion of subsequent ejecta and its own gravity to flow on the surface, followed by gradual chilling and lithification. Effusive facies is commonly found in acidic, neutral, and basic volcanic rocks, characterized by rocks derived from various magma with vesicular and rhyolitic structures. In this lithofacies, the typical textures are spherulitic, aplitic, and vitreous, and typical structures are rhyolitic, vesicular, amygdaloidal, and lithophysa (Figure 4.6). The effusive facies comprises four subfacies, including, in ascending order, the lower, middle, upper, and top subfacies (Table 4.3).

FIGURE 4.5

Volcanic breccia in fallout subfacies of explosive facies.

Table 4.2 Subfacies and Geological Marks of Explosive Facies

Facies	Subfacies	Genesis	Typical Lithology	Typical Textures	Typical Structures
Explosive facies	Splash subfacies	Formed as a result of local retraction and accumulation of magmatic melt and surrounding rocks carried by the lava during its upwelling near the crater	Breccia lava, tuff lava	Welded breccia texture, welded tuff texture	Deformed rhyolitic structure
	Hot debris flow subfacies	Formed through mixture of hot debris and magma debris with volatiles flowing on surface, driven by subsequent ejecta and gravity, through chilling, cementation, and compaction	Welded tuff containing crystal fragments, vitric fragments, magmatic fragments, and lithic fragments	Welded tuff texture	Graded beddings, volcanic glass lineation, and matrix support
	Hot base surge subfacies	Formed by gas-solid-liquid multiphase system through degassing driven by gravity, transported as suspension near the surface and then lithification via gravity deposition and compaction	Tuff containing crystal fragments, vitric fragments, and magmatic fragments	Crystalline tuff texture	Parallel beddings, cross beddings, regressive sand wave beddings
	Fallout subfacies	Formed by solid volcanic debris and plastic ejecta, driven by volcanic degassing and free fallback to surface and compaction	Agglomerate, volcanic breccia, volcanic tuff	Agglomerate texture, breccia texture, tuff texture	Particle supported normal graded beddings, ballistic trajectory drop pebbles

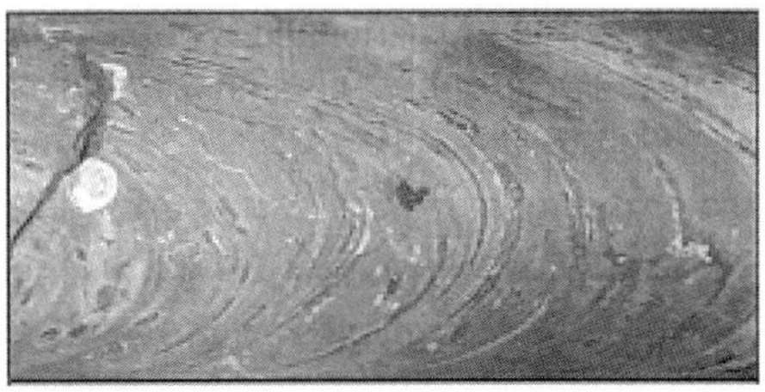

FIGURE 4.6

Rhyolitic structure in middle subfacies of effusive facies.

Table 4.3 Subfacies Types and Geologic Marks of Effusive Facies

Facies	Subfacies	Genesis	Typical Lithology	Typical Textures	Typical Structures
Effusive facies	Top subfacies	Formed during magmatic melt flow; in contact with air; rapid lithification and chill; compression and crushing by subsequent magmatic melt	Breccia lava	Autoclastic texture, welded breccia texture	Deformed rhyolitic structure
	Upper subfacies	Formed during magmatic melt flow through chilling, cementation, and compaction of upper magmatic melt, with best-developed primary vesicles	Vesicular structure lava	Spherulitic texture, aplitic texture	Vesicular structure, amygdaloidal structure, and lithophysa structure
	Middle subfacies	Formed during magmatic melt flow through chilling, cementation, and compaction of middle magmatic melt; this is a relatively dense lithofacies belt	Rhyolitic structure lava	Aplitic texture, porphyritic texture	Rhyolitic structure
	Lower subfacies	Formed during magmatic melt flow through chilling, cementation, and compaction of lower magmatic melt	Aplitic rhyolitic structure and syngenetic breccia bearing lava	Vitreous texture, aplitic texture, porphyritic texture	Blocky or discontinuous deformed rhyolitic structure

3) Extrusive facies

Extrusive facies forms in the later period of volcanic activity cycle and comprises blocky volcanic massifs formed near the crater when high-viscosity acidic and neutral magma is quenched by water or quickly chilled in the air while being compressed by internal forces to flow out of the surface. The extrusive facies is mainly domed in shape, with perlite being the predominant lithology. Its typical rock textures are perlitic and vitreous, and the typical structures are dominated by nodules and pillows (Figure 4.7). From inside out, the extrusive facies can be divided into three subfacies, namely the inner zone, mesozone, and outer zone (Table 4.4).

4) Volcanic conduit facies

A volcanic conduit facies is located in the lower and near the center parts of the volcanic edifice, comprising volcanic rocks retained and backfilled in the volcanic conduit in the process of the upward transportation of magma to the surface. This facies accumulates throughout the volcanic cycle, but only the products of later activities are preserved. In a volcanic conduit facies, cryptoexplosive breccia is the lithological marker (Figure 4.8), characterized by cryptoexplosive breccia, autoclastic, porphyritic, and cataclastic textures and a dendritic vein structure. From inside out, the volcanic conduit facies can be identified into two subfacies: the volcanic neck subfacies and cryptoexplosive breccia subfacies (Table 4.5).

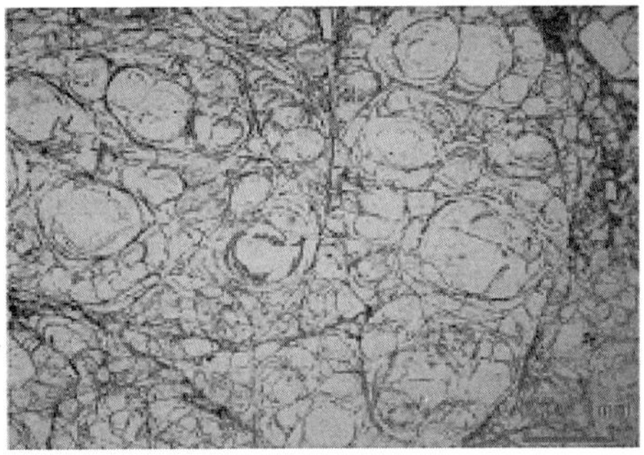

FIGURE 4.7

Perlitic texture in mesozone.

Table 4.4 Subfacies Types and Geological Marks of Extrusive Facies

Facies	Subfacies	Genesis	Typical Lithology	Typical Textures	Typical Structures
Extrusive facies	Outer zone subfacies	Outside dome-like massif; formed through cooling, deformation of linguoid front of magma that scrapes and wraps newly formed rock blocks and initial rock blocks; flow under compression of internal forces	Breccia lava, tuff lava	Welded breccia texture, welded tuff texture	Deformed rhyolitic structure
	Mesozone subfacies	Formed by high-viscous magma in the middle of the dome of extrusive facies compressed by internal force to flow; quenched when in contact with water; chilling and lithification near crater	Tight massive perlite and aplitic rhyolite	Vitreous texture, Perlitic texture, oligophyric texture, porphyroclastic texture	Blocky structure, with rock mass in bedded and lenticular shapes
	Inner zone subfacies	Formed by high-viscosity magma inside lava dome of extrusive facies compressed by internal force to flow; slow chilling and lithification near crater	Pillow and spherical perlite		Nodule structure and pillow structure

5) Subvolcanic facies

Subvolcanic facies forms in the same stage or the later stage of volcanic eruption cycles and is the result of slow chilling and crystallization of the lava intruding into the surrounding rocks. This facies is often located near the volcanic conduit in the form of sills, laccoliths, stocks, dikes, and veins. A subvolcanic facies typically has porphyritic and xenolith structures (Figure 4.9) and can be classified into three subfacies from inside out: the inner zone, mesozone, and outer zone subfacies (Table 4.6).

FIGURE 4.8

Cryptoexplosive breccia in subfacies of extrusive facies and volcanic conduit facies.

Table 4.5 Subfacies and Geologic Marks of Volcanic Conduit Facies

Facies	Subfacies	Genesis	Typical Lithology	Typical Textures	Typical Structures
Volcanic conduit facies	Cryptoexplosive breccia subfacies	Formed when volatile-enriched magma intrudes fractured rock belts, and explodes in subsurface due to pressure release; or magma encounters ground water, leading to cryptoexplosion due to release of volatiles and pressure	Cryptoexplosive breccia	Cryptoexplosive breccia texture, Autoclastic porphyritic texture, Cataclastic texture	Cylinder, laminar, vein, dendritic, and fracture filling shaped
	Volcanic neck subfacies	Formed in late stage of volcanic eruption; magma fails to erupt due to decrease of internal pressure, and becomes chilled inside volcanic conduit; rock layers near crater collapse due to thermal subsidence; cementation by continuously surging and cooling magma	Lava and breccia / tuff lava and welded breccia / tuff	Porphyritic texture, Welded texture, Breccia tuff texture	Ring or radial joint

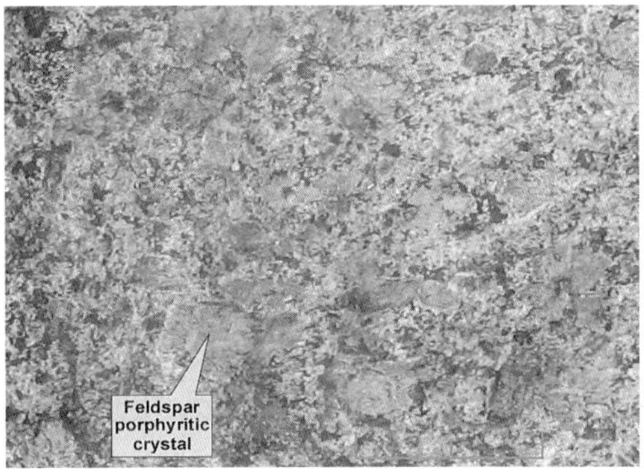

FIGURE 4.9

Porphyritic texture of orthophyre in subvolcanic facies.

Table 4.6 Subfacies and Geological Marks of Subvolcanic Facies

Facies	Subfacies	Genesis	Typical Lithology	Typical Textures	Typical Structures
Subvolcanic facies	Outer zone subfacies	Formed in the same stage or late stage of the volcanic eruption cycle, when the linguoid front of intruding magma chills; earlier or neogenic rocks become deformed, scaped, and wrapped; flow under compression of internal forces	Xenolith-bearing subvolcanic rock	Welded breccia texture, welded tuff texture	Xenolith structure
	Mesozone subfacies	Formed in the same stage or later stage of the volcanic eruption cycle; high-viscous intruding magma compressed by internal force to flow; some parts of magma undergo cooing, crystallization, and compaction within surrounding rocks	Subvolcanic rock	Porphyritic texture	Chilled margin structure, platy flow structure, linear flow structure
	Inner zone subfacies			Crystalline texture	Columnar joint, tabular joint

6) Volcanic sedimentary facies

Volcanic sedimentary facies consists mainly of clastic deposits between volcanic rocks. It often co-occurs with volcanic rocks but may not be present in every phase of volcanic activities. The clastic components of volcanic sedimentary facies contain a large amount of volcanic rock debris, and various types of sedimentary volcanic rocks and volcanic sedimentary rocks are developed in this facies with terrigenous clastic textures, rhythmic beddings, and planar beddings (Figure 4.10). This facies can be divided into three subfacies: xenoclast-bearing, retransported, and coal-bearing tuff subfacies (Table 4.7).

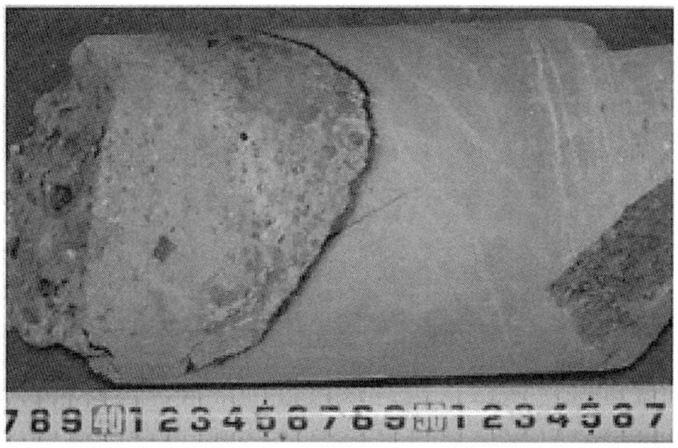

FIGURE 4.10

Horizontal bedding in volcanic sedimentary facies.

Table 4.7 Subfacies and Geologic Marks of Volcanic Sedimentary Facies

Facies	Subfacies	Genesis	Typical Lithology	Typical Textures	Typical Structures
Volcanic sedimentary facies	Coal- bearing tuff subfacies	Tuffaceous volcanic debris between volcanic rocks interbedded with peat-enriched beds; coal-forming swamp environment.	Volcanic tuff interbedded with coal beds or intercalated with coal lines	Terrigenous clastic texture	Rhythmic beddings, horizontal beddings
	Retransportation subfacies	Formed by volcaniclastic deposits reworked by water current.	Bedded volcaniclastic rock/tuff	Terrigenous clastic texture	Cross beddings, trough beddings, graded beddings, block structure
	Xenoclast- bearing subfacies	Formed primarily by volcanic debris, probably with additional terrigenous clastics	Volcanic tuffaceous glutenite containing xenoclast	Terrigenous clastic texture	Cross beddings, trough beddings, graded beddings, blocky structure

2 Logging response characteristics

1) Explosive facies

The logging curves of a complete explosive facies assemblage is characterized by an overall composite pattern of serrated or microserrated box shapes or bell shapes ~ funnel shapes. The log curve of different subfacies in the explosive facies are different: (1) the fallout subfacies is commonly characterized by high interval transit time, low density and low resistivity, and its logging curve is often serrated in shape; (2) the hot base surge subfacies is characterized by mid-low interval transit time, medium-high density and medium-high resistivity, with its log curve having a microserrated shape; (3) the hot clastic flow subfacies typically shows low interval transit time, high density and high resistivity, and its

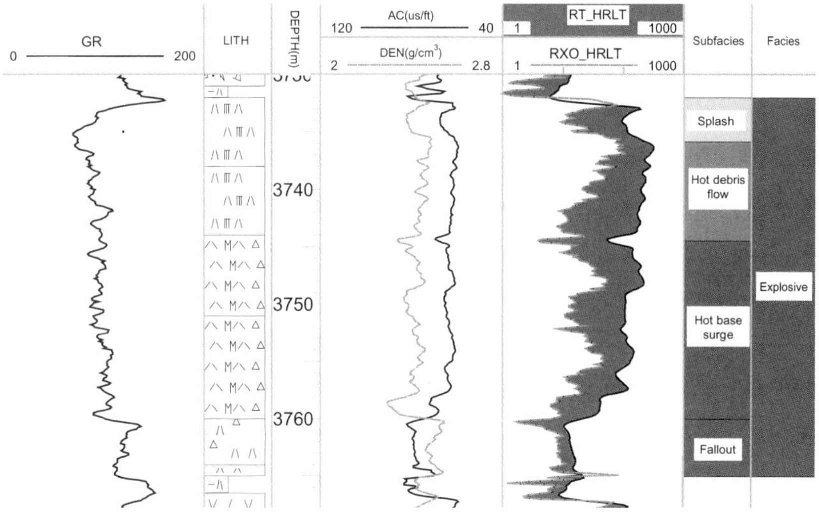

FIGURE 4.11

Diagram of log response characteristics of explosive facies (Well DD1413).

log curve is generally smooth, microserrated in shape; and (4) the splash subfacies is characterized by high interval transit time, low density and low resistivity, and its log curve often has a serrated shape (Figure 4.11).

2) Effusive facies

The logging curve of effusive facies generally has a smooth, from microserrated box to bell shape. The log response characteristics of different subfacies of the effusive facies are as follows: (1) the lower subfacies is characterized by high interval transit time, low density and low resistivity, and its log curve is often microserrated in shape; (2) the middle subfacies is characterized by low interval transit time, high density and high resistivity, and its log curve is relatively smooth, microserrated in shape; (3) the upper subfacies has an increased porosity, characterized by high interval transit time, low density and low resistivity, and its logging curve often has a micro-serrated shape; and (4) the top subfacies is characterized by high interval transit time, low density and low resistivity with various pores and fractures, and its logging curve ranges from microserrated ~ serrated bell to digitate (Figure 4.12).

3) Extrusive facies

The logging curve of extrusive facies generally has a serrated box shape, and the log response characteristics of its different subfacies are quite different: (1) the outer zone subfacies, usually contains xenolith and volcanic debris, characterized by high interval transit time, low density and low resistivity, with its log curve having a serrated shape; (2) the mesozone subfacies is characterized by high interval transit time, low density and low resistivity, and its log curve is often micro-serrated; and (3) the inner zone subfacies has a relatively consistent composition and texture, as well as a fairly dense lithology, characterized by low interval transit time, high-low density and high resistivity, and a typically micro-serrated logging curve (Figure 4.13).

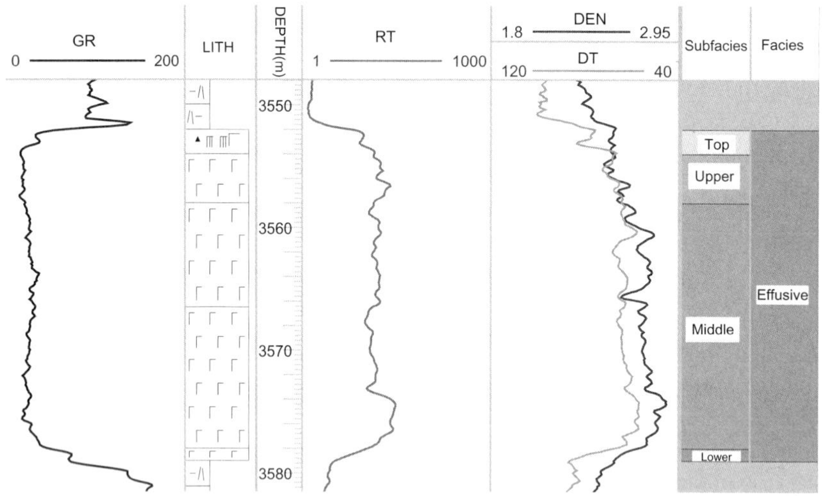

FIGURE 4.12

Diagram of log response characteristics of effusive facies (Well DD172).

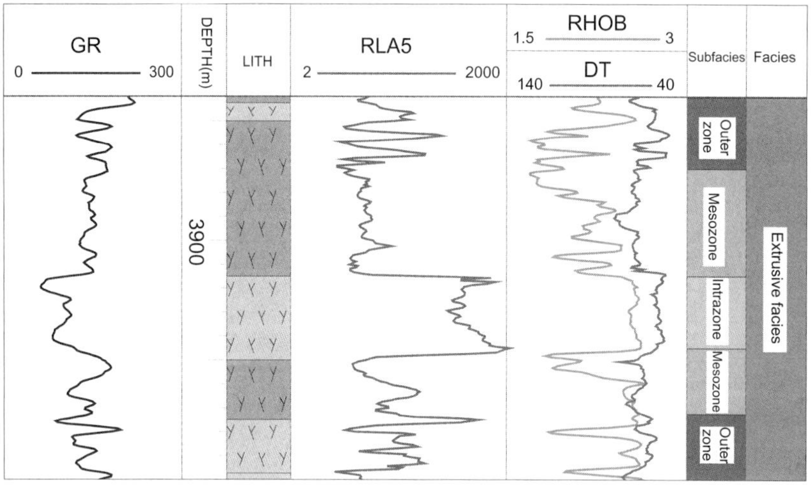

FIGURE 4.13

Diagram of log response characteristics of extrusive facies (Well XX2).

4) Volcanic conduit facies

The lithologies, rock textures, and rock structures of different subfacies of the volcanic conduit facies are different, as reflected by their logging response characteristics. The constituents of the cryptoexplosive breccia subfacies are mixed, characterized by medium-high density, medium-high resistivity and medium-low interval transit time, with a serrated logging curve. In comparison, the volcanic neck subfacies with a consistent lithology is characterized by low density, low resistivity and high interval transit time, and its log curve usually has a smooth box shape (Figure 4.14).

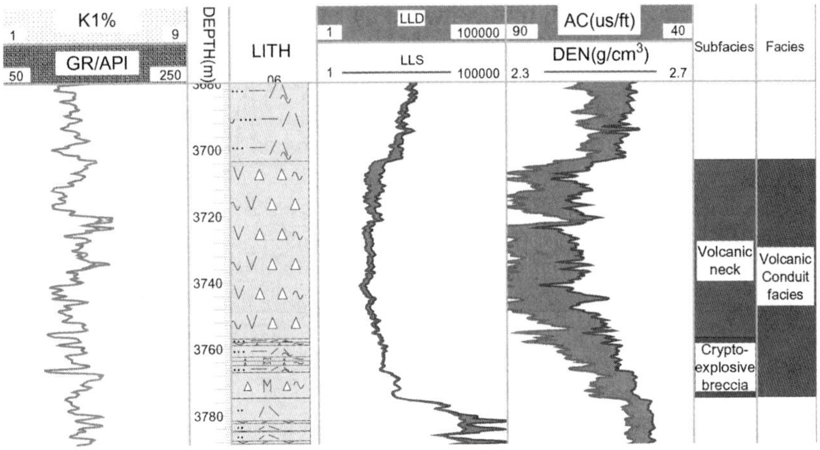

FIGURE 4.14

Diagram of log response characteristics of volcanic conduit facies (Well XX8).

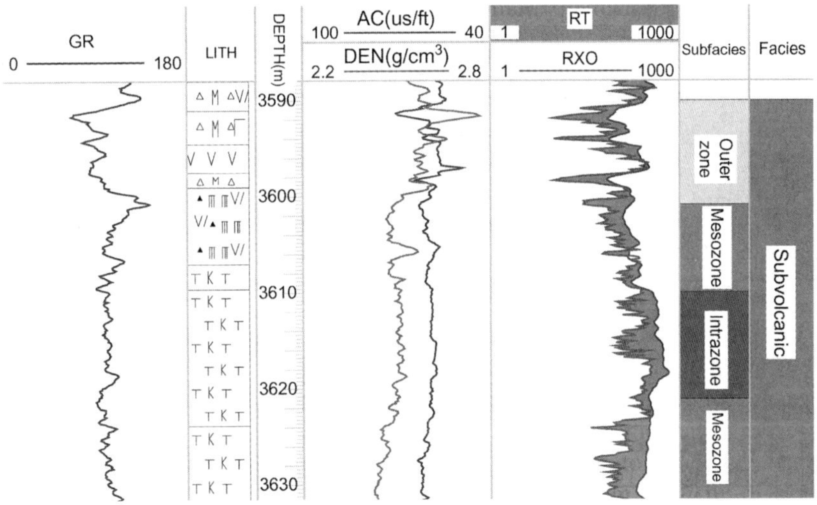

FIGURE 4.15

Diagram of log response characteristics of subvolcanic facies (Well DD1813).

5) Subvolcanic facies

The logging curve of subvolcanic facies is generally microserrated box or bell shaped. The outer zone subfacies is often characterized by low density, high interval transit time and low resistivity, with serrated logging curve; the mesozone subfacies shows a fairly high density and resistivity as well as low interval transit time, and its logging curve is micro-serrated ~ smooth in shape; and the inner zone subfacies is characterized by high density, low interval transit time and high resistivity, and a smooth logging curve (Figure 4.15).

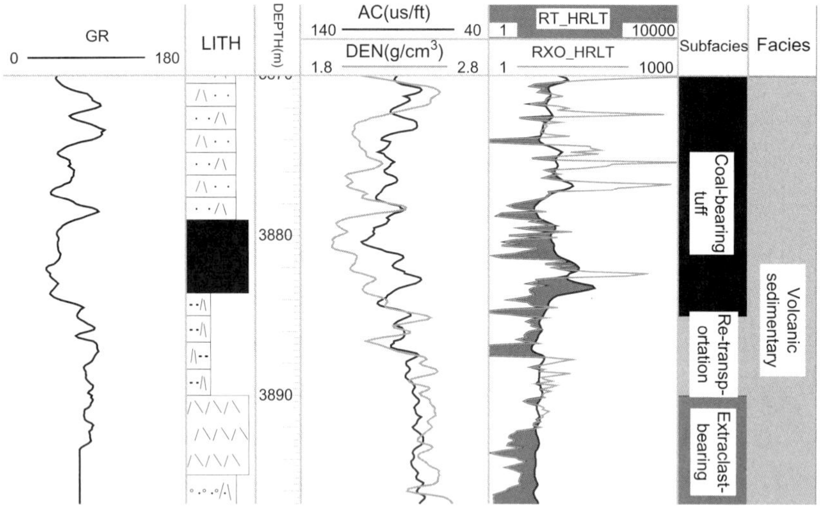

FIGURE 4.16

Diagram of log response characteristics of volcanic sedimentary facies (DD1414).

6) Volcanic sedimentary facies

Volcanic sedimentary facies often shows a composite pattern of serrated ~ serrated-digitate shape or funnel shape ~ bell shape in log curves. The xenoclast-bearing subfacies is characterized by high interval transit time, low density and low resistivity, and a microserrated logging curve; the retransported subfacies typically has high interval transit time, low density and low resistivity, and serrated logging curve; and the coal-bearing tuff subfacies is characterized by high interval transit time, low density and high resistivity, and a serrated-digitate logging curve (Figure 4.16).

4.3.2 Single-well facies identification

Lithofacies identification in single wells are based on the geological settings of volcanic edifices penetrated by the drill hole, in combination with single-well lithological identifications using geological and logging markers, to establish a foundation for profile facies identification and lithofacies prediction.

1. Determination of lithofacies settings in single wells based on volcanic edifice distribution

The types of lithofacies in different parts of the volcanic edifice are different. The analysis of lithofacies settings based on the position of the volcanic edifice penetrated in a single well is the foundation for the appropriate identification of lithofacies. The volcanic conduit facies or subvolcanic facies, for example, is only developed near the conduit of the volcanic edifice; the extrusive facies is often seen near the crater; the explosive facies occurs primarily near the crater, the effusive facies is mostly developed in the limb of the periclinal structure of the volcanic edifice; and the volcanic sedimentary facies is usually distributed in a topographic low between volcanic edifices. The volcanic facies settings can be ascertained according to the position of the volcanic edifice where the well is located. As shown in Figure 4.17, Well XX231 is located near the volcanic conduit of the volcanic edifice and is penetrated

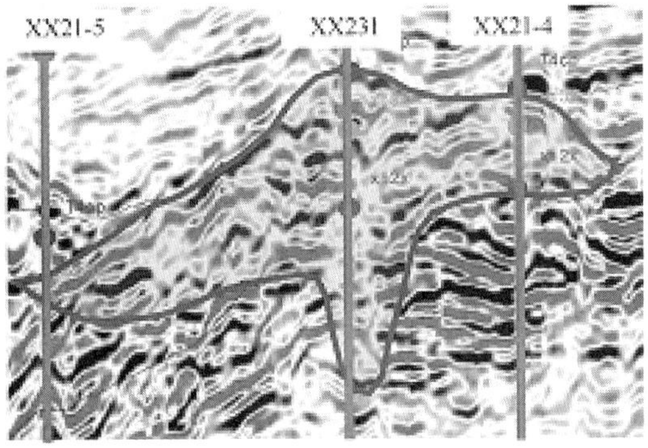

FIGURE 4.17

Estimation of the lithofacies of environment according to position of volcanic edifice.

mainly by the volcanic conduit facies; Well XX21-5 is located at the margin of the volcanic edifice and is encountered chiefly by the volcanic sedimentary facies and distal explosive facies; in comparison, Well XX21-4 is located on the limb of the volcanic edifice, where the effusive and some explosive facies are encountered.

2. Lithofacies identification and division in cored interval

Analysis of lithofacies settings in single wells can be combined with the available core data and information on rock types, texture, and structure characteristics to identify volcanic lithofacies and subfacies and to calibrate logging facies.

1) Lithofacies identification in cored intervals

One of the important steps in volcanic lithofacies study is the description of available cores, whereby the lithology, texture, and structure of volcanic rocks are investigated to identify volcanic facies. In Figure 4.17, Well XX21-4 was drilled on the limb of the volcanic edifice, and the core from interval 3847.08 to 3878.35 m can be divided into four segments: interval 3847.08 to 3851.60 m is composed of offwhite rhyolitic brecciated lava with welded breccia texture and vesicular structure, and the breccia color and composition are similar to those of the lava. This segment was formed by flowing magma that compressed, crushed, and consolidated the top rock blocks; interval 3851.60 to 3864.41 m consists of gray rhyolite with rhyolitic structure, formed during the flowing process of lava; interval 3864.41 to 3869.35 m comprises gray rhyolite-syngenetic breccia with well-defined breccia geometry, but its color is darker than that of the lava, indicating that this segment was formed by magma that captured rock blocks from early formations during the flowing process; interval 3869.35 to 3878.35 m is composed of grayish purple, crystalline, welded tuff with both volcaniclastic, and welded textures (Table 4.8).

According to the core analysis, therefore, two segments can be recognized in the lithofacies of this interval. The lower crystalline welded tuff from 3869.35 to 3878.35 m formed from volcanic eruption with medium energy and should be classified as an explosive facies; the upper rhyolite-syngenetic breccia, rhyolite with rhyolitic structure, and breccia lava in the 3847.08 to 3869.35 m interval represent an internal lithological assemblage of effusive petrogenesis and should be classified as an effusive facies.

Table 4.8 Core Characteristics of 3847.08~3878.35 m in Well XX21-4

Depth	Lithology	Color	Texture	Structure	Core Photo
3847.08–3851.60 m	Rhyolitic breccia lava	Offwhite	Welded breccia texture	Vesicular structure	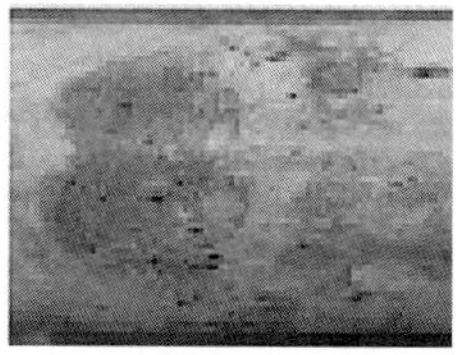
3851.60–3864.41 m	Rhyolite	Gray	Aplitic texture	Rhyolitic structure	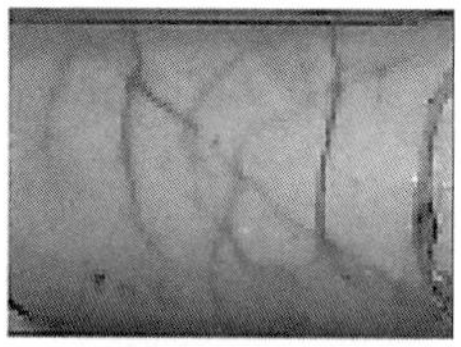
3864.41–3869.35 m	Rhyolite	Gray	Syngenetic breccia bearing texture	Deformed rhyolitic structure	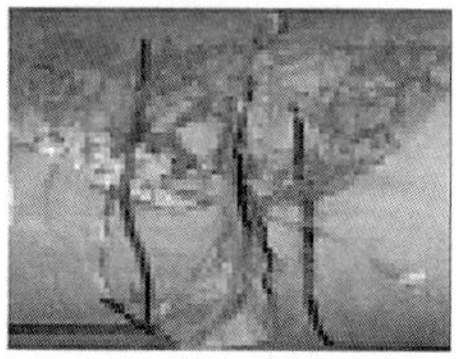
3869.35–3878.35 m	Crystal35ne welded tuff	Gray purple	Welded tuff texture, volcaniclastic texture	Volcanic glass orientational arrangement	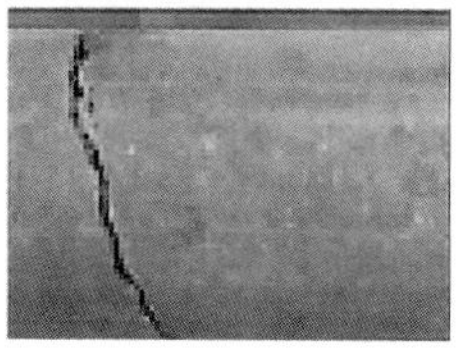

2) Subfacies classification of cored intervals

Based on lithofacies identification, subfacies can be further divided according to their lithology, rock textures and structures, and vertical assemblage relationships. For example, each subfacies of an explosive facies has different debris sizes and welded textural strengths, whereas each subfacies of an effusive facies has different vesicular and rhyolitic structures. The pillows and orbicular perlites of the extrusive facies are only developed within the mesozone and inner zone subfacies, whereas cryptoexplosive breccia of the volcanic conduit facies is only developed within the cryptoexplosive breccia subfacies. The lower portion of the 3847.08 to 3878.35 m cored interval in Well XX21-4 is crystalline welded tuff (3869.35 to 3878.35 m), with a welded textural grade between those of hot base surge and splash subfacies, which has the typical characteristics of hot clastic flow subfacies; in the upper portion, there are rhyolite and syngenetic breccia, with rhyolitic structure, which, according to the volcanic facies markers and internal subfacies assemblage characteristics, correspond to the top, middle, and lower subfacies of the effusive facies, respectively (Figure 4.18).

3) Establish logging facies patterns of different lithofacies through log calibration

Geophysical logging data are calibrated according to the classification of lithofacies and subfacies in cored intervals. Logging facies patterns for different lithofacies and subfacies are established through

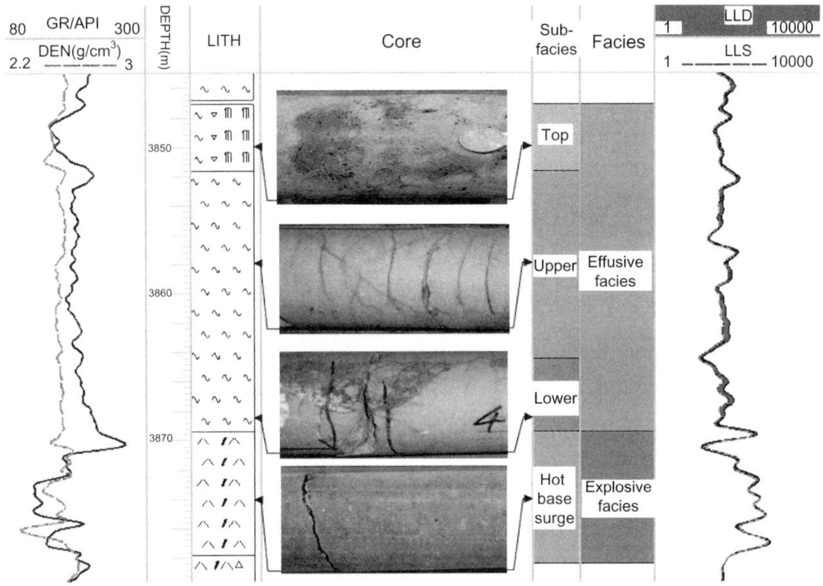

FIGURE 4.18

Lithofacies classification in cored interval of Well XX21-4.

analysis of vertical assemblage relationships and logging responses to lithofacies and subfacies, which serve as the foundation for lithofacies identification in noncored intervals.

3. Identification and classification of lithofacies in noncored intervals

Volcanic facies and subfacies in noncored intervals are identified on the basis of lithofacies setting analysis and lithological identification, using mud log and wireline log data, as well as lithological characteristics and logging facies patterns of volcanic lithofacies.

1) Lithofacies identification

Volcanic facies are determined based on the results of lithofacies setting analysis and lithological interpretations. Because volcanic rocks are characterized by rapid facies change and complex facies sequence combinations [20], it is difficult to achieve an accurate division of lithofacies based solely on wireline logs. It will be necessary, therefore, to ascertain the types of volcanic facies based on the results of lithofacies setting analysis and lithological interpretations. For example, Well DD172 encountered the limb of the volcanic edifice, and the rock massif occurrence here has a predominantly wedged shape. According to the lithofacies distribution model, therefore, it is determined tentatively that this part probably represents an effusive facies or explosive facies. The mud log and lithological interpretations from wireline logs suggest that this well encountered mainly basalt and vesicular basalt, then basaltic breccia lava, basaltic welded breccia, sedimentary tuff, and mudstone, which point to the development of a primary effusive facies, and a secondary volcanic sedimentary facies, with minor explosive facies, probably developed locally (Figure 4.19).

Lithofacies are classified and associated according to logging curve shapes and vertical changes. The lithological identification of volcanic rocks is ambiguous, and the corresponding relationship

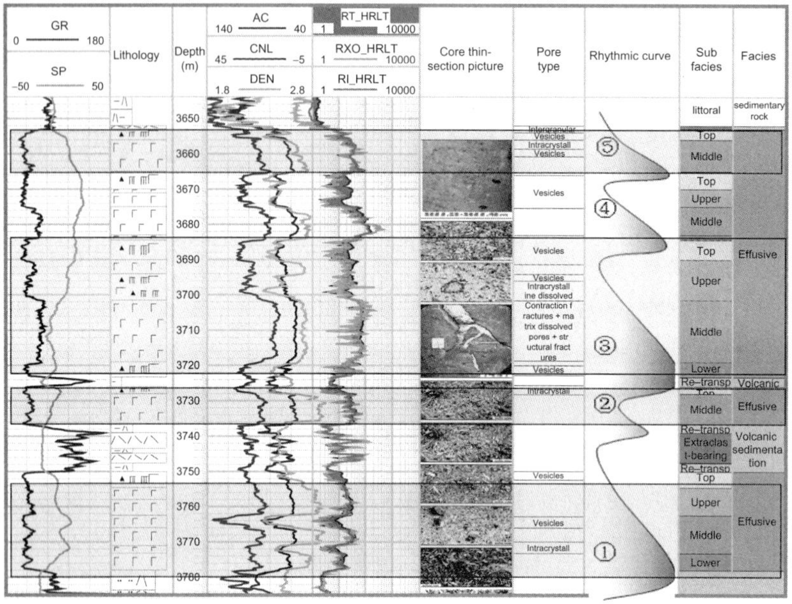

FIGURE 4.19

Single-well lithofacies of volcanic rocks.

between lithology and lithofacies is not unique. Therefore, single-well lithofacies need to be classified and associated according to the amplitudes, shapes, smoothness, and vertical changes of log curves. For example, the log curve for Well DD172 generally has a minor-serrated box or bell shape, characteristic of effusive facies, whereas the basaltic breccia lava and basaltic welded breccia generally correspond to the upper part of a box-shaped or bell-shaped curve and form part of the lithology association in the top subfacies of an effusive facies.

Therefore, according to the lithological and logging facies characteristics, seven lithofacies belts can be identified in Well DD172, including five minor-serrated box-shaped or bell-shaped curve segments corresponding to the effusive facies and two serrated curve segments corresponding to the volcanic sedimentary facies (see Figure 4.19). Among the five effusive facies belts, the log curve characteristics of the effusive facies at different segments are different: segments ① and ② largely show a box shape, segments ④ and ⑤ a bell shape, and segment ③ is transitional between box and bell shapes.

2) Subfacies division

Within the volcanic facies, subfacies are identified on the basis of the shape, variation of smoothness, and vertical association relationships of log curves.

In segment ① of Figure 4.19, for example, the log curve is divided into four subsegments: the lower segment has low density and low resistivity values, the middle segment has high density and high resistivity values, the upper segment has slightly lower density and resistivity values than the middle, whereas the top segment has low density and low resistivity, and the primary lithologies corresponding to these four subsegments are basalt-syngenetic breccia, tight basalt, vesicular basalt, and basaltic breccia lava, respectively. For this reason, Segment ① in Well DD172 can be divided into four subfacies: the lower, middle, upper, and top.

4.4 PROFILE IDENTIFICATION

Profile facies are the basis for investigating lithofacies paleogeography [21]. Based on single-well facies identification, seismic data can be calibrated using logging data, and seismic facies patterns of different lithofacies of volcanic rocks can be established. On the basis of the analysis of the volcanic edifice and massif and the seismic facies patterns of different lithofacies, profile facies of volcanic rocks are identified according to volcanic facies interpretation in single wells and the seismic reflection characteristics between wells.

4.4.1 Seismic response characteristics

Different lithofacies have different internal architectures and different seismic responses. Seismic data can be calibrated based on logging data, and then seismic response models of different lithofacies can be established.

1. Explosive facies

The overall sorting of volcanic debris in the explosive facies is relatively poor with mixed grain sizes, with rapid facies change along a plane. The seismic profiles mostly show chaotic reflection characteristics of medium-weak amplitudes, relatively high frequencies, medium-poor continuity, and hummocky or wedge shapes (Figure 4.20).

2. Effusive facies

The lava within an effusive facies has good continuity and fairly consistent distribution. The seismic profiles mostly show seismic reflection characteristics of medium-strong amplitudes, medium-low frequencies, and relatively good continuity, parallel-subparallel seismic reflection configuration, with predominantly bedded and lenticular shapes (Figure 4.21).

3. Extrusive facies

Located in the upper part of the volcanic conduit, this facies comprises the extruded bodies in regular mound, dome, cone, or bell shapes, formed by highly viscous lava that gushes out of relatively narrow conduits or fractures. The seismic profiles show chaotic seismic reflection characteristics of medium-weak amplitudes, medium-high frequencies, and poor continuity, with an umbrella-like pattern radiating from the center in the bottom (Figure 4.22).

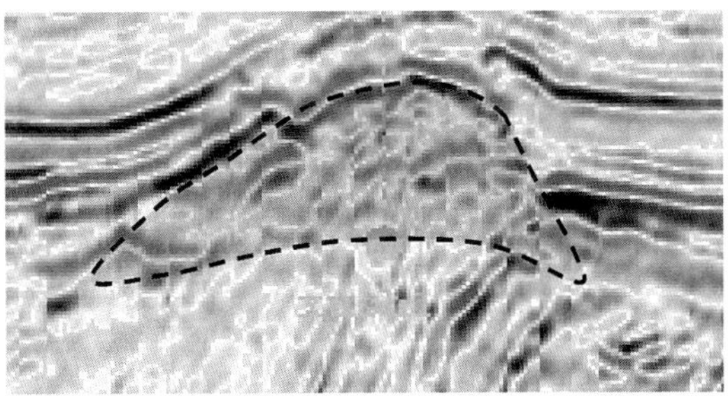

FIGURE 4.20

Seismic reflection characteristics of explosive facies.

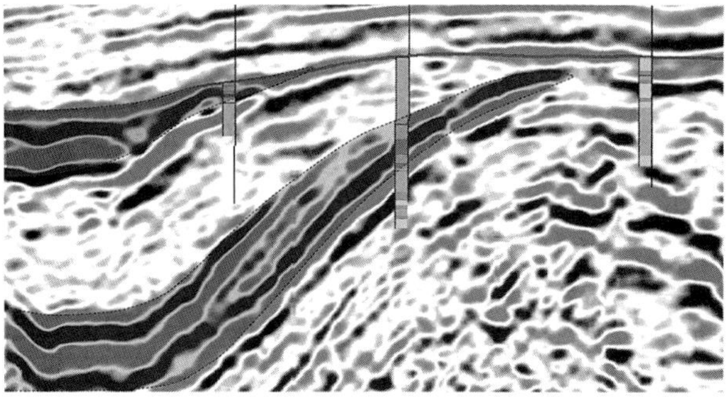

FIGURE 4.21

Seismic reflection characteristics of effusive facies.

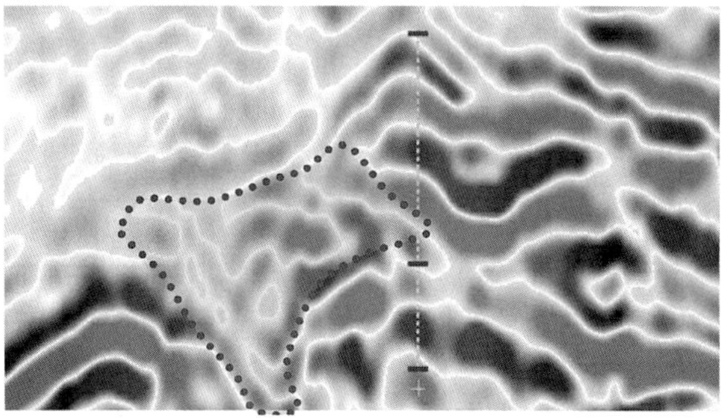

FIGURE 4.22

Seismic reflection characteristics of extrusive facies.

4. *Volcanic conduit facies*

This facies is formed by magma retained and backfilled in the volcanic conduit, and mostly in subvertical columnar shape with a mixture of volcanic rock constituents. The seismic profiles mostly show chaotic reflection characteristics of weak amplitudes, high frequencies, and poor continuity, and in nearly upright columnar shape (Figure 4.23).

5. *Subvolcanic facies*

This facies comprises mostly sills, laccoliths, stocks, dikes, and veins. The constituents of its outer zone subfacies are mixed, whereas its mesozone and inner zone subfacies have consistent constituents and structures. The seismic profiles typically show chaotic reflection characteristics of strong reflection in the top and bottom parts, and weak amplitudes, medium-high frequencies, and discontinuity inside, and have irregular lenticular reflection shapes (Figure 4.24).

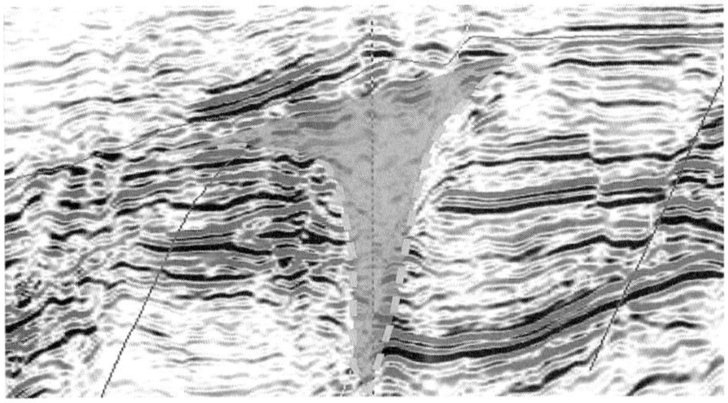

FIGURE 4.23

Seismic reflection characteristics of volcanic conduit facies.

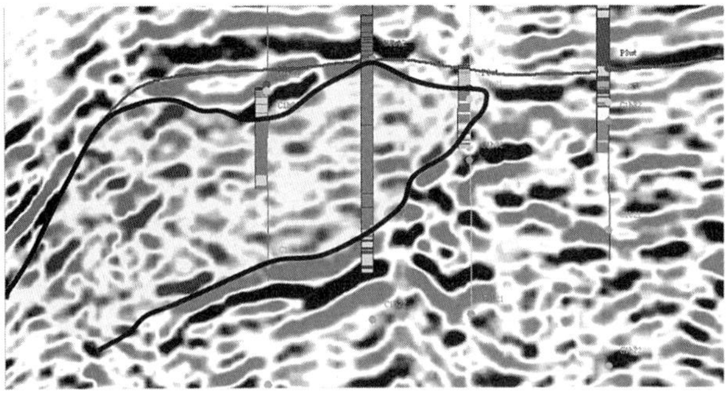

FIGURE 4.24

Seismic reflection characteristics of subvolcanic facies.

6. Volcanic sedimentary facies

This facies is represented mainly by overlap fills, often in bedded shapes, and with relatively good sorting of internal volcanic debris. The seismic profiles are characterized by parallel-subparallel reflection characteristics of strong reflection on the top and bottom, and medium-weak amplitudes, medium-high frequencies, and moderate continuity inside (Figure 4.25).

4.4.2 Profile facies identification

Profile facies of volcanic rocks are identified by tracing their top and bottom boundaries based on single-well facies identifications, well-seismic calibration, changes in seismic reflection characteristics, and interwell events of different lithofacies.

1. Calibrating profile facies based single-well facies

The resolution of single-well lithofacies identification is usually high, but the resolution of seismic data is low. As a result, when a single-well lithofacies is used for seismic calibration, the single-well data need to be properly coarsened based on the dominant lithofacies of the target layer. As shown

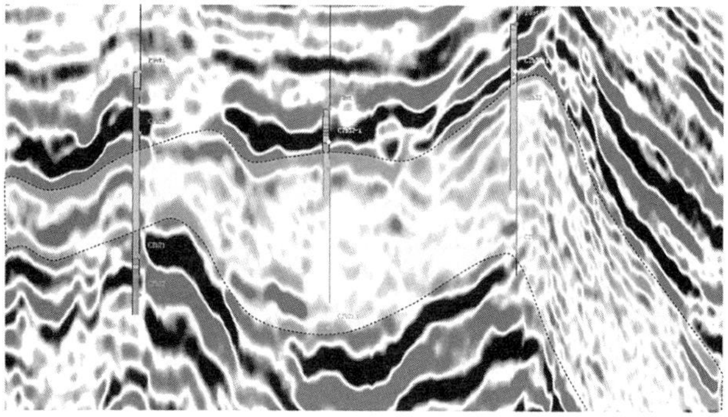

FIGURE 4.25

Seismic reflection characteristics of volcanic sedimentary facies.

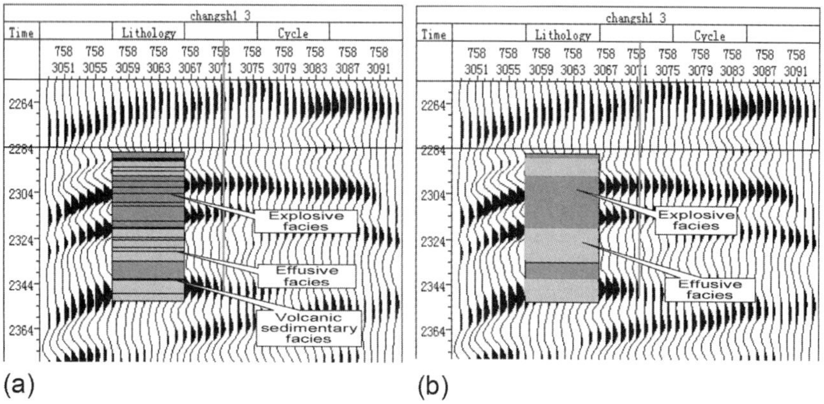

FIGURE 4.26

Calibrated seismic profiles of volcanic facies before and after coarsening (Well CC1-3). (a) Before coarsening. (b) After coarsening.

in Figure 4.26, a large number of thin interbeds of volcanic sedimentary facies are developed in the volcanic formation encountered by Well CC1-3, but these thin interbeds are unidentifiable in seismic profiles. After coarsening treatment, six facies become recognizable in this well from top to bottom, hence good corresponding relationships are established with seismic profile reflection characteristics.

2. Profile facies identification

The types, shapes, distributions, and superposition relationships of volcanic lithofacies in profiles are analyzed and the traceable scopes of different lithofacies are determined by applying seismic facies patterns based on the volcanic facies divisions in single wells and constrained by the changes in seismic reflection characteristics between wells.

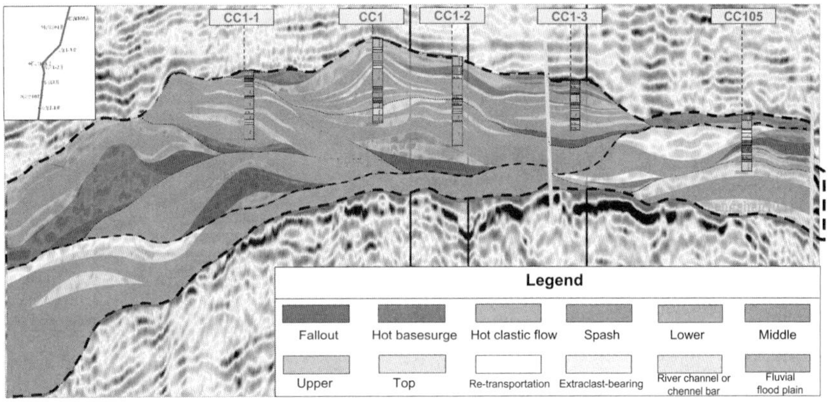

FIGURE 4.27

Profile facies map of volcanic rocks.

3. Profile facies tracing

Constrained by the internal structure of volcanic rocks and following the technical approach of boundary tracing (e.g., from big boundaries to small boundaries, from distinct boundaries to blurry boundaries), the volcanic facies boundaries are traced in succession, and volcanic facies in profiles are identified (Figure 4.27).

4.5 PLANAR PREDICTION

Volcanic facies can reveal the spatial distribution of volcanic rocks and the petrogenetic relationships between different lithological assemblages. Based on the identification of single-well facies and profile facies, therefore, seismic facies can be predicted through the correlation and extrapolation of single-well facies, skeletal network control of profile facies, and well-seismic combination techniques; ultimately, the planar distribution of volcanic facies can be revealed.

4.5.1 Planar facies prediction based on single-well facies

The planar distribution of volcanic facies can be predicted reliably through well correlation and extrapolation between wells using the results from single-well facies identification, based on sufficient data from closely spaced wells.

1. Analysis of interwell distribution of lithofacies through interwell correlation

Constrained by the internal architecture of volcanic rocks, the lithofacies characteristics and interwell changes in each eruption period are ascertained and the range of lithofacies distribution in each eruption period is predicted through interwell correlations.

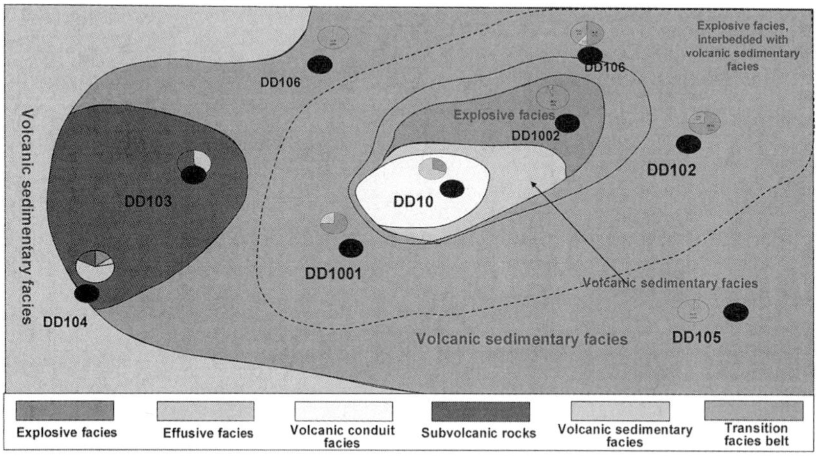

FIGURE 4.28

Plane facies map of volcanic rocks predicted on the basis of facies identification for single wells.

2. Analysis of planar distribution of lithofacies by extracting single-well lithofacies information for each eruption period

The lithofacies information (e.g., types, thicknesses) of volcanic rocks in each eruption period is extracted from wells and projected onto a plane, and the characteristics of lithofacies change are analyzed on the plane. As shown in Figure 4.28, the lithofacies in each eruption period are proportionally projected onto the planes, and the planar distribution characteristics and trends of lithofacies variation are analyzed. The results will serve as the foundation for establishing a planar lithofacies distribution model.

3. Establishing the planar lithofacies distribution model

Guided by the volcanic facies pattern, a planar lithofacies distribution model is established according to the shape of the volcanic edifice, the position of the volcanic conduit, and the characteristics of planar lithofacies distribution, which will provide guidance for predicting lithofacies distribution in different subzones.

4. Determining lithofacies boundaries to predict their planar distribution

Lithofacies boundaries are determined by extracting single-well subzone lithofacies thicknesses, with reference to the interwell and planar distribution models, to further predict the planar distribution of volcanic facies. For the superposed parts of lithofacies, relevant transition lithofacies belts, such as the eruptive-volcanic sedimentary facies transition belt, should be classified based on the lithofacies characteristics of their surrounding rocks (see Figure 4.28).

4.5.2 Predicting planar facies patterns based on profile facies

Based on seismic reflection profile identification and the established skeletal profile network that can effectively control the study area, as well as the constrained distribution pattern of volcanic edifices and volcanic massifs, the thickness variation of lithofacies is analyzed, the lithofacies boundaries are determined, and the planar distribution of volcanic facies is predicted. The prediction of planar facies distribution patterns based on profile facies can be realized through the following three steps.

1. Constructing a skeletal profile network that covers the entire study area

To effectively predict the lithofacies distribution across the study area, 22 skeletal profiles are constructed within Well Block DD17, including 11 profiles in the NE direction, 7 profiles in the EW direction, and 4 profiles in the NE-SW direction, which effectively control the planar distribution of volcanic facies in the study area (Figure 4.29).

2. Identifying profile facies, analyzing the distribution of lithofacies, and estimating the thickness variation of lithofacies

The variation characteristics of volcanic facies are analyzed in succession on the profile, and the traceable extent and interwell changes of different lithofacies are determined based on single-well lithofacies division for well points or seismic response models and seismic event variations of volcanic facies between wells. As shown in Figure 4.29, the distribution and variation trends of volcanic facies in profile are ascertained through profile facies identification.

3. Determining lithofacies boundaries to predict their planar distribution

By projecting the profile lithofacies distribution onto the plane, the lithofacies boundaries are determined and the lithofacies distribution is predicted based on lithofacies patterns, interwell lithofacies changes, and the variation trend of lithofacies thickness.

4.5.3 Predicting planar facies distribution based on seismic facies

Apparently different lithofacies have different seismic responses and seismic attributes. Based on single-well identification and profile identification, therefore, the planar distribution of volcanic massifs is ascertained by systematic analysis of seismic attributes and especially lithofacies-sensitive attributes, planar identification, and comprehensive prediction.

1. Analysis of sensitive attributes of lithofacies

Seismic attributes include a wealth of information on subsurface rock formations such as lithology, petrophysical properties, and fluid properties. In different eruption areas, the types of seismic attribute sensitive to volcanic facies are different. In volcanic facies analysis using seismic attributes, therefore, it is necessary to correlate and prioritize different seismic attributes and determine the sensitive seismic attributes according to their geological characteristics and seismic responses.

In the DD gas field case study, attributes such as waveform classification and root mean square (RMS) amplitude can be considered good reflections of volcanic facies characteristics. The explosive facies is typified by relatively tight lithology, poorly developed pores and fractures, characterized by seismic reflections of strong amplitude, low frequency, and high RMS amplitude; the effusive facies shows chaotic reflection characteristics of weak-strong amplitudes, medium-high frequencies, discontinuity, and medium RMS amplitude values; and the volcanic sedimentary facies seismically shows seismic reflections of medium-weak amplitudes, medium-high frequencies, good continuity, and low RMS amplitude.

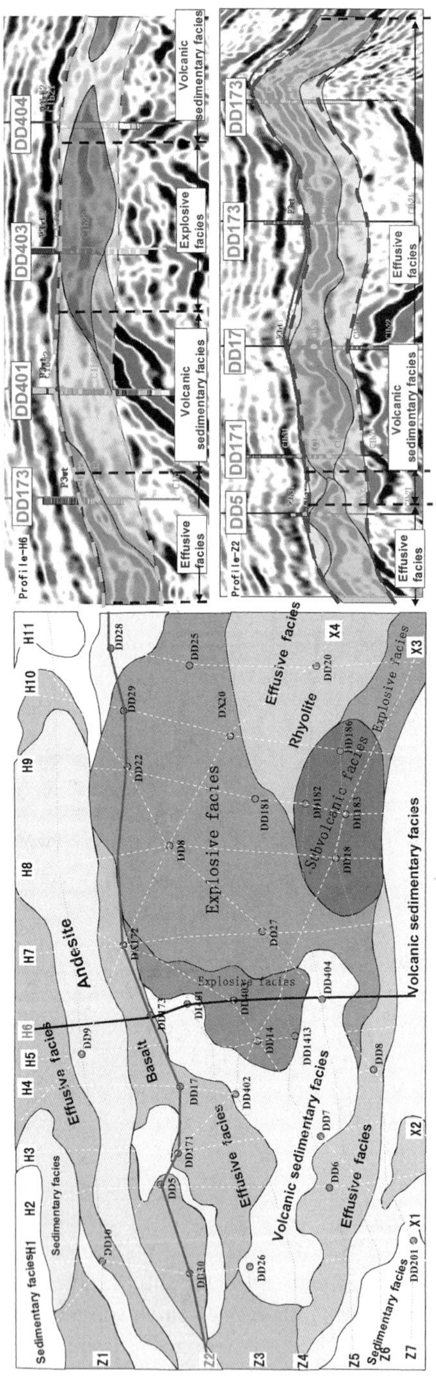

FIGURE 4.29

Skeletal profile network and corresponding lithofacies profiles for Well Block DD17.

2. *Extraction of attributes*

Volcanic lithofacies prediction has stringent requirements on seismic attributes, especially on the selection of the reference surface and time window. The basic principles are (1) select the geological interface with well-defined geological significance as the reference surface; (2) determine the time window size according to the position and thickness of the target layer. When studying the distribution of volcanic facies in various periods, for example, it is preferable to use the top and bottom contacts of different volcanic periods as the boundaries to extract seismic attributes; when studying the lithofacies distribution in gas zones, it is necessary to determine the boundaries based on the distribution and thicknesses of gas zones.

The volcanic gas zones in Well Block DD are located mainly within the range of 100 m depth below the top of volcanic rocks, and the seismic attributes are generally extracted from the 40- to 50-ms time window beneath the top of the volcanic rock body. In comparison, the volcanic gas zones in Well Block XX are located mainly within the range of 150 m in the top part of each volcanic edifice, and the seismic attributes are generally extracted by opening a 60- to 70-ms time window downward from the top of volcanic edifices. Figure 4.30 shows a case study on the extent of planar distribution and changes of the explosive facies within 50 m in the upper part of Well DD14. It is determined that the target layer in the seismic profile corresponds to the formation of 23 ms downward from the top of the volcanic edifice; therefore, seismic attributes are extracted by opening a 25-ms time window downward from the top of the volcanic edifice, and the characteristics of planar distribution and changes of the explosive facies are thus ascertained.

3. *Planar prediction*

1) Analysis of single attributes

This technique refers to the use of a single sensitive seismic attribute to predict the planar changes of lithofacies. Guided by the volcanic facies distribution model, lithofacies distribution is predicted according to the changes of attribute values and by calibrating the single-well lithofacies division against seismic attributes.

Figure 4.31 illustrates the planar distribution of lithofacies predicted through waveform classification attributes in CC gas field. The red, yellow, orange, pink, dark brown, and blue areas stand for different waveform characteristics and reflect different lithofacies types, respectively. In particular, the

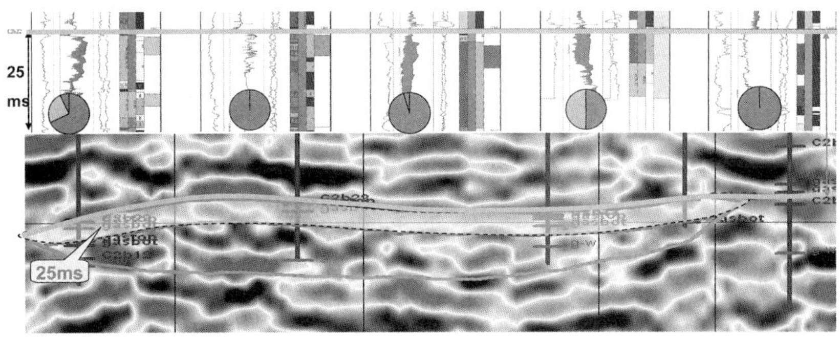

FIGURE 4.30

Selecting the seismic time window based on the research target layer.

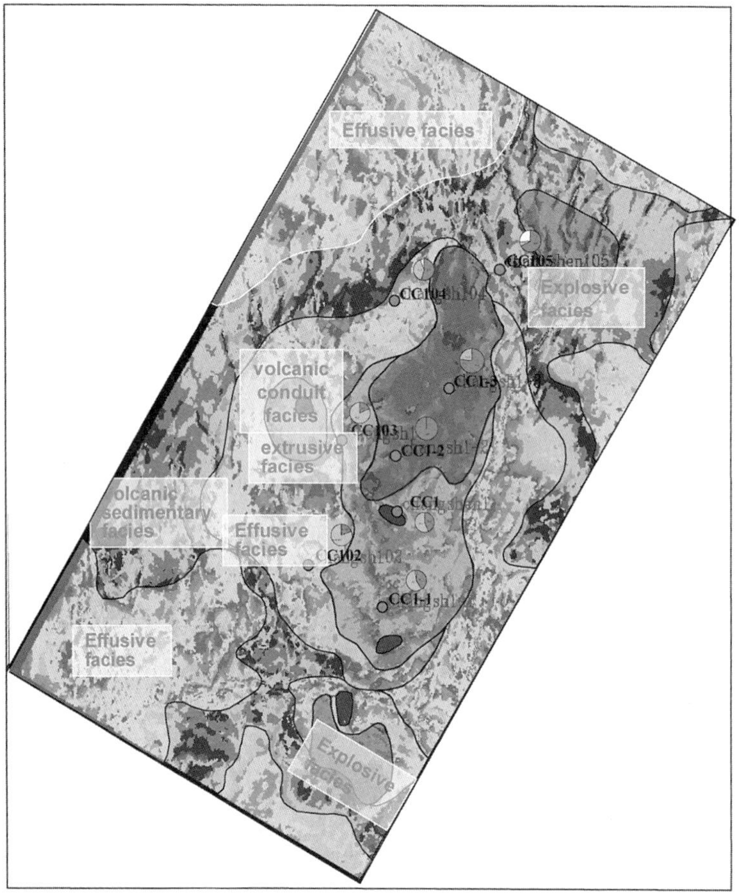

FIGURE 4.31

Seismic attributes map by waveform classification.

red areas stand for explosive facies regions, the yellow areas for effusive facies regions, the orange areas for mixed explosive and effusive facies, the pink areas for extrusive facies, the dark brown areas for volcanic conduit facies, and the blue areas for volcanic sedimentary facies. In Figure 4.32, different RMS amplitude values also reflect the distribution of different lithofacies.

2) Combined prediction of multiple attributes

Different seismic attributes reflect different geological formations and, therefore, the prediction of volcanic facies distribution from a single attribute can lead to multiple interpretations. The combined analysis determined by integrating multiple attributes can effectively reduce the uncertainty in prediction results. The principle is to give different weight values to different attributes according to their corresponding relationships to volcanic facies, and the weight values are converged to obtain a comprehensive attribute distribution map.

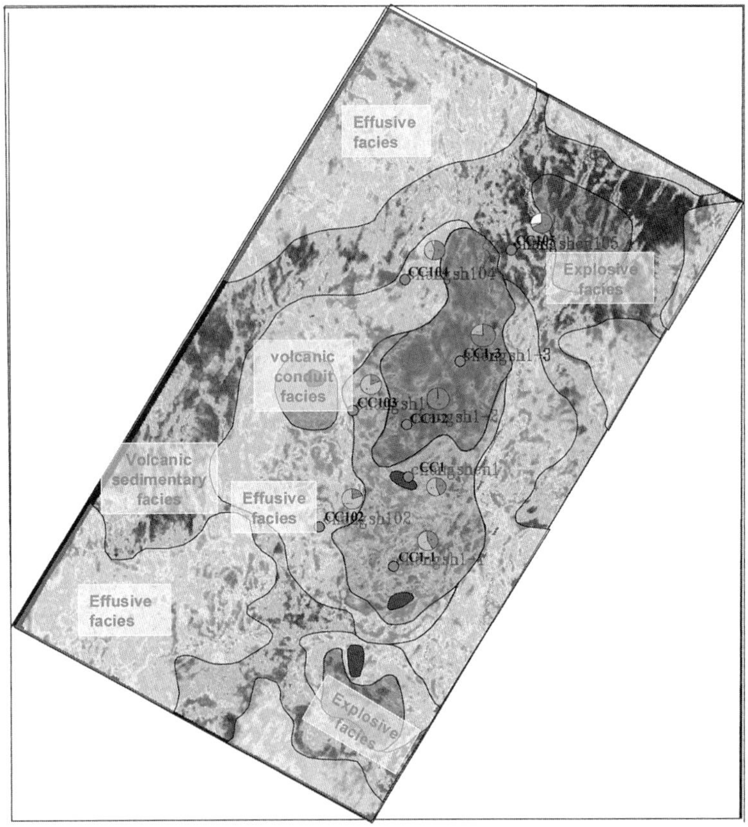

FIGURE 4.32

Seismic attributes map by RMS amplitudes.

Figure 4.33 is the lithofacies planar distribution map of the CC gas field derived from comprehensive prediction from waveform classification, RMS amplitude, and instantaneous phase attributes. Subsequently drilled wells have verified these prediction results, whereby the planar distribution characteristics of lithofacies in the CC gas field have been identified reliably.

4.6 SPATIAL PREDICTION

4.6.1 Profile tracing closure technique

To predict the 3D spatial distribution of lithofacies, the tracing and closing of volcanic facies boundaries are carried out based on the seismic facies patterns, by establishing a skeletal profile network that can effectively control the study area, with constraints by volcanic edifices and massifs. When applying this technique, the usually rapid change of volcanic facies and significant lateral changes in their thicknesses pose challenges in seismic identification and tracing of each lithofacies.

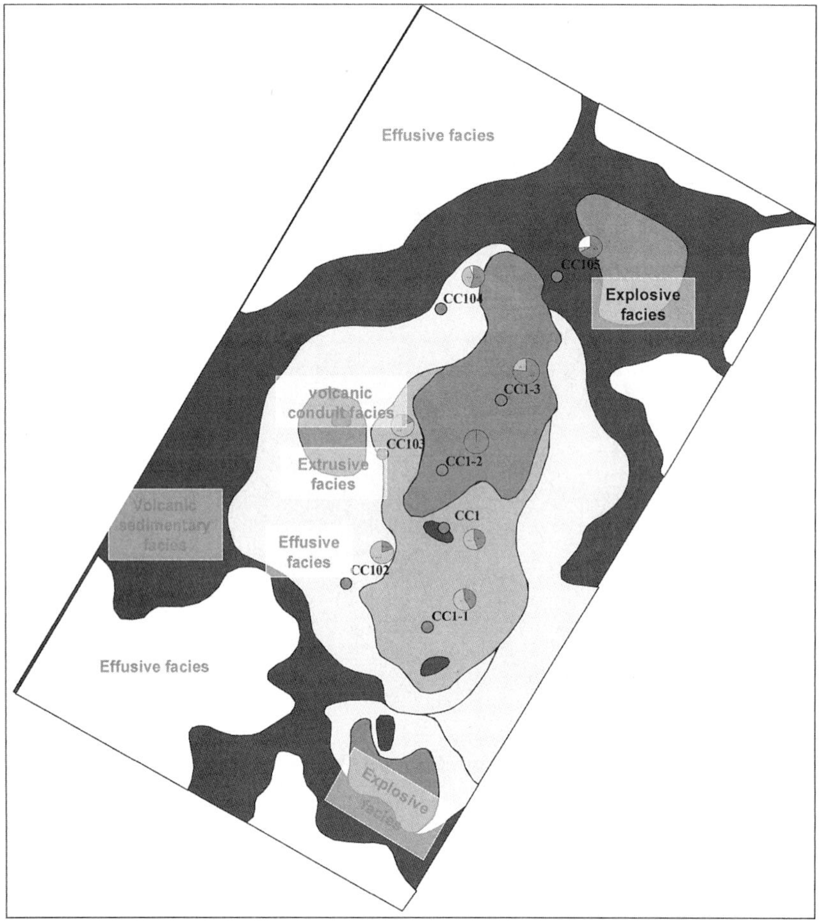

FIGURE 4.33

Lithofacies planar distribution map based on systematic prediction of multiple attributes.

4.6.2 Integrative technique for lithofacies analysis by stratigraphic slicing

Volcanic rocks are often formed in multiple periods and accordingly seismic attributes within volcanic massifs are extracted in terms of the eruption periods. On the basis of the planar prediction of lithofacies at individual periods, the spatial distribution characteristics of lithofacies are revealed according to the results of single-well lithofacies and profile facies identifications.

This technique is explained herein using the case study in the #YC_2^5 volcanic massif in Block CC1. Vertically, the YC_2^5volcanic massif encountered in Well CC1 can be separated into three vertically stacked volcanic eruption periods. Interval 3550 to 3587.6 m is composed mainly of tuff and crystalline tuff, formed as a hot debris flow subfacies and fallout subfacies of the explosive facies; interval 3587.6 to 3811.0 m consists of various rhyolite and autoclastic brecciated lava, formed as the top, upper, middle, and lower subfacies of the effusive facies; interval 3811.6 to 3852.5 m comprises welded tuff, rhyolite, and welded breccia, interbedded with volcanic breccia, accumulated as alternating effusive facies and explosive facies.

Within the #YC_2^5 volcanic massif, the planar distribution of lithofacies in each of the three eruption periods is predicted using the attribute analysis technique. As shown in Figure 4.34, the explosive facies are mainly developed in the top part of the #YC_2^5 volcanic massif, and the effusive facies in a nearly south-north (SN)-trending area to the east of Well CC1 and to the west of Well CC1-3; the effusive facies are mainly developed in the middle part, and the lithologies in Wells CC1 and CC1-2 include rhyolite, vesicular rhyolite, and autoclastic brecciated lava, with multiple effusive facies; the lower part is characterized by alternating explosive and effusive facies. The drilling results have since revealed multiple lithologies such as rhyolite, welded tuff, volcanic breccia, autoclastic brecciated lava, and so on.

The spatial changes of volcanic facies are determined systematically on the basis of planar distribution predictions of lithofacies in each period, in combination with information on vertical changes of lithofacies in wells and in different periods. The #YC_2^5 volcanic massif corresponds to a crater between Well CC1 and Well CC1-2; the lithofacies analysis shows that neither the explosive nor the effusive facies of the #YC_2^5 volcanic massif has the characteristics of inherited development vertically, its upper and lower parts being composed mainly of explosive facies and the middle part of effusive facies. The favorable facies belt is the effusive facies in the middle of the volcanic massif, and the favorable zone is the effusive facies regions to the east of Well CC1-2 and the region near the volcanic conduit between Well CC1 and Well CC1-2.

4.6.3 Technique for 3D attribute volume analysis

This technique extracts the spatial attribute parameters of lithofacies and reveals their distribution patterns by applying 3D attribute volumes sensitive to volcanic facies and establishing attribute classification criteria for different lithofacies. There are primarily four steps in this technique: (1) establishing the volcanic facies classification criteria for seismic attributes through sensitivity analysis, (2) extracting various sensitive attributes within the 3D space using the volume attribute extraction technique, (3) extracting 3D space data points of various volcanic facies according to the classification criteria and combining them based on their spatial locations, and (4) performing superposition and systematic analysis based on the prediction results of multiple attributes to achieve a comprehensive prediction of the spatial distribution of volcanic facies.

4.7 CHARACTERIZATION OF VOLCANIC FACIES

Characterization of the shapes, scales, and superposition relationships of volcanic facies and delineation of the distribution of volcanic facies are conducted based on volcanic facies identification, using outcrop, seismic, and dense well grid data.

4.7.1 Characterization of volcanic facies geometry (shape)

1. Characterizing volcanic facies geometry from outcrop data

Field outcrop investigation is an important step for volcanic facies study, as it is the most intuitive, accurate, and reliable approach to acquire data. The process of characterization consists of the following four steps.

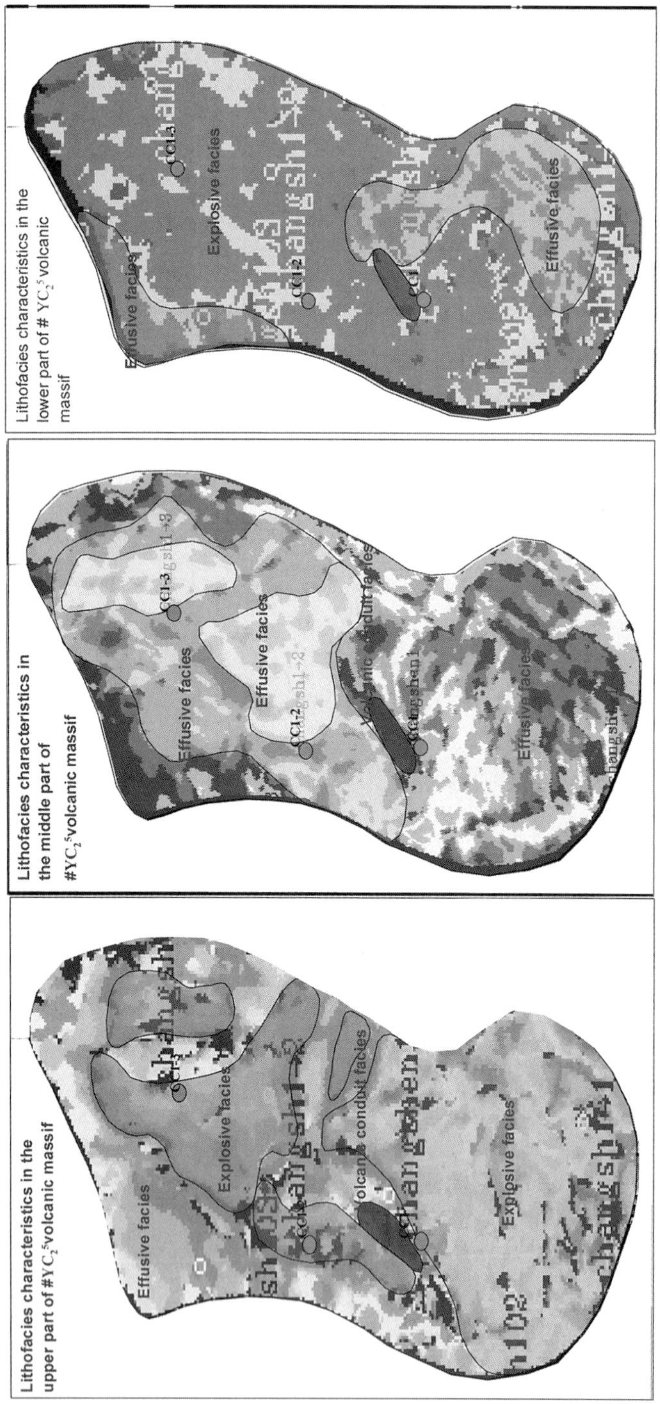

FIGURE 4.34

Planar distribution of lithofacies in each eruption period within the #YC$_2^5$ volcanic massif.

1) Profile selection

Proper profiles are selected through geological surveys and regional geological analysis. The volcanic gas reservoirs of the Songliao basin are located mainly in the volcanic rocks of the Yingcheng Formation; 37 volcanic outcrop profiles (approximately 3 km long, covering an area of approximately 13 km^2) of the formation located at Shichangcun, Huangtu'aizi, in the Yingcheng Coal Mine and the Santai regions in the east of the basin have been selected for this study.

2) Lithofacies identification

On the basis of the selected lithofacies profiles, the lithofacies of the type localities are identified and the profiles calibrated with reference to the typical rock types, textures, and structure marks of outcrop profiles.

The study involves 1886 field photos and 878 samples, with the following results: at the outcrops, the Yingcheng Formation consists of vesicular rhyolite, volcanic breccia, agglomerate, perlite, and cryptoexplosive breccia, with perlitic and cryptoexplosive breccia textures, and rhyolitic to lithophysa structures. This indicates the presence of volcanic neck subfacies and cryptoexplosive breccia subfacies of the volcanic conduit facies; fallout subfacies and hot debris flow subfacies of the explosive facies; lower, middle, and upper subfacies of the effusive facies; inner zone subfacies of the extrusive facies; and xenoclast-bearing subfacies of the volcanic sedimentary facies.

3) Mapping profile and planar facies

Identification of volcanic profile facies and planar mapping of volcanic facies distribution are based on volcanic facies markers, in combination with data on crater position, volcanic edifice distribution, change in volcanic sequence boundaries, and pinchout patterns of volcanic massifs, as well as ground-truthing measurements, trenching, and drilling data (Figure 4.35).

4) Characterizing volcanic facies shape/geometry

The volcanic lithofacies of the Yingcheng Formation in the Songliao basin consists mainly of explosive and effusive facies, with different facies showing different shapes and other characteristics.

1. *Shape of explosive facies.* The explosive facies comprises mainly volcanic breccia (fallout subfacies), crystal welded tuff (hot clastic flow subfacies) and bentonite layers (fallout subfacies), and occasionally drop-pebble volcanic bombs. This facies typically has a wedge-shaped geometry

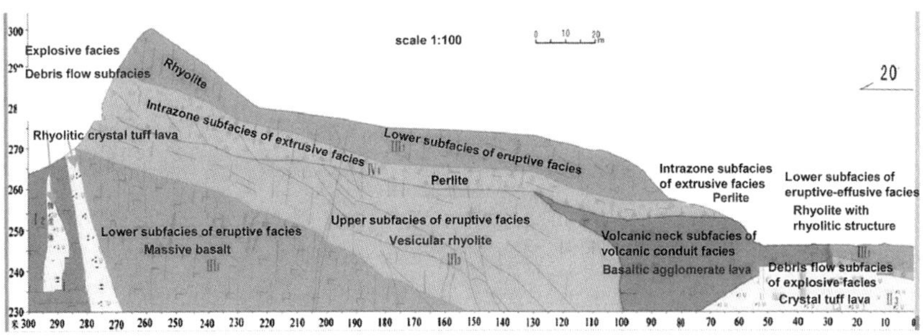

FIGURE 4.35

Volcanic profile facies of field outcrops.

but can be mound-shaped locally. The outcrops at the Yingcheng Coal Mine are the most typical of such geometry.

2. *Shape of effusive facies.* The effusive facies is characterized by tight massive basalt (lower subfacies), vesicular basalt (upper subfacies), and tight massive dacite (lower subfacies). This facies mainly has a bedded geometry (Figure 4.36), as is typified by the volcanic outcrops in the Santai area.
3. *Shape of volcanic conduit facies.* This facies is represented primarily by mauve basaltic agglomeratic lava (volcanic neck subfacies) and cryptoexplosive breccia (cryptoexplosive breccia subfacies). This facies is unique in having a columnar or dendritic morphology (Figure 4.37). It is characterized by the main profile in Guanmashan, Shichangcun.

FIGURE 4.36

Geometry of effusive facies in field outcrops.

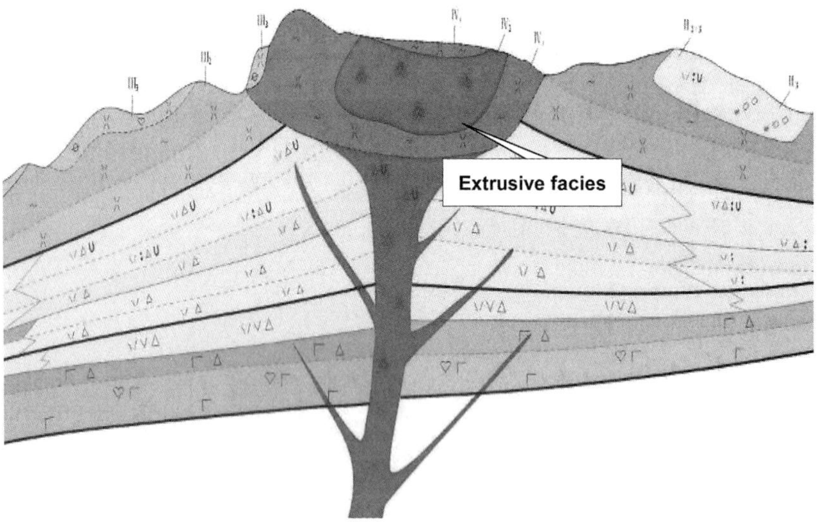

FIGURE 4.37

Shape of volcanic conduit facies and extrusive facies patterns in field outcrops.

4. *Shape of volcanic sedimentary facies.* Member Ying-I shows polygenetic conglomerates of volcanic sedimentary facies with good sphericity and poor sorting, comprising mainly debris of rhyolite, basalt, quartzite, and metamorphic rocks, as well as quartz (chert) and feldspar phenocrysts. The matrix is mostly sedimentary rocks. The volcanic sedimentary facies in this region is composed mainly of sedimentary tuff (xenoclast-bearing subfacies of volcanic sedimentary facies), having a bedded morphology. This facies is typified by the main profile in Guanmashan, Shichangcun.
5. *Shape of extrusive facies.* The extrusive facies outcrop at Huangtu'aizi is composed mainly of perlite (inner zone subfacies), but its morphology is not shown incompletely. The extrusive facies in the outcrop at Guanmashan contains perlite and brecciated lava, forming a cut-drape relationship with the explosive and effusive facies sequences. It has a flat top and steep sides, showing an umbrella- or mushroom-like geometry (Figure 4.37). This facies geometry is best represented in the Huangtu'aizi region.

2. Characterizing volcanic facies geometry using seismic data

The shapes of volcanic facies are effectively characterized through seismic profile identification as well as planar and spatial prediction, guided by the geological models derived from outcrops.

1) Volcanic facies geometry in profile

In the studied profiles, the explosive facies is primarily mound shaped, the effusive facies wedge shaped or bedded, the volcanic sedimentary facies bedded, the extrusive facies umbrella-like, the volcanic conduit facies subvertically columnar, and the subvolcanic facies usually lenticular or wedge shaped (Figure 4.38).

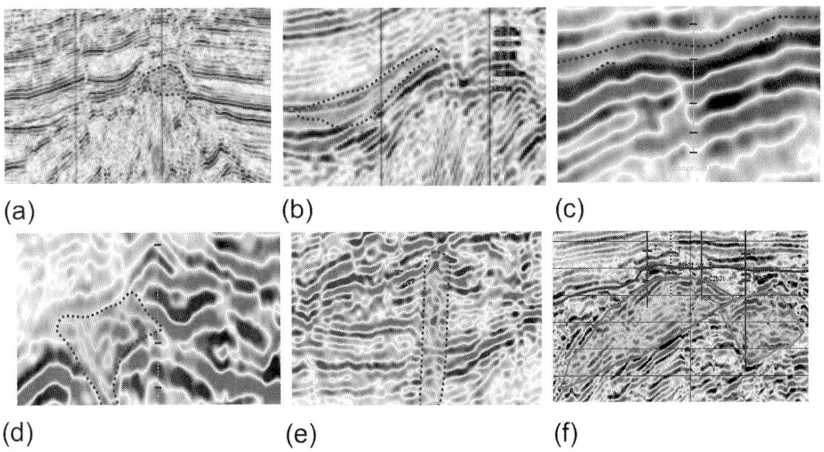

FIGURE 4.38

Characteristics of seismic profiles of volcanic facies. (a) Explosive facies: mound shape. (b) Effusive facies: wedge or bedded shape. (c) Volcanic sedimentary facies: bedded shape. (d) Extrusive facies: umbrella shape. (e) Volcanic conduit facies: columnar shape. (f) Subvolcanic facies: lenticular shape.

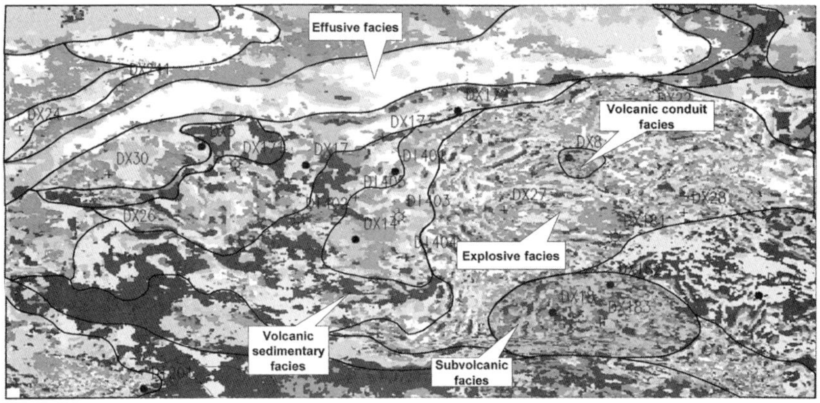

FIGURE 4.39

Characteristics of volcanic facies in plan view.

2) Volcanic facies shapes in plan view

The explosive facies are distributed around the crater and appear mostly in sheet or potato shapes in plan view; effusive facies are distributed outside the explosive facies and appear mostly in straps; the volcanic sedimentary facies occur in topographic lows and commonly appear as embayment or straps; the volcanic conduit facies is primarily rounded or elliptical; and the subvolcanic facies primarily has an elliptical outline (Figure 4.39).

4.7.2 Characterizing volcanic facies size

1. Characterization of lithofacies size using outcrop data

Field surveys revealed rapid lateral and vertical changes in the volcanic facies of the Yingcheng Formation in the Songliao basin. The explosive and effusive facies have the largest size, whereas the extrusive facies is the smallest. The explosive facies has a planar extension of 1000 to 4000 m and a vertical thickness of 15 to 200 m; the effusive facies has a planar stretch of 400 to 2700 m and a vertical thickness of 3 to 130 m; the extrusive facies has a planar extension of 3 to 600 m and a vertical thickness of 1 to 85 m; and the volcanic conduit facies has a planar extension of 110 to 1000 m and a vertical thickness of 3 to 70 m (Figures 4.40 and 4.41).

2. Characterization of lithofacies size using seismic data

Guided by the geological models derived from outcrops, the planar extent and vertical thicknesses of volcanic facies can be effectively characterized using seismic profile identification as well as planar and spatial prediction.

In the case study of Well Block XSS2, the volcanic facies in Well Block XSS changes rapidly with significant variations of thicknesses, with the explosive and effusive facies having the largest size (Figure 4.42).

The lateral extent of explosive facies has a range of 700 to 3100 m and a vertical thickness of 70 to 140 m; the lateral extent of effusive facies ranges from 500 to 4000 m and has a vertical thickness of 70 to 400 m; the lateral extent of extrusive facies ranges from 400 to 1200 m and has a vertical thickness of

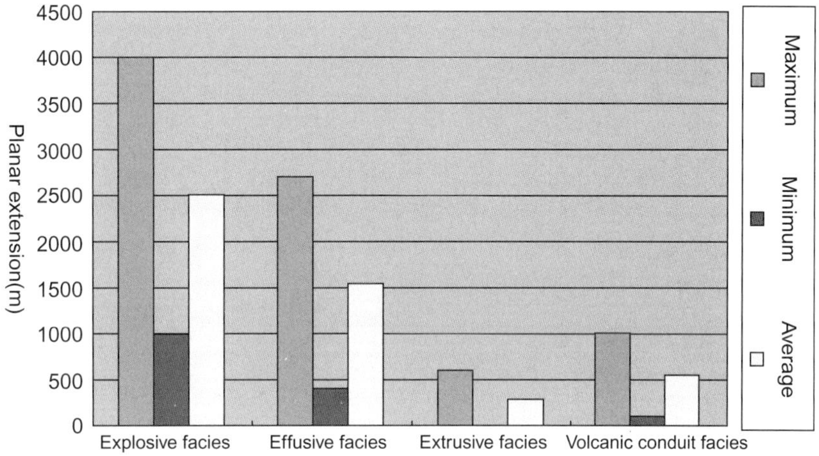

FIGURE 4.40

Planar extension of different volcanic facies in field outcrops.

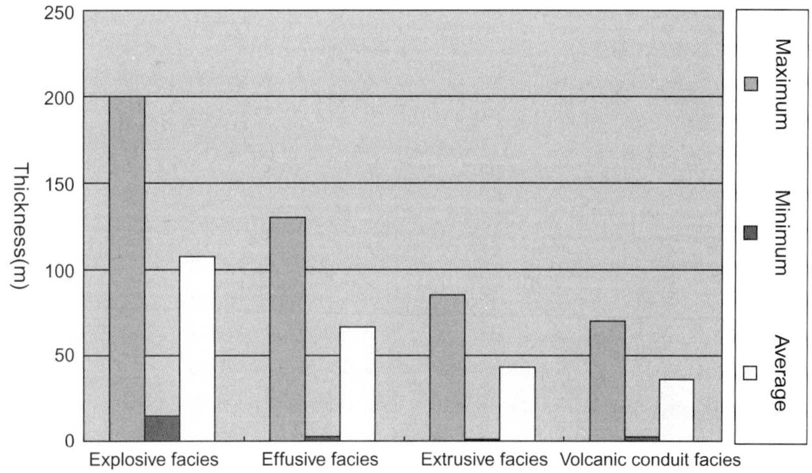

FIGURE 4.41

Thickness statistics of different volcanic facies in field outcrops.

50 to 150 m; and the lateral extent of volcanic conduit facies ranges from 700 to 1300 m and has a vertical thickness of 50 to 150 m (Figures 4.43 and 4.44).

Dense well grid data contain abundant geological, logging, and performance information and are important for volcanic facies investigations. The main steps of characterizing volcanic facies size with dense well grid data include the following:

1. Select a well block with close well spacing, and divide volcanic facies in all single wells based on lithofacies markers.

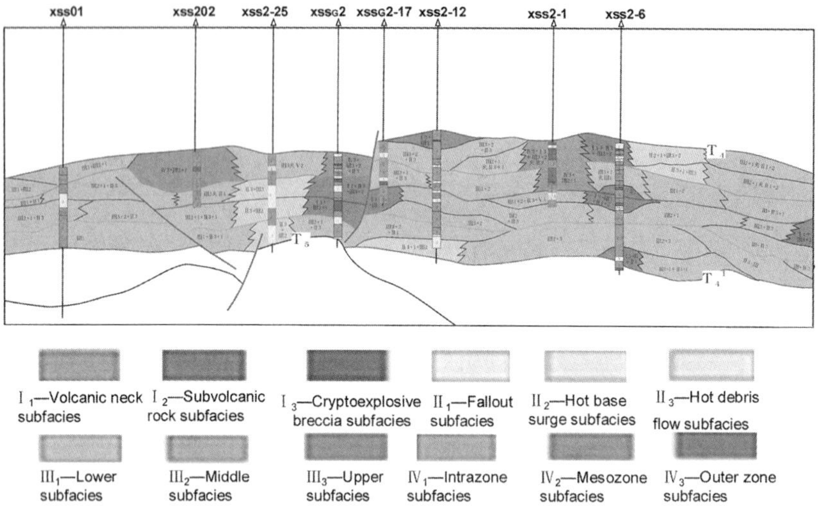

FIGURE 4.42

Identification of volcanic profile facies based on 3D seismic data.

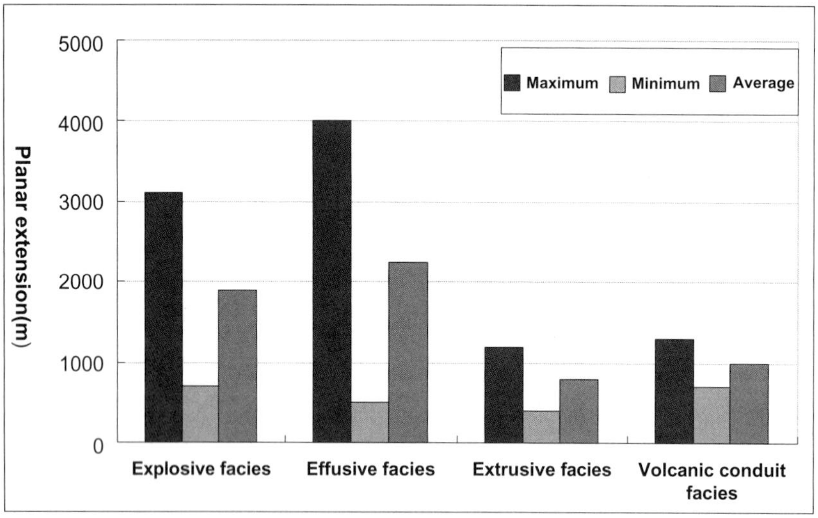

FIGURE 4.43

Planar extension of different volcanic facies underground.

2. Combine the positions of volcanic edifices and massifs in the well block through interwell correlation to analyze the lithofacies distribution between wells.
3. Calibrate seismic data using geophysical logging data to trace profile facies, extract the extent and thicknesses of different lithofacies, and determine the association relationships between lithofacies under the constraint of seismic event changes, interwell reference of seismic reflection characteristics, and lithofacies variation trends.

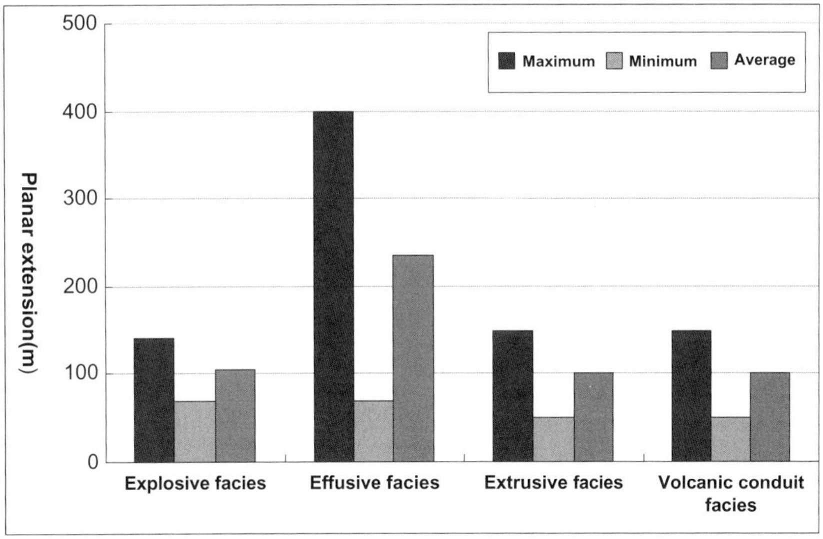

FIGURE 4.44

Thickness statistics of different volcanic facies underground.

4. Predict planar distribution of lithofacies based on single-well lithofacies division at well points or constrained by seismic attributes, as well as distribution of volcanic edifices and massifs between wells.

Volcanic massifs in Well Block XX1 are formed from multistage eruptions with multiple craters, the sizes of volcanic massifs being relatively small, and of mainly explosive facies and minor effusive facies. The lithofacies are relatively small in scale. The lateral extent of explosive facies is 200 to 800 m, and vertical thickness is 60 to 120 m; and the lateral extent of effusive facies is 300 to 1200 m, and vertical thickness is 45 to 100 m (Figure 4.45).

4.7.3 Characterizing superposition relationships of volcanic facies

The superposition relationships among volcanic facies are controlled by factors such as volcanic eruption mode, paleogeomorphology, lithofacies type, and magmatic melt volume. These relationships reflect the superposition, migration and change of volcanic rocks during its formation and evolution. The spatial superposition patterns of volcanic facies are generally classified into four types: alternating, imbricated, parallel, and isolated.

1. Alternating superposition

This superposition relationship results from frequent changes in volcanic facies and is the predominant superposition relationship of volcanic facies. The alternation occurs mainly near the crater, reflecting the development of volcanic facies during different periods stacking on one another. It is characterized by a single lithofacies forming multiple, superimposed lithofacies belts (Figure 4.46).

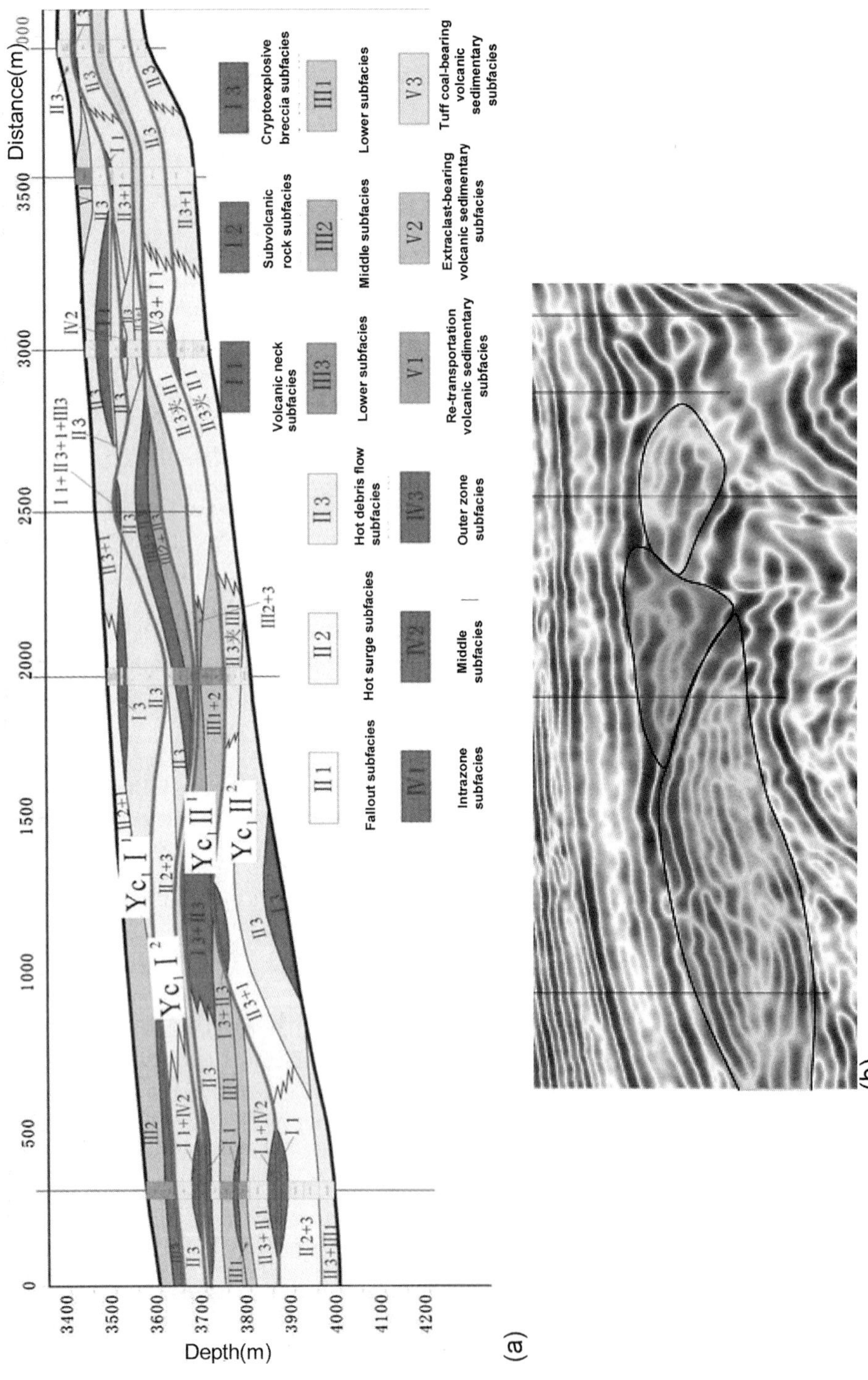

FIGURE 4.45

Volcanic profile facies identification and reflection characteristics of corresponding seismic profile in dense well grid areas. (a) Identification of volcanic profile facies in dense well grid areas. (b) Reflection characteristics of corresponding seismic profile.

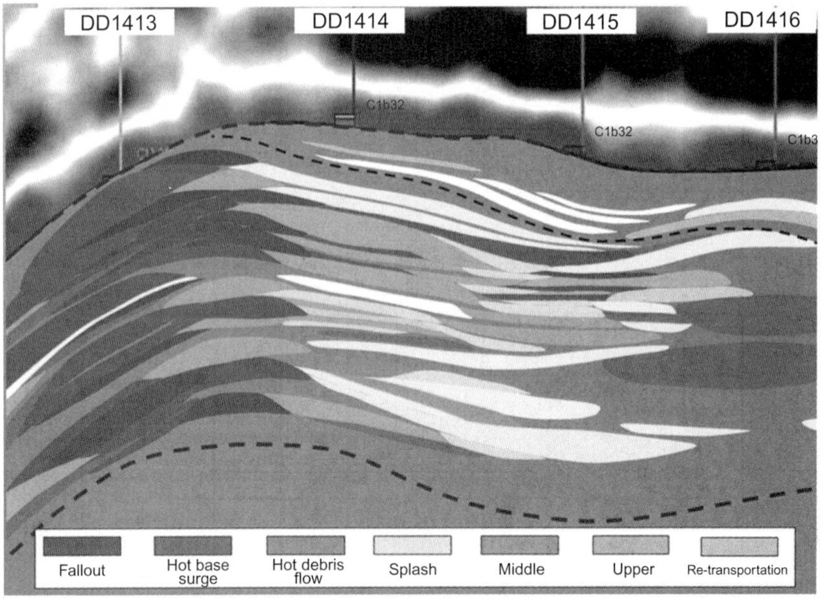

FIGURE 4.46

Alternating lithofacies pattern.

2. Imbricated (unilateral) superposition

This type is developed mainly in explosive and effusive facies formed by acidic lava, with erupted material accumulating near the crater. The explosive facies initially accumulates near the crater. As the accumulation space decreases, it migrates gradually away from the crater zone and shows a reversed imbricating superposition relationship, with the low parts superposed on high parts (Figure 4.47). The effusive facies begins by accumulating in the low parts due to magmatic melt filling. As the filling raises up, it migrates gradually closer to the crater zone, resulting in a normal imbricating superposition relationship, with the high parts superposed on low parts (Figure 4.48).

3. Parallel superposition

This relationship occurs mainly in effusive facies of neutral and basic lava and volcanic sedimentary facies. After volcanic lava surges out of the surface from the crater, it accumulates near the crater and shows a nearly bedded morphology. The lava of various eruption periods is superposed on one another vertically to form a nearly parallel superposition relationship (see Figure 4.48).

The volcanic debris of volcanic sedimentary facies from various eruption periods is transported and sorted by water currents, showing parallel and subparallel distribution characteristics in the stratigraphic succession, with a predominantly parallel superposition (Figure 4.49).

4. Isolated superposition

This type is developed mainly in subvolcanic, volcanic conduits, and extrusive facies and is characterized by unconformable contact between the lithofacies and surrounding rocks that lack associated lithofacies. In addition, isolated superposition may also occur in the limb of a volcanic edifice, where unstable slope may slide and collapse due to an excessive accumulation of volcanic deposits (Figure 4.50).

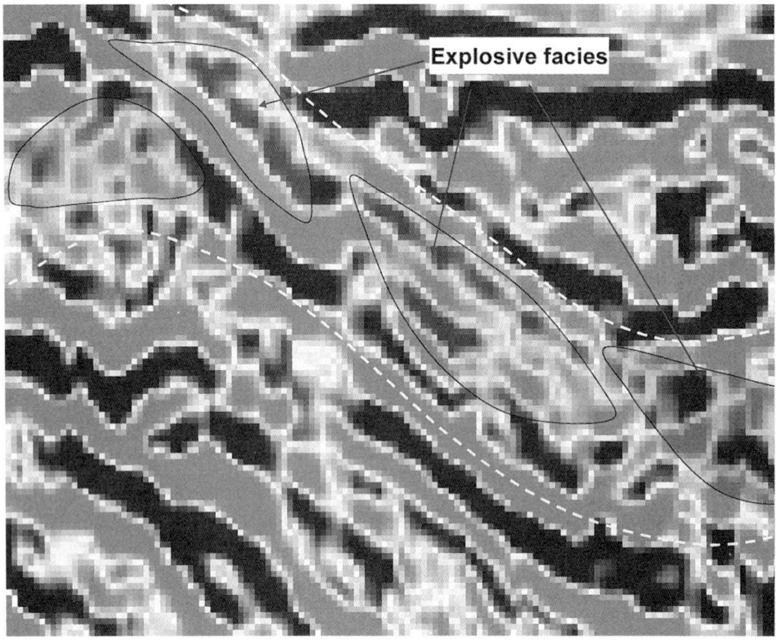

FIGURE 4.47

Imbricated lithofacies superposition pattern of explosive facies.

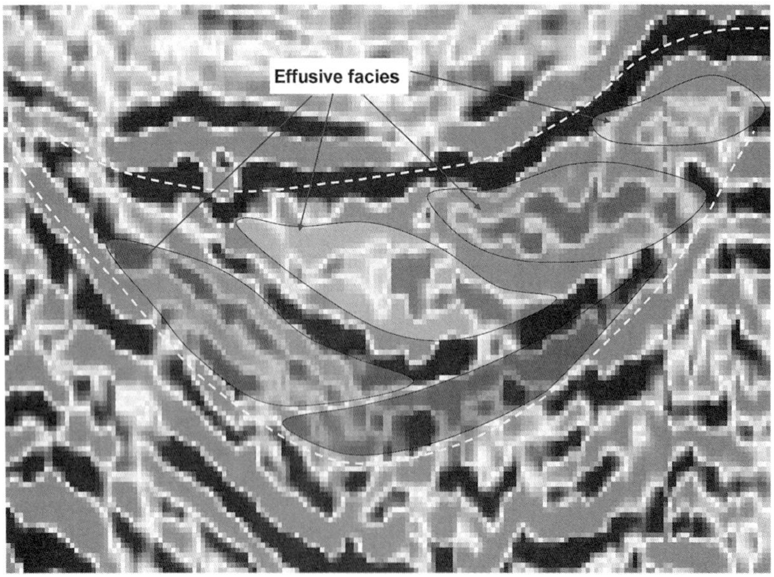

FIGURE 4.48

Imbricated lithofacies superposition pattern of effusive facies.

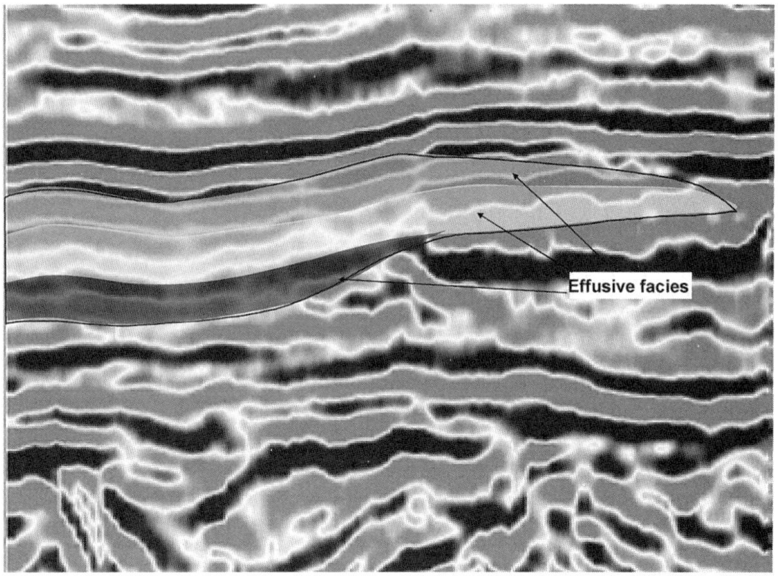

FIGURE 4.49

Parallel lithofacies superimposition.

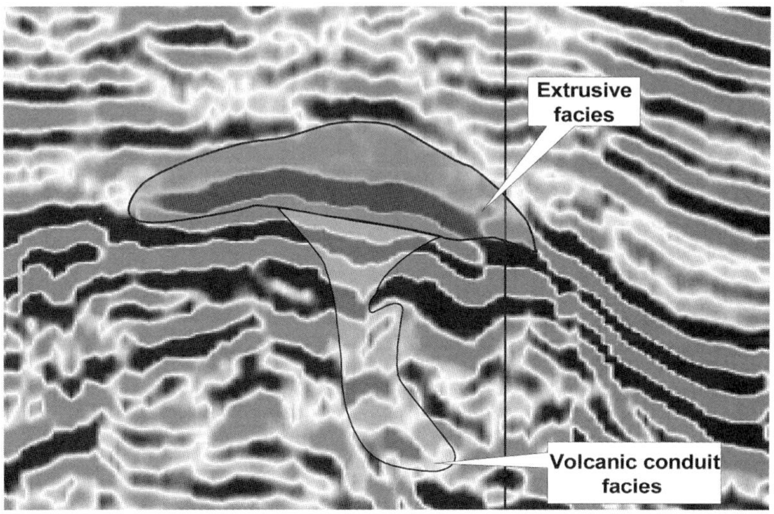

FIGURE 4.50

Isolated volcanic facies.

4.8 TECHNOLOGICAL APPLICATIONS AND EFFECTS

The volcanic facies identification and prediction technology has been applied to determine the shapes, scales, and superposition relationships of volcanic facies; to reveal the physical properties of different lithofacies; to guide reservoir prediction; and to provide facies control constraints for geological model building.

4.8.1 Revealing lithofacies distribution and predicting favorable facies

Different volcanic facies have different diagenetic mechanisms so that different pore-fracture associations are developed during their diagenetic process, ultimately leading to a different accumulation capacity for different lithofacies [22]. Accordingly, favorable petrophysical properties are found in the top and upper subfacies of the effusive facies, the splash and fallout subfacies of the explosive facies, and the outer zone and mesozone subfacies of the subvolcanic facies (Figure 4.51). With the volcanic facies distribution pattern as a guidance, the distribution characteristics of favorable lithofacies belts can be determined through lithofacies identification and prediction, which will facilitate the prediction of effective reservoirs and improve the accuracy of reservoir prediction.

In the case study on the volcanic gas reservoir in Well Block DD17, for example, the basalt, basaltic breccia lava, basaltic welded breccia, and sedimentary tuff of effusive facies are interbedded with each other, and the reservoirs are developed mainly in the top and upper subfacies of effusive facies. Because the wave impedance of the reservoir in this gas field is similar to that of sedimentary tuff, the inversion method cannot be used to identify reservoirs effectively. Planar prediction of volcanic facies revealed the distribution characteristics of favorable facies belts (effusive facies) in this well block (Figure 4.52) and provided a basis for predicting favorable reservoirs. These interpretations have since been validated by drilling results.

4.8.2 Ascertaining the spatial distribution of lithofacies and providing facies control constraints for geological model building

Volcanic lithology and lithofacies change rapidly, with correspondingly strong reservoir heterogeneity. Therefore, it is difficult to depict the complex architecture of volcanic rocks using the modeling

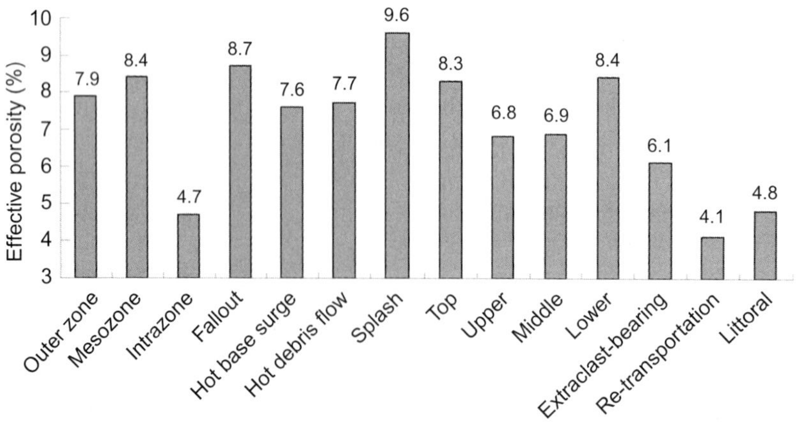

FIGURE 4.51

Porosity characteristics of different subfacies.

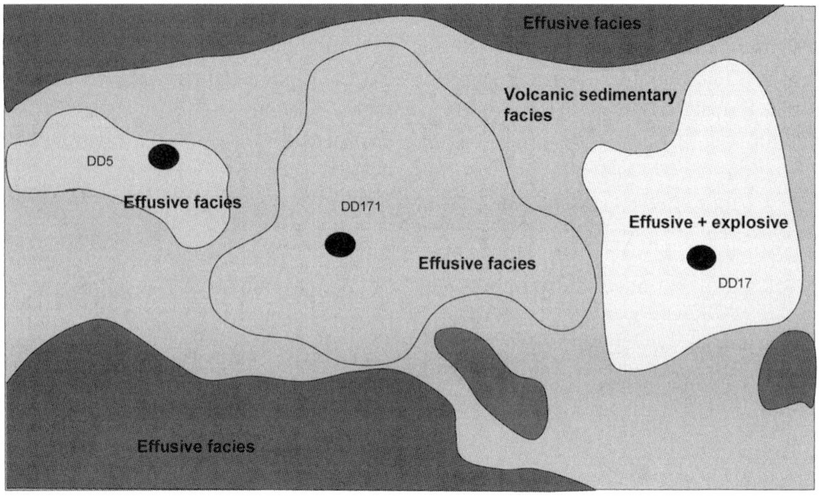

FIGURE 4.52

Lithofacies distribution in the gas field.

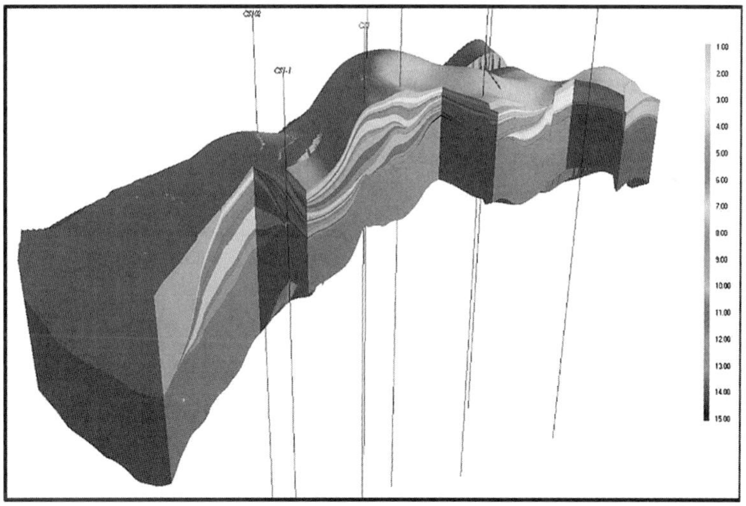

FIGURE 4.53

Framework model controlled by volcanic massifs and volcanic facies.

methods based on conventional structural constraints. In the CC1 gas field case study, 3D framework and lithofacies modeling has been achieved based on architectural analysis, lithofacies identification, and prediction of volcanic rocks, constrained by volcanic edifices, volcanic massifs, and volcanic facies (Figure 4.53). These models have helped improve the validity of the volcanic geological model and depict the architectural characteristics of volcanic rocks objectively.

References

[1] Y Lin. Study on controlling factors and distribution of volcanic gas reservoirs. Beijing: PetroChina Research Institute of Petroleum Exploration & Development; 2007.
[2] Wang P, Feng Z. Volcanic rocks in basins: lithology, lithofacies, reservoirs, gas reservoir, exploration. Beijing: the Science Press; 2008, p. 34.
[3] Wang D, Zhou X. Volcanic petrology. Beijing: Science Press; 1982, p. 10.
[4] Li S, Wang T. Volcanic rocks. Beijing: Geological Publishing House; 1981, p. 52.
[5] Kepugef DBC. Volcanic rocks and related research methods. Beijing: Geological Publishing House; 1978, p. 10–82.
[6] Li S, Wang T. Volcanic rocks. Beijing: Geologic Publishing House; 1980, p. 20–40.
[7] Fisher RV, Schmincke HC. Pyroclastic rocks. Heidelberg: Springer; 1984, p. 59–265.
[8] Tao K. Volcanic lithofacies and tectonics. Nanjing: Jiangsu Science Press; 1994, p. 12–31.
[9] Jin B, Zhang X. Study on volcanic geology in Changbai Mountain. Yanji: Northeast Korean National Press; 1994, p. 40–49.
[10] Xie J, Tao K. Volcanic geologic research in Southeast-China Continent. Beijing: Geological Publishing House; 1996, p. 40–71.
[11] Liu X, Xiang T. Energy and disasters of Cenozoic volcanic and volcanic debris deposits in China's northeast region. Changchun: Jilin University Press; 1997, p. 1–8.
[12] Liu W, Sun S, Li J. Volcanic geology and lithofacies connection characteristics of Jingangtai formation of late Jurassic at Zhaolu of Dabie Mountains. Mod Geol 1997;11(6):237–43.
[13] Wang P, Wu H, Pang Y, et al. Volcanic facies in Songliao basin: quantitative relationships amongst facies sequences, facies patterns and reservoir's physical properties. Journal of Jilin University (Geosciences Edition) 2006;36(5):805–12.
[14] Qiu J, Tao K, Zhao J, et al. Volcanic rocks. Beijing: Geological Publishing House; 1996, 10.
[15] Liu X, Xu J, Zhang X. The volcano lithology characteristics and facies pattern in north-west of Jungger basin. Xinjiang Geology 2010;28(1):73–6.
[16] Xu Z, Pang Y, Wang Y, et al. Volcanic gas reservoir development technology. Beijing: Petroleum Industry Press; 2010, p. 116.
[17] Wang P, Feng Z. Volcanic rocks in basins: lithology, lithofacies, reservoirs, gas reservoir, exploration. Beijing: Science Press; 2008, p. 47.
[18] Zhang J, Chang X. Petroleum geology. Beijing: Petroleum Industry Press; 2004, p. 113.
[19] Zhang D. In: Proceedings of the 4th national petroleum geological experimental technology and laboratory management conference. Beijing: Petroleum Industry Press; 2002. p. 246.
[20] Hou Q, Zhao Z, Wang L. Volcanic gas reservoirs. Beijing: Science Press; 2009, p. 1, 6.
[21] Qiu Y, Xue S. Oil and gas reservoir evaluation technology. Beijing: Petroleum Industry Press; 1997, p. 39.
[22] Sun Y, Song X, Ran Q, et al. Lithology and lithofacies characteristics of volcanic rocks and their control on reservoirs, Changling gas field. Petrol Explor Dev 2009;36(1):68–73.

CHAPTER

Lithological Identification and Prediction of Volcanic Rock

5

Lithology consists of characteristics such as rock color, composition, texture, and structure. For volcanic reservoirs, lithology is one of the factors that control reservoir space and other properties. Based on the schemes of rock classification and nomenclature, volcanic lithology can be identified according to logging identification models for composition, texture, and the structure of various types of volcanic rocks, and by integrating elemental capture spectroscopy (ECS) element logging, acoustic, and electric image logging and conventional logging. Through calibration of seismic data based on well logs and analysis of seismic responses of different lithologies, volcanic lithology can also be predicted based on seismic reflector profile analysis, waveform classification, frequency-divided inversion, and other techniques. The lithological distribution patterns thus revealed can provide the basis for volcanic lithofacies division, reservoir parameter calculation, gas layer and aquifer identification, and effective reservoir prediction.

5.1 CHALLENGES AND SOLUTIONS

5.1.1 Challenges in lithology identification and prediction

Compared with sedimentary rocks, volcanic rocks are more varied, with more complex compositions, textures, and structures, resulting in various challenges in their lithological identification and prediction.

1. Difficulties in accurate identification of rock composition based on log data

Volcanic rocks contain almost all existing elements and cationic oxides in the earth's crust [1–3], characterized by complex chemical compositions (e.g., SiO_2 ranging from 34% to 77%, MgO 0.06% to 46%) and a great variety of rock-forming minerals rock types (> 1000 varieties) [4–6].

Conventional logs do not measure lithological compositions directly but merely provide volcanic lithological information indirectly by measuring parameters such as rock radioactivity, bulk density, acoustic wave propagation characteristics, and conductivity. ECS logs can measure the elemental contents of formations through neutron-capture gamma ray spectroscopy, but the measurements are error prone because it is easily affected by borehole environment and ionic composition and concentration in formation water, thus posing challenges in accurate identification of rock compositions using log data.

2. Difficulties in accurate identification of rock texture and structure using well log data

The texture and structure of volcanic rocks are highly variable, with complex origins. Acoustic and electric imaging logs can reflect the textural and structural features of volcanic rocks indirectly through borehole imaging that reflects the acoustic property and conductivity, but the log interpretation is often ambiguous. Furthermore, the resolution of imaging log is inversed related to the probing depth. When the resolution of full-bore micro scanner imager (FMI) logging is high, the probing radius becomes smaller, affected by borehole conditions, formation fluids, and other factors. In comparison, the probing radius of array resistivity imager (ARI) logging is large, but the resolution is low. Therefore, it is difficult to accurately identify the texture and structure of volcanic rocks using imaging logs.

3. Difficulties in predicting lithological distribution

Volcanic rocks are characterized by a rapid and prominent change in their thickness, from less than 1 m to several thousand meters. Because of the low resolution of seismic data, the lithological prediction of thin layers tends to carry large errors. In addition, various volcanic lithologies may be represented by differences in either rock composition or petrophysical properties. Some volcanic rocks (such as volcanic breccia and rhyolite) with the same composition and small differences in physical property will also have minor differences in petrophysical property. Thus, it is difficult to predict the lithological distribution because of the small differences in seismic response to different lithologies and in the seismic reflector of lithological interfaces in syngenetic volcanic rocks.

5.1.2 Technical solutions

Through integration of geological, logging, and seismic data, technical approaches can be established by incorporating technical solutions of hierarchical identification and multiple prediction methods (Figure 5.1). Following this process, the lithological identification and prediction techniques for volcanic reservoirs are developed, which provide a basis for lithofacies identification, reservoir parameter interpretation, and gas layer and aquifer identification.

1. Classification and nomenclature of volcanic rocks

Petrographic analysis is carried out first based on core data, geochemical analysis, and thin section information. This is followed by the identification and classification of volcanic rocks using the "component + texture + genesis" scheme.

2. Lithological identification of volcanic rocks based on well logs

A volcanic rock identification scheme can be established by using core data to calibrate well log data. The procedure is as follows: (1) building an identification scheme for rock composition using ESC and conventional log curves (GR, DEN, compensated neutron log (CNL)); (2) building an identification scheme for rock texture and structure with acoustic and electric imaging logs; and (3) building a well logging response model for volcanic lithologies according to the scale, shape, and smoothness of conventional log curves. Afterward, integrated lithological identification can be conducted in noncored intervals with the above identification schemes and models.

3. Distribution prediction of volcanic lithology

Seismic data can be calibrated by lithological information from single wells; the response characteristics of lithology with near-wellbore seismic traces can be analyzed, and seismic response models of different volcanic rock can be established. After these steps, the lithological distribution pattern can be predicted by seismic attribute analysis, waveform classification, and seismic inversion.

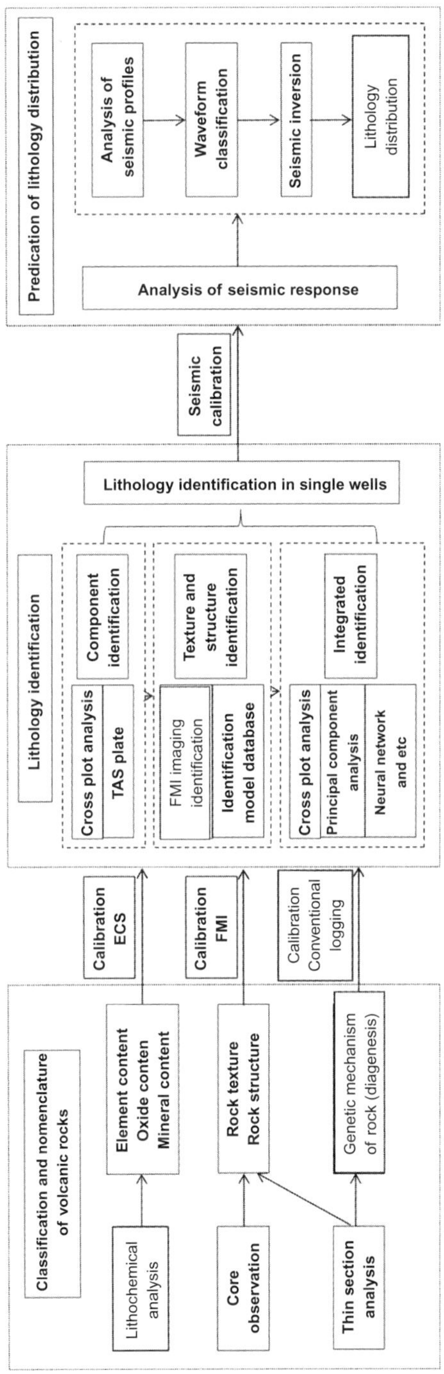

FIGURE 5.1

Flowchart of volcanic lithology identification and prediction techniques.

5.2 IDENTIFICATION OF VOLCANIC LITHOLOGY

With reference to existing research results [7–10], volcanic lithology can be identified by integrating ECS element logs, FMI imaging logs, and conventional logs, following the conventional scheme of rock classification and nomenclature (i.e., the "composition → texture → rock type" scheme) and using methods such as graphic analysis, crossplot of curves, superposition of curves, and multivariate clustering.

5.2.1 Identification of volcanic rock composition through ECS logging

ECS logging can record inelastic scatter gamma spectra and capture gamma spectra through fast neutron-induced inelastic scattering capture nuclear reaction in the formation. Based on spectral unscrambling analysis, the percentage of rock-forming elements in the formation can be determined, which provides a basis for identifying rock composition [11].

1. Identification principle

ECS logs measure the relative number (or relative yield) of the electrical pulses that were caused by the gamma ray. According to the processing of the oxide-closure model and geological calibration, the relative number (or relative yield) can be transferred to the weight percentage of the elements and oxides in rocks.

1) Determining the weight percentage of elements using the oxide-closure model

In the oxide-closure model, the total weight percentage of all elements in the minerals is 100% and the total weight percentage of oxides and carbonates is also 100%. Then the formula for a given depth is expressed by [12]

$$F\left(\sum X_i \frac{Y_i}{S_i}\right) = 1 \tag{5.2.1}$$

Where F— the normalizing factor at each depth;

X_i—mass ratio of oxides or carbonates to corresponding element i;

Y_i—relative yield of element i;

S_i—the detector's sensitivity factor to element i, corresponding to the concentration of the element and Y_i.

The key of this model lies in the determination of the elements present in the matrix but not in the fluid. The calculated results are the weight percentage of the elements.

2) Calculate the oxide contents by establishing the relationship between element contents and mineral contents

Statistical methods can be applied to analysis of the core data from various sources to summarize the relationship between the element abundance and oxide content on the premise that all the mineral elements are stable:

$$E = CM \tag{5.2.2}$$

Then,

$$M = C^{-1}E \tag{5.2.3}$$

Where M—column matrix of the weight percentage of oxides;
E—column matrix of element abundance;
C—coefficient matrix.

The percentage of various oxides can be calculated by this matrix equation. However, two aspects should be taken into consideration [11,12]: (1) the measured elements should be those that can represent the characteristics of this mineral (e.g., Si can be an indicator for quartz), and (2) the number of oxides to be determined must be less than the number of measured elements.

2. Logging response characteristics of volcanic rocks with different compositions

Based on the composition classification of volcanic rocks, ECS log data can be calibrated to analyze the responses of volcanic rocks with different compositions, so as to establish an ECS identification model for rock composition.

From ECS log curves (Figure 5.2), it can be found that Si, Al, Fe, Ti, K, and Na are the typical elements for determining the compositional variation in volcanic rocks, and Ca, S, and Gd are relatively insensitive elements. Basic volcanic rocks are characterized by "low Si, low K, low Na, high Al, high Fe, and high Ti." In contrast, acidic volcanic rocks have "high Si, high K, high Na, low Al, low Fe, and low Ti," and intermediate volcanic rocks fall between these two rock types (Table 5.1). Accordingly, oxides in basic volcanic rock are represented by "low SiO_2, low K_2O, low Na_2O, high Al_2O_3, high ($FeO + Fe_2O_3$), and high TiO_2," whereas acidic volcanic rocks are the opposite, and intermediate

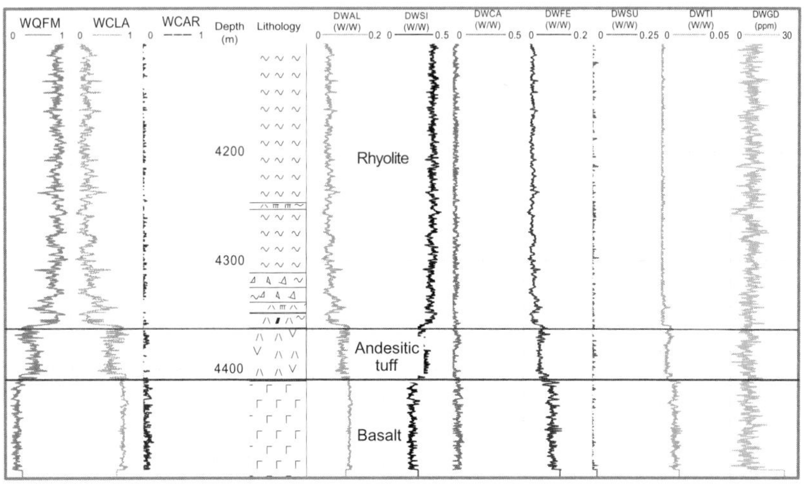

FIGURE 5.2

ECS log response characteristics of different lithologies (Well XX7).

Table 5.1 Comparison Table of ECS Element Content and Mineral Contents of Volcanic Rocks with Different Composition

Rock Type	Si/ SiO_2	Al/ Al_2O_3	Fe/ ($FeO+Fe_2O_3$)	Ti/ TiO_2	(Na+K)/ (Na_2O+K_2O)
Basic volcanic rock	Low	High	High	High	Low
Intermediate volcanic rock	Medium				
Acidic volcanic rock	High	Low	Low	Low	High

volcanic rocks fall in between. From basic to acidic volcanic rocks, therefore, there is an increase in Si content, a decrease in Al, Fe, and Ti, an increase in SiO_2 and Na_2O+K_2O, and a decrease in Al_2O_3, ($FeO+Fe_2O_3$) and TiO_2 among the major oxides (see Table 5.1).

3. Identification method

Building on the logging response analysis discussed earlier, volcanic rock compositions are determined using a crossplot analysis of element contents and total alkali versus silica (TAS) plot analysis.

1) Crossplot analysis of element contents

Basic, intermediate, and acidic volcanic rocks can be differentiated quantitatively through crossplot of element contents in volcanic rocks with different composition, as reflected by logging responses [13].

As an example from the case study of XX gas field, Figure 5.3 is a crossplot of elemental contents in volcanic rocks with different compositions in the ECS log, with Table 5.2 providing the ranges of elemental contents in volcanic rocks with different compositions in the ECS log. These data demonstrate that, on a scale of 0 to 1, the basic volcanic rocks have Si content values less than 0.24, Al more than 0.1, Fe more than 0.067, and Ti more than 0.08; acidic volcanic rocks have Si values greater than 0.3, Al less

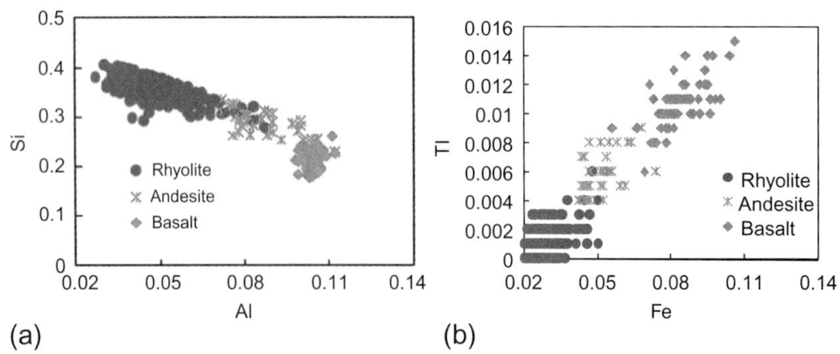

FIGURE 5.3

Crossplot of elemental contents in volcanic rocks of different compositions in the ECS log (XX gas field). (a) Crossplot of Si and Al contents. (b) Crossplot of Ti and Fe contents.

Table 5.2 The Range of ECS Elemental Contents in volcanic rock with Different Composition (XX Gas Field)

Rock Type	Si	Al	Fe	Ti
Basic volcanic rock	< 0.24	> 0.1	> 0.067	> 0.08
Intermediate volcanic rock	0.23~0.32	0.07~0.11	0.04~0.068	0.04~0.09
Acidic volcanic rock	> 0.3	< 0.07	< 0.041	< 0.042

than 0.07, Fe less than 0.041, and Ti less than 0.42; the values of elemental contents for intermediate volcanic rocks fall between those for the basic and the acidic volcanic rocks.

2) TAS plate analysis of oxides contents

TAS plate was recommended to be a part of the volcanic rock classification scheme at the 28th International Geological Congress of the International Union of Geological Science (IUGS) in 1989 [1]. In this analysis, SiO_2 content is plotted on the x-axis, representing the acidity of a volcanic rock, and $Na_2O + K_2O$ contents are plotted on the y-axis, representing the alkalinity (Figure 5.4a). On a TAS plate, 15 fundamental types of volcanic rock can be divided on the basis of geochemical analysis.

Figure 5.4 and Table 5.3 are the TAS plates and classification schemes for the volcanic rocks in XX gas field. For the basic rocks, the SiO_2 contents range from 45% to 52%, and $Na_2O + K_2O$ less than 7.2%; for the intermediate rocks, the contents of SiO_2 have a range of 52% to 63%, and $Na_2O + K_2O$ less than 11.2%; for acidic rocks, SiO_2 contents are more than 63%, and $Na_2O + K_2O$ less than 16%. In terms of the SiO_2 and $Na_2O + K_2O$ values, the volcanic rocks can be subdivided into basalt, trachybasalt, basaltic andesite, andesite, trachyandesite, dacite, trachydacite, and rhyolite. The contents of oxides can be determined using ECS log data, and the volcanic rocks can be classified on the basis of rock composition in the TAS plates.

5.2.2 Identification of volcanic rock texture and structure by imaging log

Acoustic imaging logging can scan the borehole wall and record the echo waves using a rotational ultrasonic transducer. A microresistivity scanning image log can cover the borehole wall and measure several microresistivity curves with a microelectrode array. Thus, borehole wall images can be obtained by interpreting the propagation characteristic of acoustic waves and conductivity in the formation [14,15]. Based on these borehole wall images, the texture and structure of volcanic rocks can be identified effectively.

1. Identification principle

By measuring the amplitude and travel time of the reflected waves from a borehole wall, the acoustic image logs can display the images in 360 degrees around the borehole to form high-resolution images of the entire borehole wall. In general, longer acoustic travel time and lower acoustic energy of the reflected waves are correlated with darker colors. Microresistivity scanning image logs transfer each

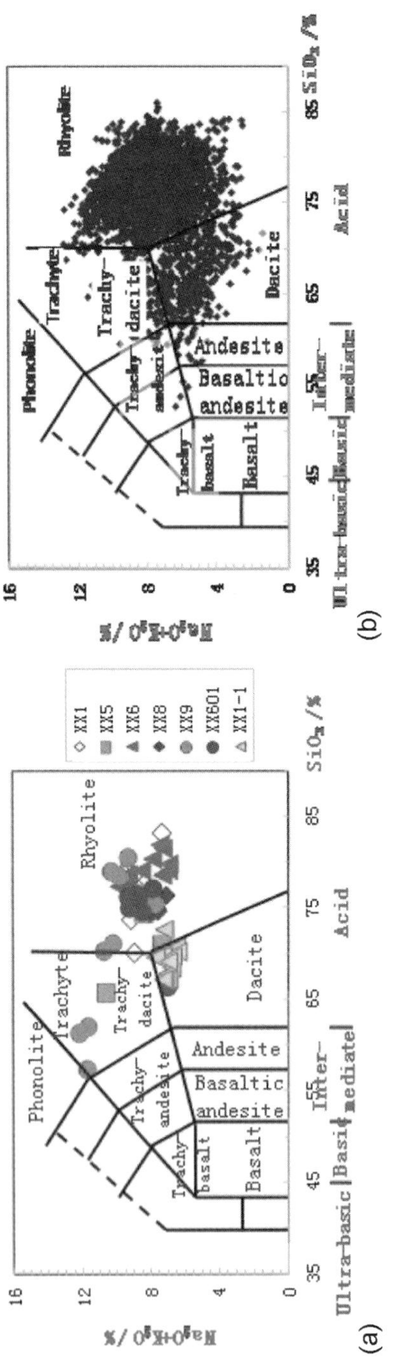

FIGURE 5.4

TAS plate of volcanic rocks with different compositions (XX gas field). (a) Core analysis data in cored wells. (b) Mineral contents in ECS log (Well XX7).

Table 5.3 TAS Plate for Classification of Volcanic Rocks with Different Compositions

Classification Basis	Ultra-Basic	Basic	Intermediate	Acidic
SiO_2 /%	< 45	45~52	52~63	> 63
Na_2O+K_2O/%	< 5	< 7.2	< 11.2	< 16

microresistivity curve into colors to form an electric image reflecting the formation conductivity. Greater conductivity and lower resistivity of a formation are rendered into a darker color.

Acoustic and electric image logs can clearly reflect various geological features of subsurface rock formations. However, the images are false-color images of the borehole, and it is necessary to first establish an interpretation model of acoustic characteristics and conductivity on the basis of core calibration before the texture and structure of volcanic rock can be identified effectively.

Among resistivity imaging logs, FMI imaging logs have a relatively high resolution, as is reflected by the wide use of FMI today for volcanic rock imaging. Therefore, this book focuses on the technique of texture and structure identification for volcanic rocks using FMI logging.

2. *Identification model of volcanic rock texture and structure*

Based on the determination of texture and structure by core data, imaging log data can be calibrated to analyze the FMI logging response of different rock textures and structures and to further establish an image identification model for rock textures and structures. The maximum resolution of imaging log is 0.2 inch (about 5 mm), implying that only relatively large-scale rock textures and structures can be identified, such as rhyolitic structure, agglomeratic texture, brecciated texture, and welded texture (Figure 5.5, Table 5.4).

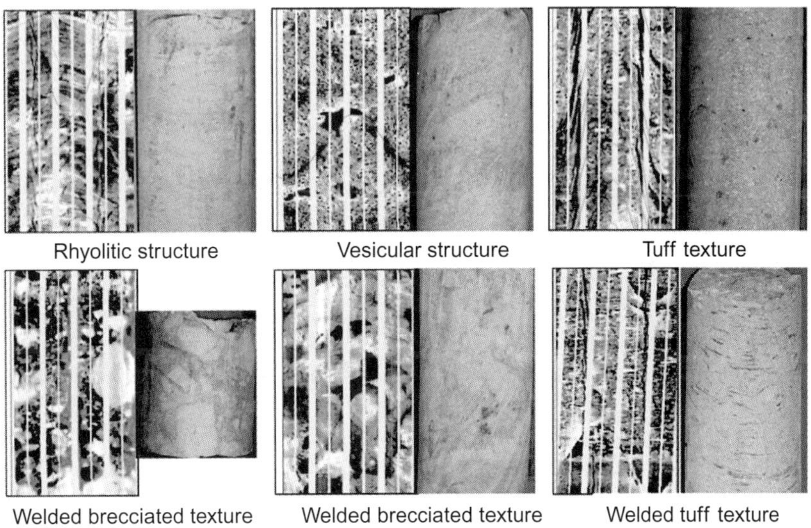

FIGURE 5.5

Identification model for volcanic rock textures and structures.

Table 5.4 FMI Imaging Log Response Characteristics of Typical Textures and Structures in Volcanic Rocks

Type of Texture and Structure		Geological Origin	Response Characteristics in FMI Imaging Log		
			Color	*Shape*	*Image Texture*
Volcanic lava	Vesicular structure	Volatile constituents escape during volcanic eruption	Black	Spot	Uniform clumpy
	Rhyolitic structure	Flow of magmatic melt causing oriented arrangement of vesicules	Dark or light color	Oval spot	Laminar
Volcaniclastic lava or welded volcaniclastic rock	Welded tuff texture	Magmatic debris being squashed and elongated, arranged in nearly identical direction	Yellow-white	Deformed band	Laminar -like
	Welded brecciated texture	Breccia is composed of rhyolite, filled with volcanic ash or volcanic breccia welded by melt	Yellow-white	Mottling	Irregular clumpy
Volcaniclastic rock	Agglomeratic texture	Volcanic debris sizes more than 64 mm	Yellow-white	Angular mottling	Irregular clumpy
	Brecciated texture	Volcanic debris sizes 2-64 mm	Light yellow	Angular grain	Irregular clumpy
	Tuff texture	Volcanic debris sizes less than 2 mm	Red	Spot	Clumpy

1) Structures in volcanic lava

Typical structures in volcanic lava are vesicular and rhyolitic.

1. *Vesicular structure*. This is formed by the escaping volatile constituents during volcanic eruption, manifesting as relatively regular black spots (with an overall clumpy appearance) in FMI images (Figure 5.6, Table 5.4).
2. *Rhyolitic structure*. This is formed by the oriented vesicular and almond-shaped structure caused by magmatic flow. Usually some degree of deformation will occur when the magma bypasses quartz and orthoclase phenocrysts. Individual vesicles or almond structures appear as oval spots in FMI images, with an overall alternation of dark and light bands, resembling laminar stratification (see Figure 5.6, Table 5.4).

2) Texture of volcaniclastic lava or welded volcaniclastic rock

Welded Tuff Texture. This is the typical texture of tuff lava and welded tuff, formed by squashing and elongation of vitric fragments and magmatic fragments with plastic deformation along the flowing direction of magma, appearing as tiny spots with deformed bands in a light color (yellow-white) and an overall laminar appearance in FMI images (see Figure 5.5, Table 5.4).

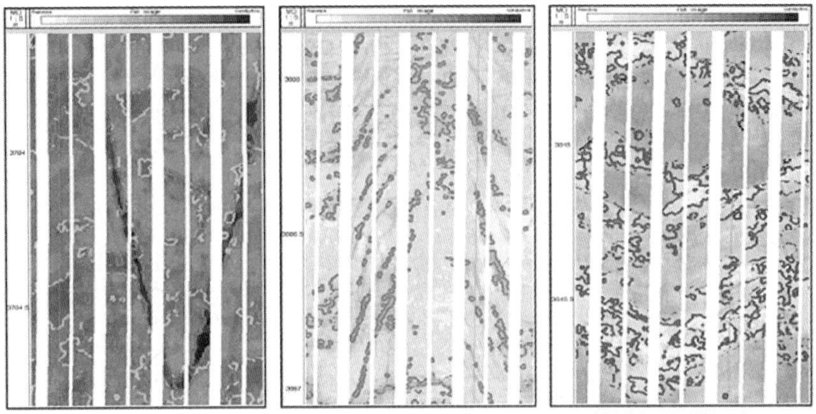

FIGURE 5.6

Identification model for volcanic rock textures and structures.

Welded brecciated texture. This is typical of breccia lava and welded breccia of two petrogeneses: (1) breccia that consists of rhyolite and is infilled with volcanic ash; and (2) volcanic breccia welded by plastic magma with blurred edges. In FMI images, the volcanic breccia appears as light-colored (yellow-white) spots with soft edges, the infilled volcanic ash as minute spots (dark colors alternating with light colors), and the plastic magma as irregular bands (which were elongated and deformed along the flowing direction). In general, volcanic rocks with welded brecciated texture have an irregular clumpy appearance in FMI images (see Figure 5.5, Table 5.4).

3) Texture of volcaniclastic rocks

Characteristic volcaniclastic rock textures are agglomeratic, brecciated, or tuffaceous (see Figure 5.5, Table 5.4).

1. *Agglomeratic texture*. Volcanic clasts are more than 64 mm in size; individual agglomerates appear as yellow-white angular mottling in FMI images, with an overall irregular clumpy look.
2. *Brecciated texture*. Volcanic clasts are 2 to 64 mm in size; individual clasts appear as yellow-white angular grains in FMI images, also with an overall irregular clumpy appearance.
3. *Tuffaceous texture*. Volcanic grain size is less than 2 mm; this type of texture is represented by minutes, black and white spots in FMI images, but in general it is only a relatively coarse grained (e.g., crystalline-vitric) tuff and plastic deformation characteristics can be recognized.

3. Identification method

Based on the preceding identification model, volcanic rock texture and structure can be identified in the following steps:

1. Analyze the response characteristics of the imaging log, and divide the interpreted horizons according to the similarity principle.

2. Calculate the indicative curve of resistivity changes (represented by the changes in corresponding colors), and determine a cutoff threshold to depict the abnormal area in the image (see Figure 5.6).
3. Analyze the petrogenesis of abnormal areas, and match the identification model to confirm the texture and structure (see Figure 5.6).

5.2.3 Identification of volcanic rock types by conventional logging

Conventional logs include caliper, spontaneous potential (SP), gamma ray, three porosity logs (density, neutron, and acoustic travel time logging), and three resistivity logs (deep lateral, shallow lateral, and microspherical focused logging, or deep induction, medium induction, and laterolog 8 logging), with the spectral gamma log and dipmeter log as the supplements for complex strata [15,16]. Based on the principle of logging identification and petrogenesis, the lithological logging response model can be established by analyzing the scale, shape, and assemblage relationships of conventional log curves for different lithologies to identify volcanic rock types.

1. Identification principle

1) Gamma ray and spectral gamma logs

Gamma ray logs can measure the total natural radioactivity of the formation. In general, radioactivity increases from basic to acidic rocks; and the radioactivity of subaqueous eruptive volcanic rock is stronger than that of other volcanic rocks with the same composition.

Spectral gamma logs can be used further to determine the relative contents of U, Th, and K. The K content is controlled mainly by the potassium feldspar content, and hence the rock composition; the U content is related to the organic absorbency of the rocks; and the Th content is mainly absorbed and retained in clay minerals, which can represent the contents of clay minerals. In general, U and Th contents in volcanic reservoirs decrease from basic rocks to acidic rocks, which are poor in the early stage and relatively abundant in the later stage of magmatic evolution.

2) Three porosity logs

Based on Compton scatter principle between gamma ray and the formation, density logging can measure the bulk density of the formation by measuring the electron density index; neutron logging can survey the hydrogen contents of the formation based on the thermal-neutron capture principle [17]; acoustic travel time logging can measure the propagation velocity of longitudinal waves in the formation based on the principles of slide wave propagation along borehole wall [18]. Porosity logging curves are related to changes in lithology, physical property, fractures, and fluid properties. In general, basic volcanic rocks have higher density and neutron responses than acidic volcanic rocks, with intermediate volcanic rocks in between.

3) Resistivity log

Rock conductivity is related to many factors, such as conductive mineral content in rock matrix, clay types and contents, development of pores and fractures, formation fluid types, and so on. Because volcanic rocks have a low clay mineral content and relatively low porosity, the values and curve shapes of the resistivity log can reflect the rock types relatively well.

2. *Well logging response characteristics of volcanic rocks with different lithologies*

Volcanic rocks with different compositions have different response characteristics in conventional logs

1) Response characteristics of volcanic rocks with different compositions in conventional logs

Basic volcanic rocks are characterized by "low radioactivity, low Th content, low K content, high density, and high neutron porosity" (Figure 5.7, Table 5.5), and acidic volcanic rocks by "high radioactivity, high Th content, high K content, low density, and low neutron porosity," whereas intermediate volcanic rocks fall in between. From basic to acidic rocks, there are general tendencies of "radioactivity increase, Th and K content increase, and density and neutron porosity decrease."

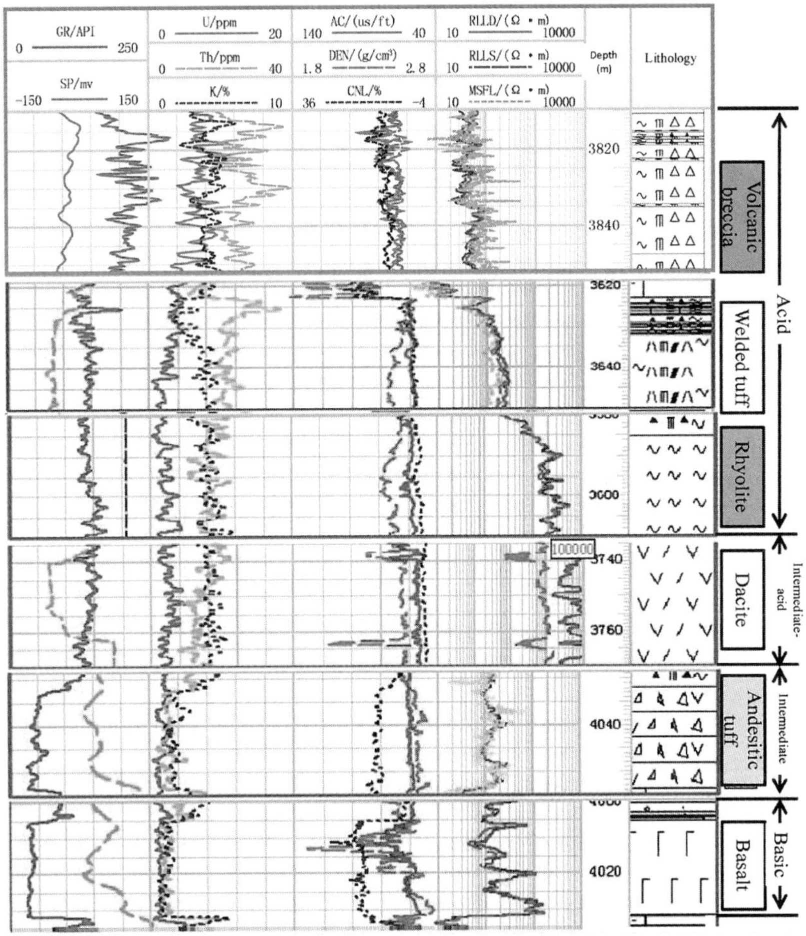

FIGURE 5.7

Well logging characteristics of different volcanic rocks.

Table 5.5 Conventional log Response Characteristics of Volcanic Rocks with Different Compositions

Rock Type	GR	Th Content	K Content	Rock Density	Neutron Porosity
Basic volcanic rock	Low	Low	Low	High	High
Intermediate volcanic rock	Between basic and acid				
Acidic volcanic rock	High	High	High	Low	Low

2) Conventional log response characteristics of volcanic rock of different petrogenses

In terms of their petrogenesis, volcanic rocks with the same composition can be divided into volcanic lava, volcaniclastic rock, and volcaniclastic lava. Volcaniclastic rocks can be subdivided into volcanic agglomerate, volcanic breccia, and volcanic tuff by grain size. Taking volcanic gas reservoir in the CC gas field as an example, conventional log response characteristics of volcanic rocks of different petrogeneses have been analyzed from three aspects (amplitude, shape, and smoothness of the curves; Table 5.6).

Value range. Rock types can be distinguished by curves of gamma ray, density, and resistivity logs, as well as the combined TH/K curve of (see Table 5.6). Rhyolite has medium-high GR, DEN, and resistivity, low TH/K; in contrast, breccia lava has medium-low GR, DEN and resistivity, medium TH/K; welded breccia is characterized by "high GR, high DEN, medium resistivity and high TH/K"; welded tuff by medium GR, DEN, resistivity, and TH/K; crystal tuff by "low GR, medium-low DEN, low resistivity, and low TH/K," and volcanic breccia by "high GR, medium DEN, low resistivity and high TH/K."

Curve shapes. The three major curve shapes are cylinder, funnel, and bell. According to the statistical results in the CC gas reservoir, the cylinder shape is most typical of rhyolite, crystalline tuff, and volcanic breccia, which accounts for 55.6%, 66.7%, and 66.7%, respectively, and the funnel shape ranks second for these lithologies, accounting for about 33.3%. Brecciated lava is dominated by the funnel shape (accounting for about 50%), and the bell shape is second (about 20%). Welded breccia is

Table 5.6 Conventional Log Response Characteristics of Volcanic Rock (CC Gas Field)

Rock Type	Amplitude				Shape			Smoothness		
	GR	*DEN*	*Resistivity*	*TH/K*	*Box*	*Funnel*	*Bell*	*Smooth*	*Microserrated*	*Serrated*
Rhyolite	Medium-high	Medium-high	High	Low	55.6	33.3	11.1	22.2	33.3	44.4
Brecciated lava	Medium-low	Low	Low	Medium	20	50	30	50	40	10
Welded breccia	High	High	Medium	High	50	0	50	0	0	100
Welded tuff	Medium	Medium	Medium	High	20	20	60	40	60	0
Crystalline tuff	Low	Medium-low	Low	Low	66.7	33.3	0	33.3	66.7	0
Volcanic breccia	High	Medium	Low	High	66.7	33.3	0	0	33.3	66.7

dominated by cylinder and bell shapes, accounting for about 50% each. Welded tuff is dominated by the bell shape, accounting for about 60%, with about 20% consisting of cylinder and funnel shapes.

Curve smoothness. Volcanic reservoirs have strong heterogeneity, and the corresponding logging curves are dominated by microserrated shapes (about 44%) and serrated shapes (about 32%). Rhyolite is dominated by serrated shapes (about 44.4%) and microserrated shapes (about 33.3%); brecciated lava is dominated by smooth (about 50%) and microserrated (about 44%) shapes; welded breccia is dominated by serrated shapes; welded tuff and crystalline tuff are dominated by microserrated shapes (about 60% and 66.7%, respectively) and smooth shapes (about 40% and 33.3%, respectively); and volcanic breccia is dominated by serrated shapes (about 66.7%) and microserrated shapes (about 33.3%).

3. Identification method

The identification method for volcanic rock types using conventional logs includes the bivariate crossplot, principal component analysis, neural network, fuzzy clustering, and the like.

1) Bivariate crossplot analysis

Based on the crossplot of two variables, the differences in logging response of different rock types can be determined to establish the identification criteria. Based on conventional log data, volcanic rock types can be identified in two steps following the technical approach of "compositions first and structure-petrogenesis second."

Type identification based on differences in rock composition. The variety of volcanic rock composition is distinguished mainly by differences in radioactivity, density, and hydrogen contents of the rocks [19]. As is shown in Figure 5.8 and Table 5.7, the GR values of basic volcanic rocks are usually less than 65 American Petroleum Institute unit (API), Th content is less than 5.2 ppm, K less than 2.3%, rock density more than 2.6 g/cm^3, and neutron porosity more than 6%; the GR values of intermediate volcanic rocks are 60 to 120 API, Th content is 5 to 12 ppm, K 1.1% to 5.5%, rock density 2.55 to 2.7 g/cm^3, and neutron porosity 3% to 12%; and the GR values of acidic volcanic rocks are more than 110 API, Th content is more then 10 ppm, K more than 3.7%, rock density less than 2.66 g/cm^3, and neutron porosity less than 10%.

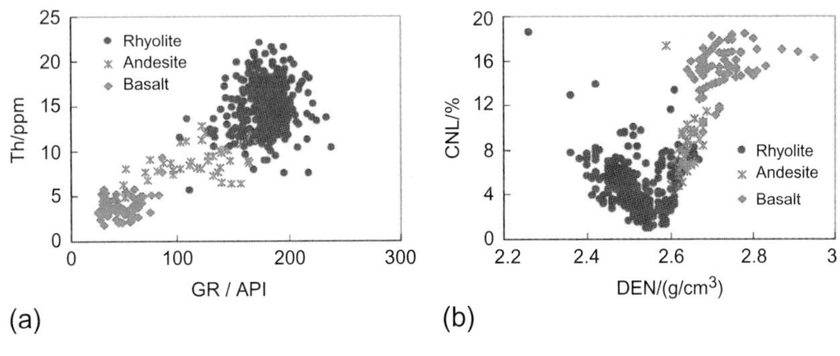

FIGURE 5.8

Bivariate crossplot of volcanic rocks with different compositions (XX gas field). (a) Crossplot of GR-Th contents. (b) Crossplot of neutron-density.

Table 5.7 Value Range of Conventional Log Curves for Volcanic Rocks with Different Compositions (XX Gas Field)

Response Characteristics of Conventional Log		Basic Volcanic Rock	Intermediate Volcanic Rock	Acid Volcanic Rock
Radioactivity	GP/API	< 65	60-120	> 110
	Th/ppm	< 5.2	5-12	> 10
	K/%	< 2.3	1.1-5.5	> 3.7
Density/(g/cm3)		> 2.6	2.55-2.7	< 2.66
Neutron porosity/%		> 6	3-12	< 10

Lithological identification based on differences in rock texture and petrogenesis. This process is illustrated by the case study of the acidic volcanic rocks in Well Block XX1 (Figure 5.9, Table 5.8).

Acidic lava is characterized by high GR, low U, low Th, high K, low neutron, high density, and high resistivity, as are indicated by GR values of 131 to 170 API, photoelectric absorption cross-section index (PE) 2.9 to 3.5, U content 0.9 to 3.9 ppm, Th content 11.8 to 21.2 ppm, K content 3.7% to 5.1%, CNL 0.07% to 4.6%, DEN 2.42 to 2.65 g/cm^3, and recorded value of laterolog deep (RLLD) 800 to 4600 Ω m.

Welded breccia has the characteristics of high GR, high U, high Th, low K, low neutron, medium density, and medium resistivity, as are shown by GR values of 175 to 212 API, PE 3.9 to 4.9, U content 7 to 13.3 ppm, Th 20 to 28 ppm, K 2.7% to 5.2%, CNL 2.56% to 4%, DEN 2.49 to 2.6 g/cm^3, and RLLD 200 to 650 Ω m.

Among the volcaniclastic rocks, tuff typically has low GR, medium U, high Th, high K, high neutron, low density, and low resistivity, as evidenced by GR values of 100 to 140 API, PE 2.9 to 3.1, U content 1.7 to 5 ppm, Th 13.6 to 16.5 ppm, K 3.6% to 4.7%, CNL 8% to 10.8%, DEN 2.38 to 2.6 g/cm^3, and RLLD 20 to 24 Ω m.

In general, volcanic lava can be diagnosed by high radioactivity, high K content, high density, low neutron, and high resistivity. In contrast, volcaniclastic rocks have high U, high Th, high neutron, low

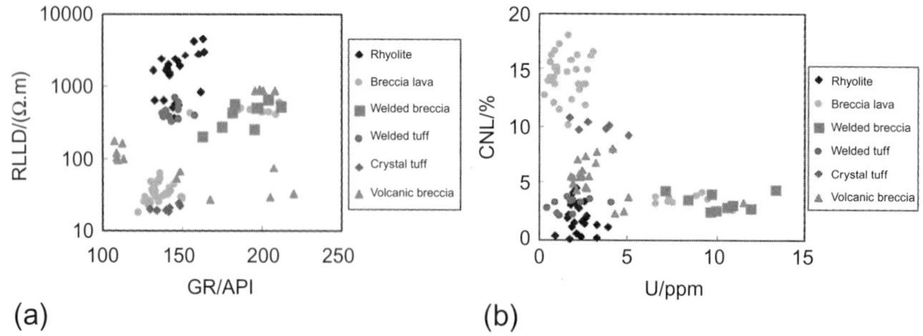

FIGURE 5.9

Bivariate crossplot of volcanic rocks of different petrogeneses (Well Block XX1). (a) Crossplot of GR-LLD. (b) Crossplot of U content-CNL.

Table 5.8 Value Range of Log Curves for Volcanic Rocks of Different Petrogenesis (Well Block XX1)

Lithology	GR/API	PE/(b/e)	U/ppm	TH/ppm	K/%	CNL/%	DEN/ (g/cm3)	RLLD/ (Ω.m)
Rhyolite	131-170	2.9-3.5	0.9-3.9	11.8-21.2	3.7-5.1	0.07-4.6	2.42-2.65	800-4600
Brecciated lava	122-150	3.3-4.3	0.3-3.6	16.5-29	2.9-5.6	3.3-18	2.3-2.5	25-210
Welded breccia	175-212	3.9-4.9	7-13.3	20-28	2.7-5.2	2.56-4	2.49-2.6	200-650
Welded tuff	138-158	3.1-3.6	0.4-2.9	13-20	3-4.8	2.2-3.7	2.4-2.52	320-700
Crystalline tuff	100-140	2.9-3.1	1.7-5	13.6-16.5	3.6-4.7	8-10.8	2.38-2.44	20-24
Volcanic breccia	150-220	3.1-4.1	2-11.4	14-20	2-4.8	4-8	2.42-2.59	26-176

density, and low resistivity, although they also have high radioactivity. These values for welded volcaniclastic rocks fall between the two rock types discussed earlier. Among volcaniclastic rocks, volcanic breccia has the characteristics of high radioactivity, low K content, low neutron, and low density, in comparison to volcanic tuff.

2) Principal component analysis

Principal component analysis is also called R-factor analysis, which can generalize and classify log data in terms of petrogenesis by using a relatively few variables with the least information loss. Principal component analysis can transfer a set of relevant initial variables into a set of mutually orthogonal variables by orthogonal transformation:

$$Y_i = \omega_i^{\mathrm{T}} X_k \tag{5.2.4}$$

Where

$$X_k = (x_1, x_2, \ldots, x_m)^{\mathrm{T}} \quad (k = 1,2,\ldots,n) \tag{5.2.5}$$

$$Y_i = (y_1, y_2, \ldots, y_m)^{\mathrm{T}} \quad (i = 1,2,\ldots,n) \tag{5.2.6}$$

Where X_k—initial variable;

Y_i—principal component variables;

ω_i—unit orthogonal eigenvector, of which the eigenvalue represents variance, usually ranked in the order of size;

m—number of initial variable, represents the number of well log curves that were adopted for lithology identification;

n—Number of samples for model building.

The first principal component among these new m variables constitutes the largest variance and maximum interpreted information; the second principal component has the second largest variance, and so forth; when the sum of the variance of the front p principal components is much more than the sum of the latter m-p principal components, the effect of the latter m-p principal components can be ignored, so as to achieve an effective interpretation and reduce the space dimensionality at the same time [8,20].

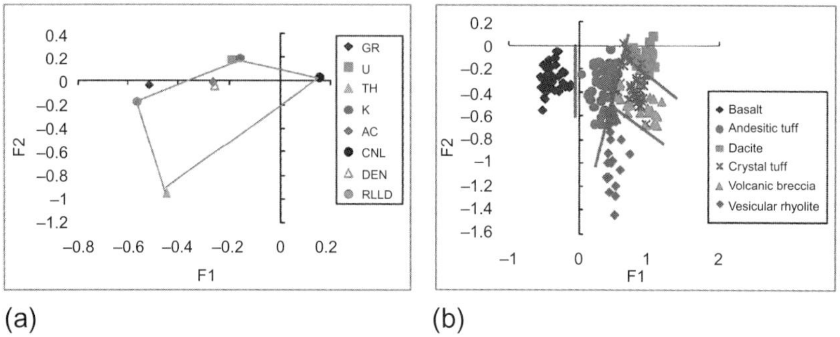

FIGURE 5.10

Map of principal component analysis. (a) Variable loading diagram. (b) Rock sample loading diagram.

Five steps of principal component analysis.

1. Selecting initial variables. According to the integrated analysis of well logs and geological data, the logs selected for volcanic rock identification include gamma ray, spectral gamma, density, neutron, sonic, and deep probing resistivity (deep laterolog or deep induction) logs.
2. Data normalization. After normalization, all logging values are converted to a scale of 0 to 1 to eliminate the weighting effects of logging dimensions and absolute values.
3. Calculating the covariance matrix on the basis of the relevant relationships among logging curves.
4. Calculating the eigenvalue of the covariance matrix and corresponding unit eigenvector by the Jacobi method.
5. Calculating the contributions of each principal component to establish the loading diagram of variables and rock samples.

Figure 5.10 shows the principal component analysis of volcanic rocks in Well Block XX9, XX gas field, where (a) and (b) are the variable loading diagram and the rock sample loading diagram, respectively. From this figure, the following can be determined:

1. The eigenvalue of the first principal component (F1) is 4.06, accounting for 55.7% of total variance. The left side of F1 is resistivity and GR, and the right side is neutron porosity, reflecting the changes in physical property and total intensity of radioactivity. The eigenvalue of the second principal component (F2) is 1.85, accounting for 28.1% of total variance and reflecting the changes in U, Th, and K contents.
2. The total eigenvalue of F1 and F2 accounts for 83.8% of total information, so that the crossplot of F1-F2 can be used to identify the basalt, andesitic tuff, dacite, vesicular rhyolite, and volcanic breccia effectively.

5.2.4 Integrated lithological identification

The lithology in single wells can be identified based on the identification model of composition, texture, and structure, using ECS element logging, FMI imaging, and conventional logs to determine the vertical characteristics of volcanic lithology, thus providing a basis for lithology prediction.

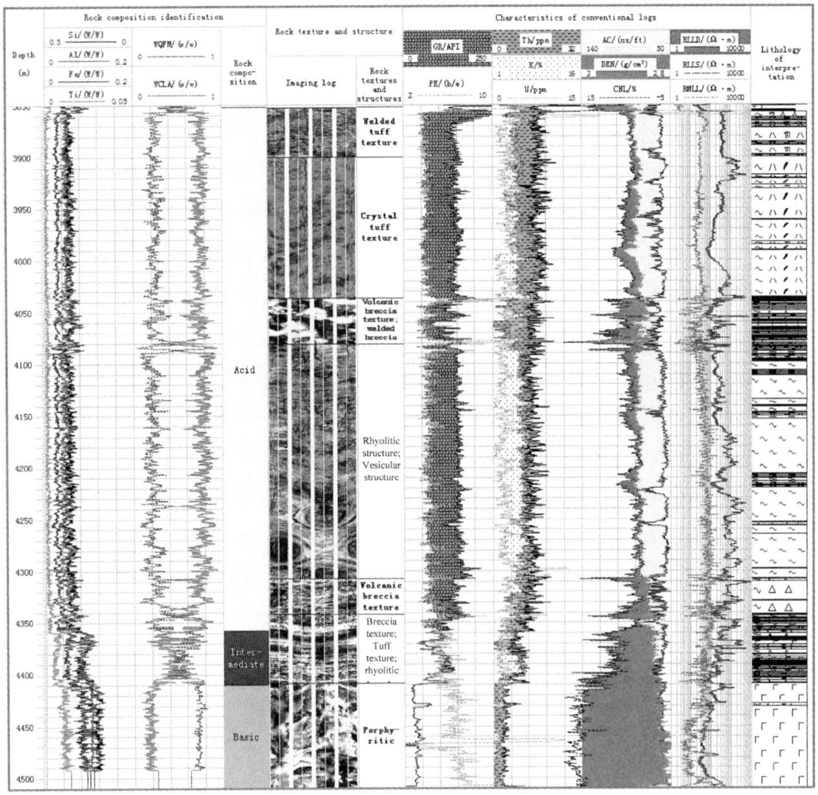

FIGURE 5.11

Comprehensive lithological identification in single wells (Well XX7).

The technique and methods can be illustrated using Well XX7 as an example for integrated identification of volcanic lithology using multiple data sources (Figure 5.11).

1. Identification of volcanic rock composition

Volcanic rock compositions can be determined by integrating ECS element logging, conventional gamma ray logging, gamma spectrometry, and compensated neutron logging as proxies for elemental content, oxide content, radioactivity, and the separation between density and neutron porosity logs in volcanic rocks with different compositions.

According to the response characteristics of ECS and conventional logs, three types of volcanic rocks with different compositions are recognized in Well XX7 (see Figure 5.11, the fourth trace). In the upper part (3850 to 4358 m), the average Si content is 36%, Al 5%, Fe 3%, Ti 0.2%, and SiO_2 77%; and the average GR value is 154 API, rock density 2.5 g/cm^3, and the average CNL 4.7%; gamma ray spectrometry (GRS) shows the characteristics of low U, high K, and high Th. These values indicate acidic volcanic rocks in this part of the well. In the middle part (4358 to 4410 m), the average Si content is 28%, Al 8.6%, Fe 5%, Ti 0.6%, and SiO_2 41%; the average GR is 100 API, rock density 2.62 g/cm^3, and average CNL 8%; GRS is characterized by medium U, medium K, and medium Th. These values indicate an intermediate type of volcanic rock in the middle part. In the lower part (4410 to 4510 m), the average Si content is 20%, Al 10.5%, Fe 8.4%, Ti 1.1%, and SiO_2 17.7%; the average GR is 35 API, rock density 2.74 g/cm^3, and average CNL 16%. GRS is represented by low U, low K, and low Th. These values confirm the presence of basic volcanic rocks in this interval.

2. *Identification of volcanic rock texture and structure*

Volcanic rock texture and structure can be identified using FMI imaging and conventional logs, based on the imaging identification model and response characteristics of conventional logs.

According to the characteristics of imaging log and conventional log curves, seven intervals with different textures were recognized in Well XX7 (see Figure 5.11, the sixth trace), including welded tuff texture in 3850 to 3900 m, crystalline tuff texture in 3900 to 4036 m, volcanic breccia texture and welded breccia texture in 4036 to 4081 m, vesicular and rhyolitic structures in 4081 to 4307 m, volcanic breccia texture in 4307 to 4342 m, volcanic breccia and tuff textures and rhyolitic structure in 4542 to 4408 m, and porphyritic texture in 4408 to 4510 m.

3. *Comprehensive identification of volcanic lithology*

Based on a rock classification scheme of "composition + texture and structure + petrogenesis," the volcanic rock types in noncored intervals in single wells can be determined (see Figure 5.11, the 11th trace) by cluster analysis such as the back-propagation (BP) neural network (Figure 5.12) in terms of rock composition, texture, and structure.

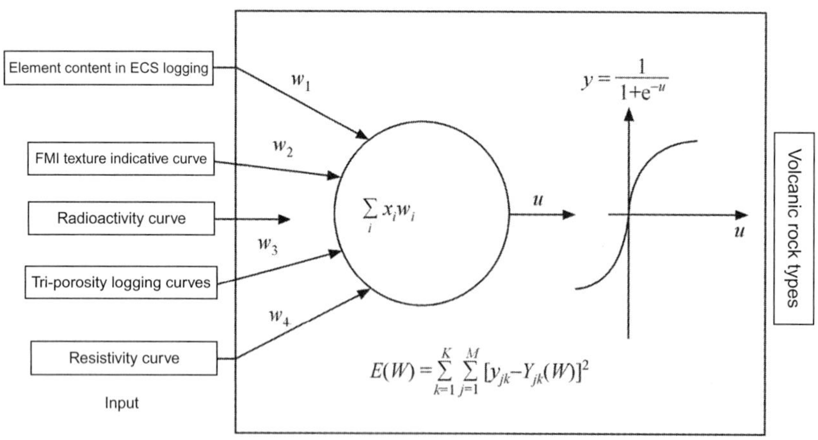

FIGURE 5.12

Schematic diagram of lithological identification for volcanic rocks using the BP neural network.

5.3 PREDICTION OF VOLCANIC LITHOLOGICAL DISTRIBUTION

After the seismic response characteristics of various volcanic rocks are determined, the planar distribution of different volcanic lithologies can be predicted by seismic profile analysis, waveform classification, and seismic inversion.

5.3.1 Seismic response characteristics of different lithologies

Seismic response models and interpretation criteria for different volcanic rocks can be established by forward modeling and response characteristics analysis on the basis of near-wellbore traces and seismic data calibrated by single-well lithology. The analyses are demonstrated by the case study of the DD gas field (Figure 5.13, Table 5.9), in which the seismic response models for basalt, volcanic breccia, orthophyre, rhyolite, and andesite are established [21].

1. Orthophyre

Orthophyre is developed mainly in the subvolcanic facies belt in Well Block DD18, characterized by high GR and high impedance. The seismic reflector of orthophyre shows a distinctly high frequency, medium-low amplitude, and discontinuity, and it often appears bosslike with an irregular wedge shape. In forward modeling analysis, the bottom and top surfaces show relatively strong reflections, which is consistent with the characteristics of unconformity surfaces in the seismic profile (Figure 5.13a).

2. Volcanic breccia

Volcanic breccia is developed mainly in the explosive facies belt in Well Block DD14, characterized by high GR and low impedance. The seismic reflector of volcanic breccia is marked by low frequency, medium-low amplitude, and discontinuity, often having a mound shape. In forward modeling analysis, the top surface has strong reflections, whereas the bottom surface has relatively weak reflections, as is typical of the unconformity surface at the top boundary and the conformable basal boundary in the seismic profile (Figure 5.13b).

3. Basalt

Formed by the eruption of basic magma, basalt is developed mainly in the effusive facies of Well Block DD17, marked by low GR and high impedance. The seismic reflector of basalt is characterized by high frequency, high amplitude, and relatively good continuity, manifesting often as a sheet or wedge with a prograding pattern. Forward modeling analysis reveals strong reflections at the bottom and top surfaces, which is consistent with unconformable surfaces in the seismic profile (Figure 5.13c).

4. Rhyolite and andesite

Rhyolite and andesite are developed mainly in the effusive facies of Well Blocks DD14 and DD10. Rhyolite is formed by the eruption of acidic magma, characterized by high GR and medium impedance, whereas andesite is formed by the eruption of intermediate magma, characterized by low GR and high impedance. Seismic reflectors of rhyolite and andesite have low frequency, high amplitude, and good

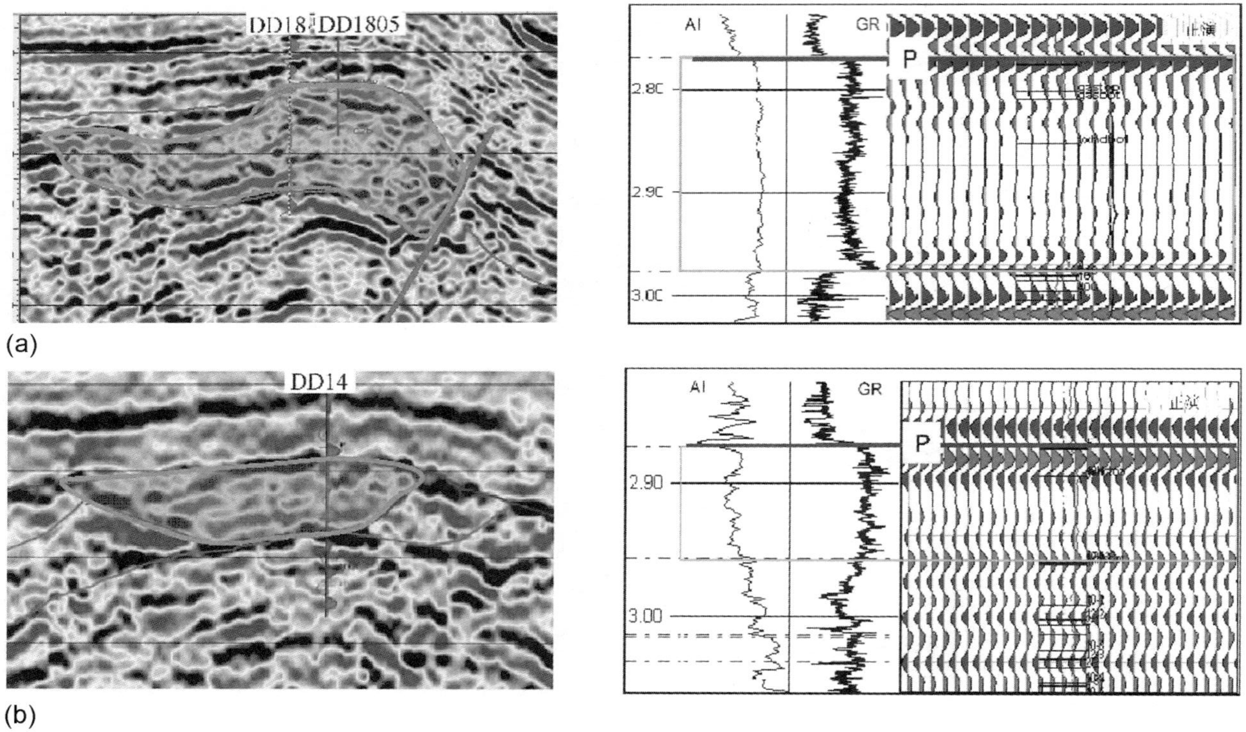

FIGURE 5.13

Seismic response characteristics and forward modeling of different lithologies (DD gas field). (a) Orthophyre. (b) Volcanic breccia.

Continued

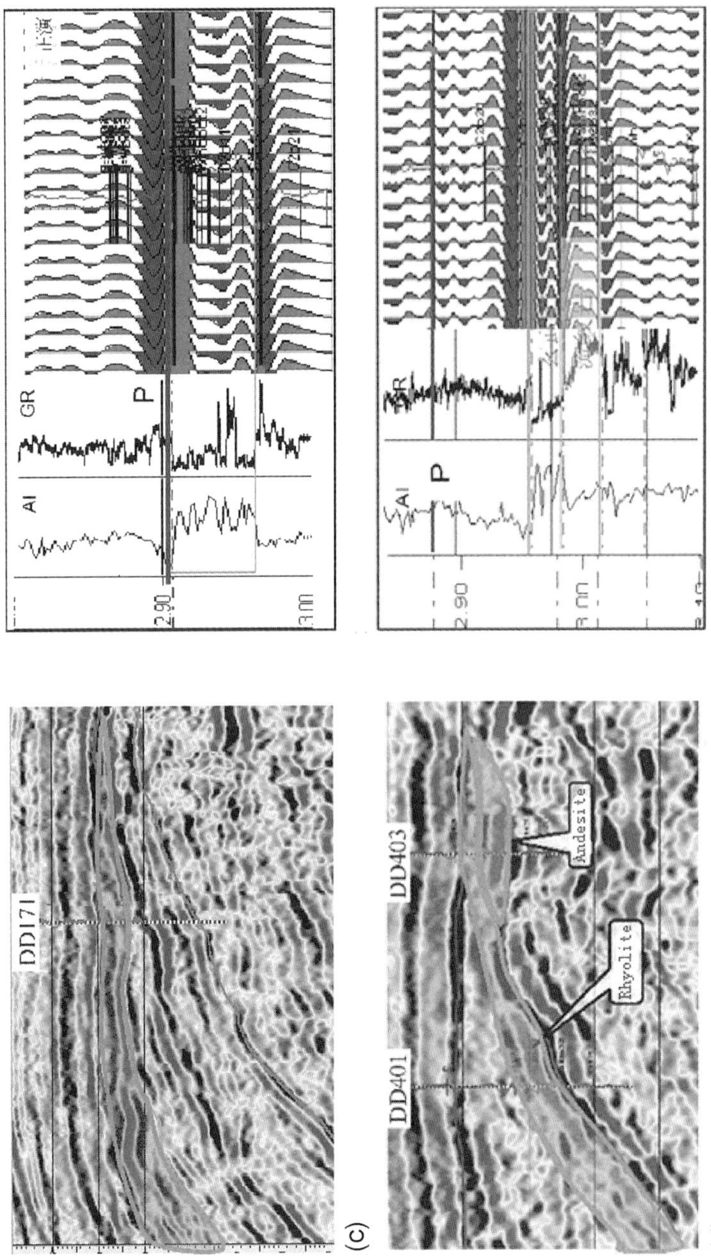

FIGURE 5.13—Cont'd

(c) Basalt. (d) Rhyolite, andesite.

Table 5.9 Seismic Response Characteristics of Different Volcanic Rocks

Lithology	Lithofacies	Seismic Response Characteristics		Representative Well
		Interior Reflection Characteristics	*Shape*	
Volcanic breccia	Explosive facies	Low frequency, medium-low amplitude, discontinuity	Mound	DD14, DD403
Basalt	Effusive facies	High frequency, high amplitude, good continuity	Sheet, wedge	DD171, DD401
Andesite		Low frequency, high amplitude, good continuity	Laminar	DD403
Rhyolite		Low frequency, high amplitude, good continuity	Laminar-like	DD403, DD401
Orthophyre	Subvolcanic facies	High frequency, medium-low amplitude, discontinuity	Bosslike, wedge	DD18

continuity, with a sheet or wedge morphology. Forward modeling analysis shows strong reflections at the bottom and top surfaces, which is consistent with unconformable surfaces in the seismic profile (Figure 5.13d).

These seismic response characteristics demonstrate that orthophyre, basalt, rhyolite, andesite, and volcanic breccia can be identified using seismic data. It should be noted that differences in acquisition method, subsurface, and surface conditions will result in discrepancies in seismic data acquired in different well blocks and at different times, which in turn will lead to discrepancies in seismic responses to different volcanic rocks.

5.3.2 Prediction of lithological distribution through seismic profile analysis

The lithological distribution of volcanic rocks can be predicted through seismic reflector profiles, based on well-seismic calibration and response characteristics of near-wellbore tracing. The methods and technical procedures are illustrated next, using Well Block DD14 as an example.

1. Preliminary prediction of lithological distribution under the constraint of volcanic rock architecture

According to the characteristics of lithology and lithofacies in Well Block DD14, a framework model of a volcanic massif can be established through dissection of the volcanic rock architecture, based on stratal occurrence, lithofacies assemblage, and fault features (Figure 5.14).

Constrained by the distribution model of a volcanic massif, the distribution range and interwell pattern (Figure 5.16) of volcanic lithology can be determined (Figure 5.15) by well integration, based on the results of lithological interpretation in single wells. It can be concluded from the figure that two sets of volcanic rocks are developed in Well Block DD14:

1. The upper volcanic rocks (the first reservoir interval). The eruption center is at the location of Well DD1424, and the typical volcanic breccia is developed along Wells DD14, DD1424, and DD402; from the center to the northwest, the lithology is dominated by rhyolite, andesite, and basalt.

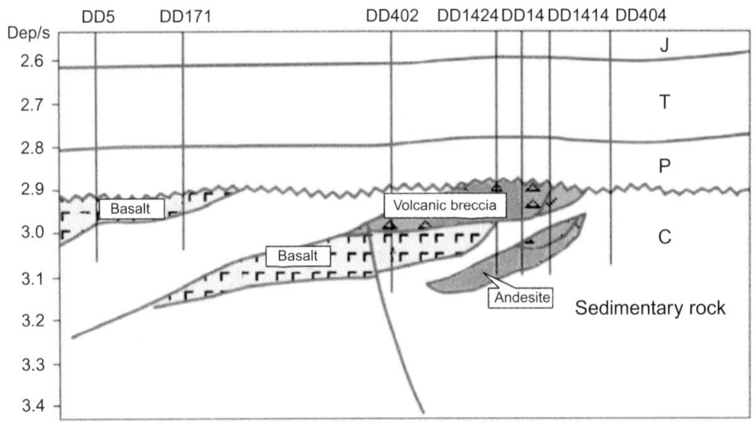

FIGURE 5.14

Distribution model of a volcanic massif (Well Block DD14).

The distribution range is extended to Wells DD401 and DD402 in the west and to Wells DD1413 and DD403 in the east.

2. The lower volcanic rocks (the second reservoir interval) are developed mainly in a belt from Well DD1415 to Well DD403; a set of andesitic volcanic breccia is present on the top, and an assemblage of interbedded volcanic breccia and andesite is developed in the lower part; this set of volcanic rock occupies a relatively small area, represented by a single thin layer.

2. Predicting profile distribution of lithology through analysis of seismic response characteristics of different rocks

The analysis of seismic response characteristics of different lithologies is conducted on the basis of the lithological distribution range and well association pattern. Multiple interwell interpretation and profile tracing are carried out to predict the profile distribution pattern of volcanic rocks (Figure 5.17).

3. Predicting planar distribution of lithology by establishing skeleton profile for seismic interpretation

The lithological interpretation on profiles is used to establish the framework profiles that can control the study area effectively, followed by prediction of planar lithological distribution by profile tracing and closing.

Volcanic breccia is the predominant lithology in reservoirs of Well Block DD14, developed mainly on structural highs in the south, as shown by the prediction results (Figure 5.18). The volcanic breccia in the upper part covers an area of $\sim 5\ km^2$, where the maximum thickness is about 160 m; the volcanic breccia in the lower part is smaller in scale, with an area of $4\ km^2$ and maximum thickness of 90 m. The favorable lithological area for development determined by the prediction results provides a basis for well location optimization.

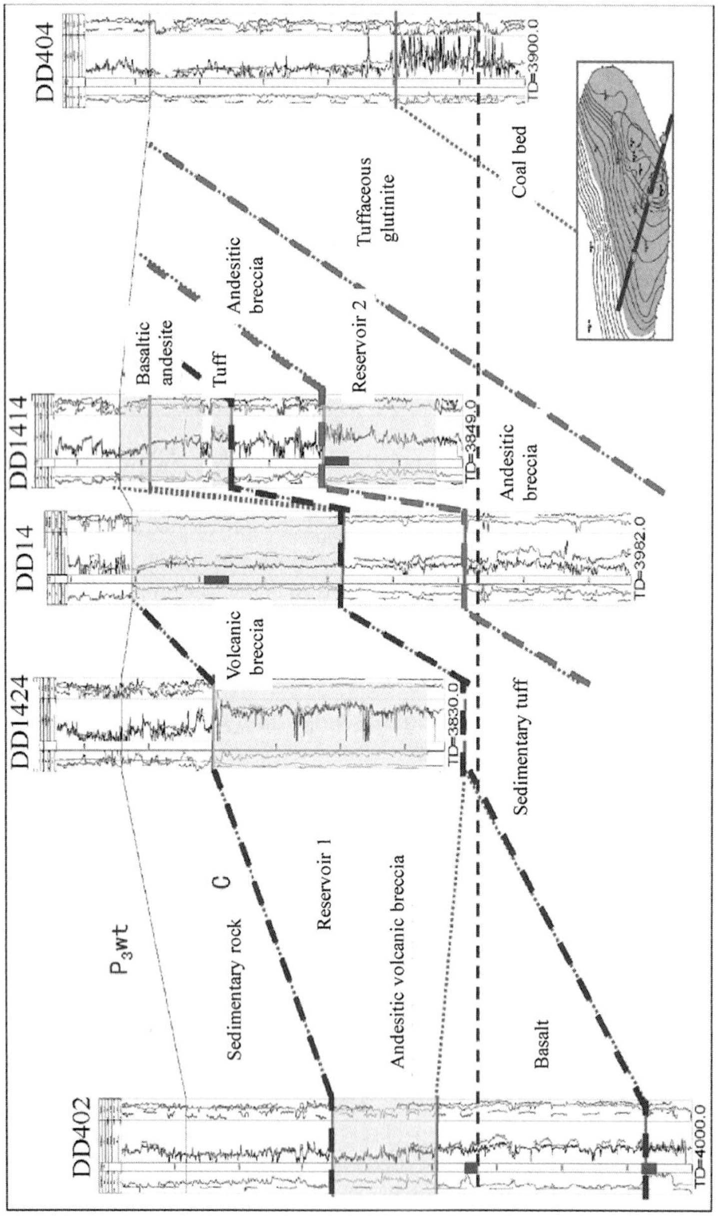

FIGURE 5.15

Multiple interwell correlation in Wells DD402-DD1424-DD14-DD1414-DD404.

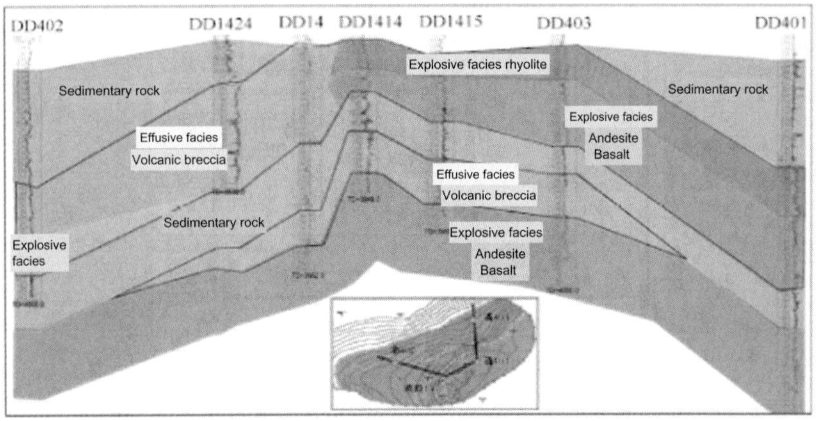

FIGURE 5.16

Lithological distribution model (Well Block DD14).

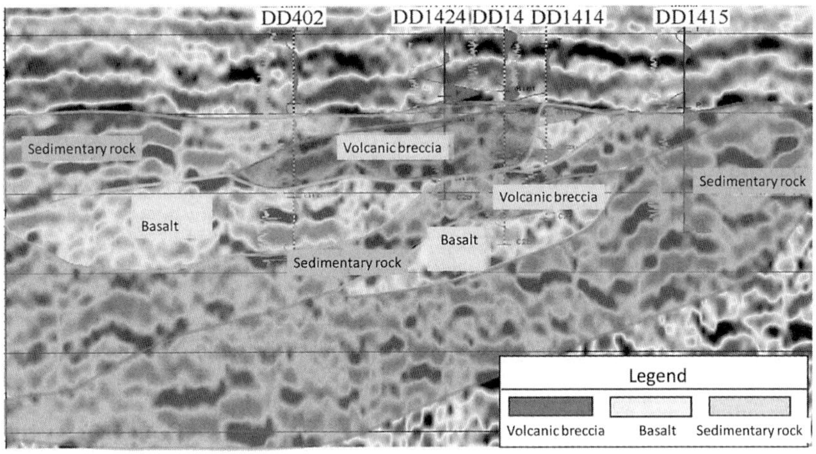

FIGURE 5.17

Profile of lithological distribution in volcanic rock (Well Block DD14).

5.3.3 Prediction of lithological distribution by seismic waveform classification

Seismic waveform classification is a technique for seismic facies analysis, usually starting from well points. Based on the analysis, generalization, and classification of actual waveforms in seismic reflectors, seismic waveform classification can be used to analyze the differences in lithology, physical property, and fluid of the target horizon by geological interpretation of waveform changes.

The major steps of predicting lithological distribution through seismic waveform classification include (1) calibrating the seismic and geological horizons in boreholes; (2) starting from individual wells, tracing the seismic reflectors of different lithological intervals in the target horizons;

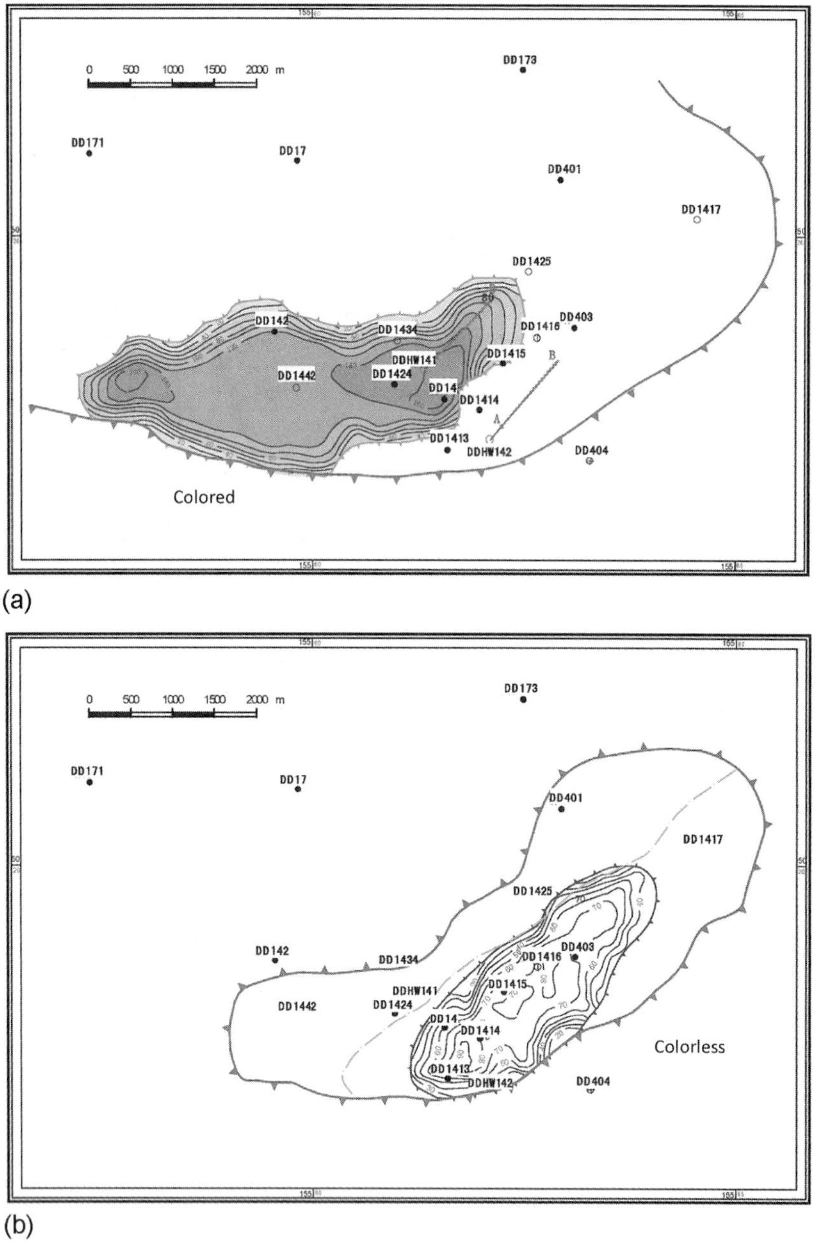

FIGURE 5.18

Planar distribution of volcanic breccia in Well Block DD14. (a) Distribution of volcanic breccia in the upper part. (b) Distribution of volcanic breccia in the lower part.

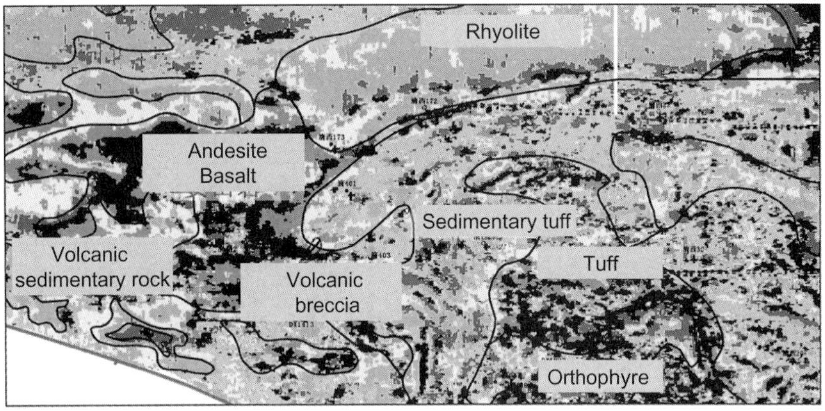

FIGURE 5.19

Planar map of waveform classification of volcanic lithology (DD gas field).

(3) according to lithological differences, dividing the seismic facies by event characteristics such as waveform, amplitude, frequency, and phase, and determining the waveform types and characteristics; (4) using the neural network technique to build the pickup model by training each waveform; (5) determining the pickup range in the target horizons and selecting the waveforms by the picking model to determine the planar distribution of different waveforms; and (6) integrating single-well and inter-well distribution characteristics of waveforms to determine the planar distribution of different lithologies.

The reservoirs of the DD gas field are developed mainly in the volcanic rocks 160 m below the top of the weathered crust, and the time thickness is about 70 ms. Thus, bounded by the top volcanic rock, the seismic waveform of 70 ms downward can be picked up (Figure 5.19), and the lithological distribution can be predicted (Figure 5.20). It can be concluded that the sedimentary tuff and sand shale along Wells DD404 to DD27 in this area are widespread and trend in approximately a NS direction. The volcanic rock can be divided into two parts (east and west). The east part is dominated by orthophyre in Well Block DD18; the north part by tuff and sedimentary tuff; and the west part by basalt and volcanic breccia, with regularly distributed sedimentary rocks, where volcanic breccia occurs mainly in Well Block DD14 and basalt in Well Block DD17.

5.3.4 Prediction of lithological distribution by frequency-divided inversion

Frequency-divided inversion is a nonlinear attribute inversion technique, which is based on amplitude versus frequency (AVF) and neural network technique. It can be adopted for lithology inversion directly and plays an important role in predicting the lithological distribution of volcanic rocks [20].

1. Principles of frequency-divided attribute inversion

Frequency-divided inversion consists of the following steps.

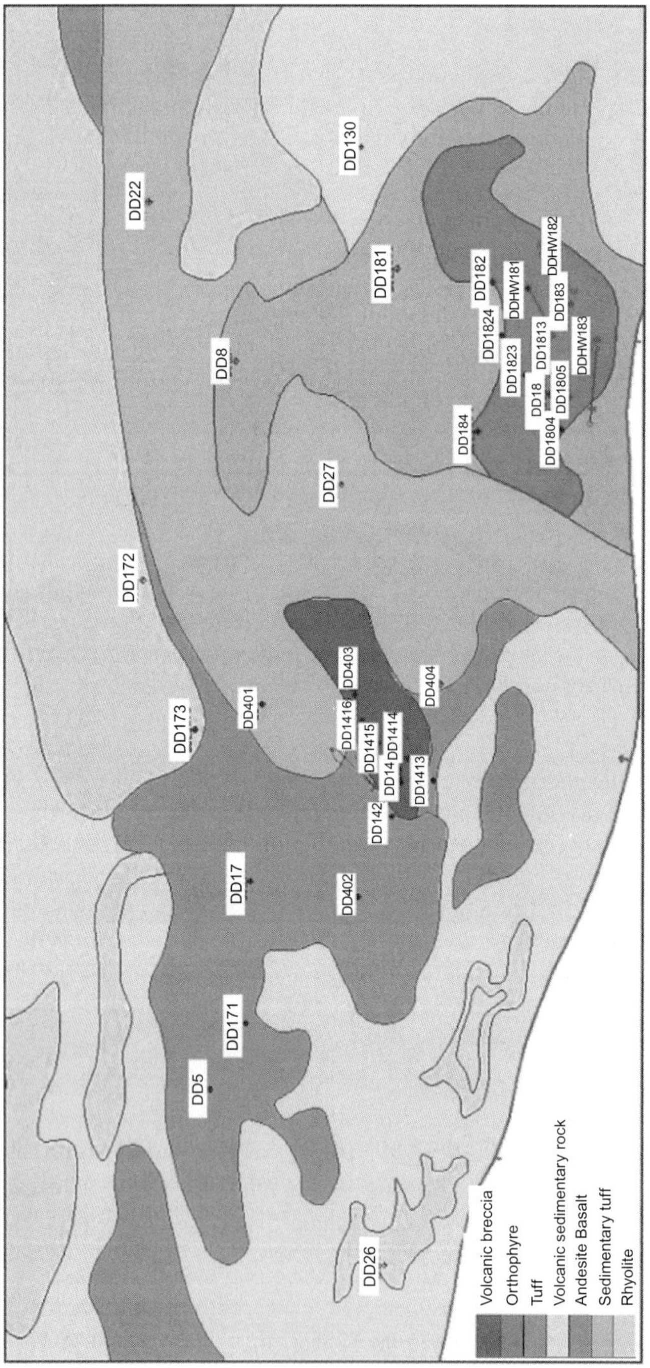

FIGURE 5.20

Planar distribution map of identified volcanic rocks (DD gas field).

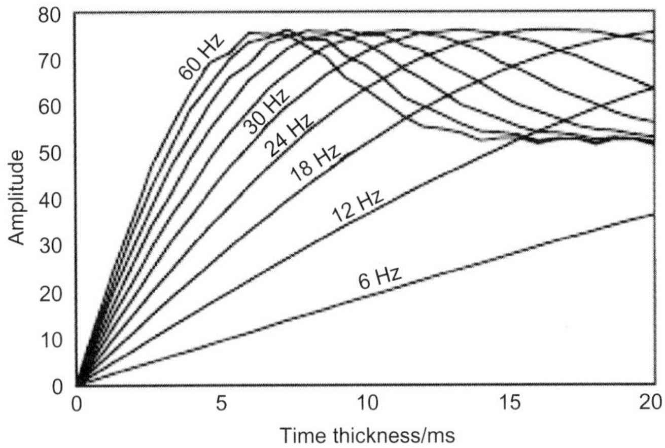

FIGURE 5.21

Relationship between amplitude and time thickness.

1) Frequency-divided processing of seismic data

This process begins with a spectrum analysis of the seismic data. According to the effective frequency band, suitable sizes can be designed. After this, the data volume of different frequency bands can be obtained in accordance with frequency-divided processing.

2) AVF relationship analysis

For a wedge model, a series of synthetic seismic profiles and tuning curves with amplitude and thickness in different frequencies (Figure 5.21) can be obtained by Ricker wavelet and its convolution with different dominant frequencies. The AVF relationship is very complex and is difficult to express by explicit functions. Therefore, tuning curves should be transferred by neural network technique to obtain the relationship curve between amplitude and frequency at different time thicknesses (i.e., AVF, Figure 5.22).

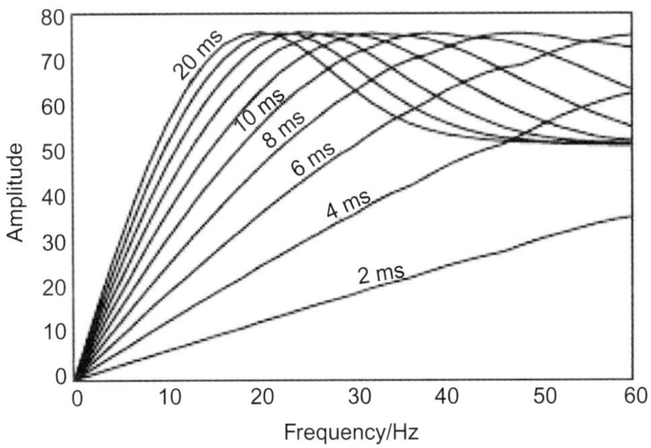

FIGURE 5.22

Relationship between amplitude and frequency.

3) Frequency-divided attribute inversion

The AVF relationship is introduced into the inversion to establish the nonlinear mapping relationship between impedance and seismic waveform to obtain inversion results.

2. Lithological distribution prediction by frequency-divided attribute inversion

Based on the frequency-divided attribute inversion technique, lithological distribution can be predicted effectively using impedance and GR inversion. The major steps are as follows.

1) Petrophysical analysis

Sensitive parameters, which can be used to determine lithology, are selected through the relationship analysis of rock elastic parameters and reservoir lithology parameters. Figure 5.23 shows a crossplot of impedance-GR and impedance-density of different lithologies. This plot demonstrates that volcanic breccia is characterized by low impedance, medium-high GR, and low density; orthophyre by medium-high impedance, high GR, and medium-high density; basalt by high impedance, low GR, and high density; andesite by high impedance, low-medium GR, and medium-high density; rhyolite by medium impedance, high GR, and medium-low density; and sedimentary rock by medium-low impedance, medium-low GR, and medium-low density. Hence, impedance and GR are the sensitive parameters for lithological prediction in this area.

2) Wave impedance inversion

The AVF relationship is introduced into wave impedance inversion. The wave impedance data volume of strata can be obtained after special processing of seismic data with the BP neural network technique.

Figure 5.24 shows a learning sample of wave impedance prediction, in which the wave impedance curve of well logging is in red, and the wave impedance curve of near-wellbore inversion is in blue. The correlation coefficient of these two curves has a range of 0.81 to 0.89, indicating a good learning result. The result of wave impedance inversion using the BP neural network technique (Figure 5.25) is basically consistent with the wave impedance of single wells and near-wellbore tracing.

3) Gamma inversion

Gamma inversion data volume can be obtained by the same method. Figure 5.26 is a learning sample of gamma prediction, in which the gamma log curve is in red and the gamma curve of near-wellbore inversion is in blue. The correlation coefficient of these two curves has a high range of 0.78 to 0.96, indicating a very desirable learning effect. The results of gamma inversion using the BP neural network technique (Figure 5.27) corroborate the natural gamma of single wells.

4) Prediction of lithological distribution

Based on petrophysical analysis, the lithological distribution of volcanic rocks can be predicted effectively through single-parameter analysis and integrated multiparameter analysis using the inversion data volume.

Planar distribution of wave impedance value. Considering the time domain of 70 ms downward from the volcanic rock top as the objects (major gas layer), the planar distribution of average wave impedance (Figure 5.28) can be obtained by extracting the wave impedance inversion data volume. In this

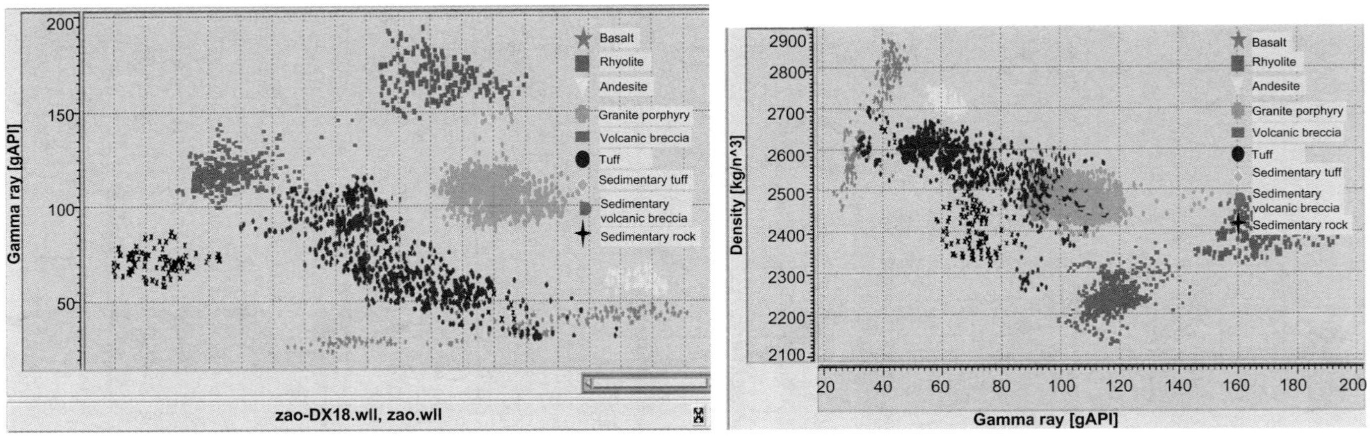

FIGURE 5.23

Crossplot of impedance-GR and impedance-density in volcanic rock (DD gas field). (a) Crossplot of impedance and GR. (b) Crossplot of impedance and density.

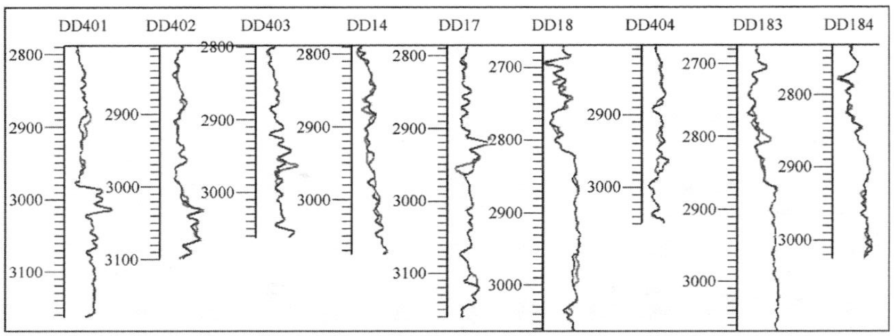

FIGURE 5.24

Learning sample for wave impedance prediction by the BP neural network technique.

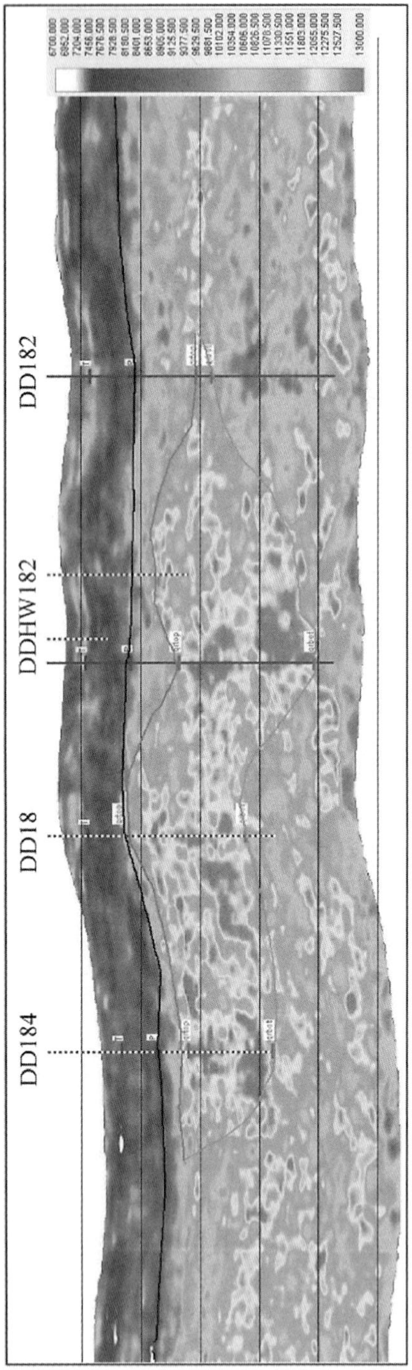

FIGURE 5.25

Results of wave impedance inversion using the BP neural network technique.

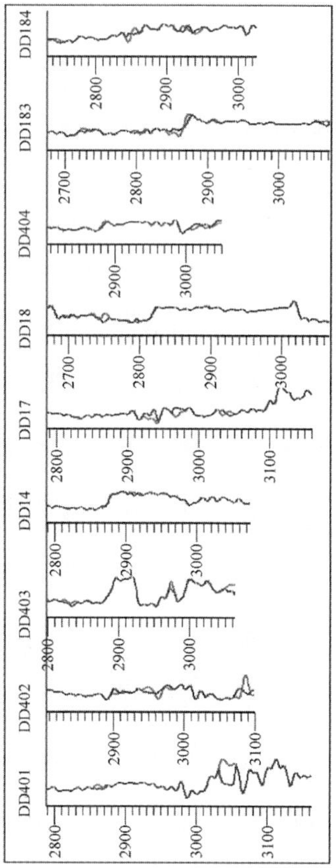

FIGURE 5.26

Learning sample for natural gamma prediction using the BP neural network technique.

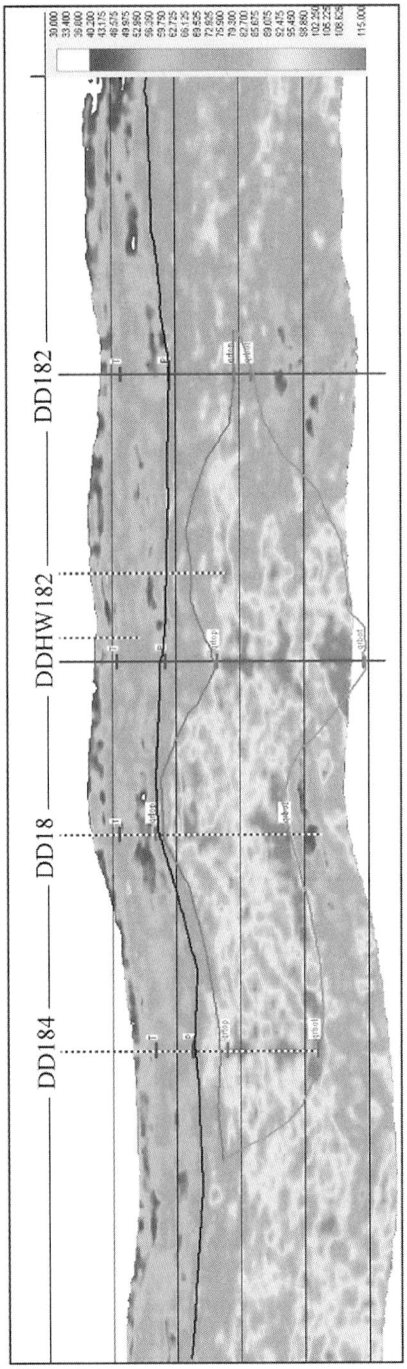

FIGURE 5.27

Natural gamma inversion results using the BP neural network technique.

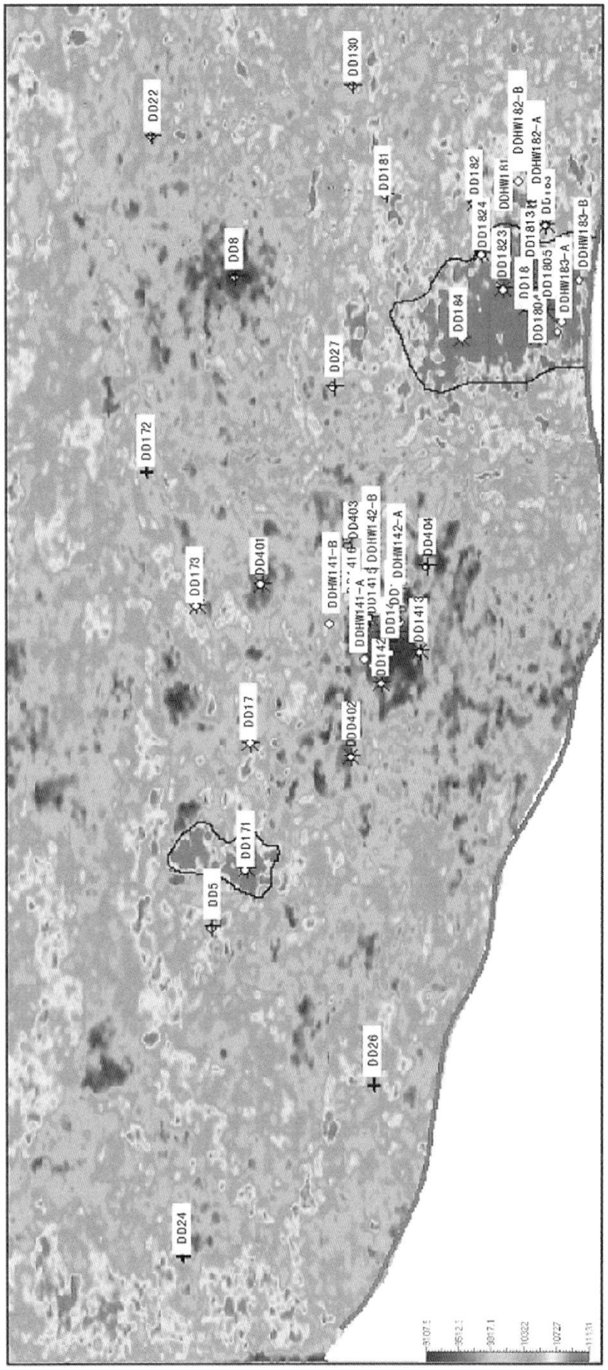

FIGURE 5.28

Planar distribution of average wave impedance (DD gas field).

figure, the highest impedance value is around Wells DD18 and DD184, the moderate impedance value is around Wells DD171 and DD17, and the lowest impedance value is around Wells DD402, DD14, and DD404.

Planar distribution of gamma value. Figure 5.29 also shows that the highest value of average gamma is around Wells DD18 and DD184, the moderate value is around Well DD14, and the lowest value is around Wells DD17, DD171, DD172, and DD173.

Planar distribution of lithology. The lithology of spatial data points can be determined through petrophysical analysis, such as multiattribute crossplot analysis with inversion data volume, so as to determine the lithological distribution of volcanic rock in the study area (see Figure 5.30). The results show that Well Block DD18 is dominated by orthophyre, Well Block DD14 by volcanic breccia, and Well Block DD17 by basalt.

5.4 APPLICATIONS

Lithological characteristics and distribution of volcanic rocks can be studied by applying various lithological identification and prediction techniques to delineate lithofacies in single wells, interpret reservoir parameters, identify gas and water layers, and predict favorable reservoir zones.

5.4.1 Identification of lithology to guide lithofacies delineation in single wells

Volcanic lithofacies have good corresponding relationships with lithology; therefore, lithological logs of single wells can be derived from lithological identification. Such lithological logs can be used in combination with lithological markers, wireline log facies markers, and the locality data of volcanic edifices in single wells to determine single-well lithofacies and to establish lithofacies profiles. This provides a basis for lithofacies prediction and characterization.

5.4.2 Building reservoir parameter interpretation models for different lithologies to improve parameter interpretation accuracy

Different volcanic rocks have different rock matrix values, different types of reservoir space, different porosity, and other physical properties. The "four-parameter" relationships of volcanic reservoirs are also variable. These differences are reflected in different slope and cutoff values in porosity-density crossplots. Therefore, lithology identification can be useful for modeling reservoir parameters for different lithologies, thereby improving the accuracy of reservoir parameter interpretation.

5.4.3 Providing a lithological basis for the establishment of gas and water layer identification models and to improve the congruence rate of identification

Different types of volcanic rocks have different conductive mechanisms and conductive pathways; gas layers in different lithotypes also have different logging response characteristics. Hence, lithological identification can provide a basis for establishing gas and water layers identification models and improving the congruence rate of identification.

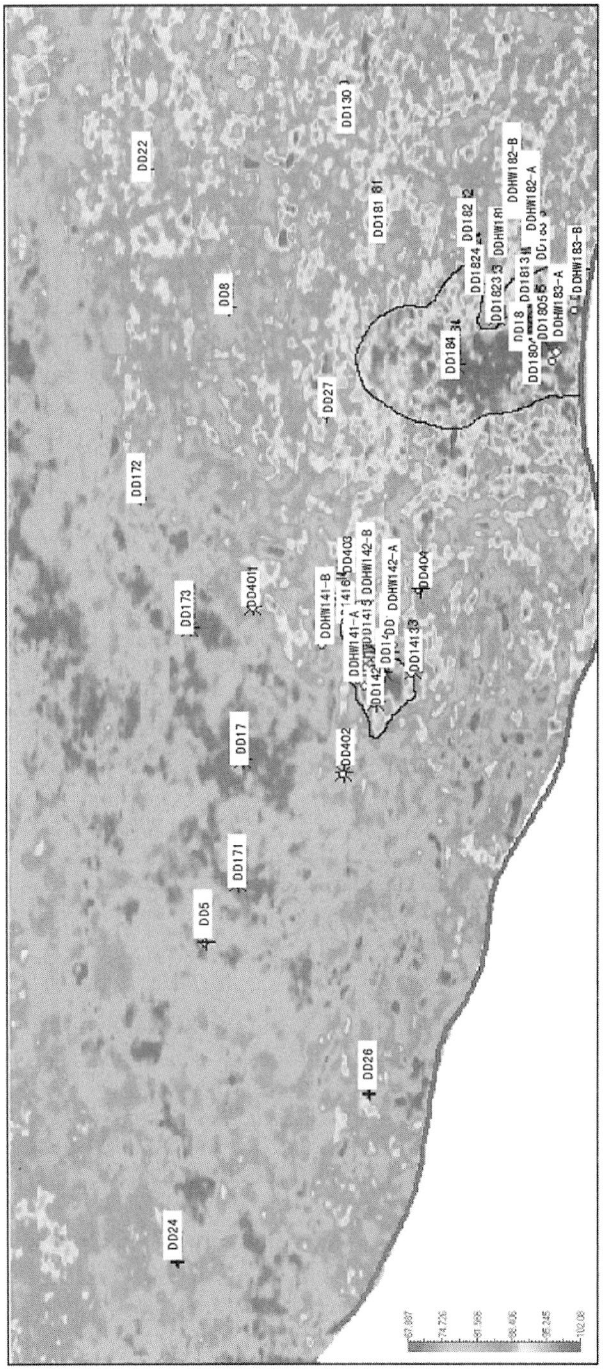

FIGURE 5.29

Planar distribution of average gamma (DD gas field).

FIGURE 5.30

Planar distribution of lithology (DD gas field).

5.4.4 Delineating lithological distribution patterns to guide the prediction of favorable reservoir zones

Lithology controls the storage space development in volcanic rocks. By predicting the spatial distribution of lithology, the spatial distribution of favorable reservoir zones can be traced and predicted by the geological rules governing storage space development, which can guide the prediction of the spatial distribution of favorable reservoir zones in volcanic rocks.

References

[1] Qiu J, Tao K, Zhao J, et al. Volcanic rock. Beijing: Geoscience Press; 1981, p. 1–6.
[2] PetroChina Exploration and Production Company. Log evaluation technology and application for volcanic oil and gas reservoirs. Beijing: Petroleum Industry Press; 2009, p. 1, 2, 57, 58, 68, 69.
[3] Li N, Tao H, Liu C. Log interpretation theory, method and application for acid volcanic rock. Beijing: Petroleum Industry Press; 2009, p. 3, 63–71.
[4] Chetverikov CD. Geochemical conversion guide. [Liu Zhixing et al., Trans.]. Beijing: Geoscience Press; 1966, p. 5–15.
[5] Nockolds SR. Average chemical compositions of some igneous rocks. The Geological Society of American; 1954, p. 1007–32.
[6] Le Maitre RW. Some problems of the projections of chemical data into mineralogical classifications. Contrib Mineral Petrol 1976;56(2):181–9.
[7] Fan Y, Huang L, Dai S, et al. Application of crossplots well log data to identifying volcanic lithology. Well Logging Technology 1999;23(1):53–6.
[8] Pan B, Yan G, Wu H. Determining lithology of igneous rocks in deep northern songliao basin using correspondence analysis. Petrol Geol Oilfield Dev Daqing 2003;22(1):7–9.
[9] Liu W. Petrophysical study on volcanic reservoir rocks. Xinjiang Petrol Geol 2003;24(5):389–91.
[10] Huang B, Pan B. The log characteristics of deep igneous rocks and lithology determination in the northern Songliao Basin. Geophys Prospect Petrol 2001;40(3):42–7.
[11] Ramamoorthy R. A look at spectroscopy. Formation Evaluation Review 2001;2(6):1–3.
[12] Liu X, Sun J, Li Z. A new generation of elemental capture spectroscopy (ECS) logging instrument and its application. World Well Logging Technology 2004;19(1):26–30.
[13] Wang Y, Li Z, Ran Q. A new logging technique in volcanic lithology identification. Faulted-Block Oil & Gas Field 2006;13(3):86–8.
[14] Zhang S, Gu C, Cao G. Imaging log technology and its application. Beijing: Petroleum Industry Press; 1997.
[15] Wang G, Guo L. Well logging geology. Beijing: Petroleum Industry Press; 2000, p. 10, 79–94.
[16] Sima L, Shu Z. Logging evaluation methodology and its application for carbonate reservoirs. Beijing: Petroleum Industry Press; 2009, p. 36–56.
[17] Huang L. Principle of radioactive logging. Beijing: Petroleum Industry Press; 1985.
[18] Chu Z. Principle of acoustic logging. Beijing: Petroleum Industry Press; 1987, p. 75–103.
[19] Wang Y, Yan L, Ran Q. Lithology identification of deep volcanic gas reservoirs in Xingcheng gas field. Journal of Southwest Petroleum University 2007;29(2):78–81.
[20] Pan B, Xue L, Li Z. Well logging evaluation methodology and its application for fractured volcanic reservoir. Beijing: Petroleum Industry Press; 2003.
[21] Ouyang Y, Zeng Q, Geng J, et al. 3D seismic processing and interpretation and reservoir prediction in Dixi 5 and Dixi 8 areas. Beijing: Research Institute of Petroleum Exploration and Development, PetroChina; 2009, p. 14–25.

CHAPTER

Identification and Prediction of Fractures in Volcanic Reservoirs

6

Fracture is a part of the volcanic reservoir space that connects pores and seepage pathways. Volcanic rock fractures and their occurrences can be investigated and evaluated based on geological study on fractures and analysis of logging response characteristics, using acoustic and electrical image logging and conventional logging. The development and effectiveness of fractures can be evaluated according to the electrical conduction mechanism of fractures, such as fracture parameters derived from micro-resistivity scanning image logs and dual laterologs. Based on fracture identification and interpretation in single wells, seismic data can be calibrated by other log data. In this chapter, we discuss the systematic integration of the analysis of near-wellbore seismic trace responses, poststack seismic attribute analysis, prestack subazimuth seismic attributes processing and analysis, and fracture parameter inversion to predict the spatial distribution of fractures in volcanic rocks, reveal the basic characteristics of fractures, and provide a basis for optimizing the designs of well grids and horizontal well locations.

6.1 RESEARCH CHALLENGES AND TECHNICAL SOLUTIONS

6.1.1 Challenges in fracture identification and prediction

Fractures in volcanic reservoirs are highly variable, with complicated origins and logging and seismic response characteristics. The challenges in their identification and prediction can be summarized into the following three aspects.

1. Difficulties in fracture identification due to various types of volcanic fractures and diverse geological characteristics

There are various natural fractures in volcanic reservoirs [1–3], and hence various challenges in their identification. Structural fractures and columnar joints have a long extension and large scale and are thus easily identifiable in outcrops, cores, and logging curves due to their easily recognizable characteristics. Contraction fractures and blast fractures, characterized by a short extension, small scale, and special shapes, are easily identified in cores and thin sections because of their distinct features but are difficult to recognize in logging curves. Weathering fractures and sutured fractures have a limited extension and small scale, and as such they are difficult to recognize in outcrops, cores, thin sections, and logging curves due to their inconspicuous characteristics. The key to identifying dissolution fractures is to look for corrosion traces. In general, dissolution fractures are easily identified in cores and thin sections, but they are difficult to recognize in logging curves.

In addition, the identification of natural fractures in volcanic rocks can be confounded by false fractures, fluid properties, and physical properties of the matrix. In general, these features pose a challenge in fracture identification.

2. Large error range in calculation of fracture parameters due to complex electrical conduction mechanisms of volcanic rocks and significant changes in log responses

The conductivity of volcanic rocks is affected by factors such as rock framework, matrix pore, fluid type, and fracture. Volcanic rocks have complex electrical conduction mechanisms and significant changes in log response characteristics. Based on rock conductivity, therefore, the fracture parameters can be calculated with resistivity imaging and dual laterologs, but the calculation error is large. Moreover, the high level of difficulty in fracture identification also increases calculation errors for fracture parameters.

3. Difficulties in seismic prediction due to various volcanic fractures and complicated stratigraphic structures

Limited by resolution, only fracture zones developed in groups can be predicted by seismic data. Among the great variety of fractures developed in volcanic rocks, structural fractures and columnar joints developed in groups can be identified easily using seismic data; blast fractures and contraction fractures are small in scale with limited extension, making them difficult to identify based on seismic data; weathering and sutured fractures are difficult to recognize using well log data and even more difficult to predict based on seismic data. In addition, volcanic rocks are characterized by rapid lithological change with various lithological boundaries, internal architectural boundaries, and complex stratigraphic structures. All of these factors pose challenges in fracture prediction using seismic data.

6.1.2 Research approaches and solutions

To meet the challenges in identification and prediction of fractures in volcanic rocks, the technical processes for volcanic fracture identification and prediction are established in this study, following the concept of integrated geological, geophysical, and seismological investigations on fractures. Through application of qualitative identification, quantitative interpretation, and multiapproach prediction (Figure 6.1), the techniques for identifying and predicting fractures in volcanic gas reservoirs can be developed.

1. Geological investigations on fractures

Fracture classification can be used as a basis to establish identification markers for lithology, lithofacies, morphology, infilling, and group characteristic of fractures in volcanic reservoirs. Outcrop, core, and thin section data are utilized to identify fractures; quantitatively characterize fracture shape, size, and accumulation and permeation capacity; and reveal the genetic mechanisms, morphological characteristics, and distribution of various fractures. These lay the foundation for log identification. Details on this subject will be discussed in Chapter 11 and will not be repeated here.

2. Log identification and evaluation

Fracture identification and the evaluation of fracture distribution can be achieved by calibration of well log data using core data, analyzing log response characteristics, and establishing log identification markers, and through the integrated use of Full-bore Micro scanner Imaging (FMI) and conventional logging. The density, width, length, surface porosity, and permeability of various fractures are calculated on the basis of fracture identification using resistivity imaging logs. Fracture width, porosity, and permeability are estimated by the application of conventional logging; thereafter, the development and effectiveness of fractures in volcanic rocks are evaluated, and the distribution characteristics of fractures in wells are ascertained.

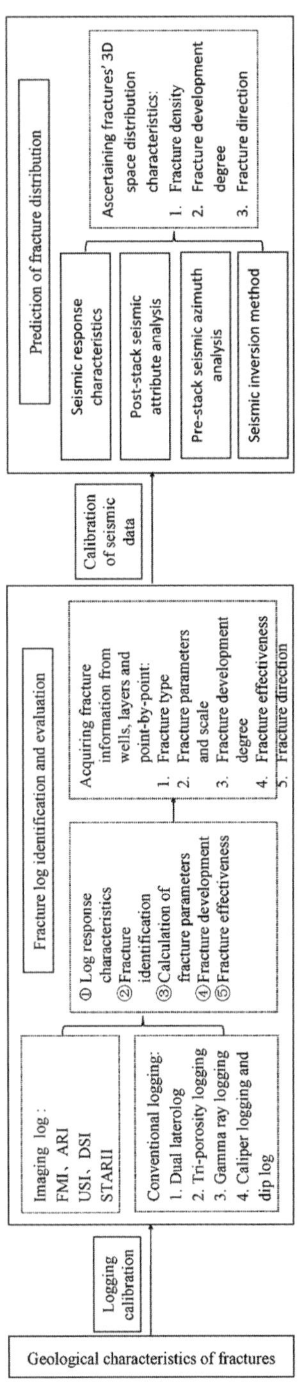

FIGURE 6.1

Technical processes for identification and prediction of fractures in volcanic rocks.

3. *Prediction of fracture distribution*

On the basis of fracture identification and evaluation in wells and analysis of seismic response characteristics of fractures, the methods of integrating poststack attribute analysis, prestack subazimuth attribute analysis, and fracture parameter inversion can be used to predict the distribution of fractures in volcanic reservoirs, reveal the distribution of fractures, and establish a basis for optimizing well grid design, as well as locations and trajectories of horizontal wells.

6.2 IDENTIFICATION OF FRACTURES IN VOLCANIC RESERVOIRS

Various types of fractures can be qualitatively identified and their degree of development evaluated on the basis of log calibration with core data by analyzing log response characteristics and establishing the imaging and conventional log identification models for fractures in volcanic reservoirs.

6.2.1 Identification of fractures based on imaging logs

The acoustic and electric imaging logs provide high-resolution sidewall images and fine depiction of geological characteristics of subsurface formation fractures, rock textures, and beddings. In volcanic reservoirs, FMI imaging logs are widely used to identify fractures.

1. *Imaging log response characteristics of fractures*

In an FMI imaging chart (Figure 6.2), open fractures have low resistivity and usually show black-colored, sinusoidal or irregular lines or narrow stripes. The characteristics of filled fractures are related to the conductivity of infilled material. If the filler conductivity is low, filled fractures will show white, sinusoidal or irregular lines or stripes. If the filler conductivity is high, filled fractures will be black sine-curves or stripes. The height difference between the peak and valley of a sine curve is related to the fracture dip angle: the greater the dip angle, the greater the height difference.

2. *FMI imaging log identification model*

Highly conductive fractures, high-resistivity fractures, microfractures, and induced fractures can be identified effectively using the sine-curve theoretical fitting method [4] to establish an FMI imaging log interpretation model for volcanic rock fractures (Figure 6.3, Table 6.1).

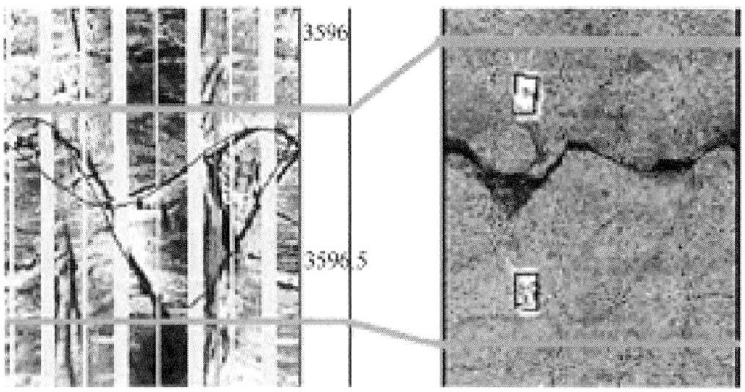

FIGURE 6.2

FMI logging response of fractures (Well XX9).

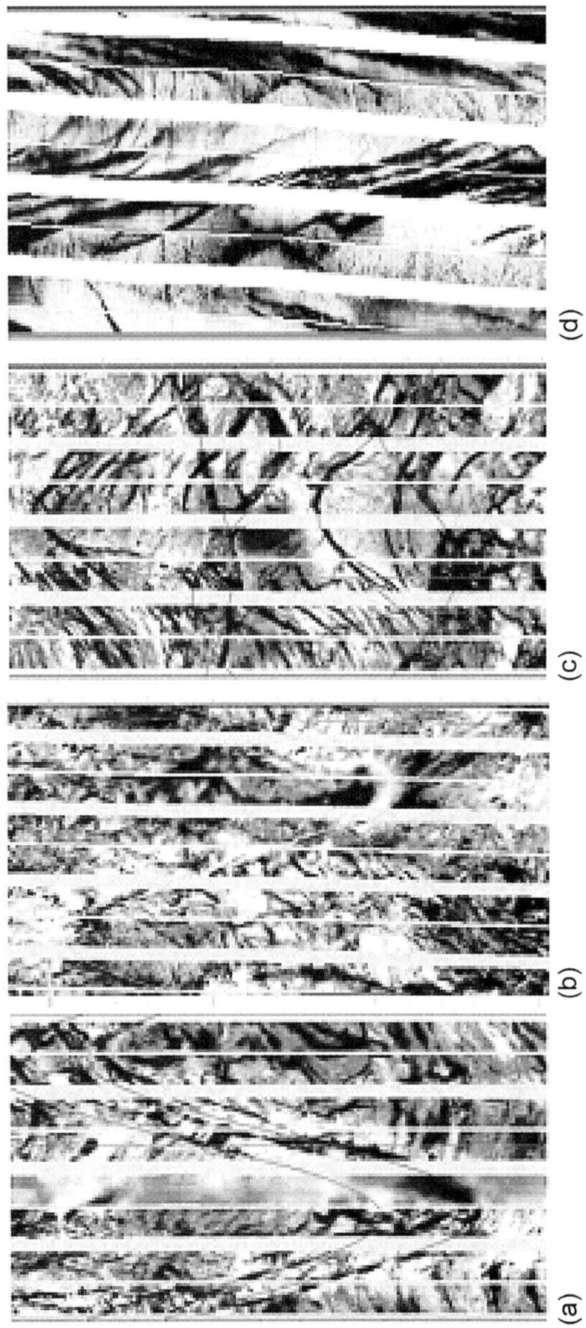

FIGURE 6.3

FMI imaging log identification model for different fractures. (a) Highly conductive fractures. (b) High-resistivity fractures. (c) Micro-fractures. (d) Induced fractures.

Table 6.1 FMI Imaging Log Response Characteristics of Different Fractures

Fracture Type	Major Geological Origin	Imaging Logging Characteristics			
		Color	*Shape*	*Grouping Characteristics*	*Extended Distance*
Highly conductive fracture	Structural fracture, columnar joint, dissolution fracture, sutured fracture	Black	Sine-curve type	Multigroup	Long
High-resistivity fracture		White	Sine-curve type	Multigroup	Long
Microfracture	Blast fracture, contraction fracture, weathered fracture	Black	Irregular	Chaotic, nongroup	Short
Induced fracture	Stress-released fracture, hydraulic fracture	Black	180-degree symmetrical feathered and chevron shapes	Single group	Long

1) Highly conductive fractures

Highly conductive fractures consist of mostly open structural fractures, columnar joints, and dissolution fractures. The direction or orientation of these fractures depends on the regional tectonic stress field, and pores and cavities are developed in as bead strings along fracture faces. In an FMI imaging chart, they show black sine curves or narrow stripes with multigroup characteristics and long extensions (Figure 6.3a, Table 6.1).

2) High-resistivity fractures

These fractures refer to early structural fractures and columnar joints, which are closed or filled with high-resistivity materials. In an imaging chart, they show white sine curves or narrow stripes with multigroup characteristics and long extensions (see Figure 6.3b, Table 6.1).

3) Microfractures

This group comprises blast fractures, contraction fractures, and weathering fractures with a limited extension, as well as structural fractures that are locally filled or closed. They serve primarily as connections between pores in the reservoir. In an FMI imaging chart, microfractures show black irregular lines or narrow stripes with a short extension, without any typical sine-curve or group characteristics. Therefore, these fractures usually do not fit the sine-curve theory (see Figure 6.3c, Table 6.1).

4) Drilling-induced fractures

These fractures are non-natural fractures generated during drilling and related to crustal stress. The fracture orientation follows the direction of the maximum horizontal principal stress. In the imaging chart, they show feathered or chevron-shaped black symmetrical curves (180 degrees) appearing along the sidewall with single-group characteristics and a long extension (see Figure 6.3d, Table 6.1).

3. Fracture identification through image logging

Natural fractures are classified and picked up based on the identification model, the result of lithology identification, and the discrimination between true and false fractures. Induced fractures are identified based on an analysis of regional structural stress. The characteristics of various fractures in wells are ascertained to establish a basis for determining fracture shapes, spatial distribution, and fracture parameters.

1) Fracture identification based on lithological constraints and identification

Different types of major fractures are developed in different lithologies. Structural fractures and dissolution fractures are developed in all lithologies; contraction fractures occur mainly in effusive volcanic rocks; blast fractures occur in explosive volcanic rocks and near volcanic conduit facies; and sutured fractures occur in the hot base surge subfacies of explosive facies. Therefore, lithological constraints for fracture identification are established through lithological identification.

2) Identification of natural fractures through image logging

Natural fractures can be ultimately identified according to petrogenetic differences and the response characteristics of image logging and by determining the differences between natural fractures and geological features (e.g., layer interfaces, fault planes, shale stripes, bedding planes, rhyolitic structures, and hydraulic fractures):

1. *Layer interface.* This refers to the interface between two sets of formations. In an image logging chart, layer interfaces are marked by a group of nearly parallel conductivity anomalies. Compared with fractures, layer interfaces have the characteristic of uniform, continuous, and integral distribution, and they do not intersect each other. Usually there is a color transition between layer interfaces and formations (Figure 6.4a).
2. *Shale stripes.* Unlike fractures, the high conductivity anomalies of shale stripes are generally parallel, regular, and have clear boundaries. Even after severe tectonic deformation, the change in their widths will not be significant (Figure 6.4b).
3. *Beddings.* Beddings appear in groups within a certain range. In general, low-angle beddings are parallel to each other and demonstrate extremely good regularity between lines (Figure 6.4c).
4. *Faults.* There is displacement in the formations at both sides of fault plane, but there is no displacement in the fracture system (Figure 6.4d).
5. *Rhyolitic structure.* In an image logging chart, rhyolitic structures are deformed sine curve in shape with a visible disturbance trace, but the line is smooth. The long axis of elongated vesicles is aligned with the direction of magmatic flow (Figure 6.4e).
6. *Perlitic structure.* As the name implies, this is a unique structure for perlite and is marked by arcuate or concentric cracks; these cracks divide acidic volcanic glass into many small balls. In imaging logs, perlitic structure shows dark-colored, arcuate, or concentric lines under a high-resistivity background. Therefore, it can be easily distinguished from fracture zones (Figure 6.4f).
7. *Heavy mud hydraulic fractures.* These are the hydraulic fractures caused by the imbalance between drilling fluid and structural stress, resembling stress-release fractures, and generally developed in two to four groups (Figure 6.4g).

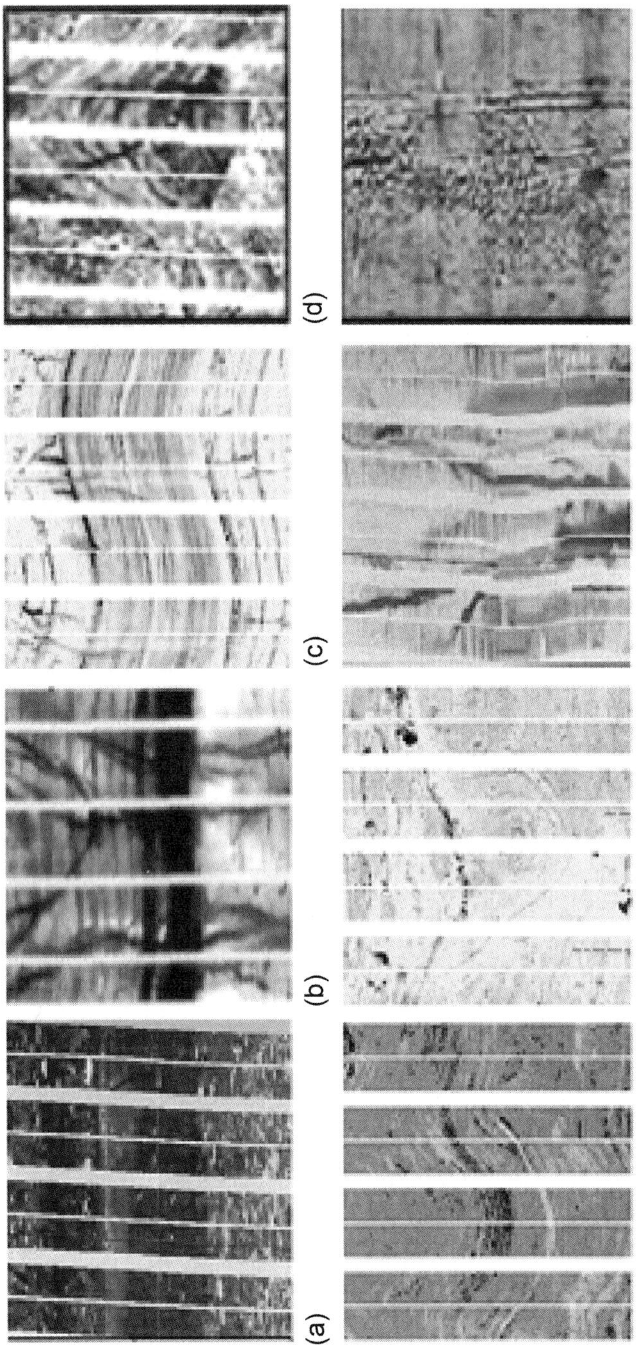

FIGURE 6.4

Diagram of FMI imaging log characteristics of different types of non-natural fractures. (a) Layer interface. (b) Shale strip. (c) Bedding. (d) Fault. (e) Rhyolitic structure. (f) Perlitic structure. (g) Heavy mud hydraulic fracture. (h) Scratches by drilling tools.

8. *Scratches of drilling tools.* These scratches are caused by the friction between drilling tools and the well sidewall, without sine-curve characteristics in the imaging chart, and their cutting depth is shallow (Figure 6.4h).

3) Classified pickup of natural fractures and induced fractures

On the basis of a clear understanding of the differences between natural fractures/induced fractures and other similar geological features, natural fractures and induced fractures are picked up in the classification process using special image logging interpretation software and the human-machine interaction approach. This lays the foundation for fracture parameter calculation and the evaluation of fracture development, fracture effectiveness, and fracture orientation.

6.2.2 Identification of fractures through conventional logging

Natural and induced fractures with different dip angles are identified by qualitative log analysis and semiquantitative fracture index calculation according to the abnormal response characteristics of fractures in conventional logging curves.

1. Conventional log response characteristics of fractures in volcanic rocks

The radioactivity, physical properties, and conductivity of a rock formation are measured through conventional logging to indirectly reveal the characteristics of fracture development in volcanic rocks. Conventional log data can be calibrated by descriptive logs of cores and analysis of the logging response characteristics of cored intervals, which will help reveal the conventional logging characteristics of fractures in volcanic rocks (Table 6.2) and establish the typical fracture log identification model.

1) Dual laterolog

A dual laterolog provides resistivity curves that reflect undisturbed formations and fracture-invaded zones.

In deep and shallow laterolog resistivity curves, open fractures result in a decrease in resistivity values and a difference between the deep and shallow laterolog resistivity values. Thus, high-angle fractures (dip angle > 75 degrees) have a positive separation (i.e., RLLD > RLLS (recorded laterolog shallow)), low-angle fractures (< 60 degrees) have a negative separation, and oblique fractures (60 ~ 75 degrees) have a small separation or no separation (see Table 6.2) [5,6].

In volcanic reservoirs, high-angle structural fractures are predominant. In dual laterolog curves, fracture-bearing intervals are characterized by *decreased resistivity value, and positive separation between deep and shallow laterologs*. Figure 6.5 is a crossplot of the surface porosity of fractures in cored intervals, which shows a difference between deep and shallow laterolog resistivity values [(RLLD−RLLS)/RLLD] in the XX gas field. This figure shows that the more developed the fractures (the greater the surface porosity), the smaller the RLLD, and the larger the RLLD−RLLS values. This implies that high-angle fractures are responsible for the decrease in resistivity and the increase in positive separation between deep and shallow laterolog resistivity values.

Table 6.2 Conventional Log Response Characteristics of Fractures in Volcanic Rocks

Fracture Type		Dual Laterolog				Acoustic					
		Log Value	*Deep/ Shallow Laterolog Separation*	*Shape*	*Smoothness*	*P-wave Interval Transit Time*	*Velocity Ratio of P wave to S wave*	*Density*	*Neutron*	*Gamma Spectrometry*	*Caliper*
Open natural fracture	High-angle fracture	Medium high	Large positive separation	Box shape	Microserrated	No response	Increase	Decrease	Increase	High uranium anomaly	Asymmetrical hole enlargement
	Oblique fracture	Medium	No separation	Bell shape, funnel shape	Microserrated	Increase, cycle wave jump					
	Low-angle fracture	Low	Negative separation	Finger shape, bayonet shape	Serrated						
	Network fracture	Low	Related to the occurrence of dominant fractures	Box shape	Serrated						
Induced fracture		High	Small positive separation	Bell shape, box shape	Smooth	No response					

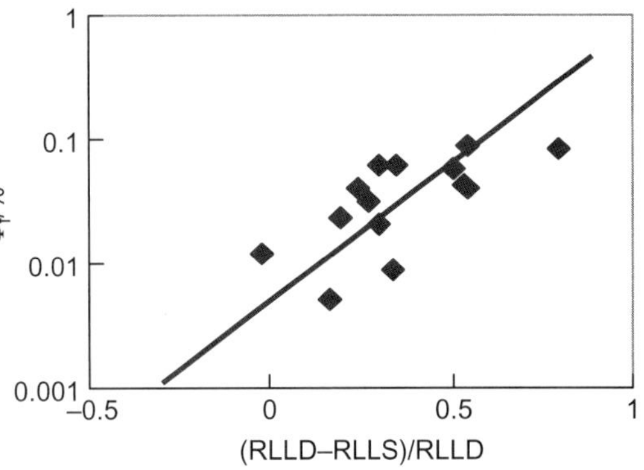

FIGURE 6.5

Crossplot of fracture surface porosity ϕ_f− dual laterolog resistivity [(RLLD−RLLS)/RLLD].

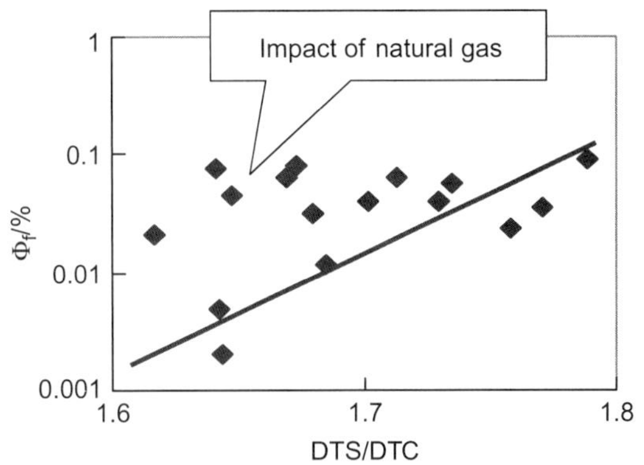

FIGURE 6.6

Crossplot of fracture surface porosity ϕ_i− P-S ratio of acoustic travel time (DTS/DTC).

2) Acoustic log

There is no response for high-angle fractures in P-wave interval transit time curves. Instead, the response increases in low-angle fracture and network fracture intervals, and even "jumping cyclic waves" occur in these intervals. At the same time, fractures cause an increase in the acoustic travel time of S waves and Stoneley waves but a decrease in their amplitude. The amount of change is greater than that of P waves [5,7], and the velocity ratio ($\Delta t_s/\Delta t_c$) of P wave to S wave increases (see Table 6.2).

Figure 6.6 is the crossplot of surface porosity of fractures in the cored interval and the P-S ratio of acoustic travel time in the XX gas field. This figure demonstrates that the P-S ratio of acoustic travel time increases as the surface porosity of fractures increases, but it is affected by fluid properties.

3) Compensated density and compensated neutron logs

A compensated density log reflects the total porosity in rocks, whereas a neutron log detects the hydrogen content in formations. Fractured intervals have reduced density and increased neutron log values, but natural gas causes a decrease in both density and neutron values (see Table 6.2). Under identical conditions, therefore, the level of fracture development is determined with the direction of relative change in density and neutron log values [5,7].

4) Gamma spectrometry log

Fluid is more active in fractures than in matrix pores, which facilitates abnormal uranium enrichment and intensifies the radioactivity in formations [5,7]. The degree of fracture development can be determined by analyzing the uranium enrichment characteristics (see Table 6.2).

5) Caliper log

In dual- (or multiple-) caliper log curves, well radius enlargement and ROP change occur when the caliper detects an open fracture. The more developed the fracture, the more significant the caliper response. Therefore, an asymmetrical hole enlargement is the primary basis for fracture identification in dual-caliper (or multiple-caliper) log curves (see Table 6.2).

2. Conventional logging identification model for fractures in volcanic rocks

A conventional logging identification model for fractures in volcanic reservoirs is established through an analysis of log response characteristics, after the conventional log is calibrated by core description logs and imaging logging identification [12].

1) High-angle fractures

In conventional log curves, high-angle fractures commonly have a smooth box shape and are characterized also by moderate amplitude values and a large separation of deep and shallow laterologs, small interval transit time, high density, and significant neutron log values.

The brecciated lava reservoir interval in 3705 to 3744 m of Well XX8 is a typical example (Figure 6.7). Core description and imaging log identification information indicate that this interval has the characteristics of abundant vesicles and dissolution pores, good matrix properties, and its fractures are mostly high-angle structural fractures. Its conventional log curve shows a smooth box shape. Its average value of deep and shallow laterolog resistivity is 38 Ω·m and 21 Ω·m, respectively. The acoustic transit time is 71 μs/ft on average, with neutron porosity of 14% on average and rock density 2.32 g/cm^3 on average. As the gas test in the studied interval shows, its natural production rate of natural gas is up to 23×10^4 m^3/d, its producing pressure differential is 8 MPa, and no water is produced. A short-period production testing for 30 days shows that the gas production rate was maintained at 15×10^4 m^3/d, the water production rate increased from 6.3 m^3/d to 98 m^3/d, and the water-gas ratio stabilized at 6.6 to 6.8 m^3/10^4 m^3, implying that the water from the lower water zone has risen up into the gas zone along high-angle fractures.

2) Low-angle fracture

Low-angle fractures have a bayonet shape in conventional log curves, with low amplitudes in resistivity curves, a large negative separation between deep and shallow laterologs, large interval transit time, small density, and large neutron values.

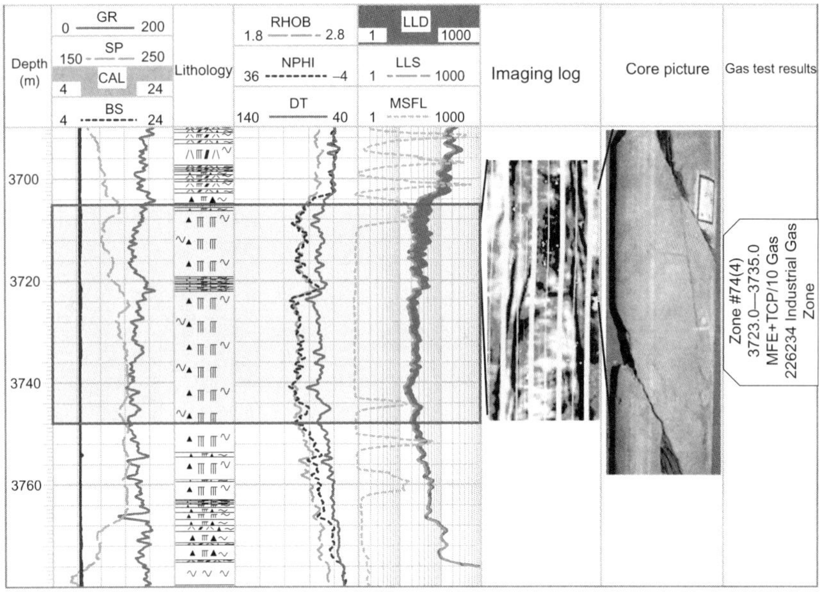

FIGURE 6.7

Log characteristics of high-angle fractures (Well XX8, brecciated lava).

The tuff-bearing volcanic brecciated reservoir in 3863 to 3866 m of Well XX5 is a typical example (Figure 6.8). Core description and imaging log interpretation indicate that a group of low-angle fractures (3 ~ 4) is developed in this interval. Conventional log curves show a bayonet shape. Their average values of deep and shallow laterolog resistivity are 44 Ω·m and 50 Ω·m, respectively, which are lower than those of the surrounding rocks. A negative separation is also observed. Their interval transit time is 64 μs/ft on average, and their neutron porosity has an average of 5%, which is slightly higher than that of the surrounding rocks. Their rock density has an average of 2.58 g/cm^3, which is slightly lower than that of the surrounding rocks.

3) Network (netlike) fractures

These are formed by intercrossing fractures with various occurrence patterns, indicating a high degree of fracture development. In conventional log curves, the intervals with network fractures have serrated and box shapes and a low-amplitude resistivity curve. Whether deep and shallow laterologs show positive or negative separation depends on the occurrence of dominant fractures. Their interval transit time is long, neutron porosity is high, and the density is small.

The brecciated lava reservoir interval in 4108 to 4122 m of Well XX9 is a typical example (Figure 6.9). The imaging log indicates that this interval is rich in predominantly high-angle network fractures. Conventional log curves are serrated or box shaped. Their average value of deep and shallow laterolog resistivity is 230 Ω·m and 95 Ω·m, respectively, and a positive separation is observed. Their interval transit time has an average of 65 μs/ft, and their neutron porosity is 4% on average—higher than that of the surrounding rocks. Their rock density is 2.52 g/cm^3 on average—lower than that of the

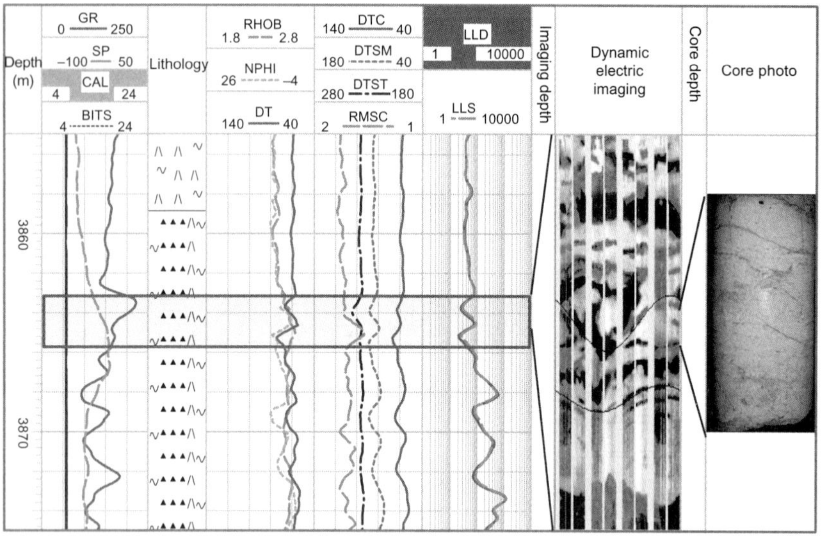

FIGURE 6.8

Log characteristics of a low-angle fracture (Well XX5, tuff-bearing volcanic breccia).

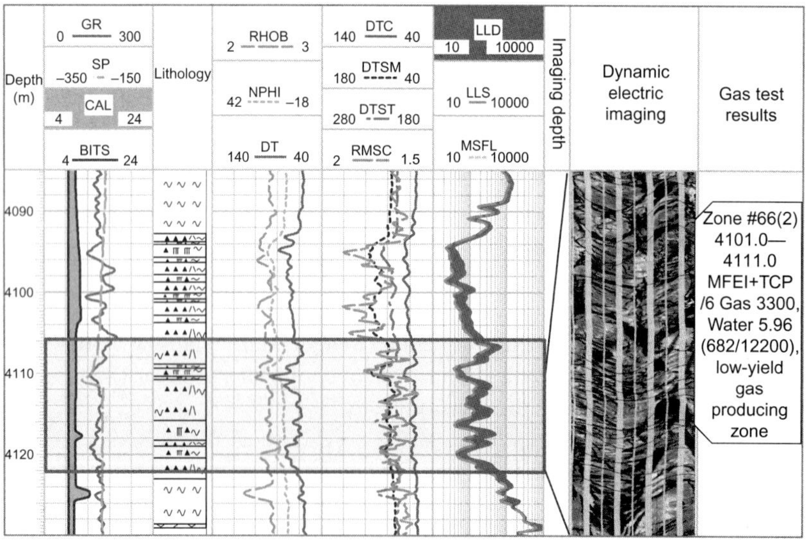

FIGURE 6.9

Log characteristics of a network (netlike) fracture (Well XX9, brecciated lava).

surrounding rocks. As shown by the gas test in this interval, the natural production rate of natural gas is 0.33 10^4 m^3/d, and the water production rate is up to 5.96 m^3/d.

4) Stress-release fractures

Stress-release fractures are developed mainly in tight intervals where tectonic stress is not easily released. In conventional log curves, they have a smooth box shape. They are characterized by a high amplitude resistivity curve, a small positive separation or no separation in deep and shallow laterolog resistivity, a short interval transit time, low neutron porosity, and high rock density values.

The dacite formation in 3980 to 3985 m of Well XX401 is a typical example (Figure 6.10). The imaging log indicates that this interval has abundant 180-degree symmetrical stress-release fractures. Its conventional log curve has a smooth bell shape. The average value of deep and shallow laterolog resistivity is 1452 Ω·m and 1244 Ω·m, respectively, and a small positive separation is observed, albeit larger than that of the surrounding rocks. Their interval transit time has an average of 52 μs/ft, and their neutron porosity is 3% on average—lower than that of the surrounding rocks. Their average rock density is 2.57 g/cm^3—higher than that of the surrounding rocks.

3. Identification of fractures through semiquantitative calculation of fracture index

According to the response characteristics of fractured intervals in conventional log curves, fractures are further identified by calculating indicative curves that reflect the degree of fracture development, using the curve combination technique.

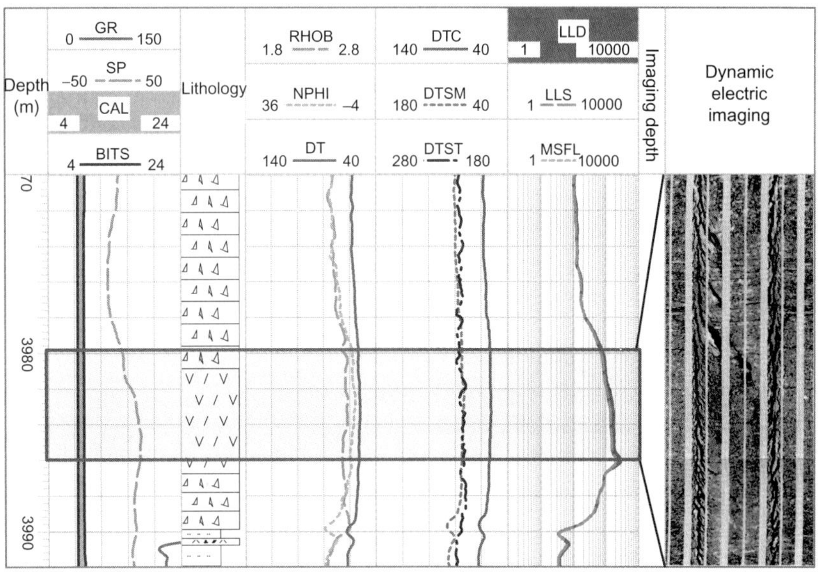

FIGURE 6.10

Log characteristics of stress-release fractures (Well XX401, dacite)

1) Calculation methods

The calculation methods commonly used are as follows:

1. *Secondary porosity method.* In reservoirs dominated by high-angle fractures, the variation of interval transit time is small, whereas the rock density and neutron porosity are high. An indicator function that reflects the degree of high-angle fracture development is thus established:

$$\mathrm{FPR2} = \phi_{\mathrm{ND}} - \phi_{\mathrm{S}} \tag{6.2.1}$$

Where

ϕ_{ND} — neutron and density crossplot porosity (%);
ϕ_{S} — acoustic porosity (%).

2. *Apparent pore structure index method.* Fractures improve the flow pathways and electric conduction paths in reservoirs, and they cause the pore structure index to decrease. An indicator function that reflects the degree of overall fracture development is thus established:

$$\mathrm{m}^* = (\log Rw - \log Rt)/\log\phi \tag{6.2.2}$$

Where

Rw — formation water resistivity (Ω·m);
Rt — original formation resistivity (Ω·m);
ϕ — matrix porosity (%).

3. *Deep and shallow laterolog amplitude difference method.* High-angle fractures cause an increase in the positive separation between deep and shallow laterolog resistivity. An indicator function that reflects high-angle fracture development degree is thus established:

$$\mathrm{FLPL} = (\mathrm{RLLD} - \mathrm{RLLS})/\mathrm{RLLD} \tag{6.2.3}$$

Where

RLLD — deep laterolog resistivity (Ω·m);
RLLS — shallow laterolog resistivity (Ω·m).

4. *Uranium anomaly index method.* Fractures provide the migration pathway for various fluids, which become more active than in the matrix. Formations with abundant fractures have higher enrichment of radioactive uranium. A fracture identification indicator function based on gamma spectrometry is established:

$$\mathrm{FPU} = U/Th \tag{6.2.4}$$

Where

U — gamma spectrometry uranium curve (ppm);
Th — gamma spectrometry thorium curve (ppm).

5. *Fracture probability function method.* A comprehensive method of fracture probability function is established based on the preceding methods. This method will further amplify the log response of fractures and improve the congruence rate of fracture identification. The fracture probability function calculation formula is

$$FIDX = \frac{W1 \cdot XFPR2 + W2 \cdot X_m + W3 \cdot XFPU + W4 \cdot XFLPL}{W1 + W2 + W3 + W4} \tag{6.2.5}$$

Where

XFPR2 — the normalized curve of FPR2;
Xm — the normalized curve of m*;
XFPU — the normalized curve of FPU;
XFLPL — the normalized curve of FLPL;
W1–W4—the weighted coefficients for XFPR2, Xm, XFPU, and XFLPL.

6. *Fracture development exponential function method.* To reduce artificial errors, a fracture development exponential function curve that semiquantitatively evaluates fracture development is further established:

$$FID2 = AFD2 \cdot FPR2/m^* \tag{6.2.6}$$

Where

AFD2—adjusting parameter, which is 100 herein.

2) Calculation results

A fracture indicative curve is established using the previously described methods. The relationship between the indicative curve and fracture parameters of FMI imaging logs is then determined. The adaptability of each method is analyzed here.

Calculation of fracture indicative curve using conventional logs. Secondary porosity (FPR2), the apparent pore structure index (m*), deep and shallow laterolog amplitude difference (FLPL), uranium anomaly index (FPU), fracture probability function (FIDX), and fracture development index (FID2) are calculated based on conventional logs using the preceding methods. Traces 8 to 13 in Figure 6.11 show that FPR2, m*, FIDX, and FID2 are generally consistent with the fracture development characteristics determined through image logging, and the correlation between FPU, FLPL, and imaging log interpretation results is poor. This implies a poor applicability or adaptability of the deep and shallow laterolog amplitude difference method and the uranium anomaly index method to the identification of fractures in volcanic reservoirs.

Applicability and adaptability. The applicability and adaptability of each method is analyzed by establishing the relationship between indicative curves and imaging log fracture parameters.

Based on calculated fracture indicative curves and the expression between each calculated curve and fracture parameters of the FMI imaging log, the adaptability of individual evaluation methods is analyzed.

In Well XX14, the thickness of volcanic rocks is approximately 400 m. Rhyolite, rhyolitic brecciated lava, rhyolitic tuffaceous lava, and rhyolitic volcanic breccia are encountered during drilling. Lithofacies include effusive facies, explosive facies, volcanic conduit facies, extrusive facies, and volcanic sedimentary facies. Therefore, this well can be regarded as highly representative of various lithologies. The calculation result of Well XX14 and analysis (Figure 6.12, Table 6.3) shows the following:

1. *The volcanic secondary porosity method.* FPR2 increases with fracture development and correlates fairly well with the hydrodynamic width of fractures determined via imaging logs (Figure 6.12c), showing a good applicability and adaptability for this method.
2. *The apparent pore structure index method.* m* decreases with the degree of fracture development and correlates fairly well with the hydrodynamic width of fractures determined via imaging logs (Figure 6.12d), also indicating a good applicability of this method.

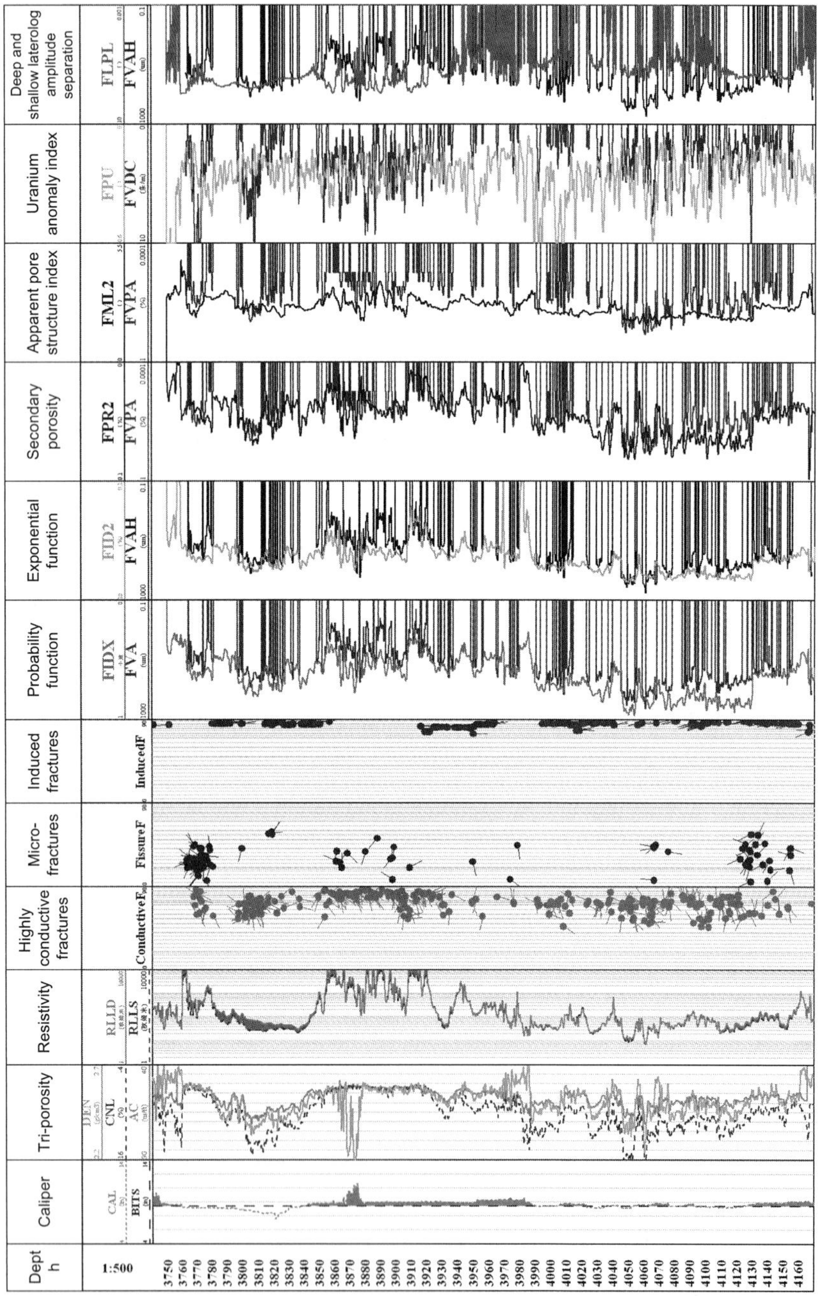

FIGURE 6.11

Correlation between fracture interpretation results of conventional logs and FMI imaging logs (Well XX14).

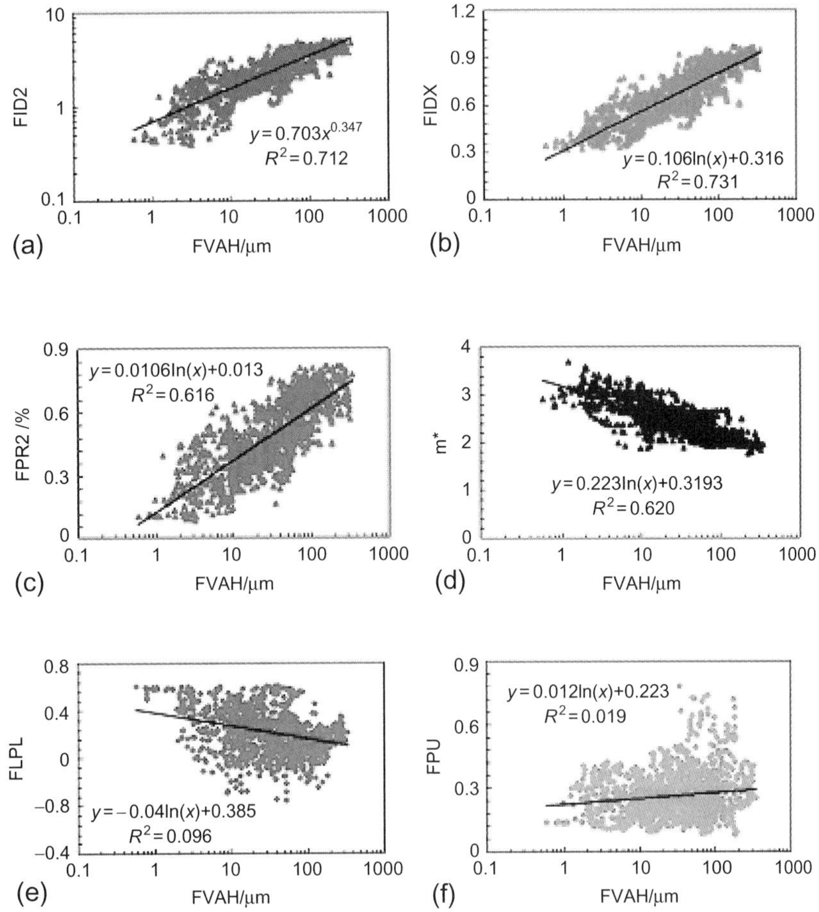

FIGURE 6.12

Crossplots between fracture indicative curves via conventional logs and fracture hydrodynamic widths via FMI imaging logs (Well XX14). (a) FID2-FVAH crossplot. (b) FIDX-FVAH crossplot. (c) FPR2-FVAH crossplot. (d) m*-FVAH crossplot. (e) FLPL-FVAH crossplot. (f) FPU-FVAH crossplot.

3. *The deep and shallow laterolog difference method.* FLPL increases with high-angle fracture development and correlates poorly with the hydrodynamic width of fractures determined on the basis imaging logs (Figure 6.12e); this method is highly affected by the variation of lithology, induced fractures, and fluid type, indicating poor applicability.
4. *The uranium anomaly index method.* FPU increases with fracture development and correlates poorly with fracture hydrodynamic width determined via FMI imaging logs (Figure 6.12f); this method is highly affected by the variation of lithology and sedimentary environment and thus is of poor applicability.

Table 6.3 Table of Adaptability Analysis of the Calculation Methods of Fracture Indicative Curves by Conventional Logs

Method	Analysis of Relationship with Fracture Hydrodynamic Width via Imaging Logs		Correlated Thickness	Lithology	Lithofacies
	Expressions	*Rt*			
Secondary porosity method	$y=0.01\ln(x)+0.013$	0.62	400 m	Rhyolite, breccia lava, tuffaceous lava, volcanic breccia	Effusive facies, explosive facies, volcanic conduit facies, extrusive facies
Pore structure index method	$y=0.22\ln(x)+3.193$	0.62			
Deep and shallow laterolog difference	$y=0.04\ln(x)+0.385$	0.1			
Uranium anomaly index method	$y=0.012\ln(x)+0.223$	0.02			
Fracture probability function method	$y=0.105\ln(x)+0.316$	0.71			
Fracture development index method	$y=0.703x0.346$	0.73			

5. *The fracture probability function method.* FIDX increases with fracture development and correlates fairly well with fracture hydrodynamic width determined via imaging logs (Figure 6.12a); it shows somewhat improved results compared with single methods and offers good applicability and adaptability.
6. *The fracture development index method.* FID2 increases with fracture development and correlates fairly well with the fracture hydrodynamic width determined from imaging logs (Figure 6.12b); it has been improved somehow compared with single methods and offers good adaptability.

Among these six evaluation techniques, the methods based on deep and shallow laterolog difference and uranium anomaly index are shown to have a poor applicability or adaptability due to the impact of such factors as lithological change, induced fractures and fluid type variation; they are mainly suitable for qualitative analysis of volcanic reservoirs with stable lithology and significant difference of fluid activity. The other four methods offer a good applicability and adaptability and can be used to semi-quantitatively evaluate the degree of fracture development.

Table 6.4 Parameter Characteristics of Fracture Indicative Curve Derived from Conventional Logs (XX Gas Field)

Indicative Curve	Distribution Range	Average Value
FPR2/%	0.009–0.082	0.048
m*	1.2–3.8	2.3
FIDX	0.32–0.97	0.67
FID2	0.3–5.0	2.5

3) Parameter characteristics of fracture indicative curve in volcanic rocks

In the XX gas field, the calculation results of several wells have been statistically analyzed, and the results can be summarized as follows (Table 6.4): secondary porosity (FPR2) is between 0.009% to 0.082% with an average value of 0.048%, which means that the secondary action in volcanic reservoirs is not strong; the apparent pore structure index m* ranges between 1.2 to 3.8 with an average value of 2.3; the fracture probability function (FIDX) ranges between 0.32 to 0.97 with an average value of 0.67; and the fracture development index (FID2) ranges between 0.3 to 5 with an average value of 2.5.

6.2.3 Comprehensive identification of volcanic fractures in wells

Comprehensive studies involving geology, imaging logs, and conventional logs are carried out by integrating core description for the cored intervals, calibrating imaging log with cores, and calibrating conventional logs with cores and imaging logs. Fractures in volcanic reservoirs are identified in single wells according to the log response characteristics and fracture identification models.

1. Describing fractures in cored intervals to calibrate imaging log data

Through fracture description for cored intervals, basic characteristics of fractures in volcanic reservoirs such as the type, configuration, and size are revealed. This helps calculate the fracture parameters, construct the volcanic reservoir fracture concept model (Figure 6.13a), and calibrate the imaging log (Figure 6.13b).

2. Identifying fractures with FMI imaging and integrated with core data to calibrate conventional logs

On the basis of the identification model for FMI imaging logs, fractures are picked up by type, piece by piece in noncored intervals, and conventional logs are jointly calibrated by core and image logging results (Figure 6.5, Figure 6.6, and Figure 6.14). As shown in Figure 6.15, traces 4 to 6 are the tadpole plots of highly conductive fractures, microfractures, and induced fractures, respectively. The figure shows that mainly oblique microfractures are developed in Well XX301, with upright induced fractures developed in its upper part.

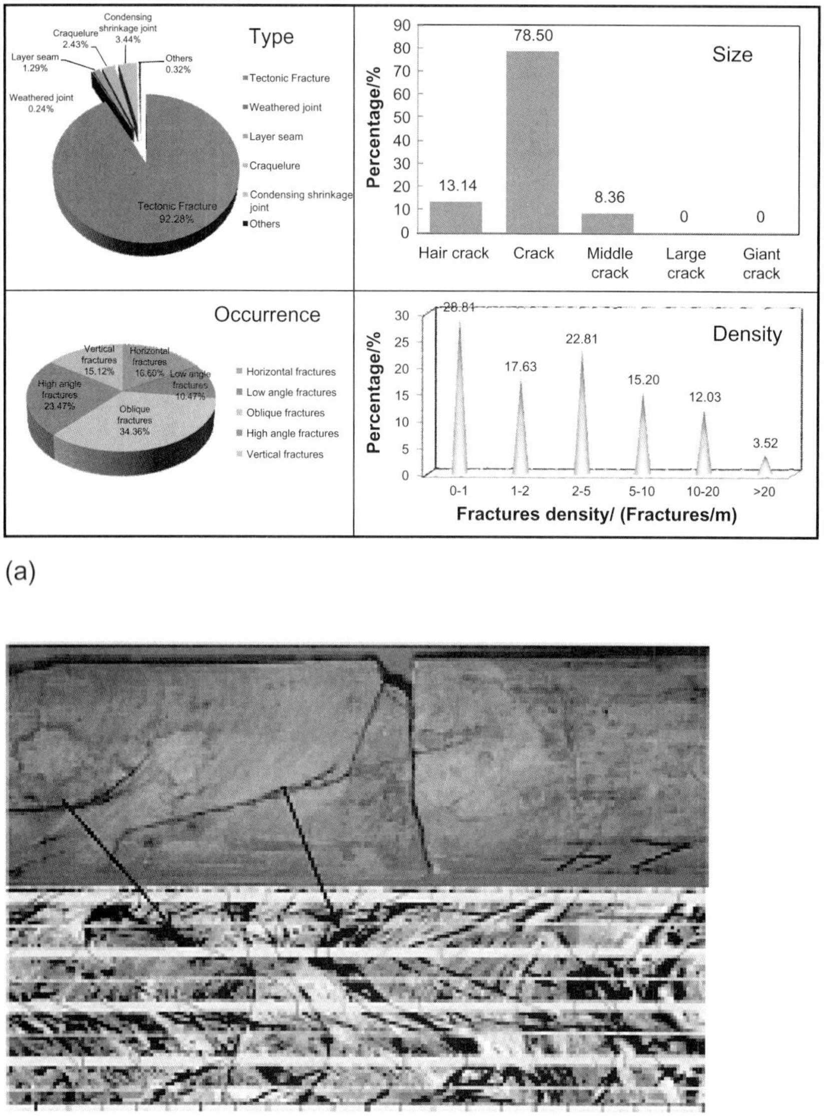

FIGURE 6.13

Research results of fractures in cored intervals. (a) Fracture description (DD gas field). (b) Core calibrating imaging log (Well XX21).

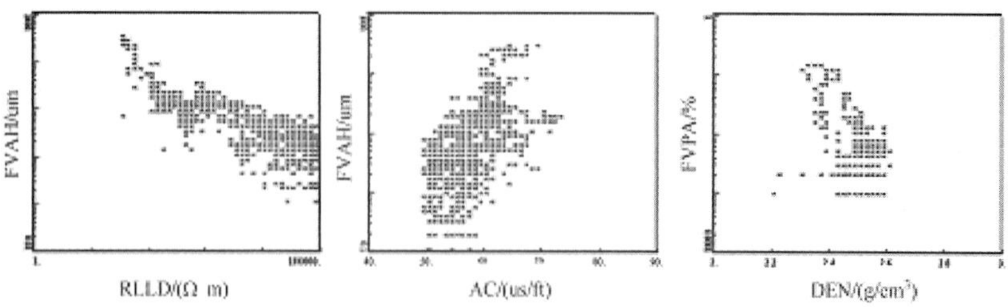

FIGURE 6.14

Conventional logs calibrated by imaging log fracture parameters (Well XX8).

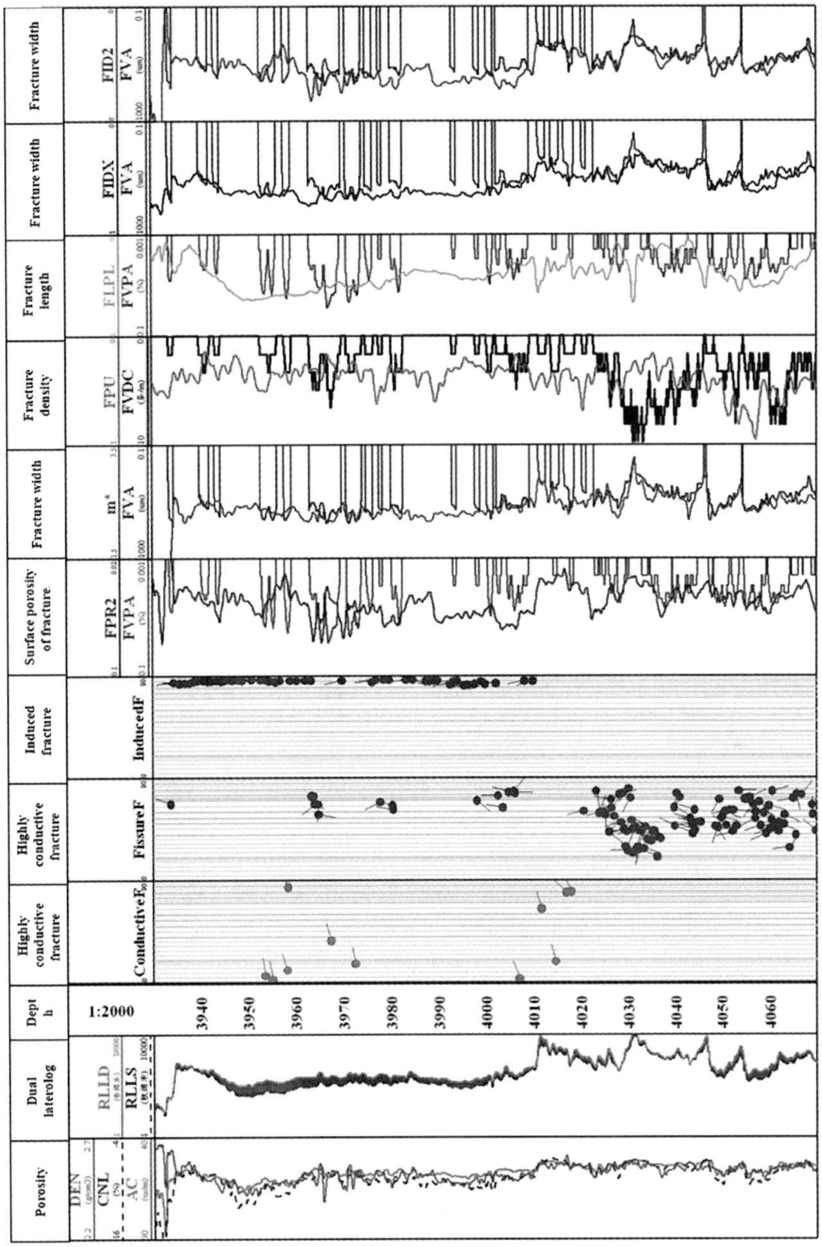

FIGURE 6.15

Diagram of fracture identification results (Well XX301).

3. Qualitative identification of different fracture occurrences based on conventional log response characteristics

The high-angle, low-angle, network, and induced fractures are qualitatively identified with conventional log data, in accordance with the log response characteristics and identification model of induced fractures and fractures of different occurrences. In Well XX301 (Figure 6.15), the upper part (3950 to 4020 m) of the log curve shows mid-low amplitude values, a box shape, and good smoothness, suggesting mostly high-angle fractures; the lower part of the log curve shows high amplitude values and box and serrated shapes, pointing to mostly network fractures. The result of conventional log identification is largely consistent with the tadpole plot results using FMI imaging logs (tracks 4 to 6).

4. Semiquantitative evaluation of the degree of fracture development based on a fracture indicative curve derived from conventional logs

The degree of fracture development is semiquantitatively evaluated on the basis of the fracture indicative curve derived from conventional log data.

In Figure 6.15, traces 7 to 12 are the integrated curve of FPR2, m*, FPU, FLPL, FIDX, and FID2. Compared with the imaging log identification results, there is a general agreement between m* and fracture width. FPR2 and FLPL are basically consistent with the surface porosity of fractures; FIDX and FID2 are in agreement with the fracture width, whereas the effectiveness of FPU is poor. Therefore, the degree of fracture development can be semiquantitatively evaluated with conventional logs [8].

6.3 INTERPRETATION AND EVALUATION OF VOLCANIC RESERVOIR FRACTURE PARAMETERS

Fracture identification, as discussed in the previous section, involves interpreting fracture parameters, evaluating the degree of fracture development, fracture orientation, and fracture effectiveness, and determining the development characteristics of fractures in volcanic reservoirs, all of which will serve as the basis for fracture prediction.

6.3.1 Interpretation of fracture parameters

This mainly includes the interpretation of fracture density, fracture width, fracture porosity, and fracture permeability.

1. Calculation methods

1) Fracture density and fracture length

In this study, fracture density (FVDC) refers to the linear density, which is defined as the number of fractures per unit length (1 m) of the sidewall (unit: 1/m). Fracture density parameters are generally calculated from the fracture pickup in the acoustic or electric imaging chart. Other methods for fracture density calculation carry various degrees of uncertainty [9].

Fracture length (FVTL) is defined as the total fracture length per unit area (m^2) of the sidewall (unit: m/m^2 or 1/m). It is statistically determined by special software based on fracture identification.

2) Fracture width

Fracture width refers to the opening width of a fracture along the normal direction. It can be determined by acoustic/electric imaging and conventional logs.

The calculation method based on FMI imaging logs. Fracture width is generally smaller than the resolution of microresistivity image logging, which cannot be read directly. Instead, it has to be calculated indirectly on the basis of fracture color representation in the imaging log chart [10], and it is calculated using the following formula:

$$\varepsilon = aAR_{xo}^{1-b}R_m^b \quad (6.3.1)$$

Where

ε — width of a single fracture (mm);
a, b — constants related to instruments, wherein b is close to zero;
A — the area of conductivity anomaly (mm^2) caused by fracture;
R_{xo}, R_m — resistivity values of invaded zone and drilling fluid (Ω·m).

The width of fractures is different in a different part of the reservoir. Fracture width can be determined by one of the two methods according to the result of point-by-point calculation:

1. The average fracture width (FVA), which is defined as the average value of all the fracture trajectory widths per unit well interval (1 m)
2. The average hydrodynamic width of the fracture (FVAH), which is defined as the cube and extraction of cubic root of fracture trajectory widths per unit well interval (1 m), which can be regarded as a fitting for the fracture hydrodynamic effect

The calculation method based on conventional logs. Before acoustic or electric imaging log was invented, Sibbit and colleagues provided a method for fracture width calculation with a dual laterolog resistivity curve [6]. This technique first requires the judgment of fracture occurrence, followed by the following parameter calculations, respectively:

$$\text{High} - \text{angle fractures}: \varepsilon = 2.50 \times 10^3 \times (C_s - C_d)/C_m \quad (6.3.2)$$

$$\text{Low} - \text{angle fractures}: \varepsilon = 8.33 \times 10^2 \times (C_d - C_b)/C_m \quad (6.3.3)$$

Network fractures: first determine the width of high-angle and low-angle fractures, respectively, and then sum them up,

Where

ε — fracture width (μm);
C_d, C_s, C_m — conductivity values of deep laterolog, shallow laterolog and mud (mS/m);
C_b— matrix conductivity (mS/m) can be read from tight intervals near the interpreted layer.

3) Fracture porosity

Fracture porosity is calculated with an acoustic/electric imaging log and a conventional dual laterolog:

1. *The calculation method based on the FMI imaging log.* For FMI imaging logs, fracture porosity (FVPA) is defined as the apparent opening area of fractures per unit area of coverage in imaging chart (unit: cm^2/cm^2), calculated using the following formula:

$$\text{FVPA} = \text{FVA} \cdot \text{FVTL} \times 400/(\pi \cdot \text{CAL}^2) \quad (6.3.4)$$

Where

FVPA — fracture porosity (cm^2/cm^2);
FVA — average fracture width (cm);
FVTL — fracture length (m);
CAL — well radius (cm).

2. *The calculation method based on conventional logs*. Fracture porosity is calculated with a dual laterolog curve using the following formulae:

$$\text{Water layer}: \phi_{\mathrm{f}} = \sqrt[m_{\mathrm{f}}]{(C_s/K_{\mathrm{r}} - C_d)/(C_m - C_{\mathrm{w}})} \tag{6.3.5}$$

$$\text{Oil/gas layer}: \phi_{\mathrm{f}} = \sqrt[m_{\mathrm{f}}]{(C_s/K_{\mathrm{r}} - C_d)/C_m} \tag{6.3.6}$$

Where

ϕ_{f}— fracture porosity (fraction);
C_{w}— formation water conductivity (mS/m);
m_{f}— fracture pore structure index, which is related to width variations and the bending of a fracture, typical range 1 to 1.3;
K_{r}— fracture's influence coefficient on shallow laterolog resistivity, which is related to fracture occurrence and development degree; typical range 1.1 to 1.3; values of 1.3 to 1.5 for reservoirs with poor fracture development.

4) Fracture permeability

Fracture permeability is divided into two types. The first type, referred to as *intrinsic fracture permeability*, only takes into account the conductivity of fracture for fluid flow, ignoring the condition of surrounding rocks. It is mainly affected by the fracture surface, the fluid flowing direction, and the intrinsic characteristics of fractures. Its typical expression is

$$K_{if} = \frac{1}{12}\left(\cos^2\alpha\sum_{i=1}^{N_\alpha} d_i^2 + \cos^2\beta\sum_{i=1}^{N_\beta} d_i^2 + \cdots\right) \tag{6.3.7}$$

For the second type, known as *fracture permeability of fractured reservoirs*, the fracture and its surrounding rocks are treated as a whole hydrodynamic unit. This permeability is related to the degree of fracture development and the connection between fracture and matrix pores, calculated using the following formula:

$$K_{\mathrm{f}} = K_{if} \times \phi_{\mathrm{f}} \tag{6.3.8}$$

Based on fracture occurrences and their combination types, the fracture permeability is calculated based the following three models:

$$\begin{cases} \text{Single group fractures}: K_{\mathrm{f}} = 8.5 \times 10^{-4} d^2 \phi_{\mathrm{f}} \\ \text{Multigroup vertical fractures}: K_{\mathrm{f}} = 4.24 \times 10^{-4} d^2 \phi_{\mathrm{f}} \\ \text{Netlike fractures}: K_{\mathrm{f}} = 5.66 \times 10^{-4} d^2 \phi_{\mathrm{f}} \end{cases} \tag{6.3.9}$$

Where

K_{if}, K_f — the intrinsic permeability of fractures, fracture permeability (mD);
d_i, d — width of the i-th fracture, average fracture width (μm);
α, β — the angle of a fracture and fluid flowing direction (°);
n_α, n_β — number of fractures in fracture group that form α and β angles with the fluid flowing direction (fractures);
ϕ_f — fracture porosity (fraction).

The calculation result of Formula 6.3.9 represents the sidewall fracture permeability [10], where the radial extension characteristics and bending degree of the fracture are not reflected. To better reflect the reservoir fracture permeability, Formula 6.3.9 is improved as follows:

$$\begin{cases} \text{Single}-\text{group fractures}: K_f = 8.5\times10^{-4}Rd^2\phi_f/m_f \\ \text{Multigroup vertical fractures}: K_f = 4.24\times10^{-4}Rd^2\phi_f/m_f \\ \text{Netlike fractures}: K_f = 5.66\times10^{-4}Rd^2\phi_f/m_f \end{cases} \quad (6.3.10)$$

Where

R — fracture radial extension coefficient.

When a the radial extension of a fracture is greater than 2 to 3 m, the fracture is approximately of infinite extension ($R=1$); when its extension is 0.5 to 2 m, it is of medium extension ($R=0.8$); when its extension is 0.3 to 0.5 m, it is of shallow extension ($R=0.4$); and when its extension is less than 0.3 m, it is of extremely shallow extension ($R=0$);

The m_f value refers to the structural characteristics index of the fracture, which reflects the fracture width and the degree of bending. For wide fractures, $m_f=1.0$; for narrow fractures, $m_f=1.3$. The greater the m_f value, the less a fracture contributes to permeability.

2. Calculation result

As an example from Well Block XX21, Figure 6.16 shows the interpretation of fracture parameters for Well XX21, with the following results: (1) with increased fracture development, there are greater values for the fracture density, width, surface porosity, and permeability; and (2) the result of conventional logging interpretation largely agrees with the FMI imaging log and is slightly better than the imaging log for the gas-bearing interval (3660 to 3720 m).

The calculation results indicate that Well Block XX21 has a maximum fracture density of 21.33 fractures/m, 4.38 fractures/m on average, but mostly (72%) in the range of 1 to 5 fractures/m; fracture length ranges between 0.01 to 22.63 m/m^2, with an average value of 4.07 m/m^2, but mostly (80.69%) 1 to 8 m/m^2; fracture width attains a maximum of 510 μm, ranging between 0 to 60 μm, with an average of 23.5 μm, and microfractures being predominant; fracture porosity is 0.537% at maximum and 0.03% on average; and fracture permeability is 103 mD at maximum and 0.9 mD on average.

6.3.2 Evaluation of fracture development

The degree of fracture development can be used to evaluate qualitatively the fracture development conditions in volcanic reservoirs. The criteria for fracture development evaluation are established on the basis of fracture identification and fracture parameter interpretation. Fractures are evaluated zone by

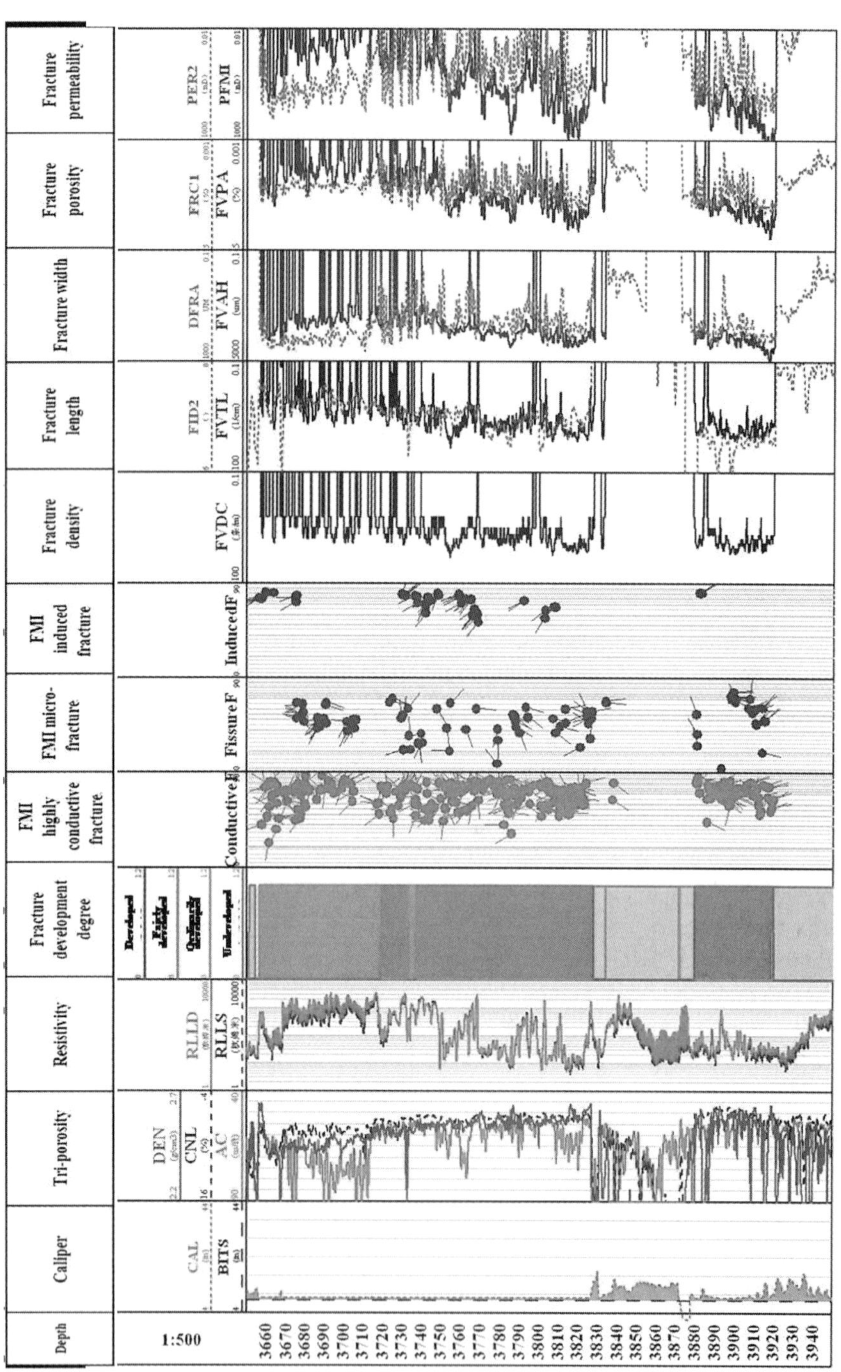

FIGURE 6.16

Diagram showing the results of fracture parameter interpretation (Well XX21).

Table 6.5 Discrimination Criteria of Fracture Development Degree (XX Gas Field)

Fracture Development	HELF / %	FID2	FVDC / (fractures/m)
Well developed	> 70	> 2.5	> 2.5
Fairly developed			1.5~2.5
Ordinarily developed			< 1.5
Fairly developed	40~70	2~2.5	> 2
Ordinarily developed			< 2
Fairly developed	10~40	1~2	> 3
Ordinarily developed			< 3
Undeveloped	< 10	< 1	< 0.5

zone, and their degree of development in different intervals of volcanic rocks in wells are ascertained, thereby providing the basis for fracture prediction.

1. Criteria for evaluating the degree of fracture development

In the XX gas field, the evaluation criteria for fracture development in volcanic reservoirs are established on the basis of parameter sensitivity analysis, through selection of three parameters, including fracture density (FVDC) interpreted based on FMI imaging logs, the fracture development index (FID2) derived from conventional logs, and the percentage of fracture intervals in reservoir thickness (HELF) [8]. The degree of fracture development in volcanic reservoirs is classified into four levels (see Table 6.5):

1. Well developed fractures: HELF > 70%, FID2 > 2.5, FVDC > 2.5 fractures/m
2. Fairly developed fractures, which are composed of three cases:
   If HELF > 70% and FID2 > 2.5, then FVDC ranges between 1.5 and 2.5 fractures/m.
   If HELF ranges between 40% to 70% and FID2 ranges between 2 and about 2.5, then FVDC > 2 fractures/m.
   If HELF ranges between 10% to 40% and FID2 ranges between 1 and 2, then FVDC > 3.0 fractures/m.
3. Ordinarily developed fractures, which are also composed of three cases:
   If HELF > 70% and FID2 > 2.5, then FVDC < 1.5 fractures/m.
   If HELF ranges between 40% and 70% and FID2 ranges between 2 and 2.5, then FVDC < 2 fractures/m.
   If HELF ranges between 10% and 40% and FID2 ranges between 1 and 2, then FVDC < 3.0 fractures/m.
4. Undeveloped fractures:
   HELF < 10%, FID2 < 1, FVDC < 0.5 fractures/m.

2. Evaluating the degree of fracture development

Fracture development is evaluated interval by interval on the basis of fracture development evaluation criteria, with fracture density (FVDC) interpreted from imaging logs, the fracture development index (FID2) from conventional logs, and the fracture thickness percentage (HELF) determined comprehensively. This lays the foundation for fracture prediction.

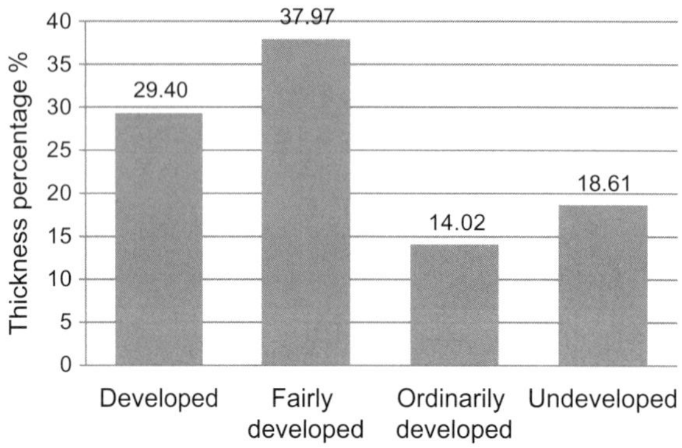

FIGURE 6.17

Evaluation results of fracture development (Well XX21).

In the case study of Well XX21 (see Figure 6.16), the results of fracture development evaluation indicate (Figure 6.17) that, according to the four-level discrimination criteria, the fracture development in volcanic rocks of this well is ranked predominantly "fairly developed," which accounts for 37.97% of total fractures; this is followed by the "well-developed" category (29.4%); "undeveloped" intervals only account for 18.6%. This suggests that the fractures in volcanic reservoirs of this well are fairly developed in general.

6.3.3 Evaluation of fracture effectiveness

Opening natural fractures are important for pore connection and fluid flow passages in volcanic reservoirs. The effectiveness of fractures in volcanic reservoirs is evaluated by geological analysis, acoustic log analysis, and comprehensive multiple-data analysis. This provides the basis for predicting effective reservoirs.

1. Geological evaluation

This line of evaluation is carried out mainly with respect to fracture width, infill characteristics, radial extension, and fluid flow capacity. Larger fracture width means longer extension, better openness, stronger fluid flow capacity for the fractures, and greater effectiveness. As is shown in the case study of Well Block XX21, comprehensive geological analysis revealed the following characteristics for the volcanic rock fractures in the study area:

1. The degree of openness is high: volcanic open fractures account for approximately 77% in cores; the highly conductive fractures and microfractures interpreted via FMI imaging log are predominantly open fractures (~ 95.7%). High-resistivity fractures account for only 4.3%.
2. The fracture width is 23.5μm on average, which is far greater than the throat diameter (0.14 μm on average).
3. The size of fractures is fairly large. Structural fractures and columnar joint fractures extend several tens of meters or even up to 100 meters. The average length of diagenetic fractures, such as blast fractures and interclast fractures, can reach up to 4.1 m/m^2.

FIGURE 6.18

Pores connected by fractures (XX23, rhyolite).

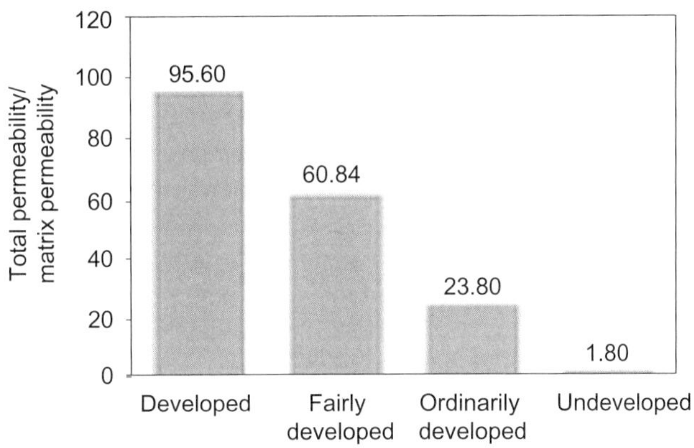

FIGURE 6.19

Reservoir permeability improved by fractures (Well Block XX21).

4. Fractures can significantly improve the fluid flow capacity of reservoirs. Diagenetic fractures mainly connect pore throats (Figure 6.18). Structural fractures are important as fluid flow channels. Thus, these fractures effectively improve the fluid flow capacity of volcanic reservoirs. In Well Block XX21 (Figure 6.19), for example, when fractures change from undeveloped to developed, the total permeability of volcanic reservoirs will increase by 1.8 to 95.6 times of the matrix permeability; therefore, fractures are crucial for improving the permeability of volcanic reservoirs.

2. Acoustic log analysis

The S-wave anisotropy of the orthogonal dipole acoustic log is one of the important methods for evaluating fracture effectiveness [11,12].

1) The fundamental principle

At a certain angle between the direction of particle vibration caused by the S wave and the fracture strike, the so-called S-wave splitting will occur—that is, the incident S wave splits into a fast S wave and a slow S wave, which have particles parallel and perpendicular to the fracture strike, respectively, vibrating and propagating upward along the borehole axis and traveling at different velocities. In fractured volcanic formations, therefore, the S-wave velocity of the dipole sonic imaging (DSI) log generally shows obvious azimuthal anisotropy. The S-wave anisotropy of the formation is measured by the orthogonal dipole technique. Stress-release fractures, other fractures, and faults are identified, and their strikes are evaluated tentatively by parameters such as energy-percentage anisotropy (OffEne), acoustic travel time anisotropy (SLOANI), and time-based anisotropy (TIMANI).

2) The evaluation method

Figure 6.20 is an example of fracture effectiveness evaluation through the integrated use of DSI anisotropy and the FMI imaging chart. In the left chart, the third trace shows the velocity and energy-percentage anisotropy of fast and slow S waves; the fourth trace shows the azimuth and error of fast and slow S waves. This chart demonstrates that this reservoir section has very strong anisotropy and a wide range of variation in fast S-wave azimuth, implying that the phenomenon causing such anisotropy has multiple distribution azimuths. In the FMI imaging log chart, the anisotropy is caused by multiple high-angle fractures of different azimuths, and both the width and extension range of fractures are quite large, suggesting that these are effective fractures. The right chart shows just the opposite. The fractures shown in the FMI imaging log chart have a small width with limited extension, thus being ineffective fractures. The anisotropic characteristics shown in the DSI log are inconspicuous.

Many factors affect rock anisotropy, including variations in lithology, directivity and changes in rock texture and porous structure, and lineation and distribution of rock constituents. On the basis of acoustic anisotropic analysis and other log data, therefore, the uncertainty of evaluation can be reduced through an integrative interpretation of fracture effectiveness.

3. *Integrative analysis based on multiple log data*

Different logging tools have different resolutions and detection depths, and different log data can be analyzed together, for example, high-resolution FMI imaging logs and other logs (e.g., Azimuthal resistivity imaging (ARI), array induction, DSI, and conventional logs) can be analyzed and compared with each other. Thus, the effectiveness of fractured reservoirs and the influence of fractures on reservoirs can be evaluated.

When fractures and the matrix are considered together, the fracture effectiveness will also depend on matrix pore development and pore-fracture combination characteristics. When matrix pores and effective fractures are developed and the pore-fracture combination is good, the fractured volcanic reservoirs will be effective. As shown in Figure 6.21, 17 fractures are picked up from the dacite formation at 4150 to 4156 m in Well XX9 based on the FMI imaging log, where the fracture density is 2.83 fractures/m. Nine fractures are picked up from the rhyolite formation at 4156 to 4162 m, where the fracture density is 1.5 fractures/m; conventional logs and DSI interval transit time curves demonstrate that the lower interval has good physical properties, has low resistivity (380 Ω·m), and contains effective fractured porous reservoirs, where fractures improved the fluid flow property of the reservoirs. The upper interval has poor physical properties and high resistivity (approximately 20000 Ω·m) and contains fractured volcanic reservoirs of poor accumulation capacity, where fractures only work as seepage channels.

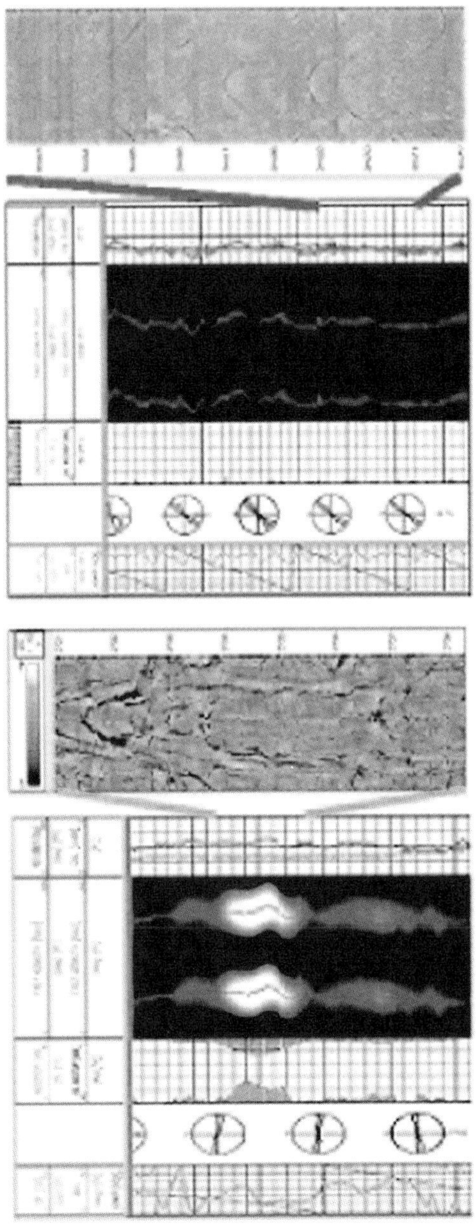

FIGURE 6.20
Evaluation of fracture effectiveness.

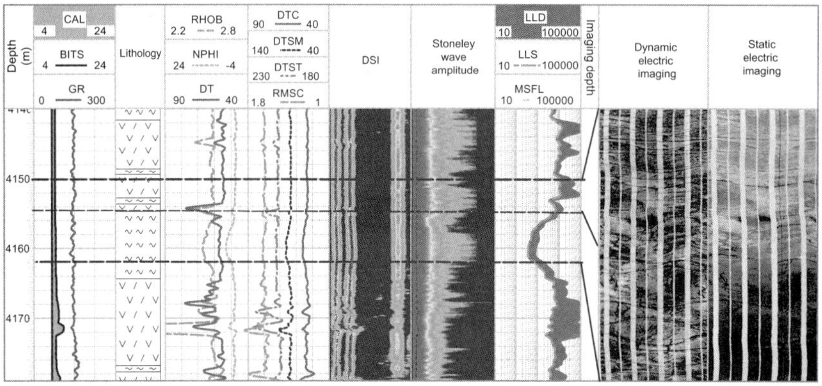

FIGURE 6.21

Fracture effectiveness evaluation via FMI, DSI, and conventional logs (Well XX9).

Methods for evaluating fracture effectiveness also include correlation analysis of FMI and ARI, Modular Dynamic Tester (MDT) testing, Stoneley wave energy attenuation analysis, dual laterolog and conventional porosity log combined analysis, and so on.

6.3.4 Evaluation of fracture occurrence (configuration)

Fracture occurrence or configuration refers to fracture dip, inclination, and strike, and it is one of the major factors for optimizing the planning of development well grid pattern and the design of horizontal well trajectory. Fracture occurrence is generally evaluated using the paleomagnetic method and imaging log method.

1. Paleomagnetic method

Rocks record the direction of the geomagnetic field when they were formed, changes of the geomagnetic field since their formation, and the direction of the latest geomagnetic field. The theory of paleomagnetism suggests that the geomagnetic field has the characteristics of a geocentric axial dipole field (GAD field), the location of the geomagnetic pole as recorded by rocks is almost always near the geographic pole, and magnetic coordinates are largely consistent with geographic coordinates. Therefore, the geographic coordinates of a core sample can be determined by measuring its direction of magnetization [13].

1) Definition of core coordinates

As shown in Figure 6.22, a standard XYZ coordinate system is established as follows: first, mark a line through the center of a circle, which is the cross section of the core column; this line is parallel with a group of fractures, and it is defined as the x-axis (the virtual north direction); second, drawing a line along the length of the core as the z-axis. Then, extract a minicore 25 mm in diameter along the direction of the marked line (x-axis), and establish a corresponding coordinate system *xyz* in this tiny core. Therefore, the corresponding relationship between these two coordinate systems is

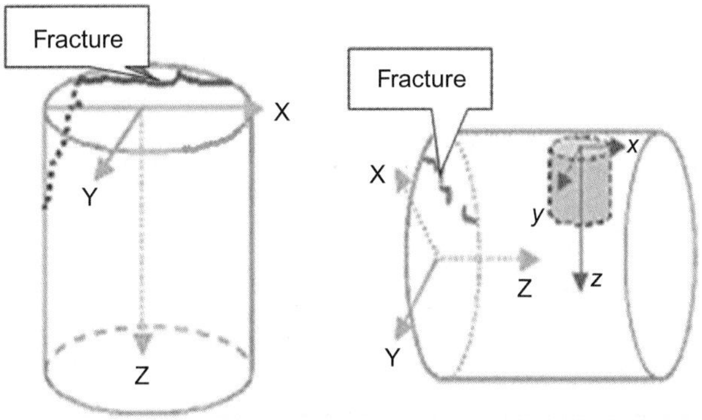

FIGURE 6.22

The coordinate system and its conversion for sampling.

$$\begin{cases} X = -z \\ Y = y \\ Z = x \end{cases} \tag{6.3.11}$$

The occurrence of the fracture face within the XYZ coordinate system is known. The geographic coordinates of the tiny core sample can also be determined through measurements. The fracture occurrences data within the geographic coordinate system can be determined using this coordinate conversion formula.

2) Measuring the direction of magnetization of minicore sample

The magnetism that volcanic rocks obtained during a rock's cooling moment is called the characteristic remanence (ChRM), which records the direction of the geomagnetic field when the rock was formed. The direction of the geomagnetic field when the volcanic rocks were formed can be determined by measuring the direction of their characteristic remanence. Once the characteristic remanence has been formed, it may decrease due to viscous demagnetization, or it may increase due to viscous remanence (VRM, which refers to the new intensity of magnetization continuously obtained below the Curie temperature). The ultimate direction of viscous remanence (which accounts for approximately 50% of the natural remanence) is consistent with the direction of the present geomagnetic field. Therefore, the direction of the present geomagnetic field (i.e., the present geographic direction) can be determined on the rock specimen through measurement of the direction of viscous remanence [14,15].

Viscous remanence can be erased under low-temperature demagnetizing conditions. Thermal demagnetization can be carried out at 12 temperature levels (i.e., 100° C, 150° C, 200° C, 250° C, 300° C, 350° C, 400° C, 450° C, 500° C, 530° C, 550° C, and 570° C). Each of the low-temperature components obtained through step-by-step thermal demagnetization below 300° C represents the direction of later viscous remanence, while each of the high-temperature components obtained above 300°C represents the direction of characteristic remanence. Figure 6.23 shows the demagnetization analysis of a core sample from the depth of 3100 m in Well XXSG2. Its natural remanent magnetization is 306 mA/m, which shows a uniform decrease from the demagnetization intensity of NRM –250° C

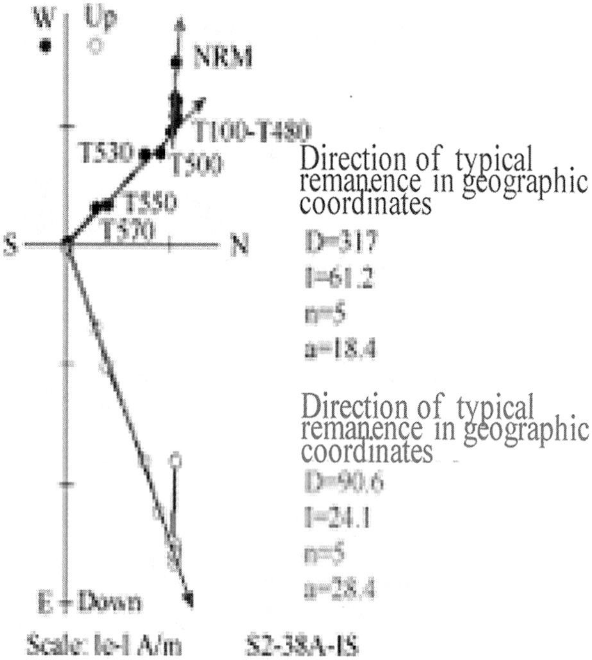

FIGURE 6.23

Thermal demagnetization diagram of a sample from Well SSG2.

to 70% of natural remanent magnetization, indicating approximately a viscous remanence demagnetization vector of NE 30 degrees. Beyond 300° C, as the demagnetization temperature rises, the remanent magnetization decreases gradually until it reaches 570° C when the remanent magnetization attenuates to 0.3 mA/m, indicating that the overall demagnetization direction of characteristic remanence remains at approximately NW 317 degrees with a dip angle of 61 degrees. This indicates that the direction of the magnetic field was NW 317 degrees when the volcanic rock was formed, whereas the direction of the present geomagnetic field is NE 30 degrees and the fracture direction is aligned largely toward the NE.

2. Imaging log method

Fracture occurrence can be determined by an image log chart. As shown in Figure 6.24, once fractures are identified, the height difference h between the peak and valley of the fracture sine curve and the average well radius d at the fractured interval are determined in the imaging chart. Thus, the dip angle θ of the fracture face can be expressed as [16]

$$\theta = \arctan(h/d) \tag{6.3.12}$$

The inclination of fracture planes can be read directly from the x-coordinate where the sine curve's valley is located, and the fracture strike is perpendicular to its inclination.

The occurrence of a single fracture can also be read from the tadpole plot interpreted from the dip log and imaging log. The x-coordinate at the circular point of the tadpole is simply the fracture dip

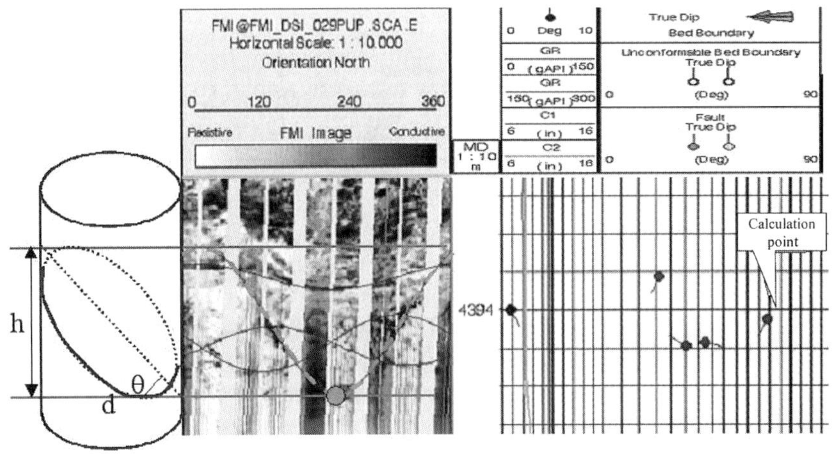

FIGURE 6.24

Sketch map of the determination of fracture occurrence based on the imaging log.

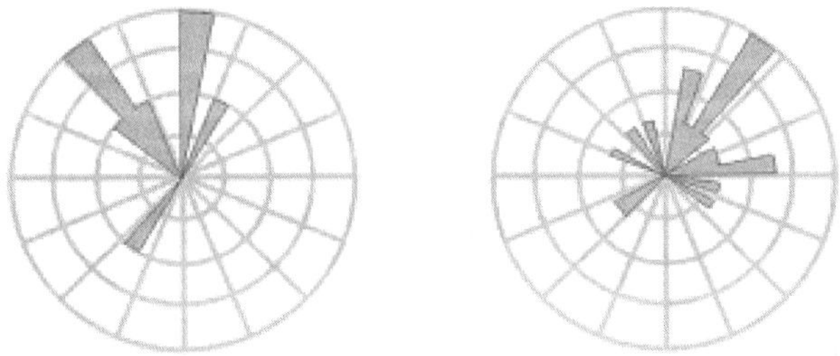

FIGURE 6.25

Azimuth frequency diagram of the fracture dip direction.

angle, whereas the azimuth at the tadpole tail point is the fracture inclination (see Figure 6.24). Azimuthal statistical analysis is performed for the target layer based on the determination of fracture occurrence. The predominant dip angle and inclination in the target layer are displayed by the azimuth frequency diagram (rose diagram; Figure 6.25).

6.4 PREDICTION OF FRACTURES IN VOLCANIC RESERVOIRS

The results of fracture prediction provide an important basis for exploring high-quality volcanic reservoirs, optimizing well spacing, and optimizing horizontal well trajectories. Seismic response characteristics are analyzed on the basis of well identification, evaluation, and seismic calibration. Planar distribution of fractures is predicted by methods such as poststack seismic attribute analysis, prestack subazimuth processing and attribute analysis, and fracture parameter inversion.

6.4.1 Seismic response characteristics of fractures

The seismic reflection, wave impedance, and azimuthal anisotropic characteristics of fractures can be determined on the basis of a feasibility study on fracture prediction in volcanic rocks, through calibration of seismic data using fracture data from wells, and based on an analysis of seismic response of fractures, which therefore will lay a foundation for fracture prediction in volcanic reservoirs.

1. Feasibility of fracture prediction

Figure 6.26 is a geological model for vertical fracture simulation. The denudated top interface in its upper part is rugged and cut by minor faults. Fractures extend from the top interface into the model. The number of fractures indicates the developmental level of the fracture zone. Figure 6.27 is the seismic profile obtained through forward geological modeling. Comparison between these two charts shows almost no response on the left of the charts, but there is obvious amplitude, frequency, and phase anomalies on the right side; from left to right, as the degree of fracture development increases, the

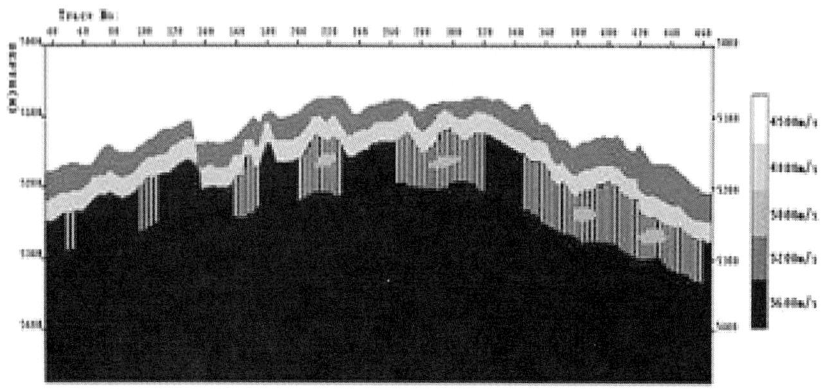

FIGURE 6.26

The geological model of a complex fracture system.

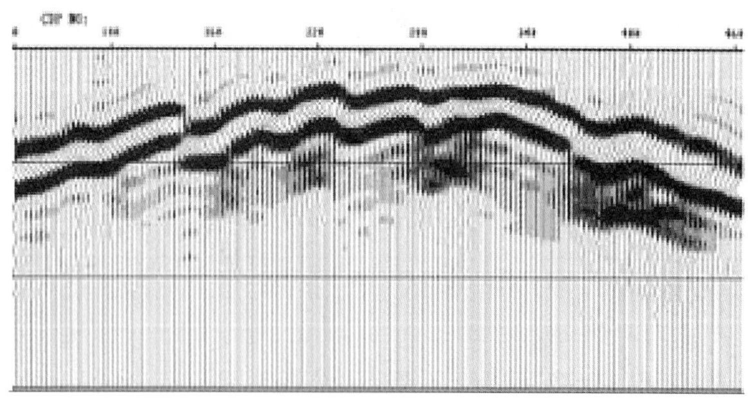

FIGURE 6.27

Forward geological modeling of a fracture system.

abnormal response in their seismic profiles becomes increasingly significant. This suggests that the seismic information cannot identify small single fractures but can identify fractured zones made up of small fractures.

2. Characteristics of seismic reflection waveform

In the forward model (see Figures 6.26 and 6.27), fractured zones have the characteristics of poor continuity, weakened energy, and reduced frequency.

In seismic profiles, fractured zones often lead to changes in seismic reflection events [17]. As shown in Figure 6.28, in the fractured interval, the near-wellbore seismic trace event of Well CC1-1 corresponds to the wave trough and shows reflection characteristics of medium-weak amplitude and poor continuity; the event of the fractured interval in Well CC102 corresponds to the wave trough and shows reflection characteristics of medium-weak amplitude and poor continuity; the event of the fractured interval in Well CC1 corresponds to both the wave crest and trough and shows reflection characteristics of medium-weak amplitude and poor continuity; the event of the fractured interval in Well CC103 corresponds to the wave crest and shows reflection characteristics of medium-weak amplitude and poor continuity; and the fractured interval in Well CC1-2 corresponds alternately to wave crests and wave troughs and shows reflection characteristics of medium-weak amplitude and poor continuity. In seismic profiles, therefore, volcanic fracture intervals show reflection characteristics of weak energy and poor continuity.

3. Wave impedance characteristics

In fractured intervals of volcanic rocks, a decrease in density and slowdown of acoustic propagation velocity are associated with a decrease in wave impedance. Figure 6.29 demonstrates the wave impedance variation characteristics of the fractured interval in Well CC103. In this figure, the fractured zone in volcanic reservoirs corresponds to a relatively low impedance, and the upper limit for the CC gas field is approximately 13200 $(g/m^3) \cdot (m/s)$.

4. Anisotropic characteristics

Fractures have obvious anisotropic characteristics. If fractures in volcanic reservoirs are symmetrically arranged, the seismic attributes of fractured intervals will have distinct anisotropic characteristics [18]. In Figures 6.30 and 6.31, the results of prestack forward modeling for Well XX5 show that its fractures are distributed along the SN direction. It can be concluded from these figures that when the incident angle is small, the reflection amplitude of various azimuths remains the same; however, when the incident angle reaches 12 degrees or higher, the reflection amplitude of various azimuths becomes different, with larger incident angles correlated to greater differences. Therefore, fracture distribution can be predicted by detecting the anisotropy in seismic attributes.

The seismic response of fractures in volcanic reservoirs is simultaneously affected by multiple factors, such as lithological boundaries, faults, and the distribution of accumulation and seepage bodies. For fracture prediction, the results of log identification and evaluation must be fully utilized. The geological constraints must be used to improve the accuracy of prediction.

6.4.2 Method of post-stack seismic attribute analysis

The poststack seismic attributes for fracture prediction mainly include coherence, dip angle, amplitude, azimuth, and curvature.

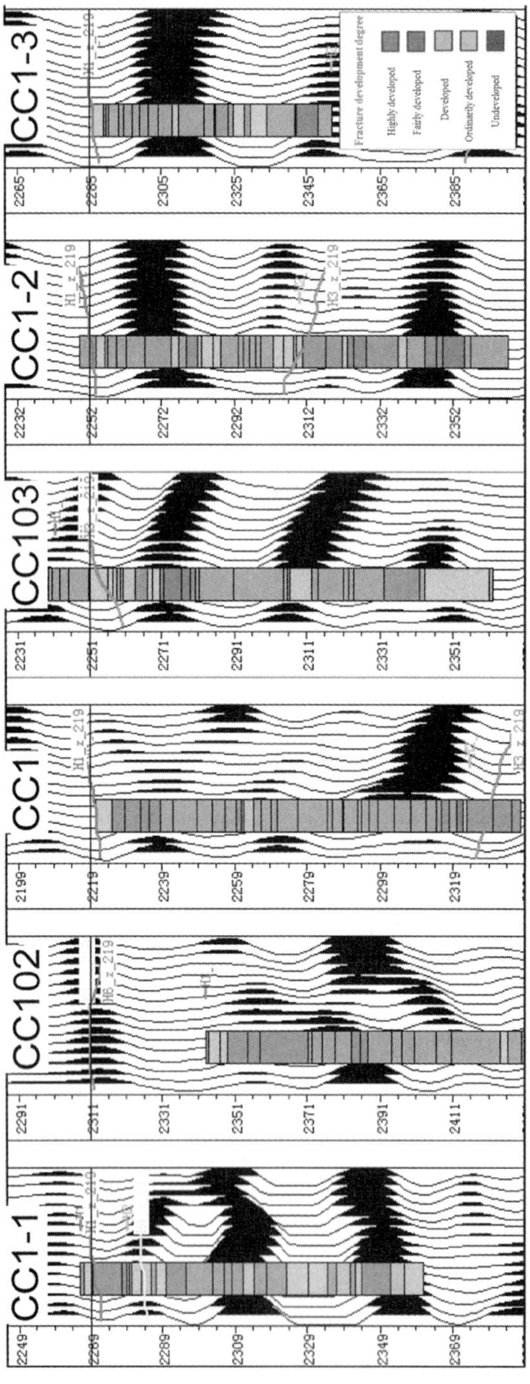

FIGURE 6.28
Seismic response characteristics of fractures in volcanic reservoirs (CC gas field).

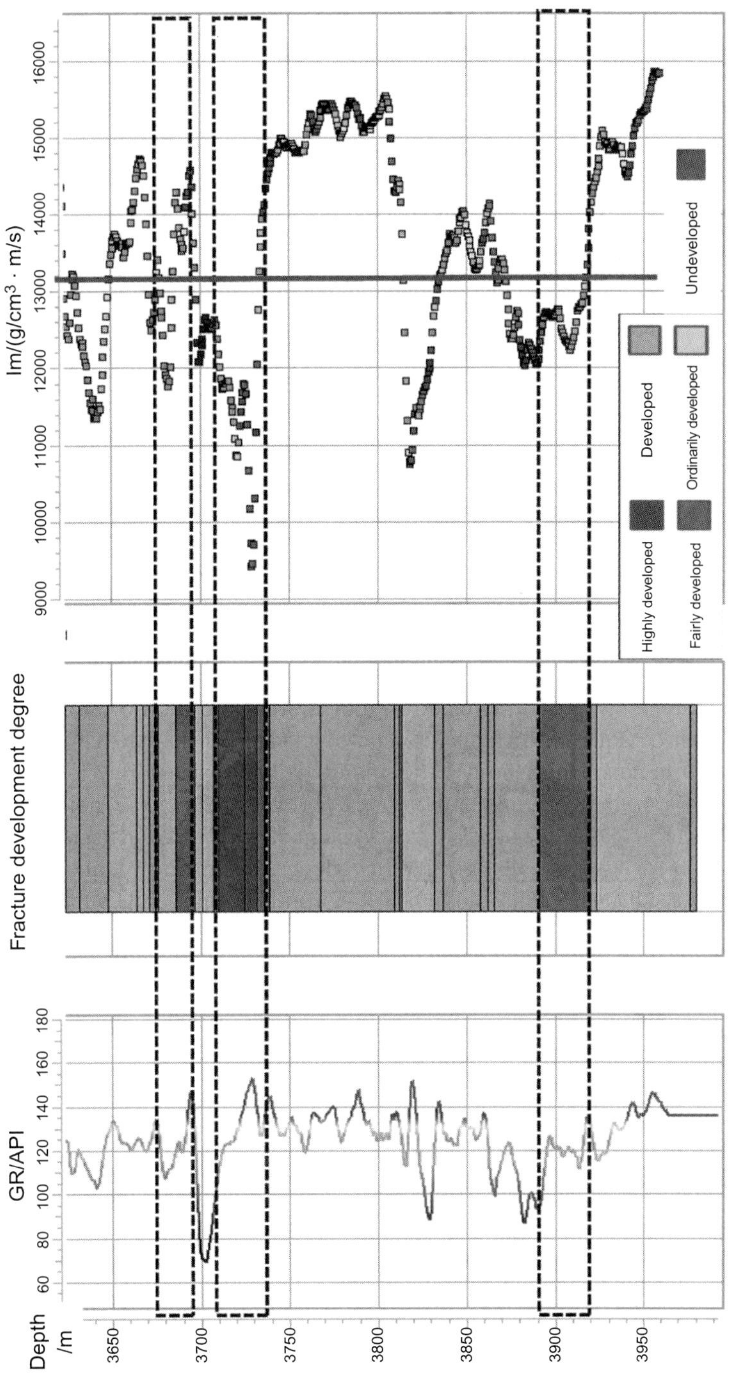

FIGURE 6.29

Wave impedance characteristics of the fractured intervals (Well CC103).

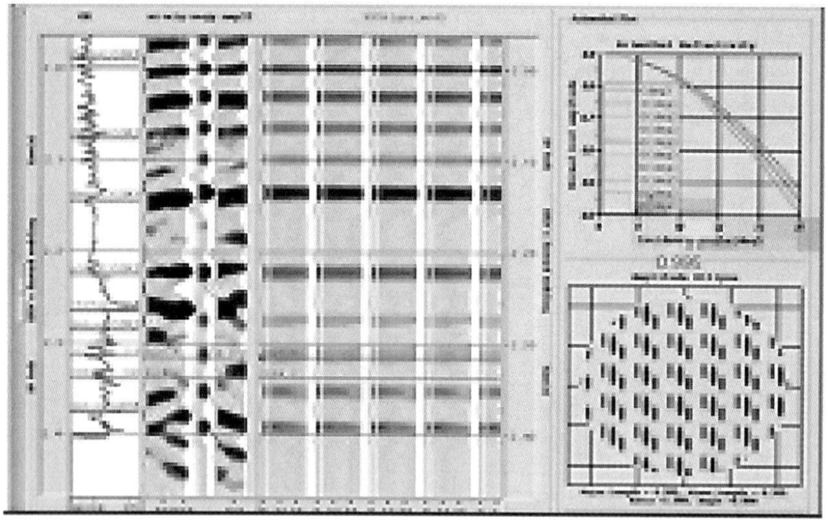

FIGURE 6.30

Results of forward modeling with an incident angle of 3 degrees.

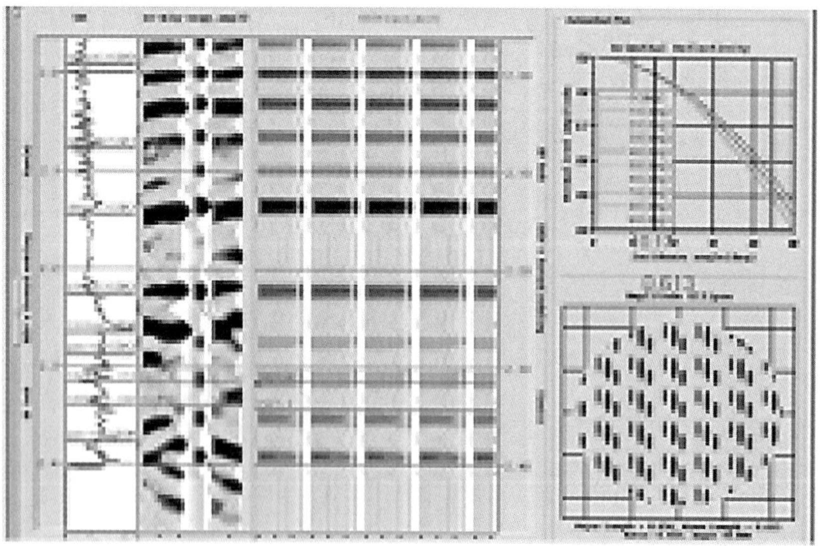

FIGURE 6.31

Results of forward modeling with an incident angle of 27 degrees.

1. Coherence analysis technique

The coherence analysis involves the calculation of dissimilarity among adjacent traces in seismic data volumes to construct coherent data volumes and to measure the similarity among multiple traces of seismic data. These data volumes are used to characterize the variation characteristics of continuity in seismic wave groups [19].

1) Coherence calculation method

The calculation method for seismic coherence is simple, using the following formula to calculate the correlation coefficient on the basis of the number of traces, the dip angle, and the time window of the given data volume:

$$\dot{R}(t,\phi_{\max}) = \frac{\sum\limits_{L=t-N/2}^{L=t+N/2} T_L T'_{L+\phi}}{\sum\limits_{L=t-N/2}^{L=t+N/2} T_L^2 T'^2_{L+\phi\max}} \tag{6.4.1}$$

Where

R — coherence coefficient, which is a function of the time of seismic trace and the dip angle between two seismic traces;
t — time;
φ — dip angle, which is affected by azimuth and is not easily to be determined;
T, T'— seismic trace data pair.

The number of coherent traces involved in the calculation includes linear 3-trace, orthogonal 3-trace, orthogonal 5-trace, and orthogonal 9-trace. A greater number of traces will lead to a greater averaging effect. Multitrace coherence is used mainly to highlight major faults. In contrast, fewer coherent traces lead to a smaller averaging effect. Coherence of fewer traces is suitable for identifying minor faults or fracture zones. Selection of the coherent time window typically depends on the apparent period t of reflection waves in the seismic profile. Its value is usually in the range of $t/2$ to 3 $t/2$.

2) Prediction of fracture distribution using coherence analysis technique

Figure 6.32 shows the T580-line migration profile, coherence profile, and T713-line coherence profile of the XX gas field. It demonstrates that the location in the migration profile with an event offset shows poor-coherence stripes in the coherence profile, whereas the location with good event continuity shows good coherence in the coherence profile. The area near Well XX602 with good coherency shows white stripes in the coherence profile, where fractures or faults are undeveloped, and the gas test confirms dry layers; the area near Well XX1 with poor coherency shows black irregular stripes in the coherence profile, where fractures or faults are developed, and the gas test shows commercial gas layers. This confirms that the coherence analysis technique can be used to identify fractured zones. However, the coherent data anomaly caused by lithological changes or poor data quality must be identified.

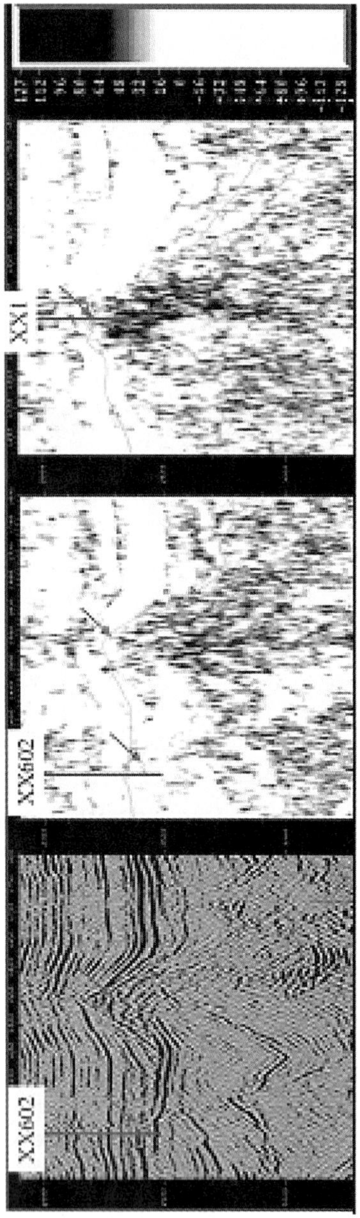

FIGURE 6.32

Migration profiles and coherence profile for lines T580 and T713 in the XX gas field.

2. Margin-detection technique

Margin refers to the end of one region and the start of another. Margin detection treats seismic data as a special image and uses margin-detection operators and the maximum gradient or the zero-crossing point of a second derivative to extract image margins (location, direction, amplitude); therefore, the configuration of an abnormal change zone can be determined to achieve fracture prediction [20,21].

Figure 6.33 shows the T580-line migration and its margin-detection profiles and the T713-line margin-detection profile from the XX gas field. In the migration profile, the location with an event offset (arrow) corresponds to high (black) values in the margin-detection profile (the high yield zone in Well XX1), whereas the location with good event continuity corresponds to low (white) values in the margin-detection profile (the dry zone in Well XX602). This confirms that the margin-detection technique can also be used to predict fracture development zones effectively, although the prediction result still has some ambiguity, with fuzzy boundaries and low definition.

3. "Three-instantaneous" attribute analysis

Petrophysical modeling experiments indicate that when the fracture opening decreases by 6.5 μm, the velocity of the seismic wave will increase by 4%, the quality factor by 70%, the amplitude by 1200%, and the dominant frequency by 44%. It is advantageous, therefore, to study fractures using the dynamical information (amplitude, frequency, and absorptive attenuation, etc.) of seismic waves. The technique of "three-instantaneous attribute analysis" predicts fractures on the basis of instantaneous amplitude, instantaneous frequency, and the instantaneous phase of seismic waves [22]. A good example for the application of this technique is found in the study of the YC3I top surface of Well Block XXS2-1 (Figures 6.34 through 6.36).

1) Instantaneous amplitude (A)

Instantaneous amplitude reflects the energy level of seismic reflection waves. The presence of pores, cavities, and fractures in reservoirs causes a decrease in reservoir velocity, reflection coefficient, and amplitude and energy of the reflection waves. Figure 6.34 demonstrates the presence of a low-amplitude areas in the southwest of Well Blocks XXS4 to XXS2-1, the east of Well XXS101, and the southeast of Well XXS2, which are in agreement with the poor-coherence area in the coherence analysis and the strong absorption area in the absorption coefficient analysis.

2) Instantaneous phase (Q)

The instantaneous phase reflects the continuity of seismic reflection waveforms. In terms of seismicity, minor faults and zones of pores, cavities, and fractures are characterized by event offsets, chaos, discontinuities, the increased number of event, and so on. In Figure 6.35, the areas of the chaotic instantaneous phase are mainly located in regions near Wells XXS4 to XXS2-1, Wells XXS to XXS2-25, west of Well XXS101, and near Well XXWS3. This is slightly different from the coherence areas and strong absorption areas.

3) Instantaneous frequency (F)

A rapid change in seismic wave frequency is related to factors such as stratigraphic pinch-outs and oil-gas-water interfaces. The presence of fractures in reservoirs causes the reservoir reflection coefficient sequence to change, which also results in abrupt changes in the instantaneous frequency, which is mainly reflected by the decrease of the instantaneous frequency. In Figure 6.36, areas of relatively

FIGURE 6.33

Migration profiles and margin-detection profile of lines T580 and T713 in XX gas field.

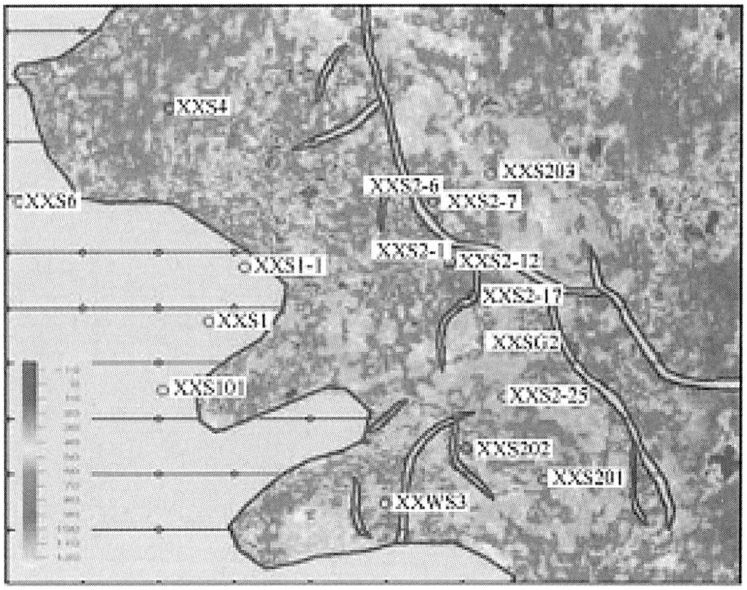

FIGURE 6.34

Instantaneous amplitude of YC3I top surface (Well Block XXS 2-1).

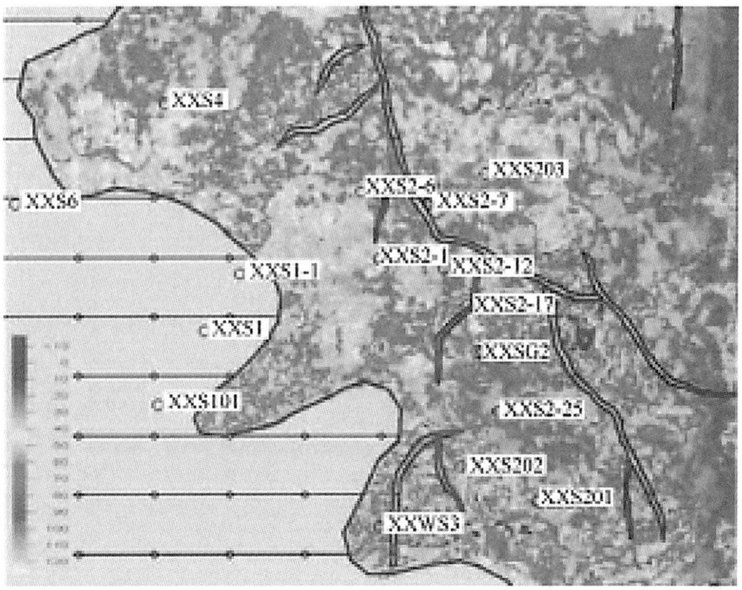

FIGURE 6.35

Instantaneous phase of the YC3I top surface (Well Block XXS 2-1).

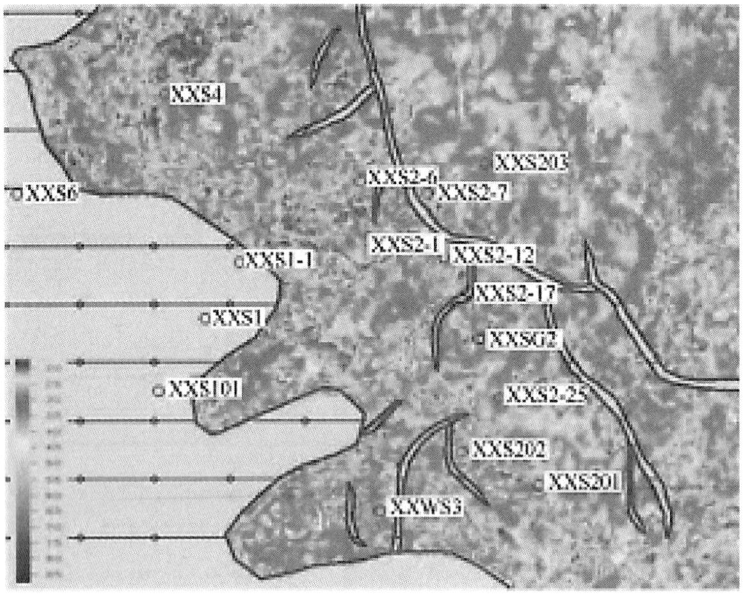

FIGURE 6.36

Instantaneous frequency of the YC3I top surface (Well Block XXS 2-1).

low instantaneous frequency are mainly located in regions near Wells XXS1-1 to XXS2-7, north of Well XXS4, east of Well XXS101, and south of Wells XXS2-25 to XXS201. These correspond well with the poor-coherence areas and strong absorption areas.

4. Absorption coefficient analysis

Seismic wave absorptive attenuation is caused by the intrinsic viscoelasticity of the rock matrix (e.g., internal friction between particles and on fracture surfaces), the relative flow of liquids in porous rocks, the local saturation effect, and geometric diffusion. Fractures and their gas-bearing property will result in changes in seismic wave attributes such as velocity, amplitude, and frequency, thereby leading to absorptive attenuation of the seismic wave and further to changes in the seismic absorption coefficient.

1) Methods for calculating absorption coefficient

Under low-frequency conditions, when the seismic wave propagates in viscoelastic media, the absorption coefficient can be expressed with the following formula:

$$\alpha \approx \frac{1}{2}\frac{\eta'\omega^2\sqrt{\rho}}{(\lambda+2\mu)^{3/2}} = \frac{1}{2}\frac{\eta'\omega^2\sqrt{\rho}}{(\kappa+4/3\mu)^{3/2}} \tag{6.4.2}$$

Where

α — absorption coefficient;
η'— viscosity coefficient of viscoelastic media;
ω — circular frequency (Hz);

ρ — rock density (g/cm^3);
κ — elastic modulus of rocks (N/m^2);
λ, μ — lame coefficient.

The absorption coefficient is related to the rock density, elastic modulus, viscosity coefficient, and the seismic wave's frequency. Assuming that c is a constant in viscosity coefficient $\eta' = c/\omega$ and that P-wave velocity v is used in place of the elastic coefficient, then the absorption coefficient can be simplified as follows:

$$\alpha_e \approx \frac{c}{2} \frac{1}{\rho v^3} \tag{6.4.3}$$

Therefore, the effective absorption coefficient is inversely related to the cube of wave velocity. In other words, the sensitivity of the absorption coefficient in reflecting seismic anomalies far exceeds that of velocity, thus making it feasible to use the absorption analysis technique to predict fractures.

2) Prediction of fractures using absorption coefficient analysis method

There are two types of absorption coefficients: absolute absorption coefficient and relative absorption coefficient. Relative absorption coefficient refers to the absolute absorption coefficient divided by the background absorption coefficient, which highlights seismic energy absorption anomaly areas to better reflect the heterogeneity of geological bodies and as such is widely used to predict fractures.

Figure 6.37 is the planar distribution diagram of the absorption coefficient on the top surface of YC3I in Well Block XXS2-1, in which strong absorption areas are located mainly near Wells

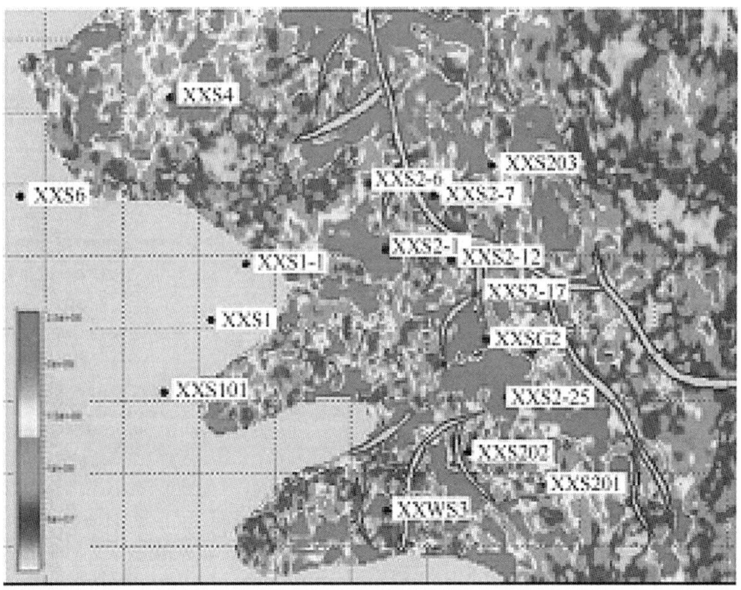

FIGURE 6.37

Absorption coefficient of the YC3I top surface (Well Block XXS2-1)

XXS2-6 to XXS2-1 and XXS2-17 to XXWS3; favorable reservoirs proven by drilling are mostly located in areas marked by high absorption coefficient values. This implies that the absorption coefficient method is suitable for predicting fractures in volcanic reservoirs.

5. *Stress analysis technique*

The tectonic strain and compressive strength of rock strata determine the occurrence and scale of faults and fractures. The formation strain in each tectonic phase is calculated using the kinematic and dynamic tectonic reconstruction method and by performing inverse and forward modeling of the structural history of rocks. By taking into consideration parameters such as formation thickness and lithology, the fractured zones and fracture direction in volcanic reservoirs can be predicted under the constraints of seismic attributes (e.g., coherence, amplitude attributes) [17,23,24].

Figures 6.38 and 6.39 show the planar distribution of formation strain and open fractures on the top surface of YC3I in Well Block XXS2-1 of the XX gas field, in which the areas of high formation strain values are mainly located near the fault zones north of Well XXS2-6 and secondarily west of Wells XXS2-1, XXS2-25, XXWS3, and XXS101. This indicates that faults and structures are important in controlling fracture development and distribution and that open fractures are mainly located near faults and in structural high areas.

6.4.3 Prestack fracture prediction method

The seismic reflection of fractured zones has anisotropic characteristics. The degree of fracture development and fracture orientation can be effectively predicted through prestack seismic subazimuth processing and attribute analysis [25–27].

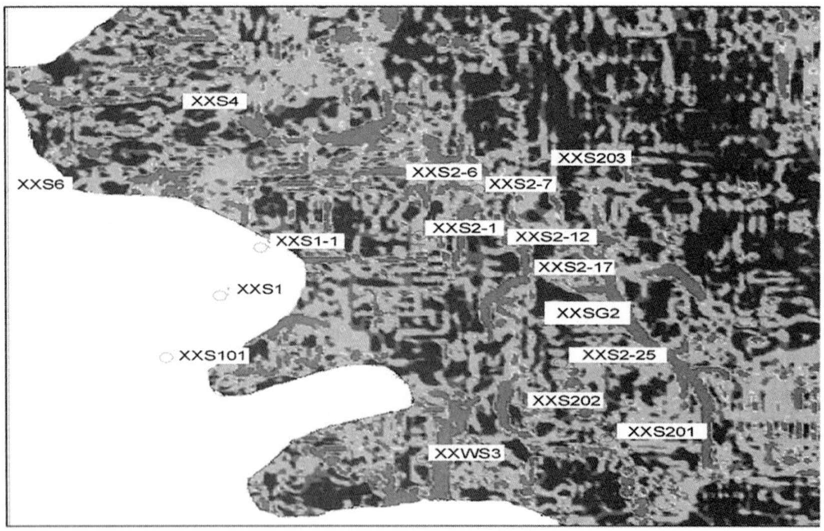

FIGURE 6.38

Formation strain of the YC3I top surface (Well Block XXS2-1)

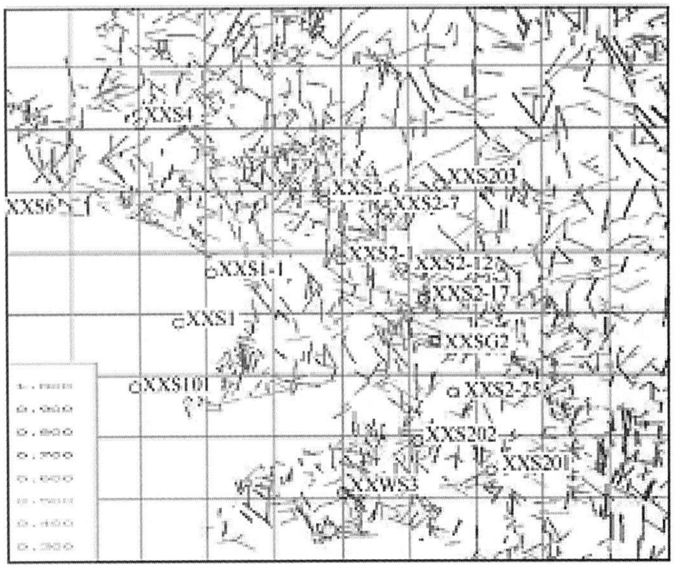

FIGURE 6.39

Fracture opening of the YC3I top surface (Well Block XXS2-1)

1. Anisotropic characteristics of volcanic reservoirs

Earth media has three types of anisotropy: intrinsic anisotropy, secondary anisotropy, and wave length anisotropy. Volcanic rocks have secondary anisotropy caused by lithological change, rhyolitic structure, fractures, and the like. In an anisotropic medium, if the elastic properties in all directions at a given plane are the same and the axial directions of all points in the perpendicular plane are all mutually parallel, this plane is called an isotropic plane, the axis perpendicular to the isotropic plane is called the axis of symmetry (Figure 6.40), and the elastic medium for the isotropic plane is called the transversally isotropic (TI) medium. When the TI medium's axis of symmetry coincides with z-axis, it is called a VTI medium, such as thin interbeds cyclically deposited in a horizontal bedded medium. When the TI medium's axis of symmetry coincides with the x-axis or y-axis, it is called an HTI medium or an

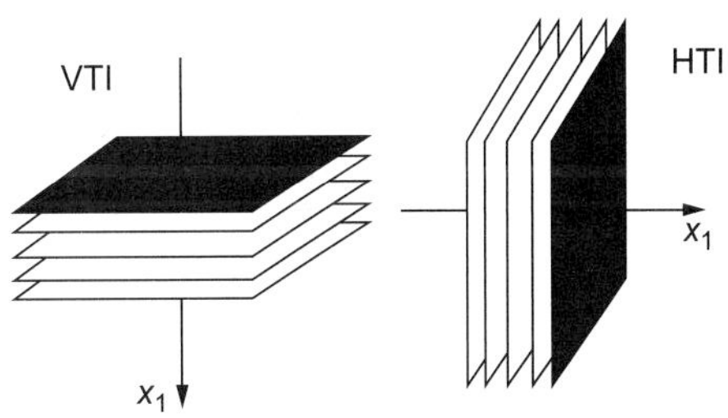

FIGURE 6.40

Illustration of the TI medium.

expansion anisotropic medium. For example, the volcanic reservoirs with vertical fractures generated by tectonic stress are isotropic longitudinally but anisotropic transversely.

2. Prediction principle

Prestack fracture detection techniques recognize fractures based on the rule of azimuthal anisotropy of 3D P-wave attributes in the HTI medium, including the VVA, AVA, and FVA techniques.

1) VVA technique

The VVA technique identifies fractures with the characteristics of reflection wave velocity that changes with the incident azimuth. As shown in Figure 6.41, the P-wave velocity of the HTI medium is [26,27]

$$V_{p0}(i,\phi) = \alpha\left[1 + \delta^{(v)}\sin^2 i\cos^2\phi + \left(\varepsilon^{(v)} - \delta^{(v)}\right)\sin^4 i\cos^4\phi\right] \tag{6.4.4}$$

Where

i — incident angle;
α — p-wave velocity during vertical incidence;
$\delta^{(v)}$, $\varepsilon^{(v)}$— Thomsen parameters of the HTI medium;
ϕ — azimuth, which is the angle between the ray plane (i.e., the source-receiver azimuth) and the HTI medium's axis of symmetry.

From Formula 6.4.4, for horizontal fractures, $\phi = 90$ degrees and $V_{p0} = \alpha$, where the V_{p0} value is the maximum.

For vertical fractures,

$$\phi = 0^\circ \text{ and } V_{p0} = \alpha\left(1 + \delta^{(v)}\sin^2 i\cos^2 i + \varepsilon^{(v)}\sin^4 i\right)$$

where the V_{p0} value is the minimum; and the azimuth and velocity between them range between these two values.

When directional vertical fractures are present in the formation, the velocity along different azimuths will show an ellipse: the velocity is the maximum when it propagates along the fracture strike. The velocity is the minimum when the propagation is perpendicular to the fracture strike. Mallick, Craft, and colleagues believe that the long axis direction of the azimuthal ellipse represents the fracture strike and the ellipticity represents the fracture density.

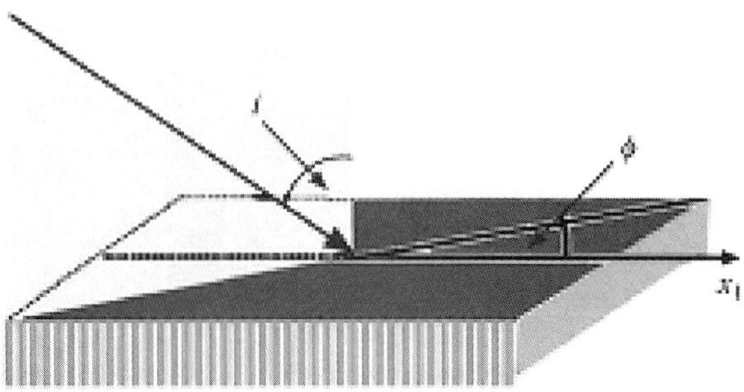

FIGURE 6.41

Illustration of seismic wave propagation in the HTI medium.

2) AVA technique

AVA refers to the change of amplitude with azimuth. The AVA technique identifies fractures based on the characteristics of reflection amplitude that change with incident azimuth. Rüger provided a P-wave reflection coefficient formula for HTI media [26,27]:

$$R_p(i,\phi)=\frac{1}{2}\frac{\Delta Z}{\bar{Z}}+\frac{1}{2}\left\{\frac{\Delta\alpha}{\bar{\alpha}}-\left(\frac{2\bar{\beta}}{\bar{\alpha}}\right)^2\frac{\Delta G}{\bar{G}}+\left[\Delta\delta^{(v)}+2\left(\frac{2\bar{\beta}}{\bar{\alpha}}\right)^2\Delta\gamma\right]\cos^2\phi\right\}\sin^2 i$$
$$+\frac{1}{2}\left[\frac{\Delta\alpha}{\bar{\alpha}}+\Delta\varepsilon^{(v)}\cos^4\phi+\Delta\varepsilon^{(v)}\sin^2\phi\cos^2\phi\right]\sin^2 i\tan^2 i \tag{6.4.5}$$

From the preceding discussion, it can be concluded that the P-wave reflection coefficient of the HTI medium is related not only to the P-wave velocity, S-wave velocity, shear modulus, and Thomsen anisotropy parameters δ and ε, but also to the incident azimuth and incident angle. Rüger believes that for a large fixed incident angle, the change of the HTI medium's P-wave reflection coefficient with incident azimuth shows elliptical characteristics, the ellipticity being proportional to fracture density, and the long axis azimuth of the ellipse representing the fracture azimuth [26,27].

3) FVA technique

The seismic scattering theory for fractured reservoirs suggests that the seismic wave's high-frequency attenuation caused by fractures results in the anisotropy of the seismic reflection frequency. The FVA technique detects fractures by the azimuthal anisotropy of 3D P-wave frequency attributes. Like VVA and AVA, the long axis direction of the ellipse of frequency changes with the azimuth and represents the fracture strike, and the ellipticity represents the anisotropic intensity of the frequency, which is related to the density of open fractures.

The frequency attributes related to attenuation mainly include the following: (1) the maximum amplitude frequency (i.e., the frequency corresponding to the maximum amplitude) is related to the fluid—the greater the seismic wave attenuation, the lower the frequency; (2) the frequency corresponding to 85% of seismic energy (i.e., the corresponding frequency when the energy in the analysis time window reaches 85% of total energy; Figure 6.42) is related to fluid properties—the greater the seismic

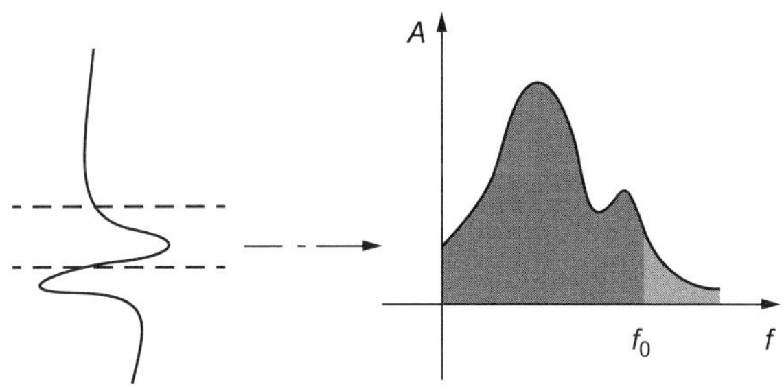

FIGURE 6.42

Illustration of the frequency corresponding to 85% of seismic energy.

wave attenuation, the smaller the frequency value; and (3) the frequency attenuation gradient refers to the intensity of frequency attenuation within unit distance (or time) and reflects the absorption of P-wave propagation by the strata—the stronger the strata's absorption, the greater the gradient value.

3. Prediction method

The fracture-sensitive seismic attributes can be optimized based on the anisotropic characteristics of the prestack seismic attributes of fractures, through subazimuth processing of prestack seismic data, and hence the degree of fracture development and azimuth can be predicted. This is discussed next using the XX gas field as an example.

1) Forward modeling

Forward modeling is an important approach for building petrophysical models and analyzing the seismic anisotropic response characteristics of the models, which is the basis for predicting fractures with prestack seismic data.

HTI media are formed in volcanic reservoirs where vertical fractures are developed, and five independent elastic coefficients are needed to describe the anisotropic characteristics. Random anisotropic media will be formed by stratum inclination caused by tectonic movement, and 21 elastic coefficients are needed to describe the anisotropy. On the basis of the fractured reservoir geological model and elastic media model, and using rock P-wave velocity, S-wave velocity, and density parameters, the prestack forward model is built to calculate the elastic tensor and anisotropy-equivalent Thomsen indices, to calibrate the response of prestack seismic reflection along individual azimuths, and to determine the relationship between seismic reflection amplitude and fractures. The calculation results provide seismic responses to fractures at near-borehole seismic traces, including prestack amplitude variation with offset (AVO) characteristics, prestack AVO characteristics of individual azimuths, and the changes in AVO characteristics with azimuth influenced by fractures.

Figure 6.43 shows the results of prestack forward modeling for Well XX5 at different incident angles. The designed fracture direction is true north, where the azimuth is 0 degrees toward the true east direction. The first trace in the diagram is the gamma curve. The second trace shows synthetic traces and seismic records. For the third trace, the upper section simulates how the reflection amplitude of volcanic reservoirs changes with the changing incident angle with different azimuths, and the lower section is the amplitude ellipse simulated by reflection amplitudes of individual azimuths at different incident angles. Figure 6.43 demonstrates the following:

1. When the incident angle is small, it is equivalent to the vertical incidence of seismic waves, where the reflection amplitude of various azimuths has no difference; the larger the incident angle, the bigger the difference. Therefore, prestack fracture detection requires data acquired using large incident angles (i.e., large offsets).
2. With a fixed azimuth, the amplitude of the reflection wave is the maximum when it is perpendicular to the fracture direction (an incident angle of 0 degrees), decreases as the incident angle increases, and becomes the minimum when it is parallel to the fracture direction (90 degrees).
3. The difference between reflection amplitudes at various azimuths increases with the incident angle, and likewise the ellipticity of reflection amplitudes increases with the incident angle. When the incident angle is ≤ 3 degrees, the ellipticity is 1. At an incident angle of 12 degrees, the ellipticity is 1.006. When the incident angle reaches 24 degrees, the ellipticity is 1.026. At an incident angle of 27 degrees, the ellipticity is 1.033.

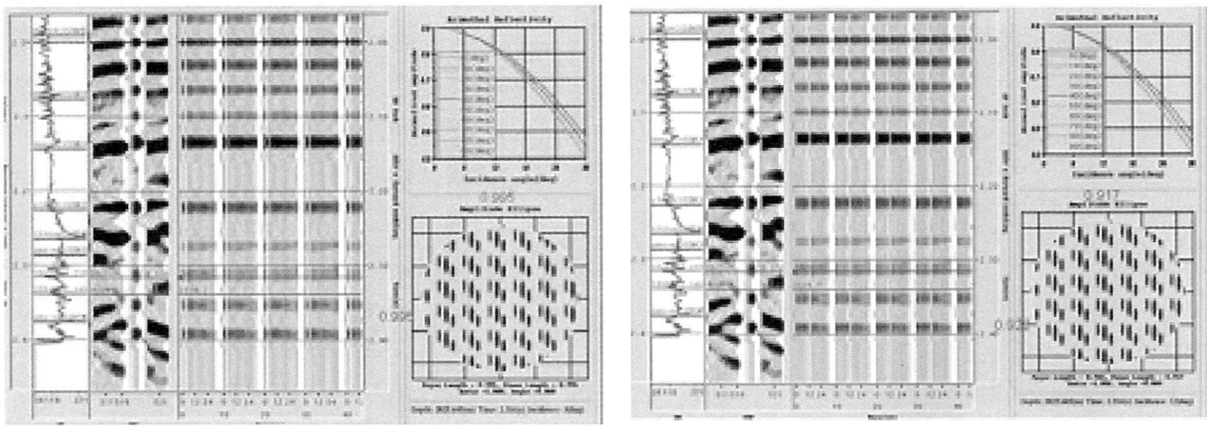

(a) (b)

FIGURE 6.43

Results of forward modeling in Well XX5 at different incident angles. (a) Incident angle at 3 degrees. (b) Incident angle at 12 degrees.

The incident angle in the target layer of the field survey system in the XX gas field is between 2 degrees and 13.5 degrees, without significant azimuthal differences. The requirements of prestack fracture detection in processing must be ensured as much as possible.

2) Prestack subazimuth processing of fractures

Prestack subazimuth processing of fractures is an important part of prestack fracture detection. The processing procedure is as follows.

Preliminary processing. This step includes denoising, deconvolution, amplitude compensation processing, and so on. The main processing parameters include common midpoint (CMP) bins 25 m × 50 m, number of record trace 1920, and sampling rate 1 ms, recorded length 6 s, and datum level 120 m, replacing speed 1600 m/s.

Prestack conventional processing. This step focuses on velocity selection and residual static correction. Velocity selection depends on whether the energy of the velocity spectrum is concentrated and whether there is an obvious trend in velocity change. Moreover, the first arrival muting and spike pulse energy depression of large amplitude must be performed, and spectrum whitening must be used to improve the data resolution.

Subazimuth processing. Subazimuth processing is the most important step in prestack fracture detection. It is affected by uneven azimuth distribution, survey system offset, and the raw data interval times. On the basis of the actual situation in the XX gas field, the azimuth division scheme is determined by the method of azimuth division with unequal spacing (Table 6.6). At the same time, the range of offset is restricted at 200 to 1600 m, for the following reasons: (1) near-offset data come from near-vertical incident seismic waves and cannot reflect the azimuthal AVO characteristics of the fracture media; (2) the azimuth distribution or seismic data volume offsets are uneven, with unequal azimuth being used for division and uniformization processing being performed in bins; and (3) the bin cover times cannot be too large because some bins will have to take a detour far away, but the cover times should not be too small either because the stack quality will deteriorate. It is generally appropriate to use the cover times of 25.

Figure 6.44 is the processed profile of different attributes at different azimuths, and the results of subazimuth processing show that the structures have a normal shape, and the signal-to-noise ratio (SNR) and resolution satisfy the fracture detection requirements.

3) Prestack fracture prediction methods

Method selection. The application of various methods on the volcanic rock reservoirs of the XX gas field within the 0 to 20 ms time window range produced four types of distribution maps of anisotropic intensities (Figure 6.45). Among these, the warm tones indicate a strong anisotropic intensity and

Table 6.6 Azimuth Division Table

Range of azimuth/ (degrees)	0~41.4	41.4~68	68~90	90~112.3	112.3~138.6	138.6~180
Average azimuth/ (degrees)	20.7	54.7	79	101.2	125.5	159.3

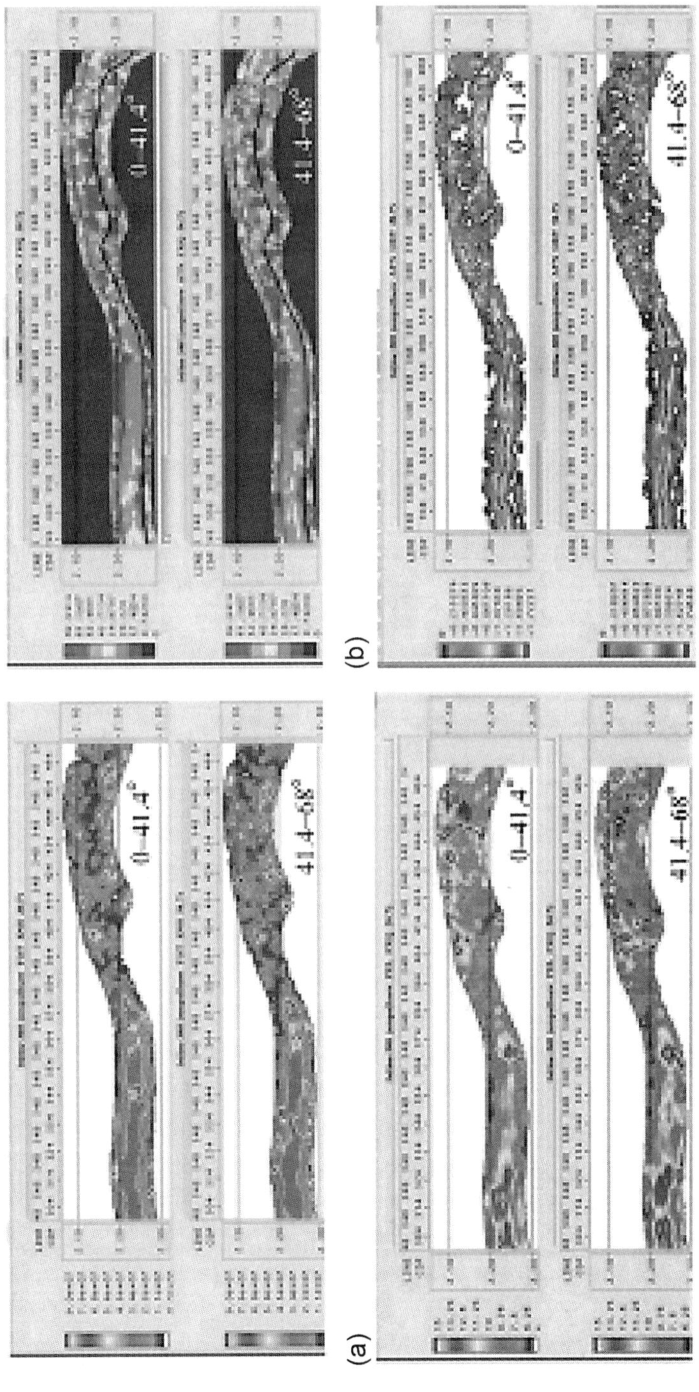

FIGURE 6.44

Processing profiles of different attributes at different azimuths. (a) Total energy. (b) Dominant frequency. (c) Quantile frequency of seismic energy. (d) Attenuation gradient.

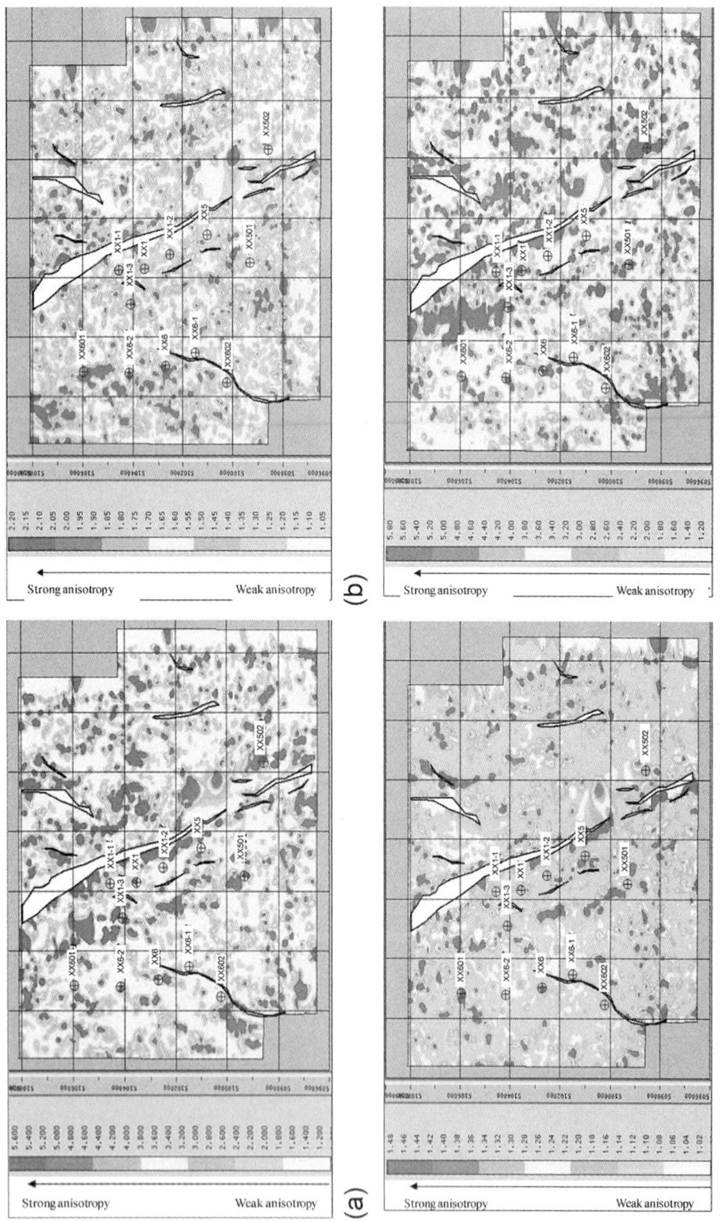

FIGURE 6.45

Distribution maps of anisotropic intensity for different attributes (XX gas field). (a) Total energy. (b) Dominant frequency. (c) Quantile frequency of seismic energy. (d) Attenuation gradient.

Table 6.7 Table of Matching Information on Fracture Development in the Upper Interval of Cycle I in Volcanic Rocks of the XX Gas Field within a 0- to 20-ms Time Window

Well Name	Fracture Density/ (Fractures/M)	Quantile Frequency of Seismic Energy	Attenuation Gradient	Dominant Frequency	Amplitude
601	4.5	Match	Unmatch	Unmatch	Unmatch
XX1	1.9	Match	Unmatch	Match	Unmatch
XX6	4.45	Match	Match	Match	Match
XX5	0	Match	Match	Match	Match
XX502	3.7	Fair match	Match	Unmatch	Match

highly developed vertical fractures, and cold tones indicate a weak anisotropic intensity and less developed vertical fractures. The correlation analysis of fracture characteristics of the drilled target layer (Table 6.7) indicates that among the four attributes, the "quantile frequency of seismic energy" offers the highest congruence rate of up to 80%, and the remaining three methods are relatively poor, with a congruence rate of 60%. Therefore, the "quantile frequency of prestack seismic energy" attribute is selected for predicting fracture development characteristics of volcanic reservoirs.

Fracture prediction. On the basis of prestack subazimuth processing of seismic data and the changes in quantile frequency of seismic energy with azimuth, the volcanic rock fracture development and orientation in target layers are predicted horizontally.

Figure 6.46 shows the following results of fracture prediction in the research area. Figure 6.46a illustrates the magnitude of ellipticity of the azimuthal ellipse for the "quantile frequency of seismic energy" attribute, where warm tones indicate a large ellipticity and a high fracture density, and cold tones depict a small ellipticity and a low fracture density. Figure 6.46b shows the result of directional rose-diagram statistics within the elliptical space of the same attribute. This result indicates the azimuth of fractures, and the size of the statistical grid is 100 m × 100 m. In this figure, the fault trends represented by black polygons are predominantly NNW and secondarily NNE; fractures are mostly developed to the left of faults along the NNW direction, and the fracture direction is mostly EW and NNW.

4) Analysis of controlling factors

The major controlling factors in prestack fracture prediction include the following.

Geological characteristics of volcanic gas reservoirs. The geological factors leading to the anisotropy in strata are complex. In addition to directionally arranged fractures, the factors also include lithological change, anisotropy inside rocks, fractures, folds, and so forth. Petrophysical research at Stanford University (in the United States) suggests that mudstone has a stronger anisotropy than sandstone. Fracture prediction based on the anisotropy in volcanic reservoirs is ambiguous. In practice, it must be combined with geological studies in order to enhance the geological guidance.

Acquisition and processing factors. Prestack fracture detection utilizes 3D data to acquire azimuthal anisotropy of P-wave attributes. Theoretically, it requires accurate seismic incident data from at least three different azimuths. Therefore, the acquisition process must satisfy wide-azimuth acquisition, as well as ensure uniform time coverage along individual azimuths and uniform distribution of offsets at individual azimuths. In actual practice, it is a challenging task to meet these requirements. During data processing, therefore, multiple methods must be used to determine processing parameters so as to minimize the negative impact of acquisition factors.

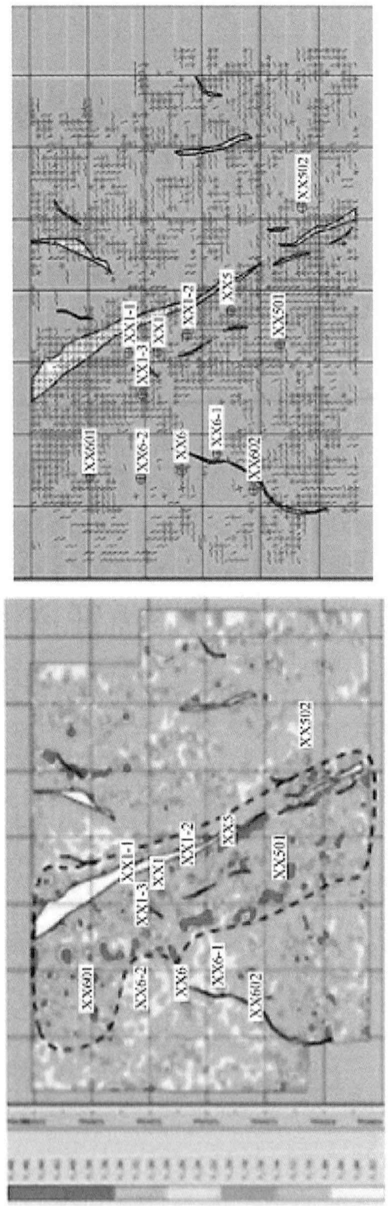

FIGURE 6.46

Planar map of fracture prediction in cycle I of volcanic rocks (Well Block XX1). (a) Fracture density. (b) Direction of fracture zone.

6.4.4 Fracture parameter inversion methods

The statistical relationship between fracture parameters and wave impedance is established on the basis of wave impedance inversion. 3D wave impedance data volumes are converted into fracture development parameter volumes with specific geological significance. These steps facilitate the effective prediction of fracture distribution in volcanic reservoirs [28].

1. Analysis of the statistical relationship between fracture parameters and wave impedance

From a petrophysical perspective, there are certain relationships among wave impedance, resistivity, and the fracture development index. The relationship between wave impedance and the fracture development index can be obtained through crossplot analysis.

1) The relationship between fracture index RTC and wave impedance

The dual laterolog resistivity is fairly sensitive to fractures. On the basis of crossplot analysis of wave impedance versus deep and shallow laterolog resistivity values (Figure 6.47), the conversion formula fitting wave impedance to resistivity is obtained:

$$RLLD = 0.043e^{0.0006\,\mathrm{Im}} \tag{6.4.6}$$

The formula converting deep laterolog to shallow laterolog is obtained through statistics:

$$RLLS = -3.853 \times 10^{-5} RLLD^2 + 0.916 RLLD + 4.234 \tag{6.4.7}$$

The formula converting deep/shallow laterolog to fracture index RTC is as follows:

$$RTC = (2.589RLLD - 1.589RLLS)/RLLS \tag{6.4.8}$$

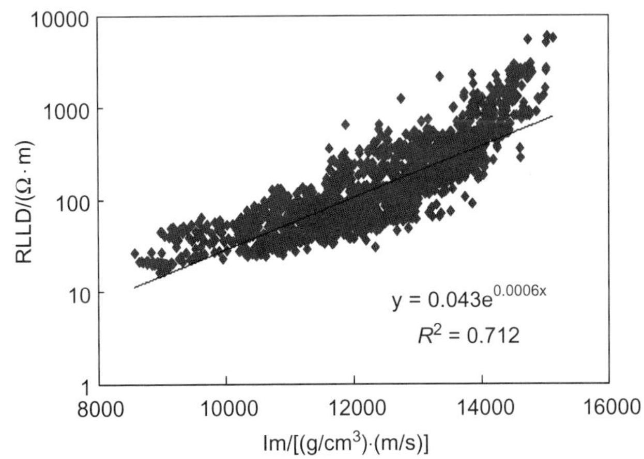

FIGURE 6.47

Crossplot between deep laterolog resistivity and wave impedance.

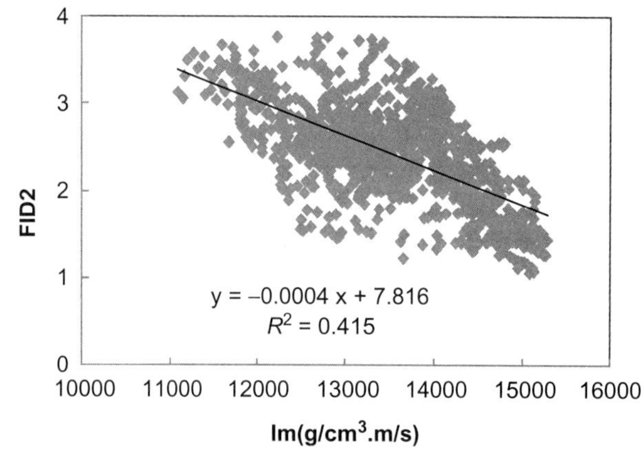

FIGURE 6.48

Crossplot of the fracture development index versus wave impedance.

Where

RLLD, RLLS — deep laterolog resistivity, shallow laterolog resistivity, (Ω·m);
Im — wave impedance [$(g/cm^3)\cdot(m/s)$];
RTC — fracture index; the larger the RTC value, the more developed the high-angle structural fractures.

2) The relationship between fracture development index FID2 and wave impedance

The degree of fracture development is positively correlated with the fracture development index but inversely correlated with wave impedance. Correlation analysis (Figure 6.48) indicates that there is a negative linear relationship between fracture development index FID2 and wave impedance:

$$FID2 = 7.82 - 0.0004\mathrm{Im} \tag{6.4.9}$$

Where

FID2 — fracture development index;
Im — wave impedance [(g/cm^3) (m/s)].

2. Wave impedance inversion

Conventional wave impedance inversion includes basic processes such as fine horizon calibration, wavelet extraction, initial model setup, and volume inversion. The inversion results yield volcanic reservoir wave impedance 3D data volumes.

3. Fracture parameter inversion

There is a good statistical relationship between RTC and wave impedance, but a poor statistical correlation between FID2 and wave impedance. Therefore, the following two types of inversion are adapted.

1) RTC inversion

This inversion is performed in three steps: (1) the previously mentioned expressions are used to convert relative wave impedance data volume obtained through sparse-spike inversion into deep and shallow laterolog resistivity data volume; (2) a geological model is built by interpolating deep and shallow laterolog resistivity curves, followed by frequency band compensation performed on the model to correct its low-frequency and high-frequency components, and finally absolute deep and shallow laterolog resistivity inversion data volumes are generated; and (3) RTC data volumes are ultimately obtained through conversion in accordance with the relationship between deep/shallow laterologs and RTC.

2) FID2 inversion

This inversion is performed in the following steps using the so-called double inversion approach (Figure 6.49): (1) an apparent acoustic curve is established by the statistical relationship between the fracture development index (FID2) and acoustic wave; (2) reservoir horizon calibration and wavelet extraction occur; (3) the inversion model is constructed; (4) the pseudo-impedance volume derived from the characteristic parameters of fractures is obtained through constrained inversion of the model; and (5) fracture development index 3D data volume is obtained through constrained inversion of the characteristic parameter curve for the fractures.

4. Fracture prediction

In the volcanic reservoir of YC_3I in Block XXS2-1, fracture distribution is predicted through RTC inversion. The results indicate (Figure 6.50) that areas with high RTC values are mainly located near Well XXS4, along the XXS2-1 to XXS2-6 to XXS2-7 to XXS203 zone, near XXS202 to XXS2-25, as well as northwest and north of the study area. Drilling operation has since confirmed that these are fracture-developed zones.

In the volcanic reservoir of YC_1I in Block XX8 and Block XX9, the results of FID2 inversion indicate (Figure 6.51) that areas marked by high FID2 values correspond to the development of horizontal strips, present in a total of five plays, including XX9 to XX902, XX3 to XX14, XX8 to XX801, XX301, and XX7 to XX3. These fractured zones in the volcanic reservoir have also been confirmed by subsequent drilling.

6.4.5 Methods for optimization of volcanic reservoir fracture prediction

Many fracture prediction methods are available. The fracture prediction method suitable for a particular volcanic reservoir is optimized through assessment of the advantages, disadvantages, contributing factors, and seismic data requirements of different prediction methods, with reference to the volcanic reservoir characteristics.

1. Characteristics of fracture prediction methods

Different fracture prediction methods are based on different principles and, therefore, their advantages and disadvantages are also different (Table 6.8).

1) Method of poststack seismic attribute analysis

1. *Coherence analysis.* This is a sophisticated attribute analysis technique. It predicts fractures based on the dissimilarity between adjacent seismic traces. Its advantages include easy extraction of attributes and simple operation. However, it is strongly affected by factors such as lithological boundaries and faults, with strong ambiguity.

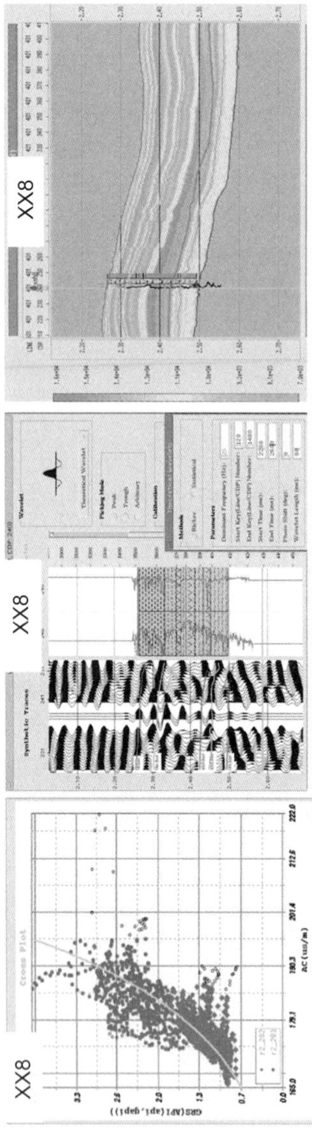

FIGURE 6.49

Processing procedure of fracture development index inversion. (a) Construction of acoustic dimensions. (b) Calibration of horizon and extraction of the wavelet. (c) Creation of the inversion model.

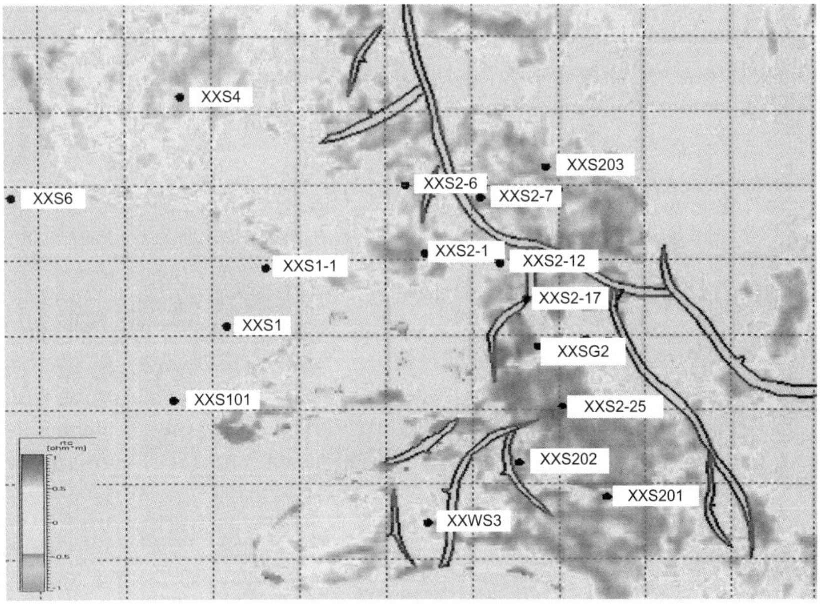

FIGURE 6.50

Planar distribution of RTC on YC_3I top surface of Block XXS2-1.

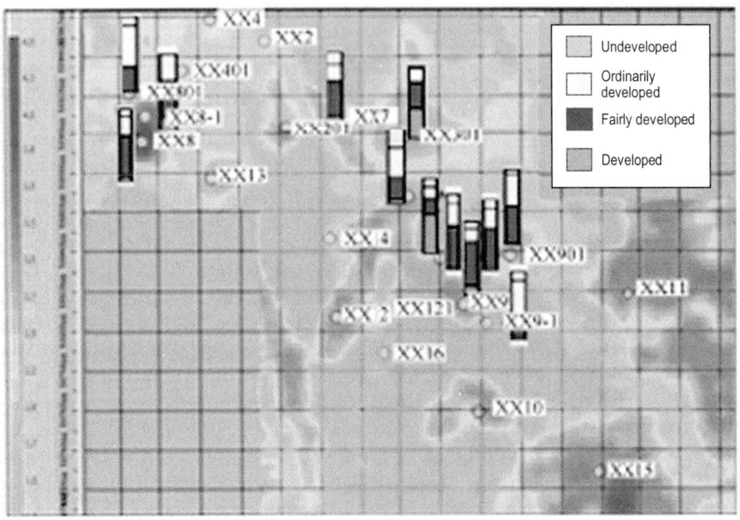

FIGURE 6.51

Planar distribution of FID2 in Reservoir YC_1I of Blocks XX8 and XX9.

Table 6.8 Comparison of Volcanic Reservoir Fracture Prediction Methods

Prediction Technique		Major Advantage	Disadvantage	Influencing Factors
Poststack seismic attribute analysis	Coherence analysis	Sophisticated technique, simple operation	Strong ambiguity	Lithological boundary, fault
	Margin detection	Intuitive image identification	Fuzzy boundary, low definition	Lithological boundary, fault boundary
	Three-instantaneous attribute	Simple, direct, and reliable	Strong ambiguity	Internal structure, fluid types
	Absorption coefficient	Sensitivity is high, result is good	Gas influence is high	Pore structure, fluid flow, local saturation effect
	Stress analysis	Theoretical basis is reliable	Paleostructure restoration too complex	Formation thickness, lithology, structure history
Prestack seismic prediction method		Subazimuth processing, ambiguity is reduced, development degree and direction are also predicted	Strict requirements for wide-azimuth acquisition of seismic data; heavy workload for long time coverage	Data acquisition and processing; other anisotropic characteristics of formation
Fracture parameter inversion method		Well-constrained, easy to implement	Dependence on level of correlation between fracture parameters and seismic information, uncertainty high	Relationship between inverted parameter and wave impedance; thickness and distribution of sedimentary rocks

2. *Margin-detection technique*. This method predicts fractures through image analysis, which is visual, intuitive, and suitable for detecting large faults. For fracture prediction, however, the boundaries tend to be fuzzy, of low definition, and easily affected by faults, lithological boundaries, and so on.
3. *Three-instantaneous-attribute method*. This method detects fractures directly with P-wave dynamical information (amplitude, frequency, phase), which is simple, direct, and reliable. However, this method is significantly affected by internal structure and fluid type in volcanic rocks, also resulting in strong ambiguity.
4. *Absorption coefficient method*. This method predicts fractures based on the principle that fractures will intensify the seismic wave absorptive attenuation. Its advantages include high sensitivity and a good prediction result. However, it is strongly affected by the gas-bearing nature of formations and is also influenced by many other factors such as porous structures, fluid flow, and local saturation effect, also leading to strong ambiguity.

5. *Stress analysis*. This method predicts fracture distribution by calculating the strain generated during tectonic activities using the kinematic tectonic reconstruction method. It has a reliable theoretical basis and offers a good prediction result. However, the tectonic reconstruction process is complicated, thus the prediction result is affected by factors such as formation thickness, lithology, and structural history, resulting in some uncertainty.

The method of poststack seismic attribute analysis requires high-quality poststack 3D seismic data.

2) Prestack seismic prediction methods

These methods predict fracture distribution by the "quantile frequency of seismic energy" attribute based on fracture anisotropy. Their advantages include the ability to reduce ambiguity through subazimuth processing. However, subazimuth processing needs a huge data volume and a long cycle time, and it requires the data volumes acquired via wide azimuth and processed via subazimuth. At the same time, the prediction result is easily affected by other anisotropic media (such as mudstone) in the formation.

3) Fracture parameter inversion method

The spatial distribution of characteristic parameters of fractures is predicted using seismic inversion techniques, which are simple to implement and offer a high well-point congruence rate. However, their prediction effectiveness depends on the level of correlation between fracture parameters and seismic information, leading to a high degree of prediction uncertainty. They are also strongly affected by the thickness and distribution pattern of sedimentary rocks, leading to further increase in uncertainty. The fracture parameter inversion methods require high-quality 3D seismic data.

2. Optimization of volcanic fracture prediction methods

Volcanic fracture prediction methods under different conditions are optimized based on the guidelines of employing the highest-quality seismic data and most advanced prediction methods available, taking into consideration characteristics of volcanic rocks such as highly variable rock types, complex internal structures, complex fluid distribution, and a great variety of fracture types:

1. If the field seismic data are collected with wide-azimuth and processed by prestack subazimuth and amplitude preserving method, the prestack fracture prediction method can be used to predict fracture-developed zones for improving prediction accuracy.
2. If only conventional poststack seismic data are available, several methods, such as stress analysis, absorption coefficient, and multiattribute coherence, can be used together to predict fracture-developed zones with reduced ambiguity.
3. Poststack seismic inversion of fracture parameters is now considered an auxiliary method for the two techniques.

In summary, when seismic methods are used to predict fracture developed zones, the seismic data must go through high-resolution, high-SNR, and high-definition processing to meet the requirements of fracture prediction.

6.5 TECHNOLOGICAL APPLICATIONS

The characteristics and distribution patterns of fractures in volcanic reservoirs can be ascertained using various fracture identification and prediction techniques. This will provide the basis for predicting favorable reservoir zones, optimizing well locations, and analyzing bottom water coning.

6.5.1 Prediction of favorable reservoir zone in volcanic gas reservoirs

In volcanic reservoirs, fractures are important for connecting pores and enhancing fluid flow pathways. As the degree of fracture development increases, the permeability of volcanic reservoirs increases by 1.8 to 95.6 folds compared to matrix permeability (see Figure 6.19). As a result, fractures constitute a major controlling factor for the permeability of volcanic reservoirs. Favorable reservoir zones in volcanic gas reservoirs can be predicted on the basis of fracture distribution and the distribution characteristics of effective reservoirs. This lays the foundation for locating high-yielding wells.

6.5.2 Well location optimization

Fractures determine the well grid pattern, well spacing, and horizontal well trajectories. For development wells in fractured gas reservoirs, irregular well patterns are generally used: (1) high-yielding wells should be placed to a maximum in fracture-developed zones; (2) wide well spacing should be adopted along the direction of fractures, and narrow well spacing along the direction perpendicular to fractures; the more developed the fractures and the longer the fracture half-length, the higher the irregularity of the development well pattern; and (3) the trajectory of horizontal wells is generally perpendicular or oblique to fractures. In the XX, CC, and DD gas fields, 12 horizontal wells were drilled on the basis of the prediction of fractures in volcanic rocks, and all of which have met their expected goals, with excellent results.

6.5.3 Determination of vertical distribution of fractures for analysis of bottom water coning

Fractures are important communicating mechanisms between gas layers and bottom water. When the production pressure differential exceeds a given value, the bottom water may migrate upward along fractures, resulting in water production from gas layers. The vertical distribution of fractures can be assessed through fracture identification and evaluation, which will provide the basis for the evaluation of bottom water coning.

References

[1] Gao F, Xu X, Zou H, et al. Reservoir space characters of Late Jurassic-Early Cretaceous volcanic rocks from Shengping area of Songliao basin and their affecting factors. Global Geology 2006;25(3):291–5.

[2] Hou Y. Characteristics and controlling factors of volcanic reservoir spaces in Xingshan-Yingshan area of Songliao basin. Special Oil & Gas Reservoirs 2001;10(1):99–105.

[3] Gao S, Li X, Song B. Features of volcanic rock reservoir spaces in Oulituozi region, Liaohe basin. Oil Gas Geol 2001;22(2):173–7.

[4] Rigby AF. Fracture identification in an igneous geothermal reservoir surprise valley. In: SPWLA, 21st annual logging symposium. Lafayette, Louisiana; 1980.
[5] Zhao L, Bu Y. Log evaluation technique of carbonate reservoirs. Beijing: Petroleum Industry Press; 1994, p. 15–30.
[6] Sibit AM, Faivre O. The dual laterolog response in fractured rocks. In: SPWLA, 26th annual symposium transaction. Dalas, Texas; 1985.
[7] Sima L, Shu Z. Log evaluation methods for carbonate reservoirs and their applications. Beijing: Petroleum Industry Press; 2009, p. 36–62, 81, 82.
[8] Wang Y, Hu Y, Ran Q. Fracture development evaluation of deep volcanic gas reservoirs. Natural Gas Industry 2007;27(8):31–4.
[9] Sun J, Wang Y. Integrated application of geophysical data. Dongying: China University of Petroleum Press; 2001, p. 173–78.
[10] Sima L. Logging geology application techniques. In: Beijing: Petroleum Industry Press; 2002 p. 40–59, 102–104.
[11] Liu Z, Dai S, Wang H, et al. Fracture effectivity evaluation of igneous rock with sonic logging. Well Logging Technology 2007;31(2):56–8.
[12] Yang T, Hu Z. Application of anisotropy in evaluating fracture effectivity. World Well Logging Technology 1999;14(3):58–60.
[13] Department of Geosciences of School of Science of Zhejiang University. A study on volcanic eruption stages in volcanic gas reservoirs at Shengping and Xingcheng area (internal research report in Chinese). p. 50–55.
[14] Li X, Zhang X, Li Q. The determination of the spatial occurrence of buried layers using DRM isolated from core. Comput Tech Geophys Geochem Explor 2001;23(3):236–9.
[15] Dong P. Orientation determination of maximum horizontal stress in reservoir formation by paleomagnetic orientation of cores. Chinese Journal of Rock Mechanics and Engineering 2004;23(4):2480–3.
[16] Zhang S, Gu C, Cao G. Imaging logging techniques and applications. Beijing: Petroleum Industry Press; 1997, p. 30–50.
[17] Yuan M, Pan M, Tong H, et al. Low-permeability fracture oil reservoir exploration. Beijing: Petroleum Industry Press; 2000, p. 199–221.
[18] Zhu C. Application effect analysis of fracture prediction techniques in Songnan prospect area. Geophysical Prospecting for Petroleum 2001;40(4):62–8.
[19] Tao Y, Zheng D, Zhang L, et al. A new correlation analytical technique and its application in complicated igneous rock region. Journal of Oil and Gas Technology 2006;28(1):58–61.
[20] He R, Huang H, Li Q, et al. Seismic fracture prediction in igneous rocks of Well Block Shang-741. Petroleum Geophysical Explorations 2005;40(6):682–7.
[21] Gou L, Peng Z. Multi-dimension marginal detection of wavelet and its application in fracture prediction. Geophysical Explorations 2005;40(3):309–13.
[22] Tao G. Integrated identification technology of multiple-factor neural network for buried hill reservoir. Petroleum Geology and Recovery Efficiency 2006;13(4):36–8.
[23] Guan W, Tan L, Li G, et al. Application of stress variation analysis to predict buried-hill fracture of ordovician in Lunnan area. Xinjiang Petroleum Geology 2009;30(2):267–8.
[24] Ren D, Zhang X, Chen S. Fissure forecasting of volcanic reservoir at Well Block Fangshen 9 in northern Songliao basin. Journal of Geomechanics 2002;8(3):279–86.
[25] Zhang M, Yao F, Han D, et al. Progress of multi-component seismic for fracture detection. Natural Gas Geoscience 2007;18(2):293–7.
[26] Du Q, Yang H. Detection method for fractures in azimuthally anisotropic media. Journal of the University of Petroleum, China (Edition of Natural Science) 2003;27(4):32–6.
[27] Gan Q, Gao Z. Application of AVA fracture detection technique. Natural Gas Industry 2005;25(5):42–3.
[28] Zheng Z, Dai Z, Zheng L, et al. Application of inversion technique for fracture prediction in Wenmingzhai oilfield. Inner Mongolia Petrochemical Industry 2008;15:135–6.

CHAPTER

7 Parameter Interpretation for Fractured Volcanic Reservoirs

Reservoir parameters constitute the basis for quantitative evaluation of the accumulation-permeation capacity and gas-bearing characteristics in volcanic gas reservoirs. The porosity, permeability, and gas saturation of volcanic rock matrices and fractures can be interpreted and the petrophysical properties and gas-bearing characteristics of volcanic reservoirs clarified on the basis of core experiments and preprocessing of log data by combining theoretical models with statistical models and through the integrated application of information derived from nuclear magnetic, acoustic-electric imaging, and conventional logs. This forms the basis for the quantitative characterization of volcanic reservoirs, the appropriate evaluation of producible reserves, and geological model building.

7.1 CHALLENGES IN PARAMETER INTERPRETATION AND TECHNICAL SOLUTIONS

7.1.1 Challenges in interpretation of volcanic reservoir parameters

Compared to sedimentary rocks, volcanic rocks have complex lithology, diverse reservoir spaces, and complex porous structures and electrical conductive mechanisms, resulting in complicated logging responses to volcanic reservoirs and increased difficulties in interpretation of reservoir parameters.

1. Difficulties in porosity interpretation due to great changes in parameters of volcanic rock matrices and the complex characteristics of porosity logging responses

Volcanic rocks are highly variable in lithology and composition, resulting in great changes in matrix parameters. In the XX gas field, for example, the density of the volcanic matrix ranges from 2.5 to 2.89 g/cm^3 with a distribution width of 0.39 g/cm^3, corresponding to a porosity variation about 20%. For acidic rhyolite, it has a matrix density distribution ranging from 2.5 g/cm^3 to 2.75 g/cm^3, with a distribution width of 0.25 g/cm^3, corresponding to a porosity variation about 13%. In the study area, the volcanic reservoirs have an average porosity of only 6% to 8%, showing a significant effect of matrix parameter variations on the interpretation accuracy for reservoir porosity.

There are various types of reservoir space in volcanic reservoirs. It may be divided into two categories based on the origin—primary porosity and secondary porosity—which can be subdivided into vesicles, interstitial pores, dissolution pores, and various fractures. Reservoir spaces can be divided into connected pores and isolated pores in terms of their connective condition. For different characteristics in terms of size, assemblage, and communication, various pores have different logging response characteristics. It is difficult to discriminate various types of reservoir space and pore-fracture assemblage with log data, making it relatively difficult to interpret porosity.

2. Difficulties in permeability interpretation due to complex porosity and permeability relationships associated with complex porous structures in volcanic reservoirs

The types of reservoir space in volcanic rocks are diverse in characteristics, size, and shape, leading to various distribution patterns and communication modes, as well as different relationships between porosity and permeability. There are multiple types of throat in volcanic reservoirs, and the characteristics of their morphology, size, tortuosity, and pore-throat configuration are different, with complex porous structures. Moreover, multiple types of accumulation-permeation association with different characteristics of fluid flow pathways lead to different permeability states in reservoirs. Therefore, the complicated relationship between porosity and permeability in volcanic rocks and the difficulties in permeability interpretation are the result of complex reservoir spaces, porous structures, and fluid flow pathways.

3. Difficulties in saturation interpretation due to complex mechanisms of electrical conductivity, fluid distribution, and gas/water relationships in volcanic rocks

A great variety of volcanic rock types are recognized corresponding to the generally complex volcanic lithology. In various types of volcanic rocks, the contents of metallic and clay mineral as well as fine-grained tuffaceous constituents are different; and the characteristics of surface conductivity caused by irreducible water film are also different, leading to different characteristics of electrical conductivity.

There are many types of porosity in volcanic rocks. The morphology, size, distribution, and connective mode of various porous spaces are different, resulting in different patterns of fluid distribution and occurrences in pores as well as distinctive effects on the three conductive media—that is, free fluid ions, irreducible water films, and fractures. These features cause differences in the response characteristics of nuclear magnetic resonance logs (NMR) and resistivity logs.

The complex porous structures and the great variety of accumulation-permeation modes in volcanic rocks are responsible for the many types and combinations of electrical conductivity modes, the complex and highly variable electrical conducting routes, and the complex response characteristics of resistivity logs.

These factors account for the high level of difficulties for interpreting the saturation states in volcanic gas reservoirs based on nuclear magnetic logs and resistivity logs.

7.1.2 Solutions and technical approaches

In view of the difficulties in interpreting volcanic reservoir parameters, the theoretical models for parameter interpretation and the statistical models derived from petrophysical experiments and logs are established for matrices and fractures in volcanic reservoirs based on core-based experiments and the preprocessing of log data as determined through the integrated use of the information acquired from nuclear magnetic, imaging, and conventional logs. Parameters such as porosity, permeability, and saturation are interpreted, and a set of parameter interpretation techniques for volcanic reservoirs is formulated (Figure 7.1).

1. Core-based experiment

From core-based experiments on conventional physical properties, petroelectrical properties, and nuclear magnetic and mercury injection, the petrophysical characteristics of volcanic reservoirs such as lithology, physical properties, electrical conductivity, and fluid mobility are ascertained, which serve a basis for constructing reservoir parameter interpretation models.

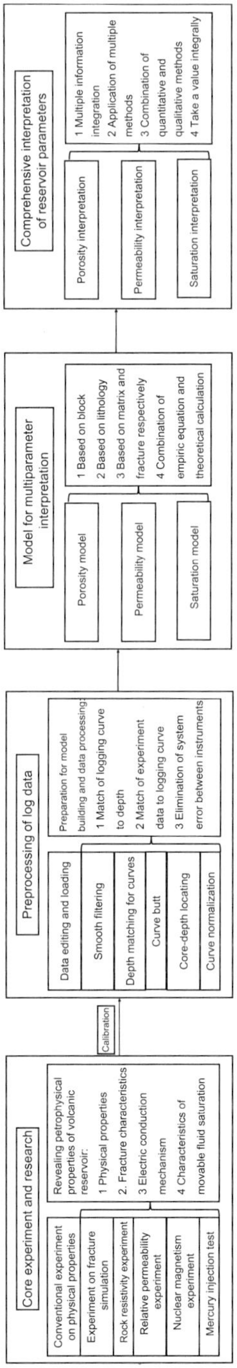

FIGURE 7.1

Technical process for interpreting the parameters in volcanic reservoirs.

2. Preprocessing of log data

By means of log data editing, loading, smooth filtering, depth matching between curves, curve-joining, core-depth reposition, and logging curve normalization, the experimental data are matched with logging curves by depths, and system errors between different instruments are corrected. These preprocessing procedures set the stage for the construction of interpretation models and data processing.

3. Establishing the parameter interpretation model

By integrating theoretical models with statistical models, the interpretation models of total porosity, matrix porosity, matrix permeability, matrix gas saturation, and fracture gas saturation are built lithologically for matrices and fractures block by block. As the interpretation of fracture parameters was described in Chapter 6, this chapter focuses mainly on the interpretation methods regarding matrix parameters.

4. Comprehensive interpretation of reservoir parameters

Reservoir parameters are calculated point by point in wells and interpreted systematically based on reservoir parameter interpretation models, by integrating quantitative and qualitative methods, and using the data derived from nuclear magnetic and conventional logs.

7.2 PREPROCESSING OF LOG DATA

The image logging system is used for most volcanic reservoirs with relatively simple processes of data editing, loading, smooth filtering, and depth matching between curves. The preprocessing mainly includes curve joining, core-depth adjustment, and normalization.

7.2.1 Curve joining

Log data in volcanic target intervals are generally acquired as a complete process; however, the acquisition may be performed interval by interval occasionally at the early stage due to insufficient knowledge. Also, all the data from the wellhead to the borehole bottom are required for refined calibration of seismic horizons. Therefore, curve joining is a rather important process, to be completed primarily according to the analogy of logging curves in the repeated intervals and taking into account the variation in lithology and physical properties (Figure 7.2).

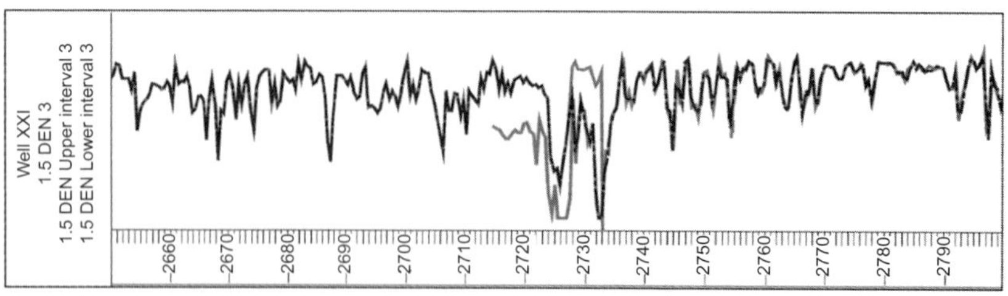

FIGURE 7.2

Curve joining of upper and lower intervals (density logging curve in Well XX1).

7.2.2 Core-depth adjustment and repositioning

When a core is collected, the depth is recorded by drilling pipes. In well log data acquisition, a cable is used to record the depth. This results in some system errors and random discrepancies in the recorded depths between these two methods. For this reason, the core depth data are calibrated and matched with logging depths on the basis of a lithological description of the cored interval by analyzing the similarity between core porosity and logging porosity, core density and logging density, and core porosity and the porosity logging curve (Figure 7.3).

7.2.3 Normalization of logging curves

System error exists between the same type of logging curves acquired in different wells because of the measuring time, instrument models, calibrating methods and operating errors. The purpose of normalization is to find and eliminate (or diminish) system errors caused by nonstratal factors by means of correlation with key layers:

1. *Selection of key layers.* Key layers in volcanic rocks are not well developed in general. Some "proxy key layers" are selected according to the criteria of pure lithology, large thickness, extension over the whole study area, regular borehole, flat curve, and minimal change. The thick glutenite in the Ying-4 interval and the thick and tight dacite in the Ying-1 interval in the XX gas field are good examples.
2. *Determination of matrix parameters.* The matrix parameters of the proxy key layers in a well are determined by logging curve crossplot analysis.
3. *Normalization of logging curves.* If the proxy key layer is a tight zone with clean lithology (e.g., the Ying-1 tight dacite), the error of normalization is determined through interwell correlation by means of histograms (Figure 7.4); if the lithology is not sufficiently pure (e.g., the Ying-4 glutenite), the error will be determined by means of crossing analysis of triple-porosity crossplots (Figure 7.5).

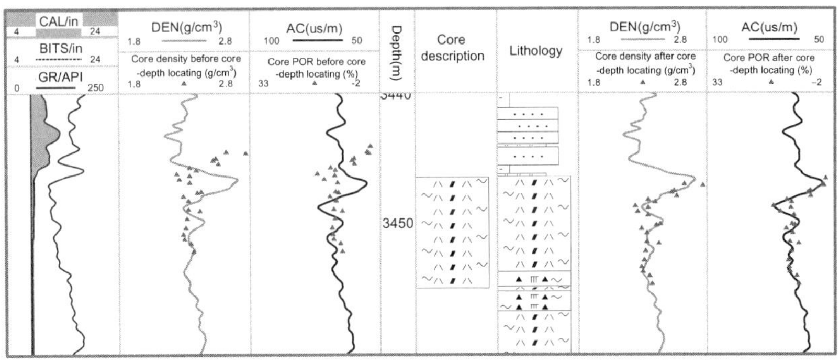

FIGURE 7.3

Core depth adjustment and repositioning (Well XX1).

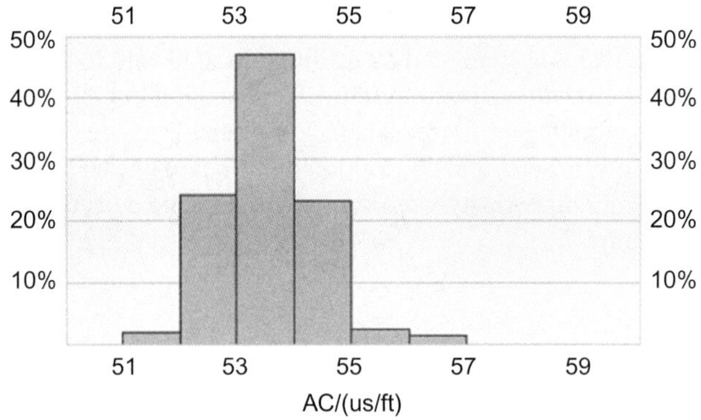

FIGURE 7.4

Histogram analysis (dacite).

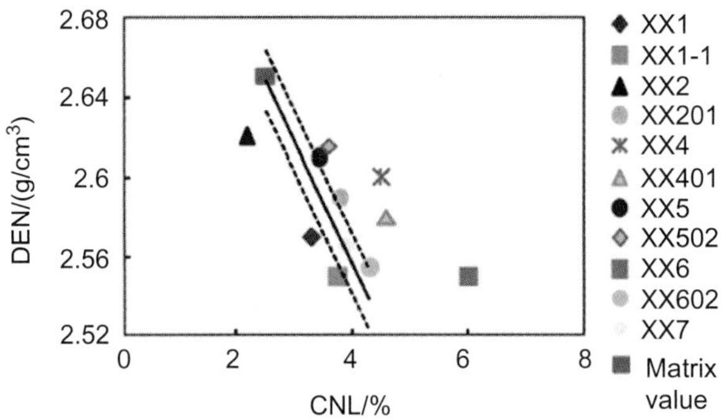

FIGURE 7.5

Crossplot analysis (glutenite).

7.3 INTERPRETATION OF THE POROSITY OF VOLCANIC RESERVOIRS

The total porosity and effective matrix porosity are interpreted on the basis of lithology, through determination of matrix parameters, integrating theoretical and statistical models, and combining conventional and nuclear magnetic logs.

7.3.1 Conceptual model for porosity calculation

The conceptual model for calculating reservoir porosity is derived from the multimedium model for volcanic rocks [1].

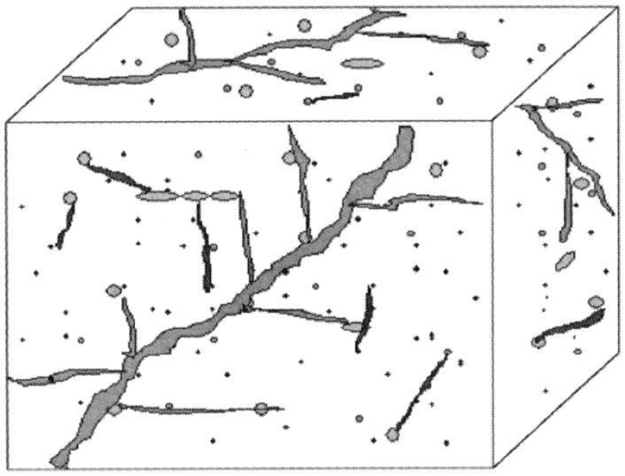

FIGURE 7.6

Multimedium model for volcanic reservoirs.

1. Multimedium model for volcanic rocks

As is shown in Figure 7.6, a multimedium model is built for volcanic rocks based on the study of characteristics of reservoir spaces, and it has been used to reveal the following basic features of volcanic reservoirs:

1. Cavities, pores, and fractures are well developed with various types of reservoir space and throat.
2. The sizes of pores and fractures vary widely, with the pore radius ranging from tens of centimeters to less than one millimeter, and the fractures including large structural fractures and small diagenetic fractures.
3. Pores connections are complicated and include three modes: throat connected, fracture connected, and unconnected.
4. In different pores, throats, and fractures of volcanic rocks, there are various fluid storage states, including mobile water, irreducible water, mobile hydrocarbon gas and residue gas, and so on.

2. Conceptual model for porosity calculation

Different principles adopted for measurement result in disparate response characteristics acquired using different logging methods for cavities, pores, and fractures. These factors must be considered when building conceptual models for calculating the porosity of volcanic reservoirs (Figure 7.7):

1. Nuclear magnetic logs are less affected by the reservoir matrix and have good responses to cavities, macropores, micropores, fractures, and irreducible water as well as responses to effectively dead pores (i.e., pores filled with fluid). According to T2 distribution, various types of porosities can also be discriminated and used to calculate the total porosity, effective matrix porosity, and the porosity of pores with various radiuses, which are effective methods based on porosity logs. However, they have the disadvantage of limited detection depth.
2. Acoustic logs mainly detect the effective porosity and irreducible water, other than the secondary porosity. Thus, they are used for calculating the effective matrix porosity.
3. Density logs have good responses to macropores, micropores, dead pores, irreducible water, and fractures, but not to the formation water and, therefore are used mainly to calculate total porosity.

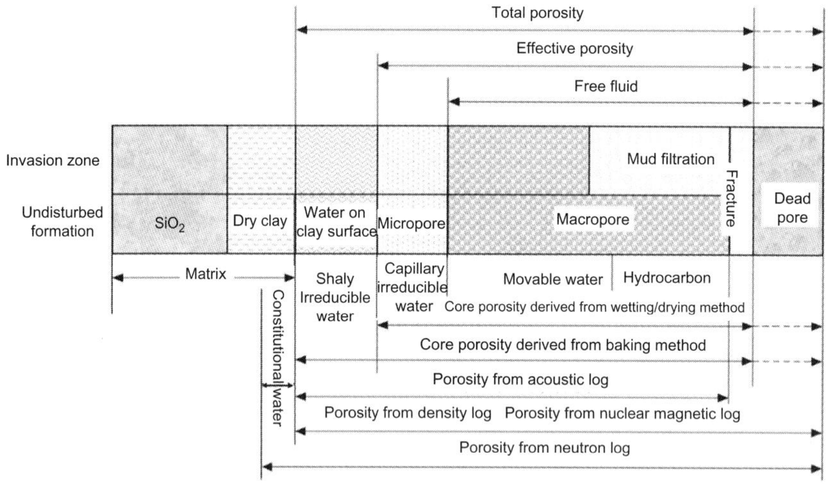

FIGURE 7.7

Conceptual model for calculating the porosity of volcanic reservoirs.

Also, through calibration, the model for calculating effective matrix porosity can be built based on lithology. Hence, statistical models can be used for calculating effective matrix porosity.

4. Neutron logs primarily detect the hydrogen contents in strata and have responses to all the porosities, which are mainly used for calculating total porosity. Neutron logs, however, are affected significantly by the formation water in the matrix and the variation in pore fluid types.

7.3.2 Determination of matrix parameters in volcanic rocks

The matrix parameters of volcanic rocks are determined by petrophysical analysis, log crossplot, and elemental capture spectroscopy (ECS) log analysis.

1. Petrophysical analysis

Rock density decreases as porosity increases, whereas interval transit time and neutron porosity increase as porosity increases. Using petrophysical data, the bivariate crossplots of rock density, interval transit time, and neutron porosity versus core porosity (Figures 7.8 and 7.9) can be obtained lithologically and regression-fitted to derive the values of density, interval transit time, and neutron porosity at a zero porosity on the regression line. Through these analytical steps, the matrix parameters of various volcanic rocks can be determined.

Table 7.1 shows the matrix parameter values in the XX and the DD gas fields. Derived from petrophysical analysis, these values indicate that (1) when rocks change from acidic to basic, the matrix density increases from 2.62 g/cm^3 to 2.84 g/cm^3 and the neutron porosity of matrix increases from 3.74% to 9.21%, but the interval transit time in matrix shows only a minimal change; (2) the matrix parameters in different types of volcanic rock are highly variable; and (3) there is a minor difference in the matrix parameters of volcanic rocks between the XX and the DD gas fields.

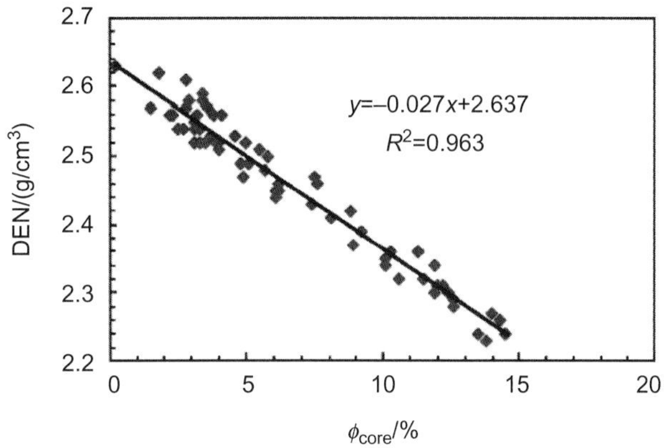

FIGURE 7.8

Crossplot of rock density versus core porosity (XX gas field).

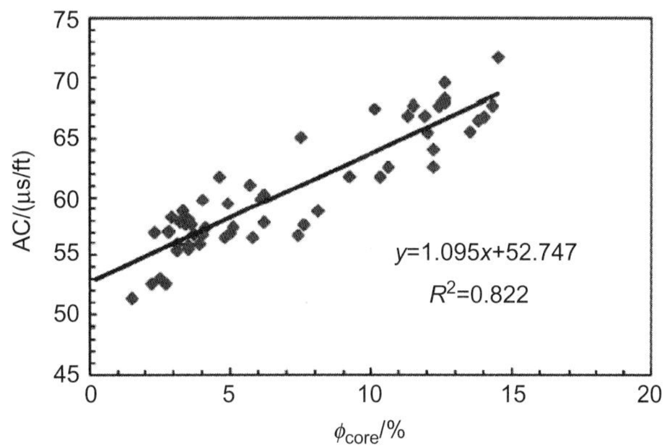

FIGURE 7.9

Crossplot of interval transit time versus core porosity (XX gas field).

2. Logging curve crossplot analysis

With lithology and fluid type being consistent, when porosity decreases, rock density tends to increase, the values of interval transit time log and porosity log decline, and rock resistivity rises. The volcanic reservoirs with the same lithology and fluid type are selected to generate the crossplots of rock density versus deep lateral resistivity (Figure 7.10), interval transit time versus deep lateral resistivity (Figure 7.11), and neutron versus deep lateral resistivity; and the density, interval transit time, and neutron porosity values at infinite resistivity are obtained and used as the values for the density, interval transit time, and neutron porosity of rock matrix in this type of volcanic reservoir, so as to determine the matrix parameters of various volcanic rocks.

Table 7.2 shows the matrix parameters in the XX gas field derived from an analysis of logging curve crossplots, which demonstrates that when rocks change from acidic to basic, the matrix density increases from 2.61 g/cm^3 to 2.80 g/cm^3 and the neutron porosity in matrix increases from 0.5% to 8%, but the interval transit times in the matrix of various types of rock differ from each other only slightly.

Table 7.1 Matrix Parameters of Volcanic Rocks Determined by Petrophysical Analysis (XX and DD Gas Fields)

Rock Type		Rock Density (g/cm^3)		Interval Transit Time (μs/ft)		Compensated Neutron %	
		XX Gas Field	DD Gas Field	XX Gas Field	DD Gas Field	XX Gas Field	DD Gas Field
Acidic volcanic rock	Lava	2.64	2.62	52.7	46.3	−1.35	−3.74
	Breccia	2.62	2.63	53.3	50.5	0.91	0.01
	Tuff	2.64	2.62	47.3	47.1	−1.74	0.02
	All acidic volcanic rock	2.64	2.63	51.2	48.9	−1.25	−0.82
Intermediate volcanic rock	Lava		2.69		46.2		3.77
	Breccia		2.65		54.4		8.00
	All intermediate volcanic rock	2.70	2.71	49.6	50.1	2.4	3.60
Basic volcanic rock	Lava		2.78		54.9		9.19
	Breccia		2.84		47.0		9.21
	All basic volcanic rock	2.80	2.79	52.7	54.7	7.5	9.15
Orthophyre			2.62		45.9		−3.23
Sedimentary volcanic rock		2.69	2.69	55.4	47.4	1.87	1.97

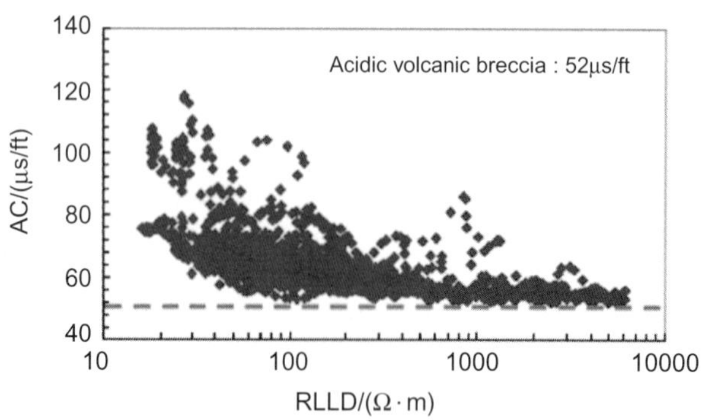

FIGURE 7.10

Crossplot of logging density versus deep lateral resistivity.

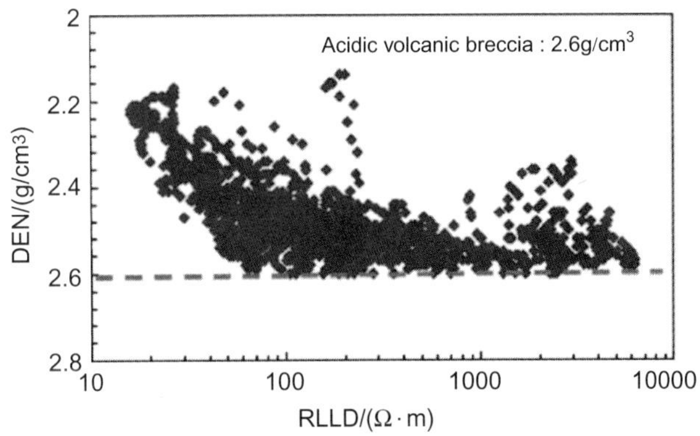

FIGURE 7.11

Crossplot of logging interval transit time versus deep lateral resistivity.

Table 7.2 Matrix Parameters of Volcanic Rocks as Determined by an Analysis of Logging Curve Crossplots (XX Gas Fields)

Rock Type		Rock Density (g/cm^3)	Interval Transit Time (μs/ft)	Compensated Neutron %
Acidic volcanic rock	Rhyolite	2.60	52.00	−0.50
	Breccia lava	2.58	51.50	−0.50
	Breccia	2.62	52.50	0.00
	Tuff	2.59	52.00	0.20
	All acidic volcanic rock	2.61	52.00	−0.50
Moderately acidic volcanic rock (dacite)		2.63	50.00	1.00
Intermediate volcanic rock (andesite)		2.72	52.00	4.00
Basic volcanic rock (basalt)		2.80	55.00	8.00

3. ECS log analysis

On the basis of experimental analysis, Schlumberger established a multivariate linear relationship between matrix parameters in volcanic rocks and elemental contents derived from ECS logging [2,3]. As a result, the continuous values of matrix parameters are obtained, which vary with depth as defined by the following equation:

$$\rho_{ma} = 3.15 - 1.10 \cdot W_{Si} - 0.98 \cdot W_{Ca} - 2.44 \cdot W_{Na} - 2.41 \cdot W_K + 1.42 \cdot W_{Fe} - 11.31 \cdot W_{Ti} \quad (7.3.1)$$

Where

ρ_{ma} — matrix density (g/cm^3);
W_{Si}, W_{Ca}, W_{Na} — weight percentages of silicon, calcium and sodium in ECS logs;
W_K, W_{Fe}, W_{Ti} — weight percentages of potassium, iron and titanium in ECS logs.

The method has a high accuracy (Figure 7.12) and can determine the matrix density values point by point (Figure 7.13).

The calculation results (see Figure 7.8) indicate that the matrix density of volcanic rocks varies with rock composition and types (see Table 7.3).

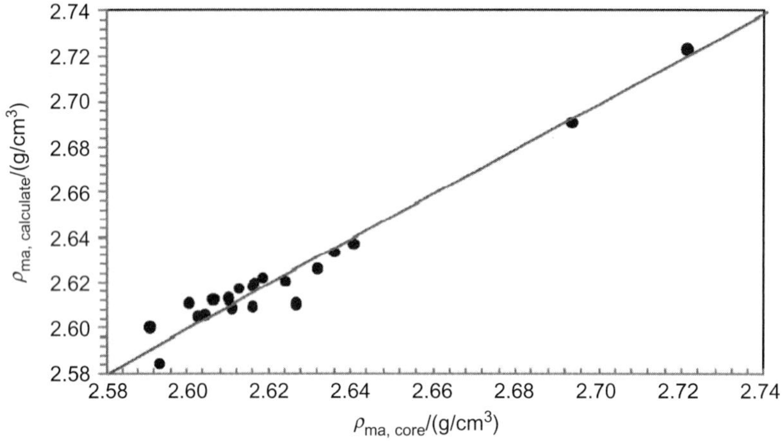

FIGURE 7.12

Matrix density of volcanic rocks as determined by the ECS method.

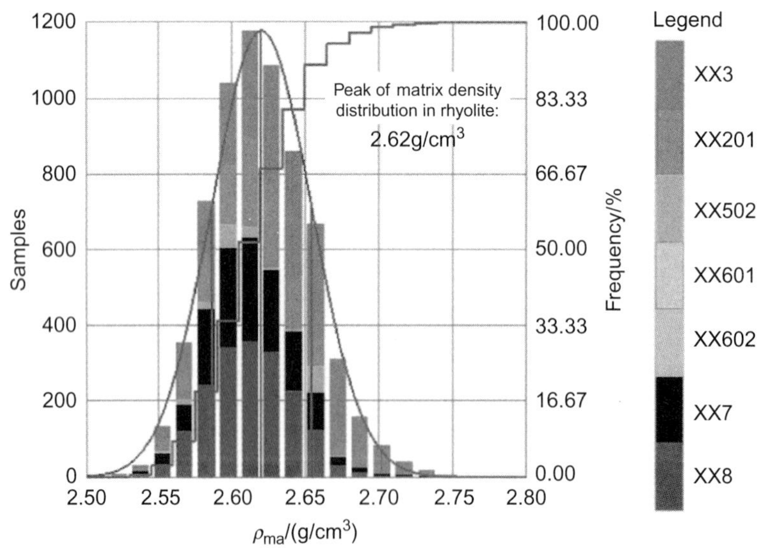

FIGURE 7.13

Density of rhyolite matrix as determined by the ECS method.

Table 7.3 Matrix Densities of Volcanic Rocks as Determined by an Analysis of ECS Logs (XX Gas Field)

Lithology		Distribution Range (g/cm^3)	Average (g/cm^3)
Acidic volcanic rocks	Rhyolite	2.52-2.72	2.62
	Agglomerate	2.55-2.71	2.64
	Breccia	2.50-2.75	2.62
	Tuff	2.53-2.71	2.62
Moderately acidic volcanic rock (dacite)		2.54-2.76	2.64
Intermediate volcanic rock (andesite)		2.60-2.80	2.70
Basic volcanic rock (basalt)		2.65-2.89	2.78

7.3.3 Effective porosity in matrix

With reference to the conceptual model for calculating the porosity of volcanic reservoirs, the theoretical models for interpreting porosity are built on the basis of nuclear magnetic logs and acoustic logs, whereas the statistical models are constructed on the basis of core experiments, acoustic logs, density logs, and neutron logs for various lithologies to systematically interpret the effective matrix porosity in volcanic reservoirs.

1. Calculation methods based on conventional logs

The theoretical and statistical models are the primary means for calculation.

1) Theoretical model based on acoustic logs

In conventional logging curves, the compensated acoustic log detects the effective matrix porosity. Thus, the theoretical acoustic model is used to calculate the effective matrix porosity in volcanic rocks using the following equation:

$$\phi_e = \left(\frac{\Delta t - \Delta t_{\mathrm{ma}}}{\Delta t_{\mathrm{f}} - \Delta t_{\mathrm{ma}}} - V_{\mathrm{sh}} * \frac{\Delta t_{\mathrm{sh}} - \Delta t_{\mathrm{ma}}}{\Delta t_{\mathrm{f}} - \Delta t_{\mathrm{ma}}} \right) * 100\% \tag{7.3.2}$$

Where

φ_e — matrix's effective porosity (%);
Δt — logging value of interval transit time (μs/ft);
Δt_{ma} — value of interval transit time in matrix (μs/ft);
V_{sh} — content of shale and fine volcanic ash;
Δt_{sh} — value of interval transit time in mudstone and fine tuff (μs/ft);
Δt_{f} — value of interval transit time in fluid (μs/ft).

Δt_{f} is in a range of 180 to 195 μs/ft, normally 189 μs/ft, which is acquired from clean mudstone and fine tuff strata that in principle should be closest to the reservoir with good borehole conditions.

2) Statistical models from experiments

In acoustic and density experiments under high temperature (139 °C) and high pressure (confining pressure at 85 MPa and internal pressure 42 MPa) with 23 pieces of acidic volcanic rock core taken from the XX gas field, methane and formation water ($NaHCO_3$ type, 10,000 ppm), respectively, were injected into these samples, and experimental data of compressional and shear wave (P&S) interval transit time and rock porosity were obtained. Through bivariate crossplot analysis, the fitting relationship of porosity versus interval transit time and rock porosity was established. Finally, the statistical models for effective matrix porosity in acidic volcanic rocks were established for analyses using density logs and interval transit time logs (Figures 7.14 through 7.16):

When the core contains methane, the equation for P-wave interval transit time is

$$\phi_e = 0.82 \times \Delta t_p - 44.93 \tag{7.3.3}$$

the equation for S-wave interval transit time is

$$\phi_e = 0.69 \times \Delta t_s - 79.10 \tag{7.3.4}$$

and the equation for rock density is

$$\phi_e = -45.05 \times \rho_b + 115.88 \tag{7.3.5}$$

When the core contains formation water, the equation for P-wave interval transit time is

$$\phi_e = 0.89 \times \Delta t_p - 46.90 \tag{7.3.6}$$

the equation for S-wave interval transit time is

$$\phi_e = 0.61 \times \Delta t_s - 70.81 \tag{7.3.7}$$

and the equation for rock density is

$$\phi_e = -69.44 \times \rho_b + 178.88 \tag{7.3.8}$$

In preceding equations, Δt_p — P-wave interval transit time (μs/ft) is

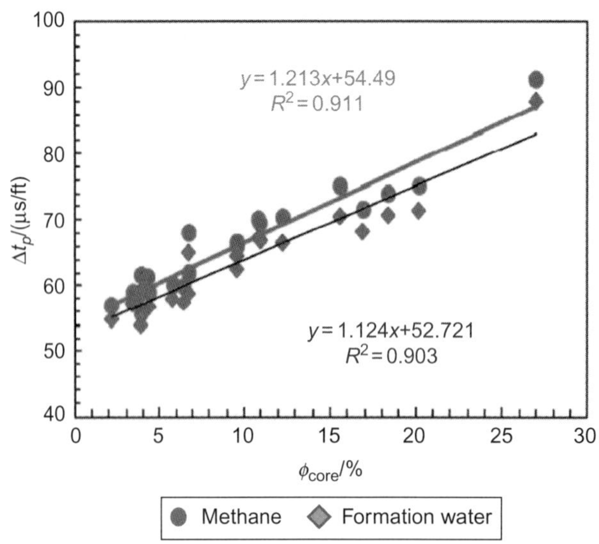

FIGURE 7.14

Crossplot of P-wave interval transit time versus rock porosity.

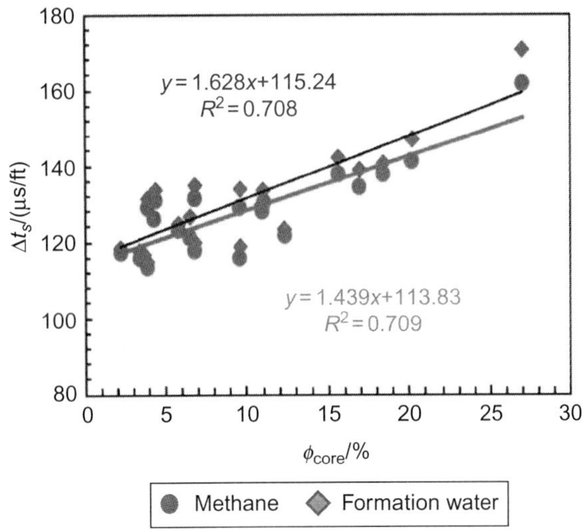

FIGURE 7.15

Crossplot of S-wave interval transit time versus rock porosity.

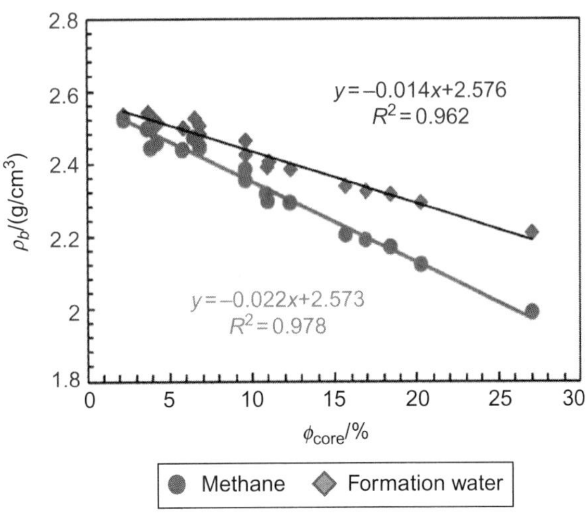

FIGURE 7.16

Crossplot of rock density versus rock porosity.

Δt_s — S-wave interval transit time (μs/ft);
ρ_b — rock density (g/cm^3).

This discussion demonstrates that the interval transit time in volcanic rocks increases as porosity increases, the rock density decreases as porosity increases, and fluid variation has a significant effect on density but only a minor effect on interval transit time.

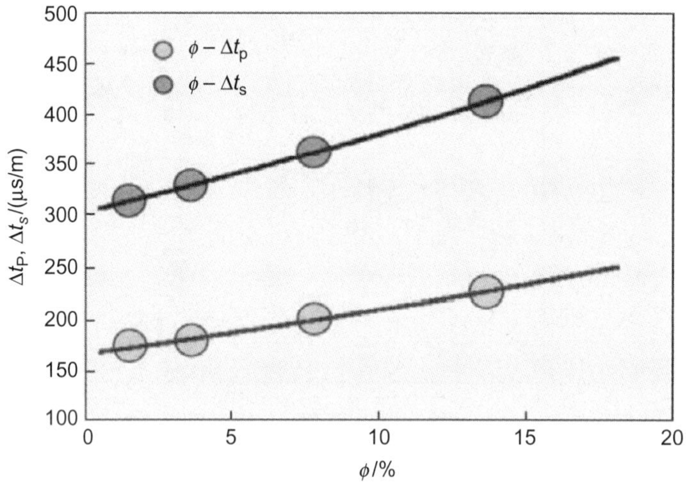

FIGURE 7.17

Crossplot of P & S wave interval transit time versus porosity.

From acoustic experiments on acidic volcanic rocks (Figure 7.17), Li established the following interpretation model for determining effective matrix porosity with P & S wave interval transit time [2]:

$$\phi_e = 0.4545 \times \ln\frac{\Delta t}{\Delta t_{\rm ma}} \tag{7.3.9}$$

For sedimentary volcanic rocks with volcanic ash and shale, the calculation using this statistical model produces an inflated value due to the effect of irreducible water and, therefore, the following equation is required to correct the effect of irreducible water:

$$\phi_e = \phi_e - V_{\rm sh} * \frac{\Delta t_{\rm sh} - \Delta t_{\rm ma}}{\Delta t_{\rm f} - \Delta t_{\rm ma}} * 100\% \tag{7.3.10}$$

3) Statistical models based on experiments and logs

Through log data preprocessing and calibrating the log data with core data, the statistical relationships of core porosity versus logging density, interval transit time, and neutron porosity are fitted to establish statistical models with experiment and logs.

In the case study of the DD gas field, more than 1000 samples of volcanic rocks were used in the petrophysical experiments. Through analysis of the relationship between core data and log data (Figures 7.18 and 7.19), the statistical models for effective matrix porosity in acidic, intermediate, and basic volcanic rocks as well as orthophyre and sedimentary volcanic rocks were established for interval transit time, rock density, and compensated neutron logging curves (Table 7.4). These models demonstrate that (1) the fitting equations for volcanic rocks with different lithologies have different slopes and cutoff values and the models are different, (2) the application of interval transit time and rock density has better results in calculating effective matrix porosity, and (3) the result of effective matrix porosity calculated with neutron porosity is unsatisfactory.

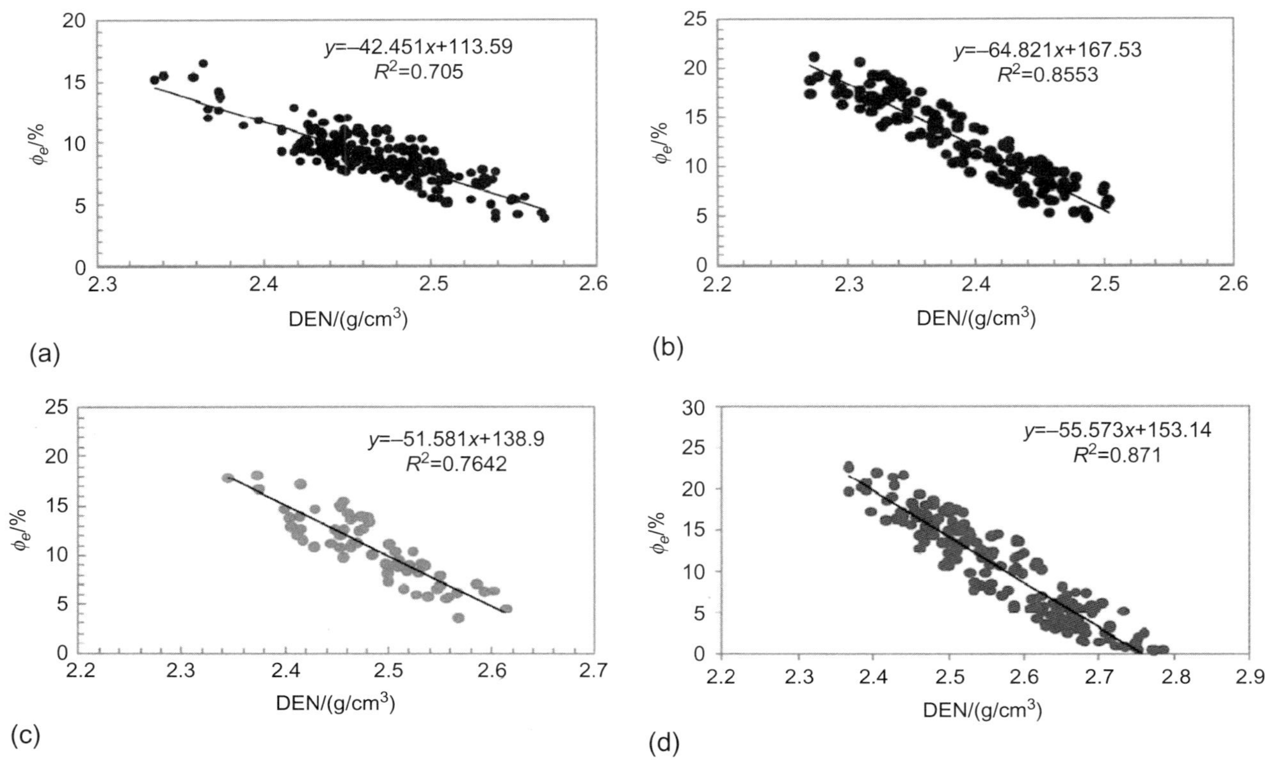

FIGURE 7.18

Crossplots of core porosity versus logging density (DD gas field). (a) Subvolcanic roks. (b) Acidic volcanic rock. (c) Intermediate volcanic rocks. (d) Basic volcanic rocks.

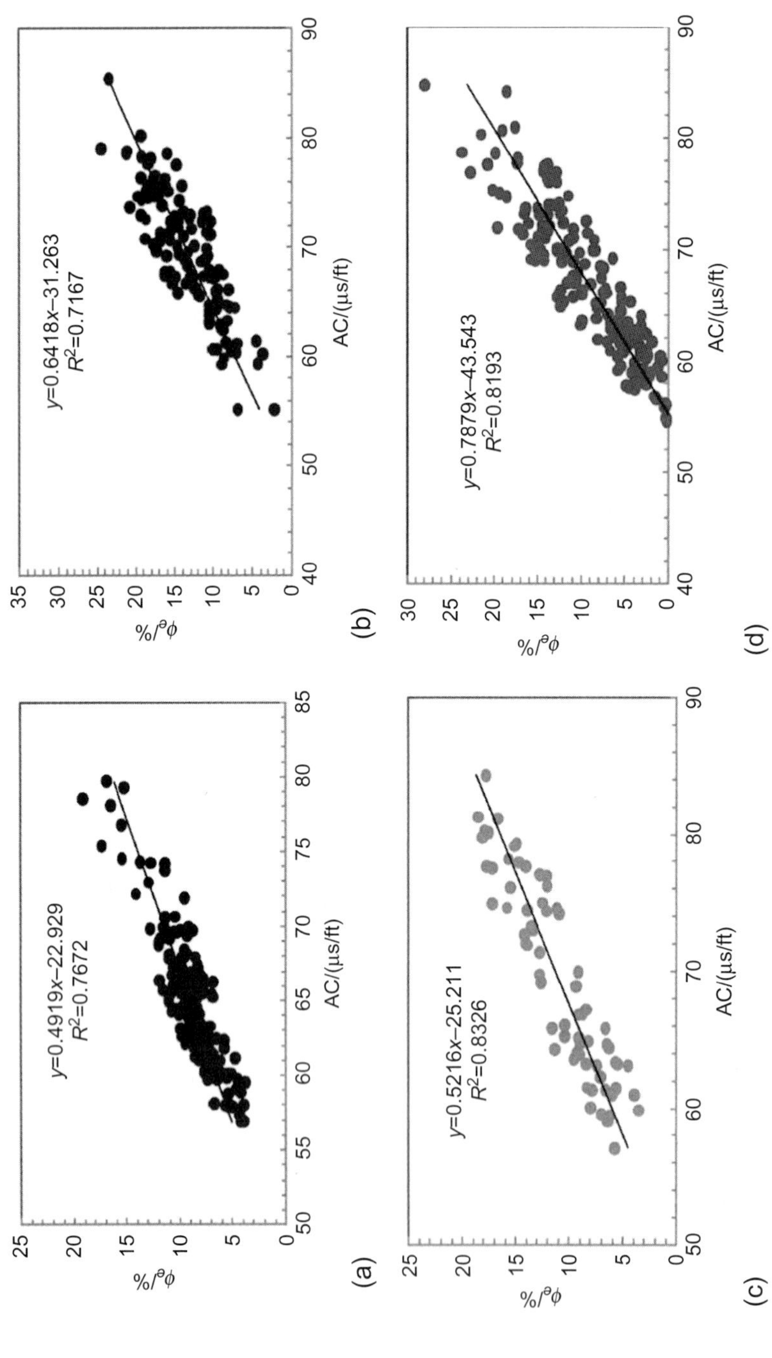

FIGURE 7.19

Crossplots of core porosity versus logging interval transit time (DD gas field). (a) Subvolcanic rocks. (b) Acidie volcanic rocks. (c) Intermediate volcanic rocks. (d) Basic volcanic rocks.

Table 7.4 Statistical Models for Effective Matrix Porosity in Volcanic Rocks Based on Experiments and Logs (DD Gas Field)

Lithology	Logging Curves	Statistical Models	Samples	R^2
Acidic volcanic rocks	Interval transit time, μs/ft	$\phi_e = 0.642$ AC – 31.26	281	0.72
	Rock density, g/cm^3	$\phi_e = -64.82$DEN + 167.53		0.85
	Compensated neutron, %	$\phi_e = 0.85$ CNL + 0.70		0.68
Intermediate volcanic rocks	Interval transit time, μs/ft	$\phi_e = 0.52$ AC-25.21	124	0.77
	Rock density, g/cm^3	$\phi_e = -51.58$ DEN +138.9		0.76
	Compensated neutron, %	$\phi_e = 0.89$CNL -3.21		0.61
Basic volcanic rocks	Interval transit time, μs/ft	$\phi_e = 0.792$ AC – 43.54	300	0.82
	Rock density, g/cm^3	$\phi_e = -55.57$DEN + 153.14		0.87
	Compensated neutron, %	$\phi_e = 1.15$ CNL + 10.51		0.75
Orthophyre	Interval transit time, μs/ft	$\phi_e = 0.52$ AC – 25.21	279	0.77
	Rock density, g/cm^3	$\phi_e = -42.45$DEN + 113.59		0.71
	Compensated neutron, %	$\phi_e = 0.61$ CNL + 1.97		0.58
Sedimentary volcanic rocks	Interval transit time, μs/ft	$\phi_e = 0.292$ AC – 13.57	79	0.73
	Rock density, g/cm^3	$\phi_e = -24.2$DEN + 67.65		0.83
	Compensated neutron, %	$\phi_e = 0.42$ CNL – 0.82		0.63

2. Theoretical model based on nuclear magnetic logs

The original echo signal of nuclear magnetic resonance logging is the superposition of the echo signals from fluid in pores of various sizes and configurations. Through inversion of the original signal, the relevant T_2 distribution [3–6] can be derived accordingly. Figure 7.20 shows the porosity distribution models for nuclear magnetic resonance logging, where the echo signals from different pores are shown in the left; corresponding T_2 distribution is shown in the middle, with red solid lines representing the ideal case and blue dotted lines representing a T_2 distribution with a widened peak due to noise. The real T_2 distribution in pores of different sizes is shown in the right [7].

If the measured signal intensity is calibrated and transformed to porosity, the sum of the signal intensities at different intervals of relaxation time will be the total porosity of the strata, and the effective porosity is the part with a porous throat greater than the lower radius limit of the total porosity. Therefore, there are two methods for effective porosity calculation based on nuclear magnetic logging [3].

1) T_2 Cutoff method (CBVI method)

Given a defined cutoff value of T_2 ($T_{2\text{cutoff}}$), the pore fluid with a T_2 value less than $T_{2\text{cutoff}}$ will stay in a irreducible state and cannot flow under formation pressure, whereas pore fluid with a value greater than $T_{2\text{cutoff}}$ is mobile (Figure 7.21). Thus, the volumetric percentages of irreducible water and free water in reservoirs will be calculated, respectively, as follows:

$$BVI = \sum_{T_2 \le T_{2\text{cutoff}}} P_i \tag{7.3.11}$$

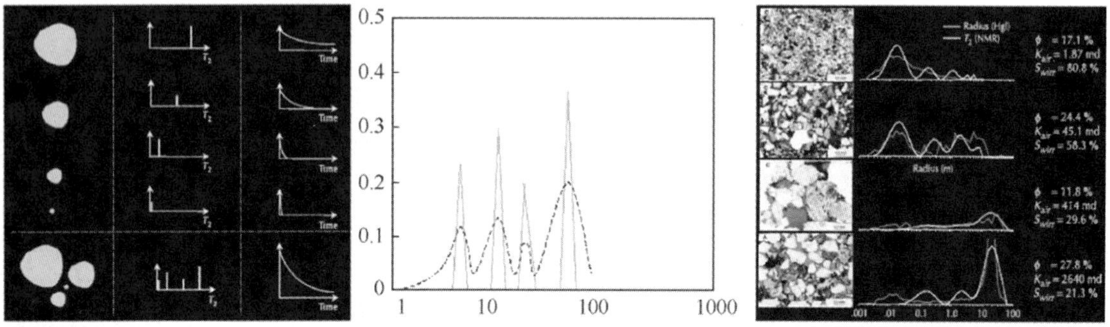

(a) (b) (c)

FIGURE 7.20

Diagrams of porosity distribution models for nuclear magnetic logging. (a) Echo signals from pores with different sizes. (b) Correlation of ideal T2 vs. measured T2. (c) Example of T2 distribution.

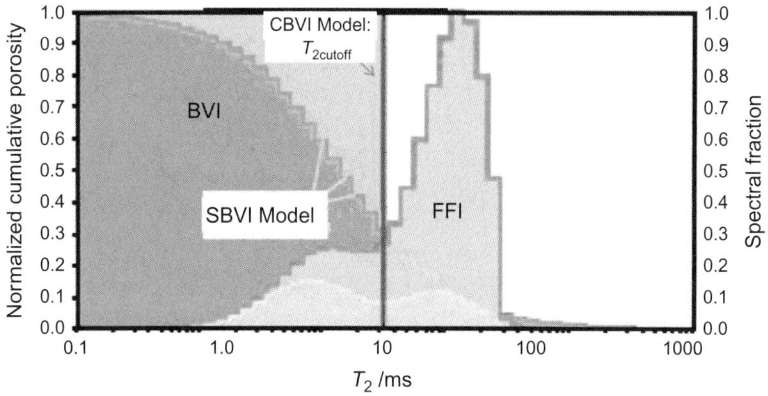

FIGURE 7.21

Model for interpretation of irreducible water.

$$\phi_e = FFI = \sum_{T_2 > T_{2\text{cutoff}}} P_i \tag{7.3.12}$$

Where

P_i — Signal intensity at various intervals of relaxation time, representing the porosity with pores of different sizes;

BVI — porous volume occupied by capillary irreducible water;

FFI — porous volume occupied by free water, or called as free fluid index.

$T_{2\text{cutoff}}$ is a parameter for characterization of reservoir characteristics, determined by experiments.

2) Method of T_2 spectral coefficient (SBVI method)

This method is based on the premise that some water is bound with the pore surfaces in every porous system regardless of the pore radius, and it cannot flow under reservoir conditions. The weight percentage of this irreducible water differs in value with different pore sizes. By determining the proportional factor of irreducible water in pores with different radii, the volumetric percentages of overall capillary irreducible water and free water can be calculated respectively (see Figure 7.21):

$$BVI = \sum_i C_i P_i \tag{7.3.13}$$

$$\phi_e = FFI = \sum_i (1 - C_i) P_i \tag{7.3.14}$$

where C_i (i = 1, 2, . . . n) is the distribution coefficient of irreducible water's T_2 spectrum, determined by experiments.

The variation in pore fluid composition has different effects on the results calculated with these two methods (Figure 7.22). When hydrocarbon is contained in the macropores (gas layer), the relaxation time of irreducible water in pores is less than $T_{2cutoff}$. In this case, better results will be obtained with the $T_{2cutoff}$ method, whereas a higher value may be obtained in the calculation using the spectral coefficient method. When only water is contained in the macropores (aquifer), irreducible water may be difficult to discriminate according to relaxation characteristics. Therefore, the result may be underestimated with the $T_{2cutoff}$ method, whereas the spectral coefficient method will provide an adequate result.

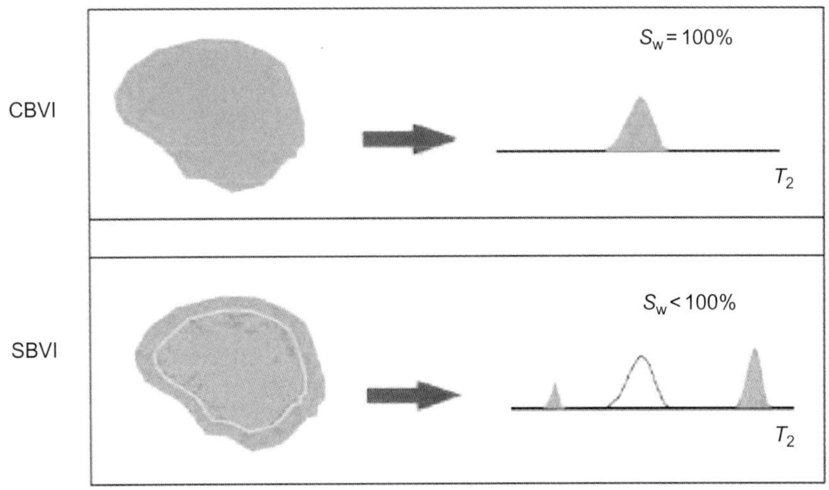

FIGURE 7.22

Effects of fluid composition on CBVI and SBVI methods.

7.3.4 Total porosity

The method of total porosity calculation—that is, to add effective matrix porosity and fracture porosity—cannot reflect the porosity in isolated and dead pores. In all porosity logging methods, nuclear magnetic, conventional density, and neutron logs can reflect the value of total porosity, but neutron logs are strongly affected by the formation water in the matrix. Therefore, nuclear magnetic logs and conventional density logs are selected for total porosity calculation in theoretical models.

1. Theoretical model based on the density log

The theoretical model used for the total porosity calculation based on density logs is a follows:

$$\phi_t = \left(\frac{\rho_b - \rho_{\mathrm{ma}}}{\rho_f - \rho_{\mathrm{ma}}} - V_{\mathrm{sh}} * \frac{\rho_{\mathrm{sh}} - \rho_{\mathrm{ma}}}{\rho_f - \rho_{\mathrm{ma}}} \right) * 100\% \tag{7.3.15}$$

Where

ϕ_t — total porosity from density log (%);
ρ_{ma} — density of rock matrix (g/cm^3);
ρ_{sh} — density of fine tuff and mudstone (g/cm^3);
ρ_f — fluid density (g/cm^3);
V_{sh} — content of shale and fine volcanic ash.

2. Theoretical model based on the nuclear magnetic Log

If the measured signal intensity of nuclear magnetic log is calibrated and converted to porosity, the total porosity in reservoirs will be equal to the sum of signal intensities at different intervals of relaxation time. Therefore, the total porosity can be calculated directly from T_2 distribution:

$$\phi_t = E(0) = \sum_i P_i \tag{7.3.16}$$

Where

ϕ_t — total porosity from nuclear magnetic log;
$E(0)$ — intensity of signal measured at zero time, which is calibrated into porosity value;
P_i — intensities of signal measured at different intervals of relaxation time, namely the porosity of pores with different sizes; the porous structures in reservoirs and reservoir quality can be known from P_i.

7.4 INTERPRETATION OF PERMEABILITY IN VOLCANIC RESERVOIRS

Permeability is an important parameter to reflect the fluid flow capacity of a reservoir. In terms of characterization objectives, it is divided into three types: matrix permeability, fracture permeability, and total permeability. This chapter discusses only the interpretation methods for matrix permeability and total permeability.

7.4.1 Matrix permeability

Matrix permeability is controlled by the type of reservoir space, the size of the matrix throat, and the configuration of the pore and throat and their level of development, and it is related to the effective porosity of reservoirs. The interpretation methods include experiment-based statistical models, the Timur equation, and theoretical models based on a nuclear magnetic log.

1. Experiment-based statistical model

For rocks with the same pore structure, permeability normally has a good correlation with porosity. From petrophysical analysis, experiment-based statistical models of matrix permeability can be built through an analysis of crossplots of porosity versus permeability. However, volcanic rocks have poor correlation between permeability and porosity for the following reasons: (1) there are many throat types in volcanic rocks, with great differences in radius, length, and behavior; (2) complex relationships exist between pore and throat configurations as well as complex pore structures; and (3) rock permeability can be improved by various types of fracture, resulting in some cores having a low porosity but high permeability. Therefore, building a model based on lithological types and reducing the fracture effect are the key issues in the establishment of statistical models for volcanic rock matrix permeability calculation.

The process can be illustrated using the acidic volcanic rocks of the XX gas field as an example. More than 400 data points derived from petrophysical experiments on volcanic rocks in this gas field were used—excluding the core samples with fractures—to establish the statistical relationships between permeability and effective porosity based on actual lithological data (Figure 7.23). This formed the basis for constructing the statistical models for matrix permeability, as shown in

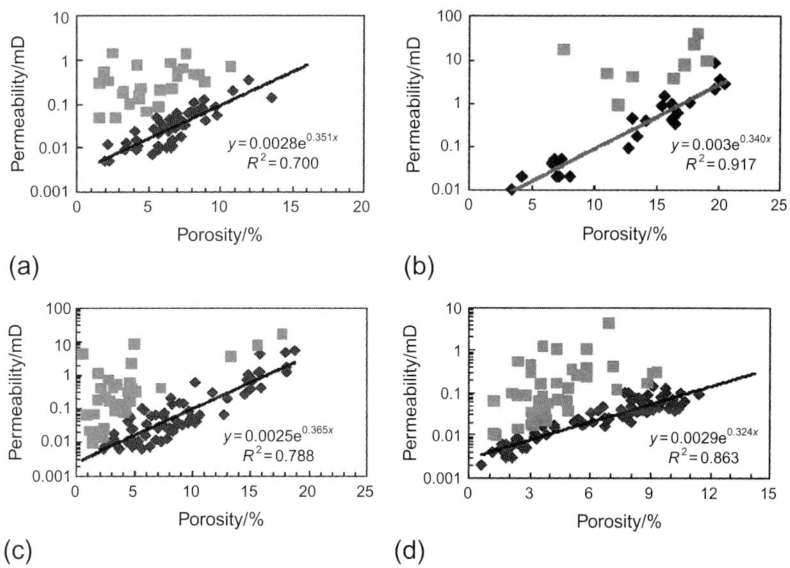

FIGURE 7.23

Models for interpreting acidic volcanic rock matrix permeability in the XX gas field. (a) Rhyolite. (b) Breccia lavel. (c) Volcanic breccia. (d) Tuff.

Table 7.5 Statistical Models of the Matrix Permeability of Acidic Volcanic Rock in the XX Gas Field

Lithology	Equations	Samples	R^2
Rhyolite	$K_b=0.0028EXP(0.351\Phi e)$	47	0.7
Breccia lava	$K_b=0.003EXP(0.34\Phi e)$	32	0.92
Breccia	$K_b=0.0025EXP(0.364\Phi e)$	69	0.79
Tuff	$K_b=0.0029EXP(0.324\Phi e)$	87	0.86

Table 7.5. The table demonstrates that (1) cores with fractures appear as abnormal points with high permeability in the charts; (2) after the fracture effect is corrected, core permeability and core porosity have a very good correlation; and (3) the slopes and cutoff values of fitting equations for various lithologies are different, indicating that different models should be used for calculation. Therefore, building lithology-based models for interpreting matrix permeability can improve the interpretation accuracy.

2. Timur equation

Saturation of irreducible water is closely related to absolute permeability [8]. The Timur equation is an empirical equation used to calculate the permeability associated with the porosity and saturation of irreducible water:

$$K_b^{1/2}=C\frac{\phi_e^x}{S_{wi}^y} \tag{7.4.1}$$

Where

ϕ_e — effective porosity;
S_{wi} — saturation of irreducible water;
K_b — matrix permeability (mD);
$C, x, y,$ — coefficients.

Using the saturation of irreducible water from experiments of NMR, relative permeability, and mercury injection, the regression relationships between permeability and porosity or irreducible water saturation are established on the basis of the correction of residual gas saturation (Figures 7.24 and 7.25). Then, three unknown parameters in Equation 7.4.1 are determined:

$$K_b^{1/2}=1.3554\left(\frac{\phi_e^{1.54}}{S_{wi}^{1.454}}\right)^{0.5437}, K_b^{1/2}=1.3554\frac{\phi_e^{0.84}}{S_{wi}^{0.79}} \tag{7.4.2}$$

Hence, $C=1.3554$, $x=0.84$, and $y=0.79$.

3. Theoretical equation of the nuclear magnetic Log

Matrix permeability is related to porosity and the specific surface of rocks, as expressed by the Kozeny equation [2,3]:

$$K=\frac{0.101\phi^3}{\Gamma(1-\phi)^2}\left(\frac{S}{V}\right)^2 \tag{7.4.3}$$

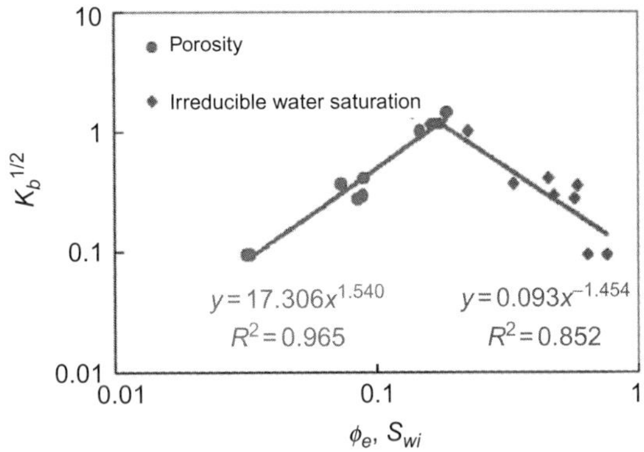

FIGURE 7.24

Crossplot of permeability versus porosity and irreducible water saturation.

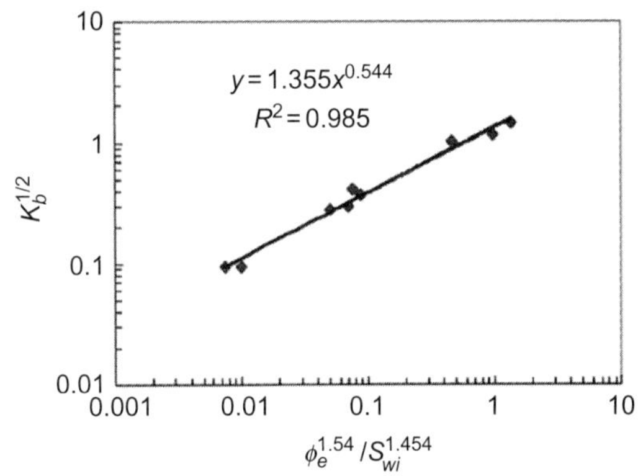

FIGURE 7.25

Model for interpreting matrix permeability.

Where

K — matrix permeability (mD);
φ — matrix porosity (%);
S/V — specific surface of pores in rocks (cm^2/cm^3);
Γ — constitutional factor, dimensionless, related to pore geometry and fluid flowing path per length in porous medium.

With the Kozeny equation, the interpretation models for rock permeability estimation are built by two methods, through analysis of the correlation between T_2 distribution and S/V (Figure 7.26):

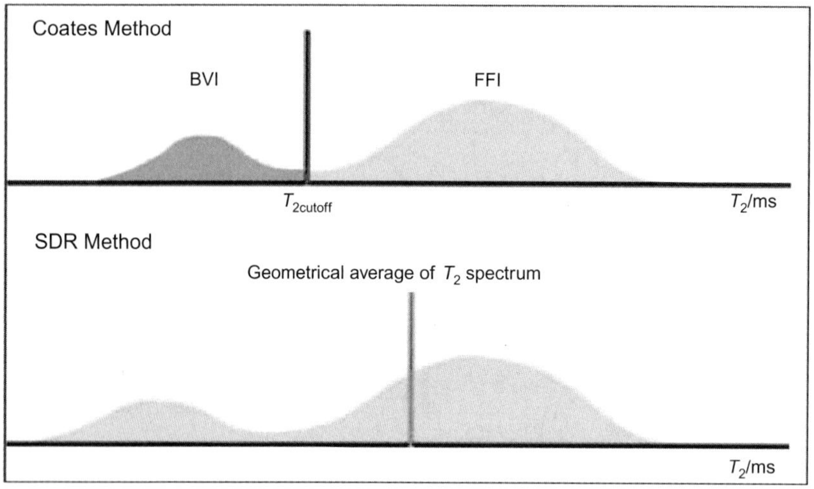

FIGURE 7.26

Schematic diagram of permeability calculation methods based on nuclear magnetic logs.

1) Coates method, also known as the free fluid model

$$K = \left(\frac{FFI}{BVI}\right)^2 \left(\frac{\phi}{C}\right)^4 \tag{7.4.4}$$

where C — coefficient, related to characteristics of the research area and reservoirs and determined from experiments; the other parameters are the same as those in the previous equations.

2) SDR method, also known as the T_2 geometric average method

$$K = C\phi^4 T_{2GM}{}^2 \tag{7.4.5}$$

Where

$T_{2\mathrm{GM}}$ — geometric average of T_2 distribution;
C — coefficient, determined from experiments.

Based on the T_2 spectrum derived from NMR logs, Schlumberger divided pores into three types: macropores, mesopores, and micropores. Two models are used for permeability calculation accordingly:

For macropores:

$$K_{\mathrm{macro}} = a'\left(\frac{V_{\mathrm{macro}}}{\phi_t - V_{\mathrm{macro}}}\right)^{b'} \phi_t^{c'} \tag{7.4.6}$$

For mesopores and micropores:

$$K_{SDR} = a(\rho * T_{2LM})^b \phi_t^c \tag{7.4.7}$$

Where

T_{2LM} — average of T_2 distribution (ms);
V_{macro} — porosity of macropores;
a, a', b, b', c, c' — coefficients, determined from experiments.

The other parameters are the same as for the previous equations.

7.4.2 Total permeability

Total permeability reflects the overall fluid flow capacity in reservoirs. Apparently, the permeability of a fractured reservoir is mainly attributable to fractures, whereas that of a porous reservoir is attributable to matrix pores and throats. In a fractured porous reservoir and a porous fractured reservoir, both pores and fractures contribute to the permeability.

The total permeability of a fractured reservoir consists of two parts: the matrix permeability K_b and the fracture permeability K_f, normally obtained from their direct addition:

$$K = K_b + K_f \tag{7.4.8}$$

where K, K_b, K_f — total permeability, matrix permeability and fracture permeability, respectively.

The total permeability interpreted from NMR logs in volcanic rocks can also be regarded as the sum of two parts: macropore permeability and micropore permeability, usually obtained from their direct addition:

$$K = K_{macro} + K_{SDR} \tag{7.4.9}$$

where K, K_{macro}, K_{SDR} — total NMR permeability, macropore permeability, and micropore permeability, respectively.

7.5 TECHNIQUES OF GAS SATURATION INTERPRETATION FOR VOLCANIC ROCKS

Gas saturation is a primary parameter for quantitative evaluation of the gas-bearing capacity in volcanic reservoirs. The calculation of gas saturation in the matrices and fractures of volcanic reservoirs are achieved based on the electrical conductive mechanism of rocks and integrated interpretation of resistivity logs and NMR logs and by taking into account the difficulties in gas saturation interpretation and fully using the data from NMR and rock resistivity experiments.

7.5.1 Mechanisms of electrical conductivity in volcanic rocks

In volcanic reservoirs, cavities, pores, and fractures are developed, resulting in complex conductive pathways. There are primarily three types of electrical conductivity: electronic conductivity from rock matrix, ionic conductivity from the mobile porous spaces, and the additional conductivity from

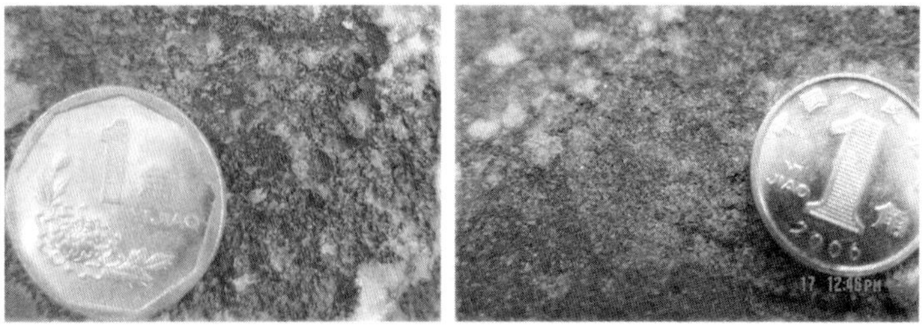

(a) (b)

FIGURE 7.27

Distribution of pyrites in volcanic rocks. (a) Nodular pyrite (Well XX13). (b) Disseminated pyrite (Well XX21-1).

irreducible water on the inner walls of pores and fractures and the electrically coupled layers of clay minerals.

1. Electronic conductivity from rock matrix

This type of conductivity refers to the electrical conductivity of rocks due to the presence of conductive minerals in their matrix [9]. Pyrites are the most common conductive minerals; they appear in a lumpy, nodular, and disseminated arrangement (Figure 7.27), and their effects are related to the distribution patterns and the concentration of pyrites, formation water salinity, and the type of resistivity logging. Generally, pyrites with a high concentration and disseminated distribution have a large effect on high frequency induction logs but a minor effect on lateral logs [10].

2. Ionic conduction in mobile porous spaces

When formation water with certain salinity is contained in the connected pores and fractures, the positive or negative salt ions dissociated in water move along or opposite to the electrical field under the action of an external electrical field, and the movements of these ions cause the rocks to exhibit electrical conductivity [9]. Ionic conduction is the primary mode of volcanic rock conductivity:

1. Ionic conductivity in matrix pores. This depends mainly on the degree of pore development, the complexity of connective pathways among pores, and the ionic composition and salinity of formation water. In general, higher rock conductivity is related to better pore development, larger throat, better communication, and higher salinity. For volcanic rocks with evenly distributed pores, their characteristics of rock conductivity comply with the conductive mechanism defined by the Archie equation.
2. Ionic conductivity in fractures. Zhao and colleagues [11] verified the dual lateral logging response characteristics of fractures in carbonate reservoirs with a flume tank experiment in Sichuan. Also, Li and colleagues studied the acoustic/electrical imaging logging response characteristics of fractures in acidic volcanic reservoirs through microfracture modeling on the basis of the acoustic/electrical imaging model [2]. Their investigations suggest that (1) fractures can cause a decrease in rock resistivity and increase the difference between deep and shallow resistivity values; and (2) logging response characteristics are related to many factors, including the development, effectiveness, and occurrence of fractures, the angle between the fracture and the conducting path, and the salinity of formation water.

3. Resistivity logging response equation for fractured-porous reservoirs

Through their experiments, Zhao and Li deduced the resistivity logging response equation for reservoirs with both fractures and pores. On this basis, a resistivity logging response equation for fractured-porous volcanic reservoirs has been derived for this book:

$$\frac{1}{R_t}=\frac{\phi_b^{m_b}S_{wb}^{n_b}}{R_w}+\frac{K_f\phi_f^{m_f}S_{wf}^{n_f}}{R_f} \tag{7.5.1}$$

Where

R_t — rock resistivity of fractured reservoir (Ω•m);
R_w, R_f — fluid resistivities in matrix and fractures (Ω•m);
Φ_b, Φ_f — porosities in matrix and fractures;
S_{wb}, S_{wf} — water saturations in matrix and fractures;
m_b, m_f — pore constituent indices in matrix and fractures;
K_f — fracture distortion coefficient, which is related to fracture occurrence and logging series.

For low angle fractured, porous reservoirs, K_f falls between 0.7 and 0.8; for single-series high-angle fractured, porous reservoirs, K_f varies from 1.7 to 2; for multiseries high-angle fractured, porous reservoirs, it varies from 1.1 to 1.3.

3. Additional conductivity

This category includes the surface conductivity produced by irreducible water on the inner walls of pores and fractures as well as the additional conductivity induced by the coupled potential of clay minerals.

1) Surface conductivity

This is also known as the conductivity of the inner layer in dual electrical layers, as an additional conduction from ionic diffusion of the irreducible water film on the pore surface. In 1992, Bremer studied the relationship between the conduction of crystalline rocks and the salinity of the formation water contained in these rocks and found that as salinity was reduced, rock conductivity decreased; when the salinity dropped below a certain value (e.g., 600 ppm), rock conductivity stopped changing with the salinity drop; when the salinity approached zero, the rock still exhibited some conductivity (Figure 7.28), which was later shown to be from the surface conductivity [1]. In 1983, Mandelbrot and Kranz found that fracture conductivity was also affected by the surface conductivity of microfissures [1]. In 2002, Hao and colleagues found in their study on resistivity variation during rock crushing that surface conductivity also existed in fractures [12].

The surface conductivity in fractures and pores is expressed as follows:

$$C_q=\frac{\sigma\times u}{\delta}F_\delta \tag{7.5.2}$$

Where

σ— electrical charge density on rock surface (c/mz);
μ — migration speed of negative ion in dual electrical layer (cm/s);
δ — thickness of dual electrical layer (nm);

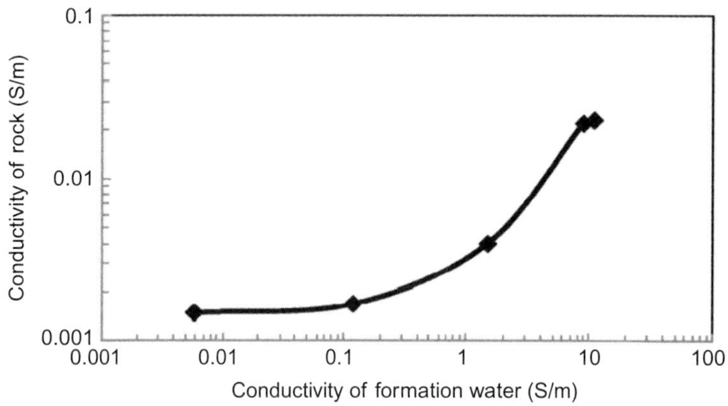

FIGURE 7.28

Relationship between the conductivity of crystalline rocks and formation water.

F_δ — formation factor with surface effect considered.

$\sigma\mu$ is about 12.6 ± 7nS; F_δ is related to the surface structures of pores and microfractures.

The associated parameters are determined from experiments.

For sedimentary rocks, surface conductivity can be ignored. In volcanic rocks, it is a macroscopic conductive property affected by microcapillary pores with microscopic structure. Surface conductivity is important for understanding fluid migration and petrophysical variation, especially for fine tuff.

2) Additional conductivity of electrically coupled layers of clay minerals

Clay minerals are composed of sheetlike particles that are stacked with two to three layers of alumino-oxygen octahedrons or silicon-oxygen tetrahedrons, with a large specific surface. Because the tetravalent silicon (Si^{4+}) in the silica tetrahedron is substituted by the trivalent aluminum (Al^{3+}) and the trivalent aluminum ion (Al^{3+}) in the silica octahedron is substituted by the divalent magnetic ion (Mg^{2+}) or ferrous ion, extra negative charges are generated on clay surfaces. To neutralize this local imbalance of electrical charges, the surface of clay particles absorb some cations as compensating ions. Hence, electrically coupled layers will be formed on the surface of clay particles, resulting in an additional conduction of clay minerals, expressed as follows:

$$C_{CLAY} = \frac{BQ_v}{F^*} = \frac{B^* CEC(1-\phi)\rho_G}{F^*\phi} \tag{7.5.3}$$

Where

C_{CLAY} — additional conductivity of clay (ms/m);

CEC, Q_v — capacity of cation exchange (mmol/cm$_3$) and concentration of cation exchange (meq/mL);

ρ_G — density of rock particles (g/cm^3);

F^* — formation factor of sedimentary volcanic rocks when C_w is high enough;

B — electrochemically equivalent conductivity for cation equilibrium [(S·cm^3)/(mmol·m)].

From Equation 7.5.2, it can be concluded that the additional conductivity of clay depends on the capability of cation exchange (CEC) and is affected by factors such as the type, concentration, and distribution of clay minerals and the ambient conditions (PH value, temperature, salinity, etc.).

7.5.2 Gas saturation in rock matrix

This is determined through integrative interpretation of conventional resistivity logs and NMR logs.

1. Interpretation method based on the resistivity Log

On the basis of the conductive mechanisms of volcanic rocks and rock resistivity experiments, the degree of gas saturation in volcanic reservoirs can be calculated from resistivity logs following the establishment of interpretation models and the calibration of fractures and surface conductivity.

1) Calculation model

According to the conductive mechanisms of volcanic rocks, the effects of electronic conductivity, surface conductivity, and fracture conductivity are eliminated and the gas saturation in the matrix of volcanic reservoirs is calculated using the theoretical model based on the Archie equation:

$$S_g = 1 - \sqrt[n]{\frac{a^* b^* R_w}{R_t^* \phi^m}} \tag{7.5.4}$$

For altered and sedimentary volcanic rocks, the W-S model is used for calculating the gas saturation in matrix:

$$S_g^{-n^*} = 1 - \frac{R_t}{F^* R_w}\left(1 + \frac{BQ_v R_w}{S_w}\right) = 1 - \frac{\phi^{m^*} R_t}{aR_w}\left(1 + \frac{BQ_v R_w}{S_w}\right) \tag{7.5.5}$$

Where

S_w — water saturation in matrix;
a, b — coefficients of lithology–electrical property relationships, determined from petroelectrical experiments;
m, n — cementation index and saturation index for rock matrix;
m^*, n^* — cementation effect index and saturation index for rock matrix after clay effect is eliminated;
R_t,R_w — formation resistivity, resistivity of formation water ($\Omega \cdot m$);
B, Q_v — same as above.

2) Specific procedures

Lithological identification and model selection for saturation calculationVolcanic rocks are less affected by clay minerals, but sedimentary and altered volcanic rocks are affected significantly. Therefore, appropriate models for saturation calculation can be selected based on lithological identification to reduce the effect of the conductivity from coupled electrical layers in clay minerals.

Correcting the effects of additional conductivityThis primarily involves correcting the effects of fracture, matrix, and surface conductivity.

1. *Correcting the effects of fracture conductivity.* For fractured reservoirs, fracture conductivity should not be ignored. Within the detection limit of logging tools, fractures are normally filled with mud, hence, $S_{wf} \approx 1$ while $R_f \approx R_m$. Therefore, the following equation is used to correct the effect of fracture conduction:

$$R_t = \left(\frac{1}{R_{\log}} - K_f \frac{\Phi_f m_f}{R_m} \right)^{-1} \tag{7.5.6}$$

   For an explanation of the symbols, see the preceding equation.
2. *Correcting the effects of matrix conductivity.* Metallic minerals such as pyrite can reduce rock resistivity. If their effects are overlooked, the calculated gas saturation will be underestimated. By taking into account the effects of the scattered pyrite of different concentrations on the resistivity logs with various frequencies, Clavier made his calibration and verification through petroelectrical experiments. His results (Figure 7.29) indicate that the correction coefficient of resistivity will increase with increased pyrite content; and under the same condition, pyrite had the greatest effect on induction logs with medium/high frequency but, less effect on lateral logs with low frequency. Normally, the chart shown in Figure 7.29 is used for correction.
3. *Correcting the effects of surface conductivity.* In general, a smaller pore or fracture size is correlated with higher rock surface conductivity. Bremer's experiment showed that the surface conductivity of crystalline rocks is about 0.0017 S/m, which is a relatively low value. In volcanic rocks, fine tuff requires Equation 7.5.7 to be used for the correction of surface conductivity, whereas no correction is necessary for other lithologies:

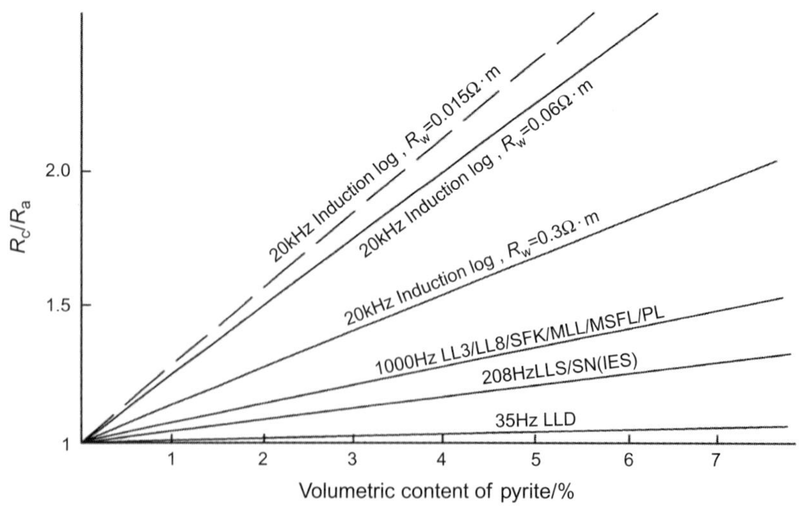

FIGURE 7.29

Chart showing the effects of different concentrations of scattered pyrite on resistivity logs with various frequencies.

$$R_t = \left(\frac{1}{R_{\log}} - C_q \right)^{-1} \tag{7.5.7}$$

where C_q is rock surface conductivity (S/m), determined by rock-based experiments.

So far, no experiment has been performed on the surface conductivity of volcanic reservoirs. As a result, a constant is taken between 0.001 and 0.01 S/m according to the real case with reference to the results of Bremer's experiment. In this book, 0.005 S/m is used as the constant.

Determination of model parameters

The values of* a, b, m, *and* n *are determined through petroelectrical experiments In the case study on the DD gas field, more than 140 data points from the rock resistivity experiments are analyzed (Figures 7.30 and 7.31 and Table 7.6) by dividing the lithologies to determine the parameters of the matrix saturation model for volcanic reservoirs.

a is a lithology correlation coefficient, reflecting the characteristics of matrix conduction [13,14]. The value of a varies from 0.94 to 3.86, with an average of 1.66, being highest for effusive rocks,

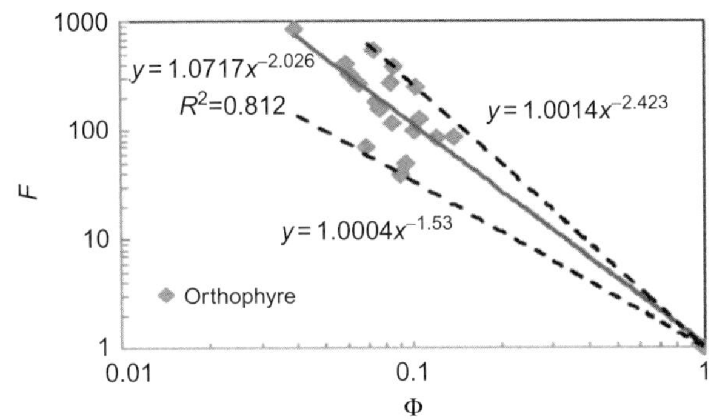

FIGURE 7.30

F-Φ crossplot of full size cores.

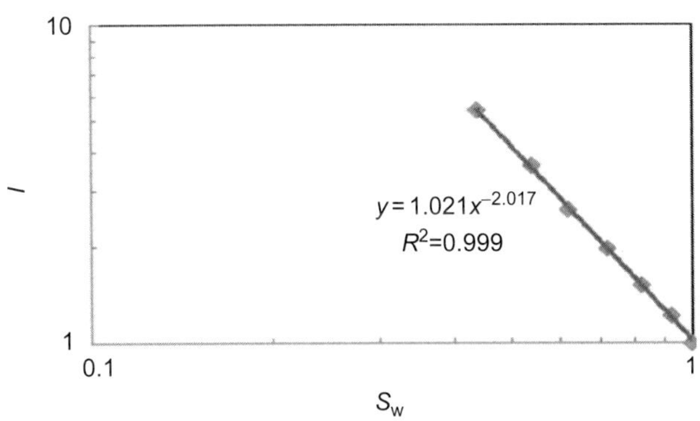

FIGURE 7.31

I-Sw crossplot of full size cores.

Table 7.6 Values of Petroelectrical Parameters for Various Volcanic Rocks (DD Gas Field)

Lithology		*m*			*n*		*b*	
		Range	*Average*	*Average*	*Range*	*Average*	*Range*	*Average*
Subvolcanic Rock		1.53-2.42	2.02	1.071	1.9-2.15	2.01	0.94-1.09	1.02
Effusive Rock	Volcanic Lava	1.49-3.57	1.44	3.86	1.87-3.64	2.04	0.95-1.34	1.05
	Breccia Lava	1.47-2.17	1.62	1.736	1.15-2.04	1.67	1-1.2	1.1
	Tuff	1.72-2.08	1.97	0.938	1.89-2.02	1.96	0.97-1.13	1.01
	Breccia	1.63-2.37	1.65	1.321	1.49-2.01	1.85	0.99-1.22	1.1
	All Effusive Rocks	1.47-3.57	1.64	2.124	1.49-2.01	1.9	0.95-1.34	1.05
Volcanic Sedimentary Rock		1.27-2.0	1.64	1.076	1.85-2.24	1.99	0.97-1.16	1.05

followed by subvolcanic rocks and volcanic sedimentary rocks. For effusive rocks, the a value is highest in lava followed by breccia and lowest in tuff.

m is a porous structure index and reflects the characteristics of conductive pathways in matrix pores and throats. The value of m varies in the range of 1.27 to 3.57, with an average of 1.73. The m value is highest in subvolcanic rocks, followed by effusive rocks. Among effusive rocks, m is highest in tuff, volcanic breccia next, and lowest in lava.

b is a heterogeneity coefficient for wetting saturation, related to lithology. It has a small range of variation, mainly from 0.94 to 1.34, with an average of 1.05.

n is a saturation index and reflects the distribution characteristics of fluid in pores. It ranges from 1.15 to 3.64, with an average of 1.92. The n value is highest in subvolcanic rocks, followed by effusive rocks, and then volcanic sedimentary rocks. Among effusive rocks, lava has the highest n value, tuff being next, volcanic breccia has the lowest value.

This shows that the model parameters can be determined on the basis of variations in lithology, types of reservoir space, and porous structures.

Resistivity of formation water R_w is determined from water analysis First, the equivalent NaCl salinity of formation water is derived. For instance, if the formation water is of the $NaHCO_3$ type, the total salinity is 9026 mg/L, including 2680 mg/L of cations (such as Na^+), 986 mg/L of Cl^-, and 5360 mg/L of other anions (e.g., $HCO_3{}^-$); this gives an equivalent NaCl salinity of 5810 mg/L according to the salinity conversion chart.

Based on this table, the resistivity of formation water at 24° C, $R_{w(24°C)}$ is calculated with a result of about 0.9 Ω·m for preceding case.

Finally, taking into account the geothermal gradient in the study area, the formation temperature T (° C) at various depths is calculated and the resistivity of formation water at formation temperature, R_w, is calculated:

$$R_w = R_{w(24°C)}\left[\frac{45.5}{T+21.5}\right] \tag{7.5.8}$$

2. Methods of NMR log interpretation

Two methods are commonly used, such as the dual T_w method and the gas column height method.

1) Dual T_w method

Considering the T_R effect, the observed echo string can be expressed as follows:

$$E(T_R,t) = \sum_i P_i S_w\left(1-e^{-\frac{T_R}{T_{1i}}}\right)e^{-\frac{t}{T_{2i}}} + \phi S_g I_{Hg}\left(1-e^{-\frac{T_R}{T_{1g}}}\right)e^{-\frac{t}{T_{2g}}} \tag{7.5.9}$$

Where

T_R — longitudinal recovery time (ms);

T_{1i},T_{2i} — longitudinal relaxation time (ms) and lateral relaxation time (ms) of water in the i porous system;

S_w, S_g — water saturation and gas saturation;

T_{1g},T_{2g} — longitudinal relaxation time (ms) and lateral relaxation time (ms) of natural gas;

I_{Hg} — index of hydrogen content in gas;
t — measure time (ms).

The two terms in the right of Equation 7.5.9 are signals of water and gas, respectively. By rational selection of T_{WL} and T_{WS} to enable that $T_{RL} \approx (3–5)\ T_{1g}$ and $T_{RS} \approx (3–5)\ T_{IW}$, the signals of water and gas should be recovered completely for T_{RL}. For T_{RS}; the signal of water is recovered completely, but that of gas is recovered only partially. Therefore, the signal of water can be totally eliminated to highlight the information on natural gas through subtraction of the two echo strings:

$$\Delta E(t) = \phi S_g I_{Hg} \left(e^{-\frac{T_{RS}}{T_{1g}}} - e^{-\frac{T_{RL}}{T_{1g}}} \right) e^{-\frac{t}{T_{2g}}} = \Delta \phi_g e^{-\frac{t}{T_{2g}}} \tag{7.5.10}$$

Given that T_{2g} is known, $\Delta\phi_g$ can be obtained through dual exponential fitting of $\Delta E(t)$, and then gas saturation S_g can be derived. The biggest difficulty with this method lies in the low S/N ratio, making the fitted energy of T_2 distribution divergent, not exhibiting a unimodal or bimodal pattern.

2) Gas column height method

The key to this method lies in the calculation of the capillary pressure curve based on NMR logs. Figure 7.32 is a schematic diagram of capillary pressure, and Figure 7.33 shows the measured result of the NMR T_2 spectrum. In combination with the principle of NMR logging, the interpretation model for calculating capillary pressure P_c from T_2 can be obtained:

$$\left\{ \begin{array}{l} P_c = 2\sigma \cos\theta / r_1 \\ 1/T_2 = \rho_2 S/V = 2\rho_2 / r_2 \end{array} \right\} \Rightarrow P_c T_2 = \sigma * r_2 * \cos\theta / (\rho_2 * r_1) \Rightarrow P_c = \frac{K_{pc}}{T_2} \tag{7.5.11}$$

Where

P_c, σ, θ — capillary pressure (MPa), interfacial tension (N/m) and wetting angle (°);
r_1 — capillary radius of pore or throat (μm);
r_2 — radius of pore (μm);
K_{PC} — conversion coefficient, related to interfacial tension, wetting angle, capillary radius, pore radius, and surface relaxation rate of rocks, determined by experiments.

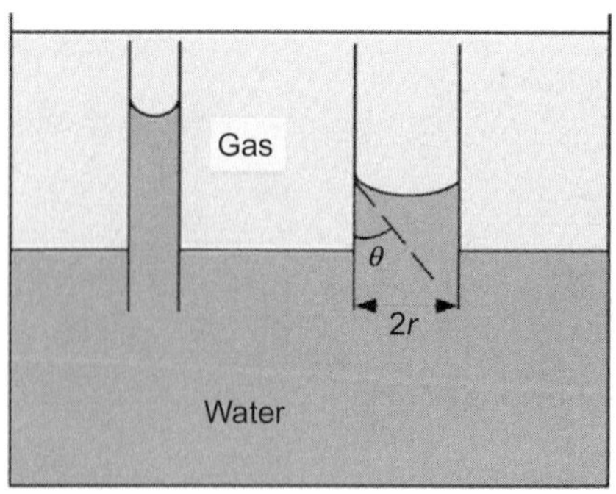

FIGURE 7.32

Schematic diagram of capillary pressure.

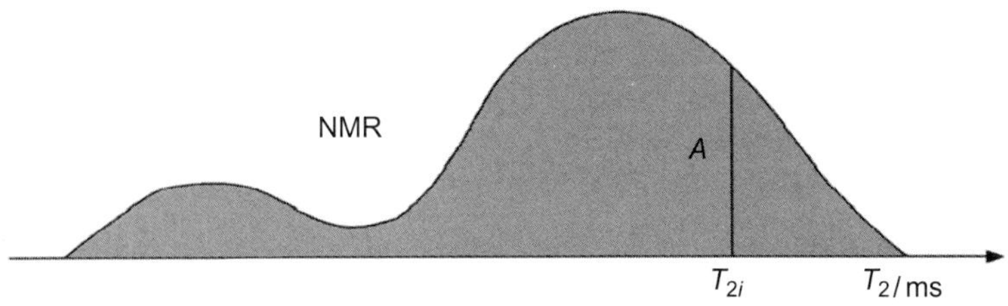

FIGURE 7.33

NMR T2 distribution spectrum.

Based on the capillary pressure curve under gas reservoir conditions, the water saturation in the gas zone above the gas-water interface is calculated with the gas column height method [15]:

$$Sw = 1 - A \cdot e^{-\left(\frac{B}{h+D}\right)^C} \text{ where, } h = \frac{100P_C}{\rho_w - \rho_g} \tag{7.5.12}$$

Where

h — gas column height (m);
$A, B, C,$ and D — calculation parameters, related to volcanic rock type, porous structure, and fluid property, determined by experiments.

7.5.3 Original gas saturation in fractures and cavities

Gas saturations in fractures and cavities are difficult to measure and interpret so they are usually treated as constants [16]. In 1974, Smihof concluded from fluid flow experiments that the thickness of water film in fractures did not exceed 0.016 μm; this conclusion is corroborated by research results showing that the average thickness of irreducible water films on the inner walls of pores and throats is 0.0113 μm in the Sinian dolomite in Weiyuan, Sichuan. Based on an average opening of 15 μm for subsurface fractures, the calculated saturation of irreducible water in fractures and cavities is only 0.2% or less. In volcanic reservoirs, the width of fractures falls in a range from several μm to several hundred μm, with an average at dozens of μm, whereas the thicknesses of water films in fractures occupy only 0.02% to 5% of the reservoir space. Therefore, the original gas saturation in cavities and fractures of volcanic gas reservoirs is normally given a value of 95%.

7.6 TECHNICAL APPLICATIONS

The technical applications for interpreting volcanic reservoir parameters include analysis of accumulation-permeation capacity and gas-bearing property, reservoir classification and evaluation, the calculation of producible reserves, and the establishment of geological models.

7.6.1 Providing parameters for reservoir classification and evaluation and reserve calculation

The interpretation of reservoir parameters can provide important information on the porosity, permeability, and saturation states in matrices and fractures in volcanic reservoirs, which in turn serve as the basis for a categorized evaluation of reservoir characteristics. For example, in Well Blocks XX8 and XX9 (Table 7.7) of the volcanic reservoir, the effective matrix porosity range is 3.6% to 16.4% with an average of 6.01%, and the matrix permeability range is 0.01 to 1.13 mD with an average of 0.07 mD. In the gas-bearing zone, the matrix gas saturation has a range of 43.5% to 79.5% with an average of 56.2%; the fracture porosity ranges from 0.01% to 0.34% with an average of 0.09%; the fracture width ranges from 0.94 to 119 μm with an average of 23.5 μm; the fracture permeability ranges from 0 to 42.6 mD with an average of 1.95 mD. This confirms that volcanic reservoirs are characterized low porosity, low permeability, and good fracture effectiveness. In reference to the classification criteria of effective reservoirs, the volcanic reservoir in these blocks is dominated by type II, accounting for 51.5%; type III is a close second, accounting for 42.7%, and type I makes up only 5.8%.

On the basis of the interpreted porosity, permeability, and saturation values for reservoirs, reservoir types can be categorized according to the cutoff criteria of effective reservoirs, and the net pay of gas reservoirs can be determined. This will provide information on the effective porosity and gas saturation in matrices and fractures as well as net pay of gas reservoirs for reserve calculations, which, in turn, serve as the basis for the proper evaluation of gas in place (GIP) and producible reserves in volcanic reservoirs.

7.6.2 Facilitating the construction of geological attribute models and fluid models

By means of the interpretation models for reservoir parameters, matrix porosity and permeability, and fracture porosity and permeability, the matrix gas saturation and fracture gas saturation states are calculated well by well, layer by layer, and point by point to provide data for establishing geological attribute models and fluid models for volcanic gas reservoirs as well as the constraints at well points.

Table 7.7 Parameter Interpretation Results of Volcanic Reservoir Based on Reservoir Classification (Well Blocks XX8 and XX9)

Reservoir Type	Total Thickness, m	Net Thickness, m	Well Count	Intervals	Total Porosity, %	Effective Porosity, %	Porosity in Fracture, %	Total Permeability, mD	Permeability of Matrix, mD	Permeability of Fracture, mD	Gas Saturation in Matrix, %	Fracture Width, μm
I	48	48	1	2	16.54	15.36	0.328	29.878	0.865	29.014	79	95.21
II	436.1	425.8	11	41	8.06	6.25	0.09	0.467	0.026	0.441	58.01	23.36
III	468.9	353.7	12	49	5.98	4.45	0.059	0.116	0.013	0.103	50.85	13.83
All reservoirs	953	827.5	12	92	7.66	6.01	0.091	2.023	0.069	1.954	56.17	23.46

References

[1] Pan B, Xue L, Li Z, et al. Logging evaluation methods and application in fractured igneous reservoirs. Beijing: Petroleum Industry Press; 2003, p. 22–25, 62–71.
[2] Li N, Tao H, Liu C. Theory, methodology and application of logging interpretation in acidic volcanic rocks. Beijing: Petroleum Industry Press; 2009, p. 3, 63–71.
[3] Xiao L, Chai X, Sun B, et al. Introduction of NMR log data interpretation and application. Beijing: Petroleum Industry Press; 2001, p. 73, 74, 144–47.
[4] Mo X, Li Z, Mei Z, et al. Methodology and application of NMR logging interpretation. Logging Technology 1997;21(6):424–31.
[5] Gao M, Gao X, An X, et al. Evaluation of porous structures in reservoirs with NMR log data. Logging Technology 2000;3(24):56–64.
[6] Tang X, He Z, Zhang C, et al. Study on regional correlation for quantitative solution of porous structure parameters by means of NMR T_2 distribution. Journal of Jianghan Petroleum Institute 2003;25(4):75–7.
[7] Robinson MA. Magnetic resonance imaging for the measurement of oil core porosity distributions. Society of Petroleum Engineers, Richardson, Texas, USA, Publication SPE23582; 1991, p. 1–32.
[8] Yong S, Zhang C. Processing and comprehensive interpretation of log data. Dongying: Press of China University of Petroleum; 1996, p. 180–300.
[9] Chen F, Yan J. Logging principles and interpretation in oil and gas fields. Beijing: Petroleum Industry Press; 1996, p. 5, 6.
[10] PetroChina Exploration & Production Company. Evaluation methods and technology of logging identification of low resistivity gas reservoirs. Beijing: Petroleum Industry Press; 2006, p. 50–55.
[11] Zhao L, Bu Y. Logging evaluation technology of carbonate reservoirs. Beijing: Petroleum Industry Press; 1994, p. 15–22.
[12] Hao J, Feng R, Zhou J. Discussion on mechanism of resistivity variation during rock breaking. Chin J Geophys 2002;23(1):426–33.
[13] Liu H, Xia H, Wang Y. Study on fractal characteristics of formation cementation factor m. Logging Technology 2001;25(1):24–7.
[14] Liu Z. Analysis of high temperature, high pressure, shaliness, wettability and experimental methodology on archie equation. Logging Technology 1998;22(4):231–6.
[15] Wang Z, Shi Z. Modern technology of reservoir description. Beijing: Petroleum Industry Press; 1999, p. 118–26, 144–47.
[16] Wang Y, Kong J, Li HG. Geology of gas reservoirs. Beijing: Petroleum Industry Press; 2004, p. 41–55.

CHAPTER

Identification of Gas and Water Zones in Volcanic Gas Reservoirs

8

As the principal means of clarifying gas and water distribution in volcanic gas reservoirs, gas and water zones in volcanic gas reservoirs are identified on the basis of mud log indicators for gas-bearing, water-bearing, and dry zones as well as formation test and logging responses as determined by combining qualitative analysis with quantitative interpretation and using individual well data. These analyses help delineate the distribution of gas and water zones and lay the foundation for identifying gas reservoir types and appropriate reserve evaluation.

8.1 RESEARCH CHALLENGES AND TECHNICAL SOLUTIONS

8.1.1 Challenges in the identification of gas and water zones in volcanic rocks

The complex relationships between gas and water in volcanic gas reservoirs result in great difficulties in the identification of gas and water zones. One of the challenges is that geological logging is a qualitative identification method, and significant uncertainty exists in its data interpretation; most log data are obtained from mud return with significant depth errors. Therefore, this method is mainly used for preliminary and qualitative evaluation of the gas-bearing properties in volcanic rocks. Due to the poor petrophysical properties of the matrix in volcanic reservoirs and the presence of fractures, it is usually difficult to obtain accurate formation test data. Moreover, hydrodynamic fracturing during production makes it difficult to assess the accurate fluid-yielding position. All of these factors cause difficulties in identifying gas and water zones accurately.

Another challenge is that the complex and highly variable lithology, composition, and structural features of volcanic rocks result in different electrical conductivity, acoustic propagation and neutron deceleration characteristics; consequently, the resistivity and porosity logging responses differ greatly. Volcanic reservoir spaces are characterized by various types and complex pore structures, resulting in great variability in reservoir permeability and electrical conductivity. The difference between the resistivity values of volcanic rocks with different reservoir spaces and pore structures can be up to several orders of magnitude. In addition, the conductive mechanisms of volcanic rocks are complex. Low-resistivity gas layers (with resistivity two times lower than that of pure water zones [1]) often exist due to matrix conductivity, irreducible water conductivity, and filtrate invasion, which contribute to the uncertainty and challenges in the identification of gas and water zones using log data.

8.1.2 Technical solutions

In light of the difficulties in gas and water zone identification, the integrated technical solutions based on geological logging, formation testing, and well log interpretation have been applied to establishing technical procedures (Figure 8.1), and the technique for identifying gas and water zones in volcanic gas reservoirs has been developed:

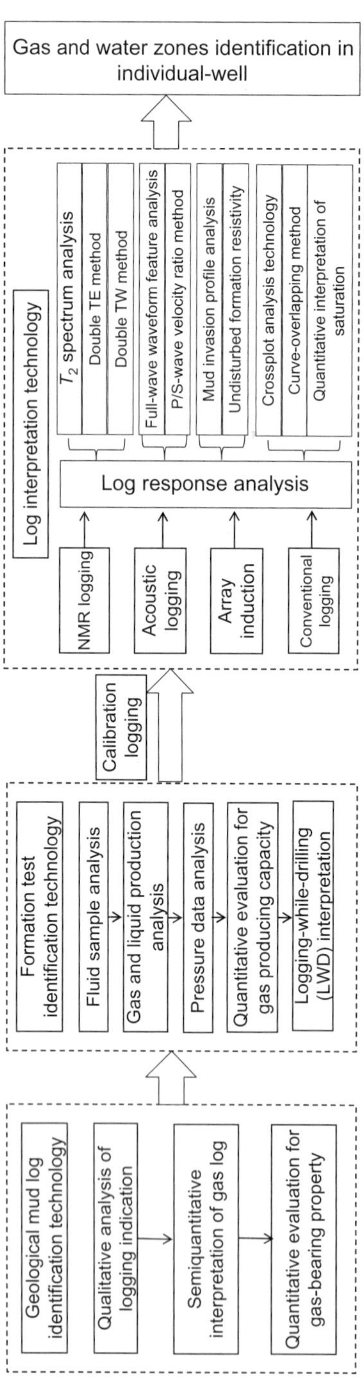

FIGURE 8.1

Technical procedures for volcanic gas and water zones identification.

1. Based on data derived from cores and cutting logs, drilling fluid logs, and gas logs, the quantitative evaluation methods for gas-bearing properties including the qualitative analysis of logging indicators, semiquantitative interpretation of gas log and sealed coring are utilized for the preliminary identification of gas and water zones in volcanic rocks.
2. Based on available data on formation fluids, pressure, fluid production, and the like, technologies such as wireline (cable) testing, drill stem testing, conventional gas testing, and gas well testing, together with fluid sample analysis, gas and fluid production analysis, pressure data analysis, and quantitative evaluation of gas producing capacity, are used to identify gas and water zones in volcanic rocks.
3. Well log calibration with geological mud logs and formation testing is used as the basis for analyzing the logging response features of typical gas zones, water zones, and dry layers. The identification modes for gas and water zones are established for different lithologies. NMR, acoustic wave, array induction, and conventional log data, in combination with methods such as T2 spectrum analysis, acoustic waveform analysis, crossplot, curve-overlapping, and saturation calculation, are used to identify gas and water zones, which lay the foundation for an evaluation of gas reservoirs.

8.2 GEOLOGICAL LOGGING AND FORMATION TEST IDENTIFICATION

Geological logging and formation tests aim to directly acquire subsurface fluid information by geological means and various testing technologies. They are important methods for identifying volcanic gas layers and aquifers and for calibrating log data. The log data on gas and water zones, fluid samples, differences in pressure features, as well as individual well gas-bearing information are used to identify gas and water zones and calibrate log data, based on the principles of geological logging and formation tests.

8.2.1 Geological logging methods for identifying gas and water zones

The geological logging methods directly acquire subsurface gas-bearing information [2] by means of core logging, cutting and drilling fluid logging, gas logging, and compound logging. This approach delineates gas and water zones by identifying the abnormalities of high gas content in log indications. It is fast, convenient, intuitive, rich in information, and easy to use.

Considering the specific characteristics of the principles and data on different geological logging technologies, a technical solution, consisting of qualitative assessment, semiquantitative identification, and quantitative evaluation, is developed to integrate log data for identifying volcanic gas and water zones.

1. Qualitative logging indicator analysis

Useful information indicating the presence of gas layers include (1) bubbles in cores, fluorescence of cuttings, and gas shows during sealed coring; (2) bubbles on the liquid surface of mud ditch during drilling fluid logging; and (3) bubbles, gas influx, well kicks, and well blowout at the mud outlet of the wellhead. More frequent and intense occurrences of the preceding gas information indicate better gas-bearing properties for the reservoirs. Therefore, through observation and qualitative analysis, it can be determined whether the drilling has encountered a gas layer, and if a gas layer is present, its gas-bearing properties can be analyzed immediately.

Well XX1 is the initial discovery well for the XX gas field. Four well kicks and one controlled well blowout took place in the Ying 1 interval of volcanic rocks when the rig drilled into the Yingcheng Formation. The well kicks (in layers 145 to 150) were so strong that the mud gushed out of the throttle pipe as high as 5 m into the air. In consequence, mud weight and viscosity dropped slightly and rice-shaped bubbles with diameters of 1 to 2 cm were observed on the trench surface (their volume making up 60% to 70%). The mud height increased by around 5 cm and the mud overflew. Well blowouts (in layers 149 and 150) were strong. The maximum distance of the gas jet reached 40 m. After ignition, the flame was orange-red, with a jet distance of 14 to 15 m and a height of 4 to 5 m. In all of the volcanic rocks in Well XX1, these drilling indicators are observed to suggest the presence of abundant natural gas within the encountered volcanic formations. Gas testing was later implemented with gas production estimated at 6.84×10^4 m^3/d. This initiated the exploration and development of the XX gas field.

2. Semiquantitative gas log interpretation

Gas log anomalies with values greater than the background values are important indicators for volcanic gas layers. Based on the information on gas, water, and dry zones determined by formation tests, together with the total-hydrocarbon data, hydrocarbon component, and nonhydrocarbon component data obtained by gas logging from the formation testing intervals, this technique relies on crossplot analysis to determine the cutoff criteria for key parameters such as total-hydrocarbon content, hydrocarbon components, and the gas log ratio of typical gas zones. It lays the foundation for interpreting gas and water zones based on gas log data.

Figure 8.2 illustrates how the volcanic gas zones were identified in the XX gas field using the gas log ratio method. It demonstrates that (1) different types of reservoirs have varied gas log ratios (GLR)

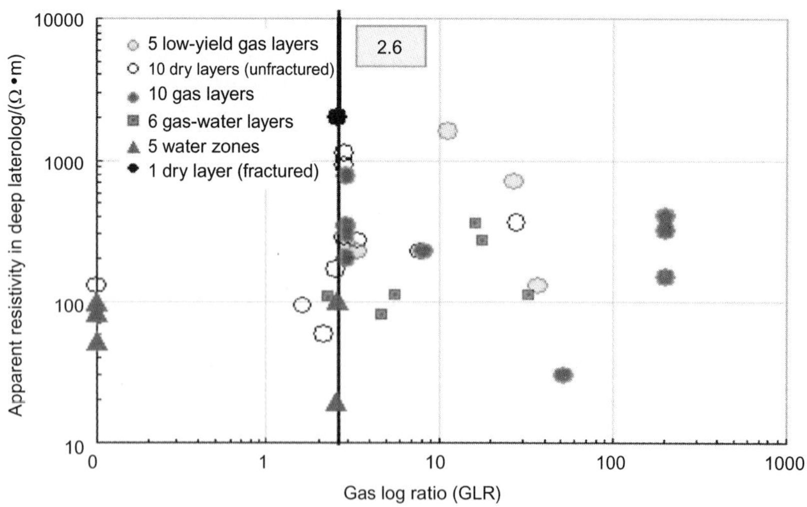

FIGURE 8.2

Gas log ratio method for identifying gas layers.

and resistivity values: high-yield gas zones have relatively high GLR and moderate resistivity values, whereas low-yield gas zones have moderate GLR but relatively high resistivity; gas-water zones have relatively low GLR and resistivity; water zones have the lowest GLR and resistivity; and dry zones have relatively low GLR but the highest resistivity; and (2) the GLR cutoff of the volcanic gas zones of this gas field is about 2.6.

3. Quantitative evaluation of gas-bearing properties

The main logging methods that can be used for the quantitative evaluation of gas-bearing properties include (1) using sealed coring data to quantitatively determine the original gas saturation in gas zones, (2) using the total-hydrocarbon analysis data from gas logging to obtain hydrocarbon gas contents, (3) using the light hydrocarbon log of headspace gas to directly obtain data on the volumetric concentration of the 29 components from C1 to C7, and (4) using porosity and permeability analysis logging to calculate the free fluid index and irreducible water saturation so as to determine the original gas saturation in volcanic gas reservoirs.

Compared with formation tests and well logging, geological logging is more susceptible to the influence of nonformation factors. Therefore, the use of the single logging technique is prone to large errors in identifying gas and water zones. Using multiple methods to validate each other can effectively reduce the interpretation uncertainties and raise the congruence rate of gas and water zone identification.

8.2.2 Formation test method for identifying gas and water zones

The formation test method uses the techniques of testing while drilling, wireline tests, drill stem tests, conventional formation tests, and gas well tests to directly collect fluid samples and formation pressure data and to estimate formation parameters so as to evaluate the commercial deliverability of reservoirs. Due to its advantageous time saving, high efficiency, and quick response, formation testing has the advantages of quick turnaround time, high efficiency, and quick response, and it is recognized as an effective and important method for identifying gas and water zones [3,4].

The following procedures have been adopted to identify gas and water zones in volcanic gas reservoirs based on the technological principles of formation testing and information acquisition.

1. Analyzing fluid samples for gas-bearing properties

Wireline tests and drill stem tests can be used to collect fluid samples at tested intervals. These samples are analyzed to extract information on the composition and density of gas and condensate and to examine the ion components and salinities of formation water. This information can then be used to verify the presence of any natural hydrocarbon gas; to determine whether the water is formation water, mud filtrate, or hydrodynamic fracturing fluid; and to estimate the water cut.

2. Identifying gas and water zones in volcanic rocks by gas and liquid production analysis

Conventional gas tests and gas well tests can be used to channel formation fluid into the borehole and then flow out to the surface. Based on the production pattern (by natural flowing or hydrodynamic fracturing) of the fluid, metering methods, choke size, and the relevant production rates of gas and water, an appropriate procedure can be adopted to calculate the daily gas production, daily water production, gas-water ratio, and cumulative gas and water production. Gas-water property analysis and pressure characteristic analysis can be combined to estimate the fluid type and the potential production capacity and thus to identify gas and water zones in volcanic rocks.

3. Identifying the gas-water interface by pressure data analysis

Pressure data is the key information derived from formation tests. Pressure data vary with depth when acquired through testing while drilling (TWD), and wireline tests and drill stem tests are used to estimate the pressure gradient and the fluid density (Figure 8.3), to identify the fluid type and two-phase fluid interface, and to identify gas and water zones.

Figure 8.3 shows an example of using Modular Dynamic Tester (MDT) data to identify the fluid type for Well CC1-2 in the CC gas field. This figure demonstrates that (1) gas zones are characterized by small pressure gradients, high resistivity, low density, and low neutron porosity. In contrast, water zones are characterized by large pressure gradients, low resistivity, high density, and high neutron porosity. Condensate layers are between the former two cases; and (2) the gas-oil interface and oil-water interface are located at about 7150 ft and 7180 ft, respectively.

4. Estimating gas production capacity by analyzing the gas production rate and pressure variations under different choke sizes

The pinpoint test method is used in conventional gas formation testing to acquire data on gas and water production and pressure variations under different choke sizes. Through gas reservoir engineering analysis, the effective permeability of the tested zone is calculated, the pressure characteristics of the tested zone are clarified, the relationship between gas production and flowing pressure is established, the gas well deliverability equation and the absolute open flow potential are determined, and the reserves controlled by wells are estimated qualitatively. Consequently, the production capacity of the tested zone in the gas well is estimated, which will provide a basis for determining an appropriate production system.

5. Interpreting gas and water zones by analyzing logging-while-drilling curves

Formation testing while drilling (TWD) offers a method to measure resistivity, acoustic waves, natural gamma ray, density, and neutron porosity, thereby generating logging curves. Based on the conventional well testing method for identifying gas and water zones, the resistivity-acoustic wave and neutron-density crossplot techniques or the sensitivity curve-overlapping method can be used to identify the types of products derived from the tested zone and hence to identify gas and water zones in volcanic gas reservoirs.

8.3 WELL LOGGING IDENTIFICATION TECHNIQUES

Well logging measures the characteristics of subsurface rocks—such as radioactivity, neutron deceleration, acoustic propagation, electrical conductivity, and proton relaxation characteristics—in order to identify gas layers and aquifers indirectly. This book presents an integrated approach to identifying volcanic gas, water, and dry zones based on the characteristics of the gas and water zones in volcanic gas reservoirs, whereby the formation test results are used to calibrate logging information. In addition, the logging response characteristics of gas, water, and dry zones are analyzed; nuclear magnetic logging, acoustic logging, array induction logging, and conventional well logging are integrated; and T_2 spectrum analysis, waveform analysis, crossplot, and curve-overlapping methods are applied.

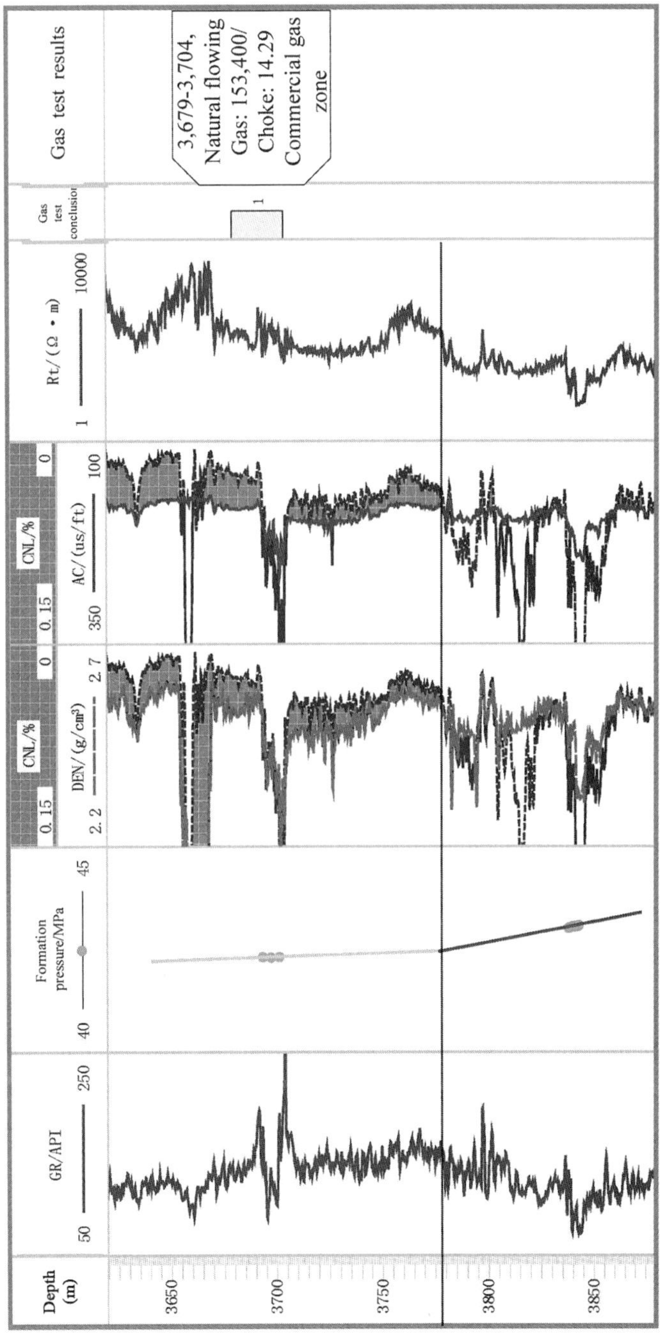

FIGURE 8.3

Pressure gradient analysis for fluid type identification (Well CC1-2).

8.3.1 Nuclear magnetic logging identification techniques

Nuclear magnetic logging is used to record the relaxation characteristics of different fluids in different states within formation pores. Several methods, such as T_2 spectral analysis, shifted spectrum analysis, and differential spectrum analysis, are used in combination to identify gas, water, and dry zones in volcanic gas reservoirs based on the principles of nuclear magnetic logging and different relaxation characteristics between gas and water.

1. Logging principles

Nuclear magnetic logging is based on the principle of nuclear magnetic resonance and measures the spin echo signal and its amplitude attenuation with time (the left side of Figure 8.4). After a multiple-factor transformation, a common T_2 spectrum is obtained (the right side of Figure 8.4). The T_1 spectrum is measured by the reversal recovery method utilizing radio frequency pulses [5,6]. Considering the requirement of wireline logging for logging speed, usually the lateral relaxation is the objective of measurement [7,8].

2. Relaxation characteristics of natural gas and water

There are three relaxation modes for protons in strata: surface relaxation, diffusive relaxation, and volumetric relaxation [9,10]. In rocks, irreducible water primarily exhibits surface relaxation; free water mainly exhibits volumetric relaxation and is also affected by diffusive relaxation. Natural gas primarily exhibits diffusive relaxation. The relaxation characteristics of formation water and natural gas differ from each other in several aspects (Table 8.1): (1) water has a high hydrogen content and a small diffusion coefficient and hence has a short longitudinal relaxation time (T_1) and long lateral relaxation time (T_2); (2) in contrast, natural gas has a low hydrogen content and a greater diffusion coefficient and thus has a long longitudinal relaxation time (T_1) and a short lateral relaxation time (T_2); and (3) there is a large difference of more than 40 times between the values of T_1 and T_2 for natural gas and formation water.

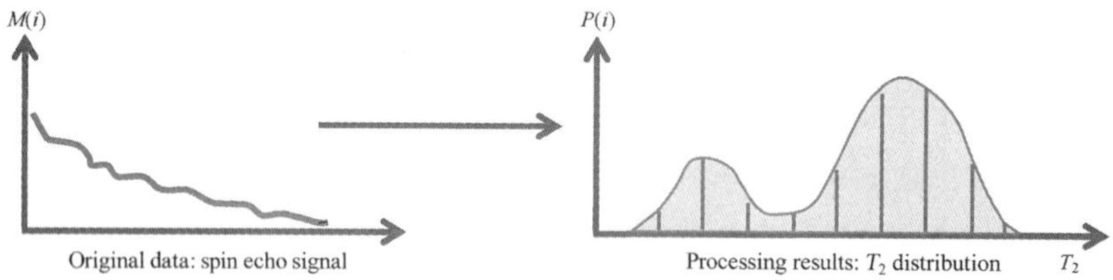

FIGURE 8.4

Multifactor transformation of the T2 spectrum in nuclear magnetic logging.

Table 8.1 Comparison between Relaxation Characteristics of Natural Gas and Water

Fluid	T_1/ms	T_2/ms	Typical T_1 /T_2	HI	η/(mPa•s)	$D/(\times 10^{-5}\ cm^2/s)$
Water	1–500	1–5000	2	1	0.2–0.8	1.8–7
Gas	2000–5000	1–60	80	0.2–0.4	0.011–0.014	80–100

Note: T_1, longitudinal relaxation time; T_2, lateral relaxation time; D, diffusion coefficient. HI, hydrogen content index; η, viscosity.

3. Logging response characteristics of gas, water, and dry zones in volcanic rocks

The pattern of T_2 in nuclear magnetic logging for volcanic reservoirs is determined by reservoir space type, pore size, and fluid type. When different types of fluids are present in a volcanic reservoir, the response characteristics of nuclear magnetic logging show the following changes (see Figure 8.4, Table 8.2): (1) for a gas layer, T_2 is short, and the peak has a large amplitude peak on the left side of the T_2 spectrum and typically exhibits an unimodal or a closely spaced bimodal feature; (2) For an water zone, T_2 is long, and the peak has a large amplitude peak on the right side of the T_2 spectrum and exhibits a more widely separated bimodal feature; and (3) for a tight dry layer, T_2 is long, and the peak has a small and wide amplitude peak and exhibits a bimodal or multimodal feature. Based on these log characteristics, nuclear magnetic logging can be used to identify gas and water zones [5].

4. Nuclear magnetic logging methods for identifying gas and water zones in volcanic rocks

T_2 spectrum analysis, shifted spectrum, and differential spectrum methods are used to identify gas and water zones in volcanic rocks based on the identification principles.

1) T_2 spectrum analysis

In accordance with the peak positions, amplitudes, and spectrum shapes of T_2 spectral distribution patterns in nuclear magnetic logging, fluid types can be qualitatively identified (Figure 8.5) following these two steps:

1. The number and amplitude of the free-fluid peaks ($T_2 > T_{2cutoff}$) of the target layer are analyzed to estimate whether the zone is a reservoir or a tight dry layer. A greater number of free-fluid peaks coupled with higher peak amplitudes usually indicate better reservoir effectiveness.
2. In a volcanic reservoir, the position and the spectral shape of the free-fluid peak are analyzed to determine the gas-bearing properties of the target layer and thereby obtain a preliminary identification of the gas and water zones. The farther left the free-fluid peak is in the spectrum, the better gas-bearing properties the reservoir has.

2) Shifted spectrum method

The shifted spectrum method is also known as the parallel TE method. Due to different diffusion coefficients, hydrocarbon gas and water zones are distributed at different positions in the T_2 spectra. The gas layer is shown approximately as a single peak, whereas the water zone usually has a wide distribution. Therefore, if acquisition time is set long enough and $T_R > 3\text{-}5T_{1h}$ (where T_R is longitudinal recovery time, and T_{1h} is the longitudinal relaxation time of natural gas) to allow complete recovery of the longitudinal relaxation in each measurement, the two echo strings can be obtained with two echo

Table 8.2 Response Characteristics of Gas, Water, and Dry Layers in Nuclear Magnetic Logging

Reservoir Type	Peak T_2 Value	Peak Amplitude	Free-Fluid Peak Position	Spectrum Shape
Gas layer	Short	Large	Left	Unimodal or closely bimodal
Water layer	Long	Large	Right	Bimodal
Dry layer	Widespread	Small	Widely spread	Bimodal or multimodal

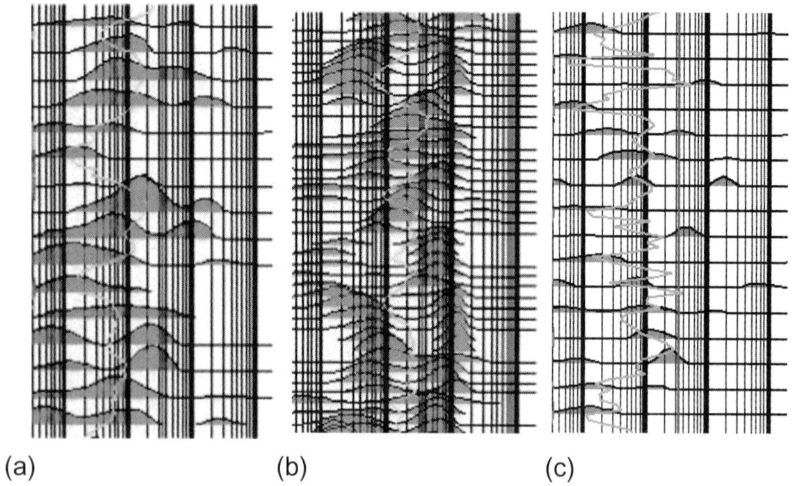

FIGURE 8.5

T_2 patterns in nuclear magnetic logs for gas, water, and dry layers in volcanic gas reservoirs. (a) Gas layer (welded tuff, Well XX601). (b) Water layer (breccia lava, Well XX8). (c) Dry layer (tight rhyolite, Well XX9).

intervals (i.e., T_{EL} and T_{ES}) and for different lengths for the measurement. In the T_2 spectra of the long echo interval (T_{EL}), T_2 for the gas layer is shorter than that for the short echo interval T_{ES}. The time variation of T_2 for the water zone is related to the pore size. T_2 would be reduced by a larger margin in a big pore than in a small pore. In addition, the displacement of signals toward T_2 for gas is much larger than that for water [5]. Hence, by calculating the T_2 distribution difference between long and short echo intervals, the signals for water can be removed and the characteristics of natural gas can be highlighted (Figure 8.6).

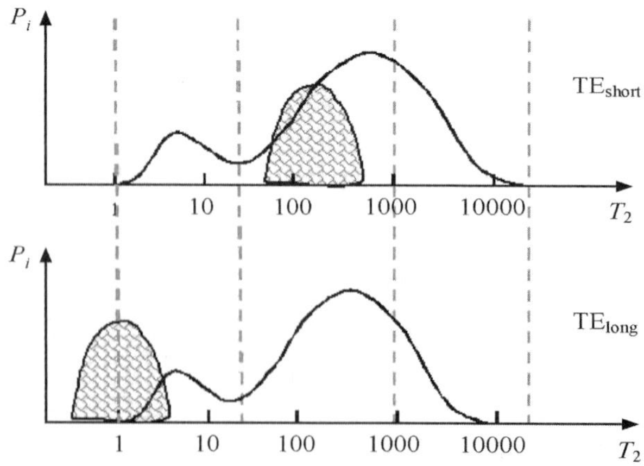

FIGURE 8.6

Schematic diagram of the parallel TE method for gas layer identification.

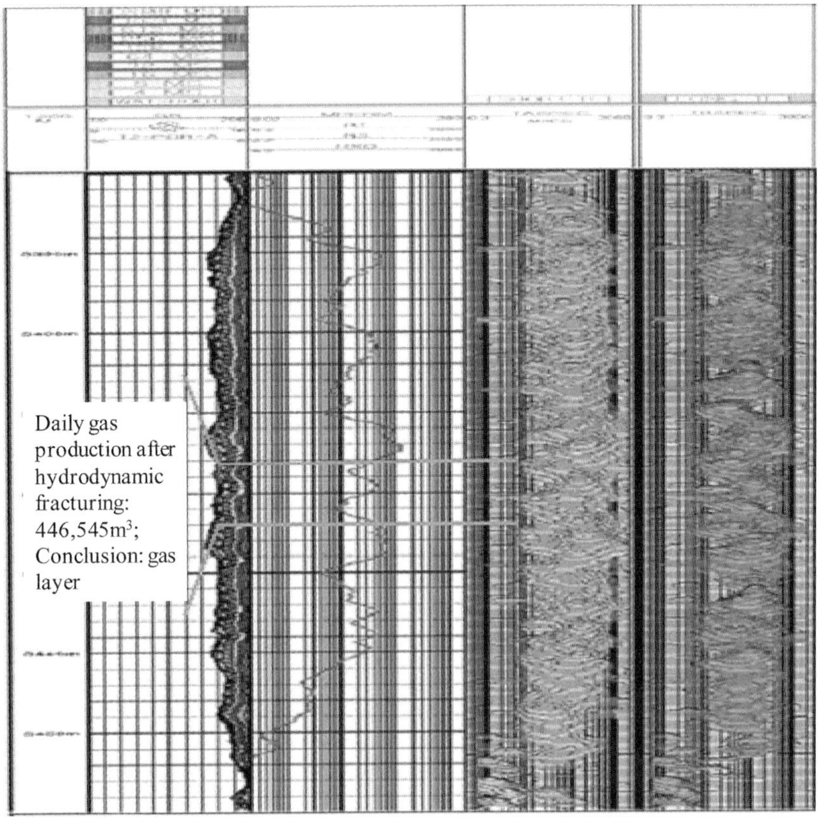

FIGURE 8.7

Example of the shifted spectrum method for identifying volcanic gas reservoirs.

Figure 8.7 shows a layer of rhyolitic welded tuff in a well of the XX gas field. The effective porosity has a range of 6% to 12%, and the permeability 1 to 3 mD. The T_2 spectra show that many long relaxation components are present in this layer, indicating the presence of well-developed macro pores. The processing results using the shifted spectrum method indicate that the leftward displacement of T_2 of the short echo interval is insignificant, whereas the leftward displacement of T_2 of the long echo interval is conspicuous, exhibiting a notable gas-bearing feature. Gas layer testing after hydrodynamic fracturing was conducted in the interval of 3416 to 3424 m. Daily gas production reached 44.65×10^4 m^3, and 14.4 m^3 of liquids were produced per day. The flow-back rate was 10.98%. Water data analysis indicates a mud filtrate. The zone was thus confirmed to be a commercial gas layer.

3) Differential spectrum method

The differential spectrum method is also known as the parallel TW method. The standard echo interval and two different lengths of acquisition time, T_{WL} (>3-5T_{1g}) and T_{WS} (>3-5T_{1w}), are used in the measurement. Because the longitudinal relaxation recovery of water is much faster than that of gas, the magnetization vectors of water and gas will recover completely in the echo information acquired with

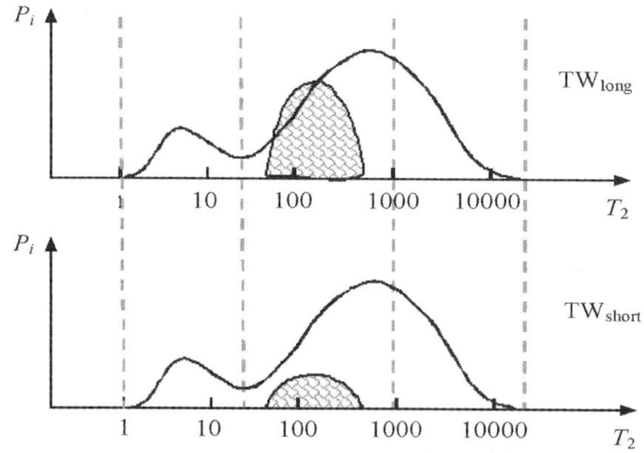

FIGURE 8.8

Schematic diagram of the differential spectrum method for identifying natural gas.

the long waiting time T_{WL}, and the T_2 spectra will contain complete gas and water information. In the echo information acquired with short waiting time T_{WS}, the T_2 spectra will include complete water information but only partial gas information. The gas information is thus obtained by subtracting one from the other (Figure 8.8) [5].

Figure 8.9 shows an example of the differential spectrum method used to identify volcanic gas layers. The T_2 signal measured with the long acquisition time (the fourth trace) is considerably greater than the T_2 signal measured with the short acquisition time (the fifth trace), and the separation between the two (the sixth trace) indicates distinct characteristics of a gas layer. After hydrodynamic fracturing, natural flowing started, with daily gas production reaching 46.14×10^4 m^3 and the open-flow potential approaching 1 MMCM (million cubic meters)/d. The formation was thus verified as a commercial gas layer.

8.3.2 Acoustic logging identification technique

1. Principle of the technique

The orthogonal dipole acoustic logging instrument consists of a controllable monopole emitter, two orthogonal dipole emitters, and eight array receivers (32 receiving units). Using six different operation modes, it can accurately acquire longitudinal waves, fast-slow shear waves, and Stoneley waves; in addition, it can obtain the time difference, reflected energy, and wave shapes of different acoustic waves.

Natural gas slows down the propagation velocity of acoustic waves in formations and strengthens the energy-absorbing capacity of the acoustic waves. When different fluids are present, the response characteristics of the acoustic log (e.g., the dipole sonic imager (DSI) log) will be different (Figure 8.10, Figure 8.11, and Table 8.3):

1. The volcanic gas layers are marked by large time lags among longitudinal waves (P wave), shear waves (S wave), and Stoneley waves, as well as rapid energy attenuation, leading to a low energy value (Figure 8.10a). On the wave interference figure, the waveform is partially missing, chaotic, and has interference fringes (Figure 8.11). In addition, as the longitudinal and shear wave velocities change at different rates, the P-S (compressive-shear wave) velocity ratio decreases by 3% to 30% [11,12].

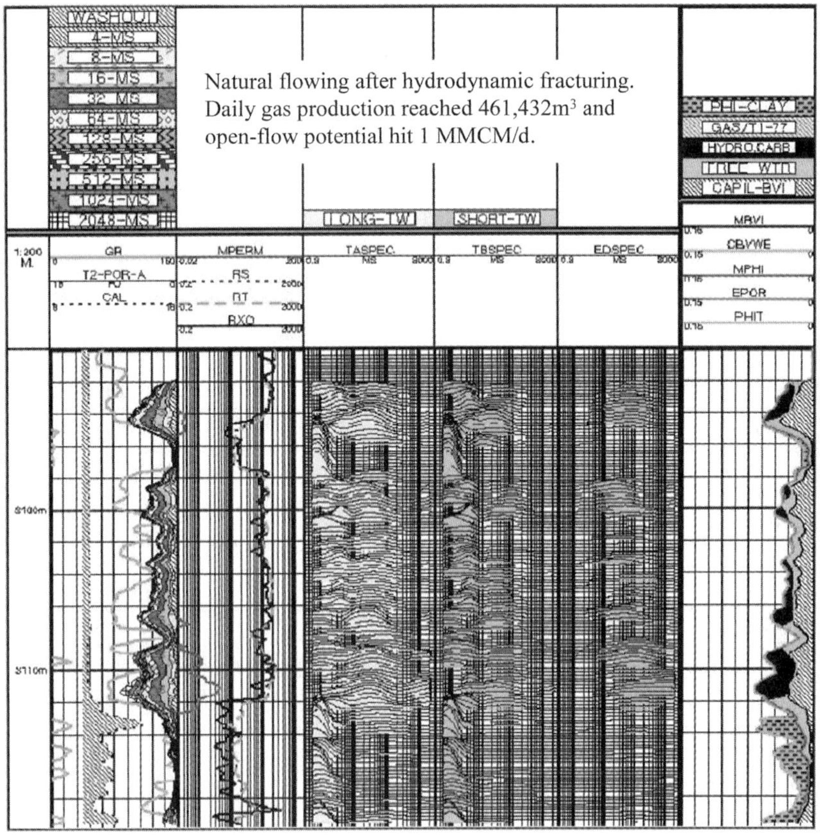

FIGURE 8.9

Example of the differential spectrum method used to identify volcanic gas reservoirs.

2. The water layers are marked by moderate time lags for longitudinal waves, shear waves, and Stoneley waves, resulting in a moderate wave energy attenuation and moderate energy (Figure 8.10b). On the wave interference figure, the waveform is complete but chaotic, and the P-S velocity ratio is relatively large.
3. For tight dry layers, the time lag of longitudinal, shear, and Stoneley waves are small, marked by minor waves energy attenuation and thus strong energy (Figure 8.10c). The waveform is complete and straight, and the P-S velocity ratio is large.

2. Identification methods

Based on the principles of acoustic log identification, two methods can be used to identify gas and water zones in volcanic rocks.

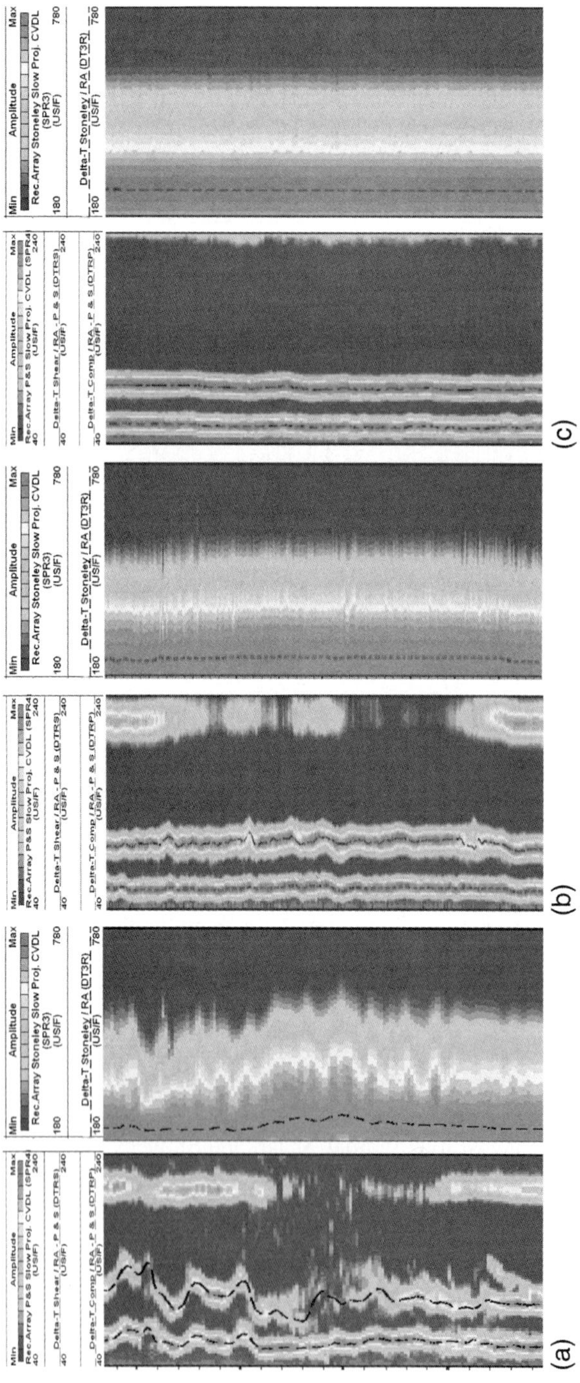

FIGURE 8.10

Interval transit time and energy characteristics in gas, water, and dry layers in volcanic gas reservoirs. (a) Gas layer (crystaltuff, XX601). (b) Water zone (breccia lava,XX401). (c) Dry layer (welded breccia, Well XX7).

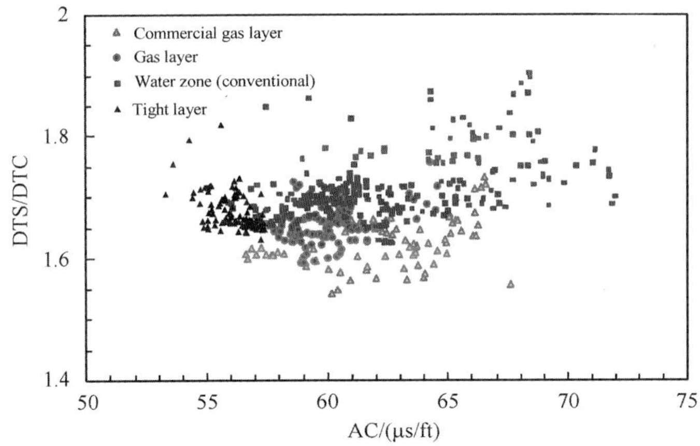

FIGURE 8.11

Crossplot of interval transit time – P/S velocity ratio for gas, water, and dry zones in volcanic rocks.

Table 8.3 Acoustic Logging Response Characteristics of Gas, Water, and Dry Zones

Reservoir Type	P-Wave Slowness	P-S Velocity Ratio	Reflected Wave Energy	Wave Interference Figure
Gas layer	Long, cycle skip	Small	Weak	Waveform partially missing, chaotic
Water layer	Moderate	Relatively large	Moderate	Waveform complete but chaotic
Dry layer	Short	Large	Strong	Waveform complete and straight

1) Analysis of full-wave oscillogram attenuation characteristics

Volcanic gas layers have a strong capacity to absorb acoustic waves, thus the acoustic wave energy would show drastic attenuation. As a result, the full wave oscillogram exhibits features such as partially missing and chaotic waveforms and interference fringes. Under the same conditions, a formation with better gas-bearing properties will have more obvious acoustic wave response characteristics. Hence, in accordance with the full wave oscillogram attenuation characteristics, the gas-bearing properties of a formation can be estimated on a preliminary basis.

2) Interval transit time and p-s velocity ratio method

Under the same conditions, a formation with better gas-bearing properties will show a larger interval transit time but smaller P-wave and S-wave velocities. Hence, a preliminary identification of gas and water zones can be achieved by establishing a crossplot of interval transit time and P-S velocity ratio. Figure 8.11 is the crossplot of the interval transit time and P-S velocity ratio for gas, water, and dry zones in the XX gas field. It shows that, in the gas layer, the interval transit time is high and the P-S velocity ratio is low, whereas in the water zone, both the interval transit time and the P-S velocity ratio have high values. For the tight dry layer, the interval transit time is low and the P-S velocity ratio is moderate.

Figure 8.12 shows an example of the acoustic logging method used to identify gas layers. In Well XX1, the 3447- to 3453-m interval is a crystalline tuff reservoir with good petrophysical properties.

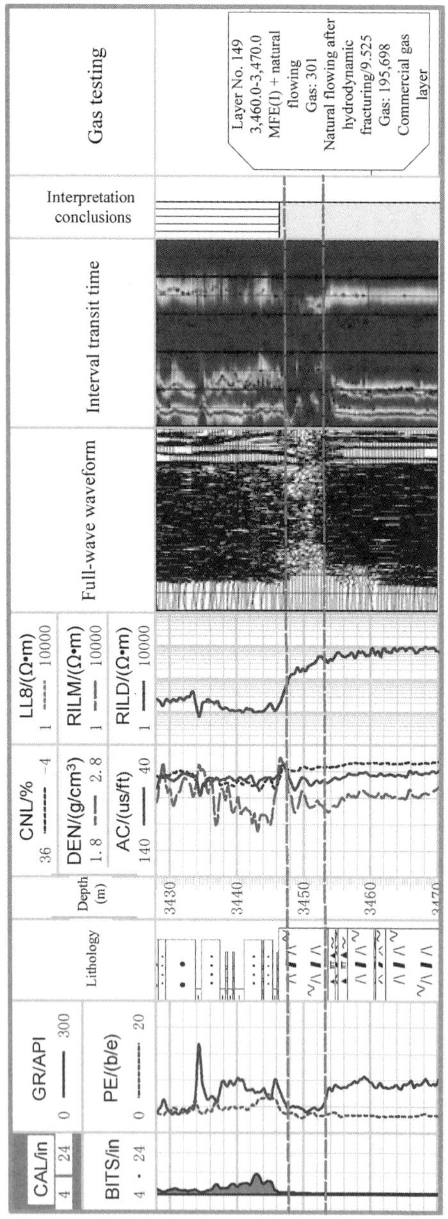

FIGURE 8.12

Full wave oscillogram and interval transit time characteristics of the volcanic gas layer (Well XX1).

The P-wave time lag is high (about 62.8 μs/ft) and cycle skip was observed. The P-S velocity ratio (V_p/V_s) is about 1.53, which is smaller than that for tight layers. The acoustic wave energy attenuates rapidly, the waveform being chaotic or even partly missing. The energies of longitudinal, shear, and Stoneley waves are weak, indicating good gas-bearing properties of the interval. Formation testing was conducted at the 3460- to 3470-m interval. Daily gas production reached 19.57×10^4 m^3, thus confirming a commercial gas layer.

8.3.3 Array induction logging identification technique

1. Principle of the technique

The high-resolution induction logger (HDIL) is equipped with an emitter coil, eight receiver coils, and eight compensating coils, which form eight array units with three working frequencies (25 kHz, 50 kHz, and 100 k Hz) and can generate hundreds of induction resistivity curves with three longitudinal resolutions (1 ft, 2 ft, and 4 ft) and six detection depths (10 in, 20 in, 30 in, 60 in, 90 in, and 120 in). Based on these curves, original formation resistivity and mud invasion profiles can be obtained through pretreatment and 1D inversion processing [13].

Due to the poor electrical conductivity of natural gas, volcanic gas layers have good petrophysical properties and high resistivity. Mud invasion will result in a decline in resistivity, causing a low mud invasion effect. Formation water has good conductivity and water zones in volcanic rocks have good petrophysical properties and low resistivity, so mud invasion will result in higher or constant resistivity, thus causing low mud invasion or no mud invasion effect. As tight volcanic dry layers have poor petrophysical properties and high to ultra high resistivity, it is difficult for mud to enter the formation. In most cases, there is no mud invasion effect (Table 8.4). Therefore, volcanic gas and water zones can be identified by comparing different original formation resistivities and analyzing the characteristics of invasion profiles from array induction logging.

2. Identification methods

Two methods can be used to identify gas and water zones in volcanic rocks based on the principles of gas layer and water zone identification in array induction logging.

1) Mud invasion profile analysis

When the formation water salinity satisfies certain conditions ($Rw < Rmf$), the mud invasion profiles of gas layers will exhibit the typical characteristics of low invasion (namely, the resistivity of the mud-invaded zone decreases). In contrast, the mud invasion profiles of water zones will exhibit the typical characteristics of high mud invasion. Tight dry layers will show no mud invasion.

In the figure of array induction processing results for Well XX5 (Figure 8.13), layer 142I (a rhyolitic volcanic breccia reservoir from 3700 to 3755 m deep) exhibits the overall characteristics of low mud

Table 8.4 Array Induction Logging Response Characteristics of Gas, Water, and Dry Zones

Reservoir Type	Petrophysical Property	Resistivity	Mud Invasion Profile
Gas layer	Good	Medium-high	Low invasion
Water layer	Good	Low	High invasion
Dry layer	Poor	High-extremely high	No invasion

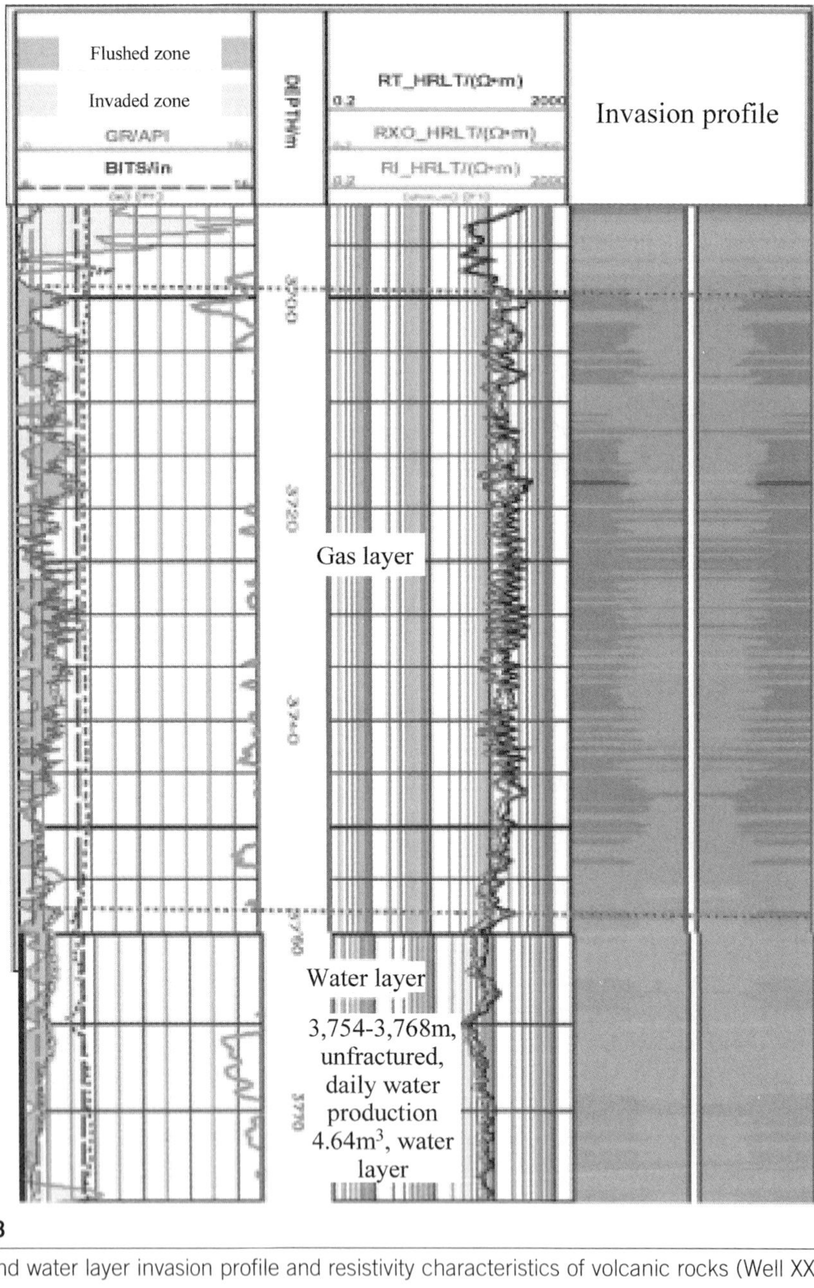

FIGURE 8.13

Gas layer and water layer invasion profile and resistivity characteristics of volcanic rocks (Well XX5).

invasion (the fourth curve) with noticeable gas-bearing features. Layer 142II (a rhyolitic crystal tuff reservoir from 3755 m to 3784 m deep) generally exhibits the characteristics of high mud invasion without noticeable gas-bearing features. A perforation test was carried out at the interval from 3754 m to 3768 m without hydrodynamic fracturing. Daily water production was maintained at 4.64 m^3. The test indicated a water layer, which validated the above assessment its gas-bearing properties.

2) Original resistivity cutoff and saturation calculation method

In general, a gas layer has medium-high resistivity values, a water layer has low resistivity, and a dry layer is characterized by high to extremely high resistivity. For this reason, gas and water zones can be identified qualitatively by comparing the resistivity values under the same conditions. Gas and water zones can be recognized semiquantitatively by examining the resistivity cutoff criteria of effective gas reservoirs. Based on saturation calculations, gas and water zones can also be distinguished and the gas-bearing properties of a reservoir can be evaluated quantitatively.

In Figure 8.13, layer 142 I shows a high resistivity. The average original resistivity of the formation interpreted by array induction logs is 300 $\Omega \cdot m$, higher than the cutoff of gas layers (at 110 $\Omega \cdot m$). The average gas saturation of reservoir matrix is 52%. The layer is interpreted as a gas layer, which is inferior to the commercial gas layers of the same type in the area in terms of gas-bearing properties. Layer 142 II is of low resistivity, with the average true resistivity of the layer interpreted by array induction logs being 60 $\Omega \cdot m$, lower than the cutoff of gas layers. The average gas saturation of reservoir matrix is 30%. The layer is interpreted as a water layer. Subsequent gas testing validated these interpretations.

8.3.4 Conventional well log-based identification technique

1. Principle of the method

The differences between natural gas and formation water in neutron deceleration characteristics, the photoelectric absorption index, acoustic propagation characteristics, and electrical conductivity are the key factors used to help identify gas and water layers in volcanic rocks (Table 8.5). The ratio of natural gas density to formation water density is about 1/8, the ratio of their compensated neutrons is less than 1/100, the ratio of their macroscopic thermal neutron capture sections is around 1/13, and the ratio of their interval transit time is greater than 4. In addition, natural gas has poor conductivity and high resistivity, whereas formation water has good conductivity and low resistivity [14].

Table 8.5 Responses of Natural Gas and Formation Water in Conventional Well Logs

Fluid Type	Compensated Density/(g/cm3)	Compensated Neutron /(p. u)	Thermal Neutron Macroscopic Capture Section Index/(p. u)	Resistivity	Interval Transit Time/ (μs/ft)
Formation water	1.19	60	157.6	Low	185
Natural gas	0.15	—	12.5	High	>800

2. Conventional well logging response characteristics of gas layers, water layers, and dry layers

Natural gas and formation water have different responses to conventional well logging; therefore, gas layers, water-bearing layers, and dry layers in volcanic rocks will have different characteristics in conventional wireline log curves (Table 8.6 and Figures 8.14 and 8.16).

1) Well log characteristics of volcanic gas layers

Volcanic gas layers are characterized by low density, medium-low neutron porosity (excavation effect), high interval transit time (cycle skip), medium-high resistivity, and positive separation between deep and shallow laterolog resistivity values (see Table 8.6).

Layer 215I in Well XX23 (depths 3892 to 3975 m; see Figure 8.14) is a rhyolite reservoir with an average matrix porosity of 9.5% and good petrophysical properties. Its average resistivity is 1175 $\Omega \cdot$m, higher than the cutoff of gas layers. Its positive difference between deep and shallow laterolog resistivities is 310 $\Omega \cdot$m. Its average neutron porosity is 3.35%, lower than the formation porosity. Its average interval transit time is relatively high, at 59.3 us/ft. Its average density is relatively low, at 2.41 g/cm^3. These parameters indicate a gas layer, which produced 27.86×10^4 m^3 of gas per day without liquids after hydrodynamic fracturing. The production lasted 30 days and production stabilized at 10×10^4 m^3/d, with stable pressure, indicating a typical commercial gas layer.

2) Well logging characteristics of volcanic water layer

Compared with a gas layer, a volcanic water layer is characterized by medium density, higher neutron porosity, medium interval transit time, lower resistivity, and negative separation between deep and shallow laterolog resistivity values (see Table 8.6).

Layer 159IV in Well XX21-1 (depths of 3760 to 3820 m; see Figure 8.15) is a breccia and welded breccia reservoir with an average matrix porosity of 7.46%. The layer has good petrophysical properties. Its average resistivity is 34 $\Omega \cdot$m, lower than the cutoff of gas layers. There is slight positive separation (at about 4 $\Omega \cdot$m) between deep and shallow laterolog resistivity values. Its average neutron porosity is up to 13.1%, higher than formation porosity. Its average interval transit time is relatively high, at 70.1 us/ft. Its average density is relatively low, at 2.49 g/cm^3. The data thus indicate a water layer. A natural production test was carried out at an interval of 3768 to 3778 m and daily gas and water production stood at 10 m^3 and 0.29 m^3, respectively, thus confirming a water layer.

Table 8.6 Conventional Well Logging Response Characteristics of Gas Layers, Water Layers, and Dry Layers in Volcanic Gas Reservoirs

Reservoir Type	Resistivity	Separation between Deep and Shallow Laterolog Resistivity	Rock Density	Neutron Porosity	Interval Transit Time
Gas layer	Medium-high	Positive separation	Low	Medium-low, excavation effect	High, cycle skip
Water layer	Low	Negative separation	Medium	High	Medium
Dry layer	High-extremely high	No separation or minor separation	High	Low	Low

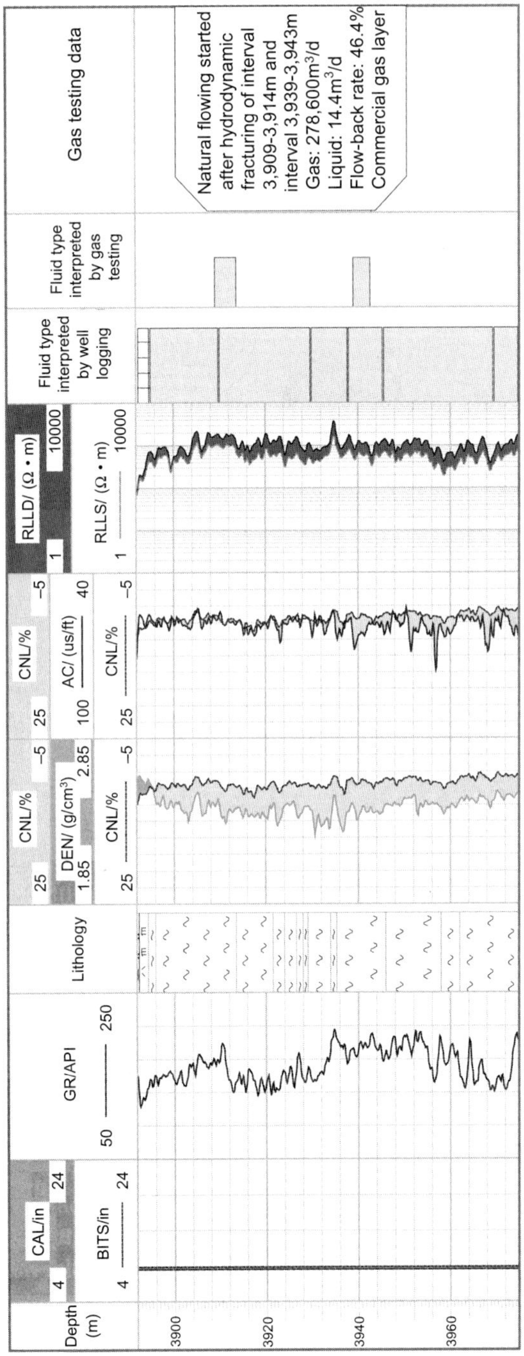

FIGURE 8.14

Conventional well log response characteristics of a typical gas layer (rhyolite, Well XX23).

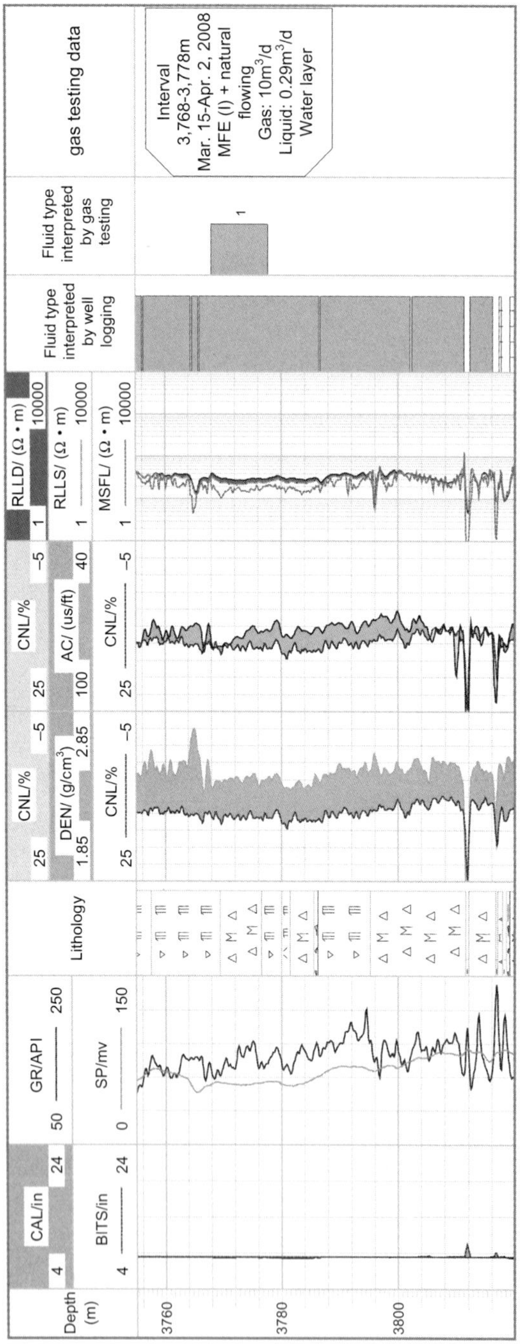

FIGURE 8.15

Conventional well log response characteristics of a typical water zone (breccia, Well XX21-1).

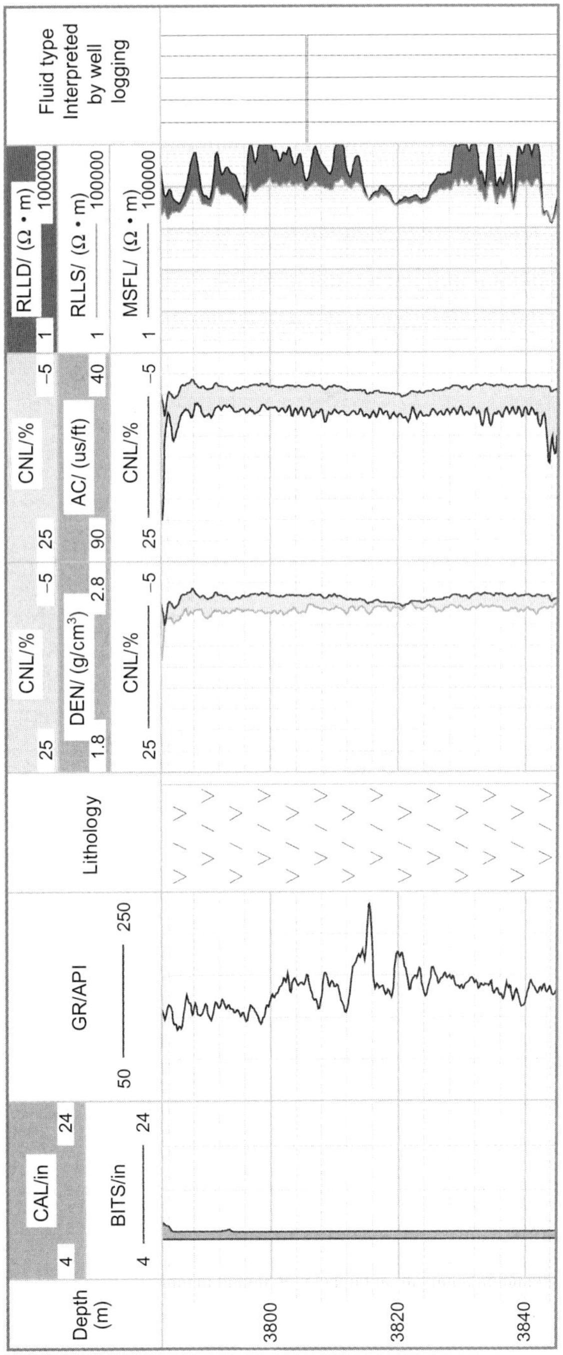

FIGURE 8.16

Conventional well log response characteristics of a tight dry layer (dacite, Well XX9).

3) Well logging characteristics of tight volcanic dry layers

In wireline logs, the poor petrophysical properties of tight volcanic dry layers are characterized by high density, low neutron porosity, low interval transit time, high to ultra-high resistivity, and a negligible separation between deep and shallow laterolog resistivity values (see Table 8.6).

Layer 60VI in Well XX9 (depths 3783 to 3845 m; see Figure 8.16) is a dacite reservoir with an average matrix porosity of 2.1%, lower than the lower limit of effective reservoirs. The formation has a very high resistivity, averaging at 33,730 $\Omega \cdot m$. There is slight positive separation between deep and shallow laterolog resistivity values, which is the result of stress-free fractures. It has an average neutron porosity is 0.28% and a low average interval transit time of 54.4 us/ft. Its average density is 2.58 g/cm^3, higher than the upper limit of effective reservoirs. These log features indicate a typical tight volcanic dry layer.

3. *Identification methods*

Through digital processing or log curve combination, the abnormal responses of logging curves in gas layers can be utilized to develop multiple methods for identifying hydrocarbon gas layers. These methods include the crossplot analysis, the curve-overlapping method, and the saturation calculation method [15,16].

1) The crossplot analysis

Based on single-curve or single-parameter identification principles, the crossplots for identifying volcanic gas layers and water layers include the neutron-density crossplot, neutron-acoustic wave crossplot, resistivity-porosity crossplot, and so forth [14]. On account of the complex volcanic reservoirs lithologies and the correspondingly variable logging responses, the crossplots for identifying volcanic gas and water layers must be established by taking the different lithologies into consideration.

As is shown in Figure 8.17, the resistivity-density log crossplot for different rocks in the gas testing interval in the DD gas field reveals the following features:

1. Volcanic gas and water layers have zoning characteristics: the water layers are located in the low-resistivity zone, the gas layers in the high-resistivity zone, and the gas-water layers in between.
2. For different volcanic rocks, gas, and water layers have different distribution characteristics in the crossplots. Subvolcanic gas layers have the highest resistivity and their distribution is the widest. Acidic volcanic gas layers have moderate resistivity and their distribution is similarly moderate. Intermediate-basic volcanic rocks have the lowest resistivity and their distribution is the narrowest.
3. For different volcanic rocks, the positions and slopes for the gas layer and water layer boundaries are different.

These results indicate that volcanic gas layers and water layers can be effectively identified by establishing crossplots for different lithologies.

2) The curve-overlapping method

Curve overlapping is a technique for evaluating reservoirs or fluid types based on curve separation [14], whereby uniform dimensions, grid scales, and baselines are used in the mapping of original logging curves or converted parameter curves, and the mapped curves are overlapped in certain direction. Based on the differences between the logging responses of gas layers and aquifers, the following curve-overlapping techniques can be used to identify volcanic gas layers and aquifers: neutron-density

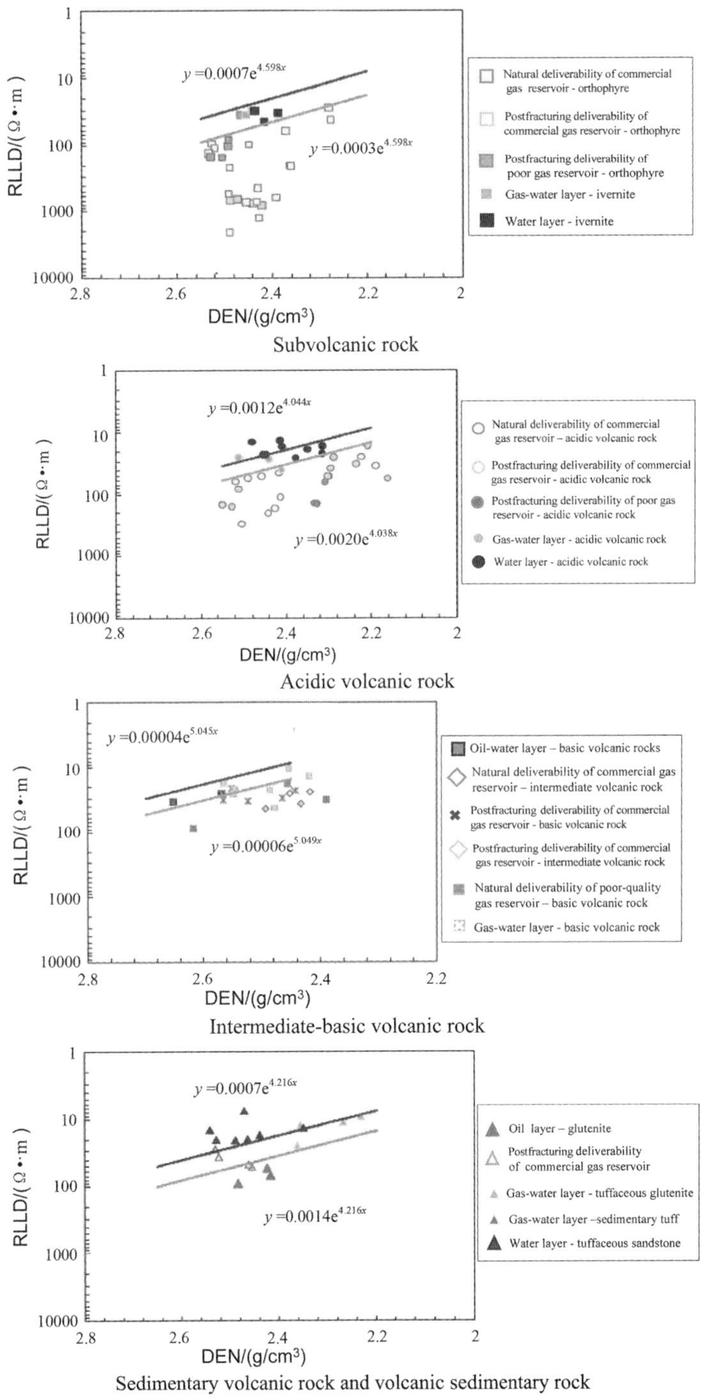

FIGURE 8.17

Resistivity-density logging crossplots for the gas tested interval (DD gas field).

curve-overlapping, neutron-acoustic wave curve-overlapping, depth resistivity curve-overlapping, P-S velocity ratio and baseline overlapping, and the overlapping of gas-water boundaries determined by the crossplot method with resistivity curves or neutron curves. Again, the lithological differences must be considered when using the curve-overlapping method to identify volcanic gas and water layers. This method can be illustrated with Well XX23 (Figure 8.18).

In Well XX23, the middle-upper interval consists of rhyolite and the lower interval of welded breccia. In Figure 8.18, seven methods are used to indicate the gas-bearing properties of the reservoir, starting with the fifth trace. These seven methods include neutron and density overlapping, neutron and interval transit time overlapping, deep and shallow laterolog resistivity overlapping, FRG1 and deep laterolog resistivity overlapping, FRG2 and deep laterolog resistivity overlapping, FRG3 and neutron overlapping, and P-S velocity ratio and its gas layer ceiling overlapping. FRG1, FRG2, and FRG3 are the gas-water boundaries determined by the RLLD-DEN, RLLD-AC, and CNL-DEN crossplots. Figure 8.18 shows that, from top to bottom, the reservoir resistivity decreases, the low mud invasion weakens, neutron increases, and the overlapping area shrinks. These characters suggest that the gas-bearing properties become poorer from top to bottom. The two turning points, located at depths of 4011 m and 4042 m, respectively, correspond to the bases of the gas layer and the gas-water layer. During the gas testing carried out at the interval of 3909 to 3943 m, daily gas production reached 27.86×10^4 m^3 after fracturing. In the 30-day pilot production, the output stabilized at 10×10^4 m^3/d and liquid production stood at 9.1 m^3/d, with a flow-back rate of 97.76%. The target layer was thus confirmed to be a commercial gas layer. The water produced subsequently came from the water body beneath the reservoir via natural or artificial fractures.

3) The quantitative saturation interpretation methods

Based on interpretations of volcanic reservoir porosity and saturation parameters, two methods can be used to identify gas and water layers.

Saturation and porosity crossplot analysis. The gas saturation of a volcanic gas reservoir is positively correlated with the petrophysical properties and gas-bearing properties of the reservoir. The crossplot of gas saturation and effective porosity can be used to establish the saturation identification model for volcanic gas layers and water zones (Figure 8.19). It is effective for identifying the types of reservoir fluid in an untested interval.

Original gas saturation cutoff method. The original gas saturation of a gas reservoir is positively correlated with the petrophysical properties of rocks. The cutoff criteria for the original gas saturation of effective reservoirs (about 40% for Block XX1) can be established based on crossplot analysis (Figure 8.20) and petrophysical property criteria. It allows engineers to use reservoir gas saturation to identify gas layers and water zones in volcanic rocks as well as to quantify the gas-bearing properties of the reservoir.

8.4 VOLCANIC GAS LAYER AND WATER ZONE IDENTIFICATION: THE COMPREHENSIVE APPROACH

To meet the challenges of identifying volcanic gas layers and aquifers, an integrative or comprehensive approach to identify volcanic gas layers and aquifers has been developed specifically for the complex geological conditions of volcanic gas reservoirs. As a principle, this approach is guided by geological

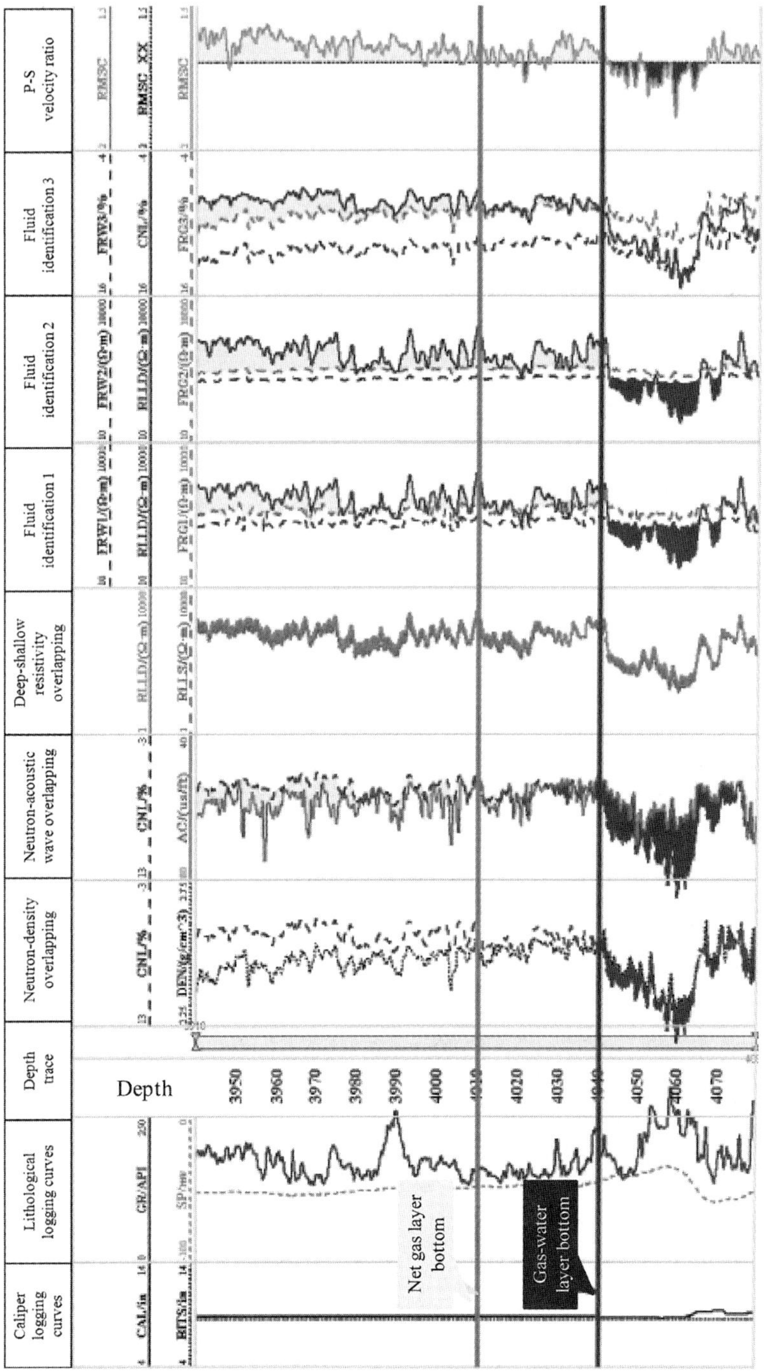

FIGURE 8.18
The curve-overlapping method for gas layer and water layer identification (Well XX23).

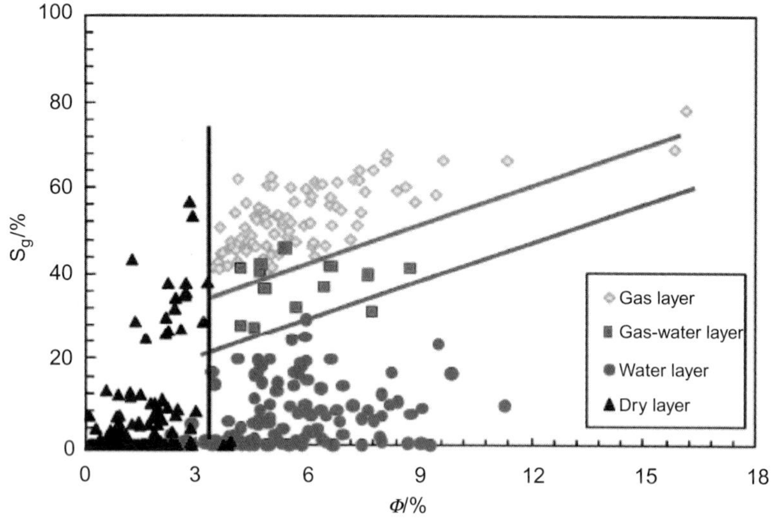

FIGURE 8.19

Sg-ф crossplot analysis for identifying volcanic gas and water layers (Block XX21).

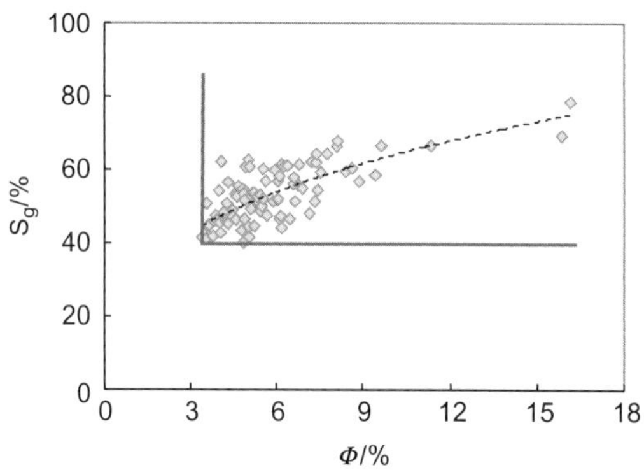

FIGURE 8.20

Identification of volcanic gas layers using the original gas saturation cut-off method (Block XX1).

concepts, based on integrated information, and validated by multiple methods, through an integration of logging, testing, and various logging interpretation techniques.

8.4.1 Establishment of gas-bearing property profiles for single wells by integrating single-well data

This approach integrates logging indicators, formation test results, and well log data to identify a reliable marker layer and establish gas-bearing property profiles for individual wells. This can be illustrated using Well XX9-1 as an example, which is a development well drilled at the second highest point in Block XX9. Prior to formation testing, the gas layers and aquifers in this well were tentatively identified on the basis of mud logging and well logging data, the gas-bearing properties were characterized qualitatively, and gas-bearing property profiles were established for individual wells (Figure 8.21).

1. Gas-bearing properties indicated by logging data

Gas logging (traces 3 to 6 in Figure 8.21) shows that layer 202I in the upper part of the borehole has the best gas-bearing properties. This layer has a maximum total-hydrocarbon content of 2.29%, a gas log ratio of 9.2 times, an absolute methane content of 5.16%, and a relative methane content of 73.5%. The calibrated logs have high amplitude and good continuity. These logging features indicate layer 202I as a gas zone.

2. Gas-bearing properties indicated by well log data

Only conventional log data were acquired from Well XX9-1, and six methods were used to characterize the gas-bearing properties of the reservoir (tracks 7 to 12 in Figure 8.21), including neutron and density overlapping, neutron and interval transit time overlapping, deep and shallow laterolog resistivity overlapping, FRG1 and deep laterolog resistivity overlapping, FRG2 and deep laterolog resistivity overlapping, and FRG3 and neutron overlapping. FRG1, FRG2, and FRG3 are the gas-water layer boundaries determined by the RLLD-DEN crossplot, the RLLD-AC crossplot, and the CNL-DEN crossplot, respectively. Figure 8.21 shows a general decline of gas-bearing properties from 3780 m downward in this well. Layer 202I in the upper part of the borehole is characterized by low resistivity, high neutron, low density, and high interval transit time. Except for the significant difference in deep and shallow laterolog resistivity values, the information of other gas-bearing properties is very weak and does not match logging indicators, resulting in great uncertainties about the gas and water layer interpretation for the reservoir.

8.4.2 Further characterization of gas-bearing properties for target layers through multiwell correlation

To resolve the identification uncertainties regarding the low-resistivity gas layers in volcanic gas reservoirs, a multiwell correlation profile can be established under the guidance of gas and water distribution patterns constrained by the internal structure of volcanic rocks. In this method, the gas-bearing properties of the target layers are characterized by means of comparative reservoir analysis to reduce the identification uncertainties about gas and water layers, as is illustrated later, again using Well XX9-1 as an example.

1. Establishment of multiwell correlation profile

The volcanic gas reservoir in the Yingcheng Formation of Block XX9 is an anticline structure, with Well XX9 situated in the high point (Figure 8.22). The volcanic rock architecture shows that Well XX9-1 and the adjacent Wells XX9, XX9-2, and XX9-3 penetrated the same volcanic massif, with their major gas zones all located in the third cycle, and the wells are interconnected (Figure 8.23). Based on these findings, a correlation profile was established for Wells XX9-3, XX9-2, XX9, and XX9-1 (Figure 8.24).

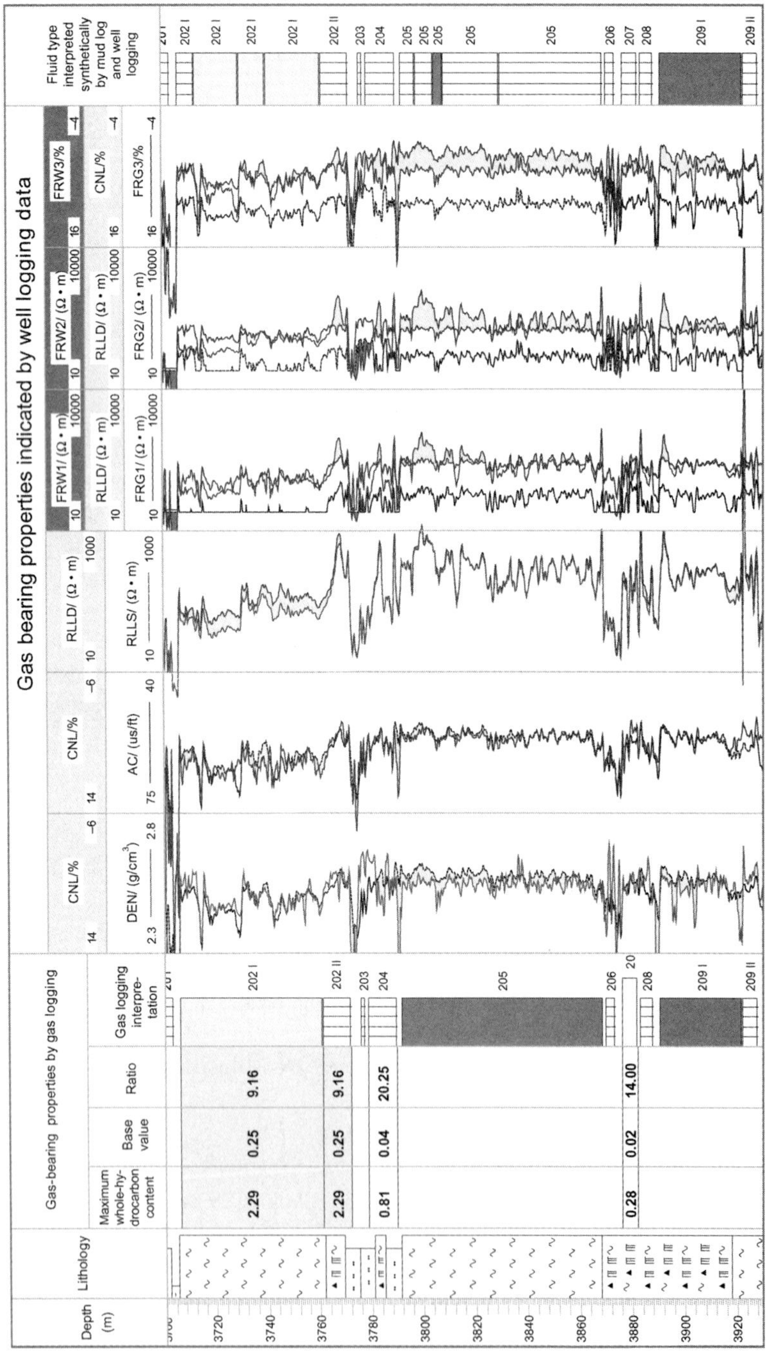

FIGURE 8.21

Characteristics of the longitudinal direction gas-bearing properties of Well XX9-1.

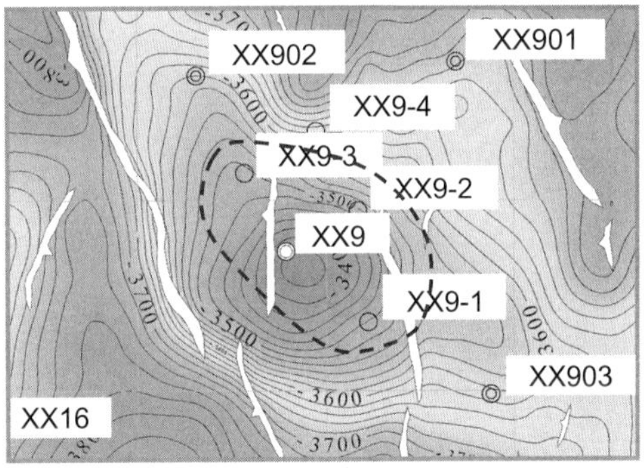

FIGURE 8.22

Structure of the gas reservoir top in Block XX9.

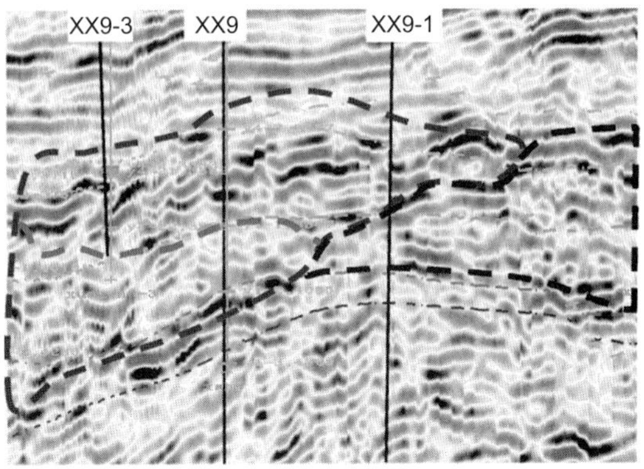

FIGURE 8.23

Through-well seismic profile for Wells XX9-3, XX9, and XX9-1.

2. Comparative reservoir interpretation

The results of gas and water layer interpretation from adjacent wells and gas testing confirm that the volcanic rock unit has a uniform water-gas contact and the target layer of Well XX9-1 is 100 m higher than the water-gas contact. Based on the gas and water distribution characteristics of the gas reservoir in this volcanic massif, namely upper gas and lower water, layer 202I in Well XX9-1 is interpreted as a gas zone.

3. Integration of multiple data to validate the characterization of low-resistivity gas layers in the target zone

The results from core observation, petrophysical property analysis, and log interpretation indicate that the target layer of Well XX9-1 is rhyolite with well-developed vesicles and fractures. The reservoir has good physical properties, with an effective porosity of 8% and deep invasion of drilling fluid.

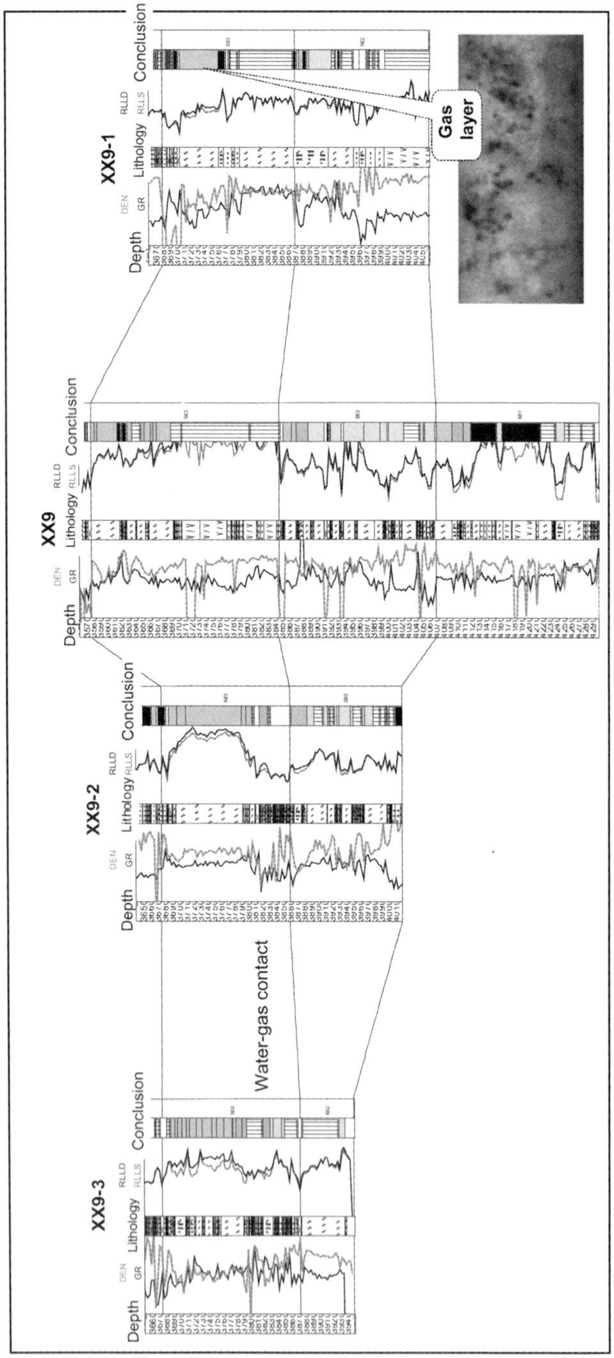

FIGURE 8.24
Correlation profile for Wells XX9-3, XX9-2, XX9, and XX9-1.

The salinity of the water sample in an adjacent well was consistent with that of the formation water in the area, and no signs of abnormal high salinity were detected.

Drilling data from that well indicate that, prior to logging, the target zone had been soaked in mud for 128 days (about 3081 hours), allowing for deep mud invasion.

After a comprehensive analysis, layer 202I in Well XX9-1 is interpreted as a low-resistivity gas layer based on deep mud invasion, which validates the results of logging indications and geological interpretation. It also explains the log indications of poor gas-bearing properties.

8.4.3 Validating interpretation results through formation tests and improving interpretation models, providing a basis for the interpretation of new wells

In June 2007, perforation gas testing was carried out in the 3717 to 3722 m and 3757 to 3761 m intervals of Well XX9-1. Under the MFE I + TCP + natural flowing system, daily gas production reached 0.09×10^4 m^3. Natural flowing started after fracturing and the daily gas production reached 10.2×10^4 m^3 with a 7.94-mm choke. The tubing pressure was 10.36 MPa. The total salinity of produced fluids was 21,000 to 27,000 ppm, which was inconsistent with the water properties in the area (7400 to 9900 ppm), and the flow-back rate of fracturing fluid was 59.69%, including fracturing fluid and filtrate. Therefore, the testing demonstrated that these two layers are commercial gas layers and no formation water is produced. This conclusion is consistent with the results of the comprehensive interpretation.

While validating the interpretation results, the formation test data from the well further enriched gas testing information for Block XX9, which facilitated the improvement of interpretation models for volcanic gas and water layers and provided a more robust model to be applied to the interpretation of new wells.

References

[1] PetroChina Exploration and Production Company. Logging evaluation methods and techniques for low-resistivity reservoirs. Beijing: Petroleum Industry Press; 2006, p. 1, 2.

[2] Chang Z. Petroleum exploration and development technologies, vol. 2. Beijing: Petroleum Industry Press; 2005, p. 945, 946, 963, 964, 978, 995, 1006-09, 1023-30.

[3] Zheng X, Gao Z, Luo D, et al. Project supervision and management of petroleum engineering. Beijing: Petroleum Industry Press; 2006, p. 34–36.

[4] Du S, Hu G, Li J, et al. Natural gas exploration. Beijing: Petroleum Industry Press; 2004, p. 1–20.

[5] Xiao L, Chai X, Sun B, et al. NMR logging interpretation and China's case studies. Beijing: Petroleum Industry Press; 2001, p. 73, 74, 144–47.

[6] Carr MB, Ehrlich R. Correlation of porosity types derived from NMR data and thin section image analysis in a carbonate reservoir. J Petrol Sci Eng 1996;14:115–31.

[7] Tan M. Observation way and analytical approach of dual waiting time by nuclear magnetic resonance logging. Logging and Perforation 2005;1:22–5.

[8] Li J, Wang B, Li X, et al. MRIL-prime NMR log data processing and application. Logging and Perforation 2004;2:1–6.

[9] Shao W, Zhuang S, Ding Y, et al. A new type of nuclear magnetic resonance logging tool—MREx. Petroleum Instrument 2004;18(2):36–9.

[10] Wang Z, Zhang C, Xiao C, et al. T2 cut-off value in low-permeability reservoirs and its application. Geophysical Prospecting for Petroleum 2004;43(5):508–10.
[11] Sima L. Geological logging application technology. Beijing: Petroleum Industry Press; 2002, p. 40, 59, 102–04.
[12] Li H. Estimating formation permeability and anisotropy using borehole Stoneley wave. Well Logging Technology Information 2004;17(4):12–8.
[13] Zhang S, Gu C, Cao G, et al. Imaging logging technology and application. Beijing: Petroleum Industry Press; 1997, p. 53–58.
[14] Yong S. Log data processing and integrated interpretation. Beijing: Petroleum Industry Press; 1989, p. 80–98, 134–39.
[15] Cao J, Liu S, Gao M, et al. Comprehensive logging data interpretation. Beijing: Petroleum Industry Press; 2002, p. 121–48.
[16] Khatchikan A. Logging evaluation of oil-bearing igneous rocks. In: SPWLA, 23rd annual logging symposium transactions; 1982.

CHAPTER

Effective Reservoir Identification and Prediction

9

As the focus of volcanic gas reservoir characterization, effective volcanic reservoirs can be identified qualitatively using a wide range of static and dynamic data and the established identification markers for different reservoirs. Effective volcanic reservoirs can be identified quantitatively by establishing cutoff criteria as well as classification and evaluation criteria. Based on well identification and seismic data calibration with log data, effective reservoirs can be classified and predicted through seismic inversion constrained by the internal structure of volcanic rocks. Furthermore, the spatial distribution patterns of effective reservoirs thus revealed will form the basis for optimizing well location and horizontal well trajectory design.

9.1 RESEARCH CHALLENGES AND TECHNICAL SOLUTIONS

9.1.1 Challenges in the identification and prediction of effective reservoirs

Effective volcanic rock reservoirs are difficult to identify and predict because of their great variability, complex origins, and complicated logging and seismic responses. The main difficulties are as follows:

1. Volcanic rocks have complex lithology and related various reservoir spaces. In terms of logging responses, the differences between various effective reservoirs are not clearly defined. It is difficult to interpret reservoir porosity and permeability and to characterize reservoir accumulation and permeation capacity. Volcanic reservoirs have complex mechanisms of electrical conductivity and contain various types of low-resistivity gas layers and high-resistivity water layers, which increase the uncertainty of fluid type identification. This makes it difficult to identify effective volcanic reservoirs.
2. Volcanic rocks have complex internal architecture and, therefore, it is hard to carry out reservoir parameter inversions based on rock architectural control. Furthermore, volcanic reservoirs have strong heterogeneity, poorly defined distribution patterns, drastic thickness changes, and low seismic resolution, which makes it difficult to predict thin effective reservoir, and even more so to classify and predict them within effective reservoirs. The effective volcanic reservoir prediction technique focuses mainly on acquiring acoustic impedance and density data volumes through body-controlling seismic inversion, applying seismic classification criteria for reservoirs to extract classified effective thickness, and predicting the spatial distribution of effective volcanic reservoirs.

9.1.2 Technical solutions

To meet the challenges posed by the identification and prediction of effective reservoirs, the technical process is developed by comprehensive multidisciplinary investigations including geology, well logging, and seismic and gas reservoir engineering (Figure 9.1), and a set of integrated techniques is established for effective volcanic reservoir identification and prediction.

1. Effective volcanic reservoir identification

Different types of reservoirs, including volcanic vesicular reservoirs, intergranular pore reservoirs, dissolution-pore reservoirs, and fractured reservoirs, are identified qualitatively by establishing identification markers for lithology and lithofacies, reservoir space, and well logging, and then integrating data from cores, thin sections, petrophysics, acoustic and resistivity imaging logs, and nuclear magnetic and conventional logs.

Based on the preceding study and core experiment, as well as deliverability tests and log interpretation, we apply the methods of water film thickness analysis, deliverability test constraint, empirical statistic analysis, and crossplot analysis to set up cutoff thresholds (or the lower limit) and classification criteria for effective volcanic gas reservoirs. Furthermore, we integrate geological data, wireline log data, reservoir parameter interpretation results, and production performance data to quantitatively identify types I, II, and III effective volcanic reservoirs.

2. Classified prediction of effective volcanic reservoirs

After calibrating seismic data with well interpretation parameters, the seismic classification criteria of volcanic reservoirs are established by analyzing near-wellbore seismic trace characteristics and the seismic response characteristics of effective reservoirs in volcanic rocks. Furthermore, the volcanic

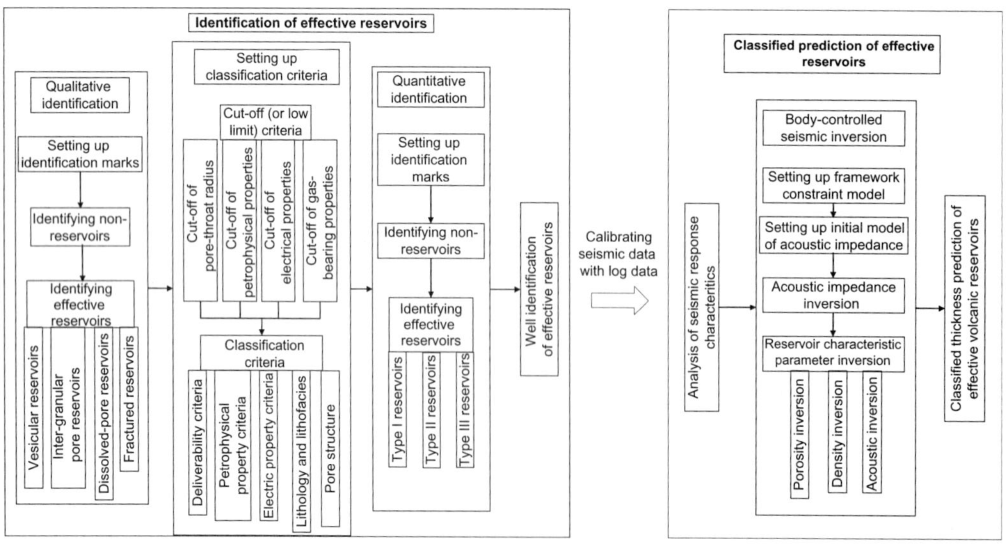

FIGURE 9.1

Process of identifying and predicting effective volcanic reservoirs.

framework model is constructed. The data volumes for 3D acoustic impedance, density, acoustics, and porosity are obtained by body-controlled seismic inversion. On the basis these studies, we extract total reservoir thickness, overall categorized (classified) reservoir thickness, and reservoir thickness above the water-gas contact in volcanic rocks according to the seismic classification criteria, so as to predict the distribution of effective volcanic reservoirs.

9.2 IDENTIFICATION OF EFFECTIVE VOLCANIC RESERVOIRS

Geological identification markers and well-logging identification markers for different types of reservoirs and nonreservoirs are established on the basis of volcanic reservoir classification and analyses of their geological and well-logging characteristics. The effective volcanic reservoirs are identified qualitatively using core, thin-section, and log data. Furthermore, the cutoff criteria and classification criteria are established, and the effective volcanic reservoirs are identified quantitatively on the basis of an integrated analysis of static and dynamic data and parameter interpretation.

9.2.1 Qualitative identification of effective reservoirs

In terms of reservoir space characteristics, volcanic reservoirs are classified into four types: vesicular, intergranular pore, dissolution pore, and fractured. Nonreservoir bodies are classified into two types: tight volcanic rocks and sedimentary volcanic rocks. After setting up the identification markers for all types of reservoirs and nonreservoirs, the effective volcanic reservoirs can be qualitatively identified using multiple data sources.

1. Identification markers

On the basis of the types and the degrees of development of volcanic pores and factures, and combining logging response characteristics, a suite of markers is set up for lithology, lithofacies, reservoir space, and well-logging response regarding each type of reservoir and nonreservoir rock body, which provides a basis for the qualitative identification of effective volcanic reservoirs (Table 9.1).

1) Vesicular reservoirs

Geological characteristics. Vesicular reservoirs are formed by effusive volcanic eruption. The reservoir lithology consists mainly of volcanic lava, brecciated lava, and tuffaceous lava. Their reservoir lithofacies mainly comprise the top subfacies, upper subfacies, and lower subfacies of the effusive facies. The reservoir space primarily consists of vesicles. Their pores are round or oval with smooth pore walls. The long axes of vesicles are usually parallel to the magma flow direction, directionally aligned, and form a rhyolitic structure.

Logging response characteristics

1. *Acoustic and resistivity imaging logs.* In general, the volcanic magma has low electrical conductivity and shows yellow (light) blocky structure in acoustic and resistivity imaging logs. On the background of this image, vesicular reservoirs show smooth round or oval black spots usually connected as alternating black-white lines or strips (i.e., vesicular rhyolitic structure) along the magma flow direction, similar to beddings.

Table 9.1 Identification Marks for Volcanic Reservoirs and Nonreservoirs

Type of Volcanic Reservoir		Geological Characteristics			Logging Response Characteristics						
						Nuclear Magnetic Logging			Conventional Logging		
		Typical Lithology	*Typical Lithofacies*	*Main Reservoir Space*	*Acoustic and Resistivity Imaging Logging*	**Spectrum Shape**	**T2 Spectrum**	**Movable Fluid Peak Amplitude**	**Curve Value**	**Shape**	**Smoothness**
Reservoir type	Vesicular reservoir	Volcanic lava, breccia lava, tuffaceous lava	Subfacies of top effusive facies, upper effusive facies and lower effusive facies	Vesicle	Yellow blocky texture, regular round and oval black spots distributed uniformly, with smooth margins	Bimodal or polymodal, scattered distribution	Wide distribution range, bigger average value	Bigger in top subfacies of effusive facies and medium in upper and lower subfacies	Medium to high Rt and DEN, medium to low AC and CN	Box-shaped, bell-shaped, funnel-shaped assemblage	Smooth-slightly denticulate
	Inter-granular pore reservoir	(Welded) volcanic breccia, crystal (welded) tuff	Fallout subfacies, hot base surge subfacies and hot clastic flow subfacies of explosive facies	Inter-granular pore	Yellow-red blocky or porphyritic texture, irregular black round dots or polygonal spots with sharp margins	Unimodal or bimodal, continuous distribution	Wider distribution range, medium average value	Bigger in breccia of fallout subfacies, medium in hot clastic flow subfacies, smaller in hot base surge subfacies	Medium to low Rt and DEN, medium to high AC and CN	Box shape, bell shape, funnel shape	Dentoid-serrated
	Dissolved-pore reservoir	All types of volcanic lithology and lithofacies		Dissolved vesicle, Inter-granular dissolved pore and matrix dissolved pore	Red-dark red blocky or porphyritic texture, with irregular black strip locally	Asymmetrically bimodal	Medium distribution range and average value	Medium-big	Medium to low Rt, low DEN, high AC, medium to high CN	Box-shaped, bell-shaped, funnel shaped assemblage	Smooth-dentoid
	Fractured reservoir	All types of volcanic rock	Middle subfacies of effusive facies, cryptoexplosive breccia subfacies of volcanic conduit facies	All types of fracture	Yellow-light yellow blocky texture, in irregular black lines with nonuniform distribution and sine line feature mostly	Unimodal or bimodal	Big distribution range	Relatively Small	High deep Rt, medium to low microresistivity, high DEN, low AC and CN	Box shape	Slightly dentate-dentoid
Nonreservoir type	Tight volcanic rock	All kinds of volcanic rock	Middle subfacies of effusive facies, hot base surge subfacies of explosive facies, and inner subfacies of subvolcanic facies	A few vesicle pores and intergranular pore, low development of fracture	White-light yellow blocky texture, with fewer nonuniformly distributed black spots and lines	Without apparent modal feature	Big distribution range	Small	High Rt, high DEN, low AC and CN	Box shape	Smooth to slightly denticulate
	Sedimentary volcanic rock	Sedimentary volcanic rock	Volcanic sedimentary facies	A few vesicle pores and intergranular pores	Black or alternating black-white porphyritic texture, with sediment beddings locally	Unimodal or asymmetrically bimodal	Small distribution range and low average value	Small	Low Rt, low (part high) DEN, high AC, high CN	Bell shape, funnel shape, finger shape, change in wave shape	Serrated

2. *T2 spectrum of nuclear magnetic logs.* The T2 value is higher for big vesicles than for small ones. Vesicular reservoirs usually show bimodal or polymodal patterns in the T2 spectrum of nuclear magnetic logs, with a wide distribution range and high average value. Their mobile fluid peak amplitudes are related to the degree of vesicle development, which is high in the top subfacies of the effusive facies and moderate in the upper and lower subfacies.
3. *Conventional logs.* Vesicular reservoirs have poor pore connectivity and conductivity, showing the characteristics of medium to high resistivity (Rt) and density (DEN), medium to low accoustic traval time (AC) and compersated nuetron (CN). Most of the log curves have a smooth and slightly denticulate box shape and occasionally have a bell-funnel-shaped association.

2) Intergranular pore reservoirs

Geological characteristics. Intergranular pore reservoirs are of volcaniclastic origin, formed by an explosive volcanic eruption. Their reservoir lithology consists mainly of volcanic breccia, welded breccia, crystalline tuff, and welded tuff. The reservoir lithofacies comprise mainly the fallout subfacies, hot base surge subfacies, and hot clastic flow subfacies of the explosive facies. Their main reservoir spaces are primary intergranular pores, such as irregular polygons with sharp margins.

Logging response characteristics

1. *Acoustic and resistivity imaging logs.* Volcaniclastic rocks have a high electric conductivity and show yellow-red blocky or porphyritic textures in acoustic and resistivity imaging logs. Intergranular pores appear as irregular black round dots or polygonal spots with sharp margins in acoustic and resistivity imaging logs.
2. *T2 spectrum of nuclear magnetic logs.* Intergranular pore reservoirs usually show continuously distributed unimodal or bimodal patterns, with a wide range of T2 value and medium average value. Their mobile fluid peak amplitudes are related to the degree of intergranular pore development, which is high in the breccia of fallout subfacies, moderate in the hot clastic flow subfacies, and small in the hot base surge subfacies.
3. *Conventional logs.* Intergranular pore reservoirs have good pore connectivity and electrical conductivity, characterized by medium to low Rt and DEN and medium to high AC and CN in conventional log curves. Most of the log curves have dentoid-serrated box, bell, and funnel shapes.

3) Dissolution-pore reservoirs

Geological characteristics. Dissolution-pore reservoirs are of secondary origin, formed mainly by dissolution during later diagenesis in primary pore reservoirs. Their reservoir lithology and lithofacies span all types of volcanic lithology and lithofacies. The main reservoir spaces are dissolution vesicles, intergranular dissolution pores, and matrix dissolution pores, with irregular pore shapes.

Logging response characteristics

1. *Acoustic and resistivity imaging logs.* The electrical conductivity of volcanic rocks improves after the dissolution of soluble components, manifesting a red to dark red and blocky or porphyritic texture, with black irregular strips appearing in acoustic and resistivity imaging logs.
2. *T2 spectrum of nuclear magnetic logs.* Dissolution-pore reservoirs usually show an asymmetrical bimodal pattern, with a moderate range of distribution of T2 values and a medium average value. Their mobile fluid peak amplitudes are medium to high.

3. *Conventional logs*. Dissolution-pore reservoirs have good pore connectivity and electrical conductivity, showing the characteristics of medium to low Rt, low DEN, high AC, and medium to high CN; most of the curves are characterized by smooth-dentoid box, bell, and funnel shapes.

4) Fractured reservoirs

Geological characteristics. Fractured reservoirs develop in all types of volcanic rocks. The reservoir lithofacies comprises mainly the middle subfacies of the effusive facies and the cryptoexplosive breccia subfacies of the volcanic conduit facies. The reservoir spaces consist of all types of fractures, with undeveloped matrix pores.

Logging response characteristics

1. *Acoustic and resistivity imaging logs*. Fractured reservoirs have poor electrical conductivity, showing a yellow to light yellow blocky texture. Under this background, fractured reservoirs appear as various irregular black lines, mostly characterized by a sine curvature, with nonuniform distribution.
2. *T2 spectrum of nuclear magnetic logs*. Fractured reservoirs usually show unimodal or bimodal patterns, with a large distribution range of T2 values. Their mobile fluid peak amplitudes are relatively small.
3. *Conventional logs*. In conventional log curves, fractured reservoirs show the characteristics of high laterolog deep resistivity (LLD), low microresistivity, high DEN, low AC, and low CN, with curves having a slightly dentate-dentoid box shape.

5) Tight volcanic nonreservoirs

Geological characteristics. Tight volcanic dry layers can develop in all types of volcanic rock, with few pores or fractures. Their lithofacies mainly include the middle subfacies of the effusive facies, the hot base surge subfacies of the explosive facies, and the inner subfacies of the subvolcanic facies.

Logging response characteristics

1. *Acoustic and resistivity imaging logs*. Tight volcanic rocks have poor electrical conductivity, showing a white to light yellow blocky texture with a minor number of unevenly distributed black spots and lines.
2. *T2 spectrum of nuclear magnetic logs*. Tight dry layers show ambiguous peak patterns, with a wide distribution range of T2 values. Their mobile fluid peak amplitudes are small.
3. *Conventional logs*. A tight dry layer is characterized by high Rt, high DEN, low AC, and low CN, with curves having a smooth to slightly denticulate box shape.

6) Sedimentary volcanic nonreservoirs

Geological characteristics. The lithology of sedimentary volcanic nonreservoirs comprises mainly the volcanic sedimentary facies. Their reservoir spaces are limited to a minor number of intergranular pores, with poor fluid mobility.

Logging response characteristics

1. *Acoustic and resistivity imaging logs*. Sedimentary volcanic nonreservoirs have additional electrical conductivity from irreducible water, showing a black or alternating black-and-white porphyritic texture in imaging logs, with local development of sediment beddings.

2. *T2 spectrum of nuclear magnetic logs*. Sedimentary volcanic nonreservoirs usually have a unimodal or asymmetrically bimodal pattern. T2 distributes in a small range, with a small average value. Their mobile fluid peak amplitudes are small.
3. *Conventional logs*. Sedimentary volcanic nonreservoirs usually show the characteristics of low Rt, low DEN (high DEN in XX gas filed), high AC, and high CN, with curves being serrated bell, funnel, and finger shaped.

2. Qualitative identification of volcanic reservoirs

Qualitative assessment of reservoir effectiveness and reservoir classification can be conducted based on the established identification markers, together with identified lithology and lithofacies, by integrating with acoustic and resistivity imaging logs, nuclear magnetic and conventional logs, and by means of imaging analysis, the correlation of free peak amplitude of the T2 spectrum and the analysis of conventional log response characteristics. The identification process is as follows.

1) Combining characteristics of lithology and lithofacies to qualitatively identify volcanic nonreservoirs

According to the identification markers (see Table 9.1), two types of nonreservoir volcanic rocks usually show two extremes in well logs: the dry mud layer images are dark colored (dark red to black), with a large irreducible fluid spectrum amplitude in the nuclear magnetic log, characterized by low Rt, low DEN (partly high DEN), high AC, and high CN in conventional logs; on the contrary, dry tight layer images are light colored (bright yellow to white), with a small irreducible fluid spectrum amplitude in nuclear magnetic logs, marked by high Rt, high DEN, low AC, and low CN in conventional logs. The common feature of these two types of dry layers is their small mobile fluid peak amplitudes. Thus, volcanic nonreservoirs can be effectively identified by integrating the characteristics of lithology and lithofacies and using various log data.

Sedimentary volcanic nonreservoirs. Intervals 3860 to 3876 m in Well XX901 are located on the top of a volcanic massif, lithologically dominated by pyromeride and mixed with mudstone (Figure 9.2). Full-bore micro scanner imager (FMI) imaging logs show an alternating black-and-white porphyritic texture. The T2 amplitude in the irreducible fluid spectrum of nuclear magnetic logs is large and lacks a mobile fluid peak. The conventional logs have a serrated finger or bell shape; the rock density is in the range of 2.58 to 2.7 g/cm^3, with a high average value of 2.62 g/cm^3; AC is in the range of 58 to 75 μs/ft with an average of 66 μs/ft, which is also high. The CN values are similarly high, with a range of 10% to 19% and an average of 14%; LLD is in the range of 15 to 40 Ω·m with an average of 25 Ω·m, which is a low value; and the P-S (compressive-shear wave) velocity ratio is in the range of 1.7 to 2 with an average of 1.8, which is a high value. These results show that most volcanic pore fluid in this interval is irreducible with low gas content, which is typical of a dry mud layer.

Tight volcanic nonreservoirs. The lithology in the 3800 to 3840 m interval of Well XX8 is dacite, representing the medium subfacies of the effusive facies (Figure 9.3). FMI imaging logs show a bright yellow blocky texture. The T2 spectrum pattern of nuclear magnetic logs is not obvious, without an irreducible fluid peak and mobile fluid peak. The conventional logs have a smooth box shape, and rock density values are high, ranging from 2.52 to 2.59 g/cm^3 with an average of 2.56 g/cm^3; AC ranges from 49 to 53 μs/ft with an average of 51 μs/ft (low value); CN values are −1.3% to 0.7% with an

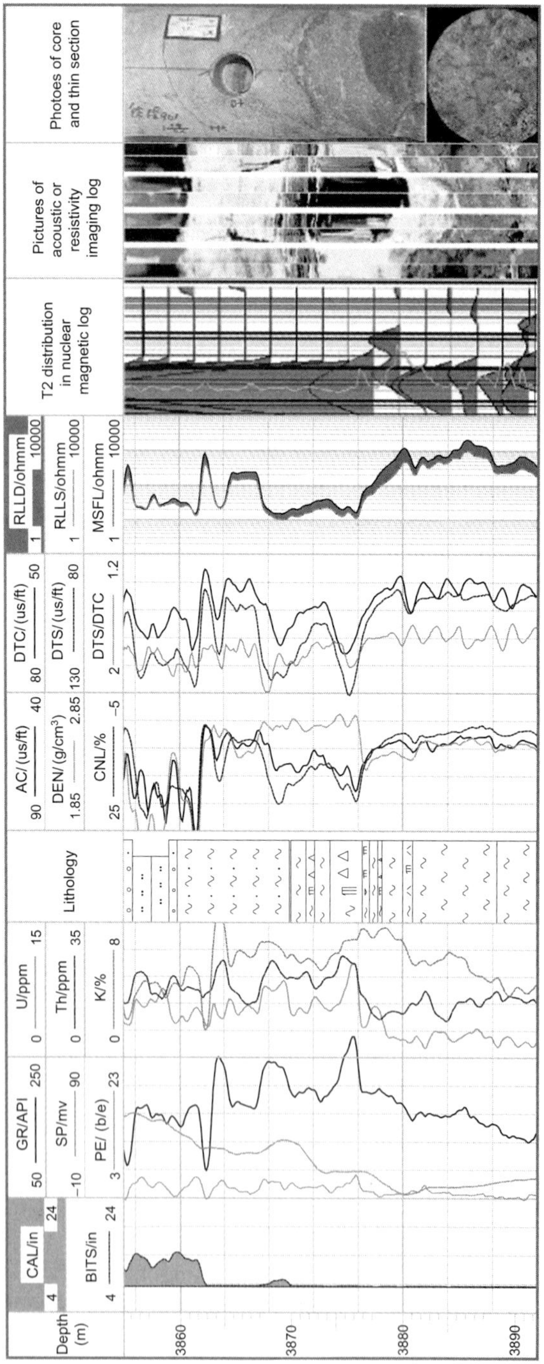

FIGURE 9.2

Integrated logs of sedimentary volcanic nonreservoirs (pyromeride mixed with mudstone, Well XX901).

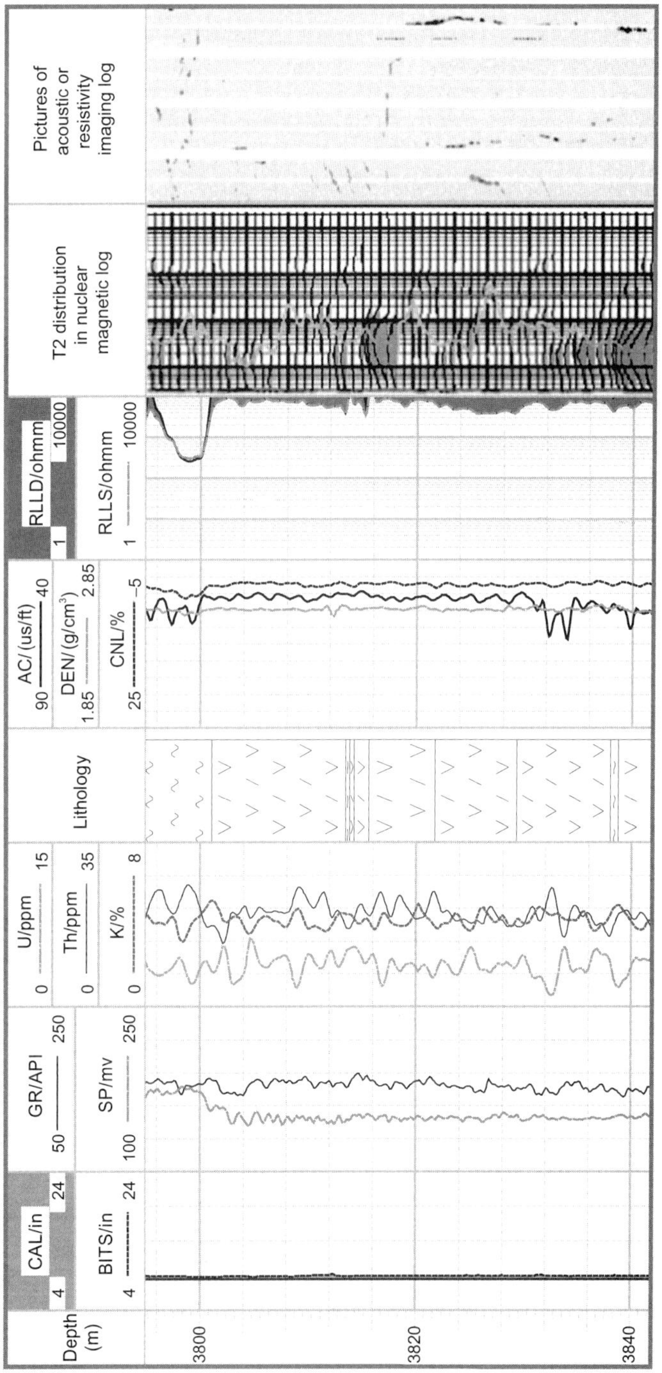

FIGURE 9.3

Integrated well logs of tight volcanic nonreservoirs (Well XX8).

average of 0.02% (low value); LLD ranges from 6428 to 104989 Ω·m with an average of 69312 Ω·m (high value). These results demonstrate poor petrophysical properties for the volcanic rock in this interval, which is a tight volcanic nonreservoir.

2) Distinguishing fractured reservoirs in tight volcanic rocks based on fracture identification

The major difference between fractured volcanic reservoirs and tight nonreservoirs lies in whether various fractures are developed. Thus, fractured reservoirs can be recognized within tight volcanic rocks through fracture identification.

The lithology in the 3644 to 3664 m interval of Well DD17 is basalt, representing the medium subfacies of the effusive facies (Figure 9.4). Fractures are observed in cores, resulting in core breakdown. FMI imaging logs show a red-yellow blocky texture with netlike black sine lines of various configurations, which indicates the development of fractures. The T2 spectrum of nuclear magnetic logs has a bimodal pattern with a wide T2 distribution range, relatively small irreducible fluid peak, and mobile fluid peak. The conventional logs have a weakly denticulate box shape, with rock density of 2.69 to 2.87 g/cm^3 and an average of 2.79 g/cm^3 (high value); AC ranges from 56 to 65 μs/ft with an average of 58 μs/ft (medium to low value); CN ranges from 15% to 23% with an average of 17% (medium to high value; high DEN and high CN are characteristics of basic volcanic rock); LLD ranges from 40 to 66 Ω·m with an average of 48 Ω·m (relatively high value for Well Block DD17). The average effective matrix porosity interpreted from well logs is 0.4%, below the reservoir cutoff; fracture porosity is 0.5%, which is of medium to high value. These results show that the volcanic matrix in this interval has poor properties and developed fractures, which are the main reservoir spaces and fluid flow channels, and as such it can be characterized as a typical fractured reservoir.

3) Assessing reservoir effectiveness by type and identifying effective volcanic reservoirs comprehensively

Reservoir effectiveness is assessed by type, based on the characteristic markers (see Table 9.1), and integration with core, thin-section, and log data.

Vesicular reservoirs. Volcanic vesicular reservoirs are marked by developed vesicles of different sizes. The lithology in the 3894 to 3958 m interval of Well XX23 is rhyolite, representing the top subfacies and upper subfacies of the effusive facies. Vesicles are developed in cores, with sizes ranging from 0.2×0.2 cm to 2×3 cm (the 11th track in Figure 9.5), which are the main reservoir spaces.

FMI imaging logs in this interval show a yellow blocky texture, with black spots of variable sizes and smooth and clear margins, and a vesicular structure (the 10th track in Figure 9.5). In the T2 spectrum of nuclear magnetic logs, there are two to four spectral peaks, with a wide T2 distribution range for each spectrum peak and a large average value; its mobile fluid peak shows a medium to high amplitude, which indicates the development of vesicles (the 9th track in Figure 9.5). The conventional logs have a weakly denticulate box shape. Rock density is 2.27 to 2.58 g/cm^3 with an average of 2.41 g/cm^3 (medium to low value); AC ranges from 52.8 to 67.8 μs/ft with an average of 58.6 μs/ft (medium to high value); CN ranges from 1.82% to 5.62% with an average of 3.47% (medium to low value); the P-S velocity ratio ranges from 1.54 to 1.74 with an average of 1.64 (medium to low value); and LLD ranges from 270 to 3722 Ω·m with an average of 1165 Ω·m (medium to high value; refer to the 6th to 8th tracks in Figure 9.5). These results show that this interval belongs to typical vesicular reservoirs, with medium physical properties and relatively good reservoir effectiveness.

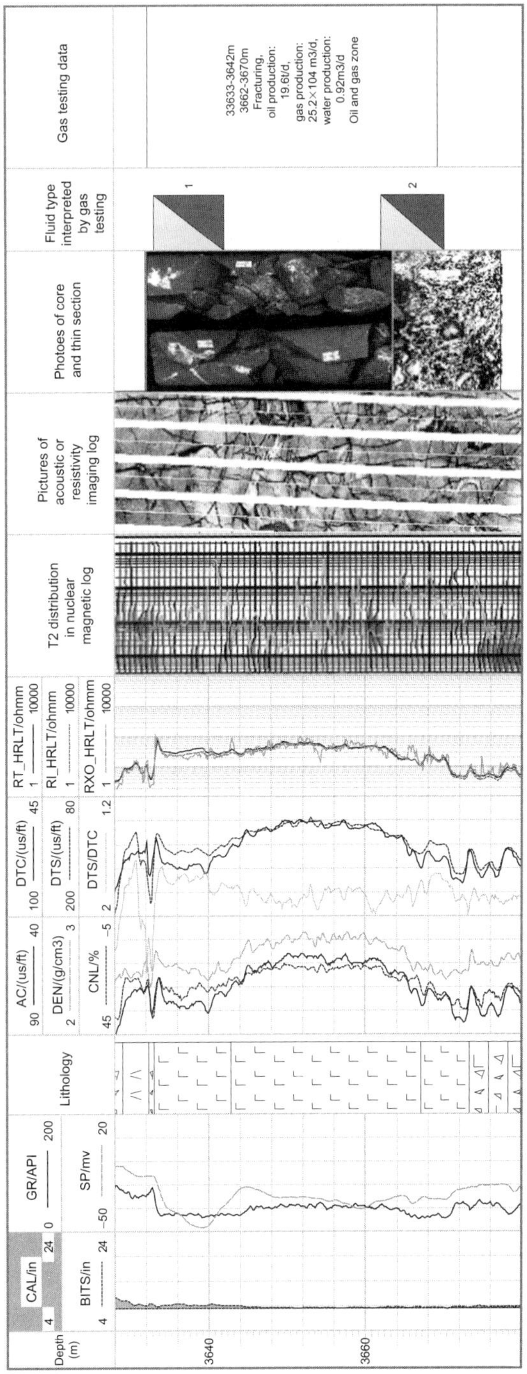

FIGURE 9.4

Integrated log map of fractured reservoirs (Well DD17).

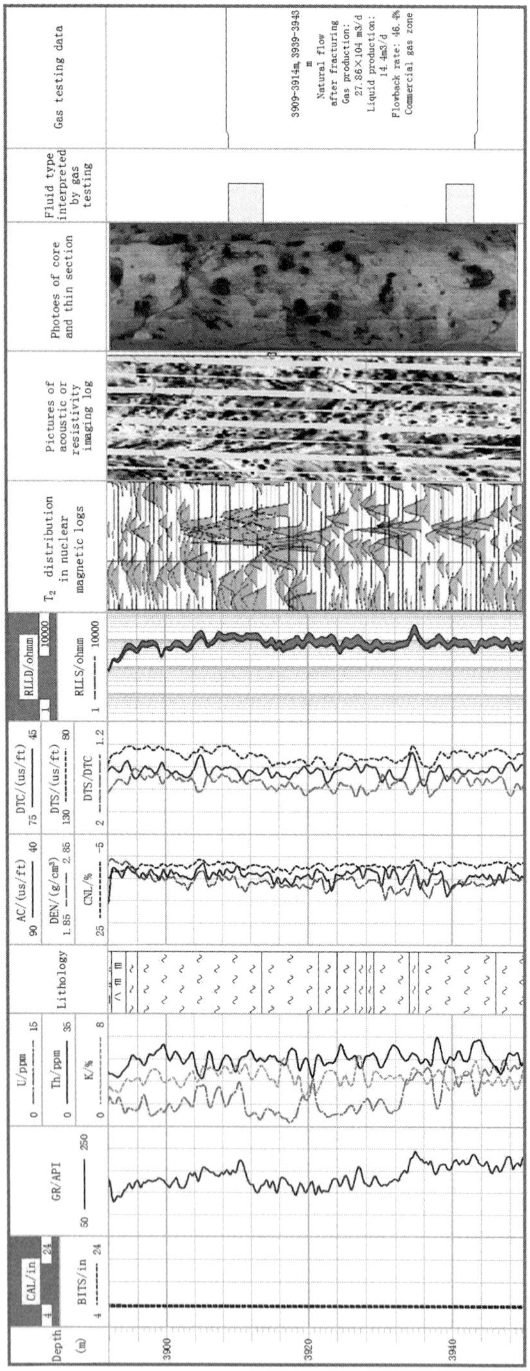

FIGURE 9.5

Integrated log map of vesicular reservoirs (Well XX23).

After gas testing in intervals of 3909 to 3814 m and 3939 to 3943 m in this well, gas and water flew naturally after fracturing, with gas production of 27.86×10^4 m^3/d, liquid production of 14.4 m^3/d, and a fracturing fluid flow-back rate of 46.4%. Gas-testing results indicate that these intervals are commercial gas zones, further proving that the 3854 to 3958 m interval in Well XX23 is a typical effective vesicular reservoir.

Intergranular pore reservoirs. Intergranular pore reservoirs are marked by the development of intergranular pores during volcanic eruption and accumulation. The 3714 to 3735 m interval of Well CC1-1 consists of rhyolitic breccia, representing the fallout subfacies of the explosive facies. There is an apparent volcanic breccia texture in cores, with mainly irregular interbreccia pores in cast thin section (the 11th track in Figure 9.6).

The FMI imaging logs for this interval show a light yellow porphyritic texture, among which the angular bright patches are volcanic breccia with poor conductivity; the irregular black patches are volcanic breccia with interbreccia pores developed, showing a high conductivity; and black spots are intergranular pores in coarse-grained tuff (the 10th track in Figure 9.6). In the T2 spectrum of nuclear magnetic logs, this interval shows a bimodal pattern, with a large T2 distribution range for each of the peaks and a medium average value; its mobile fluid peak amplitude is medium to high, indicating the development of intergranular pores (the 9th track in Figure 9.6). The conventional logs are in dentoid funnel and bell shapes, with rock density of 2.28 to 2.53 g/cm^3 and an average of 2.39 g/cm^3 (low value); AC ranges from 56.7 to 72.7 μs/ft with an average of 65.2 μs/ft (medium to high value); CN ranges from 3.6% to 14.9% with an average of 8.8% (medium to high value); and LLD ranges from 20 to 162 Ω·m with an average of 65 Ω·m (medium to low value; refer to the 6th to 8th tracks in Figure 9.6). The integrated analysis shows that this interval is typical of intergranular pore reservoirs, with good petrophysical properties and reservoir effectiveness.

Dissolution-pore reservoirs. Volcanic dissolution-pore reservoirs are formed by weathering and leaching as well as dissolution by acid water and meteoric water during late diagenesis, resulting in good petrophysical properties and connectivity. The lithology in the 3704 to 3740 m interval of Well XX8 is rhyolite breccias lava, representing the volcanic neck subfacies of the volcanic conduit facies. In cores, there are irregular volcanic breccias cemented by magma without clear margins, and various dissolution pores are developed. In cast thin section, pores show an irregular embayed shape with a high degree of development and connectivity (the 10th track in Figure 9.7).

FMI imaging logs in this interval show a dark red porphyritic texture; the light yellow to red patches are welded breccia of low electrical conductivity, and the irregular black strips are dissolution zones (the 9th track in Figure 9.7). In the T2 spectrum of nuclear magnetic logs, this interval shows an asymmetrically bimodal pattern, with a continuous T2 distribution for each peak, and a medium distribution range and average value; their mobile fluid peak amplitudes are high (the 8th track in Figure 9.7). The conventional logs have a smooth box shape, with rock density of 2.25 to 2.46 g/cm^3 and an average of 2.33 g/cm^3 (low value); AC ranges from 65 to 82 μs/ft with an average of 72 μs/ft (high value); CN ranges from 8.8% to 18.2% with an average of 14.7% (medium to high value); and LLD ranges from 24 to 63 Ω·m with an average of 39 Ω·m (low value; refer to the 6th and 7th tracks in Figure 9.7). Integrated analysis shows that this interval is typical of dissolution-pore reservoirs, with good petrophysical and good gas-bearing properties.

Gas testing was conducted for the 3723 to 3735 m interval of Well XX8 by adopting Multi-Flow Evaluator (MFEII) + Tubing Conveyed Perforating (TCP), with a 10-mm choke; it produced a natural gas flow of 22.62×10^4 m^3/d (containing 23% CO_2), thus confirming a commercial gas zone.

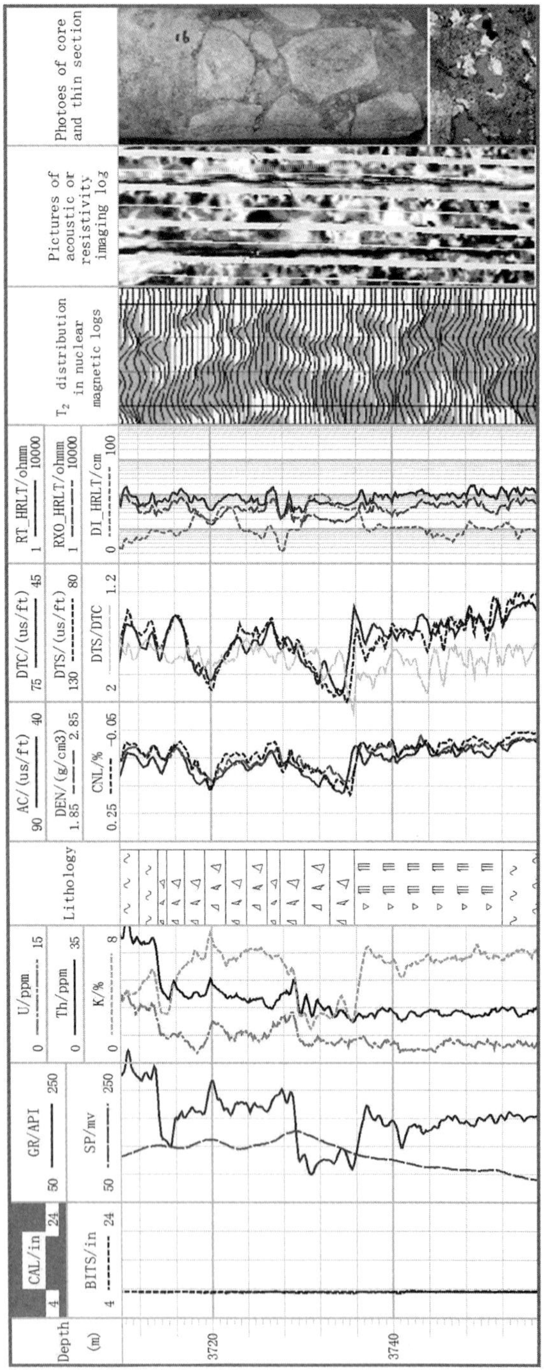

FIGURE 9.6

Integrated log map of intergranular pore reservoirs (Well CC1-1).

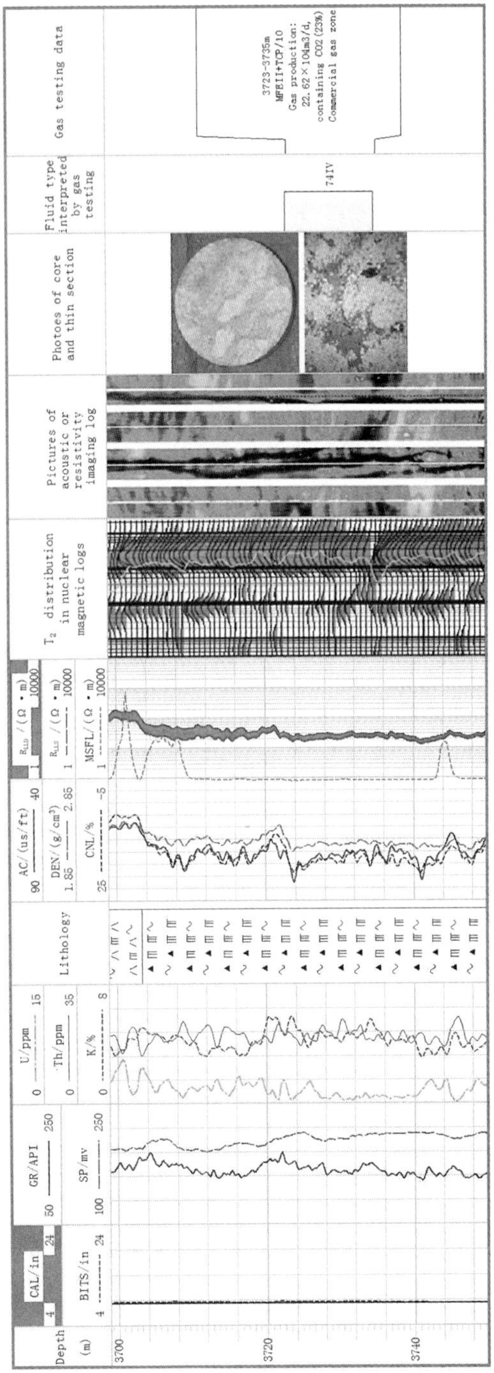

FIGURE 9.7

Integrated log map of dissolution-pore reservoirs (Well XX8).

9.2.2 Quantitative identification of effective reservoirs

On the basis of gas well deliverability, the cutoff criteria and classification and evaluation criteria of effective volcanic reservoirs are established by means of crossplot analysis and empirical statistic analysis. These criteria also facilitate the quantitative identification of various effective volcanic reservoirs using the results of fracture identification, reservoir parameter interpretation, and delineation of gas layers and aquifers.

1. Establishing cutoff criteria for effective reservoirs

The reservoir effectiveness depends mainly on lithological, petrophysical, and gas-bearing properties. The cutoff thresholds of petrophysical, electrical, and gas-bearing properties of effective volcanic reservoirs can be reasonably determined based on formation test results by integrating static data with dynamic data (such as geology, well logging, gas testing, and pilot production data) and using several analytical techniques.

1) Defining the cutoff thresholds of pore-throat radius and petrophysical properties according to water film thickness

The irreducible water film thickness experiment and the empirical equation are combined to define the irreducible water film thickness (about 0.01 m) of volcanic reservoirs. The cutoff of the pore-throat radius is defined by adding half of the diameter of a gas molecule (about 0.0005 m) to irreducible water film thickness, which is about 0.0105 m (taking 0.01 m). Based on this cutoff, the crossplot of core permeability versus pore-throat radius and crossplot of core permeability versus effective porosity are plotted with mercury injection test data, and the cutoffs of matrix permeability and effective porosity are determined according to the cutoff of the pore-throat radius.

The example in Figure 9.8 uses water film thickness to define the cutoffs of volcanic pore-throat radius and petrophysical properties in the XX gas field. The figure implies that (1) the volcanic rock permeability has a positive correlation with the pore-throat radius, with a correlation coefficient of 0.91 (Figure 9.8a); at the 0.01-μm cutoff of the pore-throat radius, the cutoff of the matrix permeability is 0.01 mD; and (2) the volcanic matrix permeability has a positive correlation with effective matrix porosity, with a correlation coefficient of 0.84 (Figure 9.8b); at the 0.01-mD cutoff of matrix porosity, the cutoff of effective matrix porosity is 3.5%.

2) Defining cutoffs of reservoir petrophysical properties and electric properties using the deliverability test method

On the basis of effective volcanic reservoirs and nonreservoirs defined by deliverability tests, the cutoff values of petrophysical and electrical properties of effective volcanic reservoirs can be determined from the correlation charts of gas well deliverability versus physical properties (porosity and permeability) and electrical properties (acoustic travel time and rock density).

Figure 9.9 shows how the deliverability test method is used to define cutoffs of the petrophysical and electrical properties of effective volcanic reservoirs in the CC gas field. The main lithology of the CC gas field consists of acidic volcanic rocks. The crossplots of permeability versus porosity and acoustic travel time versus rock density in this gas-testing interval are established based on the gas layer, water layer, and dry layer determined by gas-testing data. The correlations indicate that (1) the volcanic gas layer deliverability has a positive correlation with petrophysical properties. Being at borderline gas producibility, the cutoffs of effective reservoir petrophysical properties are defined by a permeability of 0.03 mD and a porosity of 3.5%; and (2) the volcanic gas layer deliverability

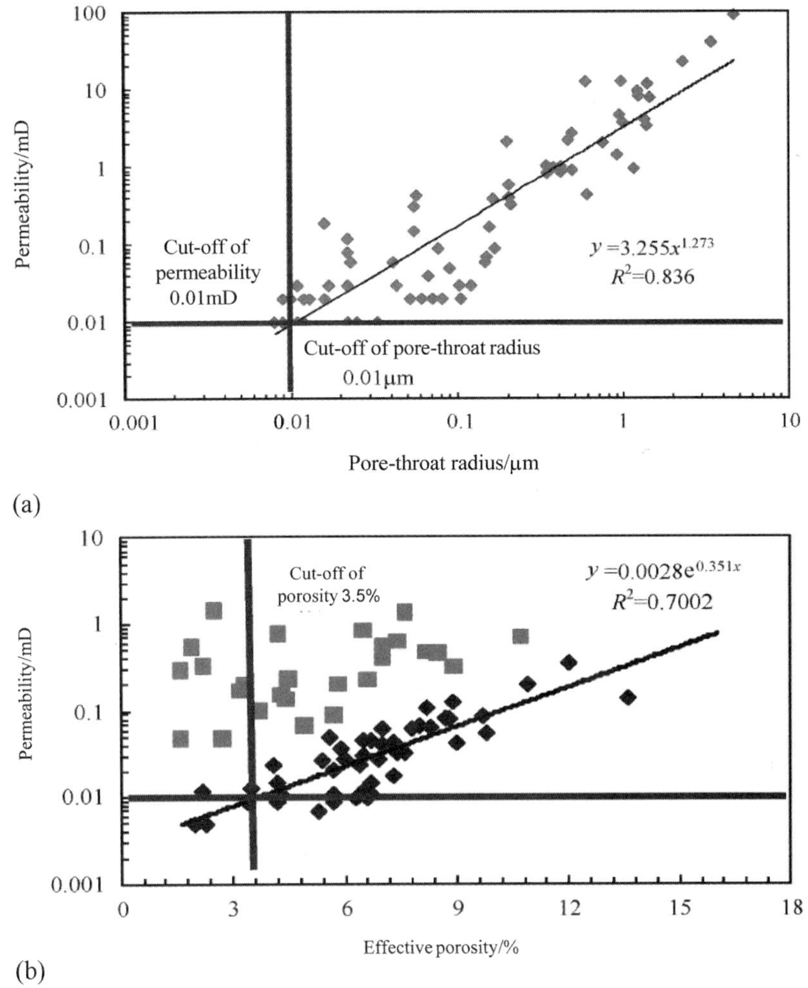

FIGURE 9.8

Defining the cutoff of petrophysical properties according to the pore-throat radius (XX gas field). (a) Crossplot of pore-throat radius versus permeability. (b) Crossplot of core porosity versus permeability.

is positively correlated with AC but inversely correlated with DEN. Based on this, the cutoffs of electrical properties in this gas field are defined by a DEN value of 2.57 g/cm^3 and an AC value of 55 μs/ft.

3) Defining cutoff thresholds of reservoir petrophysical properties using empirical statistic analysis

Experimental porosity and permeability data from core analysis provide an important basis for plotting the histograms, cumulative frequency curve of porosity and permeability, cumulative frequency curve of accumulation potential (equal to the product of porosity and thickness), and cumulative frequency

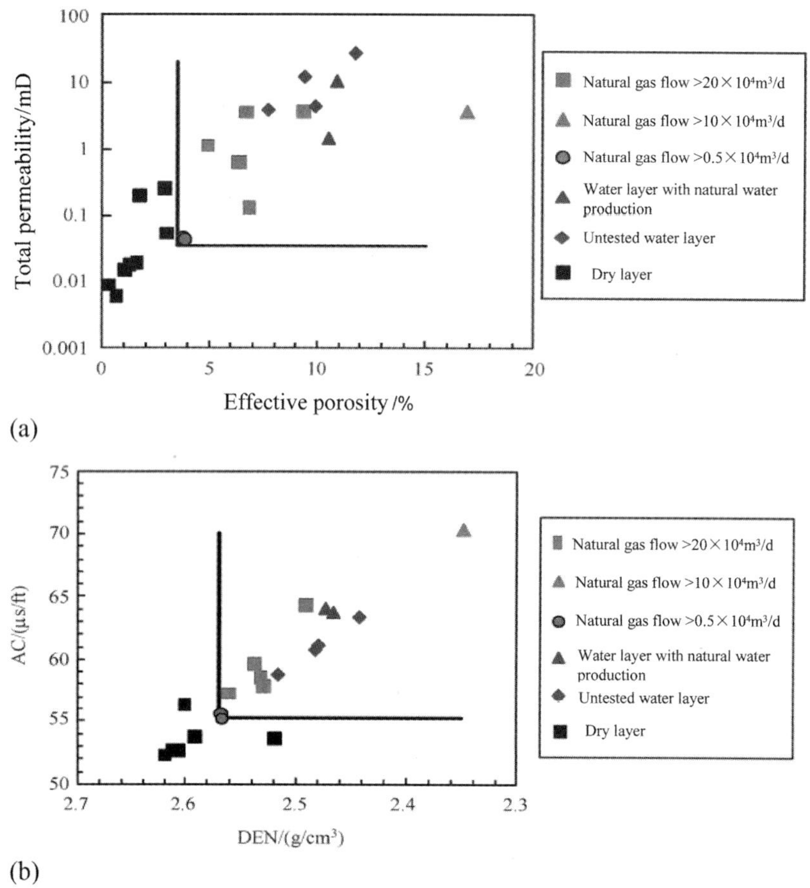

FIGURE 9.9

Defining cutoff values of petrophysical and electrical properties using the deliverability test method (CC gas field). (a) Crossplot of total permeability versus porosity in the gas-testing interval. (b) Crossplot of DEN versus AC in the gas-testing interval.

curve of deliverability (equal to the product of permeability and thickness). Based on the results of previous local studies, the percentages of effective reservoir thickness, accumulation potential, and deliverability that may have been lost during the development conditions are defined to confirm the cutoff thresholds of effective reservoir petrophysical properties [2].

Figure 9.10 uses the method of empirical statistic analysis to define the cutoffs of the petrophysical properties of volcanic lava in the DD gas field. Figure 9.10a shows the histogram of porosity distribution, cumulative frequency curve of porosity, and cumulative frequency curve of accumulation potential. Figure 9.10b shows the crossplot of porosity versus permeability at different levels of mercury intrusion saturation. According to the experience of the DD gas field, limited by 10% of accumulation potential loss and 40% of mercury intrusion saturation, the cutoffs of petrophysical properties of effective reservoirs in volcanic lava are set at a porosity value of 6.5% and a permeability value of 0.02 mD.

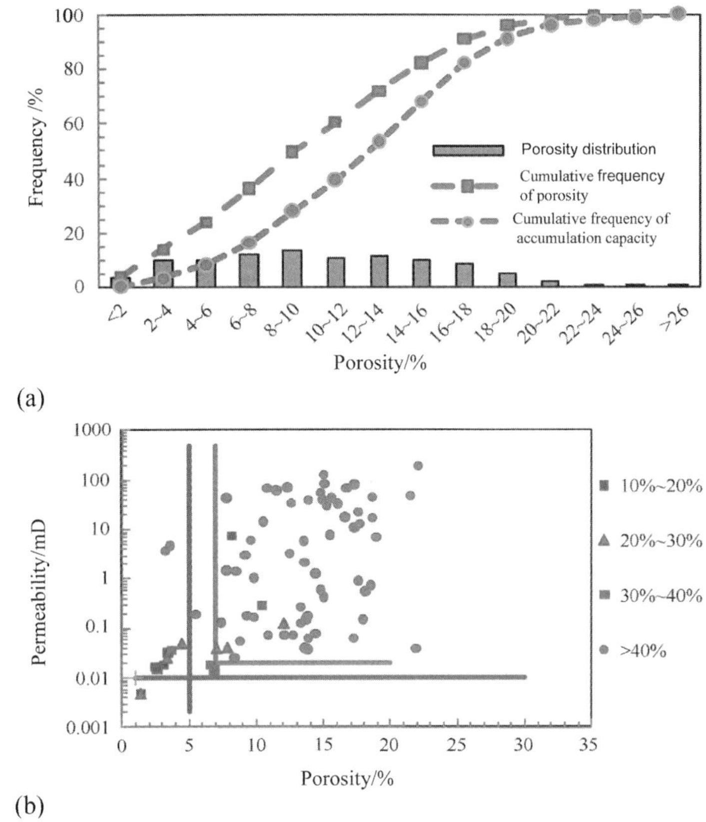

FIGURE 9.10

Defining cutoff values of physical properties using the empirical statistic method (volcanic lava in DD gas field). (a) Histogram of the porosity cumulative frequency. (b) Crossplot of porosity versus permeability at different mercury intrusion saturations.

4) Defining cutoff values of gas-bearing property with multiple approaches

Deliverability test method. The cutoff values for the resistivity and saturation of effective volcanic reservoirs are determined on the basis of effective volcanic reservoirs and noneffective reservoirs defined by employing the deliverability test, applying the crossplot analysis technique, analyzing the logging response feature, and interpreting the saturation logs.

Figure 9.11 shows how the deliverability test is used to define the cutoff values of resistivity and gas saturation of the gas layer in the CC gas field. Figure 9.11a shows the crossplot of recorded laterolog deep (RLLD) versus AC in the gas-testing interval of this gas field. Figure 9.11b shows the crossplot of gas saturation versus effective porosity in the gas-testing interval of this gas field. These analyses suggest that the resistivity cutoff of the volcanic gas layer in the CC gas field is 50 Ω·m and the gas saturation cutoff is 45%.

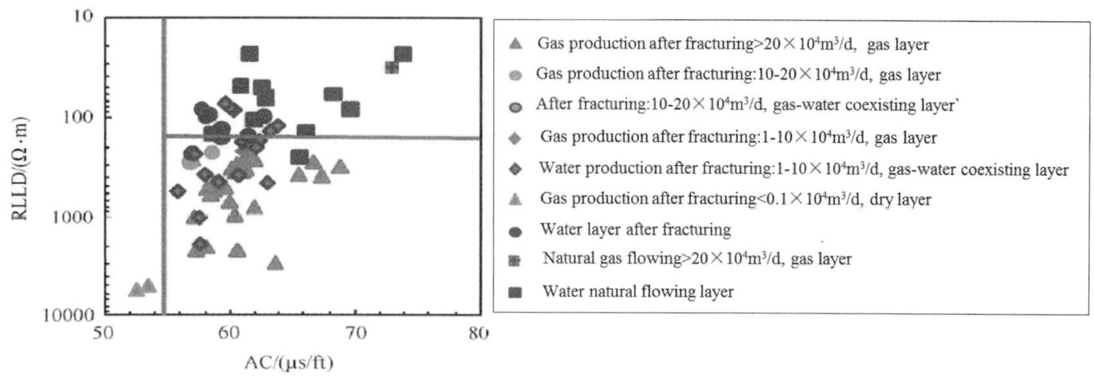

(a)

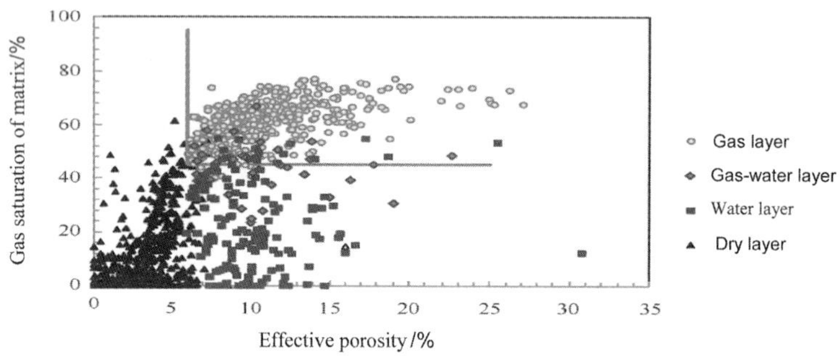

(b)

FIGURE 9.11

Defining the cutoff of gas-bearing properties using gas saturation interpretation (CC gas field). (a) Crossplot of RLLD versus AC in the gas-testing interval. (b) Crossplot of gas saturation versus effective porosity (interpreted from well logs).

Mercury intrusion method. On the basis of the porosity cutoff value and using mercury intrusion analysis data, the gas saturation cutoff of a gas layer can be defined by analyzing the relationship between mercury intrusion saturation and porosity [1].

In Figure 9.12, the mercury intrusion analysis is used to define the gas saturation cutoff of volcanic lava in the DD gas field. The crossplot of maximum mercury intrusion saturation versus porosity suggests that the initial gas saturation of volcanic reservoirs has a positive correlation with effective porosity. The gas saturation cutoff of volcanic lava in this gas field is defined at 40%, based on the changing trend and porosity cutoff of 6.5%.

2. Establishing classification criteria of effective reservoirs

Effective volcanic reservoirs are controlled by many factors and have diverse petrophysical properties. Commercial standards for classifying and evaluating effective volcanic gas reservoirs were not set up until now. In view of the complexity of volcanic gas reservoirs and the classification criteria for other oil and gas reservoirs [3–6], the classification and evaluation criteria for effective volcanic gas

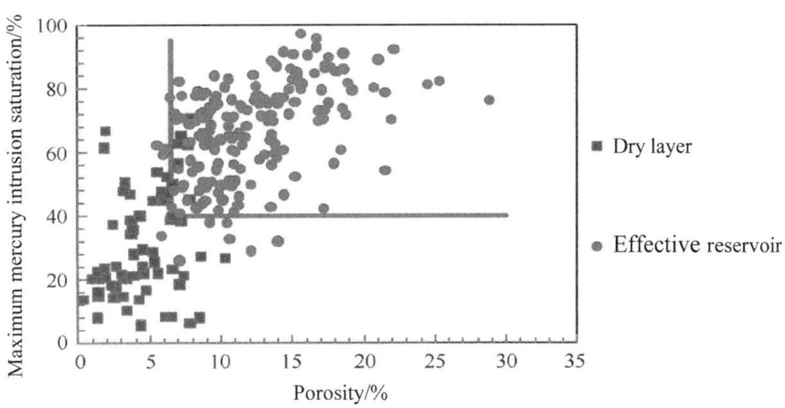

FIGURE 9.12

Defining a cutoff of gas saturation using mercury intrusion analysis (volcanic lava in the DD gas field).

Table 9.2 Classification and Evaluation Criteria for Volcanic Reservoirs in the XX Gas Field

Evaluation Item			Classification and Evaluation Criteria		
Criteria	*Parameter*	*Unit*	*Type I*	*Type II*	*Type III*
Deliverability	Gas deliverability index	$m^3/(MPa^2 \cdot d)$	> 200 (natural)	> 100 (fracturing)	10-100 (fracturing)
	Open-flow potential	$10^4\ m^3/d$	> 30 (natural)	> 20 (fracturing)	0.5-20 (fracturing)
Petrophysical property	Effective porosity	%	> 10	5-10	3.5-5
	Total permeability	mD	> 2	0.2-2	0.01-0.2
	Matrix permeability	mD	> 0.1	0.015-0.1	0.01-0.015
Gas-bearing property	Gas saturation	%	> 50	> 50	35-50
Pore structure	Pore-throat radius	μm	> 0.25	0.05-0.25	0.01-0.05
Electrical property	Density	g/cm^3	< 2.38	2.38-2.52	2.52-2.55
	AC	μs/ft	> 65	58-65	55-58

reservoirs (Table 9.2) are proposed here for the first time on the basis of gas well deliverability confirmed by formation tests and the data of core experiments, well logging, and reservoir parameter interpretation. Three types of effective volcanic reservoirs are recognized.

1) Deliverability and petrophysical property criteria

The deliverability of volcanic gas reservoirs has a positive correlation with reservoir petrophysical properties and well-controlled reserves. According to the levels of deliverability fracture development, gas layer deliverability is qualitatively classified into three types: naturally high deliverability, high deliverability after fracturing, and medium to low deliverability after fracturing. This division scheme corresponds to three types of reservoirs. On this basis, the deliverability and petrophysical property criteria for reservoir classification are set up by analyzing the relationship between gas layer deliverability and petrophysics. This is discussed here using the XX gas field as an example (Figure 9.13).

Deliverability criteria. The gas deliverability index has a positive correlation with the gas layer open-flow potential of the gas-testing interval (Figure 9.13a). In accordance with the gas-testing results, the classification criteria of the gas-testing deliverability level are set up as follows (see Table 9.2):

High-deliverability gas layer: open-flow potential $> 30 \times 10^4$ m^3/d, gas deliverability index > 200 $m^3/(MPa^2 \cdot d)$
Medium-deliverability gas layer: open-flow potential 8 to 30×10^4 m^3/d, gas deliverability index 50 to 200 $m^3/(MPa^2 \cdot d)$
Low-deliverability gas layer: open-flow potential 0.5 to 8×10^4 m^3/d, gas deliverability index 10 to 50 $m^3/(MPa^2 \cdot d)$

Gas deliverability index and permeability criteria. The gas deliverability index has a positive correlation with reservoir permeability (Figure 9.13b). The criteria for the gas deliverability index and permeability are established for three types of gas reservoirs—according to the gas-testing characteristics of naturally high deliverability, high deliverability after fracturing, and medium to low deliverability after fracturing—by analyzing the relationship between the gas deliverability index and permeability (see Table 9.2):

Type I reservoirs: gas deliverability index > 200 $m^3/(MPa^2 \cdot d)$, permeability > 2 mD
Type II reservoirs: gas deliverability index > 100 $m^3/(MPa^2 \cdot d)$, permeability > 0.2 to 2 mD
Type III reservoirs: gas deliverability index 10 to 100 $m^3/(MPa^2 \cdot d)$, permeability 0.01 to 0.2 mD

Effective porosity criteria. Reservoir permeability is positively related to effective porosity (Figure 9.13c). The effective porosity values are established for three types of reservoirs according to the gas-testing deliverability characteristics and reservoir permeability classification as determined by an analysis of the relationships between reservoir permeability and effective porosity (see Table 9.2):

Type I reservoirs: effective porosity $> 10\%$
Type II reservoirs: effective porosity 5% to 10%
Type III reservoirs: effective porosity 3.5% to 5%

Matrix permeability criteria. Volcanic matrix permeability has a positive correlation with effective matrix porosity (Figure 9.13d). According to effective porosity criteria, the matrix permeability criteria for three types of reservoirs are set up based on the relationships between matrix permeability and effective porosity (see Table 9.2):

Type I reservoirs: matrix permeability > 0.1mD
Type II reservoirs: matrix permeability 0.015 to 0.1mD
Type III reservoirs: matrix permeability 0.01 to 0.015 mD

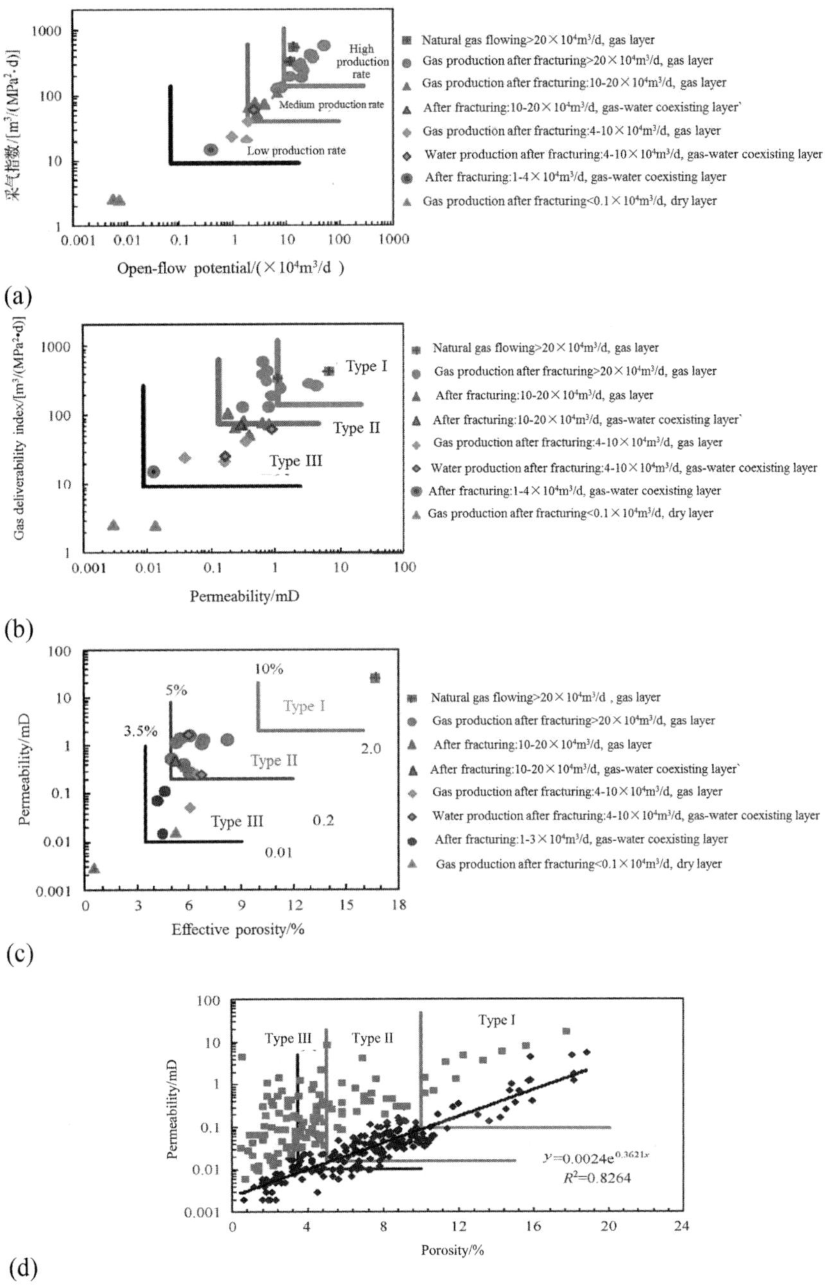

FIGURE 9.13

Deliverability and physical property criteria for volcanic reservoir classification and evaluation (XX gas field). (a) Crossplot of the gas deliverability index versus open-flow potential. (b) Crossplot of the gas deliverability index versus logging permeability. (c) Crossplot of permeability versus effective porosity in the gas-testing interval. (d) Crossplot of core permeability versus matrix porosity.

2) Electrical property criteria

Logging response characteristics of volcanic reservoirs are affected by their petrophysical properties, and there are good correlations among them: DEN decreases and AC increases with better petrophysical properties. On the basis of effective porosity criteria, therefore, the electrical property criteria for effective volcanic reservoirs are established through analysis of the relationships between core porosity and logging (Figure 9.14 and Table 9.2):

Type I reservoirs: DEN < 2.38 g/cm^3, AC > 65 μs/ft
Type II reservoirs: DEN 2.38 to 2.52 g/cm^3, AC 58 to 65 μs/ft
Type III reservoirs: DEN 2.52 to 2.55 g/cm^3, AC 55 to 58 μs/ft

3) Pore structure criteria

The pore structure parameter criteria for identifying effective reservoirs are set up based on the classification scheme of matrix permeability and effective porosity, using mercury injection data as well as the relationships among porosity, permeability, and pore-throat radius. The classification scheme of pore-throat radius for effective reservoirs of the XX gas field are as follows (Figure 9.15 and Table 9.2):

Type I reservoirs: pore-throat radius > 0.25 μm
Type II reservoirs: pore-throat radius 0.05 to 0.25 μm
Type III reservoirs: pore-throat radius 0.01 to 0.05 μm

4) Gas-bearing property criteria

According to the gas-testing deliverability, the correlation between gas deliverability index and gas saturation is set up (Figure 9.16). Based on the gas deliverability indices, the saturation classification schemes for the gas layers in the XX gas field are defined as follows (see Table 9.2):

High-deliverability gas layer: gas saturation $> 55\%$
Medium-deliverability gas layer: gas saturation 45% to 55%
Low-deliverability gas layer: gas saturation 35% to 45%

5) Criteria regarding lithology, lithofacies, and pore structures

According to these criteria, effective volcanic reservoirs are classified systematically through logging response analysis and reservoir parameter interpretation. This allows the establishment of a set of criteria in terms of lithology, lithofacies, and pore structures for the three types of effective reservoirs by combining statistical analysis with the geological characteristics of volcanic gas reservoirs (Table 9.3).

3. Quantitative identification of effective volcanic reservoirs

Various types of effective volcanic reservoirs can be identified quantitatively in accordance with the reservoir cutoff criteria and classification scheme discussed earlier, using core, experimental, and log data as well as reservoir parameter interpretation results.

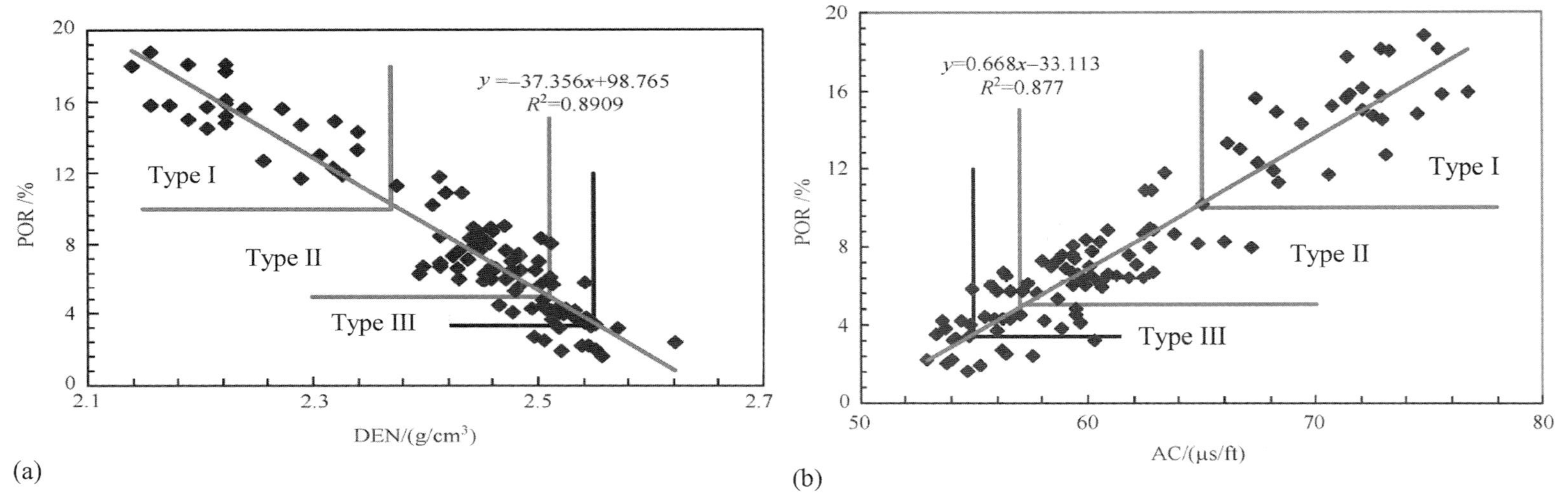

FIGURE 9.14

Electrical property criteria for volcanic reservoir classification and evaluation (XX gas field). (a) Crossplot of POR versus DEN. (b) Crossplot of POR versus AC.

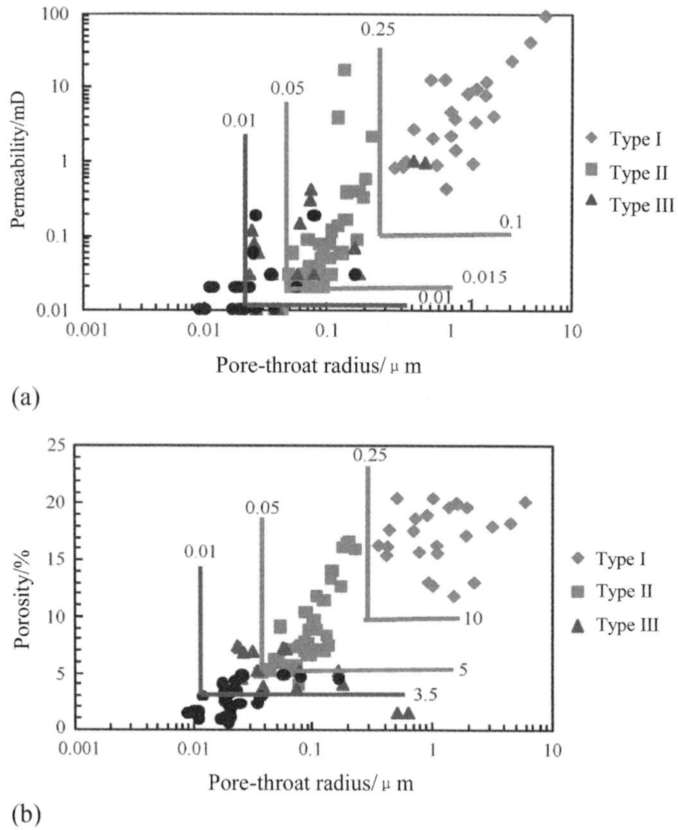

FIGURE 9.15

Pore structure parameter criteria for volcanic reservoir classification and evaluation (XX gas field). (a) Crossplot of permeability versus pore-throat radius. (b) Crossplot of porosity versus pore-throat radius.

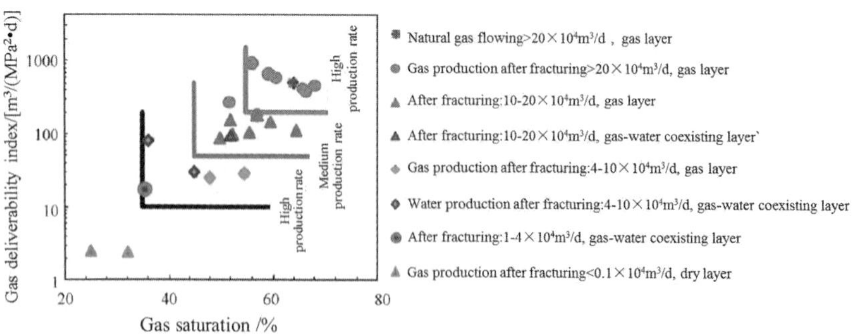

FIGURE 9.16

Crossplot of gas deliverability index versus gas saturation.

Table 9.3 Systematic Classification for Volcanic Reservoirs

		Lithology and Lithofacies Characteristics		Pore Structure Feature		
Reservoir Type	**Petrophysical Property**	*Volcanic Lithofacies*	*Rock Types*	*Accumulation-Permeation Combination*	*Pore Structure*	**Deliverability**
Type I	Good	Splash subfacies of explosive facies; top subfacies of effusive facies; volcanic neck subfacies of volcanic conduit facies	Breccia lava, vesicular rhyolite, autoclastic brecciated lava	Dissolved pore + vesicle + fracture; intergranular dissolved pore + micropore + fracture	Type I	Mainly natural high production
Type II	Medium	Hot clastic flow subfacies of explosive facies; upper subfacies of effusive facies; outerzone subfacies of subvolcanic facies	Welded tuff, crystalline tuff, vesicular rhyolite, orthophyre	Dissolved pore + vesicle + fracture; intergranular dissolved pore + micropore + fracture; vesicle + fracture; micropore + fracture	Mainly type II, a few type I and type III	Mainly high production after fracturing
Type III	Poor	Hot clastic flow subfacies and hot base surge subfacies of explosive facies; middle and lower subfacies of effusive facies; mesozone subfacies of subvolcanic facies	Welded tuff, crystalline tuff, welded breccia, vesicular rhyolite, orthophyre	Vesicle + fracture; micropore + fracture	Mainly type II and type III, a few type IV	Medium to low production after fracturing

1) Type I effective reservoirs

Type I effective reservoirs are developed in favorable facies belts, such as the splash subfacies of the explosive facies, the top subfacies of the effusive facies, and the volcanic neck subfacies of the volcanic conduit facies, lithologically dominated by brecciated lava and vesicular rhyolite. They have a reasonably good pore-fracture combination, with common type I pore structure, good petrophysical properties, good electrical property, and high natural deliverability (see Table 9.3). Their details include (see Table 9.2) (1) effective porosity > 10%, permeability > 2 mD, and matrix permeability > 0.1 mD; (2) gas saturation > 55%; (3) DEN < 2.38 g/cm^3 and AC > 65 μs/ft; (4) pore-throat radius > 0.25 μm; and (5) open-flow potential > 30×10^4 m^3/d with a gas deliverability index of > 200 m^3/(MPa^2·d).

Layer 182 II (3717 to 3735 m) in Well CC103 is an example of a type I effective reservoir (Figure 9.17). The interval consists of rhyolite and autoclastic brecciated lava. The main characteristics include (1) broken cores, with a pore structure of type I in the mercury injection curve and a pore-throat radius of 0.82 μm, which is larger than the cutoff of type I; (2) effective porosity 10% to 25.9% with an average of 16.3%; permeability 1.6 to 251 mD with an average of 46.7 mD; matrix permeability 0.13 to 2.71 mD with an average of 0.63 mD, all higher than cutoff of type I; (3) gas saturation 65% to 80% with an average of 73%, higher than the gas saturation cutoff of type I; (4) DEN 2.20 to 2.49 g/cm^3 with an average of 2.32 g/cm^3, lower than the cutoff of type I; AC 57 to 88 μs/ft, with an average of 67 μs/ft, higher than cutoff of type I; and (5) the mobile fluid peak amplitude in the T2 spectrum of nuclear magnetic logs is higher, showing a red-black porphyritic texture in FMI imaging logs and developed fractures. This interval is interpreted as a type I effective reservoir.

The test of this interval together with interval 3632 to 3648 m produced the following results: the gas production in MFE natural flowing is 51.3×10^4 m^3/d; the production pressure differential is about 15.6 MPa, indicating a natural high-production gas layer for this interval; and the open-flow potential is 74.5×10^4 m^3/d, with a gas deliverability index of 652.7 m^3/(MPa^2·d), as typical of a type I gas layer.

2) Type II effective reservoirs

Type II effective reservoirs are developed in the upper subfacies of the effusive facies, the hot clastic flow subfacies of the explosive facies, and the outerzone subfacies of the subvolcanic facies, lithologically dominated by welded tuff, rhyolite, and orthophyre. They have the characteristics of good pore-fracture combination, with mainly type II pore structure, moderate petrophysical properties, moderate electrical properties, and high natural deliverability after fracturing (see Table 9.3). Their details include (see Table 9.2) (1) effective porosity 5% to 10%, permeability 0.2 to 2 mD, and matrix permeability 0.015 to 0.1 mD; (2) gas saturation 45% to 55%; (3) DEN 2.38 to 2.52 g/cm^3, AC 58 to 65 μs/ft; (4) pore-throat radius 0.05 to 0.25 μm; and (5) after fracturing, open-flow potential > 20×10^4 m^3/d and gas deliverability index > 100 m^3/(MPa^2·d).

Layer 257 I (3887 to 3914 m) in Well XX901 is an example of type II effective reservoir (Figure 9.18), lithologically consisting of rhyolite. The characteristics (including petrophysical properties, electric properties, and gas bearing property) are all between the cutoff values for type II and type I: (1) vesicular rhyolitic structure and factures are developed in cores, without mercury injection data; (2) effective porosity is 4.7% to 8.2%, average 6.6%; permeability 0.2 to 3.3 mD, average 1.1 mD; and matrix permeability 0.01 to 0.05 mD, average 0.03 mD; (3) gas saturation is 41% to 72%, average 61%; (4) DEN is 2.36 to 2.51 g/cm^3, average 2.46 g/cm^3; AC is 56 to 63 μs/ft, average 59 μs/ft; and (5) the mobile fluid peak amplitude in the T2 spectrum of the nuclear magnetic log is

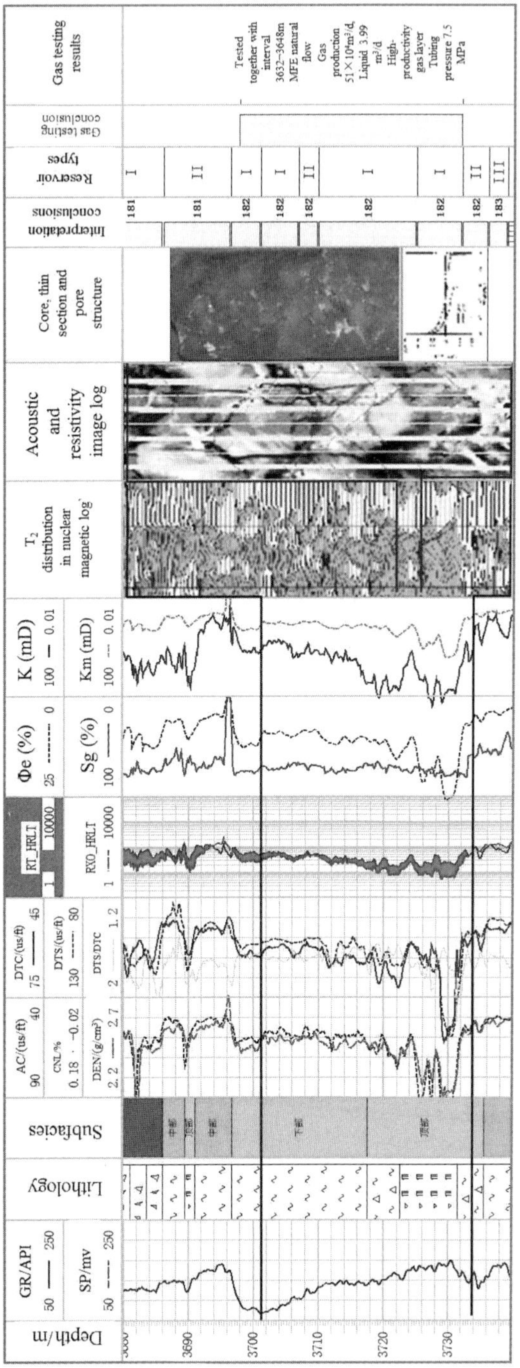

FIGURE 9.17

Type I volcanic reservoirs (Well CC103, autoclastic brecciated lava).

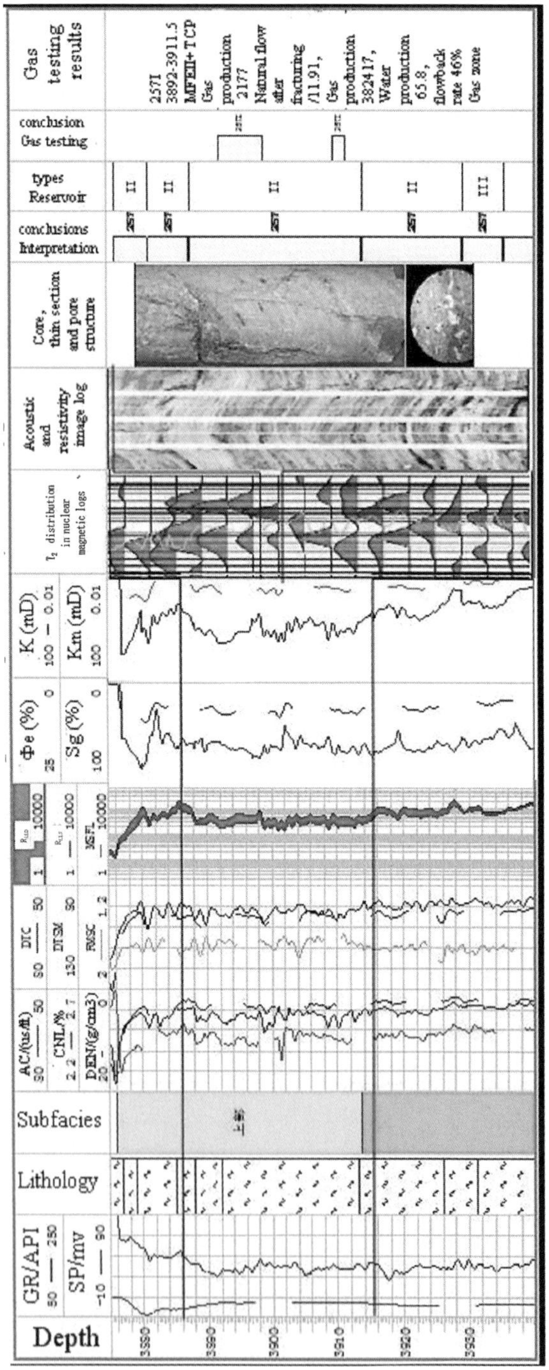

FIGURE 9.18

Type II volcanic reservoirs (Well XX901, rhyolite).

medium, showing yellow-red porphyritic texture in the FMI imaging log and developing rhyolitic structure and fractures. This interval is interpreted generally as a type II effective reservoir.

Gas-testing results show that the gas-testing deliverability after fracturing is 38.2 4 × 10^4 m^3/d; the production pressure differential is about 8.61 MPa, which is a high gas deliverability layer after fracturing; the open-flow potential is 88.2 × 10^4 m^3/d, with a gas deliverability index of 593.6 m^3/(MPa^2·d), indicating a typical type II gas layer.

3) Type III effective reservoirs

Type III effective reservoirs are developed in the upper subfacies and the lower subfacies of effusive facies, the hot clastic flow subfacies and hot base surge subfacies of the explosive facies, and the mesozone subfacies of the subvolcanic facies, lithologically dominated by welded tuff, crystal tuff, welded breccia, volcanic lava, and orthophyre. They are characterized by a poor pore-fracture combination, mainly type II and type III pore structures, and a minor type IV pore structure; poor physical properties and moderate electric properties; and moderate to low deliverability after fracturing (see Table 9.3). The details include (see Table 9.2) (1) effective porosity 3.5% to 5%, permeability 0.01 to 0.2 mD, and matrix permeability 0.01 to 0.015 mD; (2) gas saturation 35% to 50%; (3) DEN 2.52 to 2.55 g/cm^3, AC 55 to 58 μs/ft; (4) pore-throat radius 0.01 to 0.05 μm; and (5) open-flow potential after fracturing 0.5 to 20 × 10^4 m^3/d and a gas deliverability index of 10 to 100 m^3/(MPa^2·d).

Layer 161 I (3686 to 3798 m) in Well XX9-2 is an example of a type III effective reservoir (Figure 9.19), lithologically represented by rhyolite. The characteristics (including petrophysical properties, electrical properties, and gas bearing property) are (1) effective porosity 3.1% to 7%, average 4.8%; permeability 0.02 to 1.66 mD, average 0.38 mD; and matrix permeability 0.01 to 0.03 mD, average 0.02 mD; (2) gas saturation 38% to 81%, average 63%; and (3) DEN 2.45 to 2.57 g/cm^3, average 2.52 g/cm^3; AC 52.8 to 62 μs/ft, average 55.6 μs/ft; and resistivity 719 to 10946 Ω·m, average 4163 Ω·m. This interval is generally interpreted as a type III effective reservoir.

Gas-testing production on natural deliverability in this interval is 0.011 × 10^4 m^3/d; production after fracturing is 12.8 × 10^4 m^3/d; the open-flow potential is 13.85 × 10^4 m^3/d; and the gas deliverability index is 92.44 × 10^4 m^3/(MPa^2·d). These data are typical of a type III gas layer.

4. Developmental features of effective volcanic reservoirs

The developmental features of different types of effective volcanic reservoirs are basically clarified by quantitative identification. Figure 9.20 shows the classification and evaluation results of Well Block XX9. This figure implies that the main effective volcanic reservoirs in this well block are type II (about 51.5%); type III comes in second (at about 42.7%); and type I develops locally (about 5.8%). This indicates generally poor petrophysical properties for these effective volcanic reservoirs.

9.3 PREDICTION OF EFFECTIVE VOLCANIC RESERVOIRS

Based on well identification and seismic-well calibration, seismic classification criteria of effective volcanic reservoirs are set up by analyzing seismic response characteristics. The spatial distribution patterns of various effective reservoir types can be predicted by "body-controlling" reservoir inversion, which forms the basis for optimizing well locations, designing horizontal well trajectories, and improving the success rate of reservoir wells.

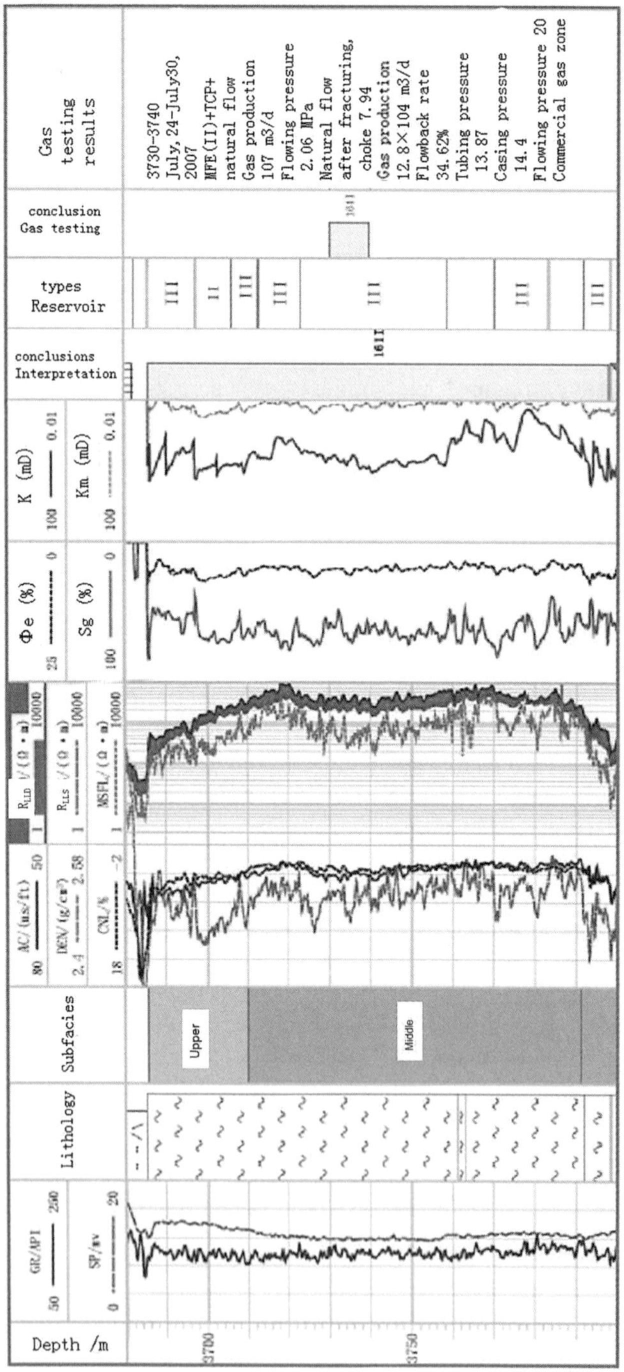

FIGURE 9.19

Type III volcanic reservoirs (Well XX9-2, rhyolite).

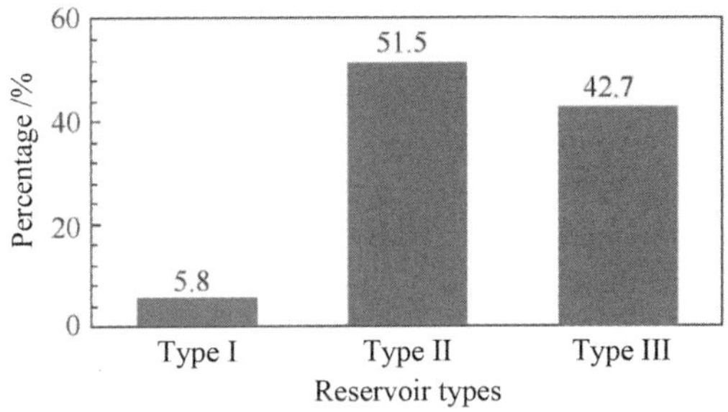

FIGURE 9.20

Developmental features of different types of effective volcanic reservoirs (Well Block XX9).

9.3.1 Seismic response characteristics of effective volcanic reservoirs

Different types of volcanic reservoirs have different seismic response characteristics. The criteria for effective reservoirs are established by calibrating seismic data against well data, analyzing response characteristics of near-wellbore seismic traces, and distilling sensitive seismic attributes, which lays a foundation for the classification and predication of effective volcanic reservoirs.

1. Seismic response characteristics of effective volcanic reservoirs

After calibrating seismic data against well identification results and analyzing reflection characteristics of near-wellbore seismic traces, the seismic facies are classified for the qualitative prediction of effective volcanic reservoirs. As is shown in the case studies of Well Blocks XX8 and XX9, four types of typical volcanic seismic response characteristics are recognized based on the differences in accumulation and permeation capacity of reservoirs (Figure 9.21).

1) Seismic response characteristics: Type A

In this reservoir type, the internal reflection is characterized by medium to low frequency, medium to weak amplitude, poor continuity, and a chaotic reflection shape, developed mainly in explosive facies. The typical intervals have an average porosity of 5.22% and an average permeability of 2.9 mD; the fractures are well developed, with an average fracture width of 19 μm, an average fracture density of 2.54 fractures/m, and an average fracture surface porosity of 0.07%. These represent the best petrophysical properties.

2) Seismic response characteristics: Type B

The seismic reflection of type B is characterized by medium to low frequency, medium to weak amplitude, good continuity, and a laminar reflection shape, developed mainly in mixed facies (volcanic lava, etc.). The typical intervals have an average porosity of 3.9% and an average permeability of 0.43 mD; both structural fractures and microfractures are well developed, with an average fracture width of 11.13 μm, an average fracture density of 2.66 fractures/m, and an average fracture surface porosity of 0.05%. These are good petrophysical properties.

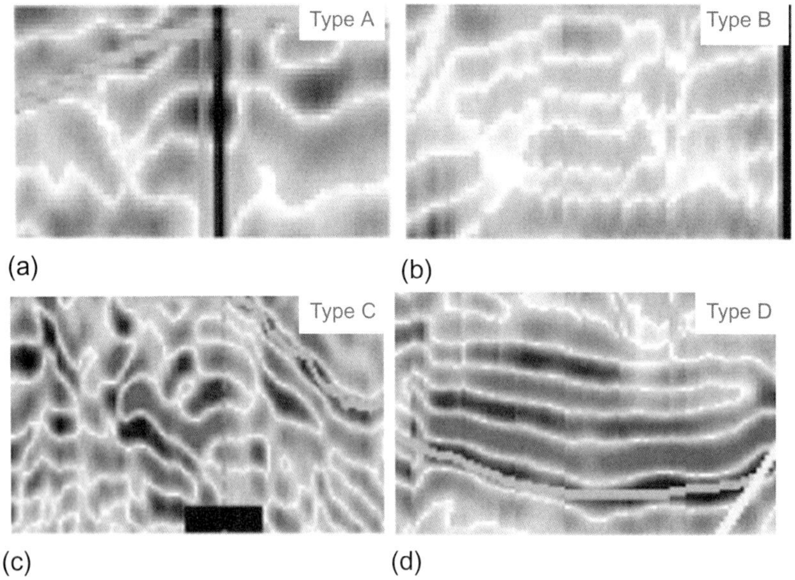

FIGURE 9.21

Seismic response characteristics of effective volcanic reservoirs.

3) Seismic response characteristics: Type C

A type C seismic reflection is characterized by medium to high frequency, medium to strong amplitude, poor continuity, and a chaotic reflection shape, developed mainly in mixed facies of alternating volcanic lava and volcaniclastic rocks. In the typical intervals, the average porosity is 3.79%, and average permeability is 0.2 mD; both structural fractures and microfractures are well developed, with an average fracture width of 15.32 μm, an average fracture density of 3.18 fractures/m, and an average fracture surface porosity of 0.04%. These are poor petrophysical properties.

4) Seismic response characteristics: Type D

A type D seismic reflection is characterized by medium to high frequency, strong amplitude, relatively good continuity, and a laminar reflection shape, developed mainly in effusive facies. The typical intervals have an average porosity of 3.61% and an average permeability of 0.2 mD; fractures are developed mainly as microfractures, with an average fracture width of 8.10 μm, an average fracture density of 3.09 fractures/m, and an average fracture surface porosity of 0.042%. These are the least favorable petrophysical properties.

2. Seismic parameter sensitivity analysis of effective reservoirs

Effective porosity is closely related to rock density in volcanic reservoirs (Figure 9.22). Acoustic impedance is the product of rock density and acoustic velocity and is also closely related to rock porosity and rock density. Therefore, rock density, acoustic velocity, and acoustic impedance are the sensitivity parameters of effective volcanic reservoirs.

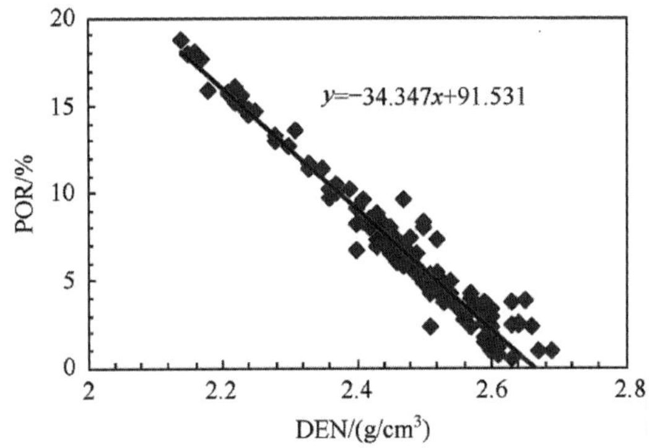

FIGURE 9.22

Crossplot of POR versus DEN.

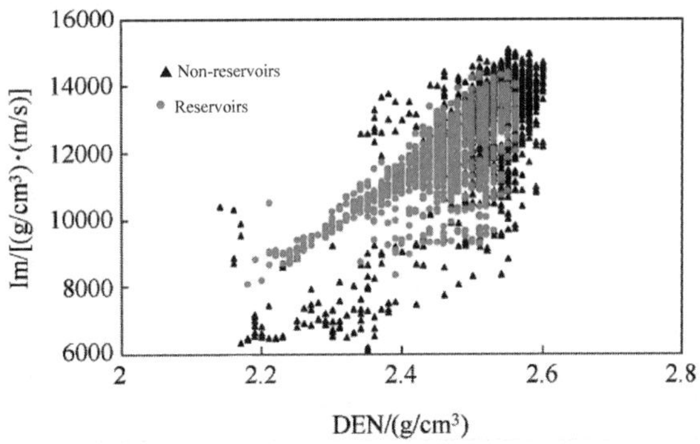

FIGURE 9.23

Crossplot of acoustic impedance versus DEN.

These relationships are shown using the CC gas field as an example. The crossplot of acoustic impedance versus well log density is established based on the results of well interpretation and the analysis of response characteristics of near-wellbore seismic traces after seismic data calibration with well log data (Figure 9.23). Figure 9.23 demonstrates that the main characteristics of effective volcanic reservoirs include low density (2.28 to 2.57 g/cm^3) and low acoustic impedance [9500 to 14500 (g/cm^3)·(m/s)].

3. Seismic classification criteria for effective volcanic reservoirs

The seismic classification criteria for effective volcanic reservoirs are set up based on the classification scheme of effective volcanic reservoir zones, using acoustic impedance, porosity, density, and acoustic data that are extracted from near-wellbore seismic traces, through bivariant crossplot analysis (Figures 9.24 and 9.25). This is shown using the CC gas field as an example:

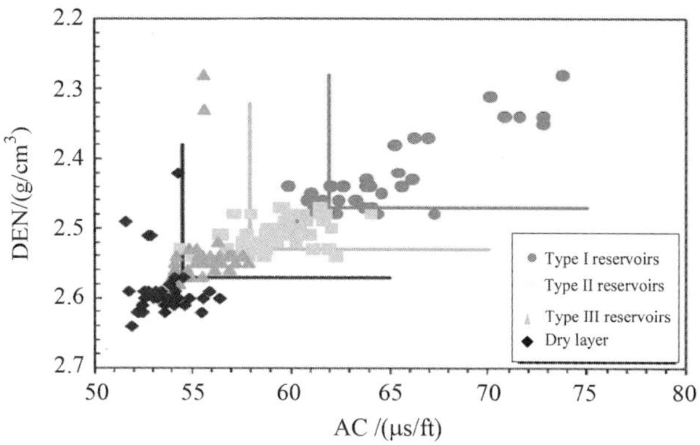

FIGURE 9.24

Crossplot of AC versus DEN.

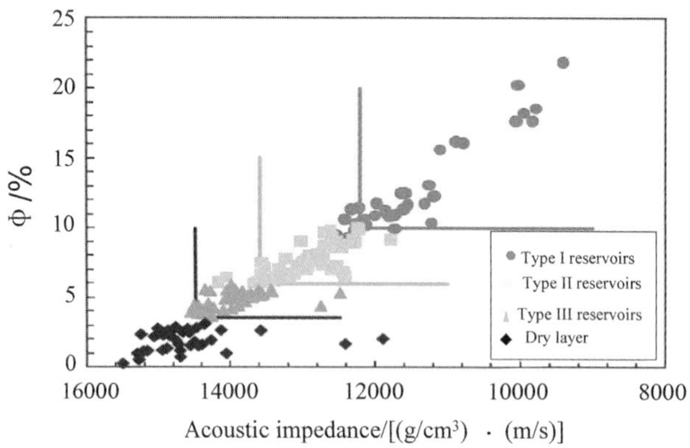

FIGURE 9.25

Crossplot of acoustic impedance versus POR.

Type I effective reservoirs: porosity > 10%, density < 2.47 g/cm^3, AC > 62 μs/ft, and acoustic impedance 9500 to 12200 (g/cm^3)·(m/s)

Type II effective reservoirs: porosity 6% to 10%, density 2.47 to 2.53 g/cm^3, AC 58 to 62 μs/ft, and acoustic impedance 12200 to 13600 (g/cm^3)·(m/s)

Type III effective reservoirs: porosity 3.5% to 6%, density 2.53 to 2.57 g/cm^3, AC 55 to 58 μs/ft, and acoustic impedance 13600 to 14500 (g/cm^3)·(m/s)

9.3.2 Classification and prediction techniques for effective volcanic reservoirs

Volcanic rock has multilevel internal structures (architectures), thus conventional reservoir inversion techniques cannot be applied to the complex geological characteristics and associated changes in volcanic gas reservoirs. Guided by the distribution models of effective volcanic reservoirs and constrained by volcanic architectures, therefore, the acoustic impedance volumes, density volumes, acoustic volumes, and porosity volumes that reflect reservoir petrophysical properties can be obtained by "body-controlling" seismic inversion. This will help extract the classified effective thickness for predicting each type of effective volcanic reservoir [7–10] by applying the seismic classification criteria for effective reservoirs.

1. Establishing the volcanic framework model to constrain seismic inversion

Volcanic reservoirs typically have a great thickness, rapid facies change, and strong heterogeneity; consequently the distribution and configuration of effective reservoirs in each architectural unit are highly variable (Figure 9.26). As a result, volcanic framework modeling based on internal structure dissection and sequence division is important for constraining seismic inversion.

On the basis of the architectural characteristics of volcanic rocks, the complete framework model includes two parts (Figure 9.27): (1) the framework model controlled by the top and bottom structural planes of volcanic reservoirs, which is used mainly for setting top and bottom constraints for seismic inversion; and (2) the framework model controlled by the internal architecture (i.e., the stratigraphic framework model for each level of internal architecture in the 3D space delineated by the gas reservoir top and bottom planes), which primarily is used to provide boundary constraints for each level of architecture for seismic inversion and provide a basis for geological control.

2. Acoustic impedance inversion

Constrained by the framework model and based on logging normalization, high-resolution reservoir calibration, wavelet extraction, and so on [11], the initial model that can reflect the volcanic architecture is constructed using seismic data and high-resolution log data. Acoustic impedance volumes can be obtained with the model-constrained sparse spike inversion method.

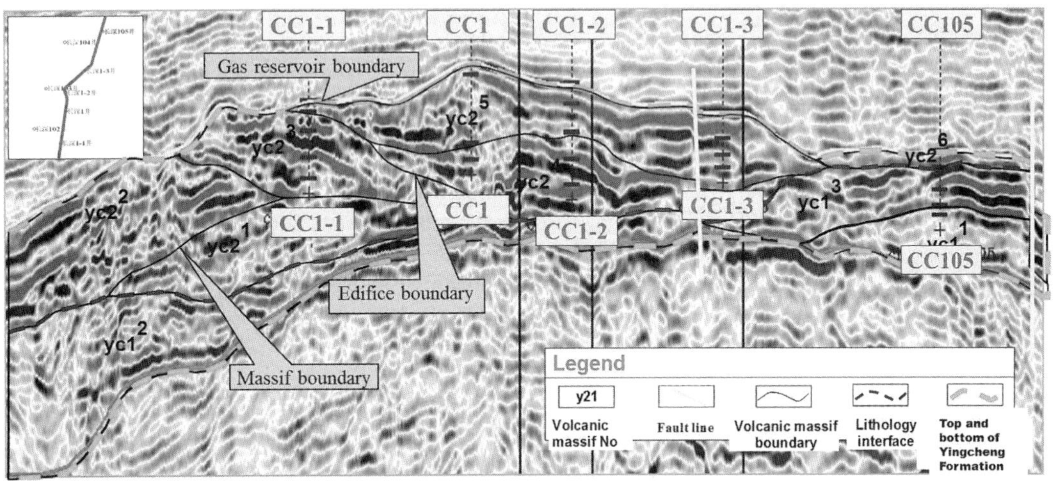

FIGURE 9.26

Architectural characteristics of volcanic rocks.

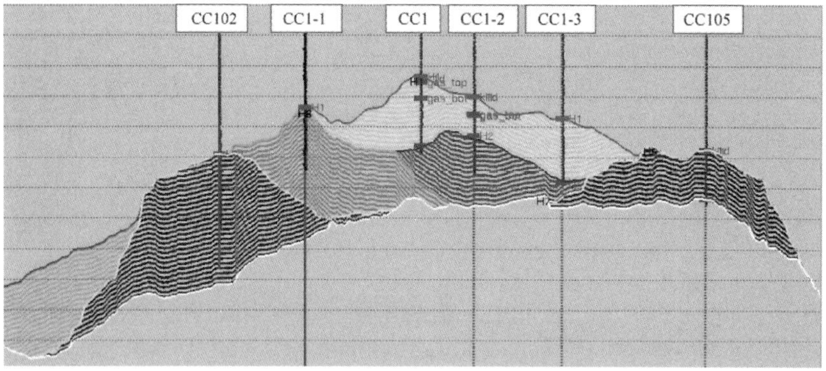

FIGURE 9.27

Reservoir framework model of acoustic impedance inversion.

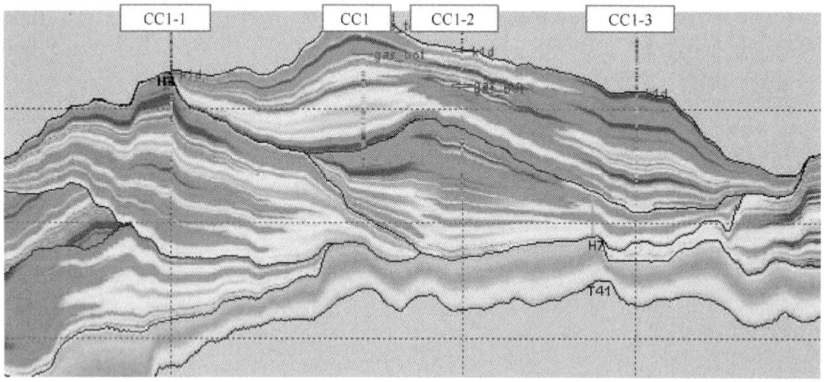

FIGURE 9.28

Initial acoustic impedance inversion model constrained by the volcanic reservoir framework model.

1) Establishing an initial acoustic impedance model

Constrained by well data, the log data are interpolated at each level of the framework model by means of linear interpolation, fractal interpolation, and waveform coherence interpolation, and the initial acoustic impedance inversion model is thus established with reference to the volcanic reservoir framework model (Figure 9.28). Compared with the initial acoustic impedance inversion model based on the conventional sedimentary reservoir concept (Figure 9.29), the initial model constrained by the framework model can reflect the internal architecture of volcanic gas reservoirs more accurately, thus forming a solid basis for acoustic impedance inversion.

2) Acoustic impedance inversion

Considering the geological features of volcanic gas reservoirs, high-quality acoustic impedance data volumes can be obtained using the model-constrained sparse spike inversion method, constrained by well data and the volcanic framework model, through iterative correction of residual error using stochastic algorithms.

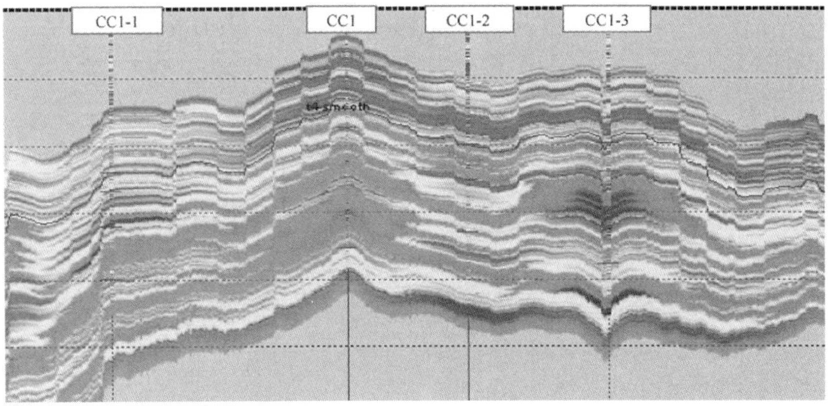

FIGURE 9.29

Initial acoustic impedance inversion model based on the conventional sediment reservoir concept.

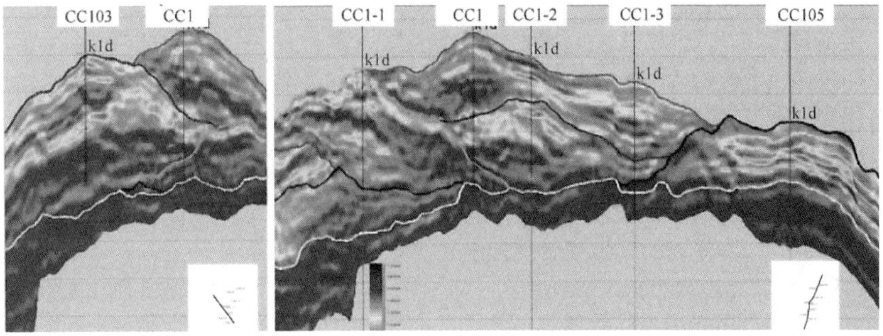

FIGURE 9.30

Acoustic impedance inversion profile of volcanic reservoirs (CC gas field).

The acoustic impedance data volumes from seismic inversion can better reflect the petrophysical properties of volcanic reservoirs (Figure 9.30) and are important for predicting effective reservoirs. As the example from the CC gas field shows, in the planar map of acoustic impedance (Figure 9.31), high-yield wells (CC103, CC1, CC1-1, etc.) with good petrophysical properties are located mainly in relatively low impedance zones; low-yield wells (CC102, CC104, CC1-2, etc.) with poor petrophysical properties are located mainly in relatively high impedance zones.

3. *Reservoir parameter inversion*

Based on acoustic impedance inversion and constrained by the volcanic framework model, reservoir parameters are inverted with the approach of "secondary inversion" (Figure 9.32): (1) constructing a pseudo-acoustic curve (using statistics and conversion methods to convert reservoir parameters to the same value range of acoustic logs); (2) on the basis of pseudo-acoustic curves, using the acoustic impedance inversion method to acquire pseudo-acoustic impedance data volumes; and (3) using waveform coherence or statistic algorithms to establish the relationship between reservoir parameters and

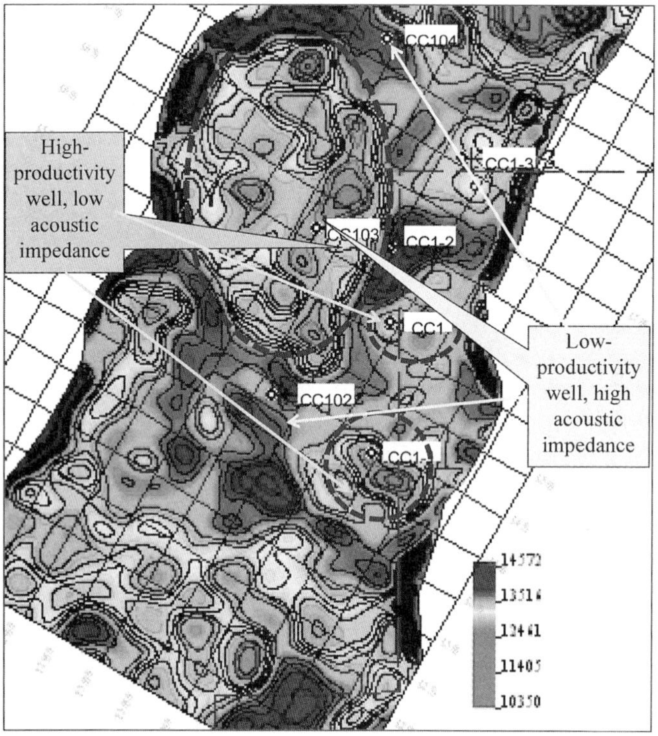

FIGURE 9.31

Planar map of acoustic impedance (CC gas field).

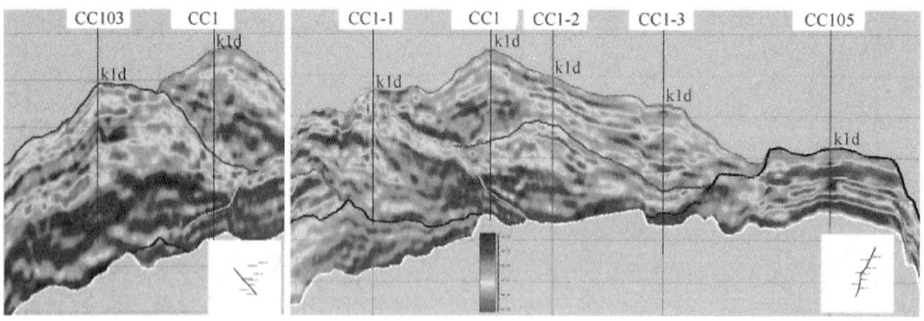

FIGURE 9.32

Density inversion profile of volcanic reservoirs (CC gas field).

pseudo-acoustic impedance data and performing a "secondary inversion" to obtain reservoir parameter volumes (including density volumes, acoustic volumes, and porosity volumes).

Data volumes from reservoir parameter inversion can more directly reflect petrophysical property changes in volcanic reservoirs and form the basis for better prediction of effective gas reservoirs. In the planar map of porosity and density in the CC gas field (Figures 9.33 and 9.34), high-production wells

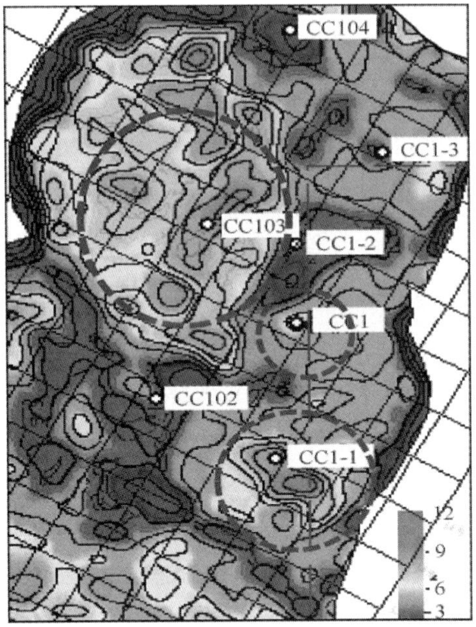

FIGURE 9.33

Planar map of inverted.

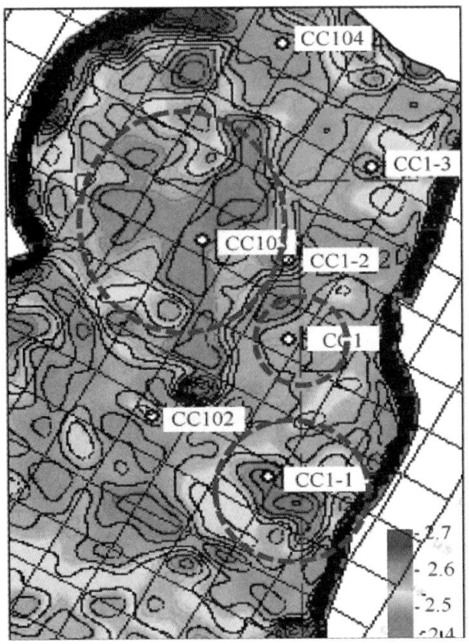

FIGURE 9.34

Planar map of inverted porosity (CC gas field) and density (CC gas field).

with good petrophysical properties are located mainly in high-porosity zones or relative low-density zones. Low-production wells with poor petrophysical properties are located mainly in low-porosity zones or relative high-density zones. The prediction results have proven to be consistent with actual production performance.

4. Extracting categories of net pay thickness

Using inverted acoustic impedance volume and reservoir parameter volumes, the categories of net pay thickness can be extracted according to seismic classification criteria of effective reservoirs: (1) extracting the reservoir gross thickness according to the seismic cutoff criteria of effective reservoirs, (2) extracting the categories of reservoir thickness according to seismic classification criteria of effective reservoirs, and (3) extracting the total net pay thickness and the categories of net pay thickness above gas-water contact according to seismic classification criteria of net pay reservoirs and constrained by gas-water contact.

By extracting the categories of reservoir thickness, the planar distribution of each type of volcanic gas reservoir is clarified. In the CC gas field, the total net pay thickness and the categorized net pay thickness above gas-water contact are extracted using inverted volumes of porosity, density, and acoustic impedance, constrained by gas-water contact (Figure 9.35). The results show that (1) in the CC gas field, the maximum total net pay thickness is 146 m with an average of 60 m, and the thickest areas are in the western and northern parts of Well Block CC1; (2) the maximum net pay thickness of type I is 56 m with an average of 15 m, mainly in areas near Wells CC103 and CC1-1; (3) the maximum net pay thickness of type II is 80 m with an average of 35 m, and the thickest area is in the northwestern part of Well Block CC1; and (4) the maximum net pay thickness of type III is 118 m with an average of 40 m, and the thickest area is near Well CC1. The new drilling data confirmed the prediction results, which provide a direct basis for well location optimization.

9.4 APPLICATIONS

The effective volcanic reservoir identification and prediction techniques discussed in this chapter are unique for volcanic gas reservoir characterization. They constitute the main approaches for clarifying the distribution characteristics of effective volcanic reservoirs. The commercial application of this technique can help optimize well location and improve the success rate of horizontal wells encountering reservoirs.

9.4.1 Optimizing well location and improving the success rate of development well placement

The planar distribution of categorized (classified) net pay zone thickness is the main guidance for development well placement in volcanic gas reservoirs. Through the prediction technique of effective volcanic reservoirs based on body-controlling seismic inversion, the categorized reservoir thickness and net pay thickness above gas-water contact can be extracted to define the planar distribution of each type of volcanic gas reservoir. Optimization of well location can effectively improve the success ratio of development well placement.

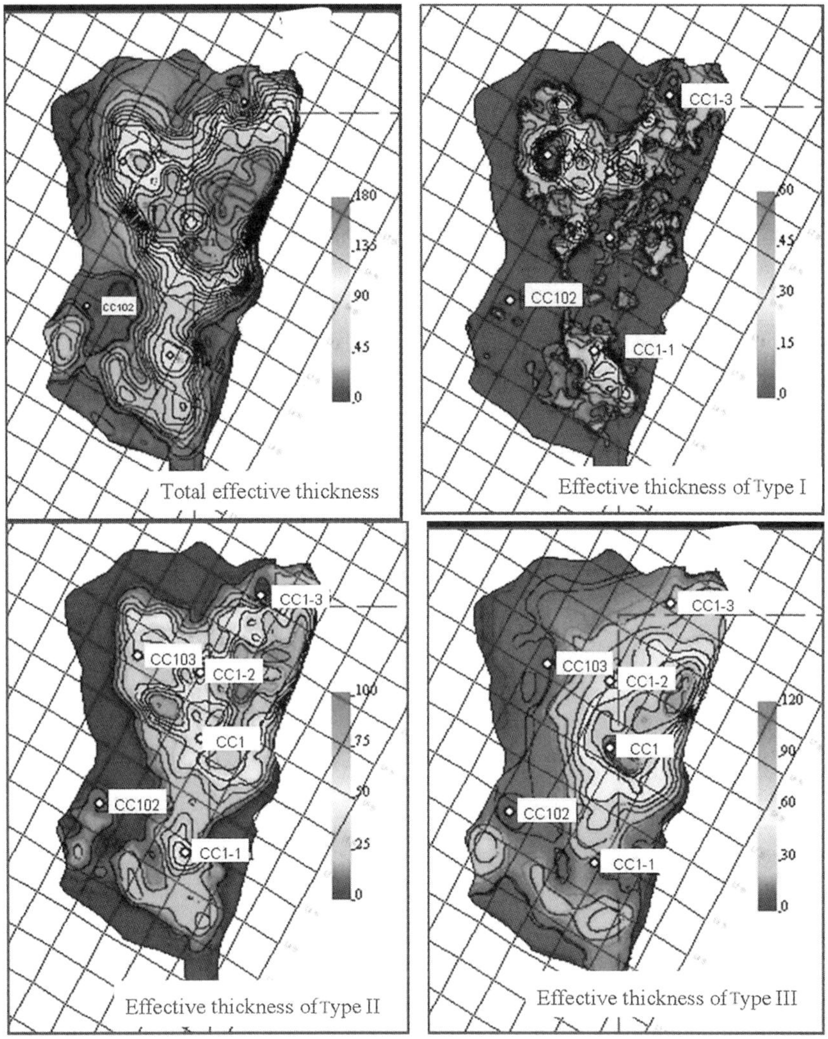

FIGURE 9.35

Planar map of categorized net pay thickness in the CC gas field.

In the CC gas field, 12 development wells were designed, including 6 vertical wells and 6 horizontal wells. After all the wells were put into production, the vertical wells reached a production of 8.7 t 34×10^4 m^3/d, and the horizontal wells reached a production of 24 to 51×10^4 m^3/d, all achieving the expected goal, with a success rate of 100%. After applying this technique in several gas fields (such as the Daqing, Jilin, and Xinjiang oil fields), the success rate of development well placement in complex volcanic gas reservoirs was over 95%, accounting for an improvement of at least 19% compared with the results derived before the application of this technique.

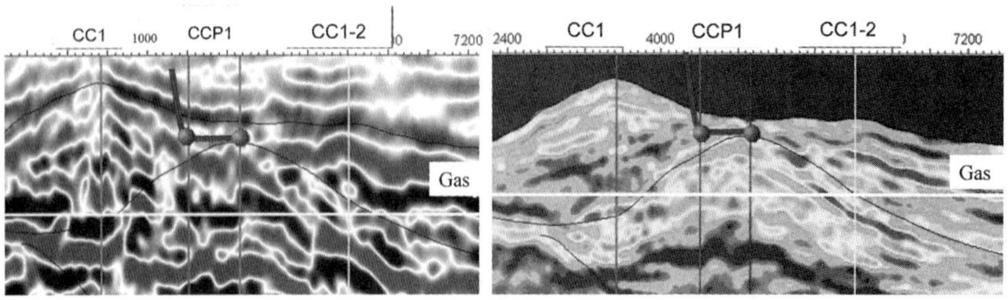

(a) (b)

FIGURE 9.36

Well trajectory design of Well CCP1. (a) Multi-well seismic profile through Well CSPI. (b) Porosity inversion profile through Well CSPI.

9.4.2 Optimizing well trajectory design and improving the reservoir encounter ratio of horizontal wells

The spatial distribution of effective volcanic reservoirs is the main basis for optimizing well trajectory design. Through identification and categorized prediction of effective volcanic reservoirs, the spatial distribution of each type of effective reservoir can be defined for optimizing the horizontal well trajectory and improving the reservoir encounter ratio of horizontal wells.

Porosity data volumes can be obtained by "body-controlling" reservoir inversion, as is the case in the design and optimization of the Well CCP1 trajectory (Figure 9.36). The actual drilled length of the horizontal interval in this well is 477 m, and the reservoir encounter ratio is up to 94.3%, including 22.2% for type I reservoirs, 40% for type II reservoirs, and 37.8% for type III reservoirs. After initializing production, the natural deliverability of this well reached up to 41.8×10^4 m^3/d, with stable pressure and production. The well is in good condition at present, with a cumulative gas production of 6890×10^4 m^3. After the application of effective volcanic reservoir identification and prediction techniques, the effective reservoir encounter ratio by horizontal wells in volcanic gas reservoirs improved by about 32%. By now, horizontal well drilling has become one of the major technologies for improving single-well production in most of the volcanic gas reservoirs.

References

[1] Wang Y, Kong J, Li H, et al. Geology of natural gas reservoirs. Beijing: Petroleum Industry Press; 2004; p. 44–56.

[2] Wang Z, Shi Z. Advanced reservoir characterization technology. Beijing: Petroleum Industry Press; 1999; p. 118–26, 128–38.

[3] PetroChina. SY/T 5830-93. Volcanic reservoir description methods; 1993.

[4] State Economic and Trade Commission. SY/T 6110-2002. The description of exploitation geology characteristics for carbonate gas reservoirs; 2002.

[5] PetroChina. SY/T 6286-1997. Detail description method for carbonate reservoirs. .

[6] State Bureau of Petroleum and Chemical Industry. SY/T 5579-2000. Fine description method for clastic reservoirs; 2000.

[7] The use of 3-D seismic data in reservoir geological model building. [Liu Z, Trans.] Foreign Petroleum Exploration 1994;6(5):600–5.

[8] Esmersoy C. Split-shear wave inversion for fracture evaluation. In: Society of Exploration Geophysicists annual meeting, 23–27 September, San Francisco, California, Abstract SN1.6, p. 1–4.
[9] Li XY. Fractured reservoir delineation using multicomponent seismic data. Geophys Prospect 1997;54:39–64.
[10] Sa L, Cao Z. Technical ideas and countermeasures for improving reservoir prediction accuracy. Petroleum Geology & Oilfield Development in Daqing 2002;21(6):6, 7.
[11] PetroChina Exploration and Production Company. Logging evaluation technique and application of volcanic hydrocarbon reservoirs. Beijing: Petroleum Industry Press; 2009, p. 68–69.

CHAPTER

Characterization of Accumulation-Permeation Units in Volcanic Gas Reservoirs

10

The *accumulation-permeation* (hereafter abbreviated as *A-P*) *unit* is the fundamental and main object in the development of volcanic gas reservoirs. The concept of A-P units in volcanic gas reservoirs is presented here for the first time by comparisons with the *flow units* in clastic rocks and *accumulation units* in carbonate rocks. By establishing identification markers, single-well data are used to identify A-P units. In terms of well identification and in combination with seismic data of single-well calibration, seismic data are used to predict the profile, planar, and spatial distribution patterns of A-P units. In these studies, the outcrop, morphology, superposition relationships, connectivity, and permeability of A-P units are characterized on the basis of outcrop and seismic data and dense well pattern data, and the distribution of A-P units is revealed. This can guide the optimization of well location, the design of horizontal well trajectory, and the preparation of development technology policy.

10.1 CONCEPT AND CHARACTERIZATION OF ACCUMULATION-PERMEATION UNITS

10.1.1 Concept of the accumulation-permeation unit

Concept is the essential cognition of objectives formed in human mind [1]. The concept of A-P units is helpful for establishing their identification models and revealing their distribution patterns. According to the characteristics of A-P units in volcanic gas reservoirs and comparisons with the flow units of clastic rocks and reservoir units of carbonate rocks, the concept of A-P unit is formed by refining their spatial distribution, petrophysical properties, and accumulation and permeability capacities.

1. Flow unit

The concept of the *flow unit* in clastic rocks was originally proposed to meet the need of detailed description and prediction of complicated reservoirs and the characterization of dynamic and static oil and gas reservoirs in studies on the formation and distribution of residual oil. In 1984, Hearn and coworkers first defined flow units as vertically and laterally continuous accumulation zones with similar permeability, porosity and bedding-plane characteristics [2]. In 1987, Ebanks presented flow units as vertically and laterally continuous reservoir rock bodies with similar geological and petrophysical rock properties, which influence fluid flow [3]. In 1993, Amacfule and colleagues defined flow units as layers with similar hydraulic characteristics within given rocks [4]. Qiu and associates considered flow units to be naturally formed fluid flow pathways that resulted when injected water displaced oil along geological structures due to reservoir heterogeneity, barrier, and bypassing conditions [5]. Mu and coworkers treated a flow unit as an accumulation unit with homogeneous permeability and waterflooding characteristics, bounded by discontinuous thin barriers, various sedimentary microinterfaces, minor faults, and permeability differences [6]. Jiao referred to flow units as building blocks within a sedimentary system, which are subdivided by subsurface hydrodynamic conditions [7], and Liu considered flow units to be special accumulation bodies that are continuous spatially, have similar petrophysical features and fluid flow characteristics, and are bounded by lithological or petrophysical barriers [8].

In summary, a flow unit has homogeneous petrophysical, hydrodynamic, and fluid flow properties, with defined barriers or boundaries, and a set of flow units can be the same time units or sedimentary units or the combination of different time units or sedimentary units.

2. Reservoir unit and fracture-cavity unit

The concept of the reservoir unit is derived from correlations of carbonate reservoirs with heterogeneous fractures and cavities. In the book *How to Understand Reservoir* compiled by Yuan (1979), the concept of the reservoir unit was defined for the first time as the basic lithological association that can form oil and gas sealing conditions in vertical profiles [9]. A reservoir unit comprises the reservoir, pay zone, cap rock, and basal bed. Among these, the pay zone and cap rock are the crucial components; the former determines the production capacity of a reservoir unit, and the latter determines its sealing capacity; the two components coexist in the entity of the reservoir unit. The concept of the reservoir unit is widely used to correlate carbonate hydrocarbon reservoirs, such as in works by Chen [10], Xiong [11], Wu [12], and Chen [13].

In 2006, Jiao and Du summarized the development results of fracture-cavity carbonate reservoirs in the Tahe oilfield and presented the concept of the fracture-cavity system and unit in order to characterize karst fracture-cavity reservoirs and their internal fluid connectivity [14,15]. A fracture-cavity system refers to the karst fracture-cavity zones or an assemblage of pores, cavities, and fractures in the same karst framework, and various systems are separated by continuous tight bodies. A fracture-cavity unit also refers to a hydrodynamic unit or connected body with a uniform pressure system, which comprises single or multiple dissolution cavities connected by a fracture network and isolated by relatively tight or low-permeability dissolution interfaces or closed fractures. A fracture-cavity system may be composed of single or multiple fracture-cavity units.

In summary, a reservoir unit comprises continuous reservoir bodies, tight dissolution or lithological boundaries, and closed fractures, developed in fracture-cavity carbonate reservoirs.

3. Accumulation permeation unit (A-P unit)

The concept of the A-P unit is proposed for the first time in this study for predicting the distribution of effective A-P bodies during the development of volcanic gas reservoirs. Similar to clastic flow units, carbonate reservoir units, and fracture-cavity units, a volcanic A-P unit comprises effective A-P bodies and peripheral flow-resistance zones, and the unit is an independent hydrodynamic system with similar fluid flow characteristics. Therefore, a volcanic A-P unit is defined as a volcanic rock unit with the same petrogenetic origin, a continuous distribution, and uniform hydrodynamic system, which is isolated by impermeable flow-resistant zones and is capable of accumulating fluid and allowing fluid to flow effectively within the unit. Under the control of volcanic rock architecture, most A-P units are developed in the same volcanic edifice, the same eruption period, and even the same lithofacies unit.

A-P units may be connected by open faults, structural fractures, or columnar joint fractures; alternatively, they form a large communicated body via vertical or lateral contact.

10.1.2 Challenges in characterizing accumulation-permeation units (A-P units)

Due to their usually small scale, complex architecture, and superposition relationships, A-P units are difficult to identify and characterize in the following two aspects.

1. Complex internal structures and small-scale A-P units pose problems for their seismic identification

Rich single-well data can reflect the change in various volcanic A-P characteristics, and it is easy to identify the A-P unit in single wells. However, the small scale, limited lateral extension, and low seismic resolution, together with complex volcanic internal structures (architecture), significantly hamper the seismic identification of an A-P unit, making it difficult to identify A-P units using seismic data.

2. Complicated shape and superposition relationships and strong internal heterogeneity make it difficult to characterize A-P units quantitatively

It is generally a challenging task to characterize the highly variable types, complex shapes, and complex superposition relationships and reservoir spaces of A-P units in a volcanic gas reservoir. In addition, due to the poorly defined superposition relationships and short production periods, the production performance data cannot clearly reflect interwell interference or coherence parameters, leading to difficulties in characterizing the connectivity of A-P units. Due to the rapid change or strong heterogeneity of volcanic reservoirs, the determination A-P capacity of A-P units outside the well points is usually fraught with high uncertainty, making it difficult to accurately characterize their accumulation-permeation capacity.

10.1.3 Technical solutions

To meet the challenges of characterizing A-P units, the technical process for identification and characterization is set up in accordance with the principle of proceeding from single-well to profile, then to plane and space, applying integrated methods of static and dynamic analyses (Figure 10.1). The complete set of A-P unit characterization techniques thus developed provides a basis for ascertaining the distribution patterns of A-P units.

1. Identification of A-P units

Lithology, lithofacies, accumulation-permeation capacity, and logging response characteristics are used for establishing identification markers of volcanic A-P layers, barriers, and interbeds, for identifying the interface of A-P layers; and for dividing A-P units in wells. With this process, the identification methods used to determine multiwell profiles, seismic reflection profiles, and inversion profiles are used to establish a profile identification model for A-P units and to reveal their profile distribution characteristics. Furthermore, sensitive attributes are extracted and analyzed to predict the planar distribution of A-P units. Finally, various methods—such as interface tracking based on seismic framework profile, spatial determination of volume attributes, and analysis of inversion data volumes—are used to predict the spatial distribution of A-P units and lay the foundation for the quantitative characterization of A-P units.

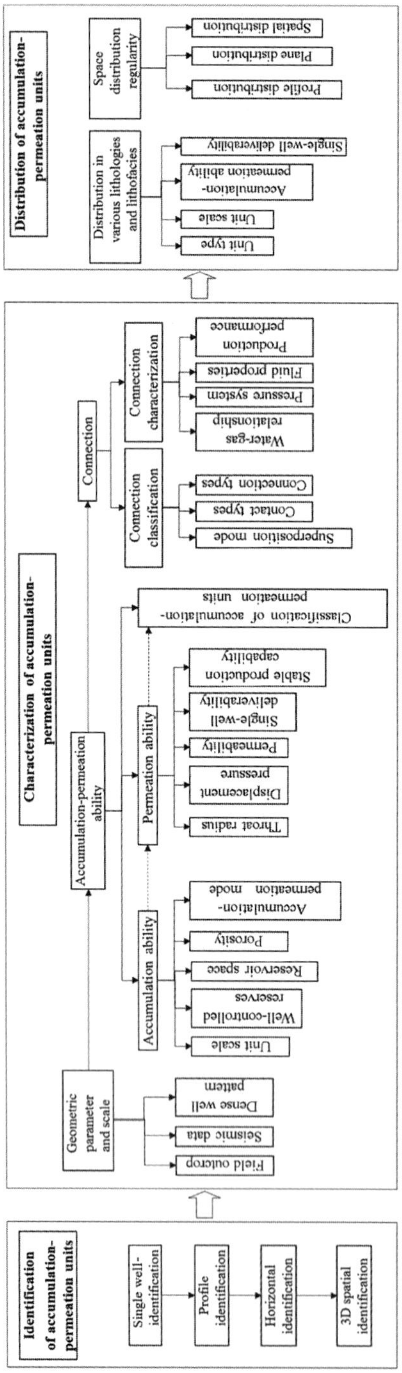

FIGURE 10.1

Technical process for characterizing A-P units in a volcanic gas reservoir.

2. Characterization of A-P units

This process includes the characterization of geometry, scale, connectivity, and the accumulation-permeation capacity of A-P units:

1. Based on outcrop surveys, seismic data interpretation, and dense well grid analysis, the geometry and superposition relationships of the A-P unit are ascertained to quantitatively characterize their geometric parameters.
2. The superposition relationships or stacking patterns of A-P units are classified on the basis of their contact relationships and connection modes. Moreover, faults, fractures, water-gas relationships, fluid properties, pressure systems, and the production performance of A-P units are analyzed comprehensively to characterize their connectivity.
3. Based on the optimization of characterization parameters of accumulation-permeation capacity, the classification criteria for A-P units of volcanic gas reservoirs are established to characterize their accumulation-permeation capacity based on various parameters.

3. Distribution of A-P units

The distribution characteristics of A-P units in various lithofacies are evaluated on the basis of their types, scale, and accumulation-permeation capacity as well as well deliverability. According to the correlation of geological profiles, identification results of seismic profiles, and planar attributes and 3D space, the distribution of A-P units in 3D space is delineated to guide effective reservoir prediction. This provides the basis for optimization of well location, trajectory design of horizontal wells, and preparation of development technology policies.

10.2 IDENTIFICATION AND PREDICTION OF ACCUMULATION-PERMEATION UNITS

To resolve the difficulties in identifying and predicting volcanic A-P units, the distribution characteristics of these units in volcanic gas reservoirs are clarified on the basis of static and dynamic data (including core, logging, seismic, gas tests; and pilot production data), in combination with single-well and profile identification, planar; and 3D spatial prediction. This provides the basis for characterizing A-P units.

10.2.1 Single-well identification

Based on the established identification marks of volcanic A-P units, single-well data are used to identify volcanic A-P layers, barriers, and interbeds and to divide A-P units in wells. This provides the basis for profile identification.

1. Identification markers of A-P units

A-P layers, along with their top and bottom barriers, are the key components of A-P units. Based on the geological characteristics of volcanic A-P units, barriers, and interbeds, the identification markers for the accumulation-permeation capacity of A-P units are established through the analysis of logging response characteristics. This provides the basis for the identifying A-P units in wells.

Table 10.1 Geological Characteristics of Volcanic A-P Layers, Barriers, and Interbeds

Stratum Type	Lithofacies Type	Typical Subfacies	Typical Lithology	Reservoir Space
A-P layer	Effusive facies	Top, upper, and lower	Vesicular lava	Vesicles
	Explosive facies	Hot clastic flow, fallout, and splash	(Welded) tuff and (welded) breccia	Intergranular pores
	Volcanic conduit facies	Volcanic neck and cryptoexplosive breccia	Vesicular lava and autoclastic brecciated lava	Vesicles and fractures
	Subvolcanic facies	Outer-zone and mesozone	Orthophyre and granite porphyry	Interstitial pores and dissolution pores
	Volcanic sedimentary facies	Xenoclast-bearing	Sedimentary volcanic breccia and sedimentary tuff	Intergranular pores
Barriers and interbed	Effusive facies	Middle	Tight volcanic lava	Underdeveloped pores and fractures
	Explosive facies	Hot base surge	Crystalline tuff	
	Volcanic conduit facies	Volcanic neck	Tight volcanic lava	
	Subvolcanic facies	Inner-zone and mesozone	Orthophyre and granite porphyry	
	Volcanic sedimentary facies	Coal-bearing tuff and retransported	Tuffaceous siltstone, tuffaceous mudstone and coal	

1) Geological characteristics of A-P units

A-P units are distributed in all lithologies and lithofacies; however, reservoir spaces in different lithologies and lithofacies are highly variable and of different petrogenetic origins, leading to varying degrees of reservoir space development and distribution characteristics of volcanic A-P layers (Table 10.1).

1. *Effusive facies A-P unit.* Vesicles are the primary reservoir space. A-P layers are developed predominantly in the top, upper, and lower subfacies, with vesicular lava being the main lithology. Barriers and interbeds are developed mainly in the middle subfacies, lithologically represented by tight volcanic lava.
2. *Explosive facies A-P unit.* Intergranular pores are the main reservoir space. A-P layers are mainly developed in the hot clastic flow, fallout, and the splash subfacies, represented by welded tuff and breccia, crystalline tuff, and volcanic breccia. Barriers and interbeds are mainly developed in the hot base surge subfacies, lithologically dominated by crystalline tuff.
3. *Volcanic conduit facies A-P unit.* Vesicles and fractures are the main reservoir space in volcanic rocks of the conduit facies. A-P layers are developed mainly in the volcanic neck and cryptoexplosive breccia subfacies, and the representative lithology is vesicular lava and autoclastic brecciated lava. Barriers and interbeds are developed mainly in the volcanic neck subfacies, lithologically typical of tight volcanic lava.
4. *Subvolcanic facies A-P unit.* Interstitial pores and dissolution pores are the primary reservoir space in volcanic rocks of subvolcanic facies. A-P layers are developed mainly in the inner-zone and mesozone subfacies, and the lithology is typically weathered and dissolution-affected orthophyre and granite porphyry. Barriers and interbeds are developed mainly in the inner-zone subfacies, and the representative lithology is tight orthophyre and granitic porphyry.

5. *Volcanic sedimentary A-P unit.* Intergranular pores are the primary reservoir space in volcanic sedimentary facies. A-P layers are developed mainly in the xenoclast-bearing subfacies, represented by sedimentary volcanic breccia and sedimentary tuff. Barriers and interbeds are developed mainly in the coal-bearing tuff and transported subfacies, dominated lithologically by tuffaceous siltstone, tuffaceous mudstone, and coal.

2) Identification markers of accumulation capacity

Accumulation capacity is used to assess the size of the effective reservoir space of an A-P unit and is related to the scale and petrophysical properties of the unit. In general, it is evaluated on the basis of effective porosity that reflects reservoir properties in wells. As a result, accumulation capacity is identified with the three-porosity curves of nuclear magnetic logging and conventional logging (Table 10.2).

Identification markers in nuclear magnetic logging. In terms of its occurrences, fluid can be divided into mobile and irreducible types by the T_2 spectrum of nuclear magnetic logging, defined by the T_2 cutoff threshold (red line in Figure 10.2). The total area of mobile fluid spectrum reflects the accumulation capacity of the A-P unit. The larger the total area, the higher the effective porosity and effective accumulation capacity of the volcanic rock (see Figure 10.2). The T_2 spectrum characteristics of nuclear magnetic logging vary with A-P layers, barriers, and interbeds (see Table 10.2). Volcanic rocks have a relatively small total area of irreducible fluid spectrum, volcanic A-P layers have a relatively large total area of mobile fluid spectrum, and volcanic barriers and interbeds have a relatively small total area of mobile fluid spectrum. Sedimentary volcanic rocks have a relatively large total area of irreducible fluid, sedimentary volcanic A-P layers have a moderate total area of mobile fluid, and sedimentary volcanic barriers and interbeds have a relatively small total area of mobile fluid spectrum.

Identification markers in conventional logging. The three-porosity curves can be used to reflect the accumulation capacity of volcanic rocks. Lithological separation results in various response characteristics of the three-porosity curves that reflect volcanic rocks and sedimentary volcanic rocks (see Table 10.2):

1. *A-P layers.* The amplitude values of density, interval transit time, and neutron porosity curves are moderate, and their curves commonly have a smooth to microdentoid box shape and a moderate to good accumulation capacity (Figure 10.3).
2. *Volcanic barriers and interbeds.* These components are characterized by high-amplitude density logging curves, low-amplitude interval transit time and neutron logging curves, and a microserrated to dentoid box shape, indicating poor accumulation capacity (see Figure 10.3).
3. *Sedimentary volcanic A-P layers.* These layers typically have moderate- to low-amplitude density logging curves, moderate- to high-amplitude interval transit time and neutron logging curves, and dentoid bell and funnel shapes, indicating moderate accumulation capacity (Figure 10.4).
4. *Sedimentary volcanic barriers and interbeds.* These components have the characteristics of low-amplitude density logging curves, moderate- to high-amplitude interval transit time and neutron logging curves, and dentoid to serrated digitate bell and funnel shapes, denoting poor accumulation capacity (see Figure 10.4).

3) Identification markers for permeation capacity

Permeation capacity is used to assess the capacity of fluid flow of an A-P unit, which is related to the degree of pore development, throat radius, and permeability. Generally, permeation capacity is evaluated on the basis of permeability parameters. Single-well permeation capacity is identified mainly

Table 10.2 Identification Markers of the Accumulation Capacity of Volcanic A-P Layers, Barriers, and Interbeds

		T_2 Spectrogram of Nuclear Magnetic Logging		Conventional Logging				
Stratum Type	**Rock Type**	*Total Area of Irreducible Fluid Spectrum*	*Total Area of Mobile Fluid Spectrum*	*Density Logging*	*Acoustic Survey*	*Neutron Logging*	*Curve Shape*	*Smooth Level*
A-P layer	Volcanic rock	Small	Large	Moderate	Moderate	Moderate	Box shape	Smooth to microserrated
	Sedimentary volcanic rock	Large	Moderate	Moderate to low	Moderate to high	Moderate to high	Bell and funnel shape	Dentoid
Barrier and interbed	Volcanic rock	Small	Small	High	Low	Low	Box shape	Microserrated to dentoid
	Sedimentary volcanic rock	Large	Small	Low	High	High	Finger, bell and funnel shape	Serrated

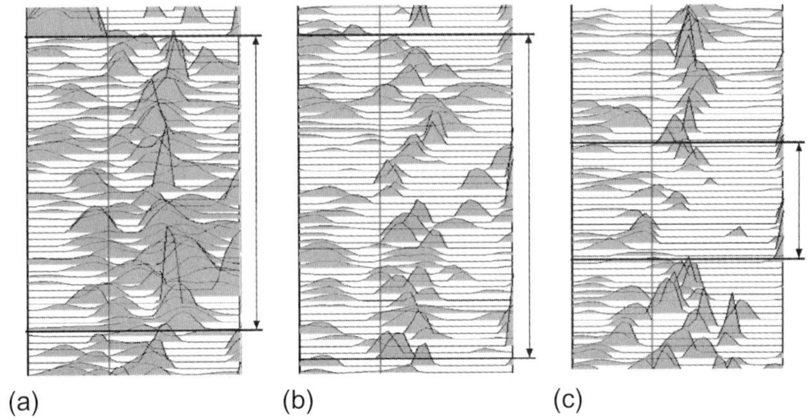

FIGURE 10.2

Response characteristics of nuclear magnetic logging for the accumulation capacity of volcanic rocks (Well XX23). (a) High permeability (4040-4065, φe=12%, breccia lava). (b) Moderate permeability (4360-4386, φe=7.5%, breccia lava). (c) Low permeability (4325-4335, φe=2%, crystalline tuff).

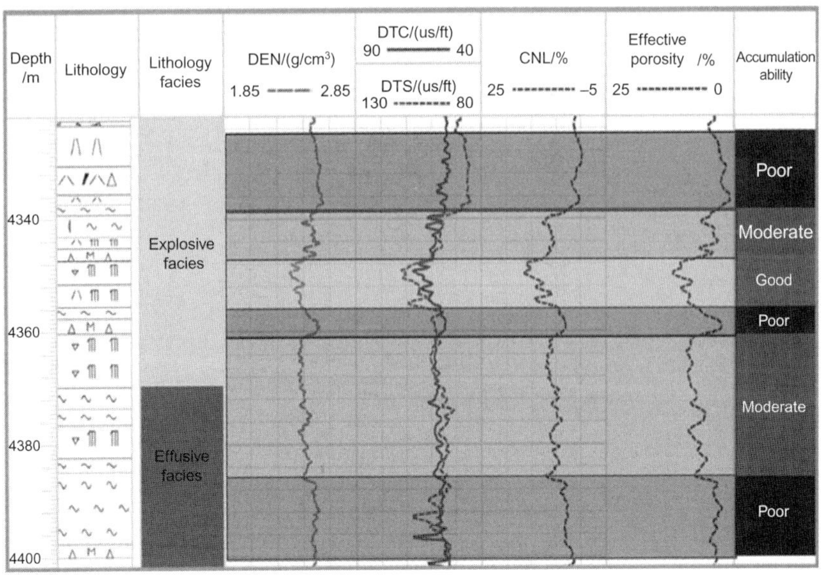

FIGURE 10.3

Response characteristics of conventional logging for the accumulation capacity of volcanic rocks (Well XX23).

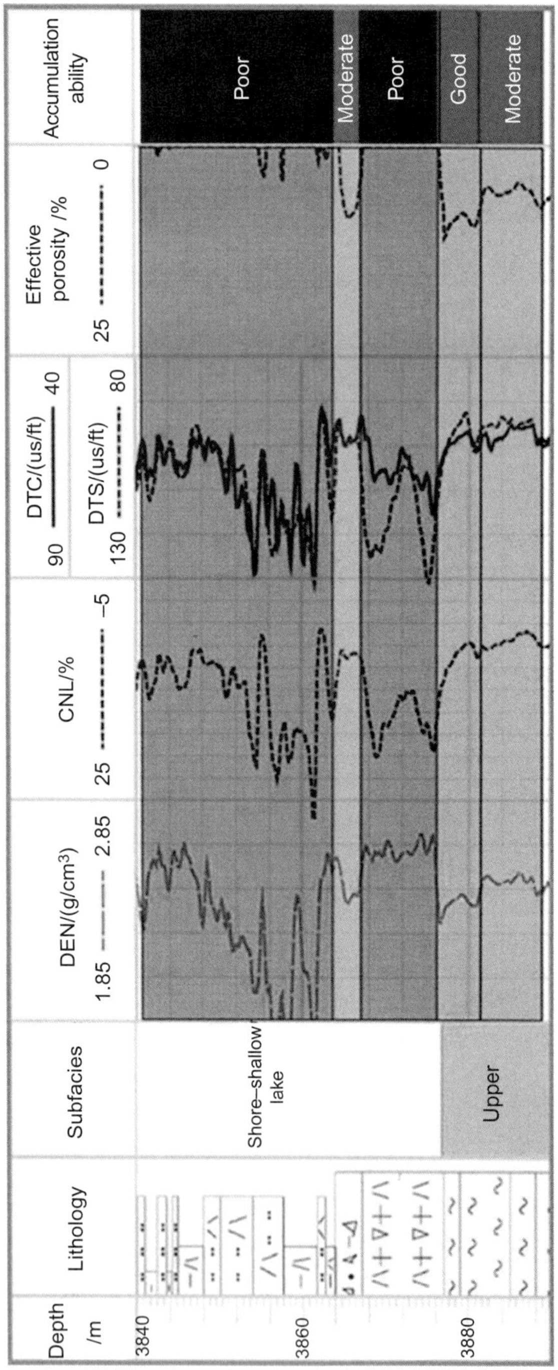

FIGURE 10.4

Response characteristics of conventional logging for the accumulation capacity of sedimentary volcanic rocks (Well XX901).

Table 10.3 Identification Markers for the Permeation Capacity of Volcanic A-P Layers, Barriers, and Interbeds

		T_2 Spectrum of Nuclear Magnetic Logging		Conventional Logging			
Stratum Type	**Rock Type**	***T_2 Peak Value of Mobile Fluid***	***Peak Area of Mobile Fluid***	***Resistivity Logging***	***Separation between Deep Resistivity and Shallow Resistivity***	***Density Logging***	***Acoustic Survey***
A-P layer	Volcanic rock	Large	Large	Moderate	Relatively high positive separation	Moderate	Moderate
	Sedimentary volcanic rock	Small	Moderate	Moderate to low	Relatively low positive separation	Moderate to low	Moderate to high
Barriers and interbed	Volcanic rock	Large	Small	High	Low	High	Low
	Sedimentary volcanic rock	Small	Small	low	Low	Low	High

through nuclear magnetic logging, three-porosity curves of conventional logging, and resistivity logging curves (Table 10.3).

Identification markers of nuclear magnetic logging. The permeation capacity of volcanic rocks is related to pore throat size, porosity, and fractures. The porous zones with various pore throats are distinguishable on the T_2 spectrum in nuclear magnetic logging curves. A higher T_2 peak value of the mobile fluid and larger area of the spectral peak indicate large-radius pores and therefore higher permeation capacity (Figure 10.5). As a result, the permeation capacity of volcanic rocks can be assessed by the T_2 time distribution and the area of mobile fluid peak in the T_2 spectrogram (see Table 10.3).

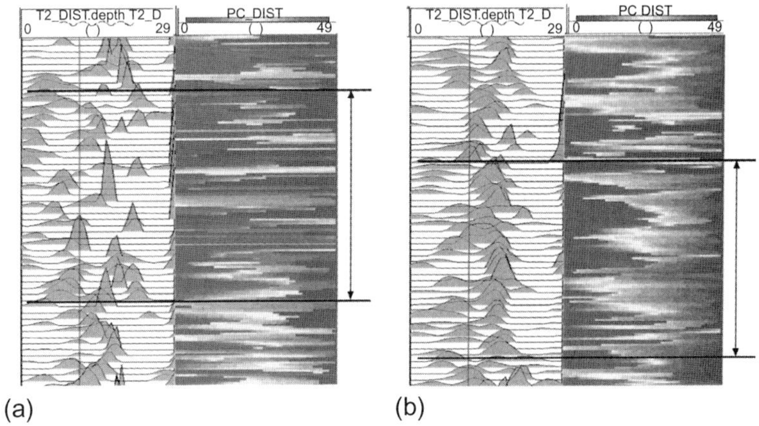

FIGURE 10.5

Response characteristics of nuclear magnetic logging for the permeation capacity of volcanic rocks (Well XX23). (a) High permeability (3942-3962, $\varphi e = 7\%$, $K = 1.4$ mD, rhyolite). (b) Moderate permeability (4233-4252, $\varphi e = 7.2\%$, $K = 0.2$ mD, crystalline tuff).

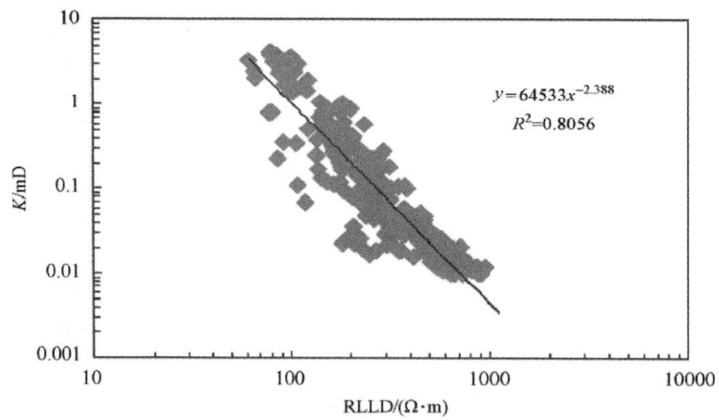

FIGURE 10.6

Crossplot of resistivity versus permeability (Well XX213).

Volcanic formations have an extensive mobile fluid peak, and overall its T2 average value is higher than that of sedimentary volcanic rocks. The trap area of the mobile fluid peak is large for volcanic A-P layers and small for barriers and interbeds (see Table 10.3).

The T2 value of the mobile fluid peak of sedimentary volcanic rocks is generally smaller than that of volcanic rocks. The trap area of the mobile fluid peak is moderate for sedimentary volcanic A-P layers and small for barriers and interbeds (see Table 10.3).

Response characteristics of conventional logging. The electrical conductivity of rocks has a good correlation with its permeability (Figure 10.6). For conventional logging, beside the three-porosity logging curves, the resistivity curve is often used to reflect the permeation capacity of volcanic rocks. Similarly, the conventional logging response characteristics reflecting the permeation capacity of volcanic rocks and sedimentary volcanic rocks are highly variable due to the variety of lithologies (see Table 10.3):

1. *Volcanic A-P layer.* Moderate-amplitude density, acoustic travel time, and resistivity logging curves and relatively large positive separation between deep and shallow laterolog resistivity curves (Figure 10.7)
2. *Volcanic barrier and interbed:* High-amplitude density logging curve, low-amplitude acoustic logging curve, and high-amplitude resistivity logging curve, as well as small separation between deep and shallow laterolog resistivity curves (see Figure 10.7)
3. *Volcanic A-P layer.* Moderate to low amplitude density logging curve, moderate to high amplitude acoustic travel time logging curve, and low-amplitude resistivity logging curve, as well as small separation between deep and shallow laterolog resistivity curves
4. *Sedimentary volcanic barrier and interbed.* Low-amplitude density logging curve, high-amplitude acoustic travel time logging curve, and low-amplitude resistivity logging curve, as well as small separation in the deep-shallow laterolog resistivity curve

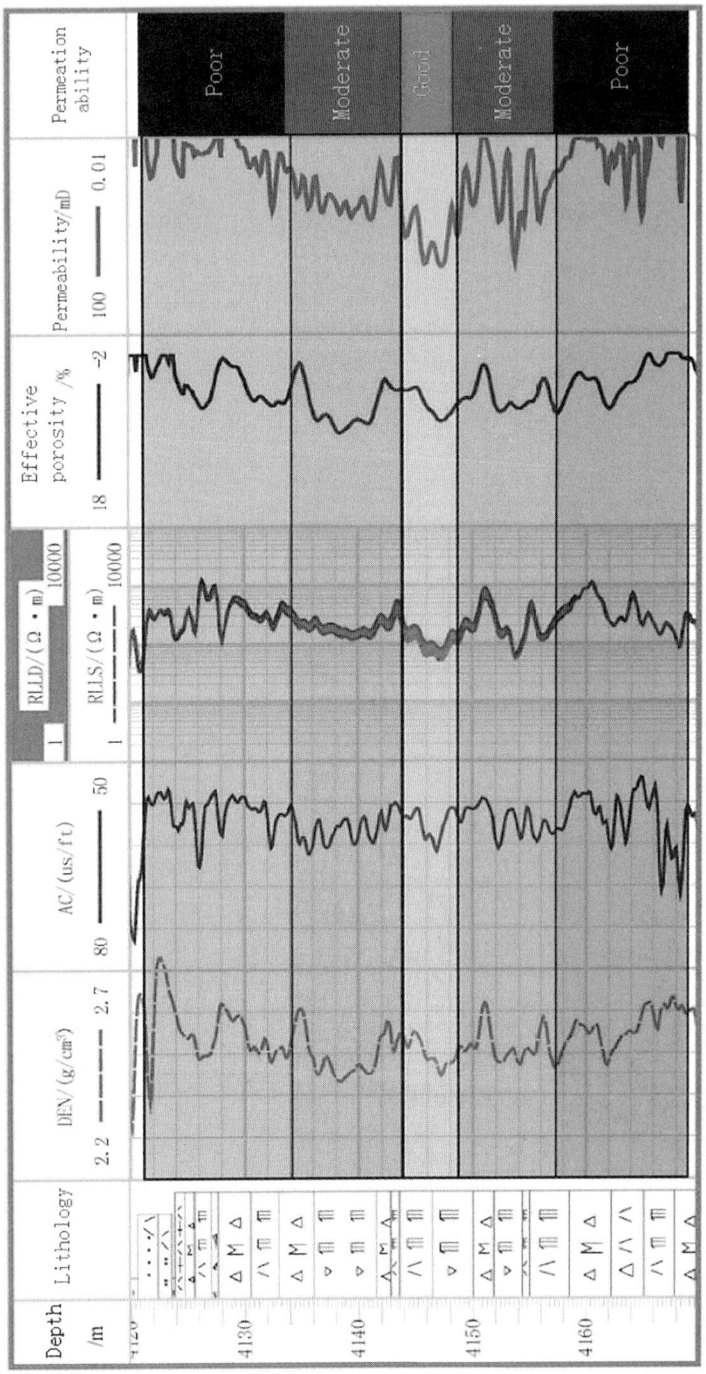

FIGURE 10.7

Response characteristics of conventional logging for the permeation capacity of volcanic rocks (Well 213).

2. Identification and division of A-P units in wells

A-P layers, barriers, and interbeds are identified using the identification markers of A-P units, static and dynamic data from single wells, and logging response analysis and reservoir parameter interpretation. Thereafter, the A-P units are classified and the interfaces or boundaries of A-P units are identified, which provide a foundation for profile identification.

1) Assessing A-P capacity and identifying A-P layers

The A-P capacity of a target reservoir can be qualitatively assessed on the basis of the identification markers of volcanic A-P layers, in combination with the identification of A-P units by lithology and lithofacies characteristics. This further facilitates the quantitative evaluation of the A-P capacity of a target reservoir through reservoir parameter interpretation and gas tests. And the identification results are verified by the cutoff criteria of effective reservoirs.

The left of Figure 10.8 is an example of the division and identification of A-P layers in Well XX9. The strata are classified, in descending order, into seven intervals.

The first interval (3520 to 3578 m) consists of sedimentary volcanic rock and volcanic rock characterized by low resistivity, low density, high interval transit time, and high neutron value. The logging curves are serrated-digitate in shape, with large amplitude of irreducible fluid peak and underdeveloped mobile fluid peak in a nuclear magnetic logging curve showing poor A-P capacity. As a result, the interval is assessed systematically as a non-accumulation-permeation (A-P) sedimentary volcanic layer. Its effective porosity is less than 1% and permeability is less than 0.1 mD, thus interpreted as a dry layer by logging characteristics.

The lithology of the second, fourth, and sixth intervals (green color) comprises rhyolite with moderate-amplitude resistivity, density, interval transit time, and neutron logging curves in microserrated box and bell shapes. The same development of irreducible fluid peak and mobile fluid peak in a nuclear magnetic logging curve suggests some A-P capacity for the intervals. As a result, the intervals are assessed systematically as effusive facies A-P layers. In logging interpretation, their porosity is 6%, 5.6%, and 5.8%, and permeability is 1.2 mD, 0.14 mD, and 0.33 mD, respectively. These values are higher than the cutoff of effective reservoirs in the XX gas field, strongly indicating that these intervals as gas layers.

The lithology of the third, fifth, and seventh intervals is rhyolite for the third and fifth intervals and dacite for the seventh interval; this is characterized by high resistivity, high density, low interval transit time, and low neutron value. The logging curves have a smooth box shape, locally serrated-digitate. The inferior development of irreducible fluid peak and underdeveloped mobile fluid peak in the nuclear magnetic log indicate poor A-P capacity for the intervals. As a result, systematic evaluation ranks them as effusive facies non-A-P layers. In logging interpretation, their porosity is 0.9%, 2.7%, and 2.3%, and permeability is 0.05 mD, 0.02 mD, and 0.01 mD, respectively. All of these values are below the cutoff for effective reservoirs, thus rendering them as dry layers.

The second and sixth intervals are perforated for gas tests. Natural production is low (0.0134×10^4 m^3/d), but the production reaches 20.9×10^4 $m^{3/}$d after hydraulic fracturing. As a result, these intervals are commercial gas layers, which verified their initial interpretation as A-P layers.

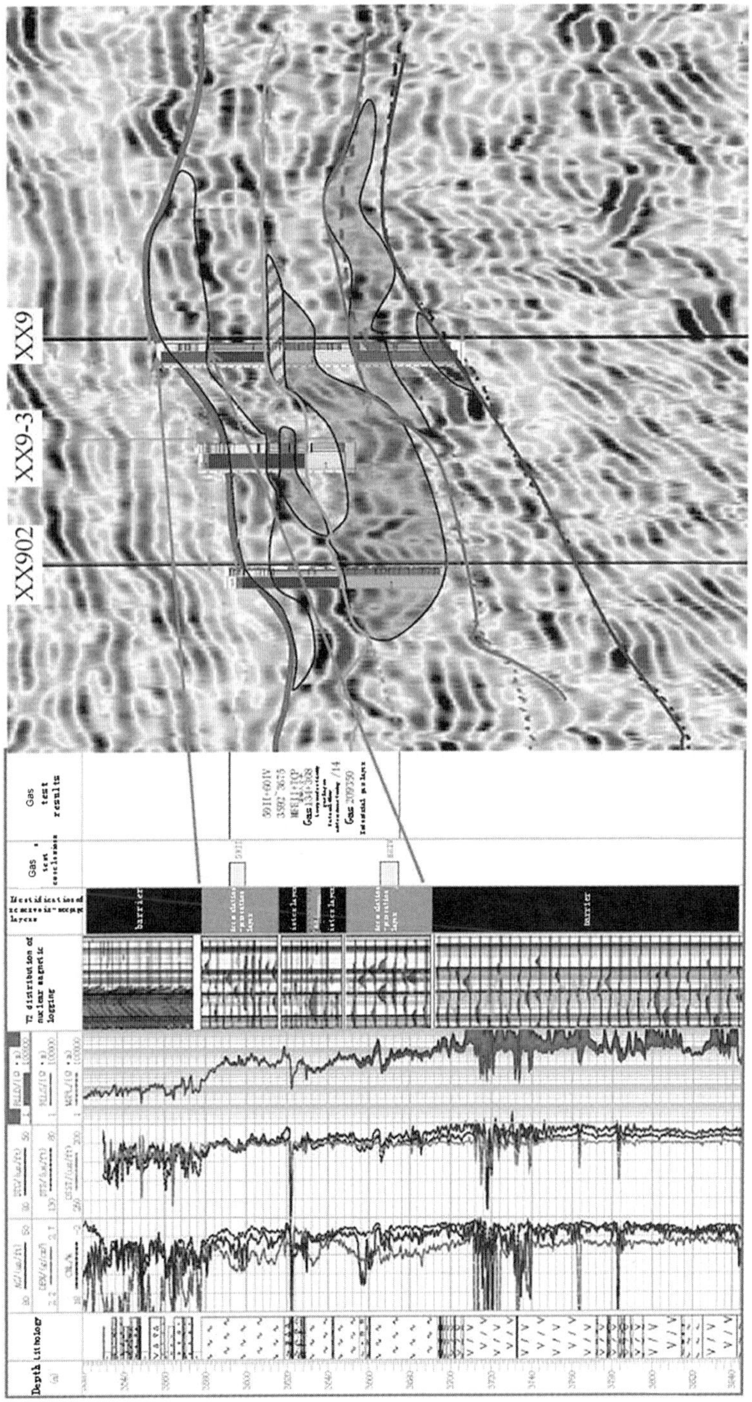

FIGURE 10.8
Identification of an A-P unit in a single well (Well XX9).

2) Tracing extension and distinguishing barriers and interbeds

Both barriers and interbeds are integral parts of A-P layers. Barriers usually have a large thickness and lateral extension, located between A-P units, and can obstruct fluid seepage. Interbeds have a small thickness and limited lateral extension, located within A-P layers, and can decrease fluid seeping. Barriers and interbeds are distinguishable based on their thickness features and the identification of A-P layers, through well-seismic calibration to track the horizontal extension of non-A-P layers in seismic profiles. This provides the basis for the division of A-P units.

In the case study of Well XX9 (see Figure 10.8), the lithology of non-A-P layers between the second, fourth, and sixth intervals is rhyolite. Due to the small thickness (about 12 m) and limited lateral extension, they cannot be identified and tracked in seismic profiles and are thus determined as interbeds with poor petrophysical properties. The first interval consists of sedimentary volcanic rock and volcanic-sedimentary rock with large thickness, which has unconformable contact with its underlying volcanic rock and is widespread in the XX gas field. As a result, the interval is determined as the cap rock of lower volcanic A-P units. The lithology of the seventh interval is dacite with a thickness of some 150 m, which appears with moderate-amplitude, low-frequency, and good-continuity reflection characteristics and a laminar-like reflection shape (on the right in Figure 10.8) and can be tracked in Well Block XX9. As a result, the interval is interpreted as a petrophysical property barrier.

3) Analyzing the relationship among cap rocks, reservoirs, and basal bed assemblage to divide A-P units

The classifications of A-P units delineation of the vertical distribution patterns of A-P units are conducted with reference to the classification scheme for carbonate reservoir units [14,15], using various static and dynamic data and taking into consideration the geological features of volcanic gas reservoirs.

Applying single-well static data to divide A-P units. A volcanic A-P unit comprises volcanic rocks of the same petrogenesis and has the assemblage characteristics of a cap rock, a basal bed, and reservoir in the middle. Therefore, the petrogenetic separation, the relative position of A-P layers, features of lower and upper barriers, and fault barriers are the key factors for classifying A-P units. In general, two adjacent A-P layers will be classified into discrete A-P units when they have the following characteristics: there is a great separation in rock composition, they are located in two disconnected structural units, they are separated by a thick barrier covering an entire volcanic edifice, or they have been isolated by a sealing fault barrier.

Two A-P layers in Well DD1001 are located within the same volcanic edifice (Figure 10.3b) and isolated by a set of sedimentary volcanic breccias with different rock compositions. The lithology of the upper layer is dacitic volcanic breccia, which is clearly from that of the lower layer comprised of andesitic volcanic breccia, thus separating the two A-P layers into different A-P units.

Using dynamic data to verify division results and determine vertical distribution modes of A-P units in wells. Fluid flow characteristics vary with A-P units. Based on the separation in produced fluid properties, the consistency of the water-gas relationship, and the relationship between the pressure system and production performance, the division results can be verified and the vertical distribution patterns of A-P units in single wells can be modified and improved. Comprehensive analyses of Well DD1001 show that the volcanic A-P unit is the similar to a carbonate reservoir unit (Figure 10.9a), characterized by the rock assemblage of an upper cap rock, middle reservoir, and lower basal bed.

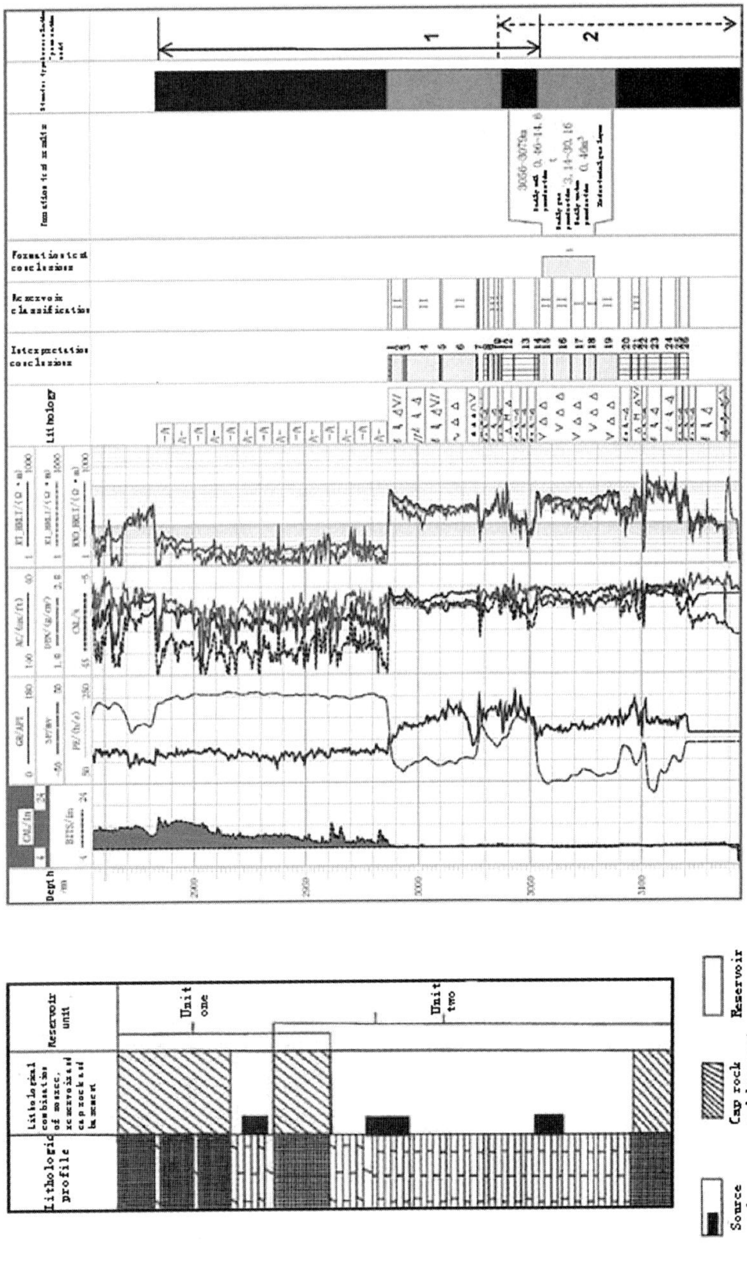

FIGURE 10.9

Comparison of a reservoir unit with an A-P unit. (a) Carbonate reservoir unit mode. (b) Volcanic A-P unit mode (Well DD1001).

4) Determining the interfaces of A-P units by comprehensive analyses

Based on the identification of A-P layers, barriers, and interbeds as well as the division results of A-P units in a single well, the type of interface between volcanic A-P units is determined through petrogenetic analysis, according to the established identification markers in single wells. This provides well point constraints for profile tracing.

The interfaces of volcanic A-P units are classified into lithological interface, petrophysical property interface, and sealing fault barrier in terms of their petrogenesis. Interface identification in single wells generally involves the former two types.

Identification of lithological interfaces. A lithological interface results from changes in petrophysical properties due to lithological variation, and can be subdivided into the following three varieties.

Abrupt interface between sedimentary and volcanic rocks This type of interface is developed mainly in the basal rock units formed at the beginning of a volcanic eruption as well as the top rock units formed at the end of a volcanic eruption. In logging curves, the interface appears as a sudden change from sedimentary rocks with high interval transit time and neutron value and low density and resistivity to volcanic rocks with low interval transit time and neutron value and high density and resistivity (Figure 10.10).

Lithofacies interfaces This results from changes in petrophysical property due to lithofacies or environmental variation. In logging curves, the interface appears as a trending change. Figure 10.11 is a typical example of Well DD14, where the lower lithology is a brecciated tuff barrier and the upper lithology is a tuffaceous volcanic breccia A-P layer. The logging curves from the bottom up are characterized by intensified radioactivity, reduced density, increased interval transit time, and increased neutron value and resistivity.

Lithological interfaces This results from poorer petrophysical properties due to variations in the magma source and petrogenesis. For example, Well DD1824 penetrated mainly orthophyre and occasionally andesitic welded breccia and sedimentary tuff; the lithological interface is formed due to relatively poor petrophysical properties. Figure 10.12 shows that the lower part is an andesitic welded breccia barrier and the upper part is an orthophyre A-P layer. The logging curves show an upward increase in radioactivity, reduced density, increased interval transit time and neutron value, and decreased resistivity.

Identification of petrophysical property interfaces. This refers to an A-P interface between the same volcanic lithologies due to relatively poor petrophysical properties of the middle subfacies of effusive facies, which clearly defines the interface between the upper and middle subfacies and between the middle and lower subfacies of the effusive facies. Also, due to the relatively poor petrophysical properties of the hot base surge subfacies, this interface is easily delineated by the borders between the hot base surge subfacies of explosive facies and other subfacies. Similar to the poor petrophysical properties of the inner-zone subfacies of subvolcanic rocks, this interface is also easily defined between the mesozone and inner-zone subfacies of subvolcanic rocks.

Relatively speaking, volcanic A-P layers with good petrophysical properties have the characteristics of low density, high neutron value, high interval transit time, and moderate-low resistivity, and non- A-P layers with poor physical properties have the characteristics of high density, low neutron value and interval transit time, and moderate resistivity in logging curves. Based on the lithology identification, logging response characteristics are analyzed to identify the petrophysical property interface of A-P units. In Figure 10.13, the map with logging curves of Well DD1804 demonstrates that

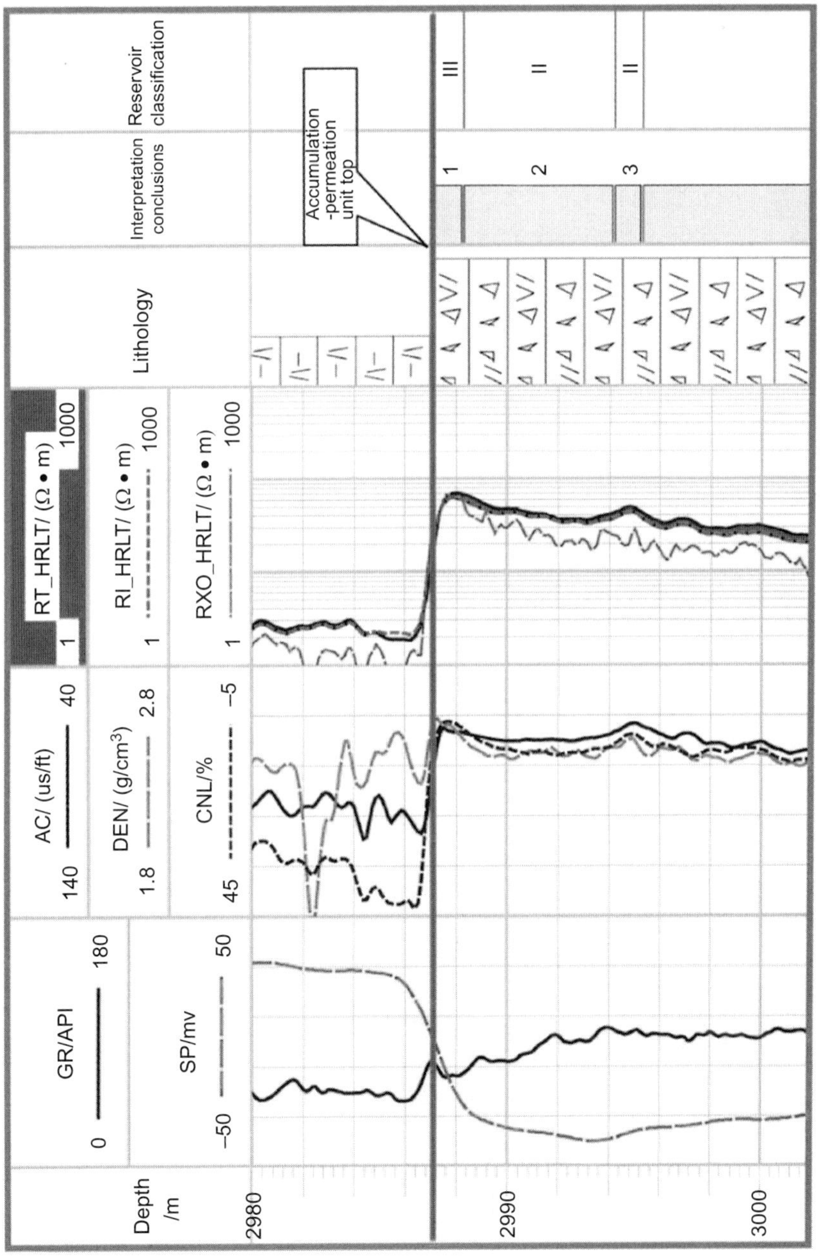

FIGURE 10.10
Abrupt interface from sedimentary rock to volcanic rock (Well DD1001).

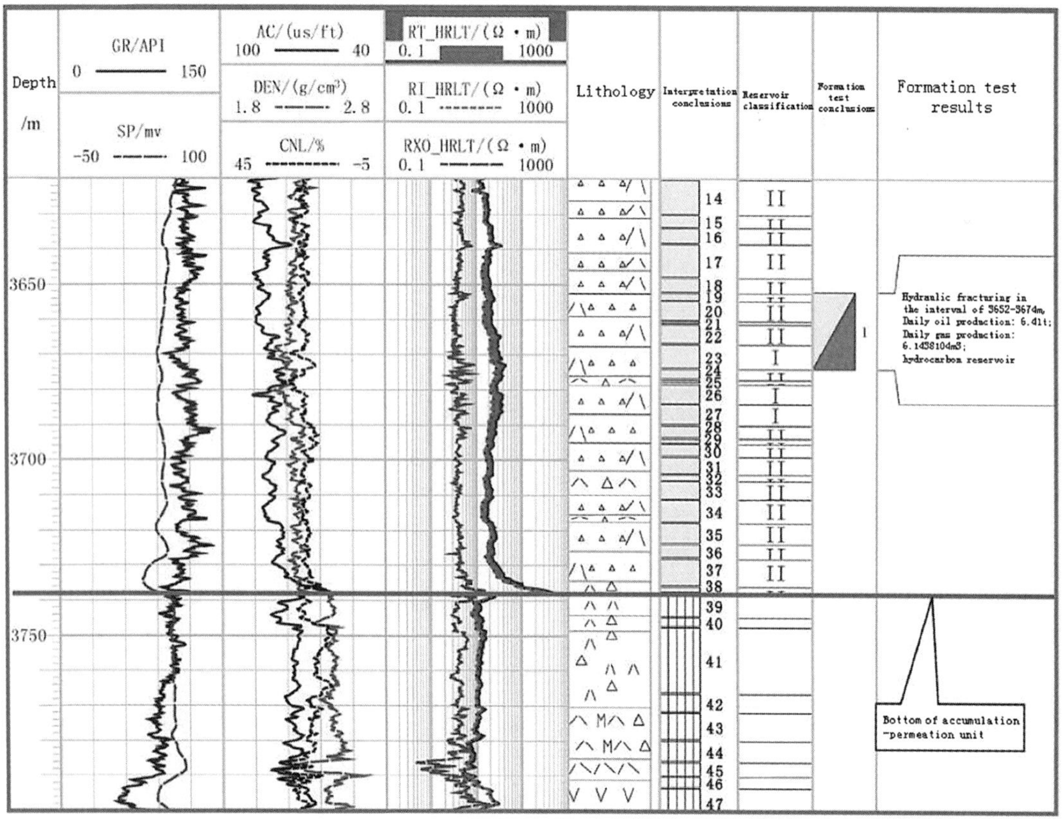

FIGURE 10.11

Lithofacies interface (Well DD14).

orthophyre has good petrophysical properties in the upper part and poor petrophysical properties in the lower part, with a petrophysical property interface located near depth 3645 m.

10.2.2 Profile identification

Multiwell profile analysis, seismic reflection profile identification, and inversion profile analysis are conducted based on single-well identifications to establish profile identification modes of volcanic A-P units. This provides the basis for 3D spatial prediction.

1. *Identification of multiwell profiles*

Identifying multiwell profiles is an effective approach for establishing the interwell distribution patterns of A-P units. In terms of the peculiarities of volcanic gas reservoirs, the principles for the interwell combination of volcanic A-P units are determined with reference to the flow–reservoir unit correlation methods [9–17] used in clastic and carbonate rocks. Constrained by the principles of interwell combination and volcanic internal architecture, various static data are used to establish multiwell profiles

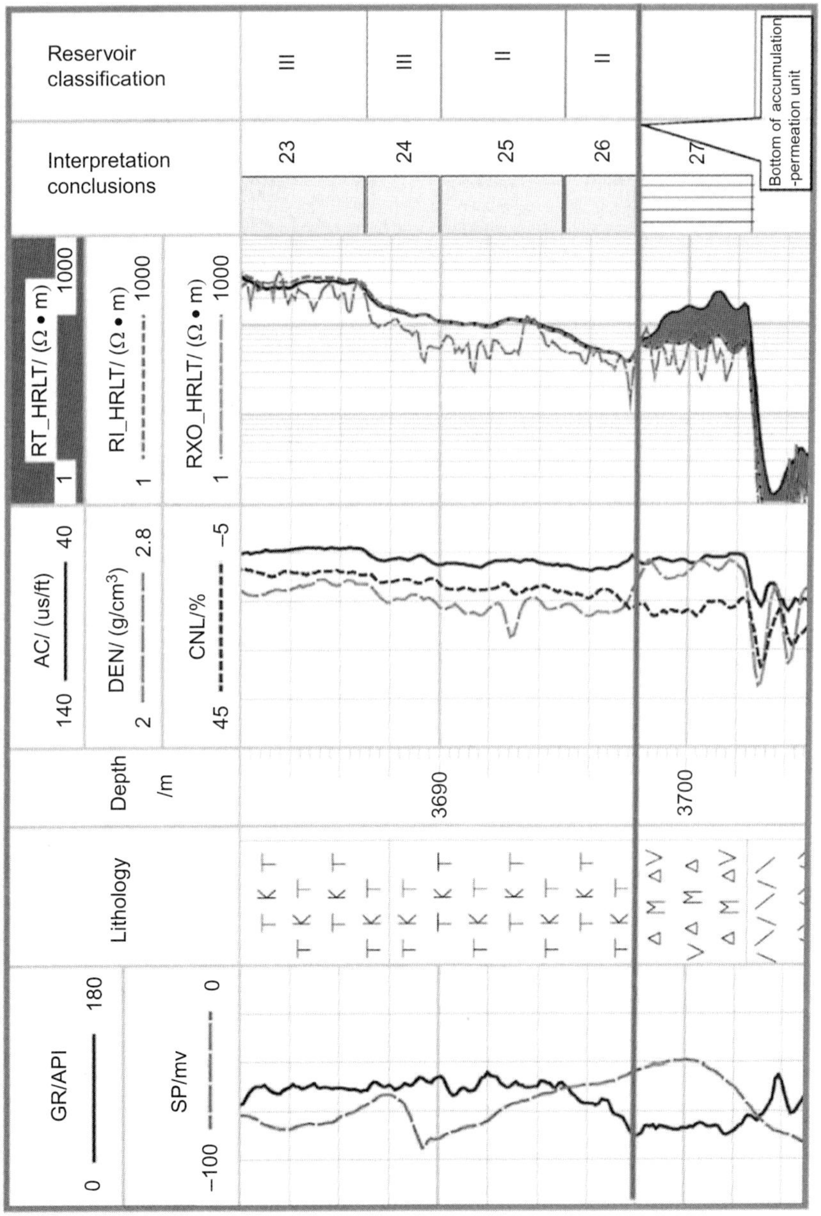

FIGURE 10.12
Interface between subvolcanic rocks and eruptive rocks (Well DD1824).

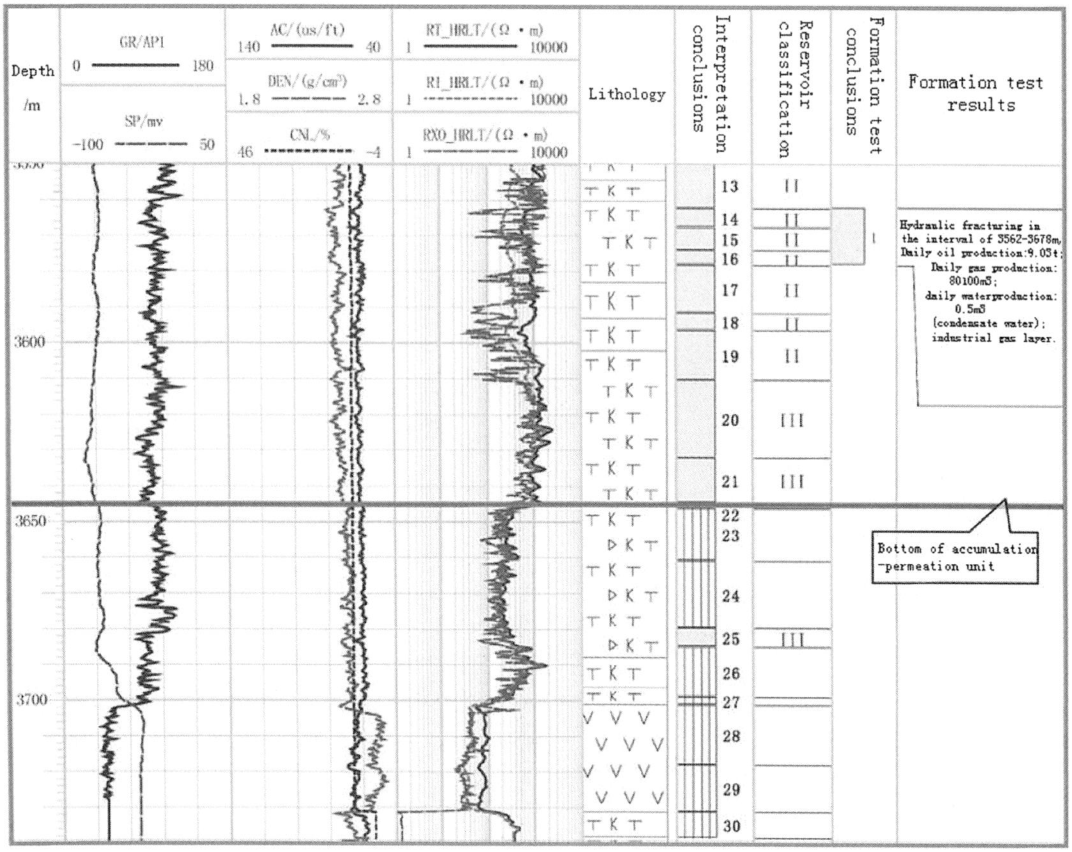

FIGURE 10.13

Physical property interface (Well DD1804).

and identify A-P units, and dynamic data are used to verify the results. The establishment of profile identification models for A-P units provides the basis for seismic profile identification.

1) Interwell combination principles

Volcanic A-P units have complex distribution patterns and multiple interwell combination modes, controlled by various mechanisms, such as modes of eruption, structures, and diagenetic processes. The principles and approaches for interwell combination and the correlation of A-P units are determined by taking into consideration the rock architecture, lithology and lithofacies, fault distribution, reservoir space type, and petrophysical characteristics of volcanic rocks:

1. Establishing the framework model of rock architectural units as well as framework constraints for the identification of A-P units based on the architectural dissection of volcanic rocks
2. Assessing fault sealing properties to provide a fault control basis for interwell combination of A-P units

3. Establishing a standard profile, dividing single-well profile, and dividing single-well A-P units by plotting the classified single-well A-P units on standard profile in the same scale
4. Connecting boundary lines of volcanic edifices, eruption cycle and period, lithology and lithofacies, and A-P units, following the approach of level-by-level control from large to small, with reference to the combination relationships between faults and associated A-P units
5. Analyzing the connectivity of A-P layers, and verifying and amending the rationality of the interwell combination mode of A-P units based on produced fluid properties, the pressure system, and the production performance of the pay zone

2) Multiwell profile identification

Based on the principles of interwell combination, the integrated multiwell profile identification methods using geological, logging, and production performance data are developed in combination with various static and dynamic data, and the interwell distribution mode of A-P units is also determined. The process is illustrated using Well DD10 as an example.

Dissecting volcanic rock architecture and establishing a constraint framework for the interwell combination of A-P units. The volcanic edifice in Well Block DD10 is cone shaped as a result of three stages of eruption from a single volcanic crater, and volcanic strata of the second and third stages are absent from the topographic highs (Well DD10) due to denudation.

The constraint framework of the interwell combination of A-P units is established (Figure 10.14) according to the results of architectural dissection and sequence division: the black lines on the top and at the bottom are the upper and lower boundaries of volcanic eruption cycles, and red lines in the middle denote the boundaries between various eruption stages of the cycles. There is no fault developed between wells.

Establishing a standard profile and dividing A-P units in single wells. Based on drilling results, the DD104–DD1001–DD10–DD102 multiwell profile from southwest to northeast is selected as the standard profile (see Figure 10.14).

Based on the interpretation results of single-well A-P layers, barriers, and interbeds, the third-stage cycle is classified into four A-P units, each of which comprises an assemblage upper cap rock and lower basal bed, with reservoir in the middle, including two gas-bearing units encountered by Well DD1001 and two water-bearing units encountered by Wells DD102 and DD104, respectively. The second-stage cycle is divided into four water-bearing A-P units (two units encountered by Wells DD102 and DD104, respectively). The first-stage cycle is divided into seven A-P units, including two gas-bearing units encountered by Well DD10 and five water-bearing units encountered by Well DD104.

Interwell combination. Following the division of single-well A-P units for well points, the interwell A-P units are combined and correlated within the framework of volcanic architectural units, constrained by seismic reflection characteristics and faults for interwell blocks, and created using the approach of level-by-level control from large to small, so as to establish the profile distribution model of A-P units in volcanic the gas reservoir of Well Block DD10 (see Figure 10.14).

3) Verification by dynamic data

Well DD10 is about 1000 m away from Well DD1001, and their major gas layers have obviously different rock compositions and lithology. In Well DD10, the lithology of two gas-tested intervals is rhyolitic welded tuff and dacitic welded breccia, respectively, and in Well DD1001, the lithology of gas-tested intervals is andesitic volcanic breccia.

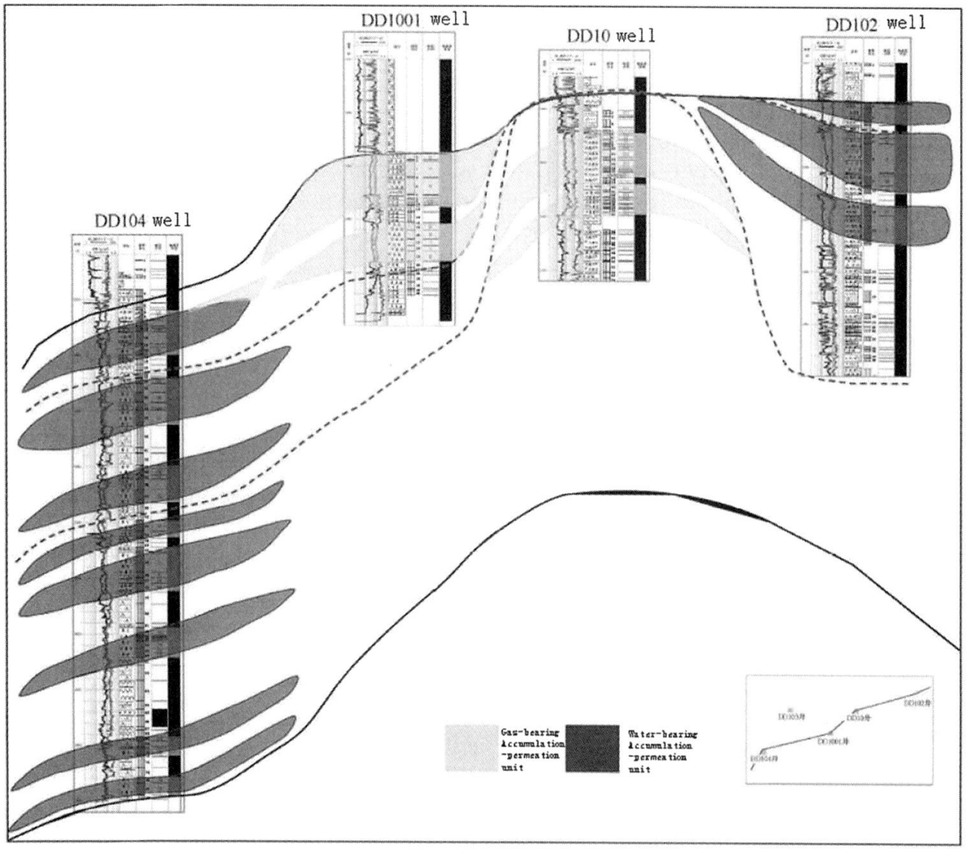

FIGURE 10.14

Distribution of multiwell profiles of A-P units (Well Block DD10).

Pressure test results show that the three gas-tested intervals belong to different pressure systems (Figure 10.15). During the early production stage, gas production accounts for 76.2% in Well DD1001, but 23.8% in Well DD10, of total gas production in this area. Compared to the relatively steady production of Well DD1001, Well DD10 has unsteady output after beginning production. The wellhead tubing pressure of Well DD10 declined from 18.5 MPa to 12.2 MPa, and daily gas production decreased from 6×10^4 m^3/d to 2.2×10^4 m^3/d within two months. Considering their quite different production performances, the three intervals are primarily interpreted as disconnected.

Furthermore, an interference well test is conducted between Wells DD10 and DD1001 to determine the connectivity of A-P units. Based on the pressure–rate history curves (Figure 10.16), no active signal of Well DD1001 is received from observation Well DD10 during the test, which indicates that the producing intervals in the wells connected poorly or disconnected. This verifies that there are multiple discrete A-P units in Well Block DD10.

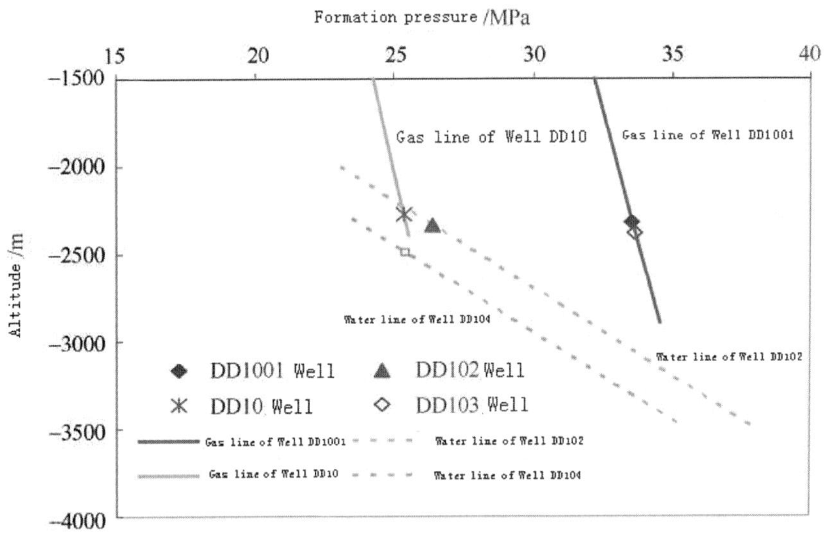

FIGURE 10.15

Pressure system of Well Block DD10.

2. Identification of seismic reflection profiles

Based on the identification of multiwell profiles, the profile distribution features of A-P units in volcanic rocks are clarified by analyzing the combined well-seismic calibration and response characteristics of near-wellbore seismic traces, establishing identification models for various types of A-P units, and conducting profile identification and interface tracing. This provides a basis for spatial prediction.

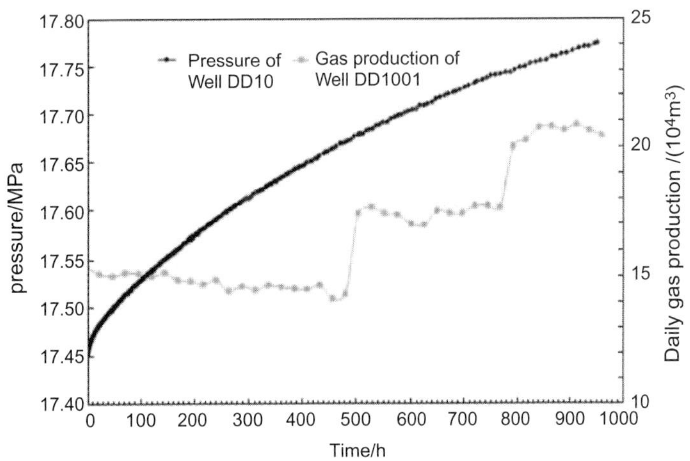

FIGURE 10.16

Interference well test of Well Block DD10.

1) Identification model

The A-P units of volcanic gas reservoirs are classified into four types in terms of the lithofacies types and characteristics. Each type of A-P unit has its own seismic reflection characteristics and configurations.

Volcanic conduit facies A-P units. The volcanic conduit facies occurs in the lower part of a volcanic edifice, in which interclast pores and internal primary pores, dissolution pores, blast fractures, contraction fractures, and columnar joint fractures are primary reservoir spaces, with good petrophysical properties. In the seismic profile (Figure 10.17), this type of A-P unit appears as umbrella- or cloud-shaped reflections with weak amplitude, high frequency, and poor continuity. Better-developed A-P units are correlated with weaker amplitude, higher frequency, and poorer continuity of the seismic reflection wave.

Explosive facies A-P units. The explosive facies mostly forms at the early stage of a volcanic eruption, in which intergranular pores, intragranular primary pores, dissolution pores, interbedded fractures, intergravel fractures, and structural fractures are primary reservoir spaces, with moderate petrophysical properties. In the seismic profile (Figure 10.18), this type of A-P unit mostly appears as a mound and

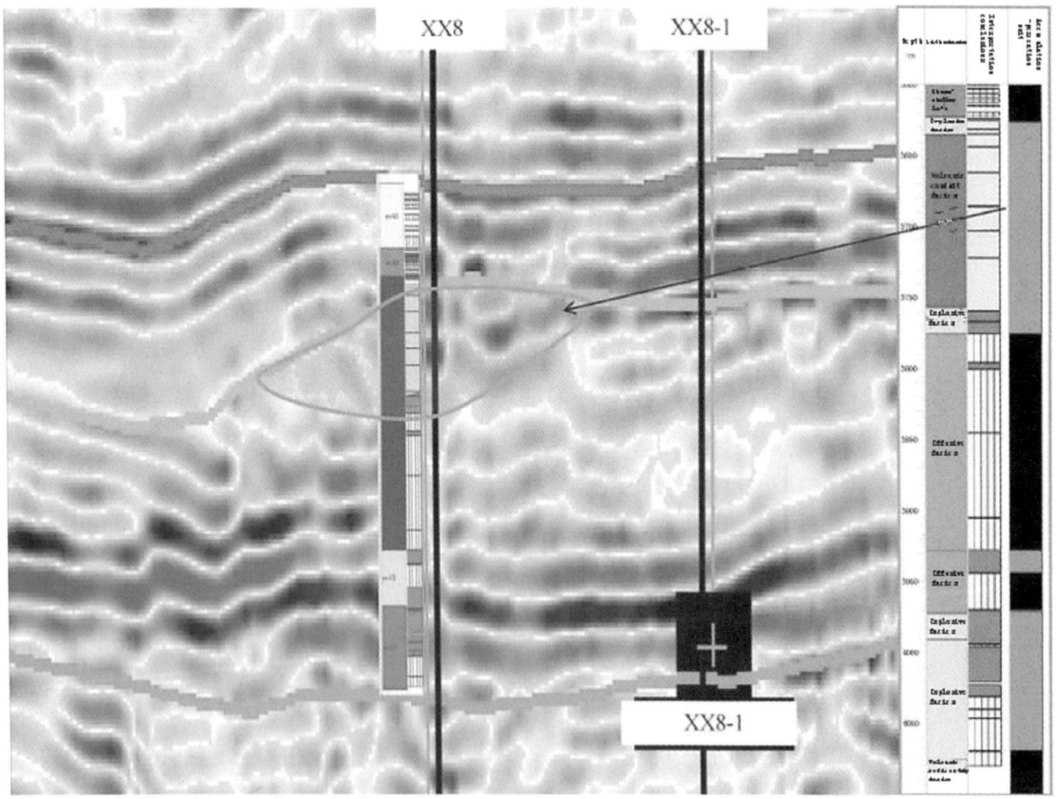

FIGURE 10.17

Seismic facies model of volcanic conduit facies A-P units (Well XX8).

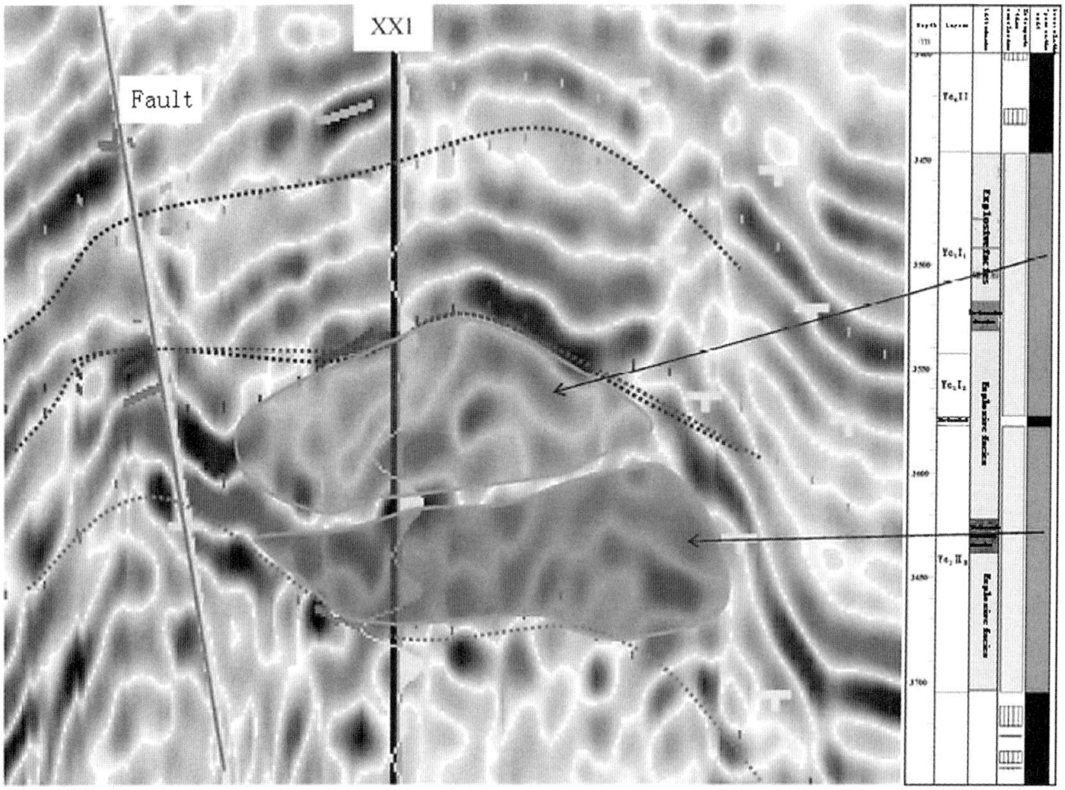

FIGURE 10.18

Seismic facies model of explosive facies A-P units (Well XX1).

shows chaotic reflections. Better-developed A-P units are represented by weaker amplitude of the seismic reflection wave. In general, discontinuous, traceable, and relatively strong reflections represent the interfaces between an A-P layer and an external flow resistance zone.

Effusive facies A-P units. The effusive facies usually forms during the middle-late stage of a volcanic eruption, in which primary pores, dissolution pores, cooling contraction fractures, and structural fractures are primary reservoir spaces and the upper and lower subfacies have relatively good petrophysical properties. In the seismic profile (Figure 10.19), this type of A-P unit mostly appears as laminoid reflections with good continuity. In the zone where A-P layers are developed, the seismic reflection has a weaker amplitude and lower frequency. If the effusive facies A-P layers are in direct contact with sedimentary rocks, a locally continuous strong reflection may occur.

Subvolcanic facies A-P units. Subvolcanic rocks are usually formed at the late stage of a volcanic eruption; they have the same source as volcanic rocks, but with intrusive occurrence, in which interstitial pores, dissolution pores, cooling contraction fractures, and structural fractures are primary reservoir spaces with relatively poor petrophysical properties. In the seismic profile (Figure 10.20), subvolcanic rocks mostly appear in an irregular massive shape and are marked by chaotic internal

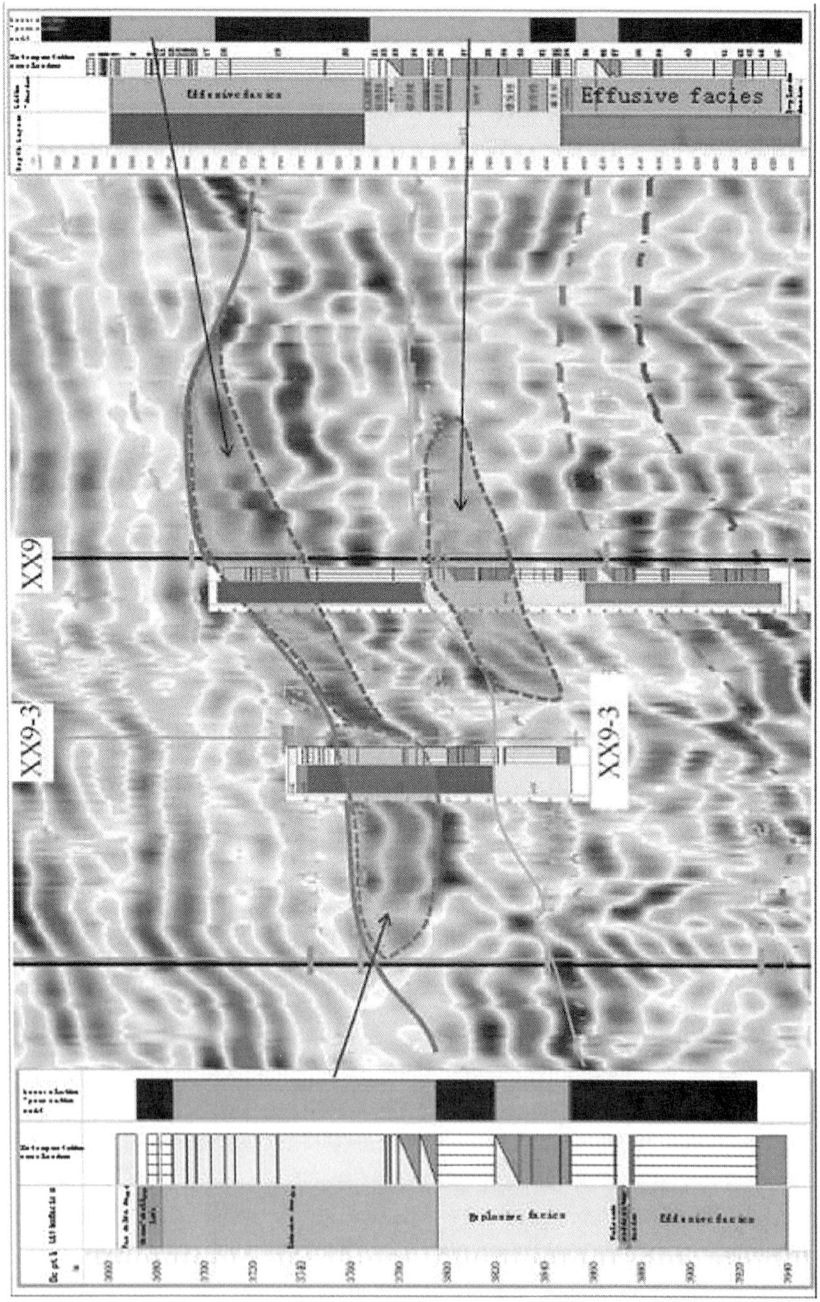

FIGURE 10.19
Seismic facies model of effusive facies A-P units (Well XX9).

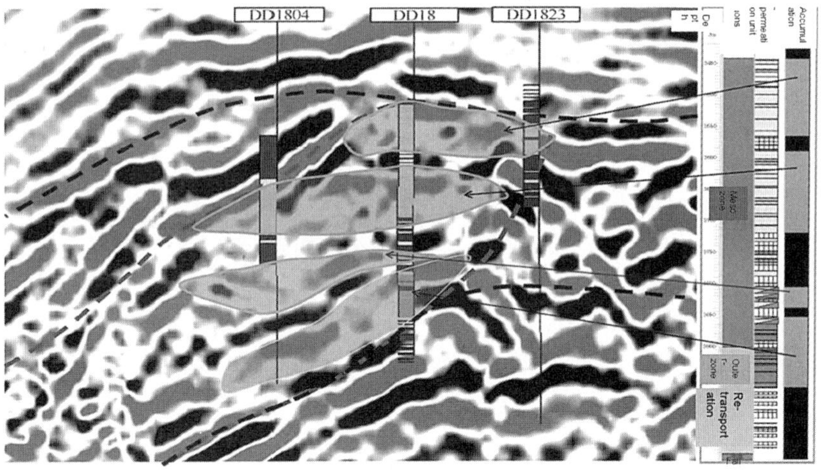

FIGURE 10.20

Seismic facies model of subvolcanic facies A-P units (Well DD18).

reflections. In the zone with A-P layers developed, the seismic reflection has a relatively weak amplitude, low frequency, and slightly better continuity. Discontinuous and traceable strong reflections usually represent the interfaces between an A-P layer and a flow resistance zone.

It is difficult to identify the A-P units directly with seismic profiles. The seismic responses presented here can provide the basis for a qualitative identification of A-P units by integrating seismic profiles with geological models.

2) Interface tracing of A-P units

Based on single-well calibration, several typical interfaces of A-P units can be traced using the identification models as a guide.

Identifiability varies with lithological interfaces. In seismic profiles, an abrupt interface from sedimentary rock to volcanic rock appears as a strong reflection of unconformity surface and is easily recognizable (Figure 10.21a); a lithofacies interface appears as a moderate reflection peak and is locally traceable (Figure 10.21b); a lithological interface resulting from variations in rock composition or grain size appears as a relatively strong reflection peak and is locally discontinuous and traceable (Figure 10.21c).

The traceability of a petrophysical property interface varies with interface types but is generally poor. An effusive facies petrophysical property interface appears as laminar reflections with relatively strong amplitudes, and locally it has a relatively good traceability (Figure 10.21c). An explosive facies petrophysical property interface appears as chaotic reflections with moderate amplitude, and its traceability is relatively poor. A subvolcanic facies petrophysical property interface appears as chaotic reflections with moderate-poor amplitudes, and it may be locally traceable (Figure 10.21d).

3. Identification of inversion profiles

Based on the volcanic body-controlling seismic inversion, the inverted data volumes of wave impedance and reservoir parameters are used to remove nonreservoirs in the inversion profile and to characterize profile distribution patterns of various types of A-P units based on the seismic cutoff for effective reservoirs.

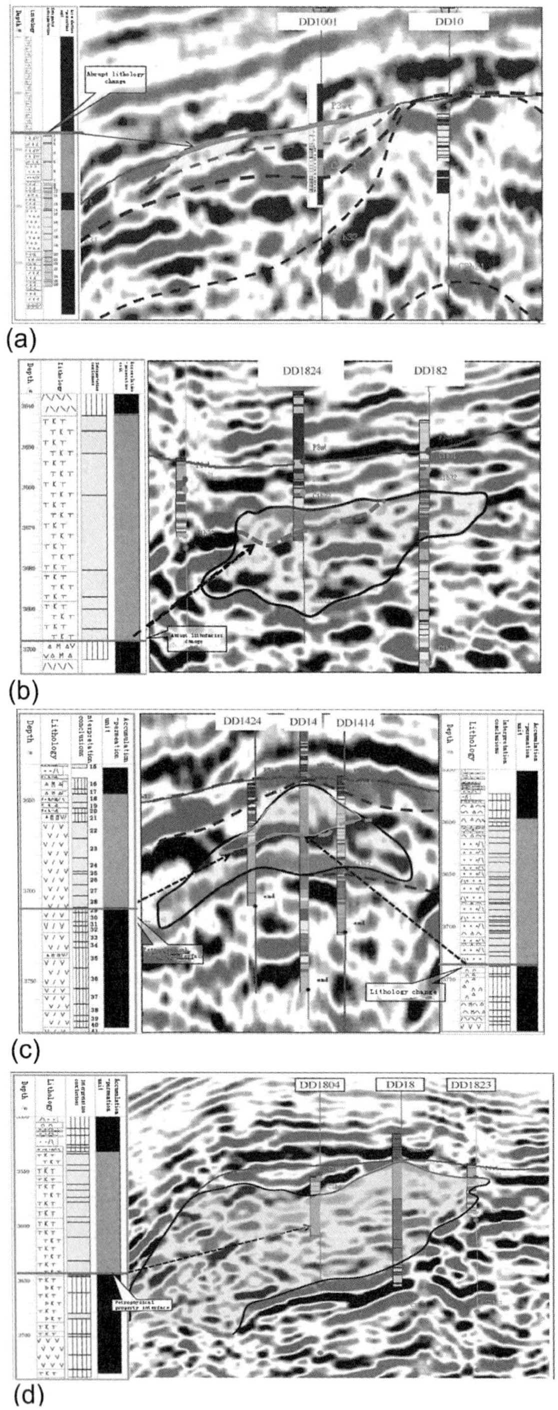

FIGURE 10.21

Interface types of an A-P unit. (a) Lithology interface (Well Block DD10). (b) Lithofacies interface (Well Block DD182). (c) Lithology interface (Well Block DD14). (d) Petrophysical property interface (Well Block DD18).

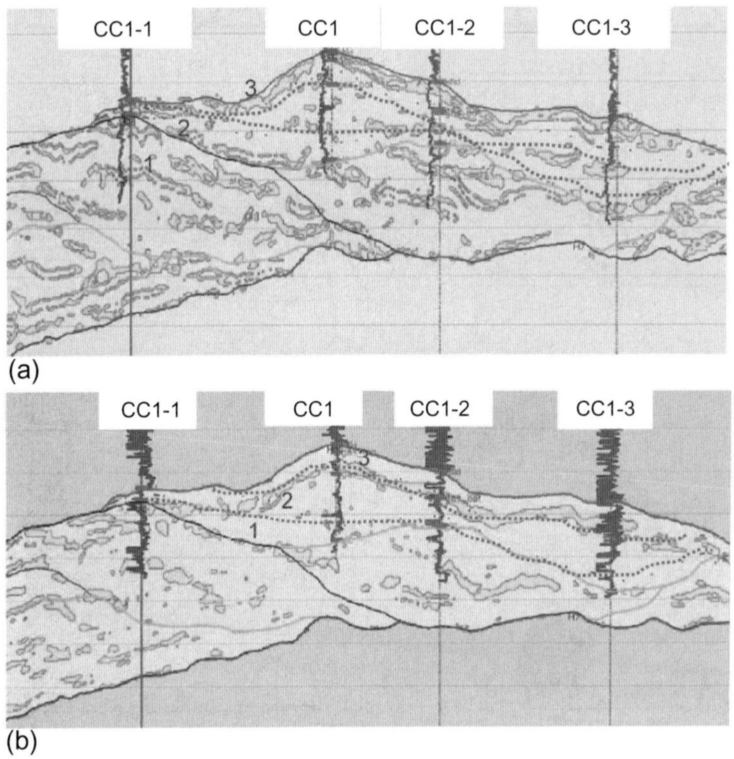

FIGURE 10.22

Distribution of A-P units in inversion profiles (CC gas reservoir). (a) Identifying A-P units in wave impedance inversion profile. (b) Identifying A-P units in density inversion profile.

For the effective reservoir in the CC gas field, the cutoff threshold for porosity is 3.5%, wave impedance ranges from 9500 to 14500 $(g/cm^3) \cdot (m/s)$, and density is 2.2 to 2.57 g/cm^3. On this basis, the data volume of wave impedance and density derived from "body-controlling" seismic inversion are applied to obtain the profile distribution of A-P units (yellow color) by changing the color of data points of the non-A-P layer as background color (blue). Figure 10.22 shows that the effusive facies A-P units mostly have a striped distribution, whereas the explosive facies A-P units mostly have an irregular subcircular distribution along the flow direction of magma.

10.2.3 Prediction of planar distribution

The prediction methods involve seismic attribute analysis and seismic inversion data analysis.

1. Seismic attribute analysis method

The useful seismic attributes are extracted to predict the planar distribution of A-P units through optimized combination analysis, according to the geological characteristics of volcanic gas reservoirs as well as the analysis of sensitive attributes of A-P layers (Figure 10.23).

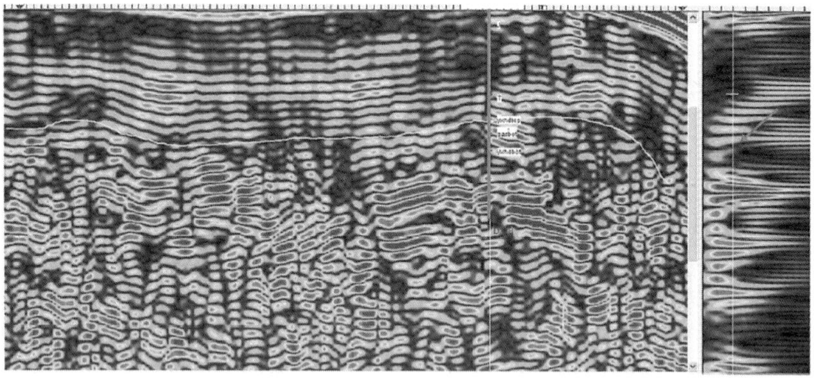

FIGURE 10.23

Frequency gathers of seismic traces near Well DD18.

1) Sensitive attributes analysis for A-P layers

Both theoretical and experimental studies show that high-frequency components are strongly absorbed when a seismic wave passes through a gas-bearing reservoir. As a result, the seismic wave of effective A-P layers is generally characterized by low energy for high-frequency amplitude and high energy for low-frequency amplitude. The case study in the DD gas reservoir shows that in the frequency acquisition of seismic trace near Well DD18, there is as an obvious trend of change from high frequency to low frequency corresponding to the transition from nonreservoir sedimentary rocks to an orthophyre gas layer (from top to bottom). According to the stratum spectral analysis of a gas-bearing property, the wavelet spectrum is marked by the phenomenon of enhancement of low frequency and attenuation of high frequency. As a result, amplitude, frequency, and attenuation are the most sensitive seismic attributes for identifying A-P units of a volcanic gas reservoir.

2) Seismic attribute extraction

The method of seismic attribute extraction involves time slice, horizon slice, and interlayer attribute extraction. The extraction targets include 2D profile and 3D data volume [18]. Both horizon slice and interlayer attribute extraction can be used to predict planar distribution.

Considering that seismic waves for volcanic A-P units have different energy levels for high-frequency and low-frequency components, the horizon slice method is used to extract high-frequency and low-frequency amplitude attributes (Figure 10.24) based on the frequency division processing of seismic data. This will render effective A-P layers as weak amplitude for high-frequency components (20 Hz) (blue area in Figure 10.24a) and relatively strong amplitude for low-frequency components (12 Hz) (red area in Figure 10.24b).

3) Optimization and combination of seismic attributes

If individual seismic attributes have a poor correlation with reservoirs, the combination and optimization of multiple attributes can be used to reduce the prediction uncertainty of the planar distribution of A-P units.

The stratum absorption coefficient obtained by calculating the change rate of wavelet spectrums can better reflect the characteristics of effective A-P layers (Figure 10.25a), although the seismic reflection

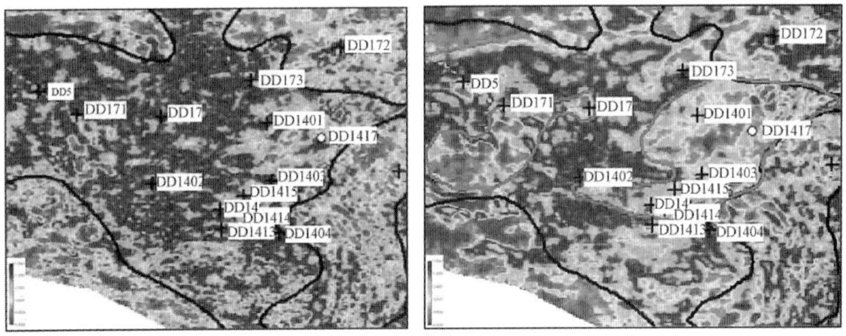

FIGURE 10.24

Predicting the planar distribution of A-P units based on the frequency-division amplitude energy attribute (DD gas reservoir). (a) High-frequency (20 Hz) amplitude energy. (b) Low-frequency (12 Hz) amplitude energy.

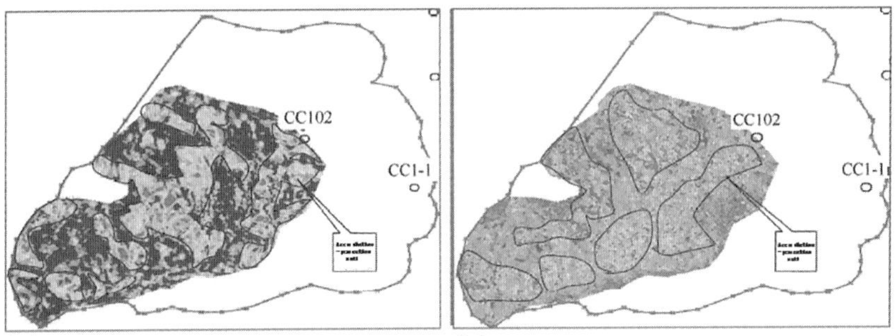

FIGURE 10.25

Predicting the planar distribution of A-P units by the absorption coefficient and the correlation dimension attribute (No. 1 volcanic massif in CC gas reservoir). (a) Absorption coefficient. (b) Correlation dimension.

frequency is related to stratum thickness, reservoir petrophysical properties, and oil-bearing properties. In general, seismic reflection signals are highly irregular. If the seismic trace is considered a random time series, the phase space reconstruction method can be used to measure its fractal dimension and obtain a correlation dimension parameter that directly reflects hydrocarbon distribution (Figure 10.25b). The analysis of the No. 1 volcanic massif in the CC gas reservoir shows that the correlation between the absorption coefficient/correlation dimension attribute and the natural gas distribution can exceed 70%.

Predicting the planar distribution of A-P units using integrated attribute analysis. A seismic attribute or attribute combination is characterized by multiple solutions under the influence of various factors.

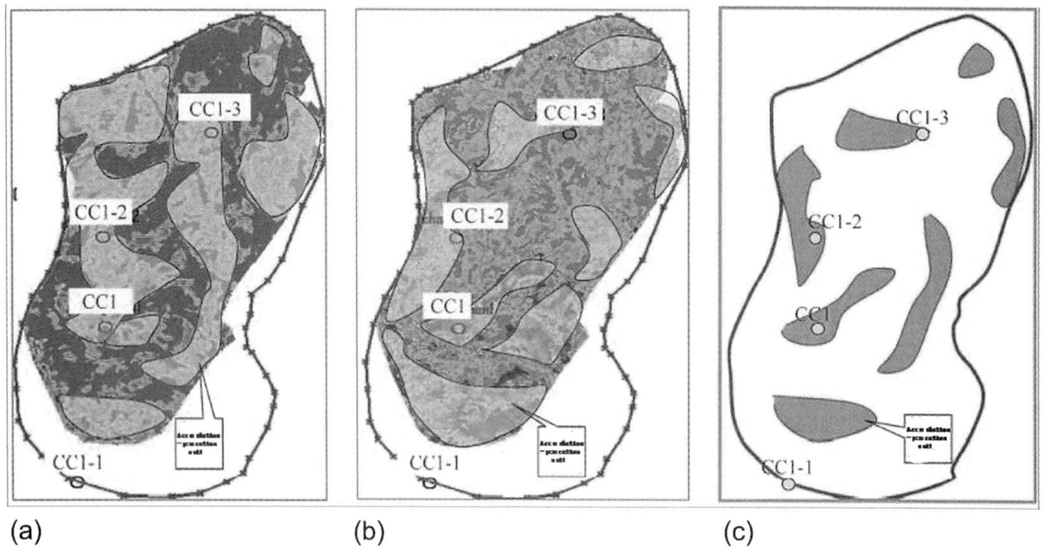

FIGURE 10.26

Predicting the planar distribution of A-P units by integrated attribute analysis (No. 6 volcanic massif in CC gas reservoir). (a) A-P unit predicted by the absorption coefficient. (b) A-P unit predicted by the correlation dimension. (c) A-P unit predicted by combining the absorption coefficient and the correlation dimension.

The method of integrated attribute analysis is effective for reducing the number of solutions and the uncertainty of prediction. The method is summarized as follows:

1. Applying single attribute to predict the possible planar distribution of A-P units (Figures 10.26a and 10.26b) based on the characteristics of individual seismic attributes or attribute combination, in combination with the analysis of static and dynamic data from the study area
2. Superposing the prediction results of all sensitive attributes and determining the confidence level for the distribution area of A-P units by superposition times (Figure 10.26c)
3. Performing a comprehensive analysis of static and dynamic data from the study area, and effectively predicting the most likely planar distribution of A-P units under the guidance of geological rules (Figure 10.26c)

2. Seismic inversion prediction method

Based on the body-controlling seismic inversion and seismic cutoff thresholds of effective A-P layers, the planar distribution of reservoir-permeation units in a volcanic gas reservoir is predicted using reservoir parameter volumes such as wave impedance, sound wave, density, and porosity:

1. Determining the extent of the planar distribution of A-P units in the main body of target strata in accordance with the planar distribution of reservoir parameters for different horizon slices

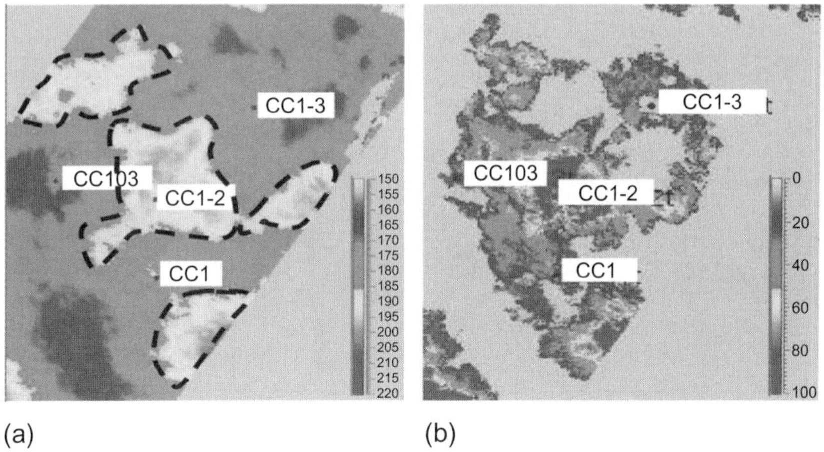

FIGURE 10.27

Planar distribution of A-P units predicted by seismic inversion (CC gas reservoir). (a) Planar distribution of A-P units predicted by mean value of interval transit time. (b) Planar distribution of A-P units shown by the thickness of A-P layers.

2. Extracting the average value of reservoir parameters within an appropriate time window based on the top structure of the target stratum (time domain) and determining the extent of the planar distribution of A-P units according to their planar distribution characteristics (Figure 10.27a)
3. Calculating the thickness value of an effective A-P layer for each point and determining the planar distribution range of an A-P unit under the constraints of top and bottom boundaries of target strata (Figure 10.27b)

10.2.4 Prediction of spatial distribution

Based on the assessment of single-well, profile, and planar distribution patterns, the following three methods are used to predict the spatial distribution of A-P units.

1. Tracing-and-closure techniques for A-P unit interfaces based on framework profiles

Based on the identification of seismic reflection profiles and the established framework profile network that effectively controls the study area, the A-P unit interface is traced and closed under the architectural constraint of the volcanic edifice in order to predict the 3D spatial distribution characteristics of A-P units.

2. 3D attribute volume assemblage technique

This technique involves the following four steps: (1) establishing the seismic attribute division criteria of A-P units by sensitivity analysis, (2) extracting the sensitive attributes from 3D space using the data volume attribute extraction technology, (3) extracting the possible data points of A-P units according to division criteria and combining the data points based on their spatial positions, and (4) predicting the spatial distribution of A-P units by superposing and optimizing the various attribute prediction results.

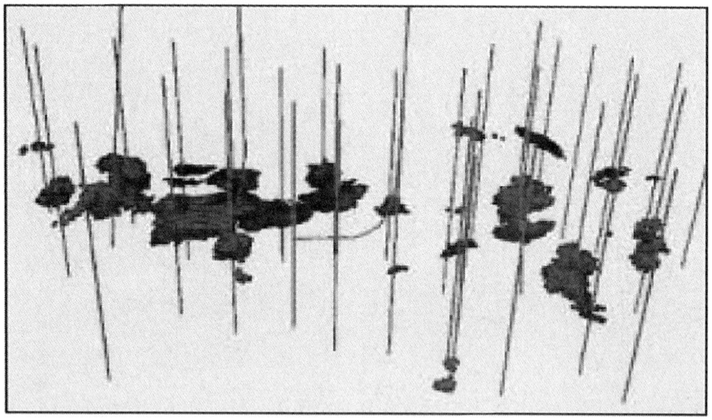

FIGURE 10.28

The 3D spatial distribution of A-P units (Well Block XX1).

3. Spatial combination technique based on inverted seismic data volume

Based on the cutoff thresholds of effective A-P layers, the possible data points of A-P units are extracted and then combined in terms of their special locations, using the inverted 3D seismic volume. On this basis, the spatial distribution of A-P units can be predicted by superposing and optimizing various inverted data volumes.

Figure 10.28 shows the 3D spatial distribution of A-P units in Well Block XX1 and demonstrates that the volcanic A-P units in 3D space are characterized by various shapes, scattered distribution, and diverse scales.

10.3 CHARACTERIZATION OF ACCUMULATION-PERMEATION UNITS

Based on the identification of accumulation-permeation (A-P) units, the geometry, connectivity, and A-P capacity of volcanic A-P units are characterized on the basis of outcrop, seismic, and dense well pattern data; accordingly, the distribution of A-P units is revealed.

10.3.1 Characterization of geometry and scale

1. Using outcrop data to characterize the geometry and scale of A-P units

Field outcrops are one of the most important means for establishing a geological prototype model for subsurface reservoirs. When an outcrop is present near the target stratum, a field outcrop survey is the most accurate and reliable method [19] for characterizing the geometry and scale of A-P units. The main procedures include the following steps.

1) Selecting the survey location

Based on the geological survey and regional geological analysis, a survey location is selected where an outcrop of the target stratum is available [20]. Interval 1 of the Yingcheng Formation is one of major gas-bearing layers in the volcanic gas reservoir of the Songliao basin. Geological surveys show that this set of volcanic strata crops out in the Jiutai area of Jilin Province. After a field survey, the Guanma Mountain-Tuanjie village area with good field exposures is selected for a detailed field outcrop survey (identified by the blue dashed line in Figure 10.29).

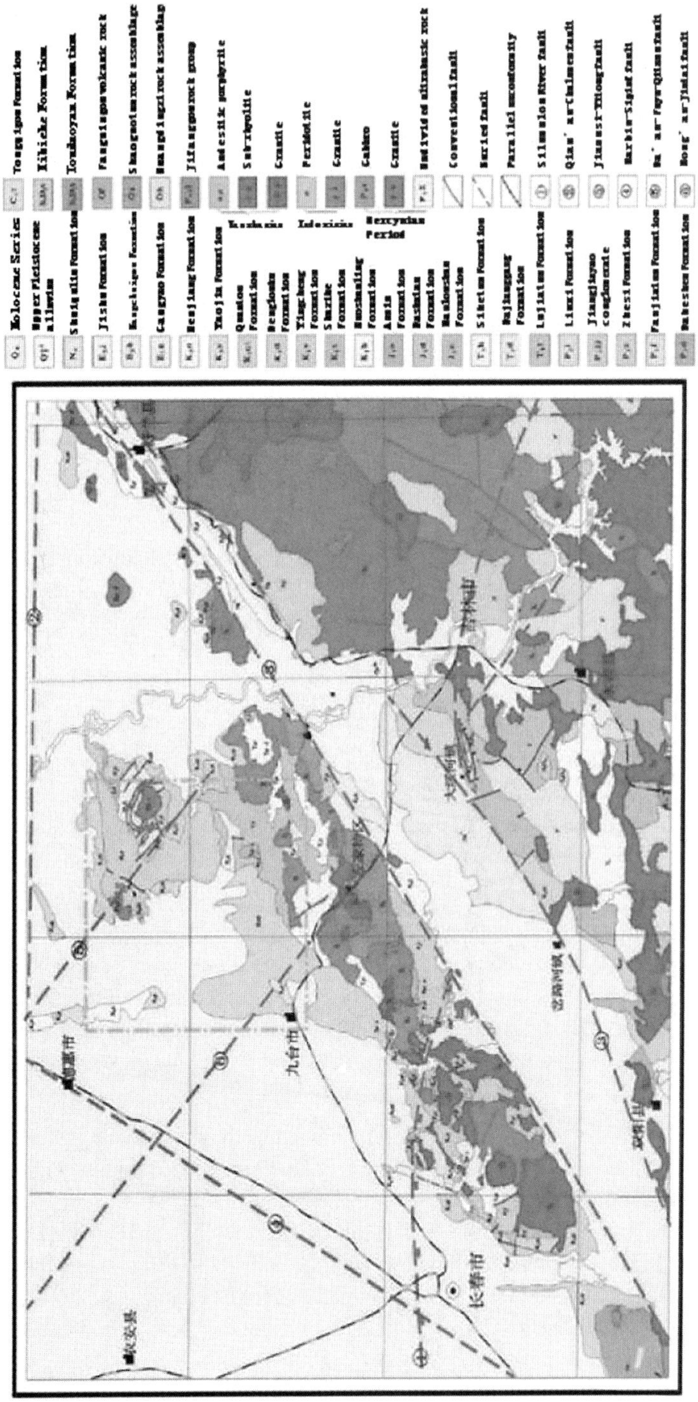

FIGURE 10.29

Regional geological map for a volcanic geological profile of the Yingcheng Formation on the eastern margin of the Songliao basin.

2) Plotting the surveyed geological profile and establishing an architectural model of volcanic edifice

Nine trench faces (1735 m) are surveyed by arranging exploratory trenches in the survey area. Based on the 20 selected samples for chemical analysis, 588 field photographs, 191 rock thin sections, and 725 micrographs, as well as the contact relationships observed from the exposed volcanic strata, the surveyed geological profile (Figure 10.30) is plotted and a geological model reflecting the architectural characteristics of the volcanic edifice is established for the target strata (Figure 10.31).

3) Obtaining drill core and dividing A-P units in a borehole

One cored well (Well Ying-D1) is drilled in a key location (i.e., the center of flank) of the volcanic edifice. Based on the lithological and lithofacies analyses as well as porosity-permeability measurement, the A-P units are classified and divided in a borehole according to the principle of upper cap rock, middle reservoir, and lower basal bed (Figure 10.32).

4) Ascertaining distribution characteristics and establishing distribution models of A-P units

Under the architectural constraints of the volcanic edifice, the spatial distribution of A-P units is predicted and the distribution modes of A-P units for this type of volcanic edifice is established on the basis of the division of A-P units in boreholes, exploratory trench measurement, and field survey of geological profiles (Figure 10.33).

5) Characterizing the shape and scale of A-P units

Outcrop survey results show that most effusive facies A-P units are bedded, explosive facies A-P units lenticular, and A-P units in mixed facies irregular. In volcanic gas reservoirs, most A-P units are 0.2 to 1.7 km long, 0.1 to 1.4 km wide, and 10 to 50 m thick, covering an area of 0.1 to 0.9 km^2, with different scales and poor continuity (Figure 10.34).

2. Using seismic data to characterize the shape and scale of A-P units

Guided by the geological models derived from outcrops, the shape, geometry parameters, and scale of A-P units in volcanic gas reservoirs can be effectively characterized through by identifying seismic profiles, as well as their planar and spatial prediction.

1) Geometry of A-P units

In profiles, A-P units in near-crater explosive facies commonly have a massive structure with large thickness and small extension; and A-P units in off-crater effusive facies are mainly bedded, with small thickness and large extension (Figure 10.35).

In plan view, volcanic A-P units have three distribution shapes (Figure 10.36): (1) a potato-shaped A-P unit, developed in explosive facies of isolated volcanic edifice, with a small amount of erupted magma and small scale; (2) a striped A-P unit, developed in effusive facies and controlled by palaeotopography, with a large planar distribution; and (3) a sheetlike A-P unit, common in both explosive and effusive facies, formed from the superposition of A-P units of multiple volcanic edifices, and relatively large in scale.

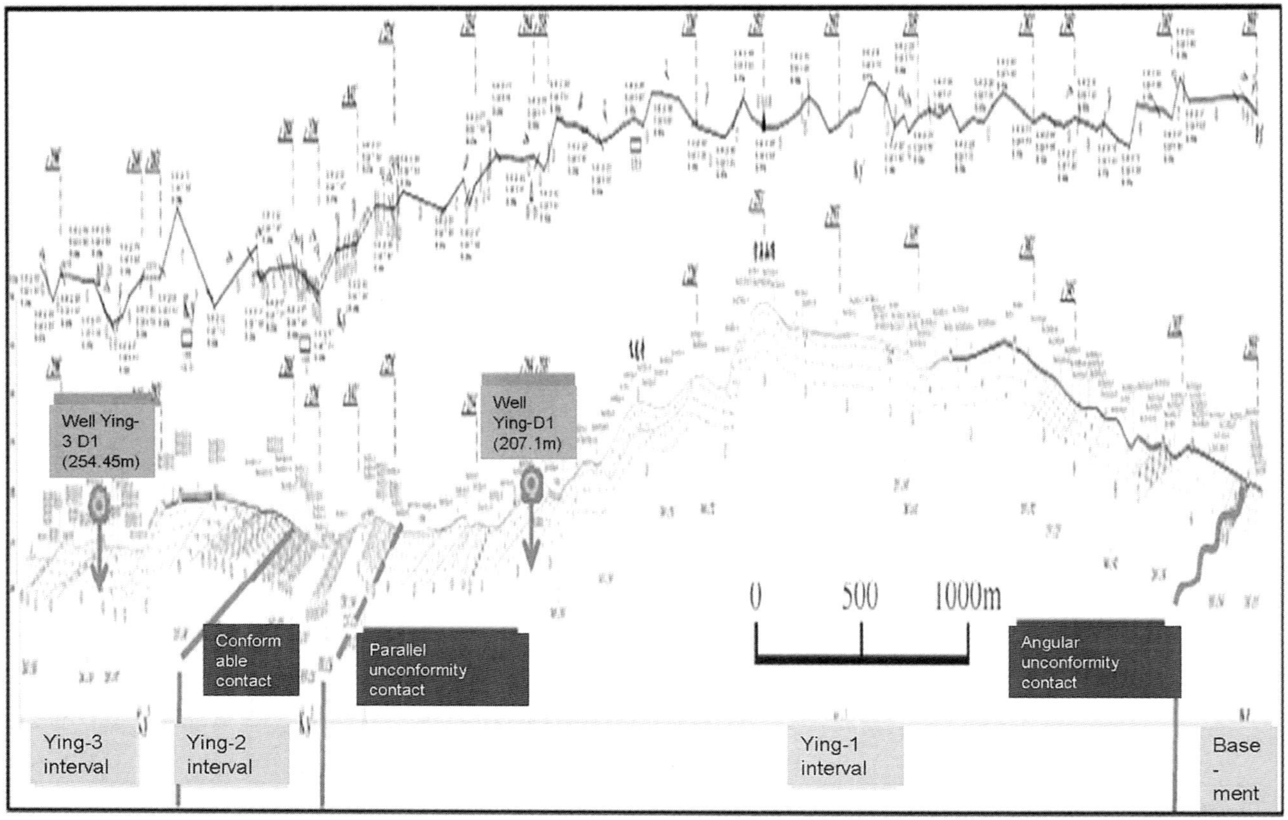

FIGURE 10.30

Surveyed geological profile for the Lower Cretaceous Yingcheng Formation, Shichang village, Jiutai City.

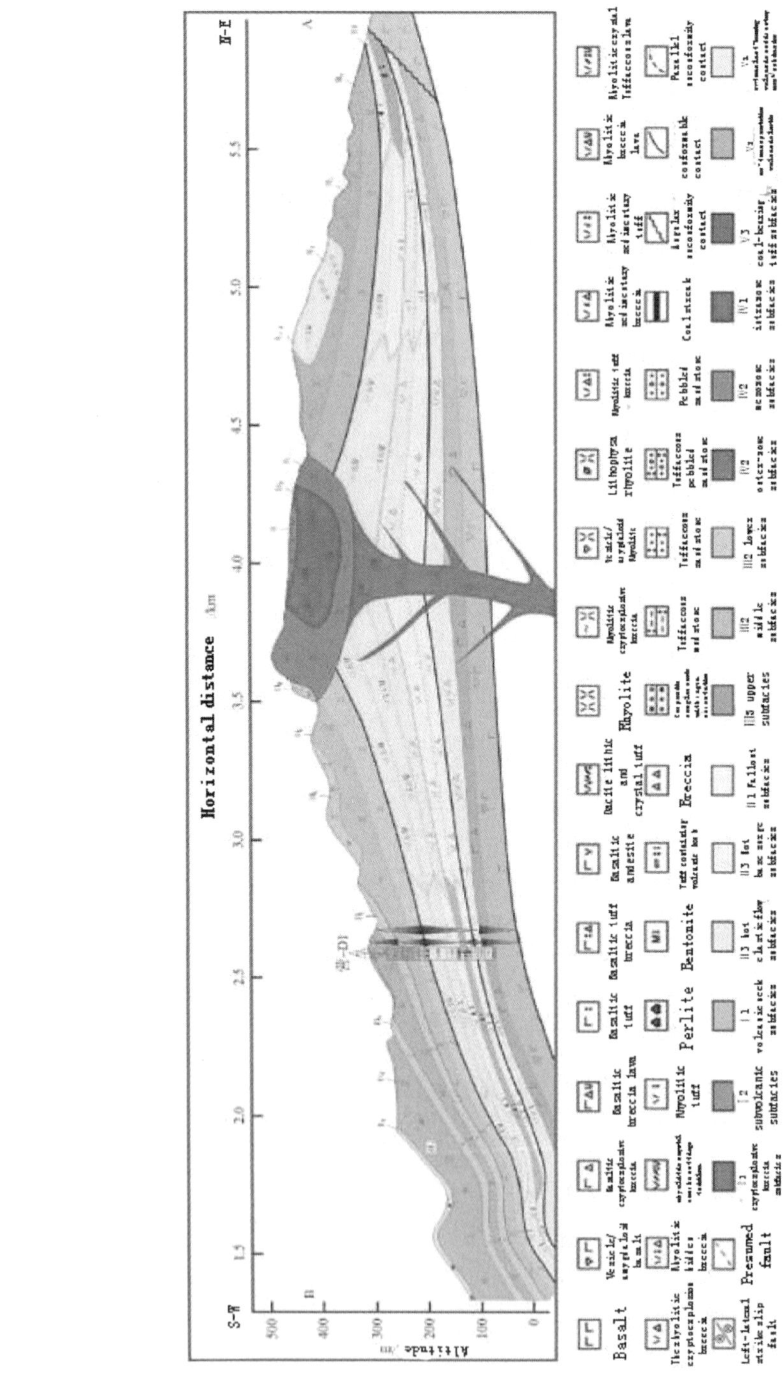

FIGURE 10.31

Architectural mode for volcanic edifice of the Yingcheng Formation in the Songliao basin.

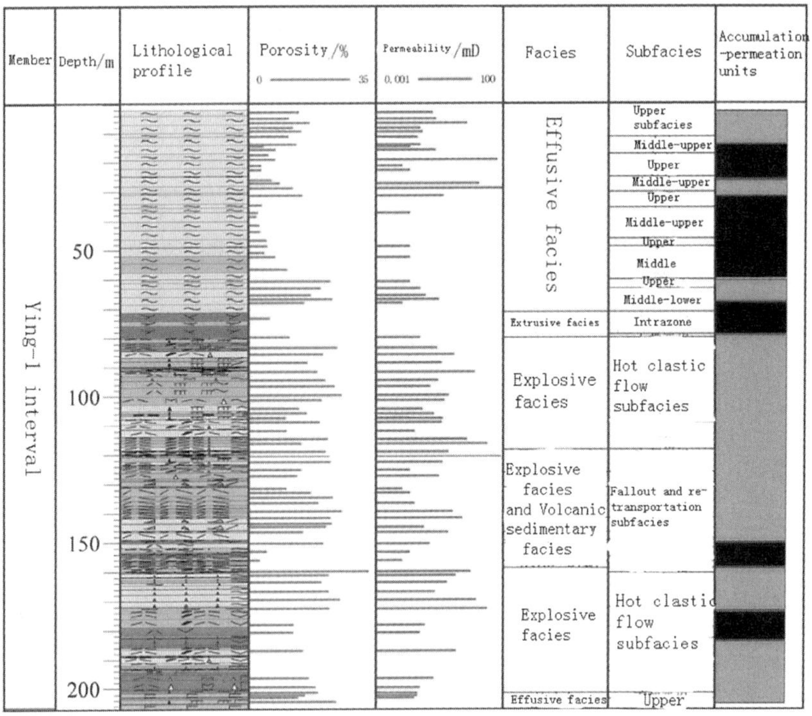

FIGURE 10.32

Distribution of borehole A-P units in Well Ying-D1.

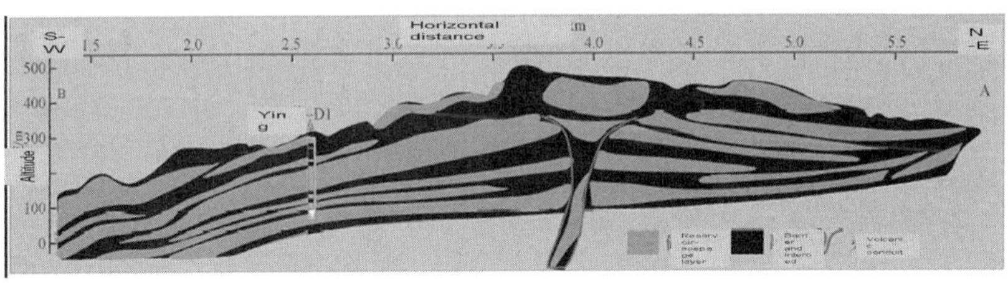

FIGURE 10.33

Distribution model of A-P units for composite volcanic edifice, Guanma Mountain.

2) Geometric parameters and scale of a-p units

As is shown by the CC gas field example (Figure 10.37), most volcanic A-P units have a discrete distribution in the gas field. Their lateral extension radius is 300 to 2100 m, with an average of 537 m, and their thickness is 9 to 124 m with an average of 20 m. As a result, they have drastically different vertical and horizontal scales.

A-P units have different scales for various lithofacies. Explosive facies A-P units are mostly 15 to 100 m thick and 200 to 800 m wide, whereas effusive facies A-P units are mostly 5 to 45 m thick and 400 to 1000 m wide.

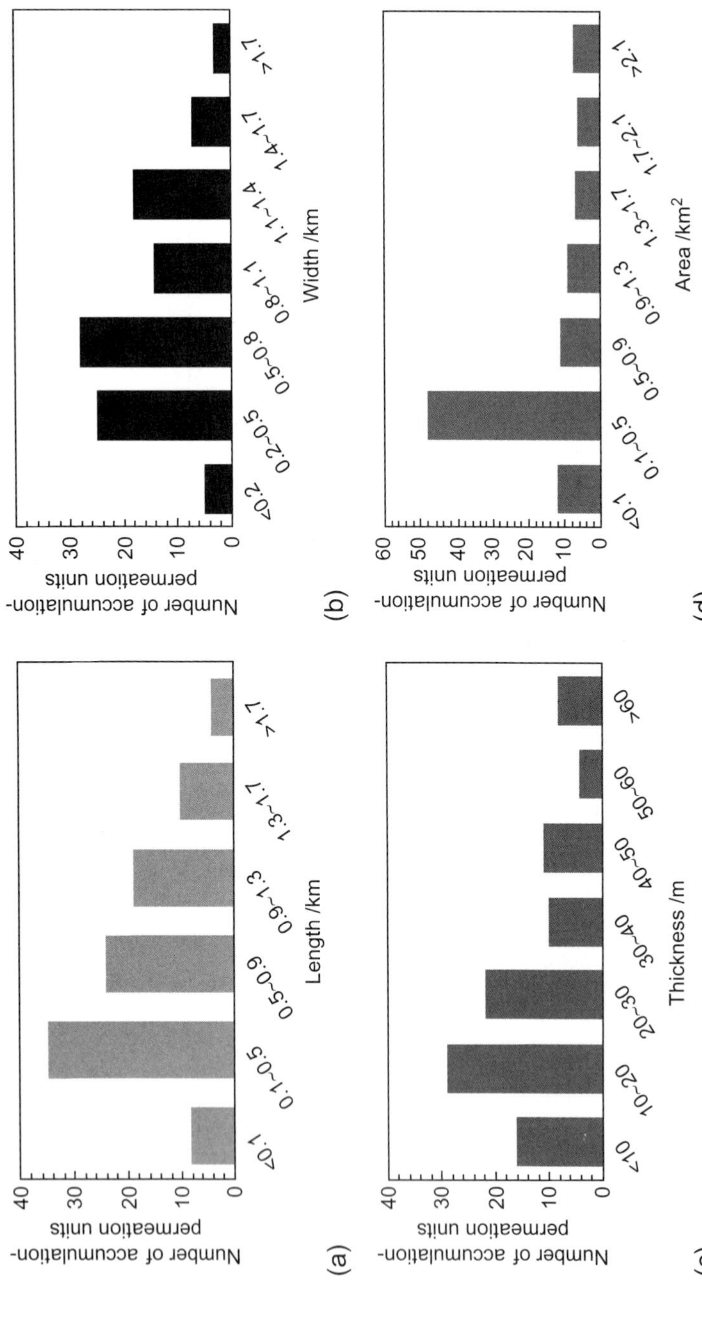

FIGURE 10.34

Geometry of volcanic A-P units (field outcrop).

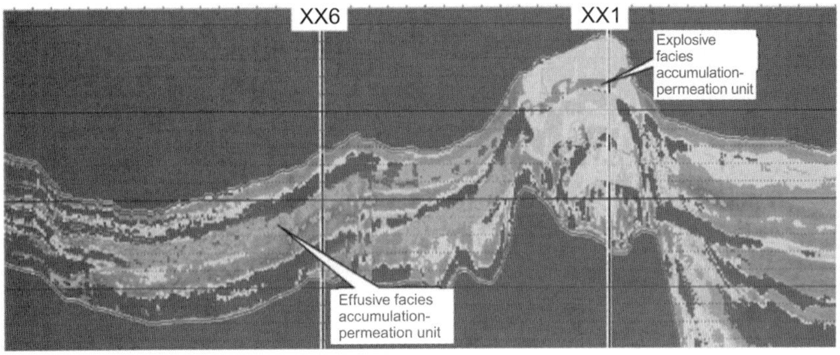

FIGURE 10.35

Profile shape of volcanic A-P units (density inversion).

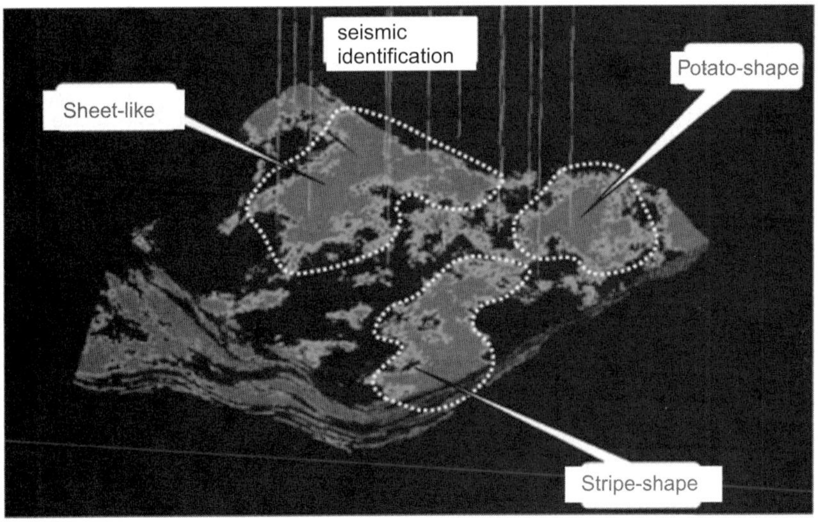

FIGURE 10.36

Planar shape of volcanic A-P units.

3. Using dense well pattern data to characterize geometric parameters and scale of A-P units

A dense well area has closely spaced wells and abundant static and dynamic data; therefore, the application of dense well pattern data is one of the most effective and direct methods to dissect subsurface volcanic architecture and to characterize the shape and scale of A-P units. The main procedures are as follows.

1) Dividing A-P units in wells

In terms of the identification markers of A-P units, coring, logging, testing, and production performance data are integrated to identify A-P layers, barriers, and interbeds and to divide A-P units in wells.

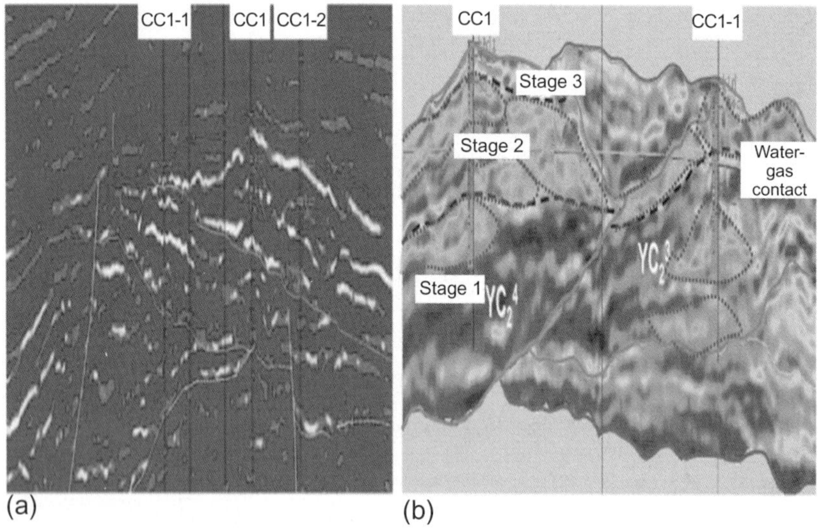

FIGURE 10.37

Profile distribution of volcanic A-P units (seismic attribute and inversion). (a) Multiwell profile of seismic energy attenuation. (b) Density inversion profile.

2) Establishing profile distribution models of A-P units

The typical cross-well skeleton or framework profile is plotted for the dense well grid area (Figure 10.38). The profile distribution model of A-P units is established through the identification of multiwell profiles and interwell combination of A-P units, taking into consideration the influences of faults, water-gas relationships, and pressure systems along with the constraints of volcanic architecture (Figure 10.39).

3) Predicting spatial distribution and characterizing A-P unit shape and scale

The spatial distribution of A-P units is predicted through the combination of techniques of A-P unit interface tracing, 3D attribute volume spatial assemblage, and seismic inversion data volume spatial assemblage. It provides a foundation for characterizing the shape and scale A-P units.

Research results show that most A-P units in the dense well area are lenticular or lamellar, bounded by lithofacies, subfacies, or petrophysical properties. They are 0.5 to 2.8 km long, 0.3 to 2.5 km wide, and 10 to 50 m thick, covering an area of 0.4 to 3 km^2 with highly variable scales (Figure 10.40).

10.3.2 Connectivity characterization

The connectivity between A-P units is characterized by the scale of connected rock bodies and determined using various static and dynamic data, based on the interpreted superposition relationships and connection modes of A-P units. This provides a basis for the characterization of A-P capacity.

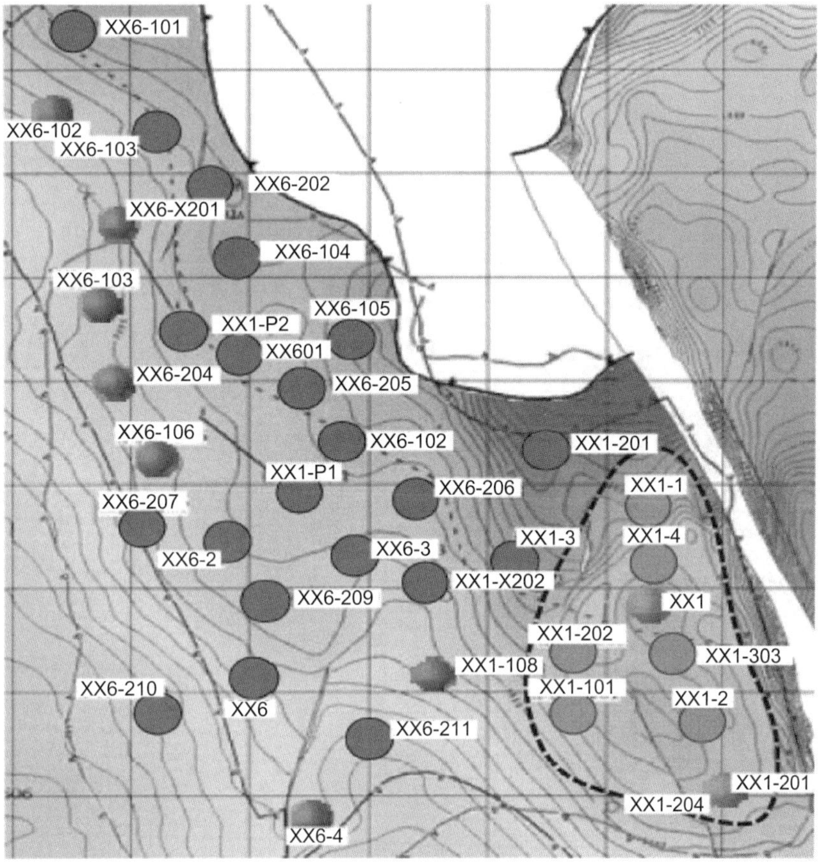

FIGURE 10.38

Horizontal well locations in a dense well grid area.

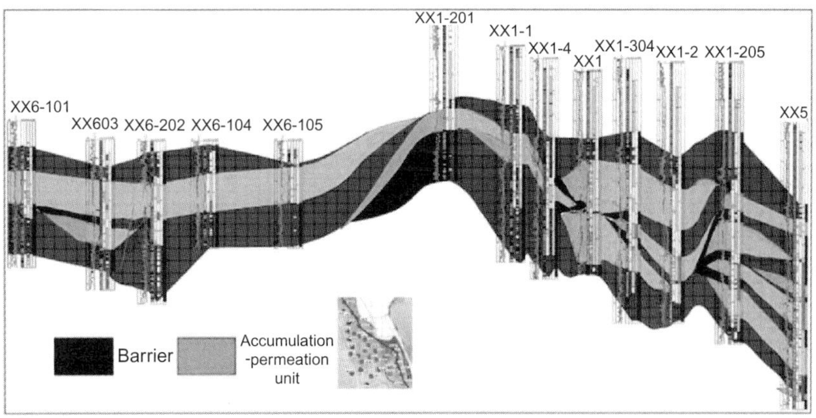

FIGURE 10.39

Multiwell profile of A-P units in a dense well pattern area.

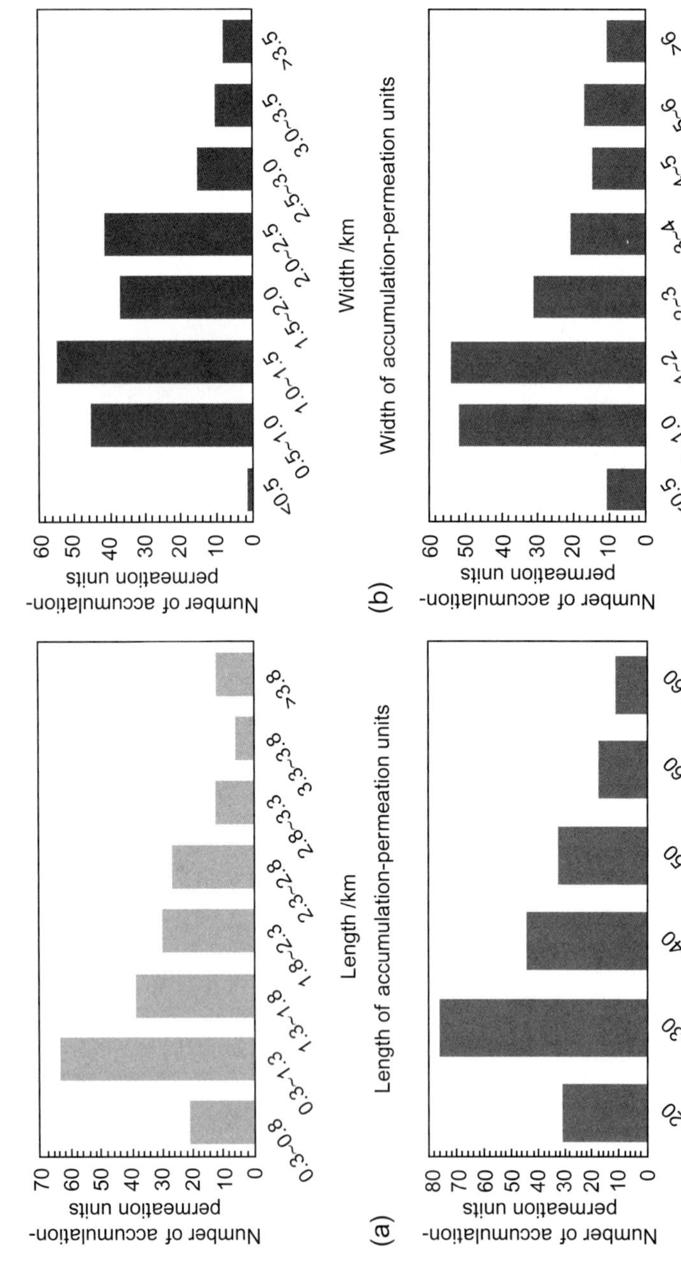

FIGURE 10.40

Geometric parameters of A-P units in a dense well grid area (Well Block XX1).

1. Connection types of A-P units

Based on the analysis of superposition relationships and connection modes, the connections between A-P units are subdivided into various types as a framework for characterizing their connectivity.

1) Superposition relationships

According to their locations, A-P units are classified into three types: vertical superposition, lateral connection, and isolated types (Figure 10.41). Vertical superposition mainly occurs between A-P units of two adjacent volcanic massifs (or stages or lithofacies) within the same volcanic edifice; lateral connection is found mainly between two volcanic edifices or volcanic massifs with lateral superposition; the isolated type, with a lenticular morphology, occurs within a discrete volcanic edifice or in a relatively independent volcanic massif within a volcanic edifice.

2) Connection modes

A-P units are connected by open fractures, structural fractures, large columnar joint fractures, and multifracture network passages. Among them, open fractures have the longest connection distance, multifracture network passages rank second, and the columnar joint has the shortest connection distance, limited within a single volcanic edifice.

3) Connection types

In accordance with the general classification principle of "superposition relationships + connection modes," the connections of A-P units are classified into 10 types, including vertical-superposition opening fracture connection type, lateral-superposition structural fracture connection type, isolated multifracture network passage, and so on. Among them, some assemblages only occur in specific settings. For example, the isolated columnar joint fracture connection type only forms among A-P units of two independent volcanic massifs within a volcanic edifice.

2. Connectivity characterization of A-P units

Based on the superposition relationships of A-P units, the connectivity of A-P units is characterized by analyzing the fault and fracture characteristics, water-gas relationships, fluid properties, and pressure systems. In the example from Well Block XX9, a total of six A-P units (Nos. I-VI) are identified in the XX9-3-XX9-XX9-1-XX903 multiwell profile (see Figure 10.41). Among them, Nos. I to III, No. IV to V, and Nos. VI A to P are developed in three separate volcanic massifs, respectively.

1. Superposition relationship: Based on their locations (Table 10.4), there is lateral superposition between Nos. I and II, and between Nos. II and III A-P units, vertical superposition between No. IV and V A-P units, and isolated type for the No. VI A-P unit.
2. *Fault communication characteristics.* Detailed structural interpretation results show that a small open fault exists between No. I and II A-P units. As a result, it is very likely that the units are connected.
3. *Fracture communication characteristics.* Logging interpretation results show that structural fractures and microfractures are developed between No. II and No. I/III A-P units, suggesting the presence of a multifracture network passage; fractures are relatively well developed between No. IV and V A-P units; but only a small quantity of microfractures are present between No. II and IV, insufficient to form a vertical connection passage.

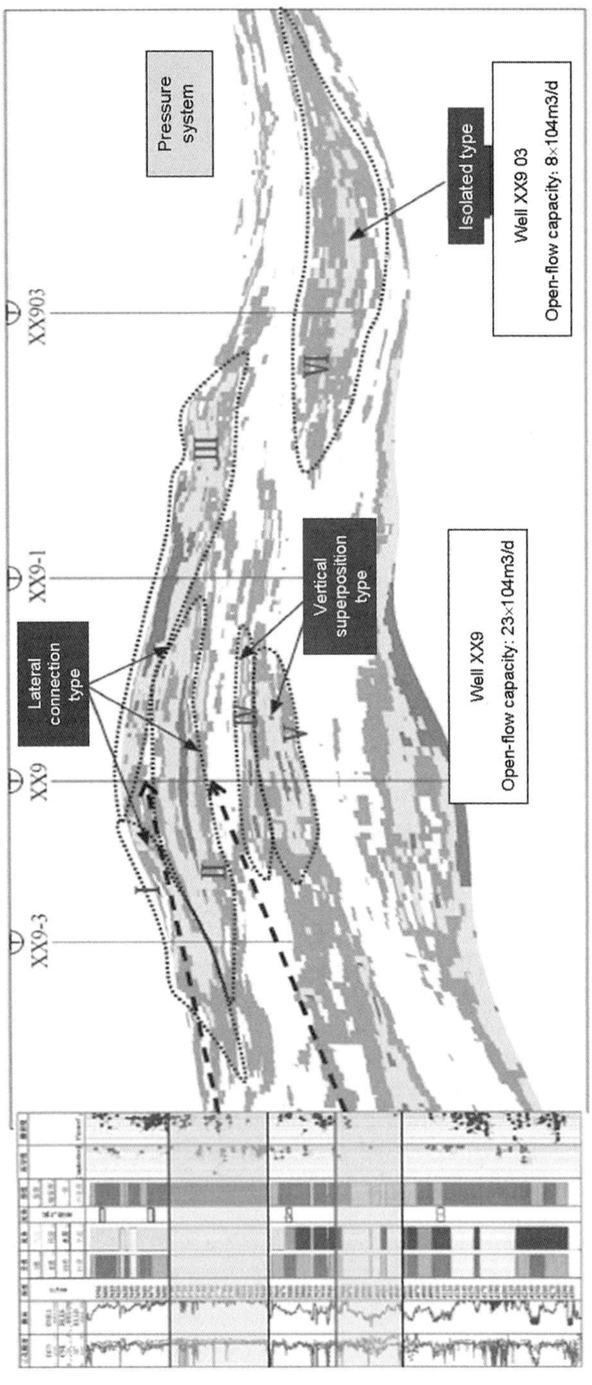

FIGURE 10.41

Superposition relationships of volcanic A-P units in Well Block XX9.

Table 10.4 Analysis of Connection Elements of A-P units in Well Block XX9

Number of A-P Units	Superposition Relationship	Fault Communication Characteristics	Fracture Communication	Water-Gas Contact (M)	Fluid Properties		Pressure System	Production Performance
					Salinity (ppm)	*CO_2 Content (%)*		
III	Lateral connection		Multifracture network passage	−3600			Scattered distribution, no obvious trend, pressure lower in basal aquifer in Well XX9 than in upper gas layer	Production of No. I to III A-P units is associated with their pressure
I		Fault connection						
II			Microfracture underdeveloped		9105	5.1		
IV	Vertical superposition			−3913.5	7501	8.1		
V					10625	9.4		
VI	Isolated type			−3704				

4. *Water-gas relationship*. No. I–III A-P units have the same water-gas contact (−3600 m), and No. IV–V A-P units also have the same water-gas contact (−3913 m). The no. VI A-P unit is an independent water-gas system with water-gas contact at −3704 m. The large differences in the three water-gas contacts indicate that these units are disconnected.
5. *Fluid properties*. Formation water salinity and CO_2 content in the three gas-water systems are significantly different (see Table 10.4), indicating that these systems are independent.
6. *Pressure system*. Scattered pressure data have no obvious trend for the six A-P units, and the pressure of the lower aquifer is lower than that of upper gas layer in Well XX9, corroborating that the upper and lower A-P units are disconnected.
7. *Production performance characteristics*. Pilot production was carried out in No. I–III A-P units in Wells XX9-3, XX9, and XX9-1, respectively. A gas test was carried out on the No. IV A-P unit in Well XX903. Pilot production and gas test results show that Well XX9-3 belongs to the type I gas well with high open-flow potential and steady production and pressure; Wells XX9 and XX9-1 belong to type IV gas well with low open-flow potential and sharp decrease in production and pressure; the production of these three wells is associated with their pressure. The low open-flow potential, rapid depletion in output and pressure when tested with large-diameter choke (12 mm), as well as slow pressure build-up after well shut-in, strongly suggest that the gas production and pressure of Well XX903 are not associated with the other three wells. This further supports the interpretation that No. I–III A-P units are connected with one other, wheteas No. VI is independent.

In summary, among the six A-P units in the study area, No. I–III A-P units belong to horizontal superposition and form a large connected body by multifracture network passages; No. IV–V A-P units are vertically stacked and form a relatively large connected body by structural fractures; and No. VI A-P unit is an independent system and forms a relatively small, independent, internally connected body.

10.3.3 Characterization of accumulation-permeation (A-P) capacity

The accumulation and permeation capacity of volcanic A-P units is quantitatively characterized in terms of scale, reservoir spaces, pore structure, petrophysical property, interbeds, and production performance, based on the classification scheme of A-P units and various selected characterization parameters. This forms a basis for preparing appropriate development technology policies. The process is discussed next using the XX gas field as an example.

1. Classification of A-P units

Based on production performance characteristics, the A-P units in volcanic gas reservoir are classified into four types (Table 10.5).

A type I A-P unit has strong accumulation-permeation capacity with large well-controlled reserves, high deliverability, and strong stable production potential. This type of A-P unit is developed in the most favorable lithology and lithofacies belts, including vesicular rhyolite and autoclastic brecciated lava of the volcanic conduit facies, breccia lava and tuffaceous lava of the splash subfacies (explosive facies), and vesicular rhyolite and brecciated lava of the top and upper subfacies (effusive facies).

A type II A-P unit has moderate accumulation-permeation capacity with moderate well-controlled reserves, moderate postfracturing deliverability, and stable production capability. This type of A-P unit occurs in relatively favorable lithology and lithofacies belts, including vesicular rhyolite and autoclastic brecciated lava of the volcanic conduit facies, welded breccia and welded tuff of the hot clastic flow subfacies, brecciated lava and tuffaceous lava of the splash subfacies, volcanic agglomerate, volcanic breccia, and tuff of the fallout subfacies (explosive facies), and vesicular rhyolite, brecciated lava, and tuffaceous lava of the top and upper subfacies (effusive facies).

A type III A-P unit has relatively poor accumulation-permeation capacity with small well-controlled reserves, low postfracturing deliverability, and weak stable production potential. This type of A-P unit occurs in unfavorable lithology and lithofacies belts with relatively poor petrophysical properties, including volcanic agglomerate, volcanic breccia and tuff of the fallout subfacies, aplitic rhyolite of the lower subfacies and tight rhyolite of the middle subfacies (effusive facies), sedimentary volcanic breccia and sedimentary tuff of the xenoclast-bearing subfacies, and tuffaceous glutenite, tuffaceous gritstone, and tuffaceous sandstone of the retransported subfacies (volcanic sedimentary facies).

A type IV A-P unit has poor accumulation-permeation capacity with small well-controlled reserves, low deliverability, and no stable production potential. This type of A-P unit is developed in relatively dense and unfavorable lithology and lithofacies belts, including crystalline, vitric, and magmatic tuff of the hot base surge subfacies (explosive facies), tight rhyolite of the middle subfacies (effusive facies), sedimentary tuff of the xenoclast-bearing subfacies, and tuffaceous pack sand and siltstone of retransported subfacies of the volcanic sedimentary facies.

2. Characterization of accumulation-permeation (A-P) capacity

A-P capacity involves accumulation capacity and permeation capacity. Various static and dynamic data can be used to quantitatively characterize both the accumulation capacity and the permeation capacity of volcanic rocks according to the classification scheme of A-P units by selecting appropriate parameters for characterization,

1) Accumulation capacity

Accumulation capacity is mainly influenced by the scale and internal effective porosity of A-P units. In general, it is characterized by the thickness and distribution area of A-P units, well-controlled reserves, the type of reservoir space, pore and fracture assemblage, A-P mode, and porosity. These can be shown using the XX gas field as an example:

1. *Type I A-P unit.* This type of A-P unit has the strongest accumulation capacity, with large scale and well-controlled reserves, multiple types of reservoir space and pore-fracture assemblage, appropriate A-P configuration, and high porosity. Its distribution area is larger than 2 km^2; net pay thickness is greater than 20 m, and well-controlled reserve is larger than 5×10^8 m^3. Reservoir space is composed mainly of intergranular pores, vesicles, dissolution pores, and fractures. The pore-fracture assemblage includes intergranular pore type, vesicle type, fracture-intergranular pore type, fracture-vesicle type, and fracture-dissolution pore type, amounting to an effective porosity higher than 10%.
2. *Type II A-P unit.* This type of A-P unit has a relatively strong accumulation capacity, with a distribution area larger than 1.5 km^2, effective thickness greater than 15 m, and well-controlled reserves ranging from 2×10^8 m^3 to 5×10^8 m^3. The reservoir space is composed mainly of intergranular pores, vesicles, dissolution pores, and fractures. The pore-fracture assemblage includes intergranular pore type, vesicle type, fracture-intergranular pore type, fracture-vesicle type, and dissolution pore type, which amount to an effective porosity greater than 5%.

Table 10.5 Classification of A-P Units in Volcanic Gas Reservoirs (XX Gas Field)

Type	Lithology/Lithofacies		Scale		Characteristics of Reservoir Space and Pore Structure		
	Lithology	*Lithofacies*	*Net Pay Thickness (m)*	*Area (m)*	*Type of Reservoir Space*	*Pore and Fracture Assemblage*	*Average Throat Radius (μm)*
I	Volcanic conduit facies	Vesicular rhyolite and autoclastic brecciated lava	> 30	> 2	Vesicles, blast fractures, and structural fractures	Fracture-vesicle type, and fracture type	> 0.3
	Splash subfacies of explosive facies	Brecciated lava and tuffaceous lava			Vesicles, intergranular pores, dissolved pores, and structural fractures	Fracture-vesicle type, fracture-intergranular pore type, and fracture-dissolved pore type	
	Top and upper subfacies of effusive facies	Vesicular rhyolite and breccia lava	> 20	> 3	Vesicles, structural fractures, and cooling contraction fractures	Vesicle type, fracture-vesicle type	
II	Volcanic conduit facies	Vesicular rhyolite and autoclastic brecciated lava	> 20	> 1.5	Vesicles, blast fractures and structural fractures	Fracture-vesicle type, fracture-intergranular pore type	> 0.1
	Hot clastic flow, splash and fallout subfacies of explosive facies	Breccia lava, tuffaceous lava, welded breccia and tuff, and volcanic breccia			Intergranular pores, vesicles, micropores, structural fractures, and blast fractures	Intergranular pore type, fracture-intergranular pore type, fracture-vesicle type, fracture-dissolved pore type	
	Upper subfacies of effusive facies	Vesicular rhyolite, brecciated lava, and tuffaceous lava	> 15	> 2	Vesicles, structural fractures, and cooling contraction fractures	Vesicle type, fracture-vesicle type	
III	Fallout subfacies of explosive facies	Volcanic agglomerate, volcanic breccia, and tuff	> 10	> 0.5	Intergranular pores, micropores, structural fractures, and blast fractures	Intergranular pore type and fracture-micropore type	> 0.01
	Lower and middle subfacies of effusive facies	Aplitic rhyolite and tight rhyolite	> 5	> 1	Vesicles, micropores, structural fractures, and cooling contraction fractures	Vesicle type, fracture-vesicle type, and fracture-micropore type	
	Extraclastbearing and transported subfacies of volcanic sedimentary facies	Sedimentary volcanic breccia, sedimentary tuff, tuffaceous glutenite, tuffaceous gritstone, and tuffaceous sandstone			Intergranular pores, matrix micropores, and fractures	Intergranular pore type, micropore-intergranular pore type, fracture-intergranular pore type	
IV	Hot base surge subfacies of explosive facies	Crystalline, vitric and magmatic tuff	< 15	< 1	Fractures + micropores + intergranular pores	Intergranular pore type, micropore type, fracture-micropore type	> 0.01
	Middle subfacies of effusive facies	Tight rhyolite	< 10	< 1.5	Vesicles, micropores, structural fractures, and cooling contraction fractures	Isolated vesicle type, micropore type, fracture-micropore type	
	Xenoclastbearing and transported subfacies of volcanic sedimentary facies	Sedimentary tuff, tuffaceous pack sand and siltite			Intergranular pores, matrix micropores, and fractures	Intergranular pore type, micropore-intergranular pore type	

	Petrophysical Property Characteristics			Interbed Characteristics		Production Performance Characteristics		
Displacement Pressure (MPa)	*Reservoir Type*	*Porosity (%)*	*Permeability (mD)*	*Density (m/ 100 m)*	*Frequency (layers/ 100 m)*	*Well-Controlled Reserves (10^8 m^3)*	*Deliverability Characteristics*	*Stable Production Capacity*
< 2	Type I	> 10	> 1	< 10	< 2	> 5	High natural flow rate, and high deliverability after hydraulic fracturing	Strong
< 6	Types I and II	> 5	> 0.1	< 15	< 3	2-5	Moderate deliverability after hydraulic fracturing	Moderate
< 12	Types I, II, and III	> 3	> 0.01	< 30	< 5	0.5-2	Low deliverability after hydraulic fracturing	Weak
< 12	Types II and III	> 3	> 0.01	< 50	< 10	< 0.5	Noncommercial gas layer	Weak

3. *Type III A-P unit.* This type of A-P unit has relatively poor accumulation capacity with a distribution area larger than 0.5 km^2, effective thickness greater than 5 m, and well-controlled reserves ranging from 0.5×10^8 m^3 to 2×10^8 m^3. The reservoir space is composed mainly of intergranular pores, vesicles, micropores, and fractures. The pore-fracture assemblage includes intergranular pore type, vesicle type, fracture-micropore type, fracture-dissolution pore type, and fracture type. These amount to an effective porosity higher than 3%.
4. *Type IV A-P unit.* This type of A-P units has poor accumulation capacity, characterized by small scale and well-controlled reserves, limited reservoir space and pore-fracture assemblage, unfavorable A-P configuration, and low porosity. Its distribution area is less than 1.5 km^2, effective thickness is less than 15 m, and well-controlled reserve is smaller than 0.5×10^8 m^3. The reservoir space is composed mainly of intergranular pores, isolated vesicles, micropores, and fractures, and the pore-fracture assemblage includes intergranular pore type, micropore type, and fracture-micropore type, amounting to an effective porosity higher than 3%.

2) Permeation capacity

Permeation capacity is influenced by matrix pore structures and the developmental characteristics of interbeds and fractures. Generally, it is characterized with throat radius, displacement pressure, permeability, single-well deliverability, and stable production capability.

1. *Type I A-P unit.* This type of A-P unit has the highest permeation capacity, referred to as the "sweet spot" of gas reservoir. It is characterized by a large throat radius, low displacement pressure, high permeability, an underdeveloped internal interbed, high single-well deliverability, and strong stable production capability, and it has a throat radius larger than 0.3 μm, displacement pressure lower than 2 MPa, permeability less than 1 mD, interbed density lower than 10 m/100 m, interbed frequency less than 2 layers/100 m, high single-well natural production or high postfracturing production, and strong stable production capability.
2. *Type II A-P unit.* This type of A-P unit has moderate permeation capacity with a throat radius larger than 0.1 μm, displacement pressure lower than 6 MPa, permeability higher than 0.1 mD, interbed density lower than 15 m/100 m, interbed frequency less than 3 layers/100 m, moderate single-well postfracturing production, and moderate stable production capability.
3. *Type III A-P unit.* This type of A-P unit has relatively poor permeation capacity with a throat radius larger than 0.01 μm, displacement pressure lower than 12 MPa, permeability higher than 0.01 mD, interbed density lower than 30 m/100 m, interbed frequency less than 5 layers/100 m, low single-well postfracturing production, and poor stable production capability.
4. *Type IV A-P unit.* This type of A-P unit has poor permeation capacity. It is characterized by a small throat radius, high displacement pressure, low permeability, well-developed internal interbed, low single-well deliverability, and weak stable production capability, and it has a throat radius larger than 0.01 μm, displacement pressure lower than 12 MPa, permeability higher than 0.01 mD, interbed density lower than 50 m/100 m, interbed frequency less than 10 layers/100 m, single-well gas production lower than the criteria of a commercial gas well, and weak stable production capability.

10.3.4 Distribution of volcanic accumulation-permeation (A-P) units

Based on the identification and characterization of A-P units, the development characteristics of A-P units in various lithofacies and locations of volcanic edifices are analyzed, and the distribution of volcanic A-P units is summarized. These analyses are discussed next using the acidic volcanic gas reservoir in the XX gas field as an example.

1. Distribution of A-P units in various lithofacies

The type, scale, A-P capacity, and single-well production characteristics for A-P units are different in various lithofacies due to various petrogenetic processes. Based on the identification and dissection of these features, the characteristics of volcanic A-P units in various lithofacies can be delineated.

1) Types of A-P units developed in various lithofacies

Volcanic conduit facies A-P units. The lithology is mainly autoclastic brecciated lava in the volcanic conduit facies, where vesicles, dissolved pores, blast fractures, and columnar joints are developed. The A-P units have a relatively small scale and good petrophysical properties. They are mainly type II, with a small quantity of types I and III.

Explosive facies A-P units. In the splash subfacies, the lithology is mainly brecciated lava and tuffaceous lava, where vesicles, intergranular pores, and fractures are developed. The A-P units are relatively small in scale and have good petrophysical properties, and they are mainly types II and I.

In the hot clastic flow subfacies, the lithology comprises mainly welded tuff and welded breccia, where intergranular pores, micropores, vesicles, and fractures are developed. The A-P units are relatively large in scale and of relatively good petrophysical properties, being mainly type II.

In the fallout subfacies, the lithology is dominated by volcanic agglomerate, volcanic breccia, and tuff, with well-developed intergranular pores, micropores, and fractures. The A-P units are of relatively large scale and moderate to relatively poor petrophysical properties, and they are mainly types II and III.

In the hot base surge subfacies, the lithology is mainly volcanic agglomerate, volcanic breccia, and tuff, with well-developed micropores and fractures. The A-P units are large in scale but have poor petrophysical properties, represented mainly by types III and IV.

Effusive facies A-P units. In the top and upper subfacies, the lithology consists mainly of vesicular rhyolite, with well-connected vesicles and well-developed fractures. The A-P units are of large scale and good petrophysical property, being mainly types I and II.

In the lower subfacies, the lithology is mainly aplitic rhyolite, with a minor number of poorly connected vesicles but well-developed fractures. The A-P units are of relatively small to moderate scale and poor to moderate petrophysical properties, represented mainly by types II and III.

In the middle subfacies, the lithology is mainly aplitic rhyolite, where fractures and minor vesicles are developed. The A-P units are of large scale and yet poor petrophysical property, being mainly type III and IV.

Volcanic sedimentary facies A-P units. The lithology comprises mainly sedimentary volcanic rocks and volcanic sedimentary rocks of the volcanic sedimentary facies, with intergranular pores developed. The A-P units are of small scale and poor petrophysical property, and they are mainly types III and IV, developed mainly in the xenoclast-bearing subfacies and retransported subfacies.

2) Scale of A-P units developed in various lithofacies (Figure 10.42)

Volcanic conduit facies A-P units. Dimensions of the AP units: length 12.2 km, average 1.4 km; width 0.8 to 2.1 km, average 1.3 km; thickness 28 to 42 m, average 35 m; area 0.8 to 2.6 km^2, average 2 km^2. With a small area and large thickness, they are relatively small in scale.

Eruptive facies A-P units. Dimensions of the AP units: length 1 to 3.2 km, average 2.8 km; width 0.9 to 3.4 km, average 2.1 km; thickness 9 to 88 m, average 35 m; area 0.9 to 7.5 km^2, average 4.2 km^2. With a relatively large area and large thickness, they are considered relatively large in scale.

Effusive facies A-P units. Dimensions of the AP units: length 1.3 to 3.5 km, average 2.5 km; width 1.3 to 3.4 km, average 2.2 km; thickness 10 to 94 m, average 25 m; area 1.9 to 9.9 km^2, average 5.8 km^2. With a large area and relatively large thickness, they are the largest in scale.

Volcanic sedimentary facies A-P units. Dimensions of the AP units: length 0.8 to 1.7 km, average 1.2 km; width 0.6 to 0.9 km, average 0.8 km; thickness 11 to 22 m, average 16 m; area 0.8 to 1 km^2, average 0.9 km^2. With a small area and small thickness, they are the smallest in scale.

In summary, the A-P units of effusive facies have the largest size, those of explosive facies rank second, and those of volcanic sedimentary facies are the smallest among various volcanic facies in the XX gas field.

3) A-P capacity and deliverability of A-P units developed in various lithofacies

Volcanic conduit facies. Type II A-P units are mainly developed in volcanic conduit facies with limited accumulation capacity but good permeation capacity, low well-controlled reserves, high deliverability, but poor stable production capability.

Explosive facies. Types II and I A-P units are mainly developed in the splash subfacies, with relatively good A-P capacity, high well-controlled reserves, high deliverability, and strong stable production capability. Type II A-P units are also common in the hot clastic flow subfacies, with relatively good A-P capacity, high well-controlled reserves, relatively high deliverability, and relatively strong stable production capability. Types II and III A-P units are predominant in the fallout subfacies, with moderate A-P capacity, well-controlled reserves, moderate deliverability, and stable production capability. Types III and IV A-P units are commonly developed in the hot base surge subfacies, with poor A-P capacity, low well-controlled reserves and deliverability, and weak stable production capability.

Effusive facies. Types I and II A-P units are commonly developed in the top and upper subfacies, with strong A-P capacity, high well-controlled reserves and production, and strong stable production capability. Types II and III A-P units are mainly developed in the lower subfacies, with relatively strong A-P capacity, relatively high well-controlled reserves and production, and relatively strong stable production capability. Types III and IV A-P units are dominant in the middle subfacies, with poor A-P capacity, low well-controlled reserves and production, and poor stable production capability.

Volcanic sedimentary facies. Types III and IV A-P units are common in the volcanic sedimentary facies with poor A-P capacity, low well-controlled reserves and production, and poor stable production capability.

2. Spatial distribution of volcanic A-P units

Based on the identification of A-P units, a preliminary assessment of the spatial distribution of volcanic A-P units can be achieved by analyzing the profile distribution (Figure 10.43), planar distribution

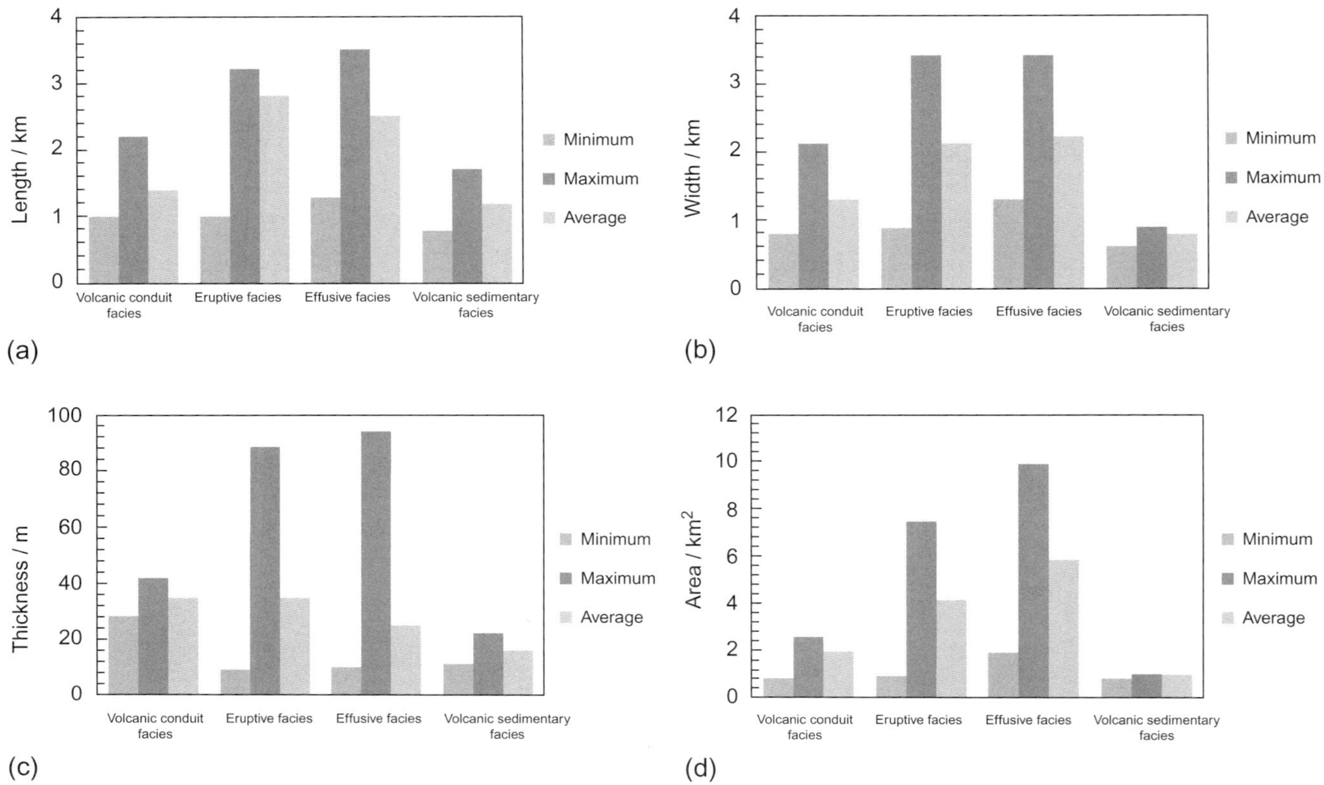

FIGURE 10.42

Scale of A-P units in various facies (acidic volcanic rock in the XX gas field). (a) Length of accumulation-permeation units /km. (b) Width of accumulation-permeation units /km. (c) Thickness of accumulation-permeation units /m. (d) Area of accumulation units /km^2.

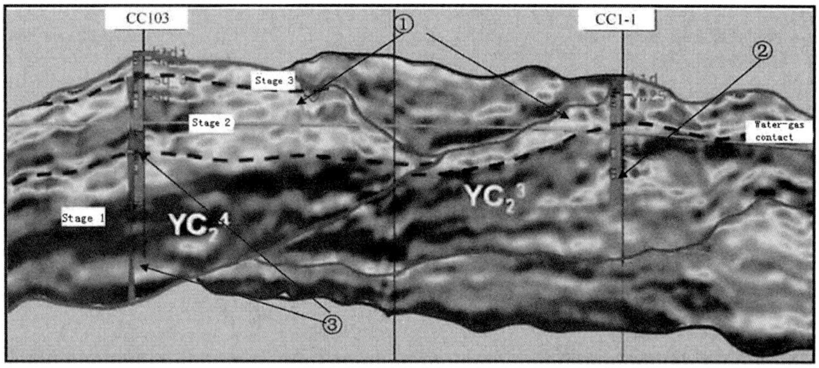

FIGURE 10.43

Profile distribution of A-P units in volcanic gas reservoir (density inversion profile).

(Figure 10.44), and spatial distribution (see Figure 10.28) characteristics of A-P units. The favorable areas for A-P unit distribution are classified into three types as follows:

1. *The top area adjacent to the volcanic massif* (see Figure 10.43). The top part of the rock body has well-developed primary pores and fractures, with strong secondary dissolution. This facilitates the formation of fracture-dissolution pore type A-P units with relatively large thickness and fracture-primary pore type A-P units.
2. *The main part of the volcanic massif.* This refers to the crater and near-crater facies belts. Due to rapid accumulation, there are various primary pores and fractures, which favors the formation of fracture-primary pore type A-P units.
3. *The upper and lower parts of effusive facies.* There are vesicles and contraction fractures with little effect of compaction, which facilitates the formation of primary laminar pore type A-P units.

10.4 APPLICATIONS

The characterization technology of A-P units discussed in this chapter can be used to delineate the shape, scale, connectivity, and accumulation-permeation capacity of various A-P units in a volcanic gas reservoir. This provides a basis for optimizing well location and horizontal well trajectory design, evaluating well-controlled dynamic reserves, and optimizing development technology policy.

10.4.1 Revealing the planar distribution of A-P units to guide the optimization of well location

The planar distribution of A-P unit is the basis of well placement. The planar prediction of A-P units helps delineate their planar distribution pattern, which in turn serves as a guide for optimizing well location and improving the success rate of developmental well placement.

In accordance with the planar distribution of volcanic massifs (Figure 10.45a) and A-P units (Figure 10.45b), a total of seven horizontal wells were placed near the crater facies belt, near the

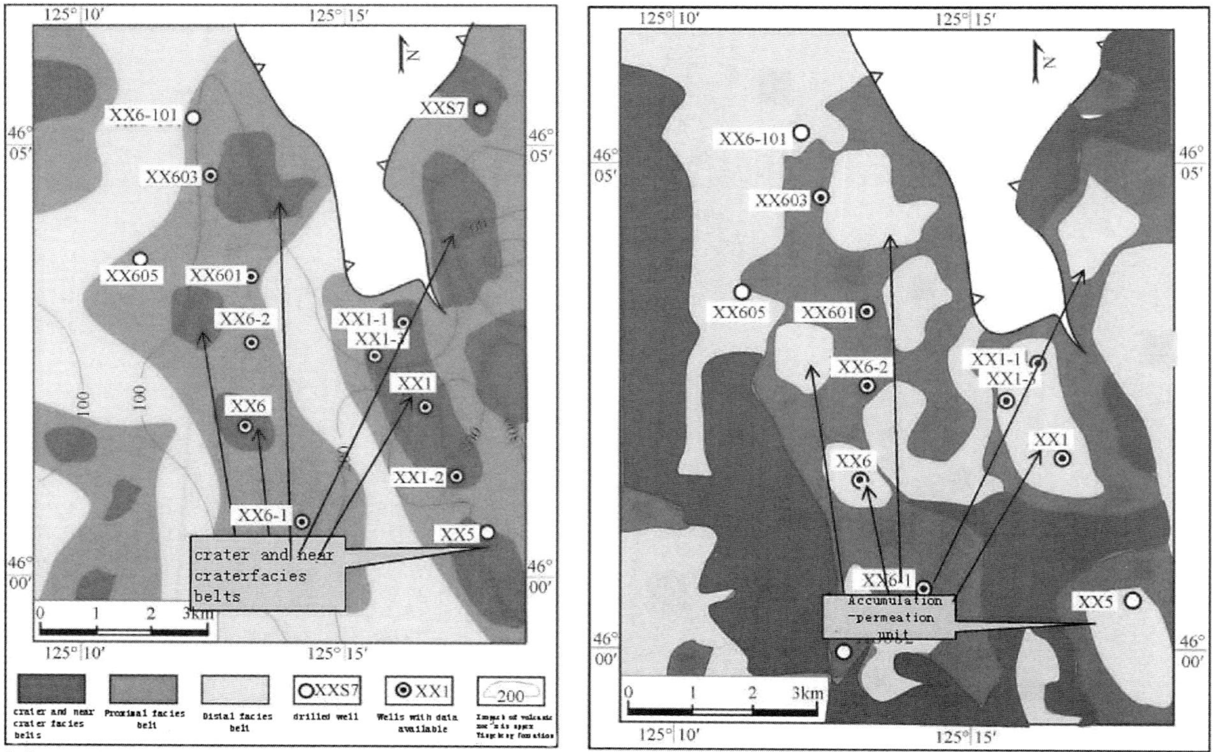

(a) (b)

FIGURE 10.44

Planar distribution of lithofacies and A-P units in a volcanic gas reservoir (Well Block XX1). (a) Planar distribution of facies belts in volcanic edifice. (b) Planar distribution of volcanic A-P units.

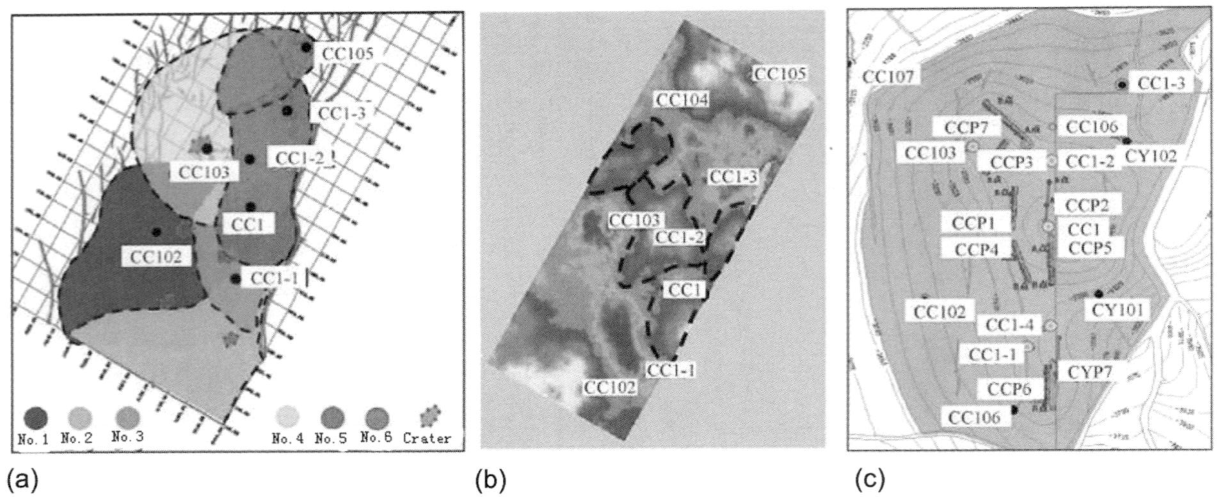

FIGURE 10.45

Planar distribution of volcanic massifs and A-P units and well placement (CC gas reservoir). (a) Planar distribution of volcanic massifs. (b) Planar distribution of A-P units. (c) Well placement.

top area of a volcanic massif and the effusive facies area in the CC gas field (Figure 10.45c). The gas test after well completion shows that single-well daily gas production ranges from 24×10^4 m^3/d to 51×10^4 m^3/d, which is a desirable verification.

10.4.2 Optimizing horizontal well trajectory design under the guidance of the spatial distribution of A-P units

The spatial distribution of A-P units serves as a guide for optimizing horizontal well trajectory design. Based on the spatial prediction results of several methods, the spatial distribution characteristics of volcanic A-P units are clarified, and they provide a basis for optimizing horizontal well trajectory design, improving the drilling success rate of horizontal wells, and enhancing single-well production.

In the case study of the XX gas field, the trajectory of horizontal Well XX1-P2 (Figure 10.46) is optimized and modified timely during drilling based on the prediction of the 3D spatial distribution of A-P units (see Figure 10.28). After the completion, the effective reservoir drilling success rate of the horizontal wells reached 98%, and the type I effective reservoir drilling success rate was up to 86%. After perforation, daily gas production reached 24×10^4 m^3/d under natural conditions.

10.4.3 Guiding well-controlled dynamic reserve evaluation based on the scale and A-P capacity of A-P units

Both the scale and the A-P capacity of A-P units usually have an effect on well-controlled dynamic reserves, as is shown by the XX gas field. Characterization of the scale and A-P capacity of A-P units is used to establish the correlation between well-controlled dynamic reserves and the volume of A-P units, as well as the relationship between well-controlled dynamic reserves and the permeability of a connected rock body (Figures 10.47 and 10.48). The figures show that larger-scale A-P units are positively correlated with higher permeability and larger single-well-controlled reserves. As a result, quantitative characterization of both the scale and the capacity of A-P units has important implications for evaluating well-controlled dynamic reserves.

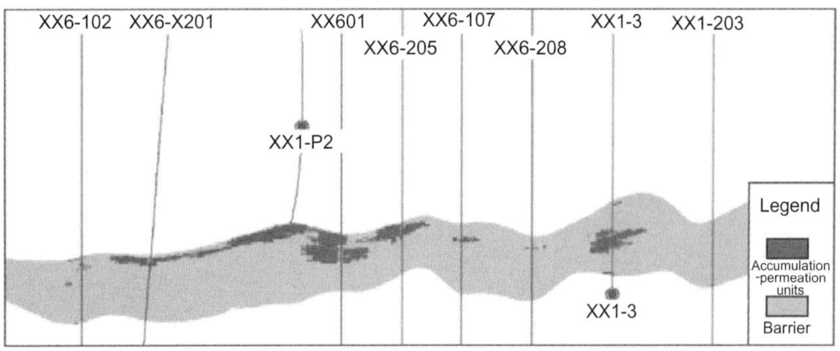

FIGURE 10.46

Optimization of horizontal well trajectory design based on the distribution of A-P units (Well XX1-P2).

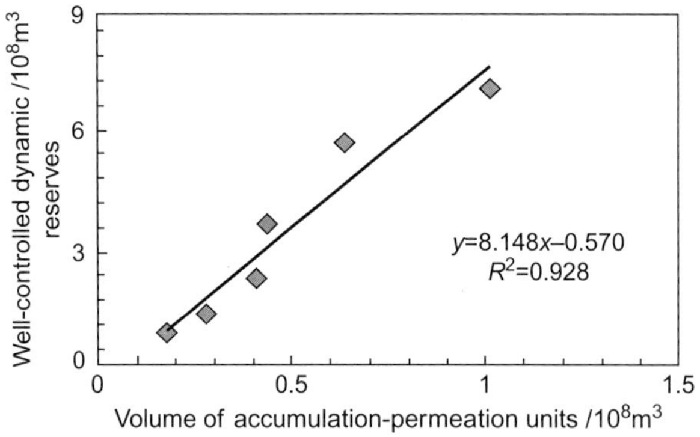

FIGURE 10.47

Crossplot of well-controlled dynamic reserves versus the volume of A-P units.

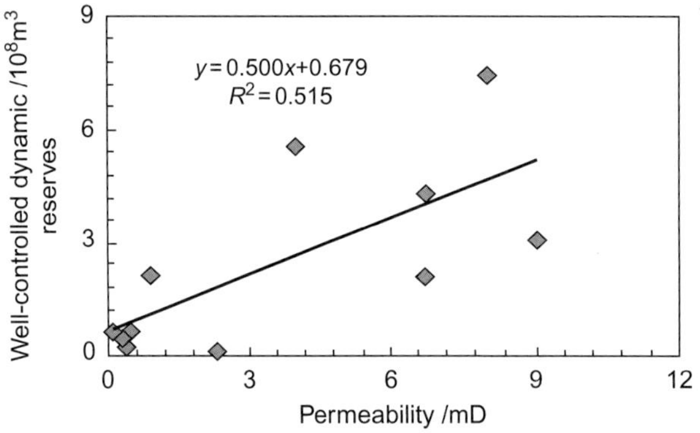

FIGURE 10.48

Crossplot of well-controlled dynamic reserves versus the permeability of the connecting body.

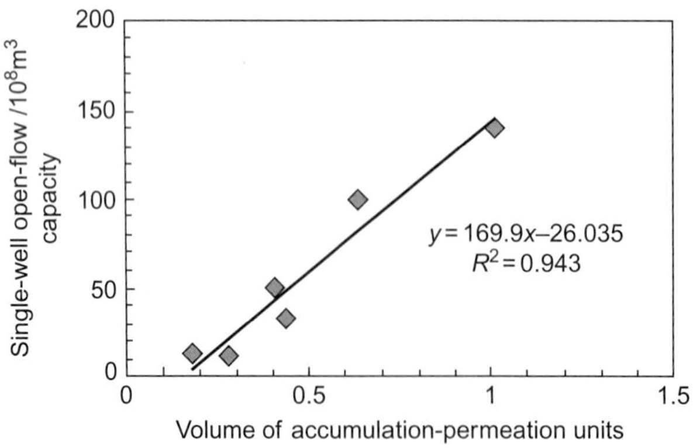

FIGURE 10.49

Crossplot of single-well open-flow potential versus the scale of A-P units.

The deliverability of gas wells is influenced by both the scale and the A-P capacity of A-P units. Larger-scale A-P units are positively correlated with higher permeability and higher single-well open-flow potential. As the XX gas field example (Figure 10.49) shows, the correlation coefficient between single-well open-flow potential and the scale of A-P units is up to 0.95. Therefore, the characterization of the scale and A-P capacity of A-P units provides a basis for assessing the stable production capability of gas wells.

References

[1] Cai C. An outline of the history of logic dialectics. Guangzhou: Jinan University Press; 1996, p. 182–186.

[2] Hearn CL, Ebanks JR, et al. Geological factors influencing reservoir performance of Hartzog Draw field. Wyoming. J Petrol Tech 1984;36(8):1335–44.

[3] Ebanks WJ. Flow unit concept-integrated approach to reservoir description for engineering projects. AAPG Bulletin 1987;71(5):551, 552.

[4] Amaerule JO, Altunbay M, Tiab D, et al. Enhanced reservoir description: using core and log data to identify hydraulic (flow) units and predict permeability in uncork intervals/wells. SPE 26436, 1993.

[5] Qiu Y, Jia A. Development of geological reservoir modeling in past decade. Acta Petrolei Sinica 2000;21 (4):101–4.

[6] Mu L, Jia W, Jia Ai. A new approach to build quantitative geological reservoir model. Petroleum Exploration & Development 1994;21(4):82–6.

[7] Jiao Y, Li S. Geologic modeling for outcrop reservoir of continental basin and the conceptual systems. Experimental Petroleum Geology 1998;20(4):346–53.

[8] Liu W. Dynamic model and remaining oil distribution for polymer flooding reservoir. Beijing: Petroleum Industry Press; 2008, p. 41–46.

[9] Yuan Y. How to understand reservoir. Beijing: Petroleum Industry Press; 1979, p. 38–40.

[10] Chen B. Subsurface geology. Beijing: Petroleum Industry Press; 1987, p. 165–68.

[11] Xiong Q. Foundations of logging geology (part 2). Beijing: Petroleum Industry Press; 1987, p. 214, 215.

[12] Wu Y, Wu S, Cai Z. Subsurface geology (the 3rd version). Beijing: Petroleum Industry Press; 2005, p. 140–42.

[13] Cheng G, Wang Y. Subsurface geology of oil and gas fields. Beijing: Petroleum Industry Press; 2007, p. 172–75.

[14] Jiao F, Dou Z. Development research & practice on fracture-pore carbonate reservoir in Tahe oilfield. Beijing: Petroleum Industry Press; 2006, p. 22–70.

[15] Jiao F, Dou Z. A collection of research papers on the development of Tahe oil and gas field. Beijing: Petroleum Industry Press; 2006, p. 17–22.

[16] Zhu Y. The Impact of formation damage upon flow units in lithologic pinchout oil and as accumulation with reference to Huachi and Maling oilfield in Ordos basin. In: Xi'an: Northwest University Press; 2005. p. 18–28.

[17] Dou Z. Research on reservoir flow units. Beijing: Petroleum Industry Press; 2000, p. 28–42.

[18] Qin T. Precise seismic description for Cainozoic reservoir of Huaguang sag in Qiongdongnan basin. Beijing: China University of Petroleum; 2009, p. 77–95.

[19] Yu X. Clastic petroleum reservoir sedimentology. Beijing: Petroleum Industry Press; 2008, p. 23.

[20] Wang P, Feng Z. Basin volcanics: lithology, lithofacies, reservoir, gas reservoir and exploration. Beijing: Science Press; 2008, p. 234–50.

CHAPTER

Characterization of Microstructures of Volcanic Gas Reservoirs 11

Reservoir microstructures refer to the geometry, size, distribution, and communication of pores and throats in reservoir rocks, which directly affect the reservoir accumulation and permeation capacity and ultimately determine the variable distribution of gas reservoir deliverability [1]. By means of core description, microscopic observation, and experimental analysis, the microstructural characterization of volcanic gas reservoirs can reveal the microscale shape, size, and accumulation capacity of volcanic reservoir spaces as well as the shape, scale, and permeation capacity of seepage pathways. It can also clarify the accumulation-permeation capacity and distribution characteristics of volcanic reservoirs with various accumulation-permeation modes, and ensure that the effectiveness and fluid-producing capacity of volcanic reservoirs are characterized quantitatively, so as to provide guidance for evaluating their microscale producing capacity and calibrating their recovery ratio.

11.1 CHALLENGES AND SOLUTIONS

11.1.1 Challenges in characterizing volcanic reservoir microstructures

Volcanic reservoirs have their specific petrogeneses and a great variety of reservoir spaces and throats as well as complex pore-throat configurations. This results in extremely complex microstructures that pose tough challenges in their identification and characterization with regard to the following aspects.

Intergranular pores are predominant in conventional sedimentary reservoirs, and reservoir spaces in volcanic reservoirs are highly variable, as there are up to a dozen types. Because of their complex petrogenetic origins, various influencing factors, and different preservation conditions, reservoir spaces in volcanic reservoirs are extremely complex for a variety of reasons: (1) various types of reservoir spaces differ significantly in shape, and the shapes of the same type of reservoir space vary greatly under different conditions of occurrence and preservation; (2) the radius of a reservoir space can range from several micrometers to tens of millimeters, with a wide range of uneven distribution; (3) the variety in the shapes and sizes of reservoir spaces contributes to the large differences in accumulation capacity. In volcanic reservoirs, porosity varies from several thousandths to over 20%, and the maximum mercury intrusion saturation ranges from a few percentages to over 95%. Therefore, it is difficult to characterize the shape, size, and accumulation capacity of reservoir spaces in volcanic rocks.

Compared to conventional sedimentary reservoirs, the throats in volcanic reservoirs have more complex petrogenetic origins and are represented by a variety of types (contracted, dissolved, and fractured), a wide range of sizes (with throat radii that can range from 10^{-3} μm to several μm), and distinctly different permeation capacities (with a permeability differential up to several thousands). Currently, the research methods are relatively simple. Specifically, the throat shape is observed qualitatively in cast thin sections, and the throat size is determined quantitatively from the data derived from mercury injections and constant-rate mercury injections. Therefore, it is difficult to characterize the shape, size, and permeation capacity of throats in volcanic reservoirs.

The complexity of reservoir spaces and throats leads to complicated and variable accumulation-permeation modes in volcanic reservoirs. Based on the assemblage modes of reservoir spaces and seepage pathways, eight basic accumulation-permeation (A-P) patterns are recognized in volcanic reservoirs. Currently, the categories of A-P patterns are identified qualitatively by examining cores and thin sections; the accumulation and permeation capacities of various volcanic reservoirs are evaluated by log interpretation and dynamic testing. Due to the variation in geological characteristics of the tested horizon, the reliability of the testing method, and the information quality and quantity of testing points, it is difficult to quantitatively evaluate the accumulation-permeation capacity and deliverability of volcanic reservoirs with different A-P patterns.

Multiple modes of pore-fracture assemblage and the extremely strong heterogeneity of volcanic rock reservoirs result in complex microporous structures in volcanic reservoirs in 3D space. Currently, mercury injection and nuclear magnetic resonance are the primary means for determining the effective pore volumes and mobile (movable) fluid volumes under the control of throats, permeability, and displacement pressures at various levels and for evaluating the producing capacity of different reservoir fluids. Constrained by many factors such as insufficient quantity and representation of samples, it is difficult to precisely evaluate the effectiveness and producing capacity of volcanic reservoirs.

11.1.2 Technical solutions and approaches

Most previous studies on the microscopic characteristics of volcanic reservoirs provided qualitative descriptions. In this book, the microstructures are characterized in detail step by step according to a technical approach that progresses from qualitative to semiquantitative and then to quantitative study, taking into consideration the complexity of volcanic reservoir microstructures, by systematically integrating data from cores, thin sections, and experiments such as mercury injection, constant-rate mercury injection, and nuclear magnetic resonance. This study focuses on the four key topics: reservoir space, the fluid flow pathway, the accumulation-permeation pattern, and the microscopic producing capacity (Figure 11.1).

1. Characterization of reservoir spaces

The volcanic reservoir spaces are usually characterized in three steps. (1) In terms of their origins, shape, and size, a logical classification scheme is established to divide the reservoir spaces in volcanic gas reservoirs. (2) The classification scheme provides identification markers for various types of reservoir spaces; then, the types of reservoir spaces in volcanic rocks are identified, and the distribution and accumulation characteristics of various types of reservoir spaces are described based on the shape, scale, configuration, and dissolution traces, in combination with the characteristics of lithology and lithofacies, rock structure, and texture that are associated with the reservoir spaces. (3) The morphology of various types of reservoir spaces is characterized qualitatively based on an observation of cores and thin sections; the size and accumulation capacity are characterized quantitatively by means of experiments such as porosity and permeability tests, conventional mercury injection, and constant-rate mercury injection.

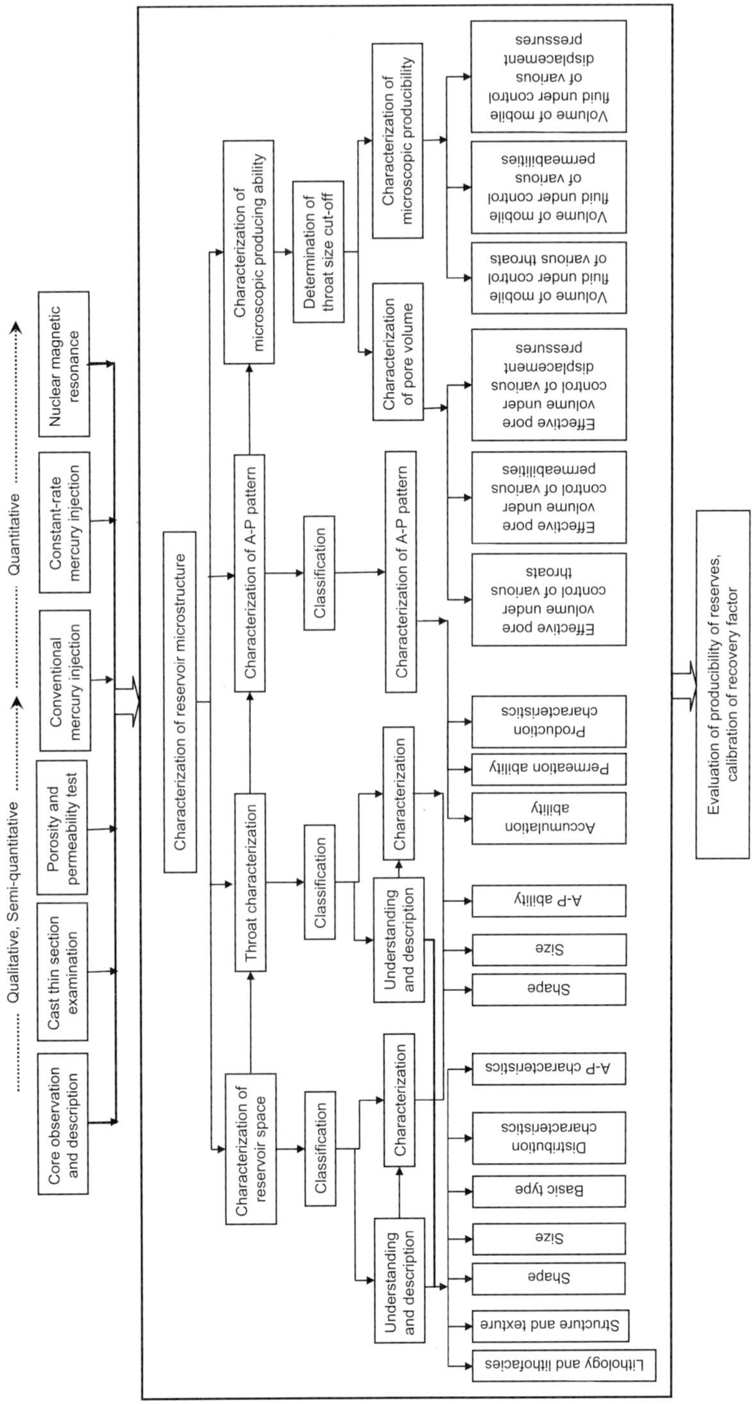

FIGURE 11.1
Technical process of characterization of microstructures in volcanic gas reservoirs.

2. Characterization of throats

Similar to the characterization of reservoir spaces, throats in volcanic reservoirs are also characterized in three steps. (1) In terms of their origins, shape, and size, the throats in volcanic reservoirs are classified. (2) Identification markers of various types of throat are set up; then the throat types in volcanic rocks are identified and the distribution and seepage characteristics of various types of throats are described based on throat size, shape, and dissolution traces, in combination with the characteristics of lithology and lithofacies, rock texture, and structure that are associated with the throats. (3) The shape characteristics of various types of throats are characterized qualitatively using core observation and thin section examination, and the sizes and permeation capacities are characterized quantitatively by means of experiments such as porosity and permeability tests, conventional mercury injection, and constant-rate mercury injection.

3. Characterization of A-P patterns

On the basis of characterized reservoir spaces and throats, the A-P patterns in volcanic gas reservoirs are established according to the modes of pore-throat-fracture assemblage. For various A-P patterns, the accumulation-permeation capacity and production characteristics are characterized respectively and the control of the A-P pattern on production characteristics is clarified by integrated analysis of static and dynamic data, so as to guide the formulation and optimization of technical policy for development.

4. Characterization of reservoir microscopic producing capacity

The cutoff threshold of throat size in volcanic gas reservoirs is determined by various methods. On this basis, the permeability cutoff of volcanic reservoirs is determined, the classification criteria of throat and permeability are set up, the effective pore volumes and mobile fluid volumes are evaluated under the control of various throat, permeability, and displacement pressure parameters, and the characteristics of the microscale fluid-producing capacity of volcanic rocks under different controlling conditions are clarified. These criteria provide a basis for evaluating reserve producing capacity and calibrating recovery parameters for volcanic gas reservoirs.

11.2 CHARACTERIZATION OF RESERVOIR SPACES IN VOLCANIC GAS RESERVOIRS

11.2.1 Classification of reservoir spaces in volcanic gas reservoirs

The classification scheme of reservoir spaces in conventional sedimentary reservoirs is not suited for volcanic reservoirs because of the fundamental difference in the origin of volcanic reservoirs compared with normal sedimentary reservoirs. By now, a great deal of work has been done on the classification of reservoir spaces in volcanic rocks [2–7]. However, a unified classification scheme is still unavailable due to the complexity of volcanic reservoir spaces. To solve this practical problem, a classification scheme for reservoir spaces in volcanic rocks is established in this study according to the features of volcanic reservoir spaces, with references to the previous research results.

1. Research results from previous work

Zhao and colleagues [2–4] divided the reservoir spaces in volcanic rocks into two categories: pores and fractures. They subdivided pores into 10 categories: vesicles, pores within amygdaloids, interporphyrocrystic pores, contraction pores, intermicritic pores, interhyalinocrystalline pores, intracrystal pores, dissolution pores, expanded pores, and plastic flow pores. They subdivided fractures into six categories: structural fractures, cryptoexplosive fractures, diagenetic fractures, weathering fractures, vertical joints, and columnar joints.

Ren and coworkers [5] examined in detail the categories and characteristics of reservoir spaces in volcanic reservoirs in their study on the volcanic reservoir in Well Block 609 of the Liaohe depression. As a result, these authors divided the volcanic reservoir spaces into a primary category and a secondary category; the primary pores include primary vesicles, residual vesicles, interporphyrocrystic pores, intercrystal pores, pores within amygdaloids, and contraction pores; the primary fractures include cooling contraction fractures, contraction joints, and interclast fractures. The secondary pores include porphyrocrystic dissolution pores, intraporphyrocrystic and intramatrix dissolution pores, matrix dissolution pores, devitrified dissolution pores, dissolution pores in amygdaloids, dissolution pores in altered rocks, postfilling redissolution pores, and dissolution pores in metasomatic deposits; the secondary fractures include structural fractures and weathering fractures.

Yu and colleagues [6] divided the reservoir spaces in volcanic rocks into primary and secondary categories in their study on the volcanic reservoirs of the Wucaiwan depression (Junggar basin). The primary reservoir spaces include (1) primary pores (primary vesicles, residual vesicles, intercrystal pores and intra-crystal pores); (2) primary fractures (cooling contraction fractures); (3) secondary pores, including porphyrocrystic dissolution pores and dissolution pores in amygdaloids; and (4) secondary fractures, mainly including structural fractures and weathering fractures.

Liu and coworkers [7] divided the reservoir spaces in volcanic rocks into pores and fractures based on the morphology in their study on the evolution of the volcanic reservoir space in the Yingcheng Formation of the Xujiaweizi faulted depression in the Songliao basin. These authors recognized 15 types in four categories in terms of their origins: (1) primary pores, including vesicles, amygdaloidal pores, interclast pores, and intercrystal pores; (2) secondary pores, including interclast dissolution pores, intercrystal dissolution pores, intracrystal dissolution pores, devitrified dissolution pores, and dissolution pores in altered deposits; (3) primary fractures, including contraction fractures, contraction cleavages, and grain explosion cracks; and (4) secondary fractures, including structural fractures, dissolution fractures, and weathering fractures.

As this discussion implies, the classification schemes of reservoir spaces have been proposed from different perspectives, which partly reflects the complicated origins and diverse types of reservoir spaces in volcanic reservoirs. However, the shapes and origins of reservoir spaces have been taken into account when volcanic geologists have proposed classification schemes.

2. Classification scheme used in this book

With regard to the features of reservoir spaces in volcanic gas reservoirs such as their complex origins and their highly variable shapes and scales, the reservoir spaces in volcanic rocks are divided into two categories based on shape first: pores and fractures. Pores refer to a type of reservoir space that has an approximately equidimensional morphology; fractures refer to sheetlike reservoir spaces in 3D, in which two of the dimensions are far larger than the third (i.e., the ratio is greater than 10:1). Then, in terms of their origins, pores are subdivided into three categories: vesicles, intergranular pores, and dissolution pores, and fractures are subdivided into five categories: contraction fractures, structure fractures, blast fractures, dissolution fractures, and sutured fractures. Finally, according to the scale of reservoir space, the pores of different origins are further subdivided into 11 types at four levels: cavities, macropores, small pores, and micropores; and fractures are divided into four types: macro-, medium, small, and microfractures (Table 11.1).

1) Vesicles

When a volcano erupts, the magma containing a great amount of volatiles is degassed upon reaching the surface, and the volatiles escape from this magma to leave behind primary pores (e.g., vesicles), which are the products of effusion. The radius of vesicles in volcanic rocks varies significantly (up to 100 mm and over). In terms of pore radius [8], vesicles can be divided into four types: vesicular pores (those with a radius > 2 mm), microvesicle (with a radius from 0.5 mm to 2 mm), small vesicles (with a radius from 0.01 mm to 0.5 mm), and microvesicles (with a radius < 0.01 mm).

2) Intergranular pores

Intergranular pores refer to the primary pores formed between clasts during accumulation and compaction after a great amount of volcanic clastic mass erupted out of surface as the product of explosion. In terms of their radius and genesis, intergranular pores can be divided into four types: intergranular cavities (with a radius > 2 mm), macro-intergranular pores (with a radius from 0.5 mm to 2 mm), small intergranular pores (with a radius from 0.01 mm to 0.5 mm), and matrix micropores (with a radius < 0.01 mm).

3) Dissolution pores

Dissolution pores refer to the secondary pores originating from the formation of vesicles, intergranular pores, and matrixes as a result of partial or complete dissolution of soluble components such as feldspar phenocryst, and volcanic ash in rocks by meteoric water, organic acids, and acidic water with CO_2 retained during burial and diagenetic evolution after volcanic rocks and primary pores formed. In terms of pore radius, dissolution pores can be divided into three types: dissolution cavities (with a radius > 2 mm), dissolution pores (with a radius from 0.5 mm to 2 mm), and matrix dissolution pores (with a radius < 0.5 mm).

4) Fractures

Fracture refers to the sheetlike space in volcanic rocks. In terms of their origins, fractures are divided into five types: (1) contraction fractures, which are primary fractures formed by irregular contraction of magma due to an uneven decrease in internal temperature during volcanic action; (2) blast fractures, primary fractures formed in volcanic rocks when gas-liquid exploded in a volcano; (3) structural fractures, secondary fractures generated by tectonic activities after volcanic rocks are formed; (4) dissolution fractures, secondary fractures formed by enlargement of original fractures due to the dissolution of soluble components when meteoric water or underground water seeps along fractures; and (5) sutured fractures, serrated contact fractures between adjacent rock beds or two parts of the same rock bed, generated by a pressure solution. With reference to the classification criteria of fracture width in carbonate rocks [8], fractures in volcanic rocks are divided into four categories: macrofractures (those with a width > 10 mm), medium fractures (with a width from 1 mm to 10 mm), small fractures (with a width from 0.1 mm to 1 mm), and microfractures (with a width < 0.1 mm). Generally speaking, structural fractures in volcanic rocks have a large width and length and are mostly small to macrofractures; dissolution fractures have complex origins and their width and original length vary greatly before dissolution enlarges the width further, but they are mostly small to macrofractures; sutured fractures have a large extension but are usually closed; contraction and blast fractures have a small width and extension and are normally micro to small fractures. In comparison, the large columnar joints developed near volcanic conduits are of contraction origin, dominated by medium fractures.

11.2.2 Identification of reservoir spaces in volcanic gas reservoirs

The identification of reservoir spaces is crucial for understanding the developmental characteristics of reservoir spaces in volcanic rocks. It provides a basis for characterizing the shape, scale, and accumulation capacity of reservoir spaces. By taking into consideration the complexity of reservoir spaces in volcanic rocks, the identification markers for various types of reservoir spaces are set up (Table 11.2) based on the classification, and the identification is conducted by integrating outcrop observation, core description, and thin section examination, according to the technical approach that progresses from macroscopic to microscopic analysis. The procedures mainly include (1) distinguishing pores and fractures according to shape markers; (2) identifying dissolution pores and fractures according to markers of dissolution traces; and (3) identifying effective reservoir spaces in terms of the types of reservoir spaces (such as pores and fractures), respectively, on the basis of lithology, lithofacies, rock structure, and texture as well as the identification markers such as shape, scale, and configuration of reservoir spaces.

1. Identification of vesicles

Vesicles are the product of effusion, with different characteristics of size, shape, and configuration from other types of pores. According to their origins and characteristics, therefore, vesicles in volcanic rocks can be identified effectively by taking an integrated analysis of their characteristics in lithology, lithofacies, texture, and structure of vesicle-bearing volcanic rocks.

1) Characteristics of lithology and lithofacies

Volcanic lithofacies represent an important controlling factor for the type and development of primary pores [9]. Effusion and degassing of volatiles are key factors controlling the formation of vesicles. In volcanic rocks, therefore, vesicles are typically developed in the effusive facies (see Table 11.2), especially in lithotypes such as rhyolite, andesite, and basalt. In addition, vesicles can also be developed in other volcanic lithofacies with effusion involved, such as the splash subfacies and hot clastic flow subfacies of the explosive facies, with greater effusion correlated with better development of vesicles.

2) Characteristics of rock structure and texture

As the markers for lithology and lithofacies characteristics, typical rock textures and structures can serve as indirect guides for identifying reservoir spaces. The effusion of magma usually results in the directional arrangement of vesicles and magma constituents; therefore, typical rhyolitic structure and welded texture are formed, and spherulitic textures are usually formed during the cooling of magma. Hence, rhyolitic structure, spherulitic texture, and welded texture are indirect identification markers for vesicles (Figure 11.2b and c).

Table 11.1 Classification and Characteristics of Reservoir Spaces in Volcanic Gas Reservoirs

Category of Reservoir Spaces			**Basic Characteristics of Reservoir Spaces**			
Shape Type	***Genetic Classification/ Distribution in XX Gas Reservoir***	***Scale Classification***	***Pore Diameter, mm***	***Distribution Percentage in XX Gas Reservoir***	***Lithology***	***Lithofacies/Subfacies***
Pore	Vesicle, 45%	Vesicular cavities	> 2	3	Vesicular lava, breccia lava	Top subfacies of effusive facies, splash subfacies of explosive facies
		Macrovesicles	0.5-2	12	Vesicular lava	Upper subfacies of effusive facies, lower subfacies of effusive facies
		Small vesicles	0.01-0.5	20	Tight lava, welded tuff, welded breccia	Middle subfacies of effusive facies, hot clastic flow subfacies of explosive facies
		Microvesicles	< 0.01	10	Tight lava	Middle subfacies of effusive facies
	Inte-granular pore, 35%	Intergranular cavities	> 2	2	Volcanic agglomerate, volcanic breccia	Fallout subfacies of explosive facies
		Macro-intergranular pores	0.5-2	8	Volcanic breccia, welded breccia, autoclastic breccia lava	Fallout subfacies of explosive facies, hot clastic flow subfacies of explosive facies, volcanic conduit facies
		Small intergranular pores	0.01-0.5	15	Volcanic breccia, welded breccia, welded tuff	Fallout subfacies of explosive facies, hot clastic flow subfacies of explosive facies
		Matrix micropores	< 0.01	10	Crystalline tuff, welded tuff	Hot base surge subfacies of explosive facies, hot clastic flow subfacies of explosive facies
Pore	Dissolution pore, 20%	Dissolution cavities	> 2	3	Vesicular lava, breccia lava, volcanic agglomerate, volcanic breccia	Top subfacies of effusive facies, upper subfacies of effusive facies, splash subfacies of explosive facies, fallout subfacies of explosive facies,

Basic Characteristics of Reservoir Spaces							
Pore Shape	*Accumulation Ability*	*Throat Type*	*Throat Shape*	*Throat Radius, μm*	*Perm-eation Ability*	*Pattern of Accumulation-Permeation*	*Deliverability*
Nearly round, elliptic, rhombic	Strong	Pore-contraction (large throat), fractured	Thick-short tubular, sheetlike	> 0.5 dominantly	Strong	Vesicular	High
Sub-rounded, elliptic, rhombic	Fairly strong	Pore-contraction (large throat, medium throat), fractured	Thick-short tubular, small-short tubular, sheetlike	0.25-0.5 dominantly	Strong, medium	Vesicular, fractured-vesicular	High, medium
Nearly round, elliptic	Fairly weak	Pore-contraction (medium throat, small throat), fractured	Small-short tubular, sheetlike	0.1-0.25 dominantly	Weak	Fractured-vesicular	Medium, low
Nearly round	Weak	Pore-contraction (microthroat), fractured	Small-short tubular, sheetlike	< 0.1 dominantly	Weak	Microporous, fractured-microporous	Low
Stripelike, dendritic	Strong	Pore-contraction (large throat), fractured	Thick-short tubular, thick-long tubular, sheetlike	> 0.5 dominantly	Strong	Intergranular porous	High
Rhombic, striped, dendritic	Fairly strong	Pore-contraction (large throat, medium throat), fractured	Thick-short tubular, thick-long tubular, small-long tubular, sheetlike	0.25-0.5 dominantly	Medium	Intergranular porous, fractured–intergranular porous	Medium
Rhombic, stripelike	Fairly weak	Pore-contraction (medium throat, small throat), fractured	Small-short tubular, small-long tubular, sheetlike	0.1-0.25 dominantly	Medium, weak	Intergranular porous, fractured–intergranular porous	Medium, low
Rhombic	Weak	Pore-contraction (microthroat), fractured	Small-short tubular, sheetlike	< 0.1 dominantly	Weak	Microporous, fractured-microporous	Low
Nearly round, stripe like, regular phenocrystic	Strong	Dissolution throat (large throat), fractured	(Ir)regular thick-short tubular, (ir)regular thick-long tubular, irregular sheetlike	> 0.5 mainly	Strong	Dissolution-porous, fractured–dissolution-porous	High

Continued

Table 11.1 Classification and Characteristics of Reservoir Spaces in Volcanic Gas Reservoirs—cont'd

Category of Reservoir Spaces			**Basic Characteristics of Reservoir Spaces**			
Shape Type	***Genetic Classification/ Distribution in XX Gas Reservoir***	***Scale Classification***	***Pore Diameter, mm***	***Distribution Percentage in XX Gas Reservoir***	***Lithology***	***Lithofacies/Subfacies***
		Dissolution pores	0.5-2	10	Vesicular lava, breccia lava, welded breccia, volcanic breccia, autoclastic breccia lava	Upper subfacies of effusive facies, splash subfacies of explosive facies, hot clastic flow subfacies of explosive facies, fallout subfacies of explosive facies, volcanic conduit facies
		Matrix dissolution pores	< 0.5	7	Vesicular lava, tuffaceous lava, welded tuff, crystalline tuff, tuff	Lower subfacies of effusive facies, splash subfacies of explosive facies, hot clastic flow subfacies of explosive facies, hot base surge subfacies of explosive facies, fallout subfacies of explosive facies
fracture	Structural fracture	Small-macrofractures	> 0.1	70	Various volcanic rocks	Various lithofacies/subfacies
	Dissolution fracture	Small-macrofractures	> 0.1	5	Various volcanic rocks	Various lithofacies/subfacies
	Contraction fracture	Micro- medium fractures	Various sizes	12	Vesicular lava	Upper subfacies of effusive facies, lower subfacies of effusive facies, volcanic conduit facies
	Blast fracture	Micro- small fractures	< 0.1	8	Crystal tuff, cryptoexplosive breccia	Hot base surge subfacies of explosive facies, cryptoexplosive breccia subfacies of volcanic conduit facies
	Sutured fracture	Micro fractures	< 0.1, basically closed	5	Volcanic breccia, crystalline tuff	Fallout subfacies of explosive facies, hot base surge subfacies of explosive facies

Basic Characteristics of Reservoir Spaces							
Pore Shape	*Accumulation Ability*	*Throat Type*	*Throat Shape*	*Throat Radius, μm*	*Permeation Ability*	*Pattern of Accumulation-Permeation*	*Deliverability*
Near elliptic, rhombic, harbor like, regular phenocrystic	Strong	Dissolution throat (medium and small throats), fractured	(Ir)regular thick-short tubular, (ir)regular thick-long tubular, (Ir)regular fine-long tubular, irregular sheetlike	0.25-0.5 mainly	Medium, strong	Dissolution-porous, fractured–dissolution-porous	Medium
Regular pheno-crystic, dissemi-nated	Fairly strong	Dissolution throat (thin and micro-throat), fractured	(Ir)regular thick-short tubular, irregular sheetlike	Mostly <0.25	Medium	Fractured–dissolution-porous	Medium
Developed in group (flat, slightly curved) linear, stripelike	Weak				Strong	Fractured	Low yield, without stable production capability
(Flat, slightly curved, highly curved) linear, sheetlike	Weak				Strong		
Crevasse cracklike, concentric circular	Very weak				Fairly strong		
(Flat, straight, slightly curved) linear, netlike	Very weak				Fairly strong		
Highly curved linear	Very weak				Weak		

Table 11.2 Identification Markers of Reservoir Spaces in Volcanic Rocks

Type of Reservoir Space		Identification Marks of Reservoir Space						
		Lithology	*Lithofacies/ Subfacies*	*Structures/ Textures*	*Shape*	*Size or Scale*	*Occurrence*	*Dissolution Trace*
Pore	Vesicle	Volcanic lava, volcanic clastic lava, welded clastic rock	Effusive facies, splash subfacies of explosive facies, hot clastic flow subfacies of explosive facies	Rhyolitic structure, spherulitic texture, welded texture	Nearly rounded, elliptic, rhombic	Commonly cavity and macropore	Usually in directional arrangement and rhyolitic texture is formed	None
	Intergranular pore	Volcanic clastic rock, welded clastic rock, autoclastic breccia lava	Explosive facies, volcanic conduit facies	Agglomerate texture, volcanic breccia texture, tuffaceous texture, autoclastic texture	Rhombic, striped, dendritic	Small pore and micropore mainly	Non-directional	None
	Dissolution pore	Various volcanic rock	Various lithofacies	Various rock textures	Various shapes, more irregular	Commonly cavity and macropore	Consistent to primary pores	Dissolution residues, dissolution dents and pitches, dissolution embayment
Fracture	Contraction fracture	Volcanic lava	Effusive facies, volcanic conduit facies	Perlitic structure, columnar joint	Crevasse crack like, concentric circular	Small to large sized pores are all developed	Joint fractures spread along passages and contraction fractures spread disorderly	None

Blast fracture	Volcanic clastic rock	Explosive facies, volcanic conduit facies	Agglomerate texture, volcanic breccia texture, tuffaceous texture, cryptoexplosive texture	(Flat, slightly curved) linear, netlike	Small width, short extension, small scale	Scattered and chaotic	None
Structural fracture	Various volcanic rock	Various lithofacies	Various rock textures	Developed in group (flat, slightly curved) linear, striped	Fairly large width, long extension, large scale	With low to high angle, developed in group	None
Dissolution fracture				Under control of primary fractures			Dissolution residues, dissolution dents and pitches, dissolution embayment
Sutured fracture	Volcanic clastic rock	Hot base surge subfacies of eruptive facies, fallout subfacies of explosive facies	Volcanic breccia texture, crystalline tuffaceous texture	Highly curved linear (serrated)	Small width, usually closed	Consistent with rock bed interface	None

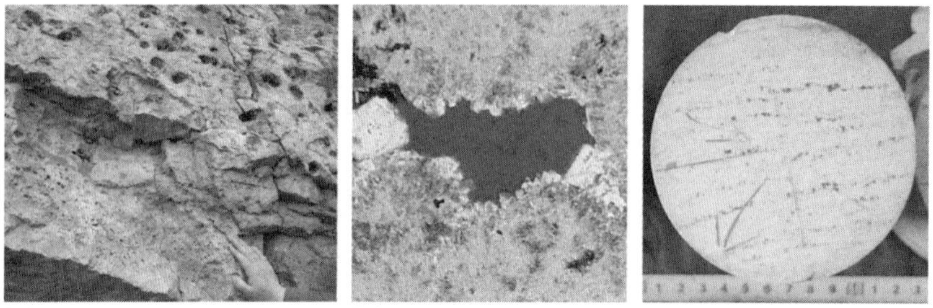

(a) (b) (c)

FIGURE 11.2

Identification markers of vesicles. (a) Vesicle shape and size (rhyolite from Jiutai, Jilin). (b) Spherulitic texture (rhyolite from Well XX901). (c) Directional arrangement of vesicles (rhyolite from Well XX901).

3) Characteristics of shape and configuration (occurrence)

Vesicles are formed by the free escape of volatile constituents in the form of bubbles along the direction toward lower pressure after degassing from the magma. The vesicle shape is affected by many factors, such as pressure distribution in the magma, bubble escape pathways, and magma flow dynamics. In general, well-preserved vesicles are characterized by a simple and regular porous shape and a smooth pore wall, mostly in sub-rounded or ovoid shapes (Figure 11.2a); when the vesicles are filled by secondary minerals after formation, their shapes become complicated and may result in a rhombic shape (Figure 11.2b).

As affected by magma flow, the important difference between vesicles and intergranular pores lies in the directional arrangement of vesicles along the direction of magma flow.

4) Characteristics of size

The vesicle size and degree of vesicle development are affected by many factors, such as the composition, gathering speed, and escaping speed of volatile constituents, as well as magma viscosity and external pressure. Higher volatile concentration, faster gathering and escaping speeds, lower magma viscosity, and lower external pressure correspond to larger vesicles and better-developed vesicles. In the top subfacies of the effusive facies, vesicles are typically large (fist sized) and are well developed like honeycombs; in the middle subfacies of the effusive facies, vesicles are small (pinhole sized), poorly developed, and scattered; in the lower subfacies of the effusive facies, the vesicles sizes are in between those of the upper and middle subfacies (Figure 11.4). Volcanic lava is more compaction resistant than volcaniclastic rocks and thus can better preserve the vesicle porous spaces. The large pores observed in outcrops and cores are mostly vesicles.

2. Identification of intergranular pores

Intergranular pores are the product of explosive volcanic eruption. They are affected by the process of volcanic clastic formation, accumulation, and compaction, and their size, shape, and occurrence differ from those of vesicles. Intergranular pores in volcanic rocks are identified systematically by integrating the characteristics of lithology, lithofacies, structure, and texture of the rocks associated with intergranular pore development.

1) Characteristics of lithology and lithofacies

Intergranular pores are developed mainly in rocks of the explosive lithofacies (see Table 11.2), such as volcaniclastic rocks. Stronger explosions will produce better-developed intergranular pores. In the explosive facies, for example, the intergranular pores of the hot clastic flow subfacies are better developed than those of the splash subfacies. Additionally, in volcanic conduit facies, some intergranular pores are developed in the autoclastic brecciated lava formed by cryptoexplosion.

2) Characteristics of rock structure and texture

Volcanic explosion breaks the magma into volcanic debris of various sizes and ejects them out of volcanic conduits, and volcanic clastic rocks and their unique intergranular pores are formed by accumulation and compaction of the ejecta. Therefore, the structure and texture of volcaniclastic rocks may be used as indirect identification markers for intergranular pores. In volcanic agglomerates, the agglomeratic textures are developed and their reservoir spaces are dominated by large interclast pores. In volcanic breccia, breccia textures are developed. In autoclastic breccia lava, autoclastic textures are developed and their reservoir spaces are dominated by intergranular pores. In tuff, tuffaceous textures are developed and matrix micropores are dominant (Figure 11.3).

3) Shape characteristics

The shape of an intergranular pore is affected mainly by the shape, sorting, and textural maturity of volcanic debris. Because of the irregular shape of volcanic debris, poor sorting, and low textural maturity, the well-preserved intergranular pores usually have features of simple porous shape with distinct angles (see Figure 11.3), with larger intergranular pores having more irregular shapes. The interclast pores developed in volcanic agglomerate and volcanic breccia are mostly rhombic and stripe shaped, whereas the matrix micropores developed in tuff usually have a simple rhombic shape. Different from vesicles, intergranular pores normally do not show any directional pattern.

(a) (b)

FIGURE 11.3

Identification markers of intergranular pores. (a) Shape of intergranular pores and volcanic breccia (agglomerate) texture. (b) Micropores and tuffaceous texture.

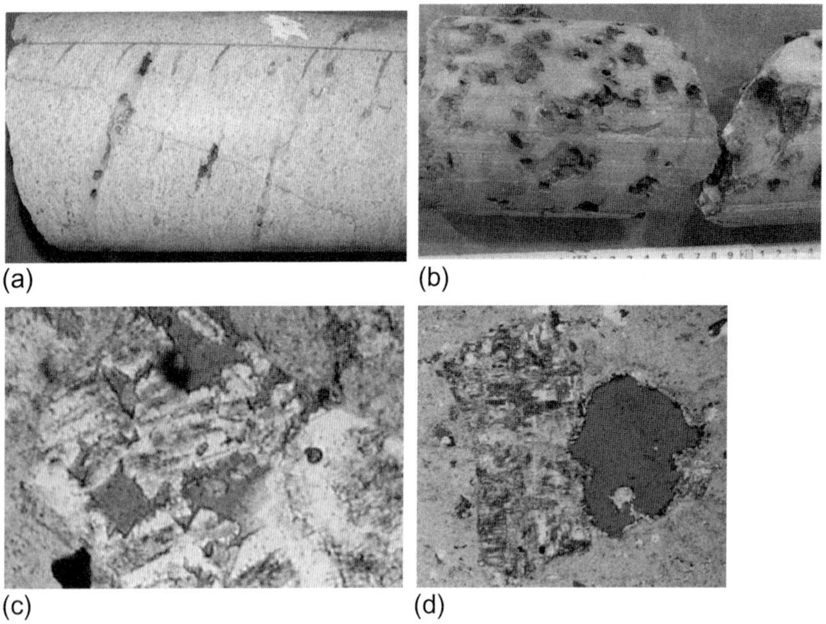

FIGURE 11.4

Identification markers of dissolution pores. (a) Dissolution pores distributed along fractures (welded tuff from Well XX1-2). (b) Irregular rounded dissolution pores (rhyolite from Well XX23). (c) Dissolution embayment (crystal tuff from Well 801). (d) Regular porphyritic dissolution pores (rhyolite from Well XX9).

4) Size characteristics

The size of intergranular pores is affected by the type, size, sorting, and filling material of volcanic clastic rocks. Volcanic agglomerate and breccia with good sorting tend to be clast supported, forming large interclast pores. In contrast, poorly sorted agglomerates and breccia result in small grains or matrix filling in between the large clasts and hence smaller interclast pores. The matrix micropores developed in tuff are fairly small (see Table 11.1 and Figure 11.3).

3. Identification of dissolution pores

Dissolution pores are of secondary origin, formed by the dissolution of soluble components such as feldspar phenocrysts and volcanic ash in vesicles, intergranular pores, and matrix mass. Compared with vesicles and intergranular pores, the shapes of dissolution pores are fairly diverse and occur in a wide range of lithology and lithofacies. Dissolution pores can be identified systematically by integrating dissolution trace markers with features associated with changes in pore shape and size.

1) Characteristics of lithology and lithofacies

Dissolution occurs normally in places where primary pores and fractures are developed and high concentrations of solvable material are present, typically in the proximal location of volcanic conduits or near major deep faults. Therefore, the typical lithofacies for the development of dissolution pores include volcanic neck subfacies of the volcanic conduit facies, splash subfacies of the explosive facies, and top subfacies of the effusive facies, typified by volcanic clastic lava and volcanic lava, with volcanic clastic lava of the splash subfacies having mostly well developed dissolution features (see Table 11.2). Dissolution pores are developed also in the top subfacies of the effusive facies, lower subfacies of the effusive facies, hot clastic flow subfacies, fallout subfacies, and hot base surge subfacies of the explosive facies, and the degree of pore development depends on the burial depth, property, and activity intensity of formation water.

2) Markers of dissolution traces

Because of the differences in soluble components or incomplete dissolution, traces such as dissolution residues, dents, and pitches as well as dissolution embayment are commonly formed in dissolution pores (see Figure 11.4), providing important identification markers. From outcrops and cores from the volcanic gas reservoirs of the XX gas field, dissolution pores distributed along structural fractures can be observed, and black or grayish black dissolution residues can be found in pores (Figure 11.4a). Additionally, due to the dissolution of soluble components, the radius of dissolution pores tends to be enlarged and its shape becomes more irregular (Figure 11.4b). In cast thin sections, dissolution embayment formed by the incomplete dissolution of soluble components can be seen (Figure 11.4c), and regular porphyritic dissolution pores originating from complete dissolution of phenocrysts are visible (Figure 11.4d).

3) Shape characteristics

The shape of dissolution pores is affected mainly by the shape of primary pores, dissolution pathways, dissolution intensity, and distribution of soluble components. When a phenocryst dissolves completely, the cast pore thus formed has a fairly regular outline (Figure 11.4d), whereas that formed by partial dissolution of soluble components appears irregular, complicated, and varied (Figure 11.4a to c). Therefore, the shapes of dissolution pores are mostly sub-rounded, rhombic, embayed, and disseminated, rarely assuming a regular phenocryst shape.

4) Size features

The sizes of dissolution pores are affected by the sizes of primary pores and dissolution intensity. In general, their radiuses are larger than those of primary vesicles and intergranular pores, and they are more widespread. At present, the largest radius of dissolution pores observed in cores from the XX volcanic gas reservoir is up to 10×30 cm.

4. Fracture identification

Fractures refer to the sheetlike spaces in rocks, distinctly different from pores with an approximately equidimensional shape. Based on their origins, fractures in volcanic rocks can be divided into five types: contraction fractures, blast fractures, structural fractures, dissolution fractures, and sutured fractures. They have different sizes, shapes, and occurrence and are developed in different lithological and lithofacies settings. On the basis of these characteristics, various types of fractures in volcanic rocks can be effectively identified according to fracture-bearing lithology and lithofacies as well as their rock textures and structures.

1) Characteristics of lithology and lithofacies

Contraction fractures are produced from uneven magma cooling, especially in volcanic lava of the upper and lower subfacies of the effusive and volcanic conduit facies. Blast fractures are formed by gas-liquid explosion, developed mainly in volcanic clastic rocks of the explosive facies and cryptoexplosive breccia of the volcanic conduit facies. Structural fractures and dissolution fractures may be developed in various types of lithologies and lithofacies. Sutured fractures are found mainly in crystalline tuff in the hot base surge subfacies of the explosive facies.

2) Shape characteristics

The shapes of fractures in volcanic rocks are diverse, with each fracture type having its own distinct morphological characteristics. Contraction fractures are affected by the distribution and variation of temperature during magma cooling, and they appear mostly as concentric circular and crevasse cracks (Figure 11.5a); large columnar joints near a volcanic conduit is regularly polygonal in cross section and flat linear in longitudinal section (Figure 11.5b). The shape of blast fractures is influenced by the intensity of gas-liquid explosion and the size and shape of phenocrysts, being generally netlike, with single fractures showing a flat or slightly curved linear pattern (Figure 11.5c). Structural fractures are formed from tectonic activities and the differential in the stress resistance of rocks, appearing in a linear pattern or grouped stripes, and a single fracture shows a flat or slightly curved linear pattern with a flat and regular fracture surface (Figure 11.5d). The shape of dissolution fractures is affected by the geometry of primary fractures and the dissolution pathways and intensity, and they manifest mostly netlike, crevasse-crack, flat to highly curved linear or stripe patterns, with various widths of fracture surfaces and irregular edges (Figure 11.5e). A sutured fracture is associated with the shape of its interface with the rock bed and pressure solution surfaces, mostly having a serrated or highly curved linear pattern with uneven fracture surfaces (Figure 11.5f).

FIGURE 11.5

Characteristics of various types of fractures in volcanic rocks. (a) Contraction fracture in perlite. (b) Large columnar joints in basalt. (c) Blast fracture in breccia welded tuff. (d) Structural fracture in rhyolite. (e) Dissolution fracture in rhyolitic volcanic breccia. (f) Sutured fracture in rhyolitic crystalline tuff.

3) Characteristics of fracture scales

Contraction fractures are normally small in scale, with a small width and short extension. However, contraction joint fractures near volcanic conduits may have large scales, with vertical extension up to tens of meters. Blast fractures are small in scale, with a small width and short extension. Structural fractures tend to be large in scale, wide, with a long extension. The size of dissolution fractures is related to the size of the primary fractures and the dissolution activity, characterized by a large scale, with a great width (wider than that of the primary one) and long extension. Sutured fractures may extend longer but are usually closed with a small scale; thus, they are of little significance in terms of accumulation and permeation capacity.

4) Occurrence characteristics

Primary fractures include small contraction fractures distributed stochastically, as well as large columnar joint fractures spreading in the direction of passages. The strike and trend of blast fractures are highly random and their occurrence has no obvious patterns. Structural fractures are controlled by the paleostress field, and their strikes and trends are regular as developed mostly in groups. In the volcanic reservoirs of the XX gas field, for example, high-angle structural fractures have a predominantly north-north-west (NNW) strike. The occurrence of dissolution fractures is influenced by the occurrence of primary ones. Most sutured fractures have low angles, with their occurrence basically consistent with that of the strata or lithofacies (Figure 11.5f).

5) Markers of dissolution traces

The dissolution traces—shown on fracture surfaces such as dissolution residues, dents and pitches, and embayments—constitute the main basis for the identification of dissolution fractures in volcanic rocks (Figure 11.5e).

5. *Distribution characteristics of reservoir spaces*

The identification results of volcanic gas reservoir spaces in the XX gas field revealed the following characteristics of classified reservoir spaces in volcanic gas reservoirs (see Table 11.1). (1) Pores are predominant in volcanic reservoir spaces, with a share of matrix porosity over 95% of the total porosity; fractures are well developed but make up less than 5% of the total porosity. (2) In these gas reservoirs, volcanic lava formed by center-fissured eruption is predominant and the effect of later dissolution is limited, so vesicles are mostly well developed in volcanic rocks, accounting for about 45% (Figure 11.6); intergranular pores rank second, accounting for about 35%; and dissolution pores are poorly developed, accounting for about 20%. (3) Among fractures in volcanic rocks, structural fractures are most common, accounting for about 70% (Figure 11.7); contraction fractures account for about 12%; blast, dissolution, and sutured fractures account for about 8%, 5%, and 5%, respectively.

11.2.3 Characterization of reservoir spaces in volcanic gas reservoirs

On the basis of the identification of classified reservoir spaces in volcanic gas reservoirs, the shape, scales, and accumulation capacity of these reservoir spaces are characterized further using methods of core observation, thin section examination, and experimental analysis.

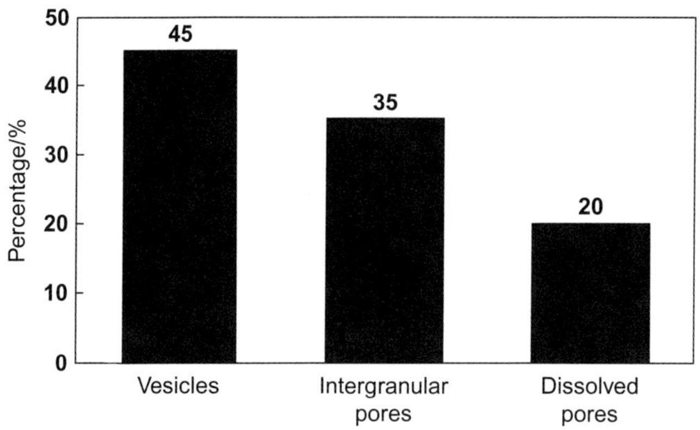

FIGURE 11.6

Characteristics of classified reservoir spaces in volcanic rocks (XX gas field).

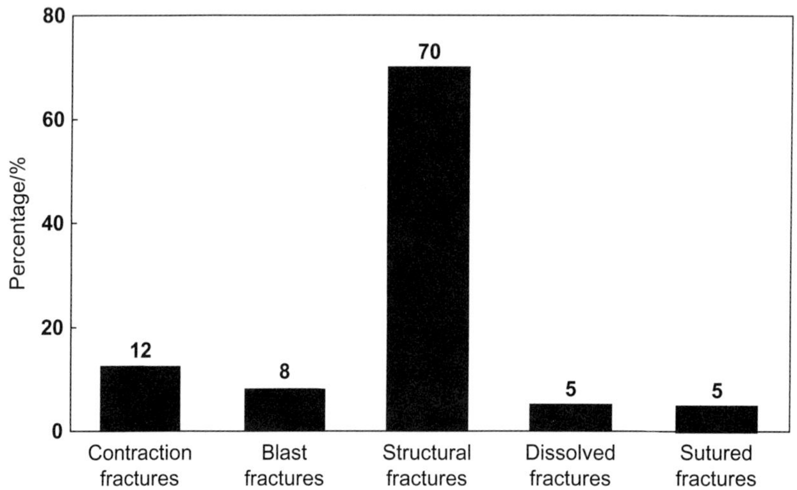

FIGURE 11.7

Characteristics of classified fractures in volcanic rocks (XX gas field).

1. Shape characterization of reservoir spaces

Cores provide important firsthand information for understanding reservoirs in the subsurface [10]. From core observation and description, the shapes of large reservoir spaces can be identified and confirmed visually. As to the small reservoir spaces, which cannot be observed directly and visually, Scanning Electric Microscope (SEM) uses cast thin sections to observe the shape of pores. For shape characterization of reservoir spaces, therefore, the two methods described earlier are integrated, and numerous cores and cast thin sections have been examined and analyzed to explain the characteristics of shape and distribution of various types of reservoir spaces.

Table 11.3 Classification of Vesicle Shapes in Volcanic Gas Reservoirs

Reservoir Space	Origin	Shape Classification		
Vesicle	Formed by volatile degassing during eruption and surface flow	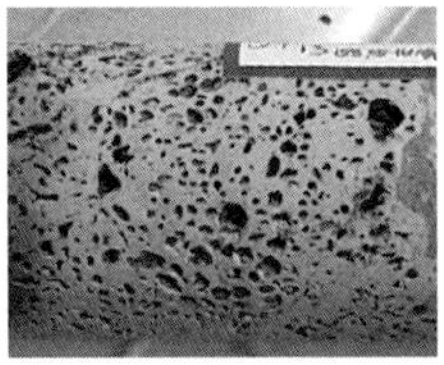		
		Sub-rounded	Ovoid	Rhombic

1) Vesicles

Normally, volatile gas is expelled from magma as bubbles, which is affected by the flow of magma. As a result, vesicles show elongation and deformation along the flowing direction of magma, mostly having sub-rounded and ovoid shapes. After vesicles are formed, an irregular rhombic morphology would be produced if matrix filling occurred or filling and replacement by secondary minerals took place during later diagenetic processes (Table 11.3). In general, the nearly rounded vesicles are distributed mostly in the middle subfacies of the effusive facies, which is rarely affected by magma flow. Ovoid vesicles are distributed mainly in the top, upper, and lower subfacies of the effusive facies, and rhombic vesicles are usually distributed in the top and lower subfacies. In the XX gas field, ovoid vesicles are most common, accounting for about 50%, sub-rounded ones account for about 30%; and rhombic vesicles account for 20%.

2) Intergranular pores

Normally, volcanic debris tends to accumulate in situ, and therefore they have poor roundedness and sorting, leading to irregular-shaped intergranular pores. In comparison, well-sorted volcanic debris lead to rhombic intergranular pores with granular support; for moderate sorting, a striped pattern supported by multiple grains tends to be developed in intergranular pores; poor sorting is associated with a complicated dendritic pattern jointly supported by grains and matrix (Table 11.4). The rhombic and

Table 11.4 Shape Classification of Intergranular Pores in Volcanic Gas Reservoirs

Reservoir Space	Origin	Shape Classification		
Intergranular pore	Primary pores, formed by accumulation and compaction after volcanic debris ejected onto surface	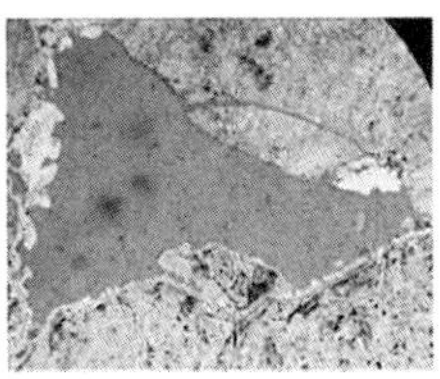	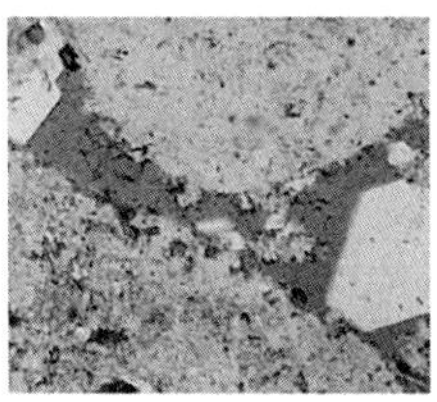	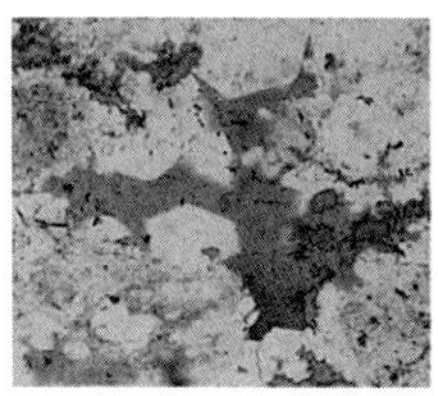
		Rhombic	Striped	Dendritic

striped intergranular pores are usually developed in volcanic breccia of the fallout subfacies and in welded breccia of the hot clastic flow subfacies, both of the explosive facies; the dendritic intergranular pores tend to be developed in the tuffaceous breccia or brecciated tuff of the fallout subfacies; and the shape of micropores is relatively simple, most common in welded tuff of the hot clastic flow subfacies, crystalline tuff of the hot base surge subfacies, and tuff of the fallout subfacies. In the XX volcanic gas reservoir, rhombic intergranular pores are predominant, accounting for about 60%; dendritic ones account for about 25%; and striped ones account for only about 15%.

3) Dissolution pores

The shape of dissolution pores is affected mainly by dissolution and the property of soluble components. When phenocrysts are dissolved completely, the dissolution pores assume the regular phenocryst morphology; when secondary minerals in vesicles or intergranular pores are locally dissolved or the phenocrysts and matrix are partially dissolved, most of the dissolution pores appear as dissolution embayments; when volcanic ash is partially dissolved through weak dissolution activities, the dissolution pores usually have a disseminated and complex pattern (Table 11.5). Regular phenocryst dissolution pores are mostly developed in porphyrocrystallic tuff and rhyolite; dissolution embayments are widespread in various volcanic lava and volcanic clastic rocks; disseminated dissolution pores are most common in various volcanic tuff. In the XX volcanic gas reservoir, dissolution embayments are predominant, accounting for about 50%; disseminated ones account for about 35%; and regular porphyrocrystallic ones are poorly developed, accounting for only about 15%.

4) Fractures

Outcrop observation, core description, and thin section examination indicate that the fracture shapes in volcanic reservoirs are closely related to their petrogenetic origins. Overall they can be divided into three types: concentric circular, crevasse cracklike, and netlike (or net shaped). Individual fractures manifest in three patterns: flat linear, slightly curved linear, and highly curved linear (or serrated) (Table 11.6). Significant differences exist among various fracture shapes. (1) The shape of contraction

Table 11.5 Shape Classification of Dissolution Pores in Volcanic Gas Reservoirs

Reservoir Space	Origin	Shape Classification		
Dissolution pore	Secondary pores, formed by dissolution of soluble constituents such as feldspar phenocrysts and volcanic ash in vesicles, intergranular pores and matrix mass.	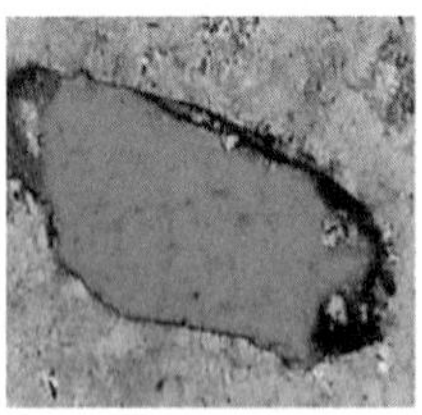 Regular porphyrocrystallic	 Embayment	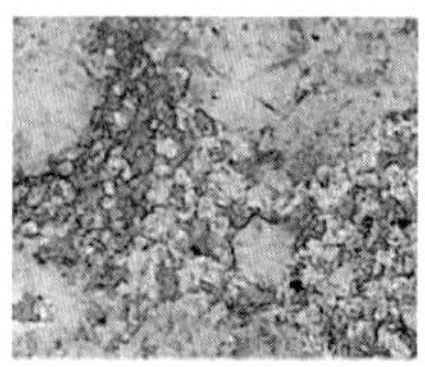 Disseminated

Table 11.6 Shape Classification of Fractures in Volcanic Gas Reservoirs

Reservoir Space	Origin	Shape Classification		
Fracture	Sheetlike space, formed by fracturing of volcanic rocks due to loss of cohesion	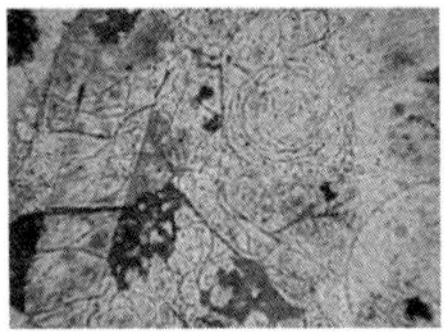 Concentric circular	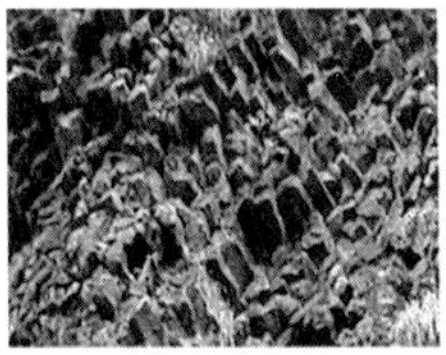 Crevasse crack like	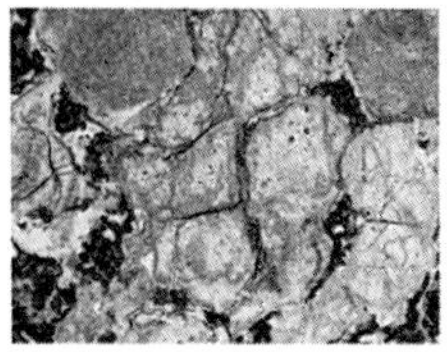 Netlike (net shaped)
		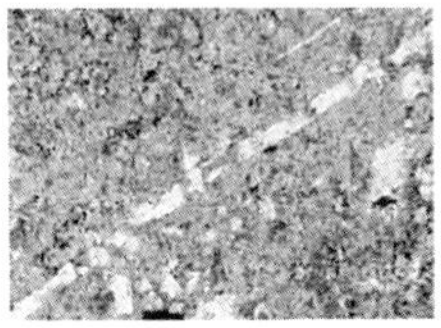 Flat linear pattern or striped	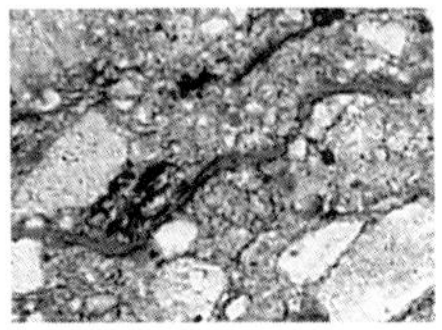 Slightly curved linear pattern or striped	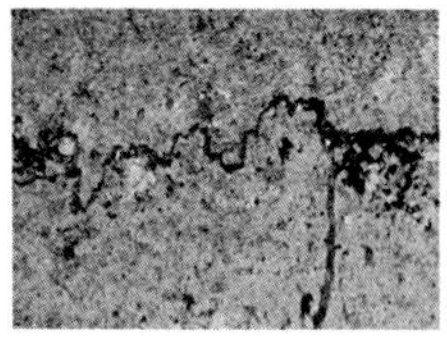 Highly curved linear pattern (serrated) or striped

fractures follows the evolution pattern in crystallo-mineralogy, appearing mostly as concentric rings and crevasse cracks, whereas the large columnar joints near volcanic conduits are regular net shaped. (2) Blast fractures are formed by cryptoexplosions, manifesting as crevasse cracks and networks, and the individual fractures have a flat to slightly curved linear pattern. (3) Structural fractures exhibit flat to slightly curved linear and striped patterns mostly developed in groups. (4) The shape of dissolution fractures is complex and usually appears as crevasse cracks and networks, with individual fractures having slightly to highly curved linear or striped patterns. (5) Finally, sutured fractures are formed by compaction and pressure solution controlled by multiple factors, including rock bedding, tectonic stress field, and formation compressibility, and characterized by a highly curved linear (or serrated) pattern. In the XX volcanic gas reservoir, the predominance of structural fractures is associated with flat to slightly curved linear patterns, accounting for about 80%; the crevasse crack and netlike patterns account for about 12%; the highly curved linear and concentric circular patterns are poorly developed, accounting only for about 5% and 3%, respectively.

2. Size characterization of reservoir spaces

1) Pore sizes

Overall features. Pore sizes in volcanic gas reservoirs are characterized by integrating core observation, thin section examination, and constant-rate mercury injection.

Core observation Cores offer important firsthand information. Through visual observation, the characteristics of pore size, development, and distribution of the larger pores can be identified, so as to qualitatively characterize the features of pore size in volcanic rocks. In the XX gas field (Table 11.7), the pore radius in volcanic rocks ranges mainly from 0.1 mm to 1 mm up to a maximum of 300 mm. Small pores are mostly well developed, accounting for about 57%; macropores account for

Table 11.7 Characteristics of Pore Size in Volcanic Reservoirs (XX Gas Field)

Characterization Method	Maximum Pore Radius, mm	Main Range, mm	Various Sizes of Pores, %			
			Cavity (> 2 mm)	*Macropore (0.5-2 mm)*	*Small Pore (0.01-0.5 mm)*	*Micropore (< 0.01 mm)*
Core observation	300	0.1-1	8	35	57	0
Cast thin section	10	0.01-0.5	2	20	40	38
Constant-rate mercury injection	0.46	0.1-0.2	0	0	100	0

about 35%; cavities are poorly developed, accounting for only about 8%; and micropores cannot be observed visually.

Thin section examination For the microscopic pores that cannot be observed visually, SEM examination is applied by using cast thin sections, allowing the characteristics of pore size, development, and distribution to be determined so as to semiquantitatively characterize the features of pore size in volcanic rocks. In the XX gas field (see Table 11.7), thin sections demonstrate that the pore radius in the volcanic rocks usually ranges from 0.01 mm to 0.5 mm, reaching a maximum of 10 mm. Small pores are mostly well developed, accounting for about 40%; micropores account for about 38%; macropores account for about 20%, and cavities account for about 2%.

Constant-rate mercury injection method In this method, mercury is injected into pores at a constant low rate (normally as low as 0.00005 mL/min), allowing them to approach a quasi-static state. The moment mercury breaks through the throat restriction and enters into a pore space, pressure drops instantaneously and then builds up again when the entire pore is being filled, until mercury enters into the next pore (Figure 11.8). The measurements of pressure drop and rise during mercury injection are used to interpret pores and throats within the rock and to generate the capillary pressure curves of pores and throats as well as the characteristic parameters of microscopic porous structure such as distributions of pore radius, throat radius, and the ratio of pore radius to throat radius [11]. The analysis of 24 volcanic rock samples used for constant-rate mercury injection in the XX gas field (see Table 11.7)

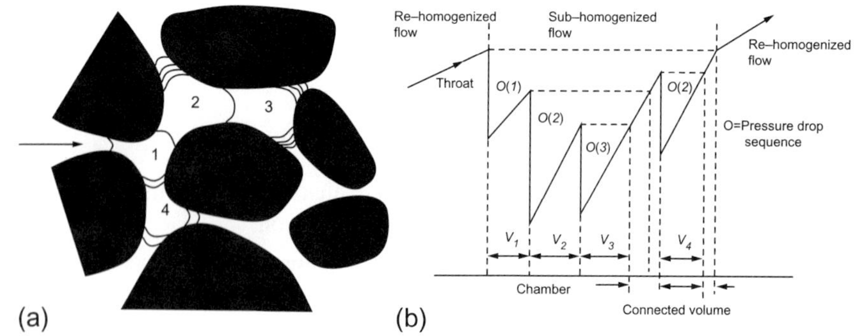

FIGURE 11.8

Principles of the constant-rate mercury injection test.

Table 11.8 Size Features of Various Pore Types in Volcanic Reservoirs (XX Gas Field)

Pore Type	Maximum Pore Radius, mm	Main Range, mm	Different Sizes of Pores, %			
			Cavity (> 2 mm)	*Macropore (0.5-2 mm)*	*Small Pore (0.01-0.5 mm)*	*Micropore (< 0.01 mm)*
Vesicles	200	0.1-1	10	45	40	5
Intergranular pores	50	0.01-0.5	6	22	43	29
Dissolution pores	300	0.1-0.2	10	50	32	8

shows that the pore radius in the volcanic rock falls mainly in the range of 0.1 mm to 0.2 mm, up to a maximum of 0.46 mm, all being in the class of small pores.

In the three methods discussed, cores can provide information only on the distribution of large, visually recognizable pores. Thin sections can provide a better characterization of pore sizes in volcanic rocks, but the representative samples are inadequate. In constant-rate mercury injection, mercury is injected into pores in a quasi-static process, but a problem arises if mercury fails to enter some micropores and throats or if the representative samples are inadequate. Therefore, it has been necessary to integrate the three methods for the characterization. The analytical results indicate a wide range of pore sizes in volcanic rocks, but overall small pores are predominant, macropores and micropores rank the second, and cavities are developed only locally (see Table 11.7).

Sizes of various pore types. Pore sizes are highly variable in volcanic rocks due to their different origins (Table 11.8).

Vesicles Volatile constituents are most prone to escape from the top of the flowing magma but are most stable in the middle. Therefore, vesicular cavities are mainly developed in volcanic lava of the top and upper subfacies of the effusive facies as well as breccia lava and tuffaceous lava in the splash subfacies of the explosive facies. Macrovesicles are predominant in volcanic lava in the upper and lower subfacies of the effusive facies. Small vesicles are common in tight volcanic lava in the middle subfacies of the effusive facies as well as welded tuff and welded breccia in the hot clastic flow subfacies of the explosive facies. Microvesicles are mainly formed in massive tight volcanic lava of the middle subfacies (see Table 11.1). In the XX gas field (see Table 11.8), the pore radius in volcanic rocks mainly ranges from 0.1 mm to 5 mm, reaching a maximum of 300 mm. On the whole, macropores are most common, accounting for about 45%; small pores account for about 40%; and vesicular cavities and micropores are poorly developed, accounting for about 10% and 5%, respectively.

Intergranular pores A stronger explosion is correlated with larger volcanic debris and a larger radius and scale of intergranular pores. Therefore, intergranular pores are mainly distributed in volcanic agglomerate and volcanic breccia of fallout subfacies of the explosive facies. Macro-intergranular pores are distributed primarily in volcanic breccia of the fallout subfacies, welded breccia in hot clastic flow subfacies of the explosive facies, and autoclastic breccia lava of the volcanic conduit facies. Most small intergranular pores occur in volcanic breccia of the fallout subfacies as well as welded breccia and welded tuff of the hot clastic flow subfacies. Matrix micropores are found mostly in crystalline tuff of the fallout subfacies and welded tuff of the hot clastic flow subfacies (see Table 11.1). In the XX gas field (see Table 11.8), the intergranular pore radius in the volcanic rocks ranges primarily from

0.01 mm to 0.5 mm, reaching a maximum of 50 mm. In general, small intergranular pores are the most common, accounting for about 43%; matrix micropores and intergranular macropores rank second, accounting for about 29% and 22%, respectively; and intergranular cavities are poorly developed, accounting for only about 6%.

Dissolution pores Larger primary pores and higher soluble contents reflect the stronger dissolution activity and larger radius and scale of dissolution pores. Dissolution cavities are normally developed in vesicular lava of top and upper subfacies of the effusive facies, brecciated lava of splash subfacies of the explosive facies, and volcanic breccia of fallout subfacies of the explosive facies. Most dissolution pores are distributed in vesicular lava of the upper subfacies, breccia lava of the splash subfacies, welded breccia of hot clastic flow subfacies of the explosive facies, volcanic breccia of the fallout subfacies, and autoclastic breccia lava of the volcanic conduit facies. Matrix dissolution pores are usually found in vesicular lava of lower subfacies of the effusive facies, tuffaceous lava of the splash subfacies, welded tuff of the hot clastic flow subfacies, crystalline tuff of hot base surge subfacies of the explosive facies, and fallout subfacies tuff (see Table 11.1 and Figure 11.9). In the XX gas field, the dissolution pore radius in the volcanic rocks usually ranges from 0.01 mm to 10 mm, reaching a maximum of 300 mm; macropores are mostly well developed, accounting for about 50%; small pores account for about 32%; and cavities and micropores are poorly developed, accounting for about 10% and 8%, respectively.

2) Fracture sizes

Core observation, cast thin section examination, and logging interpretation are usually used to characterize fracture sizes in volcanic reservoirs, taking into consideration the specific origin and morphology of various fractures (see Table 11.9 and Figure 11.10):

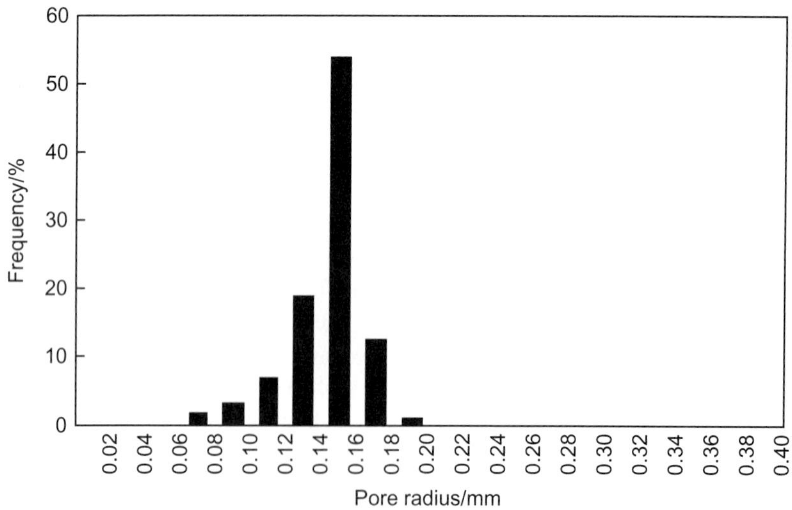

FIGURE 11.9

Distribution of pore radii as determined from constant-rate mercury injection.

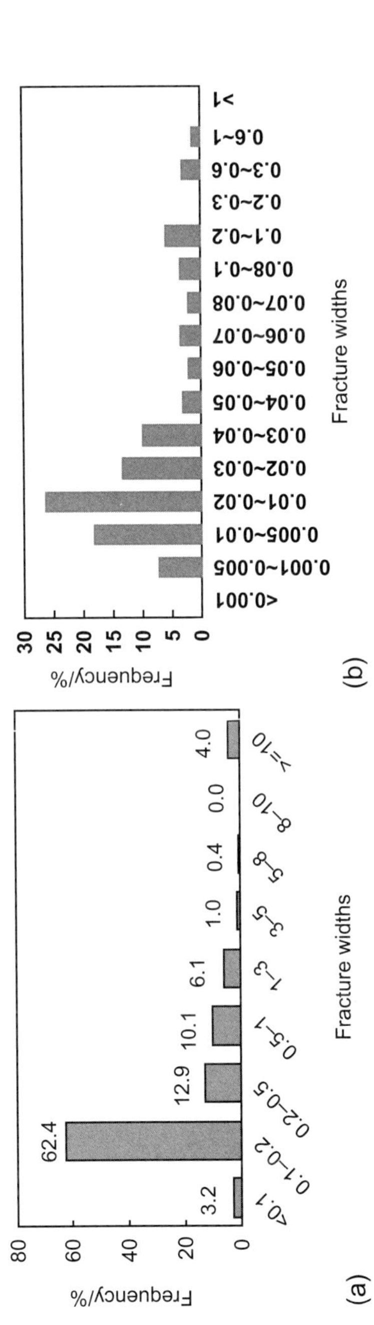

FIGURE 11.10

Characteristics of fracture size distribution in volcanic reservoirs. (a) Distribution of fracture width determined from core observation (XX gas field). (b) Distribution of fracture width determined from log interpretation (XX gas field).

1. *Core observation.* In cores, the maximum fracture width in volcanic rocks of the XX gas field is 12.5 mm, and the fracture width ranges from 0.1 mm to 1 mm. Most fractures are in the small category, accounting for about 78%; medium factures rank second, accounting for about 12%; and micro- and macrofractures are poorly developed, accounting for 6% and 4%, respectively.
2. *Cast thin section analysis.* In cast thin sections, the maximum fracture width in volcanic rocks of the XX gas field is 2 mm, and the fracture width ranges from 0.01 to 0.5 mm. Most of fractures are of the small category, accounting for about 50%; microfractures are the next, accounting for about 40%; medium fractures are poorly developed, accounting for about 10%; and macrofractures are not observed.
3. *Log interpretation.* Fracture width is calculated through electrical imaging log color analysis and statistical models of conventional dual laterologs (see Chapter 6 for details). The log interpretation indicates that the maximum fracture width in volcanic rocks of the XX gas field is about 1.5 mm, and the fracture width ranges between 0.005 and 0.55 mm. Fractures of the micro category are dominant, accounting for about 75%; small fractures rank second, accounting for about 20%; medium fractures are poorly developed (about 5%); and macro fractures are not developed.

In the three methods discussed, core observation and cast thin section analysis are used as the chief means for characterizing fracture sizes in volcanic rocks under surface conditions, and log interpretation is applied for the characterization of subsurface data. Therefore, the integrated analysis shows that small fractures are predominant on the surface, whereas microfractures are mostly well developed in subsurface volcanic rocks.

3. Characterization of the accumulation capacity in reservoir spaces

Accumulation capacity is an important indicator for evaluating the sizes of various types of reservoir space and their fluid-bearing capacity. Two parameters—porosity and the maximum mercury intrusion saturation—are usually used to characterize the accumulation capacity of reservoir spaces in volcanic rocks.

1) Accumulation capacity of pores

Porosity and permeability tests, mercury injection experiments, and log interpretation information are used to characterize the overall accumulation capacity of pores in volcanic rocks and the accumulation capacity of various types of pores.

Overall accumulation capacity of pores. Because of their complex origins and variable shapes and sizes, the accumulation capacity of pores varies significantly. In the XX volcanic gas reservoir (Table 11.10, Figures 11.11 and 11.12), the porosity as revealed by whole core experiments of volcanic rocks ranges from 0.44% to 18.8%, mainly between 3% and 8%, with an average of 7.10%. The log interpretation indicates that the maximum effective porosity is 16.6%, the average is 6.5%, and the maximum mercury intrusion saturation has total range of 11% to 99%, a main range of either 15% to 40% or 70% to 100%, and an average of 65%, suggesting a generally strong but highly variable accumulation capacity of pores.

Accumulation capacity of pores with various origins. Pores with different volcanic petrogeneses show great variation in accumulation capacity because of their different shapes and sizes (see Table 11.10). Dissolution pores have the highest values in effective log porosity (average 7.45%), experimental

Table 11.9 Characteristics of Fracture Size in Volcanic Reservoirs (XX Gas Field)

Characterization Methods	Maximum Fracture Width, mm	Main Range, mm	Different Fracture Sizes, %			
			Macro Fracture (> 10 mm)	*Medium Fracture (1-10 mm)*	*Small Fracture (0.1-1 mm)*	*Micro Fracture (< 0.1 mm)*
Core observation	12.5	0.1-1	4	12	78	6
Cast thin section	2	0.01-0.5	0	10	50	40
Log interpretation	1.5	0.005-0.05	0	5	20	75

Table 11.10 Characteristics of the Accumulation Ability of Pores in Volcanic Reservoirs (XX Gas Field)

Pore Type	Typical Rock	Logging Interpreted Effective Porosity, %	Experimental Porosity, %	Maximum Mercury Intrusion Saturation, %
Vesicle	Rhyolite	5.14	6.75	65.42
Intergranular pores	Rhyolitic volcanic breccia	4.5	5.58	60.21
Dissolution pores	Rhyolitic brecciated lava	7.45	8.4	73.42
All pores	Volcanic rocks	6.5	7.1	65

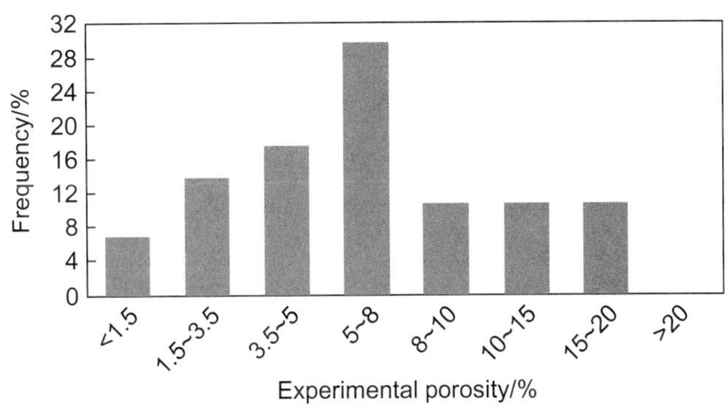

FIGURE 11.11

Frequency distribution of whole core porosity.

porosity (average 8.4%), and maximum mercury intrusion saturation (average 73.42%), thus they have the strongest accumulation capacity. Vesicles have moderate values in effective log porosity (average 5.14%), experimental porosity (average 6.75%), and maximum mercury intrusion saturation (average 65.42%), hence they have a moderate accumulation capacity. Intergranular pores have the lowest

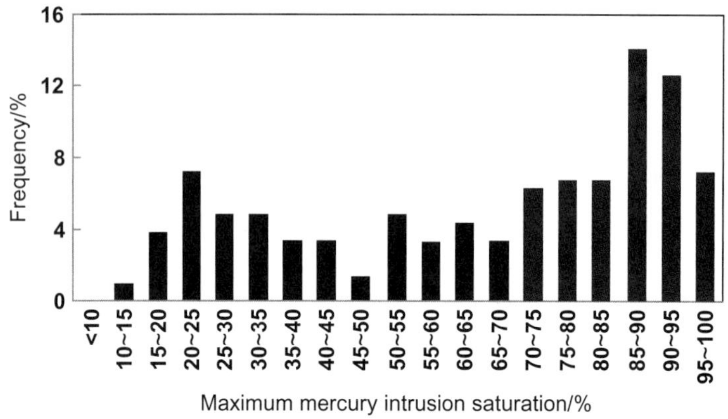

FIGURE 11.12

Frequency distribution of mercury intrusion saturation.

values in effective log porosity (average 4.5%), experimental porosity (average 5.58%), and maximum mercury intrusion saturation (average 60.21%), and thus they have the lowest accumulation capacity.

Accumulation capacity of pores with different sizes. Pores with different sizes in volcanic rocks have different levels of accumulation capacity (Table 11.11).

Vesicles Macrovesicles (including vesicular cavities) are mostly developed in top, upper, and lower subfacies of the effusive facies and splash subfacies of the explosive facies. The complete escape of volatiles results in large sizes and well-developed vesicles, and thus favorable petrophysical properties and accumulation capacity. Most small vesicles (including micropores) are developed in the middle subfacies (effusive facies) and hot clastic flow subfacies (explosive facies). The incomplete escape of volatiles results in small sizes and poorly developed vesicles, and consequently rather poor petrophysical properties and accumulation capacity. In the XX volcanic gas reservoir, macrovesicles have relatively high values in effective log porosity, experimental porosity, and maximum mercury intrusion, at 5.16%, 7.14%, and 65.02%, respectively. Small vesicles have relatively low values in effective log porosity, experimental porosity, and maximum mercury intrusion, at 3.17%, 5.56%, and 51.69%, respectively, indicating that macrovesicles have a notably stronger accumulation capacity than small vesicles.

Intergranular pores Macro-intergranular pores (including intergranular cavities) are mainly developed in fallout and hot clastic flow subfacies of the explosive facies. In these grain-supported volcanic breccia rocks, the intergranular pores are well developed, with a large radius, good petrophysical property, and strong accumulation capacity. Most small intergranular pores (including matrix micropores) are common in the fallout and hot base surge subfacies. Because smaller breccia grains and matrix usually provide support among breccia bodies, intergranular pores have a small radius and are poorly developed, characterized by poor petrophysical properties and a relatively weak accumulation capacity. In the XX gas field, the macro-intergranular pores have relatively high values of effective log porosity, experimental porosity, and maximum mercury intrusion, at 5.35%, 5.87%, and 69.81%, respectively. The small intergranular pores have relatively low values of effective log porosity, experimental porosity, and maximum mercury intrusion, at 3.87%, 5.29%, and 46.66%, respectively, indicating that macro-intergranular pores have a stronger accumulation capacity.

Table 11.11 Statistics on the Accumulation Ability of Various Types of Pores in Volcanic Gas Reservoirs (XX Gas Field)

Pore Type		Subfacies	Effective Log Porosity, %	Experimental Porosity,%	Maximum Mercury Intrusion Saturation, %
Vesicle	Macro vesicles	Top subfacies of effusive facies Upper subfacies of effusive facies Lower subfacies of effusive facies Splash subfacies of explosive facies	5.16	7.14	65.02
	Small vesicles	Middle subfacies of effusive facies Hot clastic flow subfacies of explosive facies	3.17	5.56	51.69
Intergranular pore	Macro intergranular pores	Fallout subfacies of explosive facies Hot clastic flow subfacies of explosive facies	5.35	5.87	69.81
	Small intergranular pores	Fallout subfacies of explosive facies Hot base surge subfacies of explosive facies	3.87	5.29	46.66
Dissolution pores	Dissolution pores	Top subfacies of effusive facies Upper subfacies of effusive facies Splash subfacies of explosive facies Fallout subfacies of explosive facies	6.63	8.62	75.06
	Matrix dissolution pores	Lower subfacies of effusive facies Hot clastic flow subfacies of explosive facies Hot base surge subfacies of explosive facies Volcanic conduit facies	5.31	6.96	65.49

Dissolution pores Dissolution pores with a large radius (including dissolution cavities) are formed mainly in top and upper subfacies of the effusive facies, as well as splash and fallout subfacies of the explosive facies. Because primary pores and fractures are well developed through the actions of surface water and underground water, dissolution pores tend to have good petrophysical properties and strong accumulation capacity. The matrix dissolution pores (including some dissolution pores) with a small radius are distributed mainly in the volcanic conduit facies, lower subfacies of the effusive facies, and hot clastic flow and hot base surge subfacies of the explosive facies. Because of the poorly developed primary pores and fractures, limited fluid and dissolution activity, and small radius, matrix dissolution pores tend to have relatively poor petrophysical properties low accumulation capacity. In the XX gas field, the dissolution pores in volcanic rocks have relatively high values of effective log porosity, experimental porosity, and maximum mercury intrusion, at 6.63%, 8.62%, and 75.06%, respectively. These values are lower for matrix dissolution pores, at 5.31%, 6.96%, and 65.49%, respectively. This indicates that dissolution pores with a large radius have a higher accumulation capacity.

After further comparative analysis (see Table 11.11), it can be observed that dissolution pores have the strongest accumulation capacity among the six types of pores in three categories, macrovesicles and matrix dissolution pores rank second, and the accumulation capacity of small intergranular pores is the weakest.

2) Accumulation capacity of fractures

Fracture accumulation capacity is characterized through core description and log interpretation:

1. *Core description.* With cores, the fracture width and length are measured to calculate the ratio of fracture area to core surface area so as to obtain the surface porosity of fractures. In the XX gas field, fracture description for over 1400 m cores shows that (Figure 11.13) the maximum surface porosity of fractures in the volcanic gas reservoirs is 2.7%, and the surface porosity of fractures ranges from 0.02% to 0.6%, with an average of 0.28%, accounting for about 3.9% of the total reservoir porosity.

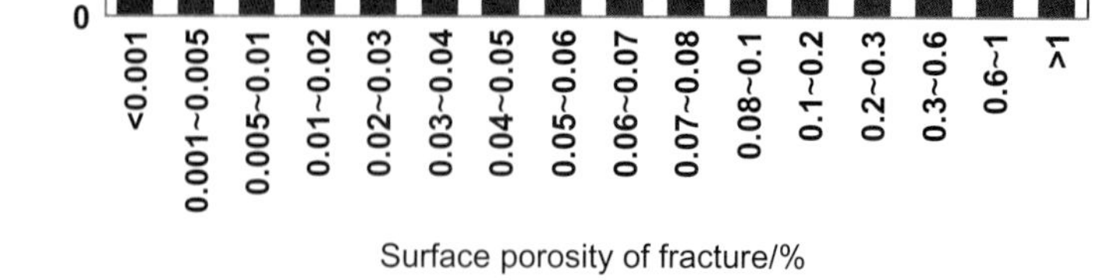

FIGURE 11.13

Distribution of surface porosity of fractures determined by the core description (XX gas field).

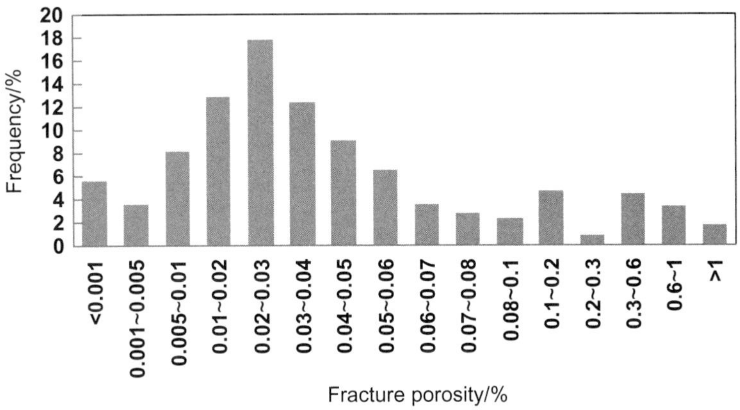

FIGURE 11.14

Distribution of fracture porosity determined by log interpretation (XX gas field).

2. *Log interpretation.* Fracture porosity can be derived from electrical imaging log analysis and the statistical models of conventional dual laterologs (see Chapter 6 for details). In the XX volcanic gas reservoirs, log interpretations based on more than 50 wells are used to determine fracture porosity. The results (Figure 11.14) indicate that the maximum fracture porosity in volcanic gas reservoirs is 1.9%, and the fracture porosity ranges from 0.01% to 0.07%, averaging 0.06%, thereby accounting for about 0.8% of the total reservoir porosity.

From the preceding discussion, one can conclude that (1) in volcanic reservoirs, the overall accumulation capacity of fractures is limited, and the subsurface accumulation ability of fractures is lower because of their low compaction resistance; (2) the fracture accumulation capacity varies greatly because of the complex origins and highly variable types and sizes, and the fracture accumulation ability varies greatly; (3) among various types of fractures, structural fractures tend to be well developed, with a large width, and they have the strongest accumulation capacity; the accumulation capacity of dissolution fractures ranks second; the accumulation capacity of contraction fractures and blast fractures are relatively weak; most sutured fractures are closed and have no effect on the accumulation-permeation.

11.3 CHARACTERIZATION OF THROATS IN VOLCANIC GAS RESERVOIRS

Throats refer to the narrow passages connecting various types of reservoir spaces, which control the permeation capacity of volcanic reservoirs. Following the throat classification scheme, throats in volcanic reservoirs are identified and the shape, size, and permeation capacity are characterized for each throat type by means of thin section examination, mercury injection analysis, and constant-rate mercury injection analysis.

11.3.1 Classification of throats in volcanic gas reservoirs

Little research has been conducted on throats in volcanic reservoirs, thus there is no commonly accepted classification scheme for throats in volcanic reservoirs so far. In this book, a classification scheme is proposed for throats in volcanic reservoirs (see Tables 11.1 and 11.12) by taking into account

Table 11.12 Classification and Characteristics of Throats in Volcanic Gas Reservoirs (XX Gas Field)

Throat Type							Throat Characteristics	
			Shape Classification	*Size Classification*				
Genetic Classification				*Radius, μm*	*Length*	*Type*	*Throat Permeation Ability*	*Typical Lithology*
Matrix throat	Pore-contraction		Tubular	≥ 0.25	Long	Large-long	Fairly strong	Volcanic agglomerate, volcanic breccia
					Short	Large-short	Strong	Vesicular lava, volcanic breccia
				< 0.25	Long	Small-long	Weak	Volcanic breccia
					Short	Small-short	Fairly weak	Vesicular lava, welded tuff, crystalline tuff
	Dissolution		Irregular	≥ 0.25	Long	Large-long	Fairly strong	Vesicular lava, breccia lava, volcanic breccia
					Short	Large-short	Strong	Vesicular lava, welded tuff, crystalline tuff
Fractured throat	Structural fractures		Sheetlike space, linear in group	Normally ≥ 0.25	Long	Large-long	Strong	Various volcanic rocks
	Dissolution fractures		Sheetlike space, crevasse crack or linear pattern					
	Contraction fracture	Columnar joints	Sheetlike space, crevasse crack pattern					
		Contraction fractures	Sheetlike space, concentric circular or crevasse crack pattern					Volcanic lava
	Blast fractures		Sheetlike space, flat-slightly curved linear pattern	Normally <0.25		Small-long	Fairly strong	Volcanic lava
								Volcanic breccia, autoclastic breccia lava

Throat Characteristics					
	Reservoir Space				
Typical Lithofacies	***Type***	***Shape***	***Accumulation Ability***	***Accumulation -Permeation Pattern***	***Deliverability***
Fallout subfacies of explosive facies	Intergranular cavity, macro intergranular pore	Rhombic, stripe shaped	Fairly strong	Intergranular porous	Medium, high
Upper subfacies of effusive facies, fallout subfacies of explosive facies	Vesicular cavity, macro vesicle, macro intergranular pore	Elliptic, rhombic	Strong	Vesicular, intergranular porous	High
Fallout subfacies of explosive facies	Small intergranular pore	Rhombic, stripe shaped	Medium, weak	Intergranular porous	Low
Lower subfacies of effusive facies, hot clastic flow subfacies of explosive facies, hot base surge subfacies of explosive facies	Small vesicle, microvesicle, matrix micropore	Nearly round, elliptic, rhombic	Fairly weak	Vesicular, intergranular porous, microporous	Medium, low
Top subfacies of effusive facies, splash subfacies of explosive facies, fallout subfacies of explosive facies	Dissolution cavity, dissolution pore	Elliptic, porphyrocrystic, rhombic	Fairly strong	Vesicular	Medium, high
Upper subfacies of effusive facies, hot clastic flow subfacies of explosive facies, hot base surge subfacies of explosive facies	Dissolution pore, matrix dissolution pore	Rhombic, harbor shaped, disseminated	Strong	Intergranular porous	High
Various lithofacies	Various pores	Various shapes	Weak-strong	Fractured-porous	Medium, high
					High
			Medium, strong		
Volcanic conduit facies	Macro vesicle, small vesicle	Nearly round, elliptic	Medium, strong	Fractured-vesicular	Medium, high
Effusive facies	Vesicle	Elliptic, rhombic	Medium, strong	Fractured-vesicular	Medium, high
Explosive facies, volcanic conduit facies	Intergranular pore	Rhombic, stripe shaped, dendritic	Medium	Fractured-intergranular porous	Medium, low

the peculiarity and complexity of throats, as well as their origins, shape, radius, and length. First, throats in volcanic reservoirs are divided into three basic types in terms of origin: pore-contraction throats, dissolution throats, and fractured throats. They are divided into two types in terms of shape: tubular throats and sheetlike throats. Then, in terms of radius [12], throats are divided into large throats ($\geq$ 0.5 μm), medium throats (0.25 to 0.5 μm), small throats (0.1 to 0.25 μm), and microthroats (< 0.1 μm), and into two types in terms of length: long and short throats. Finally, the large and medium throats are grouped as the large category ($\geq$ 0.25 μm) and the small and microthroats are grouped as small throats (< 0.25 μm) for convenience. As a result, this study recognizes eight types of throats (see Table 11.12).

1. Pore-contraction throats

Pore-contraction throats refer to the throats formed by pore contraction between vesicles or intergranular pores, and they are widespread in volcanic reservoirs. Throats between vesicles have two origins: (1) the throat is formed at the contracted part of a vesicle, and (2) the throat is developed within a vesicle or between vesicles when contraction occurs due to secondary mineral growth (e.g., quartz). For the pore-contraction throats between intergranular pores, they are formed from the accumulation and compaction of volcanic debris. Pore-contraction throats usually have a tubular shape. Most throats that originate from vesicular contraction are regular in shape, whereas those that originate from debris accumulation and compaction and crystal growth are more complicated. The radius and length of pore-contraction throats vary greatly; they can be divided into four types: large long, large short, small long, and small short (Figure 11.15).

2. Dissolution throats

Dissolution throats are formed by dissolution after primary throats and matrix mass. A dissolution throat is a major type between dissolution pores, with two origins: (1) the soluble components in a pore-contraction throat are dissolved and the original throat is restored; and (2) the soluble components in the matrix are dissolved and a new throat is generated. The dissolution throat usually follows the shape of its primary throat but may become irregularly shaped due to dissolution. The radius of dissolution

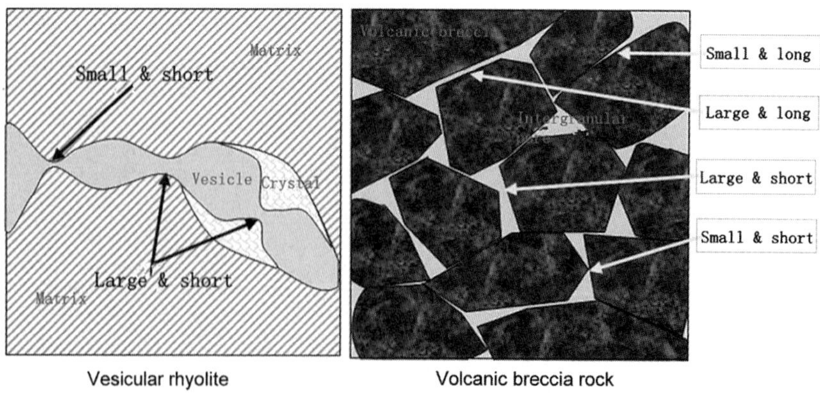

FIGURE 11.15

Patterns of pore-contraction throats.

throats tends to be larger than that of the original ones, and most are large throats. Therefore, dissolution throats mostly fall into the large and long type and large and short type based on their radius and length.

3. Fractured throats

Fractured throats refer to fractures that become pathways among vesicles, intergranular pores, and dissolution pores, playing the role of pore communication. Fractured throats are formed by rock contraction, gas-liquid explosion, tectonic activities, and dissolution, and they usually have a sheetlike configuration. Because a fracture has two dimension that are far greater than the third (with a ratio greater than 10), fractured throats are divided into two types according to their width and extension length: the large and long type, and the small and long type. Throats that originate from structural fractures, dissolution fractures, and large columnar joint fractures are mostly of the large-long type, whereas those that originate from contraction fractures and blast fractures are generally of the small-long type (see Table 11.12).

11.3.2 Identification of throats in volcanic gas reservoirs

The throat classification scheme is used as a basis for establishing the identification markers of various throat types (Table 11.13). By integrated use of cast thin section examination, mercury injection, and constant-rate mercury injection, the throats in volcanic reservoirs are identified level by level in combination with core observation and description. This enhances our understanding of the basic geological characteristics of throats in volcanic reservoirs and provides a basis for characterizing the shape, size, and permeation capacity of throats.

1. Identification of pore-contraction throats

Pore-contraction throats, formed variously from volcanic debris accumulation and compaction, degassing of volatiles in magma, and secondary crystal growth, are the constricted part of vesicles or intergranular pores. Therefore, these types of throats have distinct shapes and sizes and are developed in specific reservoir spaces and specific lithologies and lithofacies. Pore-contraction throats can be effectively identified based on their origin, shape, and size and by taking into account the characteristics of reservoir space, lithology, lithofacies, and the structure and texture of throat-bearing rocks.

1) Characteristics of lithology, lithofacies, rock structures, and textures

Pore-contraction throats are mainly developed in volcanic lava of the effusive facies and volcanic clastic rock of the explosive facies, and various throat types are developed in different lithology, lithofacies, rock structures, and textures (see Table 11.13). Large-long throats are developed mainly in volcanic agglomerate and volcanic breccia of fallout subfacies of the explosive facies, where the rocks typically have volcanic an agglomerate texture and a volcanic breccia texture. Large-short throats are developed mainly in vesicular lava of upper subfacies of the effusive facies as well as volcanic breccia of the fallout subfacies, with common spherulitic texture, volcanic breccia texture, and rhyolitic structure. Small-long throats are most common in volcanic breccia of the fallout subfacies, with a typical volcanic breccia structure. Small-short throats are developed in aplitic lava of the lower subfacies, tight volcanic lava of the middle subfacies, welded tuff of the hot clastic flow subfacies, and crystalline tuff

Table 11.13 Identification Markers of Throats in Volcanic Reservoirs

Type of Reservoir Space		Identification Markers of Throats in Reservoirs							
		Lithology	*Lithofacies*	*Rock Structure/ Texture*	*Pore Type*	*Shape*	*Throat Radius, μm*	*Throat Length*	*Dissolution Trace*
Pore-contraction	Large-long	Volcanic agglomerate, volcanic breccia	Fallout subfacies of explosive facies	Volcanic agglomerate texture, volcanic breccia texture	Intergranular cavities, macro-intergranular pores	Large-long tubular, sandglass-like outline	≥ 0.25	Long	N/A
	Large-short	Vesicular lava, volcanic breccia	Upper subfacies of effusive facies, fallout subfacies of explosive facies	Rhyolitic structure, spherulitic texture, volcanic breccia texture	Vesicular cavities, macrovesicles, macro-intergranular pores	Large-short tubular, dumbbell-like outline		Short	N/A
	Small-long	Vesicular lava, volcanic breccia	Upper subfacies of effusive facies, fallout subfacies of explosive facies	Rhyolitic structure, volcanic breccia texture	Macrovesicles, small intergranular pores	Small-long tubular, sandglass-like outline	< 0.25	Long	N/A
	Small-short	Fine crystal lava, tight volcanic lava, welded tuff, crystal tuff	Lower subfacies of effusive facies, middle subfacies of effusive facies, hot clastic flow subfacies of explosive facies, hot base surge subfacies of explosive facies	Aplitic texture, massive structure, welded texture, tuffaceous texture	Small vesicles, microvesicles, matrix micropores	Small-short tubular, dumbbell-like outline		Short	N/A
Dissolution	Large-long	Vesicular lava, breccia lava, volcanic breccia	Top subfacies of effusive facies, splash subfacies of explosive facies, fallout subfacies of explosive facies	Rhyolitic structure, welded texture, volcanic breccia texture	Dissolution cavities, dissolution pores	Irregular tubular, sandglass-like outline, dumbbell-like outline	≥ 0.25	Long	Dissolution residues, dissolution dents, pitches, and embayment

	Large-short	Vesicular lava, welded tuff, crystalline tuff	Upper subfacies of effusive facies, hot clastic flow subfacies of explosive facies, hot base surge subfacies of explosive facies	Rhyolitic structure, welded texture, tuffaceous texture	Dissolution pores, matrix dissolution pores	Irregular tubular, thorny outline		Short	
Large-long	Structural fractures	Various volcanic rocks	Various lithofacies	Various structures and textures	Various pores	Sheetlike space, linear pattern	≥ 0.25	Long	N/A
	Dissolution fractures					Under control of primary fracture			Available
	Columnar joint fractures	Volcanic lava	Volcanic conduit facies	Rhyolitic structure, columnar joint	Macrovesicle, small vesicle	Crevasse crack like			N/A
Small-long	Contraction fractures	Volcanic lava	Effusive facies	Rhyolitic structure, pellet structure, felsitic texture	Vesicles	Concentric circular, crevasse crack like	< 0.25		N/A
	Blast fractures	Volcanic clastic, autoclastic breccia lava	Explosive facies, volcanic conduit facies	Volcanic clastic texture, cryptobreccia texture	Intergranular pores	Flat- slightly curved linear shape			N/A

of the hot base surge subfacies, with a typical aplitic texture, welded texture, and tuffaceous texture, as well as a massive structure.

2) Characteristics of pore types

Pore-contraction throats occur mainly between vesicles or intergranular pores and their radiuses are associated with the pore size (see Table 11.13). Large-long throats are developed primarily between intergranular cavities or macro-intergranular pores; large-short throats are most common among vesicular cavities, macrovesicles, and macro-intergranular pores; small-long throats are found mainly between small intergranular pores; and small-short throats are found among small vesicles, microvesicles, and matrix micropores.

3) Characteristics of throat shape and size

In terms of their complex origins and variable radii and length, pore-contraction throats can be divided into four types: large long, large short, small long, and small short (Figure 11.16a and b). Large-long throats are primarily developed between intergranular pores with a large radius and are in the shape of

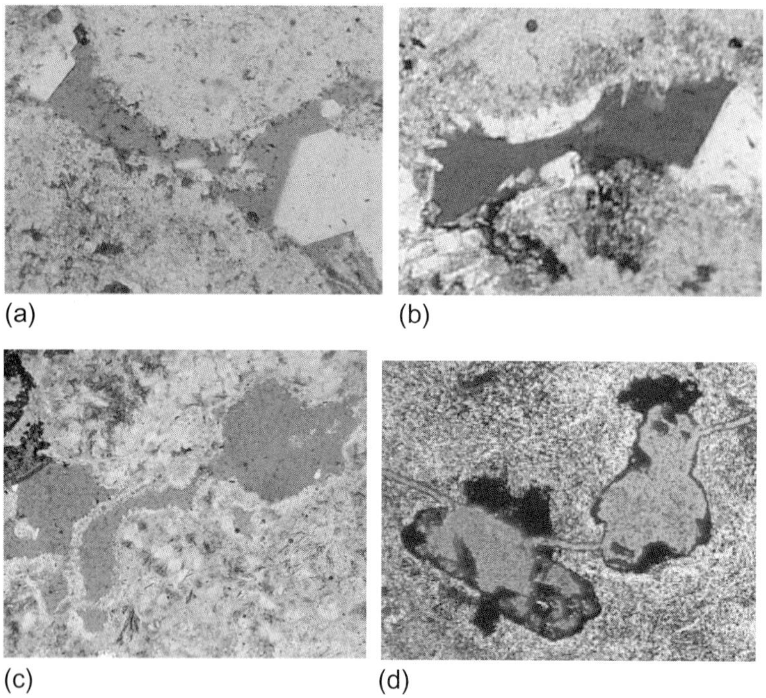

FIGURE 11.16

Typical characteristics of various types of throats. (a) Pore-contraction throats (large-long) in volcanic breccia, Well XX401, 3797.55 m. (b) Pore-contraction throats (large-short) in pelletal rhyolite, Well XX9, 3597.78 m. (c) Dissolution throats in rhyolitic tuffaceous lava, Well XX901, 3864.2 m. (d) Fractured throats in rhyolite, Well CC4, 4403 m.

sandglass-like tubes (Figure 11.16a). Large-short throats are most common between vesicles or intergranular pores with a large radius and are in the shape of dumbbells (Figure 11.16b). Small-long throats are found commonly between intergranular pores with a small radius and are in the shape of slender sandglasses. Small-short throats are developed between vesicles and matrix micropores with a small radius and are in the shape of minute dumbbells.

2. Identification of dissolution throats

Compared with pore-contraction throats, dissolution throats are of complex origins and variable morphology, developed in a wide range of lithologies and lithofacies. Therefore, dissolution throats can be effectively identified based on throat shape and size as well as dissolution trace markers in combination with the lithology, lithofacies, structure, and texture of the rocks in which the throats are developed.

1) Characteristics of lithology, lithofacies, rock structures, and textures

Dissolution throats are developed in the dissolution-prone part of the rock. They can be divided into a large-long type and a large-short type. The former is most commonly developed in vesicular lava of top subfacies of the effusive facies, breccia lava of splash subfacies of the explosive facies, and volcanic breccia of fallout subfacies of the explosive facies, with typical welded and volcanic breccia textures and a rhyolitic structure. Large-short throats are primarily developed in vesicular lava of upper subfacies of the effusive facies, welded tuff of hot clastic flow subfacies of the explosive facies, and crystalline tuff of hot base surge subfacies of the explosive facies and are characterized by welded and tuffaceous textures and a rhyolitic structure.

2) Characteristics of pore types

Dissolution throats are developed between dissolution pores, and their radii are related to the sizes of dissolution pores (see Table 11.13). Large-long throats are usually developed between dissolution cavities and dissolution pores with a large radius, whereas large-short ones are usually developed between dissolution pores and matrix dissolution pores with a large radius.

3) Characteristics of throat shape

Dissolution tends to enlarge the radius of the original throat by selectively dissolving the soluble components such as feldspars and volcanic ash. As a result, dissolution throats are characterized by a large throat, rough inner wall, and irregular shape (Figure 11.16c). Large-long throats usually have irregular, tubular, sandglass, or dumbbell shapes, whereas large-short throats have irregular, tubular, or thorny shapes.

4) Markers of dissolution traces

Dissolution traces such as dissolution residues (mostly secondary minerals such as kaolinite), dissolution dents, pitches, and embayments are important identification markers for dissolution throats.

3. Identification of fractured throats

Fractured throats are formed from rock contraction, cryptoexplosion, and tectonic activities, and they differ from matrix throats in that they have their own distinct origins and shapes. Therefore, fractured throats can be identified systematically based on their origins, shapes, and sizes, in combination with the characteristics of lithology, lithofacies, structure, and texture of the rocks in which fractures are developed.

1) Characteristics of lithologies, lithofacies, rock structures, and textures

Volcanic rock fractures may be developed in various lithology types, but different types of fractures correspond to different characteristics of lithologies, lithofacies, structures, and textures (see Table 11.13). The throats originating from structural or dissolution fractures occur in a variety of volcanic lithofacies. Those originating from columnar joint fractures are mainly developed in volcanic lava of the volcanic conduit facies with typical a rhyolitic structure and columnar joint. Throats originating from contraction fractures are found mostly in volcanic lava of the effusive facies, with typical spherulitic and felsitic textures and a rhyolitic structure. Those originating from blast fractures are most common in volcanic clastic rocks of the explosive facies and autoclastic breccia lava of the volcanic conduit facies, and they have volcanic clastic textures (e.g., agglomerate texture, volcanic breccia texture, and tuffaceous texture) and a cryptoexplosive breccia texture.

2) Characteristics of pore types

Throats originating from structural and dissolution fractures may be developed between various pores. Those originating from columnar joint fractures are primarily developed between macrovesicles or microvesicles. Throats originating from contraction fractures usually occur between various types of vesicles. Those formed by blast fractures are common between various types of intergranular pores.

3) Characteristics of throat shape

Most fractured throats appear as sheetlike spaces with a linear outline, so they are long-fractured types. In the three types of large-long throats, throats associated with structural fractures are sheetlike spaces with a linear outline. The shapes of throats in dissolution fractures are influenced by primary fractures, but the smooth walls become rough and the outlines change to irregular. Throats originating from columnar joint fractures are represented by crevasse cracks. In the two kinds of small-long throats, those originating from contraction fractures appear as concentric circles and crevasse cracks, whereas those in blast fractures mostly have a flat to slightly curved linear pattern.

4) Markers of dissolution traces

In all fractured throats, the typical markers of dissolution traces are present in the large-long throats of dissolution fracture origin, but these markers are absent from other fractured throats.

11.3.3 Characterization of throats in volcanic reservoirs

On the basis of classified identification, using cast thin section examination, rock porosity and permeability tests, conventional mercury injection; and constant-rate mercury injection analysis, the shapes, sizes, and permeation capacity of throats in volcanic reservoirs are characterized following the technical process that progresses from qualitative to quantitative characterization.

1. Characterization of throat shape

Because throat has a small radius, cast thin section examination is mainly used for SEM observation of the shape and distribution of various types of throats in volcanic reservoirs. This provides a basis for the characterization of microstructures in volcanic reservoirs.

1) Pore-contraction throats

These throats are the constricted parts between vesicles or intergranular pores and normally have a tubular shape. The vesicle contraction throats usually have the large-short and small-short tubular features, with a sandglass-like outline (Table 11.14b and d). Crystal growth throats are characterized by a rough inner wall and an irregular shape, usually of the small-long tubular category, with a sandglass outline. Most contracted throats in intergranular pores are of the large-long and small-long types, with a dumbbell outline (Table 11.15a and c). In the XX gas field, where volcanic lava is the major lithology in the volcanic reservoir, vesicles with a small radius are predominant in pores. Small-short throats are usually the well-developed type of pore-contraction throat, accounting for about 55%; the large-short type accounts for about 20%; the small-long type accounts for about 15%; and the large-long type accounts for about 10%.

2) Dissolution throats

These throats are formed by dissolution that reshapes the pore-contraction throats or by dissolution of the soluble components in a matrix. Due to selective dissolution of feldspar and volcanic ash, most dissolution throats have an irregular tubular shape, but completion dissolution of phenocrysts may result in regular-shaped throats. In the two types of dissolution throats, the large-long type is sandglass shaped or dumbbell-like in outline, whereas the large-short type tends to have a thornlike outline. In the

Table 11.14 Shape Classification of Pore-Contraction Throats in Volcanic Reservoirs

Throat Type	Origin	Shape Type	
Pore-contraction throat	Formed by contraction of intergranular pores or vesicles	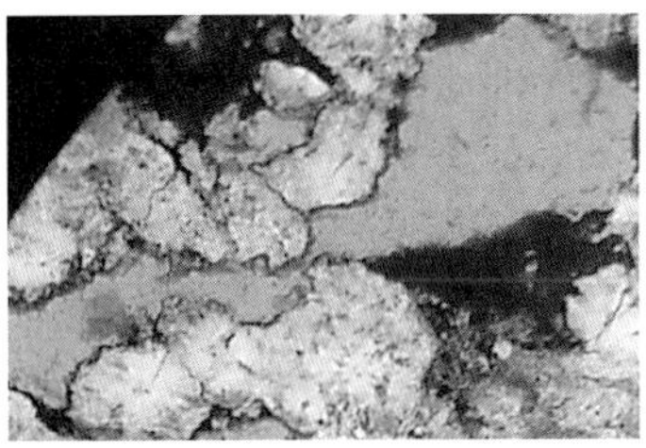 (a) Large-long tubular, sandglass like (volcanic breccia).	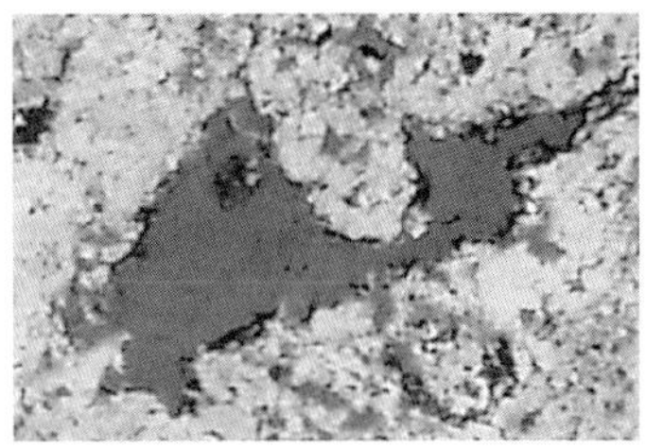 (b) Large-short tubular and dumbbell like (rhyolite).
		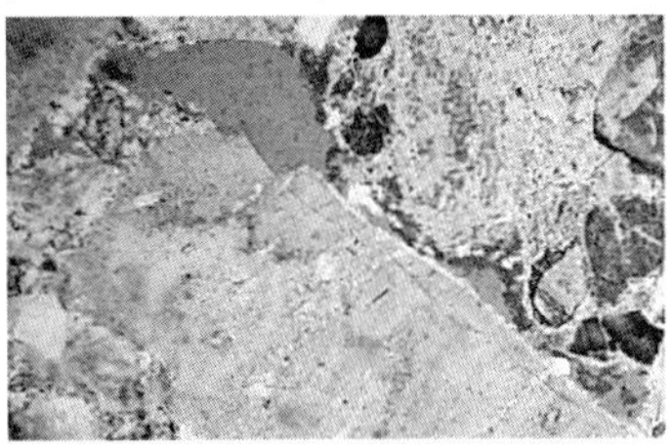 (c) Small-long tubular, sandglass like (volcanic breccia).	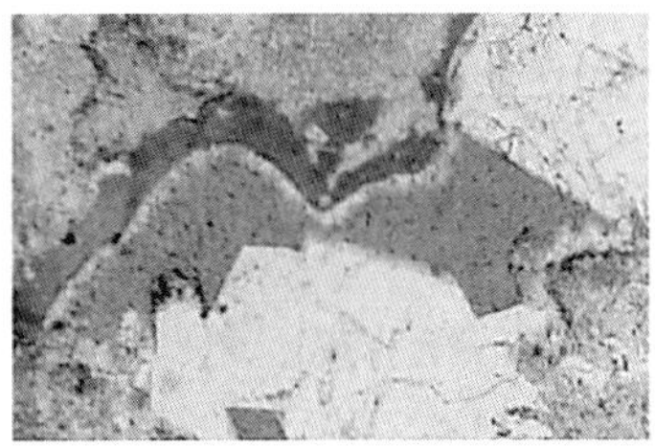 (d) Small-short tubular and dumbbell like (rhyolite).

Table 11.15 Shape Classification of Dissolution Throats in Volcanic Reservoirs

Throat Type	Origin	Shape Type (Irregular Tubular)		
Dissolution throat	Through dissolution, pore-contraction throats are reformed, or the soluble components are dissolved in matrix	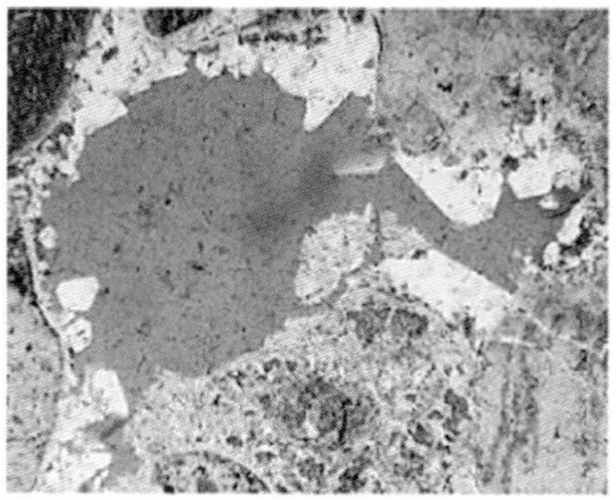 (a) Large-long type, sandglass-like in outline.	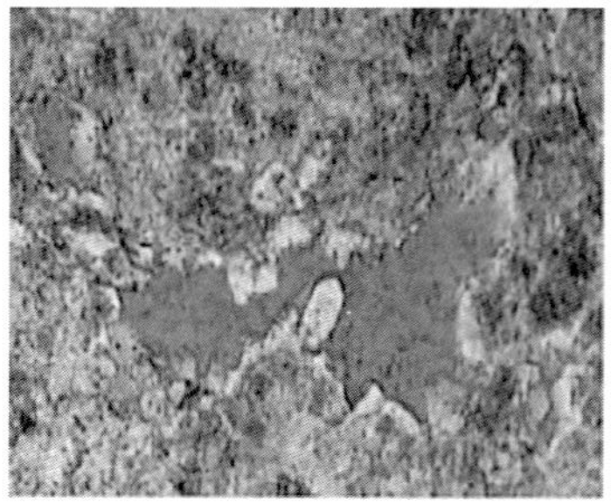 (b) Large-short type, dumbbell-like in outline.	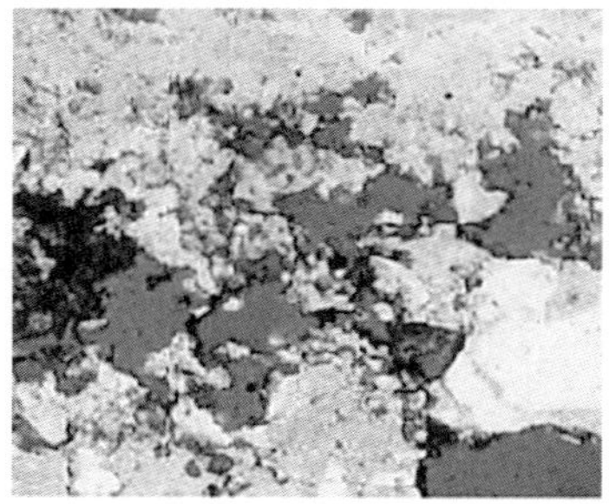 (c) Large-short type, thorny in outline.

XX gas field, vesicles with a small radius, intergranular pores, and matrix micropores are predominant in the volcanic reservoir and the dissolution is relatively weak; as a result, the large-short dissolution throats are dominant, accounting for about 70%, whereas the large-long type accounts for 30%.

3) Fractured throats

These are largely sheetlike throats. In the large-long type formed by large fractures, throats originating from structural fractures have a large size and a linear pattern in plan view and may be developed in various types of pores (Table 11.16a). Those associated with dissolution fractures have larger sizes than those of precursory fractures, as their shape is generally controlled by the precursory fractures, but the inner wall tends to become rough with an irregular the outline, and they may occur in various types of pores. The throats originating from columnar joint fractures are large in size, manifesting as crevasse cracks in the cross section and a linear pattern in the longitudinal section, and they develop mainly between macrovesicles or small vesicles in volcanic lava of the volcanic conduit facies. In the small-long throats originating from small fractures, those originating from contraction fractures are characterized by concentric rings and crevasse cracks, and they develop between various types of vesicles in volcanic lava of the effusive facies (Table 11.16b); those originating from blast fractures have the smallest size, usually showing a flat to slightly curved linear pattern, and they are found commonly between various types of intergranular pores in volcanic clastic rocks of the explosive facies and autoclastic breccia lava of the volcanic conduit facies (Table 11.16c).

2. Characterization of throat size in reservoirs

Throat size determines the fluid flow capacity of a reservoir. In terms of throat features in volcanic reservoirs, the size characteristics of various types of throats can be clarified by integrating core observation, cast thin section examination, capillary pressure curve analysis, constant-rate mercury injection analysis, and log interpretation. This provides a basis for characterizing the accumulation-permeation capacity of throats.

1) Core observation and cast thin section examination

Core observation is an effective method for examining large-sized pores and throats. Because of their generally small size, matrix throats are difficult to observe visually, so core observation is used mainly to determine the size and distribution characteristics of fractures. In the XX gas field, the width of open fractures connecting pores can reach a maximum of 6000 μm, but it mainly ranges from 60 to 600 μm with an average of 300 μm. Therefore, the largest radius of a fractured throat (large-long type) is 3000 μm, primarily ranging from 30 to 300 μm with an average of 150 μm, all of which are macrothroats (> 0.5 μm).

SEM examination of cast thin sections is important for studying the evolution of pores and fractures in reservoirs and for evaluating the types and characteristics of pores, throats, and fractures. Through SEM observation and measurement, pore-contraction throats, dissolution throats, and fractured throats can be effectively distinguished in cast thin sections, and the radii of larger throats can be determined semiquantitatively. In the XX gas field, more than 50 pieces of cast thin sections of volcanic rocks were examined and analyzed semiquantitatively to achieve a preliminary understanding of the characteristics of throat size in the volcanic reservoir: (1) the throat radius in reservoir matrix (including pore-contraction throats and dissolution throats) is normally 0.1 to 6 μm; (2) the radius of fractured throat (0.5 to 15 μm) is much larger than that of matrix throat.

Table 11.16 Shape Classification of Fractured Throats in Volcanic Reservoirs

Throat Type	Origin	Shape Type (Sheetlike Space)		
Fractured throat	Fracture plays a role in pore communication and results in fractured throats	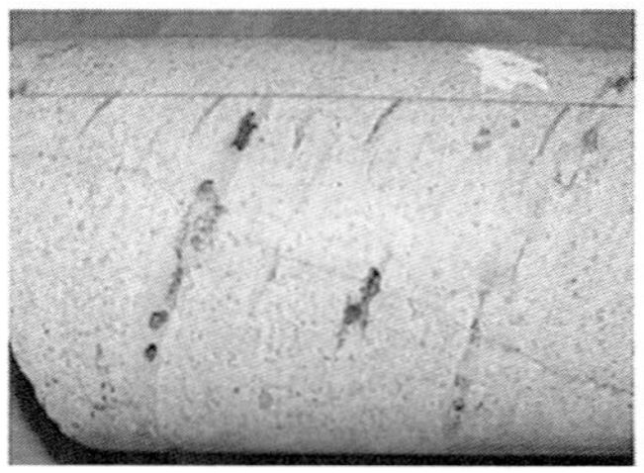 (a) Large-long type (structural fracture), linear pattern.	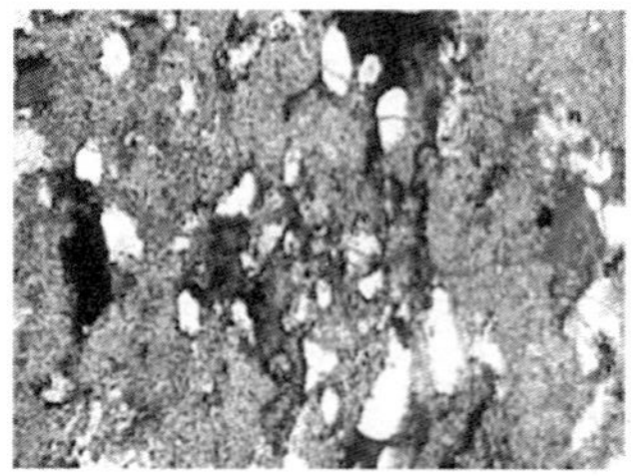 (b) Small-long type (contraction fracture), linear pattern.	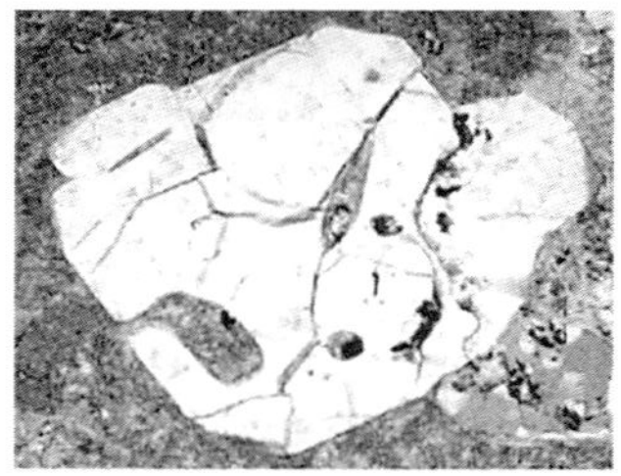 (c) Small-long type, as crevasse cracks.

2) Mercury injection

Mercury injection experiments measure the mercury volumes injected into a core plug under different pressures; the capillary pressure curves are derived and then the throat size and pore volume controlled by the throats are calculated using the J function [13]. Through processing and statistical analysis of a large number of capillary pressure curves, the size and distribution characteristics of various types of throats in a volcanic reservoir matrix can be assessed.

In the XX gas field, capillary pressure curves were analyzed with conventional mercury injection information derived from 206 pieces of volcanic rock core samples, and the radius distributions of various types of matrix throat were obtained (Table 11.17). The largest matrix throat radius is 0.64 μm and the radii range from 0.005 to 0.4 μm with an average of 0.07 μm (Figure 11.17a). In terms of size, microthroats are dominant (accounting for about 76.6%), small throats rank second (accounting for about 18%), and medium throats and large throats are poorly developed (accounting for about 4.9% and 0.5%, respectively) (Figure 11.17b). In matrix throats, the pore-contraction throats have the largest radius at 0.38 μm, with the radius values ranging from 0.008 to 0.2 μm and averaging 0.06 μm; in terms of size, microthroats are dominant (accounting for about 83.9%), small throats account for about 13.4%, medium throats account for only about 2.7%, and large throats are not developed (Figure 11.17c). The dissolution throats have a maximum radius of 0.64 μm, with the radii ranging from 0.01 to 0.4 μm and averaging 0.11 μm; in terms of size, microthroats are dominant (accounting for about 57.9%), small throats account for about 29.8%, medium throats account for only about 10.5%, and large throats are poorly developed (accounting for about 1.8%) (Figure 11.17d).

3) Constant-rate mercury injection

In this method, mercury is injected into rock pores in a process of quasi-static state. Pore and throat are identified by the pressure drop that occurs when mercury breaks through the throat, and the characteristic parameters of microscopic porous structure in rocks, such as distributions of pore radius and throat radius controlled by large throats (≥ 0.1 μm), can be derived. In the XX gas field, 24 pieces of volcanic rock sample were analyzed in the constant-rate mercury injection experiments. The results (Figure 11.18 and Table 11.17) indicate that the maximum radius of a matrix throat in the volcanic reservoir is 5.4 μm, with the radius values distributed from 0.16 to 1.2 μm and averaging 0.6 μm; medium throats are dominant, accounting for about 41.7%; small throats rank second, accounting for about 33.3%; and very few large throats are developed (about 25%). Because mercury injection is a quasi-static process, micropores cannot be detected.

Table 11.17 Statistics on Throat Sizes in Volcanic Gas Reservoirs (XX Gas Field)

Characterization Method	Matrix Throat Radius, μm			Pore-Contraction Throat Radius, μm			Dissolution Throat Radius, μm			Fractured Throat Radius, μm		
	Max	*Range*	*Min*	*Max*	*Range*	*Min*	*Max*	*Range*	*Min*	*Max*	*Range*	*Min*
Core observation										3000	30-300	150
Mercury injection	0.64	0.008-0.4	0.07	0.38	0.008-0.2	0.06	0.64	0.01-0.4	0.11			
Constant-rate mercury injection	5.4	0.16-1.2	0.6									
Log interpretation										750	2.5-25	14

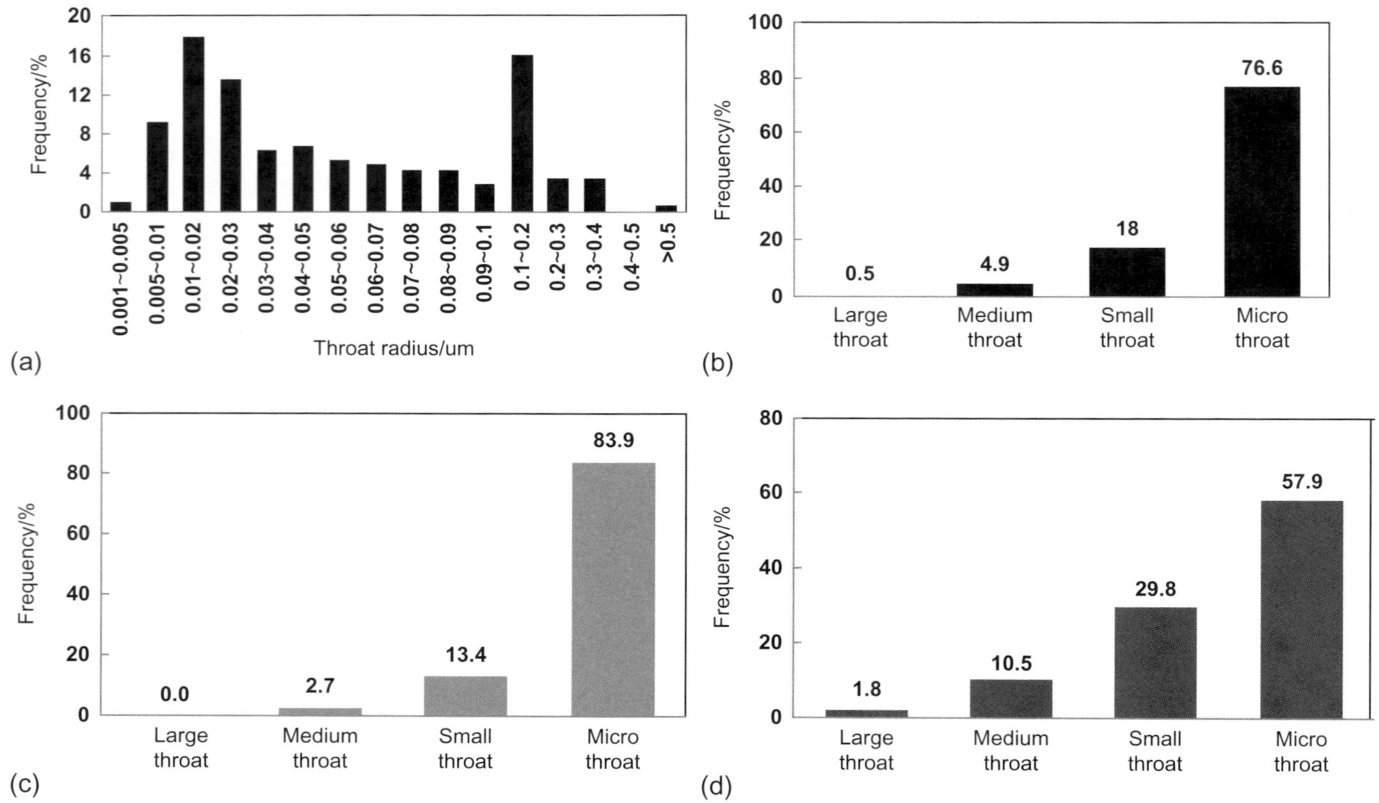

FIGURE 11.17

Characteristics of the matrix throat size of volcanic reservoirs determined by mercury injection information (XX gas field). (a) Radius distribution of matrix throats. (b) Various matrix throat sizes. (c) Various pore-contraction throat sizes. (d) Various dissolution throat sizes.

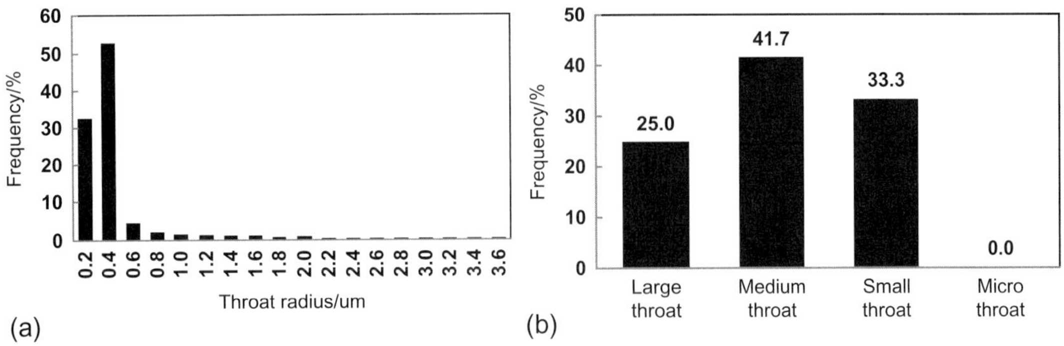

FIGURE 11.18

Characteristics of matrix throat size in volcanic reservoirs determined by constant-rate mercury injection information (XX gas field). (a) Radius distribution of matrix throats (welded tuff). (b) Types of matrix throats in terms of size.

4) Log interpretation

In this method, fracture width is calculated by means of FMI imaging logs and conventional dual laterologs, and the radius range of fractured throats is also obtained. The results (see Table 11.17) of log interpretation for the XX gas field reveal that the maximum radius of a fractured throat in the volcanic reservoir is 750 μm, with a range of 2.5 to 25 μm and an average of 14 μm, all falling under the category of large throats.

The preceding discussion suggests that (1) the throat radius in volcanic gas reservoirs has a wide range of values, including various throats developed from macro- to microscales; (2) the radius of a matrix throat is small, and microthroats and small throats are the dominant types of matrix throats; the radius of a dissolution throat is usually larger than that of a pore-contraction throat; and (3) the radius of a fractured throat is considerably larger than that of a matrix throat, and fractures play an important role for the permeability in volcanic reservoirs.

3. Characterization of reservoir permeation capacity

Permeation capacity is an important indicator for evaluating the sizes of various types of throats and their fluid flow properties. Permeability and displacement pressure are two parameters normally used to characterize the permeation capacity of throats in volcanic reservoirs.

1) Permeation capacity of matrix throats

By integrated use of core porosity and permeability tests, mercury injection experiments, and log interpretations, the overall permeation capacity of matrix throats and various individual types of matrix throats in volcanic reservoirs are characterized.

Overall permeation capacity of reservoir matrix. In volcanic reservoirs, matrix throats have complex origins and highly variable shapes and radius sizes, resulting in significant variations in their permeation capacities. In the XX gas field (Figure 11.19 and Table 11.18), the volcanic rock cores have

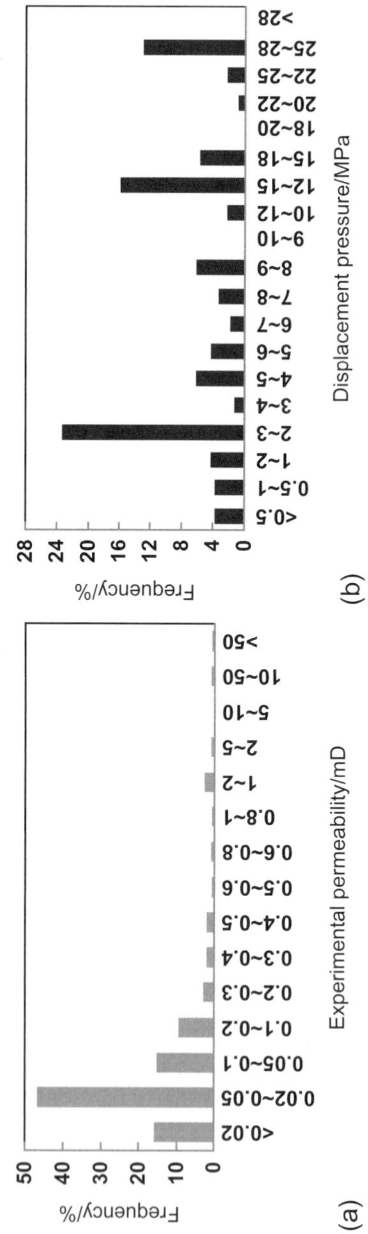

FIGURE 11.19

Characterization of permeation capacity of a volcanic reservoir matrix (XX gas field). (a) Distribution of matrix permeability in whole matrix cores. (b) Distribution of displacement pressure in a reservoir matrix.

Table 11.18 Statistics on the Permeation Capacity of Various Types of Throats in Volcanic Gas Reservoirs (XX Gas Field)

Throat Type	Typical Rock	Matrix Permeability Derived from Logging, mD	Experimental Matrix Permeability, mD	Expulsion Pressure, MPa
Pore-contraction throat	Rhyolitic volcanic breccia	0.05	0.08	10.57
Dissolution throat	Rhyolitic breccia lava	0.16	0.28	8.06
Matrix throat	All volcanic rocks	0.07	0.15	9.77

a maximum permeability of 17.3 mD, ranging from 0.02 to 0.2 mD with an average of 0.14 mD. The matrix permeability derived from logging interpretation has a maximum of 3.2 mD, with a range of 0.01 to 0.15 mD and an average 0.07 mD. The displacement pressure derived from mercury injection experiments has a wide distribution, with an average of 9.77 MPa. It is thus confirmed that the matrix permeation capacity in the volcanic reservoir is relatively weak and varies significantly.

Permeation capacity of various types of throats by origin. The two types of matrix throat—that is, the pore-contraction throats and dissolution throats—have different origins, shapes, and sizes, leading to different permeation capacities (see Table 11.18). For pore-contraction throats, the matrix permeabilities derived from logs and experiments are 0.05 mD and 0.08 mD, respectively, and the displacement pressure is 10.57 MPa; whereas these values for dissolution throats are 0.08 mD, 0.28 mD, and 8.06 MPa, respectively. This suggests a relatively good permeation capacity for dissolution throats.

Permeation capacity of various types of throats by size. Throats of different sizes and lengths have different permeation capacities. In general, a larger throat size corresponds to a stronger permeation capacity. In the pore-contraction throat example (Table 11.19) from the XX gas field, the matrix permeability of large-short throats is the highest (average 0.124 mD) and their displacement pressure is the lowest (average 9.16 MPa), indicating the strongest permeation capacity. The matrix permeability of large-long throats is fairly high (average 0.108 mD) and their displacement pressure is fairly low (average 9.86 MPa), showing a

Table 11.19 Statistics on the Permeation Ability of Various Types of Throats in the Volcanic Reservoirs of XX Gas Field (in Terms of Subfacies)

Throat Type		Typical Subfacies	Matrix Permeability, mD	Displacement Pressure, MPa
Pore-contraction throat	Large-short	Upper subfacies of effusive facies	0.124	9.16
	Large-long	Fallout subfacies of explosive facies	0.108	9.86
	Small-long	Fallout subfacies of explosive facies	0.048	12.31
	Small-short	Lower subfacies of effusive facies Hot base surge subfacies of explosive facies	0.032	16.17

relatively good permeation capacity. The matrix permeability of small-long throats is fairly low (average 0.048 mD) and their displacement pressure is high (average 12.31 MPa), showing a weak permeation capacity. Finally, the matrix permeability of small-short throats is the lowest (average 0.032 mD) and their displacement pressure the highest (average 16.17 MPa), representing the weakest permeation capacity.

2) Permeation capacity of fractured throats

In light of the fracture complexity of volcanic rocks, the permeation ability of fractured throats is characterized using core description and log interpretation, with two parameters: radius of fractured throat and fracture permeability:

1. *Radius of fractured throat.* As Table 11.17 shows, the radius of visually identifiable fractured throats ranges from 30 to 300 μm and averages 150 μm, more than 2000 times that of matrix throats; and the radius of fractured throats derived from log interpretation ranges from 2.5 to 25 μm and averages 14 μm, being 20 to 200 times that of matrix throats. In general, therefore, the radius of a fractured throat is far larger than that of a matrix throat.
2. *Fracture permeability.* Herein, it refers specifically to the permeability of fractured reservoirs. In the XX gas field, the total permeability of the volcanic reservoir falls ranges from 0.02 to 59.65 mD with an average of 2.18 mD; the matrix permeability ranges from 0.02 to 5.09 mD with an average of 0.07 mD; and the maximum fracture permeability is 59.46 mD with an average of 2.11 mD. In general, fracture permeability is 30 times that of the matrix permeability, constituting up to 97% of the total permeability of this reservoir. Therefore, fracture is a crucial factor for the enhancement of permeation capacity and therefore the production in volcanic gas reservoirs.

There are many types of fractures in volcanic reservoirs, with different fracture types having different origins and fracture sizes, and the throats that originate from different fracture types having various permeation capacities. In general, structural fractures are wide and extensive, so that the throats originating from structural fractures have a strong permeation capacity. Dissolution fractures enlarge the width of their precursory fractures and enhance the fluid flow capacity; therefore, the throats associated with dissolution fractures also tend to have a strong permeation capacity. Columnar joint fractures near volcanic conduits are fairly wide and extend far; consequently, the throats derived from columnar joint fractures have a strong permeation capacity. In contrast, contraction and blast fractures tend to have a small width, short extension, and poor pore connections, so their permeation capacity is relatively weak; sutured fractures are basically closed and are irrelevant in terms of permeation.

11.4 CHARACTERIZATION OF ACCUMULATION-PERMEATION PATTERNS IN VOLCANIC RESERVOIRS

In this study, *accumulation-permeation* (hereafter abbreviated as *A-P*) pattern refers to the dynamic combination mode of reservoir spaces and throats. The types and characteristics of A-P patterns determine the accumulation-permeation capacity of volcanic reservoirs and directly influence their deliverability and stable production rate. By taking into account the geological features of volcanic gas reservoirs, the A-P patterns are classified, and the lithology, lithofacies, pore-throat-fracture assemblage, accumulation ability, permeation ability, and deliverability characteristics of different A-P patterns are clarified on the basis of reservoir space and throat characterization. This provides a basis for assessing the control of A-P patterns on gas reservoir deliverability.

11.4.1 Classification of accumulation-permeation patterns in volcanic gas reservoirs

Attempts have been made to study accumulation-permeation (A-P) patterns in metamorphic and carbonate reservoirs. For example, Bai and colleagues [14] proposed four types of A-P patterns on the basis of their work on the A-P patterns of buried-hill oil reservoirs in the Bohai Gulf, and they determined that various A-P patterns exhibit different production characteristics. Chen and coworkers [15] recognized six A-P patterns for fractured porous oil reservoirs in their study on the Lower Ordovician fractured porous carbonate oil reservoir in the Tahe oil field of the Tarim basin, and they concluded that the recovery characteristics of such oil-producing reservoirs were closely related to the A-P patterns of the accumulation body encountered by drilling. Considering the complex origins of highly variable reservoir spaces and throats in volcanic gas reservoirs, this study recognizes eight types of A-P patterns in two categories (Table 11.20) in the XX volcanic gas reservoir in terms of the dynamic assemblage of various reservoir spaces and throats, based on the characterization of reservoir spaces and throats, with reference to the previous research results, and taking into account the formation and evolution of reservoir spaces and throats in volcanic gas reservoirs.

1. Single accumulation-permeation (A-P) pattern

A single A-P pattern denotes that the reservoir spaces are simple and that throats and reservoir spaces are related in origin. In light of the genetic classification and assemblage relationship of reservoir spaces and throats, the single pattern in volcanic gas reservoirs includes four types: vesicular, intergranular porous, microporous, and fractured. Because of the differences in pore types, throat types, and characteristics of lithology and lithofacies, different types of A-P patterns have different accumulation-permeation capacities, resulting in distinct dynamic characteristics of deliverability and stable production capability in volcanic reservoirs (see Table 11.20):

1. *Vesicular A-P pattern.* This pattern is mainly developed in vesicular lava of top and upper subfacies of the effusive facies; vesicles with a large radius are the predominant reservoir space, characterized by a bead-string-like distribution; most throats are of the large-short type formed from vesicle contraction, and the small-short type ranks second. Fractures are poorly developed. Hence, this pattern provides a high accumulation capacity and moderate permeation capacity.
2. *Intergranular porous A-P pattern.* This pattern is mainly formed in volcanic breccia of fallout subfacies of explosive facies and welded breccia of hot clastic flow subfacies of explosive facies. Intergranular pores with a large radius are predominant in reservoir spaces and most throats originate from intergranular pore contraction, and large and long, large and short, small and long, and small and short throats are all developed; fractures are poorly developed. Therefore, this pattern has features of moderate accumulation ability and moderate or poor permeation ability.

Table 11.20 Major Accumulation-Permeation (A-P) Patterns and Characteristics of Deep Volcanic

Classification of A-P Patterns					Characteristics of A-P Pattern	
No.	***Pattern Type***		***Pore Type***	***Throat Type***	***Lithology***	***Lithofacies***
1	Single pattern	Vesicular	Vesicles	Pore-contraction (large-short, small-short)	Vesicular lava	Top subfacies of effusive facies, upper subfacies of effusive facies
2		Intergranular porous	Intergranular pores	Pore-contraction (large-short, small-short, large-long, small-long)	Volcanic breccia, welded breccia	Fallout subfacies of explosive facies, hot clastic flow subfacies of explosive facies
3		Microporous	Matrix micropores	Pore-contraction (small-short)	Crystal tuff, welded tuff	Hot clastic flow subfacies of explosive facies, hot base surge subfacies of explosive facies
4		Fractured	Sporadic vesicles	Fractured	Tight volcanic rock	Lower subfacies of effusive facies, middle subfacies of effusive facies, hot base surge subfacies of explosive facies
5	Combined pattern	Fractured - vesicular	Vesicles	Pore-contraction (large-short, small-short), fractured	Vesicular lava, breccia lava	Top subfacies of effusive facies, upper subfacies of effusive facies, splash subfacies of explosive facies
6		Fractured – intergranular porous	Intergranular pores	Pore-contraction (large-short, small-short, large-long, small-long), fractured	Volcanic breccia, welded breccia	Fallout subfacies of explosive facies, hot clastic flow subfacies of explosive facies
7		Fractured - microporous	Micropores	Pore-contraction (small-short), fractured	Welded tuff, crystalline tuff	Hot clastic flow subfacies of explosive facies, hot base surge subfacies of explosive facies
8		Fractured-dissolution-porous	Primary pores and dissolution pores	Dissolution, fractured	Various volcanic rock	Various lithofacies

Reservoirs in the Songliao Basin

Characteristics of A-P Pattern				A-P Pattern	Typical Schematic A-P Pattern
Fracture Features	*Accumulation Ability*	*Permeation Ability*	*Deliverability Character*	*Representative Blocks*	
Poorly developed	High	Moderate	Natural deliverability, postfracturing high production	XX2-1	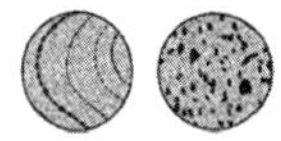
Poorly developed	Moderate	Moderate–low	Postfracturing moderate–low production	XX1	
Poorly developed	Moderate–low	Low	Postfracturing low production	XX1	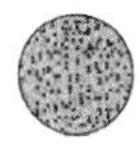
Structural fractures, contraction fractures	Low	High	Dry zone	XX9 and XX231	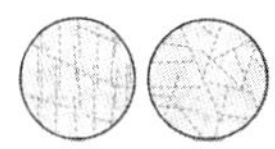
Structural fractures, contraction fractures	High	Moderate–high	Natural high production–postfracturing moderate production	XX8, XX9, CC1 and XX21	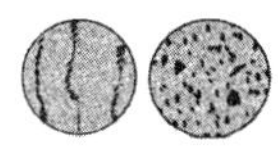
Structural fractures	Moderate	Moderate–high	Postfracturing high–moderate production	XX1 and XX21	
Structural fractures, blast fractures	Moderate–low	Moderate	Postfracturing high–moderate production	XX1	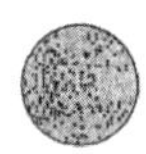
Structural fractures, contraction fractures, blast fractures	Varied greatly	Moderate–high	Postfracturing high–moderate production	XX8, CC1 and XX28	

3. *Microporous A-P pattern.* This pattern occurs primarily in welded tuff of hot clastic flow subfacies of the explosive facies and crystal tuff of hot base surge subfacies of the explosive facies. The reservoir space is dominated by matrix micropores with a small radius. The small-short types of throats that originated from matrix micropore contraction are predominant; fractures are poorly developed. Thus, this pattern usually has relatively poor accumulation and permeation capacities.
4. *Fractured A-P pattern.* This pattern is found mainly in tight volcanic rocks of lower and middle subfacies of the effusive facies as well as hot base surge subfacies of the explosive facies. Structural and contraction fractures are predominant, making up the major reservoir spaces and seepage pathways. Most of the matrix pores are scattered small vesicles, microvesicles, and matrix micropores. Fractured throats are predominant, including the large-long throat type formed from large fractures as well as the small-long throat type that originates from small fractures. Therefore, this pattern is characterized by poor accumulation capacity but good permeation capacity.

2. Combined A-P pattern

In a combined A-P pattern, various pores act as reservoir spaces and matrix throats, and fractured throats act as seepage pathways. Because the permeation capacity of fractured throats is much higher than that of matrix throats, the matrix throats in combined pattern have a smaller contribution to the permeation capacity of reservoirs. In light of the genetic types of reservoir spaces, the combined A-P patterns fall into four types: fractured-vesicular, fractured–intergranular porous, fractured-microporous, and fractured–dissolution-porous. In different types of combined A-P patterns, there are differences in pore types, fracture types, lithology, and lithofacies, and consequently there are differences in accumulation capacity, permeation capacity, and reservoir deliverability (see Table 11.20):

1. *Fractured-vesicular A-P pattern.* This A-P pattern is developed mainly in vesicular lava of the top and upper subfacies of the effusive facies. The reservoir space is dominated by vesicles with a large radius, featuring a bead-string-like distribution. Most of the throats are of the vesicle contraction and fractured types. Vesicle contraction throats include the large-short and small-short categories, whereas fractured throats include the large-long and small-long categories. Most fractures are well developed and are of the structural and contraction types. Hence, the fractured-vesicular A-P pattern has good accumulation and permeation capacities.
2. *Fractured–intergranular porous A-P pattern.* This A-P pattern is mainly formed in volcanic breccia of fallout subfacies of the explosive facies and welded breccia of hot clastic flow subfacies of the explosive facies. Most of the reservoir spaces are intergranular pores with a large radius. Intergranular pore-contraction throats and fractured throats are dominant, with large-long, large-short, small-long, and small-short types of throats developed. Structural fractures are predominant and well developed. The fractured–intergranular porous A-P pattern, therefore, has a moderate accumulation capacity and a fairly good permeation capacity.
3. *Fractured-microporous A-P pattern.* This A-P pattern is found primarily in welded tuff of hot clastic flow subfacies of the explosive facies and crystalline tuff of hot base surge subfacies of the explosive facies. Most of the reservoir spaces are matrix micropores with a small radius. Pore-contraction throats and fractured throats are predominant, usually including the large-long, small-long, and small-short types. Structural and blast fractures are common and well developed. As a result, this A-P pattern usually has a moderate-low accumulation capacity and a moderate permeation capacity.

4. *Fractured–dissolution-porous A-P pattern.* This A-P pattern occurs mainly in volcanic rocks of various lithofacies. The reservoir spaces are primarily composed of dissolution pores of variable radii, including dissolution cavities, dissolution pores, and matrix dissolution pores. Dissolution and fractured throats are predominant. Structural contraction and blast fractures are well developed. Therefore, this pattern has a moderate-strong accumulation capacity and a good permeation capacity.

11.4.2 Characterization of accumulation-permeation (A-P) patterns in volcanic gas reservoirs

The accumulation capacity of various A-P patterns is characterized in terms of porosity, probe radius, and other parameters; the permeation capacity is characterized by permeability; the deliverability is characterized by production and open flow potential; and the stable production capability of various A-P patterns is characterized by gas production per pressure drop and stable production duration. The control of A-P patterns on deliverability and the stable production capability of gas-producing reservoirs is assessed on the basis of classification; systematic analyses of static and dynamic data derived from cores, thin sections, logs, gas tests, pilot production tests, and practical productions; and by comparing the accumulation capacity, permeation capacity, and deliverability characteristics of various A-P patterns (Table 11.21). This will provide guidance for developing technical procedures regarding volcanic gas reservoirs.

1. Vesicular A-P pattern

A vesicular A-P pattern is developed mainly in volcanic lava of the effusive facies. Its accumulation capacity is primarily affected by vesicle size and the degrees of vesicle development, as well as reservoir extension, whereas the permeation capacity is influenced by the radius of vesicle contraction throats and pore-throat configuration. From the middle-lower subfacies to the upper-top subfacies of the effusive facies, the radii of vesicles and throats increase because of the increased degassing of volatile constituents, with better developed and better communicated vesicles, resulting in improved accumulation-permeation capacity, deliverability, and stable production capability for the reservoirs.

Horizon No. 93 in Well XSS2-1 has a typical vesicular A-P pattern, characterized by vesicular rhyolite of the upper subfacies of the effusive facies. Its reservoir space is dominated by macrovesicles with a bead-string-like distribution. Most of the throats are of the large-short type that originated from vesicle contraction, exhibiting good communication between pores. Fracture is not observed in cores (Figure 11.20a). In this horizon, the effective porosity is 13.2%, the average permeability is 0.74 mD, and the probe radius for the inner zone is 65 m from well test interpretation with an infinite outer zone, indicating a good accumulation capacity but a moderate permeation capacity. During gas testing, the natural deliverability was 15.1×10^4 m^3/d, with a pilot production yield of 12.15×10^4 m^3/d and open flow potential of 40.93×10^4 m^3/d. After stream injection, a stable production of 9 to 15.7×10^4 m^3/d is maintained, with a slow pressure decline and a gas production per pressure drop at 7839×10^4 m^3/MPa. So far, the cumulative gas production has been 0.6×10^8 m^3, and the stable production duration is estimated to be 10 years (Figure 11.20b and Table 11.21). Therefore, these rocks of the upper and top subfacies with a vesicular A-P pattern have good accumulation-permeation capacities, high production levels, and a strong potential for stable production.

Table 11.21 Characteristics of Volcanic Reservoirs with Different Accumulation-Permeation (A-P) Patterns

		Typical Well/Horizon				
Classification of A-P Pattern		***Well and Horizon No.***	***Lithology***	***Lithofacies***	***Reservoir Space***	***Fracture and Development***
Single pattern	Vesicular	Horizon No. 93 in Well XSS2-1	Vesicular rhyolite	Upper subfacies of effusive facies	Macrovesicles	Poor
	Intergranular porous	Horizons No. 166 and 170I in Well XX3	Welded breccia	Hot clastic flow subfacies of explosive facies	Small intergranular pores	Poor
	Microporous	Horizon No. 129 in Well XX1-4	Crystal tuff	Hot base surge subfacies of explosive facies	Matrix micropores	Poor
	Fractured	Horizon No. 224 in Well XX21	Tight rhyolite	Middle subfacies of effusive facies	Microvesicles	Structural fracture, contraction fracture, blast fracture
Combined pattern	Fractured - vesicular	Horizons No. 59II and 60 IV in Well XX9	Vesicular rhyolite	Upper subfacies of effusive facies	Small vesicles	Structural fracture, contraction fracture
	Fractured - intergranular porous	Horizon No. 150 in Well XX1	Volcanic breccia	Fallout subfacies of explosive facies	Intergranular pores	Structural fracture, blast fracture
	Fractured-microporous	Horizon No.141-I in Well XX8-107	Crystalline tuff	Hot base surge subfacies of explosive facies	Matrix micropores	Poorly
	Fractured – dissolution-porous	Horizon No. 74XI in Well XX8	Breccia lava	Volcanic neck subfacies of volcanic conduit facies	Dissolution pores	Structural fracture, dissolution fracture

Accumulation Capacity		Permeation Capacity	Deliverability Characteristics			Stable Production Potential	
Effective Porosity, %	*Radius of Investigation, m*	*Total Permeability, mD*	*Gas-Testing Production, 10^4 m^3/d*	*Production Test Yield, 10^4 m^3/d*	*Open Flow Potential, 10^4 m^3/d*	*Gas Production per Pressure Drawdown, 10^4 m^3/MPa*	*Estimated Period of Stable Production, Years*
13.2	65 m (inner), infinite (outer)	0.74	15.1 (natural)	12.15	40.93	7839	10
5.51	67.6 m	0.12	0.08 (natural), 5.27 (fracturing)	—	7.49	—	—
4.87	213.5 m	0.08	0.26 (natural), 12.86 (fracturing)	6.4	19.49	—	3
1.44	426 m	8.09	Horizon No. 223: 3.0 (natural), 40.76 (fracturing)	12.02	55.18	448.7	4
6.3	19.8 (boundary 10,27 (boundary 2)	1.19	0.05 (60 IV natural), 0.03 (59 II natural), 20.9 (commingled, fracturing)	5.24	23.16	86.3	4
8.6	205 m (inner), infinite (outer)	3.71	53 (fracturing)	14.07	118	2630.4	7
6.31	53 m	0.42	0.17 (natural), 9.7 (fracturing)	7.6	11.24	—	—
16.7	330 m	25.2	22.6 (natural)	10.11	43.81	1792.8	6

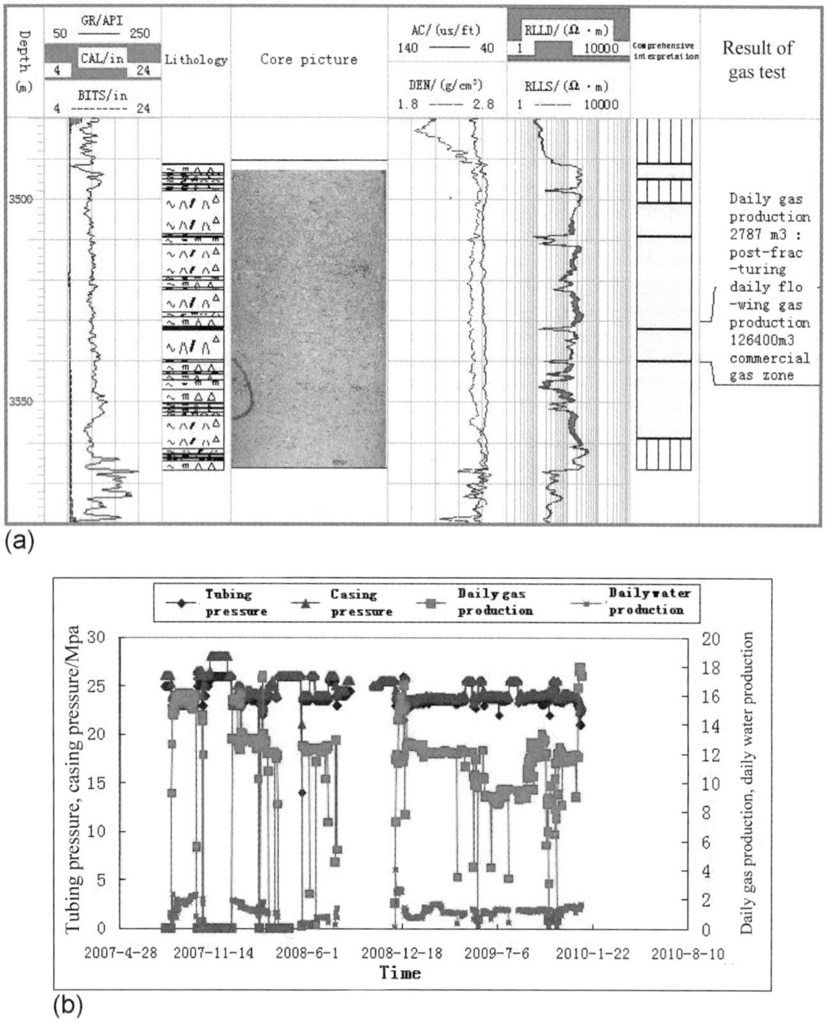

FIGURE 11.20

Vesicular accumulation-permeation pattern (Well XSS2-1). (a) Composite logging map. (b) Production curve.

2. Intergranular porous A-P pattern

This A-P pattern occurs primarily in large-grained volcanic clastic rocks of the explosive facies. Its accumulation-permeation ability is affected by many factors, such as the clastic grain size, sorting, accumulation, and compaction. Large and well-sorted grains, strong grain support, and weaker compaction usually correspond to large, well-developed, and well-connected intergranular pores and, consequently, a strong accumulation-permeation capacity. Due to weak resistance against compaction, matrix filling usually occurs within the intergranular pores. Under deep burial conditions, therefore, the intergranular porous A-P pattern tends to have poor accumulation-permeation capacities, deliverability, and stable production potential.

In Well XX3, horizons No. 165, 166, and 170-I have a typical intergranular porous A-P pattern, hosted in welded tuff of the hot clastic flow subfacies. The reservoir space is predominantly composed of intergranular pores with a small radius. Most of the throats belong to the small-long category and originated from contraction of intergranular pores. Fractures are not developed, resulting in poor communication between pores (Figure 11.21a). In this reservoir interval, the average effective porosity is 5.51%, average permeability is 0.12 mD, and the probe radius is 67.6 m, indicating a poor

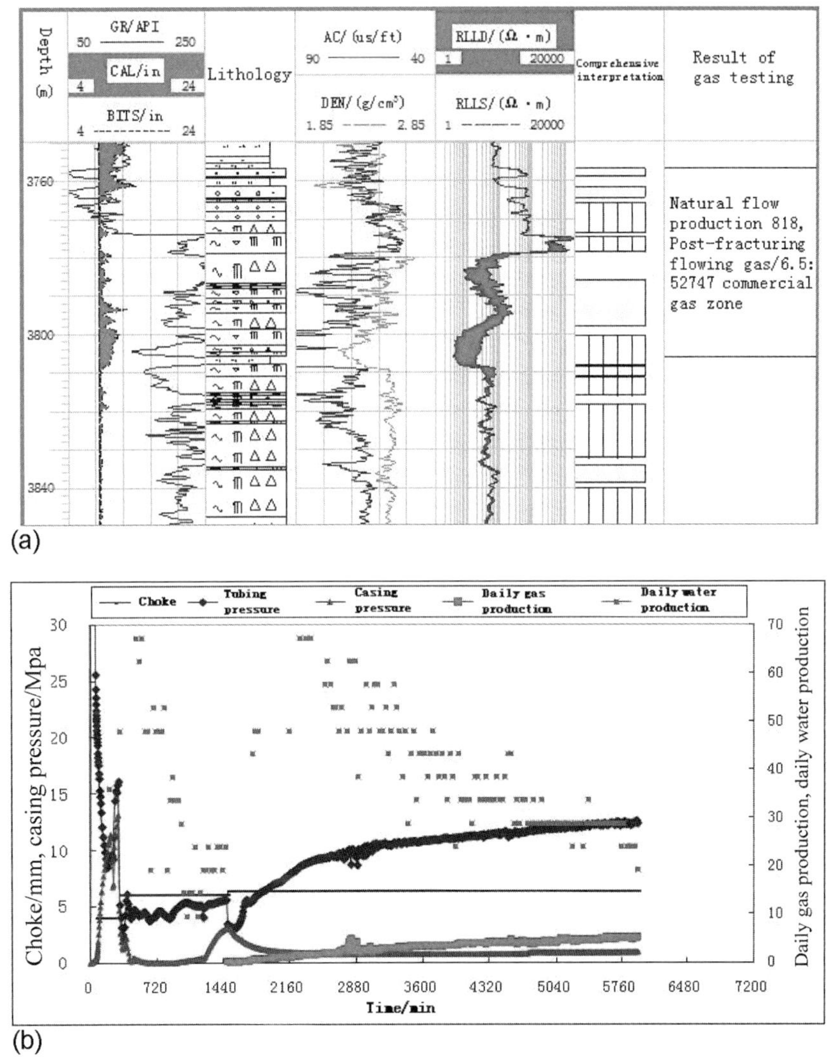

FIGURE 11.21

Intergranular porous accumulation-permeation pattern (Well XX3). (a) Comprehensive logging map. (b) Gas-testing curve.

accumulation and permeation capacity. Gas testing in the interval produced the following results: natural deliverability 0.08×10^4 m^3/d, deliverability up to 5.27×10^4 m^3/d after fracturing, and open flow potential 7.49×10^4 m^3/d (see Table 11.21). The discharge curve in production tests (Figure 11.21b) shows that this reservoir interval has a certain level of capacity for stable production, at a low yield of 2 to 3×10^4 m^3/d.

3. Microporous A-P pattern

A microporous A-P pattern occurs mainly in welded tuff of the hot clastic flow subfacies and crystalline tuff of the hot base surge subfacies. The reservoir space is dominated by matrix micropores and their associated small-short throats, resulting in poor accumulation and permeation capacities. Generally speaking, the volcanic reservoir with this A-P pattern has no natural deliverability, although a medium-low yield and relatively weak stable production can be achieved by fracturing.

Horizon No. 129 in Well XX1-4 has this type of A-P pattern, found in rhyolitic breccia and crystalline tuff of the hot base surge subfacies. The reservoir space is dominated by matrix micropores, connected by small-short throats that originated from the contraction of matrix micropores, without fractures (Figure11.22a). In this horizon, the average effective porosity is 4.87%, the average permeability is 0.08 mD, and the probe radius is 213.5 m, showing a moderate accumulation capacity but poor permeation capacity. After gas testing, natural production reached 0.28×10^4 m^3/d, and postfracturing production rose to 12.64×10^4 m^3/d; the yield during production testing was 6.4×10^4 m^3/d, and open flow potential was 19.49×10^4 m^3/d. After stream injection, it produced intermittently at a yield of 3.4 to $6.43 \times 1\ 0^4$ m^3/d, showing a rapid pressure drop when the well was open and a slow pressure buildup after shut-in. So far, the cumulative gas production has been up to 0.12×10^8 m^3 (Figure 11.22b and Table 11.21). This suggests that the microporous A-P pattern has poor accumulation-permeation capacity, low deliverability, and weak stable production potential.

4. Fractured A-P pattern

This pattern may be developed in various types of lithologies and lithofacies. Fractures are predominant in reservoir spaces and seepage pathways, characterized by poor accumulation capacity but good permeation capacity. The characteristics of the production performance of volcanic reservoirs with a fractured A-P pattern show that a certain level of production can be obtained at the initial stage when the reservoir is opened, but it declines rapidly and fails to maintain at a stable level. Therefore, a volcanic reservoir with this A-P pattern is rarely opened directly; instead, the reservoir is put into development indirectly by opening overlying and underlying pay zones.

Horizon No. 224 (3737 to 3822 m) in Well XX21 is a typical example of a fractured reservoir that occurs in tight rhyolite of the middle subfacies of the effusive facies. Vesicles are absent but structural, blast, and contraction fractures are developed (Figure 11.23a). In this horizon, the average effective porosity is 1.44 %, and the average permeability is 8.09 mD, making it a low-porosity and high-permeability reservoir. No gas tests or production tests have been conducted in this horizon. However, in its overlying formation, No. 223 (3656 to 3737 m, with an effective porosity of 4% to 8%, and permeability of 0.1 to 8 mD), a natural deliverability of 3×10^4 m^3/d was obtained during gas testing, post-fracturing production rose to 41.96×10^4 m^3/d, and the open flow capacity was 55.18×10^4 m^3/d (see Table 11.21). In horizon No. 223, the yield during production testing was stabilized at 12×10^4 m^3/d;

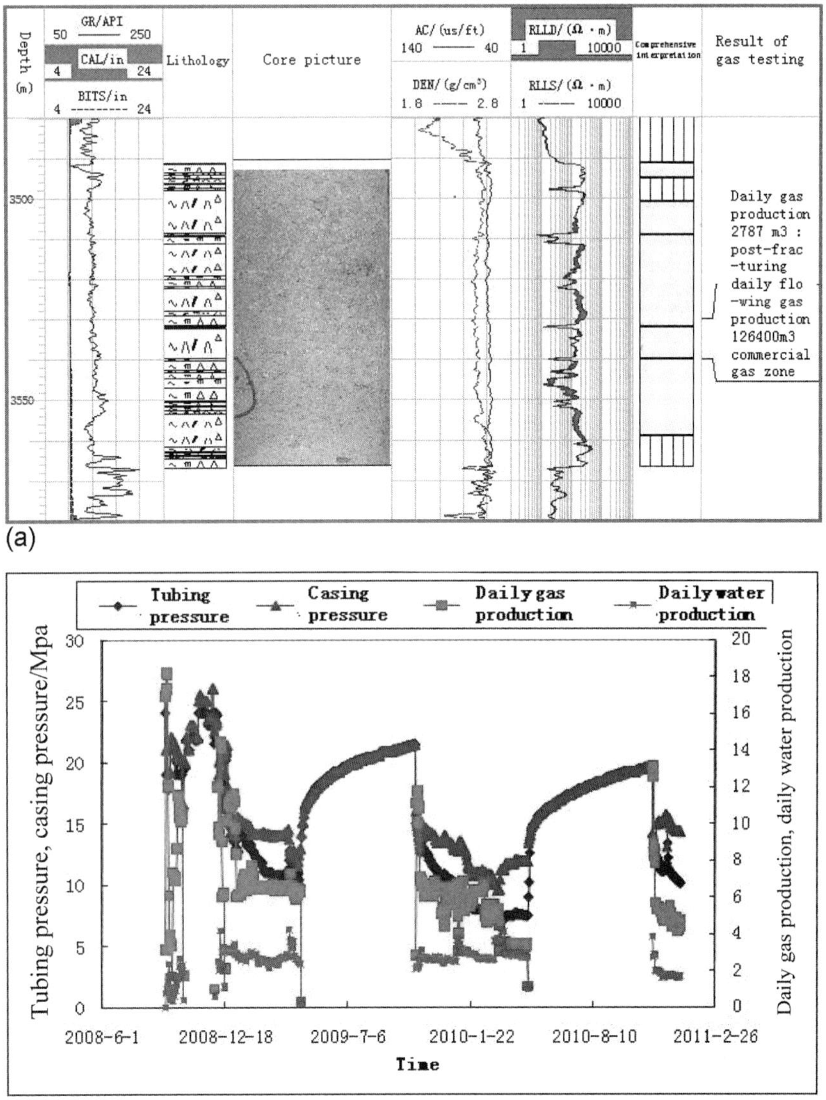

FIGURE 11.22

Microporous accumulation-permeation pattern (Well XX1-4). (a) Comprehensive logging map. (b) Production curve.

the pressure dropped rapidly but built up quickly after shut-in, and then increased gradually to approach a stable level (Figure 11.23b). This implies that some of the initial deliverability of No. 223 horizon is the contribution from the fractures in horizons No. 223 and No. 224, and the rapid production decline is consistent with the deliverability characteristics of a fractured A-P pattern.

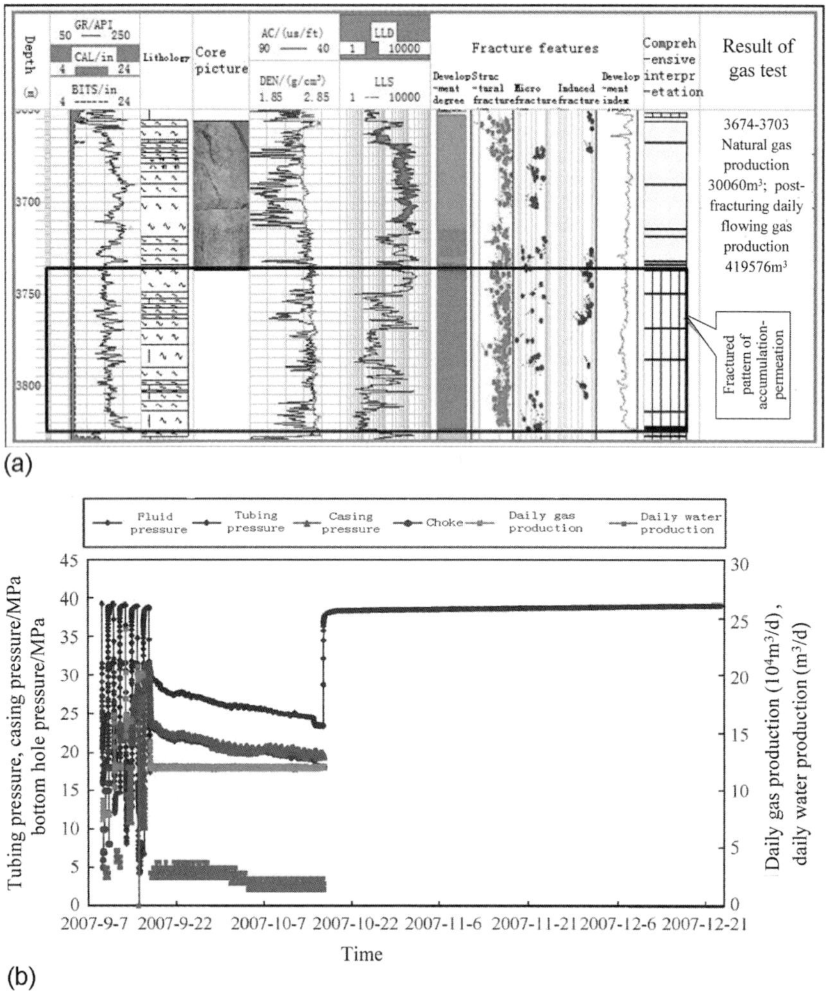

FIGURE 11.23

Fractured accumulation-permeation pattern (Well XX21). (a) Comprehensive logging map (horizon No. 224 as fractured). (b) Production test curve (horizon No. 223).

5. Fractured-vesicular A-P pattern

This A-P pattern is primarily developed in vesicular rhyolite of the top and upper subfacies of the effusive facies as well as breccia lava of the splash subfacies of the explosive facies. Its accumulation capacity is mainly controlled by the vesicular development and reservoir extension. The permeation capacity is affected by many factors such as fracture types and their degree of development, the radius of vesicle contraction throat, and the assemblage relationships of fractures and vesicles. This A-P pattern generally has a moderate accumulation-permeation capacity, and the volcanic reservoir of this pattern has a moderate deliverability and some stable production potential.

Horizons No. 59 II (3592 to 3600 m) and No. 60 IV (3665 to 3675 m) of Well XX9 consist of the upper subfacies vesicular rhyolite of the effusive facies. Reservoir spaces are predominantly the uniformly distributed small vesicles with well-developed structural and contraction fractures well connected by fractured and vesicle contraction throats, representing a typical fractured-vesicular A-P pattern (Figure 11.24a). In these formations, the average effective porosity is 6.3%, the average permeability is 1.19 mD, and the probe radius is 19.8 to 27 m, indicating a moderate accumulation and permeation capacity. Gas tests produced the following results: natural deliverability 0.05×10^4 m^3/d, postfracturing production 20.9×10^4 m^3/d, and open flow potential 23.16×10^4 m^3/d; after production commenced, the yield was maintained at 6×10^4 m^3/d, pressure dropped quickly and built up rapidly after well shut-in, and the gas production per pressure drop was 86.3×10^4 m^3/d/MPa. So far, the cumulative gas production has been 0.3×10^8 m^3, with the stable production period estimated at four years. This demonstrates that these reservoirs have a moderate accumulation-permeation capacity but a relatively low to poor potential for stable production (Figure 11.24b and Table 11.21).

6. *Fractured–intergranular porous A-P pattern*

This A-P pattern occurs mainly in volcanic breccia of the fallout subfacies and welded breccia of the hot clastic flow subfacies, both of the explosive facies. Its accumulation capacity is affected by the degree of intergranular pore development and reservoir extension, whereas the permeation capacity is affected by multiple factors such as fracture types and their development, throat radius, and the assemblage relationships of fractures and intergranular pores. When intergranular pores and fractures are well developed, the reservoir permeation capacity is enhanced by fractures. Volcanic reservoirs with this A-P pattern tend to have high deliverability, a slow production decline, and strong potential for stable production.

Horizon No. 150 in Well XX1 is a volcanic breccia of the fallout subfacies, with large grain sizes. The reservoir space is dominated by intergranular pores of medium radius, with structural and blast fractures and intergranular pore-contraction throats and fractured throats. These features point to a typical fractured–intergranular porous A-P pattern (Figure 11.25a). In this reservoir, the average effective porosity is 8.6%, average permeability is 3.71 mD, and the probe radius is 205 m, suggesting good accumulation-permeation capacity. In this horizon, a deliverability of 53×10^4 m^3/d was obtained during gas testing after fracturing, the yield of production testing was 14.07×10^4 m^3/d, and the open flow potential was 118×10^4 m^3/d. After stream injection, it reached a production level of 12 to 24×10^4 m^3/d, with a slow pressure drop; the gas production per pressure drop was up to 2630.4×10^4 m^3/MPa. So far, the cumulative gas production has reached 2.5×10^8 m^3, with the stable production period estimated at seven years (see Table 11.21).

7. *Fractured-microporous A-P pattern*

This A-P pattern is found primarily in crystalline tuff of the hot base surge subfacies and welded tuff of the hot crystal flow subfacies (both of the explosive facies). Its accumulation capacity depends on the development of matrix micropores and the reservoir extension, whereas the ability of permeation is affected by the type, development, and distribution of fractures. Because fractures can improve the reservoir permeation capacity, a volcanic reservoir with this A-P pattern tends to have a high initial production, but rapid production decline, and poor potential for stable production.

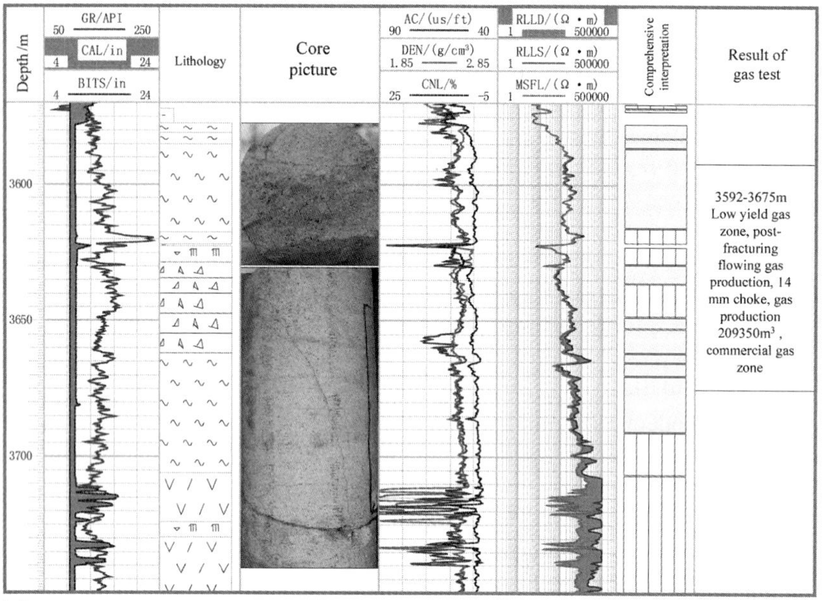

(a)

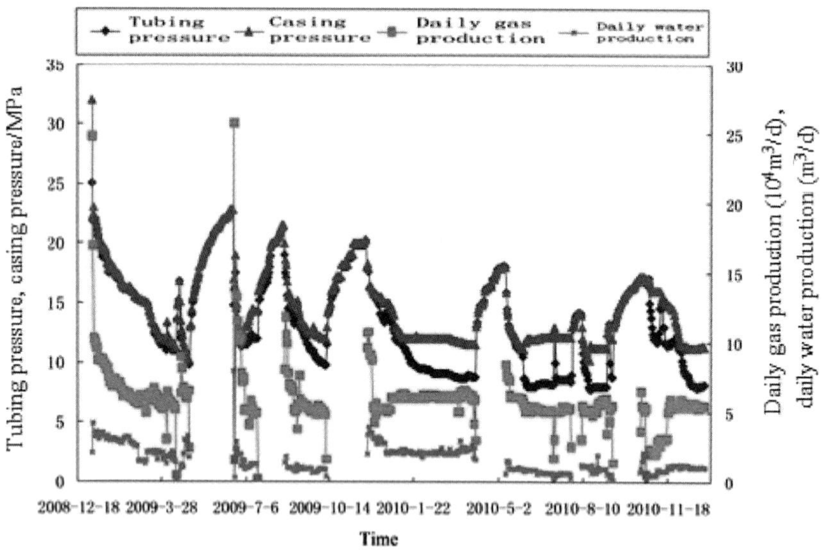

(b)

FIGURE 11.24

Fractured-vesicular accumulation-permeation pattern (Well XX9). (a) Comprehensive logging map. (b) Production curve.

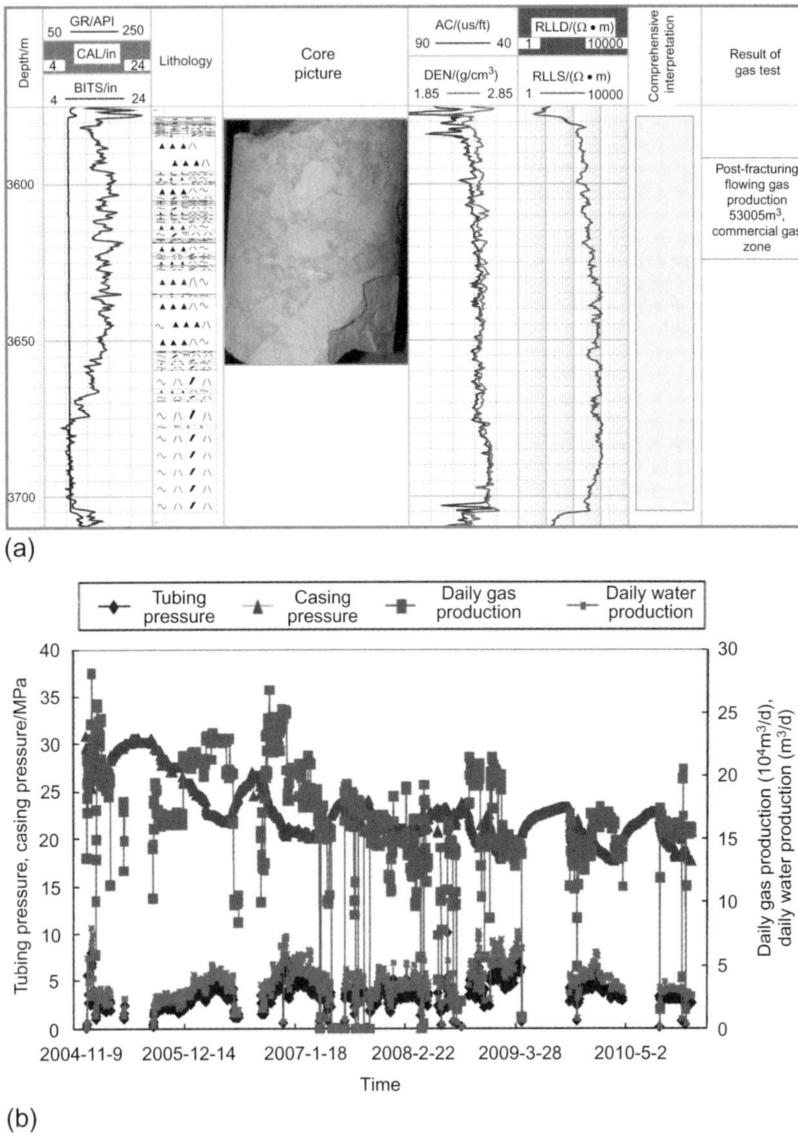

FIGURE 11.25

Fractured–intergranular porous A-P pattern (Well XX1). (a) Comprehensive logging map. (b) Production curve.

Horizon No. 141-I in Well XX6-107 is characterized by the fractured-microporous A-P pattern, comprising rhyolitic breccia and crystalline tuff of the hot base surge subfacies. Matrix micropores are predominant in the reservoir spaces, with well-developed structural and blast fractures (Figure 11.26a). In this horizon, the average effective porosity is 6.31%, average permeability is 0.42 mD, and the probing radius is 53 m, indicating a fairly poor accumulation and permeation

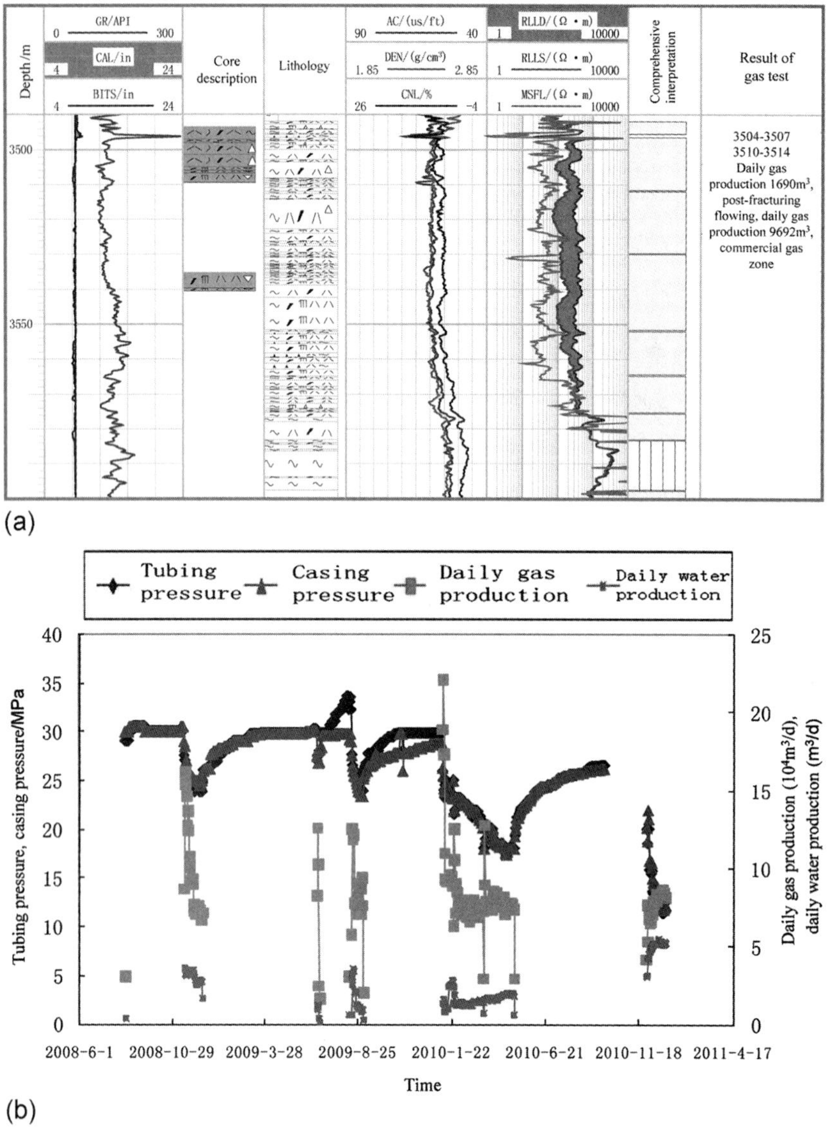

FIGURE 11.26

Fractured-microporous A-P pattern (Well XX6 - 107). (a) Comprehensive logging map. (b) Production curve.

capacity. During gas testing, a natural deliverability of 0.17×10^4 m^3/d was obtained, postfracturing gas testing production rose to 9.7×10^4 m^3/d, the pilot gas production was 7.6×10^4 m^3/d, and open flow potential was 11.24×10^4 m^3/d. After production commenced intermittently, the pressure declined rapidly and built up slowly after well shut-in, indicating insufficient gas deliverability and a very low potential for stable production (Figure 11.26b and Table 11.21).

8. Fractured–dissolution-porous A-P pattern

This A-P pattern may occur in various volcanic rocks, especially common in the splash subfacies of the explosive facies, volcanic neck subfacies of the volcanic conduit facies, and top and upper subfacies of the effusive facies. The accumulation capacity of this pattern is controlled mainly by the development of dissolution pores and some primary pores, whereas the permeation capacity is controlled by the type and development of fractures, the radius of dissolution throats, and the assemblage relationships of fractures and dissolution pores. Dissolution increases the pore and throat sizes, whereas fractures further improve the reservoir permeation capacity. As a result, the fractured–dissolution-porous A-P pattern is generally characterized by good petrophysical properties, good communication between pores, and strong accumulation-permeation capacity. The volcanic reservoir with this A-P pattern generally has a high natural deliverability.

In Well XX8, horizon No. 74 XI consists of brecciated lava of the volcanic neck subfacies. Dissolution pores among large-sized grains are predominant in reservoir spaces, with additional vesicles and minor intergranular pores. Structural fractures, contraction fractures, and dissolution fractures can be seen in cores and are well developed. Due to its favorable pore-fracture association, the fractures play a large role in pore communication. These features denote a typical fractured–dissolution-porous A-P pattern (Figure11.27a). In this horizon, the average effective porosity is 16.7%, average permeability is 25.2 mD, and the probe radius is 36.6 m, indicating a fairly good accumulation-permeation capacity. During gas testing in this horizon, the natural deliverability reached 22.6×10^4 m^3/d and the open flow potential was 43.81×10^4 m^3/d. Due to its high CO_2 content (up to 24.52%), this well has not been put on stream so far. Production testing for more than three months (Figure 11.27b) produced a yield value of 10 to 15×10^4 m^3/d, with a stable pressure and a high ratio of gas yield per pressure drop at 1792.8×10^4 m^3/MPa. These processes encountered a problem of upward breakthrough of bottom water along fractures. So far, the cumulative gas production has reached 0.12×10^8 m^3, and the stable production period is estimated to be six years (see Table 11.21).

In summary, among the eight A-P patterns discussed, the fractured–dissolution-porous pattern has the highest accumulation-permeation capacity, the highest natural deliverability, and the best stable production potential; the vesicular, fractured-vesicular, and fractured–intergranular porous A-P patterns rank next; and the microporous pattern has the poorest properties.

11.5 CHARACTERIZATION OF THE MICROPRODUCING CAPACITY FOR VOLCANIC GAS RESERVOIRS

On the basis of characterized reservoir spaces and throats, various methods are used to determine the cutoff thresholds of reservoir throats, and the grading standard for throats and permeability is established with reference to the classification scheme of effective reservoirs. The microproducing capacity of volcanic gas reservoirs is evaluated by characterizing the pore volumes and mobile fluid volumes controlled by different grades of throat radius, permeability, and displacement pressure. It will provide a basis for evaluating the producing capacity of reserves in different categories.

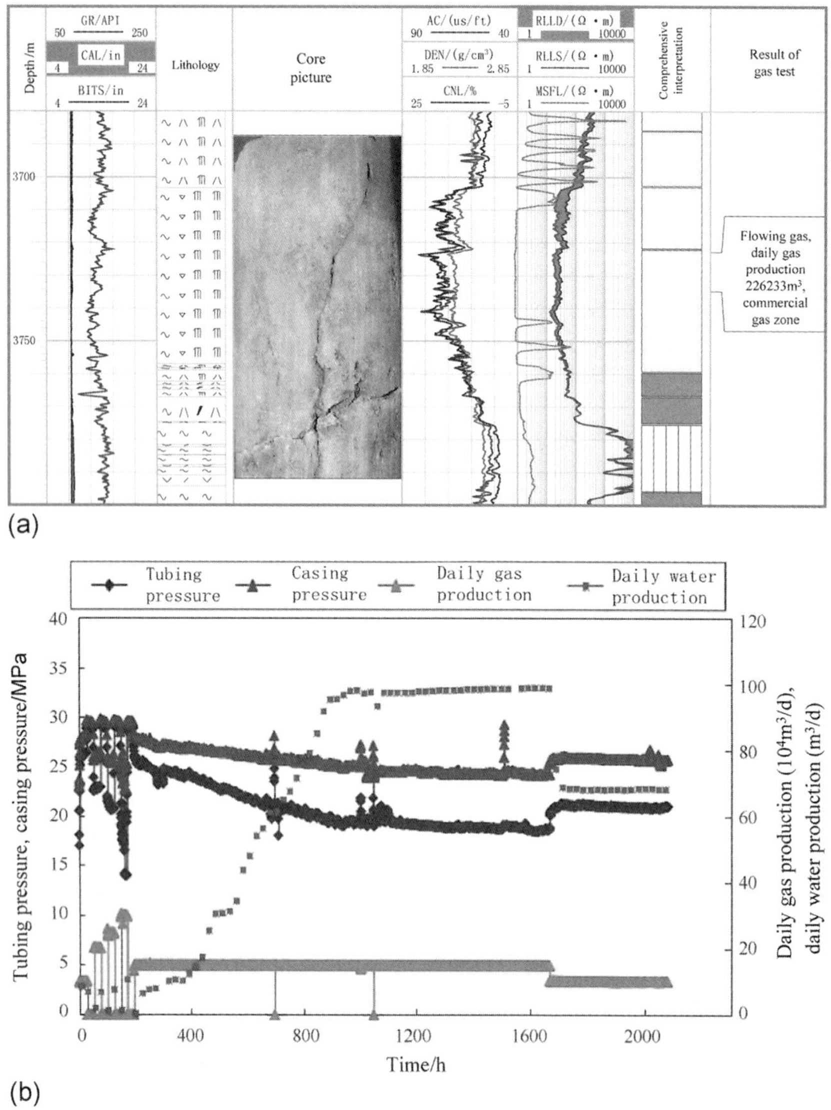

FIGURE 11.27

Fractured–dissolution-porous A-P pattern (Well XX8). (a) Comprehensive logging map. (b) Production test curve.

11.5.1 Throat cutoff thresholds of volcanic gas reservoirs

The throat cutoff threshold, also known as critical throat, refers to the minimum throat radius in a volcanic gas reservoir, at which natural gas can flow through. Below the threshold, the throats do not allow fluid flow, and this will not contribute to reservoir permeability [16]. Considering the geological features of volcanic gas reservoirs, three methods— including irreducible water film thickness analysis, the empirical formula, and the centrifuge experiment—are used to systematically determine the throat cutoff value in volcanic gas reservoirs.

1. Irreducible water film thickness analysis

A layer of stable irreducible water film is absorbed on the inner wall of pores and throats, which is difficult to remove under the pressure differential of the gas layer [17]. According to the adsorption theory, the sum of irreducible water film thickness on the inner wall and the motion diameter of the natural gas molecule (10^{-4} to 10^{-3} μm) is the throat diameter cutoff threshold for gas production.

In the laboratory, the thickness of irreducible water film is determined by the following procedures [18]. A core is taken from the net gas pay zones with oil-based drilling fluid, and then the microwave method or extraction method is used to measure the irreducible water volume in these cores. For the same core samples, high-pressure (over 400 MPa) mercury injection and adsorption methods are used to measure the specific surface of pores and throats. The irreducible water volume divided by the specific surface of pores and throats is the thickness of the irreducible water film. At present, the experiment to measure the thickness of irreducible water film is not yet performed for volcanic reservoirs. Therefore, the measurement results in carbonate reservoirs are used for reference. In the Sinian dolomite reservoirs of the Weiyuan gas field, the thickness of irreducible water film under different porosity conditions is measured. The cutoff value of the throat radius in effective reservoirs is calculated on the basis of the molecular diameter of natural gas at 0.001 μm (Table 11.22). As Table 11.22 shows, when porosity is in the range of 3% to 4%, the average thickness of irreducible water film is 0.0113 μm. Twice the water film thickness plus the molecular diameter of natural gas is 0.0235 μm—that is, when the throat diameter is less than 0.0235 μm, it will be difficult for natural gas molecule to flow through the throat. Therefore, the cutoff of the throat radius is half of the diameter—that is, about 0.0118 μm.

The porosity cutoff for volcanic gas reservoirs in the XX gas field determined by dynamic analysis and empirical statistics is about 3% to 4%. With reference to the thickness of the irreducible water film in the Sinian dolomite reservoir, the throat radius cutoff for volcanic gas reservoirs in this gas field is determined to be 0.01 μm.

Table 11.22 Statistics on the Relationships between Porosity and Irreducible Water Film Thickness in Sinian Dolomite in the Weiyuan Gas Field [15]

Porosity/%	< 1	1-1.5	1.5-2	2-3	3-4	4-5	5-6
Average thickness of irreducible water film/μm	0.0047	0.0057	0.0087	0.0106	0.0113	0.0148	0.0217
Twice the water film thickness + the natural gas molecular diameter/μm	0.0104	0.0124	0.0184	0.0220	0.0235	0.0305	0.0444
Cutoff of the throat radius/μm	0.0052	0.0062	0.0092	0.0111	0.0118	0.0153	0.0222

2. Empirical formula method

Xiang and colleagues [19] studied the water film thickness in tight sandstone gas reservoirs. Based on the theoretical equation of water film thickness in a soil grain surface, the equation to calculate the core water film thickness is derived as follows:

$$d_i = 7.142\phi_e S_{wi}/(A_{specific})\rho \tag{11.5.1}$$

Where,

d_i, water film thickness in core, μm;
ϕ_e, core porosity;
S_{wi}, irreducible water saturation in core;
$A_{specific}$, specific surface area of rock, m^2/g;
ρ, rock density, g/cm^3.

With reference to this empirical formula and the characteristics of volcanic gas reservoirs, the irreducible water film thickness in volcanic gas reservoirs is calculated to determine the throat cutoff in reservoirs.

The experimental data on relative permeability in 46 core samples from the XX volcanic gas reservoir revealed that the reservoir porosity is 2.5% to 16.9 %, irreducible water saturation is 32.61% to 53.02%, rock density is 2.13 to 2.65 g/cm^3, and the specific surface area of rock is 3.25 m^2/g [20]. Using Equation 11.5.1, the irreducible water film thickness is calculated to be 0.0106 to 0.061 μm. If the motion diameter of natural gas molecules is 0.001 μm, the throat radius cutoff will fall in the range of 0.0113 to 0.0615 μm. According to the calculation, the relationship between core porosity and water film thickness is plotted in Figure 11.28, and the cutoff distribution of the throat radius under different porosities is determined (Table 11.23). Based on these results as well as the porosity cutoff for

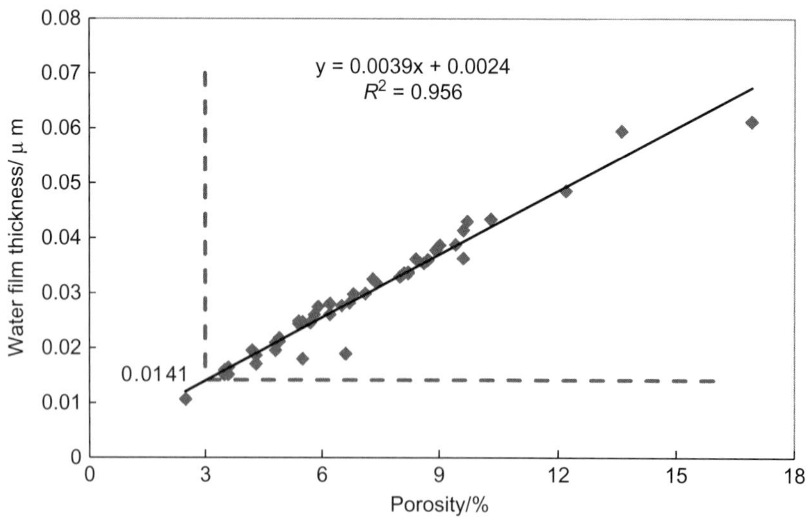

FIGURE 11.28

Crossplot of porosity versus water film thickness.

Table 11.23 Statistics on the Relationships between Porosity and Irreducible Water Film Thickness in Volcanic Rocks (XX Gas Field)

Porosity/%	2.5-3	3-5	5-8	8-10	10-12	12-15	15-17
Average thickness of irreducible water film/μm	0.0106	0.0183	0.0262	0.0367	0.0434	0.0436	0.0613
Twice of water film thickness + natural gas molecular diameter/μm	0.0222	0.0376	0.0534	0.0744	0.0878	0.0882	0.1236
Cutoff of throat radius/μm	0.0111	0.0188	0.0267	0.0372	0.0439	0.0441	0.0618

volcanic gas reservoirs in the XX gas field determined by other methods, the thickness of irreducible water film is determined to be 0.0141 to 0.0161 μm, and the cutoff of the throat radius is 0.0146 to 0.0166 μm, with the average taken at 0.015 μm.

3. Centrifuge experiment method

For different sized throats, the centrifugal force required for fluid separation in a centrifuge experiment is different. According to the relationship between centrifugal force and throat radius, the centrifuge experiment can be used to determine the throat radius cutoff threshold.

Based on the capillary force formula, the relationship between centrifuge force and throat radius for a volcanic rock core sample is expressed by the following equation:

$$p = \frac{2\sigma\cos\theta}{r} \quad r = \frac{2\sigma\cos\theta}{p} \tag{11.5.2}$$

Where:

p, centrifugal force, dsi;
r, throat radius, μm;
σ, water - gas interfacial tension, mN/m, taking 72 mN/m;
θ, wetting angle (= 0°)

Yang and coworkers [21] selected 12 volcanic core samples from the XX gas field and used six of them to perform a centrifuge experiment and to measure the residual water saturation after the centrifuge process was measured (Table 11.24). The results are used to plot the relationship between centrifugal force and residual water saturation (Figure 11.29). The plot shows that the residual water saturation decreases with the increase of centrifugal force; when the centrifugal force is more than 2.07 MPa, the range of variation in residual water saturation narrows down. In these experiments, the characteristics of mobile (movable) water saturation controlled by different throats were obtained (Figure 11.30). The mobile (movable) water saturation decreases with the decrease in throat radius. At a throat radius of 0.04 μm, the saturation of mobile water is only 1.14% on average. However, the residual water saturation (55.24% on average) is slightly more than the irreducible water saturation (52.46% on average). This indicates that the throat with a radius smaller than 0.04 μm still controls 2.78% of the mobile water. In these experiments, no data are available to describe what occurs when the centrifugal force is higher than 3.45 MPa, so the trend of change is unclear for that range. It is believed that the cutoff of the throat radius in volcanic gas reservoirs shall be less than 0.04 μm.

Table 11.24 Measurements of Residual Water Saturation in Core Samples at Different Centrifugal Forces (from Yang Zhengming)

Sample Number	**Residual Water Saturation after Different Centrifugal Forces Are Applied/%**							**Irreducible Water Saturation Derived from NMR Experiments**
	No Centrifugal Force Is Applied	*0.35 MPa*	*0.69 MPa*	*1.38 MPa*	*2.07 MPa*	*2.76 MPa*	*3.45 MPa*	
2	100	53.13	86.92	71.99	61.64	55.66	54.5	53.13
6	100	31.77	59.14	48.47	43.17	40.06	38.67	31.77
10	100	50.52	74.34	61.82	54.91	51.83	51.1	50.52
30	100	39.07	70.12	52.86	45.59	42.16	40.88	39.07
55	100	47.90	74.74	61.31	54.39	49.8	48.87	47.90
63	100	72.77	96.79	90.48	80.91	76.3	74.64	72.77
68	100	66.94	98.03	88.65	75.96	73.46	73.1	66.94
73	100	50.74	79.07	63.75	55.9	53	52.22	50.74
95	100	42.80	84.15	62	53.72	49.31	48.14	42.80
103	100	68.91	91.47	83.27	76.61	72.25	70.31	68.91

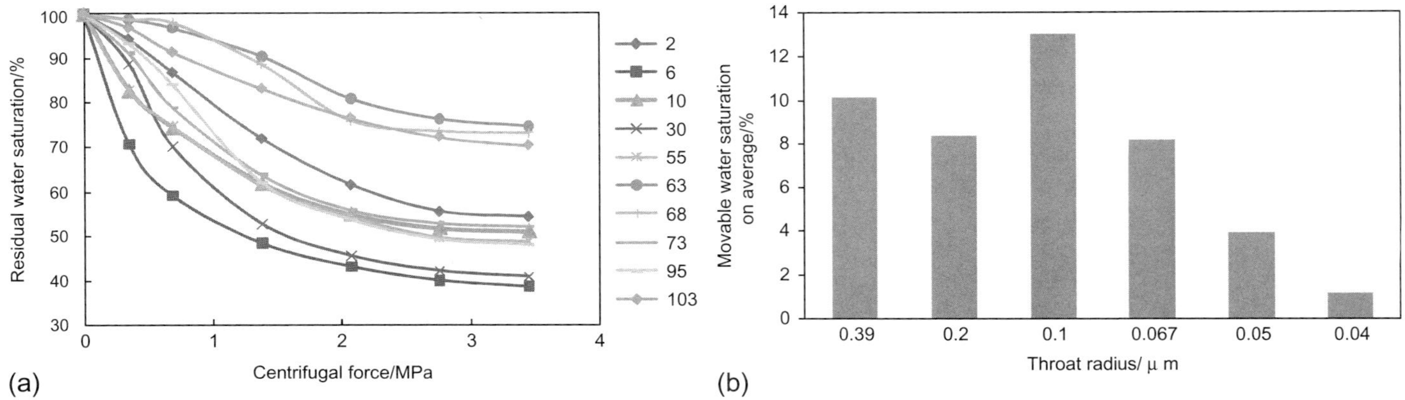

FIGURE 11.29

Throat radius cutoff threshold determined from centrifuge experiment. (a) Crossplot of centrifugal force versus residual water saturation. (b) Mobile water saturation controlled by throats with different sizes.

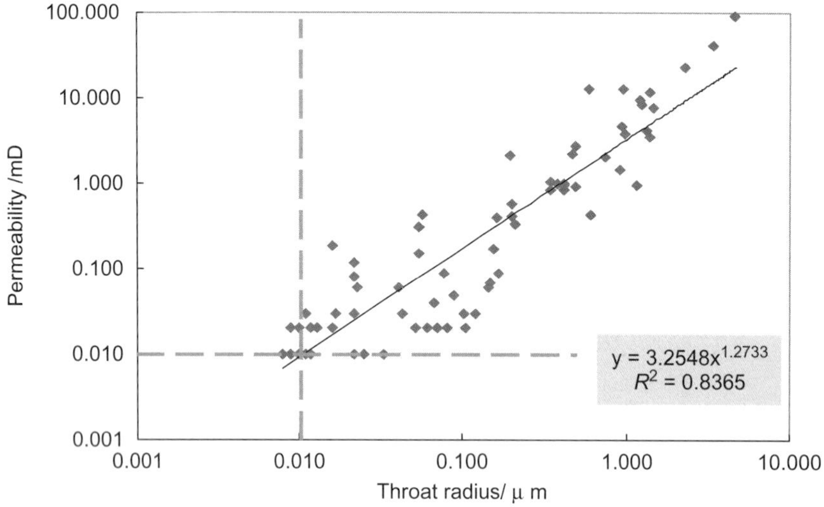

FIGURE 11.30

Throat radius cutoff and permeability.

4. Throat cutoff threshold in volcanic reservoirs

Among the three methods used to determine throat cutoff values (Table 11.25), water film thickness analysis and the empirical formula method are based on the adsorption theory, which involves integrating the method of experiment and empirical formula, respectively, to determine the irreducible water film thickness under different porosity conditions and then combining the result with the natural gas molecular diameter to determine the minimum throat radius that will allow natural gas flow. The throat cutoff values determined by these two methods are relatively close (0.01 to 0.015 μm). According to the centrifuge experiment and the capillary force equation, the residual water saturations under different centrifugal forces are measured to characterize the mobile water saturation controlled by throats of different sizes. The throat cutoff threshold is determined by comparison with the irreducible water saturation determined by Nuclear-Magnetic Resonance (NMR) experiment. The comprehensive analysis suggests that when the throat radius is more than 0.04 μm and the centrifugal force is lower than 3.45 MPa, most of the mobile fluid may be produced; when the throat radius is between 0.01 to 0.04 μm, increased centrifugal force will produce part of the mobile fluid; when the throat radius is less than 0.01 μm, no mobile fluid exists in the reservoir. Therefore, the throat radius cutoff threshold in volcanic gas reservoirs is determined to be 0.01 μm.

Table 11.25 Comparison of Throat Radius Cutoffs in Volcanic Gas Reservoirs Determined by Different Methods (XX Gas Field)

Method	Irreducible Water Film Thickness Analysis	Empirical Formula Method	Centrifuge Experiment Method	Cutoff Determined after Comprehensive Analysis
Cutoff of throat radius/μm	0.01	0.015	< 0.04	0.01

11.5.2 Permeability cutoff and permeability classification in volcanic gas reservoirs

The throat radius determines the microflow capacity. The permeability determines the macroscopic producing capacity, with the two factors related with each other. In order to characterize the effective pore volume and mobile fluid volume in volcanic gas reservoirs from different perspectives, the relationship between reservoir permeability and throat radius is established using mercury injection data, and consequently the permeability cutoff is determined. The throat radius and permeability classification standard for different types of reservoirs are established by combination with the classification standard for effective reservoirs in volcanic rocks.

Figure 11.30 shows the crossplot of reservoir permeability versus throat radius, based on the mercury injection data from the XX gas field. The plot shows that rock permeability is positively correlated to throat radius. With reference to the throat radius cutoff value (0.01 μm), the permeability cutoff is determined to be about 0.01 mD. Furthermore, the classification standard for throat radius and permeability in volcanic gas reservoirs and the corresponding relationships between them are further established according to the classification and evaluation standard for effective reservoirs in volcanic gas reservoirs (Table 11.26).

11.5.3 Characterization of effective pore volume in volcanic gas reservoirs

Effective pore volume refers to the pore volume connected by throats with a radius greater than the cutoff threshold (> 0.01 μm). It is often expressed by the percentage of effective pore volume to the total pore volume. Volcanic reservoirs have a high degree of heterogeneity at the microscopic scale. Throat, permeability, and displacement pressure at different levels control different effective pore volumes.

1. Effective pore volume controlled by different throats and permeability

In general, a larger pore volume associated with large throats corresponds to a higher reservoir permeability and better reservoir fluid-producing capacity. The following information on pore volumes controlled by different throats has been obtained using mercury injection test data of volcanic gas reservoirs in the CC gas field (Table 11.27):

1. The effective throats (radius > 0.01 μm) in volcanic rocks control 66.40% of pore volume. Among these, the throats with a radius of 0.25 to 0.05 μm control the most pore volume (accounting for 38.54%), the throats with a radius larger than 0.25 μm control 20.49%, and the throats with a radius of 0.05 to 0.01 μm control the least amount of pore volume (7.37%). This implies that most of the volcanic gas reservoir pores in this gas field are controlled by larger throats.

Table 11.26 Permeability and Throat Radius Classification

Parameter	Unit	Reservoir Type		
		I	*II*	*III*
Matrix permeability	mD	> 0.1	0.1-0.015	0.015-0.01
Average throat radius	μm	> 0.25	0.25-0.05	0.05-0.01

Table 11.27 Pore Volumes Controlled by Different Throats and Permeability (CC Gas Field)

Parameter		Range			Total
Throat radius/μm		> 0.25	0.25~0.05	0.05~0.01	
Matrix permeability/mD		> 0.1	0.1~0.015	0.015~0.01	
Pore volume controlled by different throats or permeability/%	Type I reservoir	41.97	36.73	3.83	82.53
	Type II reservoir	17.52	39.62	7.16	64.30
	Type III reservoir	1.98	39.26	11.12	52.36
	Volcanic reservoir	20.49	38.54	7.37	66.40

2. A better reservoir type has a larger pore volume controlled by effective throats. The effective throats in types I, II, and III reservoirs control 82.53%, 64.3%, and 52.36% of pore volumes, respectively. With the improvement of reservoir type, the pore volumes controlled by effective throats increase notably.
3. A better reservoir type has a larger effective pore volume controlled by large throats. For a type I reservoir, the effective pore volume controlled by throats with a radius larger than 0.25 μm is the largest (accounting for 41.97%), followed by throats with a radius of 0.25 to 0.05 μm (36.73%). For type II reservoirs, the effective pore volume controlled by throats with a radius of 0.25 to 0.05 μm is the largest (accounting for 39.62%), and throats with a radius of more than 0.25 μm account for 17.52%. For type III reservoirs, the effective pore volume controlled by throats with a radius of 0.25 to 0.05 μm is the largest (39.26%), followed by throats with a radius of 0.05 to 0.01 μm (11.12%).

Matrix permeability is correlated to throat radius (see Figure 11.30). The pore volume controlled by different grades of permeability has the same characteristics as the pore volume controlled by different throats (see Table 11.27): (1) the effective volcanic reservoirs control 66.40% of the pore volume, which is higher than that controlled by matrix permeability over 0.015 Md; (2) a better reservoir type will have a larger effective pore volume controlled by high matrix permeability.

2. *Effective pore volume controlled by different displacement pressures*

A mercury injection test is performed to characterize the pore volume controlled by different displacement pressures, which can provide a basis for determining the appropriate production pressure differential for gas reservoir development. The displacement pressure of volcanic reservoirs has a good correlation with the throat radius (Figure 11.31). The effective pore volume controlled by different displacement pressures has the same characteristics of effective pore volume controlled by different grades of throats (Figure 11.32, Table 11.28):

1. When displacement pressure is greater than 40 MPa, the effective pore volume in volcanic reservoirs accounts for 65.48% of total pore volume; when displacement pressure is less than 1, 5, 10, 15, and 25 MPa, the effective pore volumes controlled by displacement pressure are 10.91%, 27.98%, 40.67%, 49.26%, and 58.26%, respectively. This suggests that a lower displacement pressure controls the effective pore volume in volcanic reservoirs.

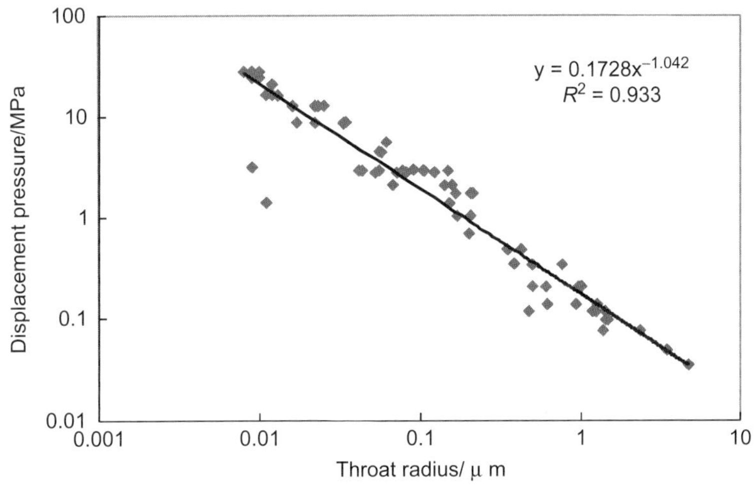

FIGURE 11.31

Crossplot of displacement pressure versus throat radius.

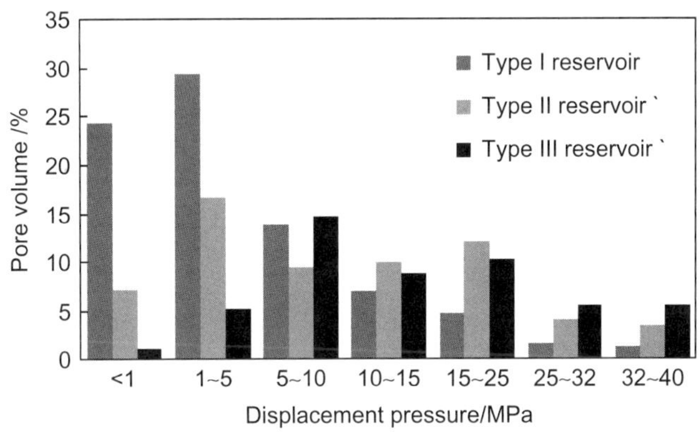

FIGURE 11.32

Pore volume controlled by the displacement pressure of different types of reservoirs.

2. A better reservoir will have a larger effective pore volume controlled by a lower displacement pressure. For a type I reservoir, more than half of the effective pore volume (53.71%) is associated with a displacement pressure < 5 MPa; for a type II reservoir, more than half of the effective pore volume (33.40%) is controlled by a displacement pressure < 10 MPa, with the main controlling pressure ranging from 1 to 25 MPa; for a type III reservoir, most of the effective pore volume (29.88%) is controlled by a displacement pressure < 15 MPa, with the main controlling pressure ranging from 5 to 25 MPa.

Table 11.28 Effective Pore Volume Controlled by Different Displacement Pressures (CC Gas Field)

Parameter		Distribution Ranges							
Displacement Pressure/MPa		*< 1*	*1-5*	*5-10*	*10-15*	*15-20*	*25-32*	*32-40*	**Total**
Pore volume controlled by different displacement pressures/%	Type I reservoir	24.34	29.37	13.83	6.99	4.70	1.59	1.38	82.20
	Type II reservoir	7.25	16.62	9.53	9.96	12.05	4.13	3.36	62.90
	Type III reservoir	1.15	5.22	14.69	8.82	10.28	5.59	5.61	51.36
	Volcanic reservoir	10.91	17.07	12.68	8.59	9.01	3.77	3.45	65.48

11.5.4 Characterization of the microproducing capacity of volcanic gas reservoirs

The producing capacity of reservoir fluid is the key for assessing the potential for natural gas reservoir development [22]. It is usually characterized by the mobile (movable) fluid volume, which refers to the pore volume occupied by mobile fluid, usually expressed by the percentage of total pore volume that is occupied by mobile fluid. Considering the complexity of volcanic reservoirs, NMR tests and mercury injection tests are combined to quantitatively characterize the mobile fluid volume controlled by different throats, different permeability values, and different displacement pressure values, so as to evaluate the microproducing capacity of volcanic gas reservoirs.

1. Mobile fluid volume controlled by different throats and permeabilities

Throats control the fluid flow capacity of reservoirs. The saturation of mobile fluid in reservoirs is positively correlated to the throat radius (Figure 11.33). In the XX gas field, NMR tests and mercury injection tests are combined to quantitatively characterize the mobile fluid volume controlled by different throats (Figure 11.34, Table 11.29). The tests show the following results:

1. The effective throats (> 0.01 μm) control 45.67% of mobile fluid volume, among which throats with a radius of 0.1 μm control 40.51% of mobile fluid volume. This indicates that the mobile fluid volume in volcanic reservoirs is small, controlled by larger throats.
2. A better reservoir corresponds to a larger mobile fluid volume controlled by effective throats. The effective throats in type I, type II, and type III reservoirs control 65.19%, 47.40%, and 24.43%, respectively, of mobile fluid volume. The mobile fluid volume controlled by effective throats increases notably with better reservoir properties.
3. A better reservoir also corresponds to a larger movable fluid volume controlled by large throats. In type I reservoirs, the mobile fluid volume controlled by throats with a radius that is > 0.75 μm and in the 0.5 to 0.25 μm range is the largest (21.38%, 22.47%, respectively), and that controlled by throats with a radius of 0.25 to 0.1 μm ranks second (12.96%). In type II reservoirs, the mobile fluid volume controlled by throats with a radius of 0.25 to 0.1 μm is the largest (26.55%), and that controlled by throats with a radius of 0.5 to 0.25 μm ranks next (15.18%). In type III reservoirs, the movable fluid volume controlled by throats with a radius of 0.1 to 0.05 μm is the largest (8.05%), and that controlled by throats with a radius of 0.25 to 0.1 μm ranks second (7.46%).

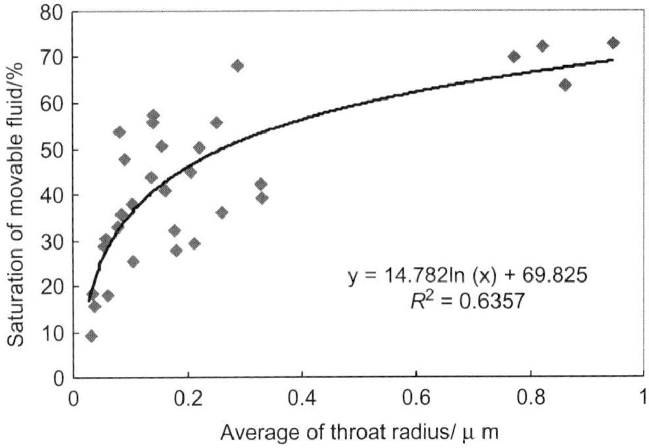

FIGURE 11.33

Crossplot of throat radius versus saturation of mobile fluid.

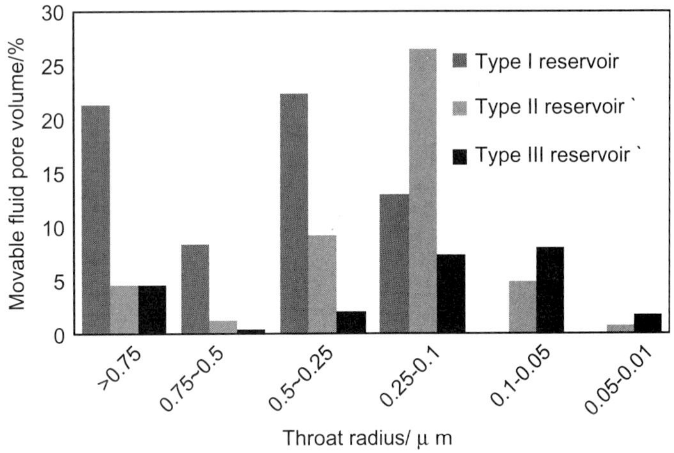

FIGURE 11.34

Mobile fluid pore volume controlled by the throat radius in different types of reservoirs.

Similarly, the mobile fluid volume controlled by different permeabilities is analogous to that controlled by different throats (see Table 11.29): (1) the effective reservoirs in volcanic rocks control 45.67% of the movable fluid volume, among which the movable fluid volume controlled by a matrix permeability higher than 0.1 mD is the largest (40.51%); and (2) the better the reservoir, the larger the mobile fluid volume controlled by a high matrix permeability.

2. Mobile fluid volume controlled by different displacement pressures

Mobile fluid saturation is well correlated to the displacement pressure (Figure 11.35). Lower displacement pressure in a volcanic reservoir corresponds to a higher movable fluid saturation. The mobile fluid volume controlled by different displacement pressures is related to the reservoir types (see Figure 11.36, Table 11.30):

Table 11.29 Comparison of Mobile Fluid Saturation Controlled by Different Throats (XX Gas Field)

Parameter		Distribution Range						Total
Throat radius/μm		> 0.75	0.75-0.5	0.5-0.25	0.25-0.1	0.1-0.05	0.05-0.01	
Matrix permeability/mD		> 2.3	2.3-1.3	1.3-0.5	0.5-0.2	0.2-0.07	0.07-0.01	
Mobile fluid volume controlled by different throats/%	Type I reservoir	21.38	8.38	22.47	12.96	0.00	0.00	65.19
	Type II reservoir	4.66	1.38	9.15	26.55	4.92	0.75	47.40
	Type III reservoir	4.56	0.52	2.06	7.46	8.05	1.78	24.43
	Volcanic reservoir	10.20	3.43	11.23	15.66	4.32	0.84	45.67

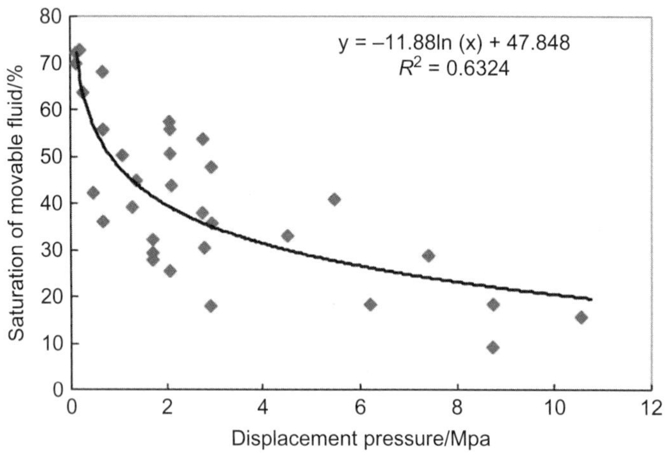

FIGURE 11.35

Crossplot of throat radius versus the saturation of mobile fluid.

1. In the XX gas field, the mobile fluid volume in volcanic reservoirs is about 45.67%, and the displacement pressure is less than 25 MPa. When the displacement pressure is less than 1, 3, 5, 10, and 15 MPa, the mobile fluid volume controlled by the displacement pressure is 10.54%, 27.3%, 36.57%, 43.69%, and 45.27%, respectively. Most movable fluid is distributed in reservoirs with displacement pressures less than 10 MPa.
2. A better reservoir has a larger mobile fluid volume controlled by low displacement pressure and better mobility of the reservoir fluid. For type I reservoirs, displacement pressures at ranges less than 2 MPa and 3 MPa correspond to 42.46% and 55.25% mobile fluid, respectively; for type II reservoirs, most mobile fluid (38.08%) corresponds to displacement pressures of 2 to 10 MPa. In type III reservoirs, most of the mobile fluid (15.83%) is related to displacement pressures of 3 to 15 MPa. Therefore, the better the reservoir, the more mobile the reservoir fluid, and the lower the production pressure differential required.

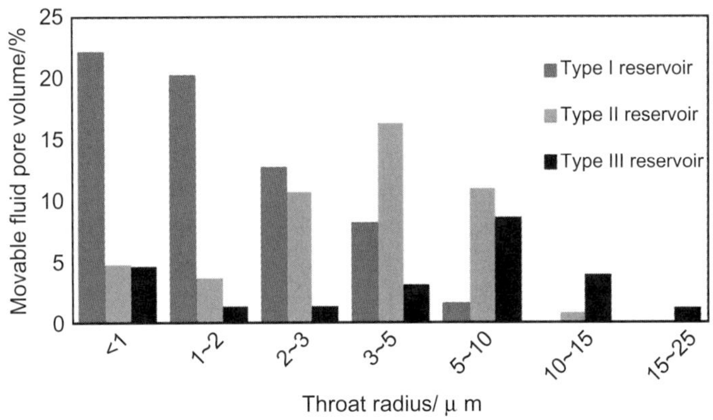

FIGURE 11.36

The mobile fluid pore volume controlled by the throat radius in different types of reservoirs.

Table 11.30 Mobile Fluid Volume Controlled by Different Displacement Pressures (XX Gas Field)

Parameter		Distribution Ranges							
Displacement Pressure/MPa		*< 1*	*1-2*	*2-3*	*3-5*	*5-10*	*10-15*	*15-25*	**Total**
Mobile fluid volume controlled by different expulsion pressures/%	Type I reservoir	22.21	20.25	12.79	8.26	1.68	0.00	0.00	65.19
	Type II reservoir	4.78	3.75	10.73	16.32	11.03	0.79	0.00	47.40
	Type III reservoir	4.63	1.37	1.39	3.22	8.66	3.95	1.20	24.43
	Volcanic reservoir	10.54	8.45	8.31	9.26	7.12	1.58	0.40	45.67

Considering the complexity of microtexture in volcanic gas reservoirs, this chapter has focused on the characterization of reservoir spaces and the shapes, sizes, and accumulation-permeation capacity of throats, using the methods of core observation, thin section examination, laboratory testing, log interpretation, and dynamic analysis. On the basis of these studies, the characteristics of effective pore volume and mobile fluid volume controlled by different throats, permeabilities, and displacement pressures are analyzed. Ultimately, the volcanic gas reservoir microtexture characterization is developed, which provides technical support for evaluating the producing capacity of different types of effective reservoirs. For the volcanic gas reservoir in the XX gas field (Table 11.31), NMR tests and mercury tests show the following features:

1. In a type I reservoir in this gas field, the effective pore volume accounts for 93.78% of the total pore volume, and the mobile fluid volume accounts for 65.19% of the total pore volume.

Table 11.31 Analysis of the Producing Capacity of Different Types of Volcanic Gas Reservoirs (XX Gas Field)

Reservoir Type and Characteristics						
Reservoir Types	*Matrix Permeability/ mD*	*Throat Radius/μm*	*Effective Pore Volume/ %*	*Movable Fluid Volume/ %*	**Reserves Proportion/%**	**Evaluation of Producing Ability**
Type I reservoir	> 0.1	> 0.25	93.78	65.19	10	Recoverable under present economic and technological conditions
Type II reservoir	0.015-0.1	0.25-0.05	82.69	47.40	30	Recoverable by applying new technology
Type III reservoir	0.01-0.015	0.05-0.01	61.75	24.43	35	Recoverable through new R and D technology
					25	Difficult to develop at present

The reservoir fluid has high seepage and good producing capacities. Its natural gas reserves (about 10% of the proven in situ gas) can be recovered effectively under present economic and technological conditions.

2. In a type II reservoir, the effective pore volume accounts for 82.69% of the total pore volume, and the mobile fluid volume accounts for 47.40% of the total pore volume. The reservoir fluid has medium seepage and relatively good producing capacities. Its natural gas reserves (about 30%) can be recovered effectively by applying new technologies such as staged fracturing in horizontal wells.
3. In a type III reservoir, the effective pore volume accounts for 61.75% of the total pore volume, and the mobile fluid volume accounts for 24.43% of the total pore volume. The reservoir fluid has poor seepage and producing capacities. For its natural gas reserves (about 60% of the proven in situ gas), about 35% can be recovered effectively by applying new research and development technologies, but the remaining 25% will be difficult to extract at present.

References

[1] Zhang Y. Experimental and theoretical study on microporous structure of reservoir rocks. Science and Technology Information 2008;27:42–3.
[2] Zhao C, Liu M, Hu A, et al. Special oil and gas reservoirs. Beijing: Petroleum Industry Press; 1997.
[3] Zhao C, Meng W, Jin C, et al. Volcanic rock and hydrocarbon in Liaohe basin. Beijing: Petroleum Industry Press; 1999.

[4] Zhao C. Forming mechanism of reservoir spaces in volcanic rocks and their hydrocarbon-bearing property. Geological Review 1996;42(Suppl.):37, 43.
[5] Ren Z, Jin C. Reservoir space characteristics of volcanic reservoirs in Wa 609 well block in Liaohe depression. Petroleum Exploration and Development 1999;26(4):54–6.
[6] Yu C, Zheng J. Reservoir properties and effecting factors of volcanic rocks of basement beneath Wucaiwan depression, Junggar basin. Earth Sci J China Univ Geosci 2004;29(3):303–8.
[7] Liu W, Zhu X. Reservoir space evolution of volcanic rocks in Yingcheng Formation of Xujiaweizi fault depression, Songliao basin. Petroleum Geology and Experiment 2005;27(1):44–9.
[8] CNPC (China National Petroleum Corp). SY/T6110–2002. The description of exploitation geologic characteristics for carbonate gas reservoirs. Beijing; 2002.
[9] Yan C, Yu F. Pores development and reservoir properties of volcanic rocks in Jianghan basin. Journal of Jianghan Petroleum Institute 1996;18(2):1–6.
[10] Zhu K. Building and management of data base for core analysis. Scientific and Technological Management of Land and Resources 2001;18(3):35–8.
[11] Yu J, Guo D, Wang X. Study of microscopic behaviors of low permeable reservoirs through constant rate mercury injection technique. Journal of Daqing Petroleum Institute 2006;30(2):22–5.
[12] CNPC (China National Petroleum Corp). SY/T5830–93, The description methods for volcanic reservoirs. Beijing; 2002.
[13] Luo Z, Wang Y. Pore structure of oil-gas reservoir. Beijing: Science Press; 1986.
[14] Bai S, Tang F. Development mode for fractured buried hill bedrock reservoir. Beijing: Petroleum Industry Press; 1997.
[15] Chen Z, Dai Y, Lang Z. Storage-percolation modes and production performance of the karst reservoirs in Tahe oil field. Petroleum Exploration and Development 2005;32(3):101–5.
[16] Luo M. Quantitative reservoir geology. Beijing: Geology Press; 1998, p. 55–58.
[17] Dai J, Pei X, Qi H. Natural gas geology in China, vol. 1. Beijing: Petroleum Industry Press; 1992, p. 160–62.
[18] Wang Y, Kong J, Li H, et al. Natural gas engineering: gas reservoir geology. Beijing: Petroleum Industry Press; 2004, p. 90–91.
[19] Xiang Y, Xiang D. Study of gas recovery and water film thickness in water drive for tight sandstone gas reservoir. Journal of Chengdu University of Technology 1999;26(4):389–91.
[20] Yang J, Kang Y. Research on diffusibility of the gas in tight sand gas reservoirs. Journal of Southwest Petroleum University (Science and Technology Edition) 2009;31(6):76–9.
[21] Yang Z, Guo H, Jiang H, et al. Experimental study on different lithology of volcanic gas reservoirs using nuclear magnetic resonance technique. Acta Petrolei Sinica 2009;30(3):400–8.
[22] Huang Y. Seepage mechanism of low-permeability oil reservoir. Beijing: Petroleum Industry Press; 1998, p. 30–35.

CHAPTER

Geological Modeling for Volcanic Gas Reservoirs 12

Geological modeling of reservoirs aims to demonstrate the final results of gas reservoir characterization [1]. A 3D geological model that reflects the complex rock architecture and dual media represents the synthesis of volcanic gas reservoir characterization results. Such a model serves as an important basis not only for reserve evaluation and well location design of volcanic gas reservoirs [2], but also for their numerical simulation and the prediction of development indices.

12.1 CHALLENGES IN MODEL BUILDING AND TECHNICAL SOLUTIONS

12.1.1 Geological characteristics and challenges in geological model building

Compared with conventional clastic rocks, volcanic rocks are characterized by complex architecture, multiple porous media (e.g., pores, cavities and fractures), and complex fluid distribution (Table 12.1). Many factors influence the reservoir development, which poses tough challenges in the construction of geological models, specifically in the following aspects.

1. Multilevel architectures

Volcanic rocks consist of multilevel architectures, with complex spatial geometry and superposition relationships at various levels, and their occurrence and surface trend are difficult to control, thus making it difficult to build structure models and reservoir framework models.

The multilevel architectures of volcanic rocks include volcanic edifices, volcanic massifs, volcanic facies, and volcanic lithology, as well as complicated stratigraphic structures and reservoir frameworks. Not only must the structure model reflect the influence of regional structures (e.g., volcanic eruption cycles) on the variation in the distribution and thickness of volcanic rocks, but it must also demonstrate how the local structures formed by volcanic edifices control the scale of gas reservoirs and how the microstructures formed by volcanic massifs affect the gas-water system in gas reservoirs. At the same time, various levels of architectural units of volcanic rocks have complicated superposition relationships and abrupt changes in thickness and occurrence; therefore, the volcanic reservoir framework must be constructed on the basis of architectural descriptions and under the multilevel constraints of volcanic edifices, volcanic massifs, volcanic facies, and accumulation-permeation units. This will enable the model builders to control the shapes and variation trends of 3D spatial frameworks of volcanic reservoirs, although building reservoir framework models is a challenging process.

Table 12.1 Geological Characteristics and Model-Building Characteristics of Volcanic Reservoirs

Model Type	Geological Characteristics		Contents and Characteristics of Volcanic Model Building	
	Sedimentary Rock	*Volcanic Rock*	*Content*	*Characteristics*
Structural model of gas reservoir	Sedimentary origin, primarily comprising bedded strata	Volcanic origin, comprising irregular massive strata that are a complex of edifices, massifs, and lithofacies	Regional cycle top-bottom structure model; edifice top-bottom structure model controlling gas reservoir distribution; massif top-bottom structure model controlling gas and water distribution	Structures of volcanic eruption cycles; structures of volcanic edifices; structural geometry of volcanic massifs
Reservoir framework model	Reservoir frameworks consisting of effective sand bodies, with reservoirs controlled by sand bodies	Controlled by multiple factors such as eruption, structure, and diagenesis to form complex reservoir framework systems	Spatial framework model of volcanic eruption cycles, volcanic edifices, volcanic massifs, volcanic facies, and A-P units	Shapes, scales, superposition relationship, and spatial distribution of cycles, edifices, massifs, lithofacies, and A-P units
Reservoir attribute model	Mainly related to changes in petrophysical properties of porous reservoirs	Distribution of reservoirs controlled by architecture; reservoirs are multiple porous media consisting of pores, cavities, and fractures	Matrix porosity/ permeability model constrained by architecture, seismic and geological origins	Reflecting characteristics and changes in petrophysical properties of the reservoir matrix controlled by architecture
			Fracture porosity/ permeability model constrained by architectures, seismic and geological origins	Reflecting characteristics and changes of reservoir fractures controlled by architecture
Fluid distribution model	Mainly related to changes in fluid distribution and saturation in pores	Fluid distribution is jointly controlled by the structure, internal structures, and reservoir attributes	Matrix gas and water distribution model constrained by the structure, architecture, and reservoir attributes	Reflecting fluid distribution in matrix as controlled by the structure, architecture, and reservoir petrophysical properties
			Fracture gas and water distribution model constrained by the structure, architecture, and reservoir attributes	Reflecting fluid distribution in fractures as controlled by the structure, architecture, and reservoir petrophysical properties

2. Complicated origins

Volcanic reservoirs have complicated origins, and their distribution is controlled by rock architectures, and multiple porous media consisting of pores, cavities, and fractures are developed in these reservoirs. This presents challenges for both the quantitative characterization of reservoir parameters and the establishment of attribute models.

In terms of their origins, volcanic gas reservoirs are affected by multiple factors such as palaeogeographical settings, volcanic eruptions modes and energy levels, tectonic activities, and diagenetic processes. Their distribution is controlled by volcanic rock architectures with many types of pores, cavities, and fractures, as well as complicated pore structures and distribution patterns. Therefore, it is difficult to quantitatively characterize these reservoirs and build reservoir attribute models. At the same time, fractures are the important feature of volcanic gas reservoirs. However, fractures in volcanic gas reservoirs are highly variable, with complex distribution patterns. Fracture identification and prediction are difficult, which thus hinders the quantitative characterization and geological modeling of these fractures. Therefore, it is difficult to build multiple porous media attribute models that can reflect the constraints of complex volcanic rock architectures.

3. Complex gas-water relationships

Volcanic gas reservoirs have complex gas-water relationships. Their gas and water distribution is controlled by the structures, architectures, and reservoir attributes. This presents a challenge in building gas saturation models.

The gas and water distribution in volcanic gas reservoirs is generally controlled by the structures characterized by "upper gas and lower water." Under the same structural background, however, the gas and water distribution in volcanic rocks is further controlled by their internal architectures, whereby the volcanic edifice controls the spatial distribution and size of gas reservoirs and the volcanic massif controls the spatial distribution and size of gas-water systems. In addition, volcanic rocks commonly contain multi-medium reservoirs, so the gas saturations in the matrix and fractures are not always the same. Factors such as the structure, architecture, and dual-medium lead to difficulties for building gas saturation models.

12.1.2 Technical solutions

Considering the geological characteristics and modeling challenges of volcanic gas reservoirs, the structural model, reservoir framework model, reservoir attribute model, and fluid distribution model are constructed based on well-point data such as core, mud log, logging, and production performance data and constrained by architectural units such as volcanic eruption cycles, volcanic edifices, volcanic massifs, volcanic facies, and accumulation-permeation (A-P) units, with seismic attribute volumes and inverted parameter volumes as the co-constraints. The geological modeling technique for complex architectures and dual-medium volcanic gas reservoirs provides support for the reserve evaluation, well location optimization, numerical simulation, and development scheme optimization of gas reservoirs (Figure 12.1). The main technical solutions are discussed next.

1. Structural model building

On the basis of stratigraphic sequence divisions and the dissection of volcanic edifices and volcanic massifs, the regional structure model can be constructed with the controls of eruption cycles, the local structure model controlled by volcanic edifices, and the microstructural model controlled by volcanic massifs.

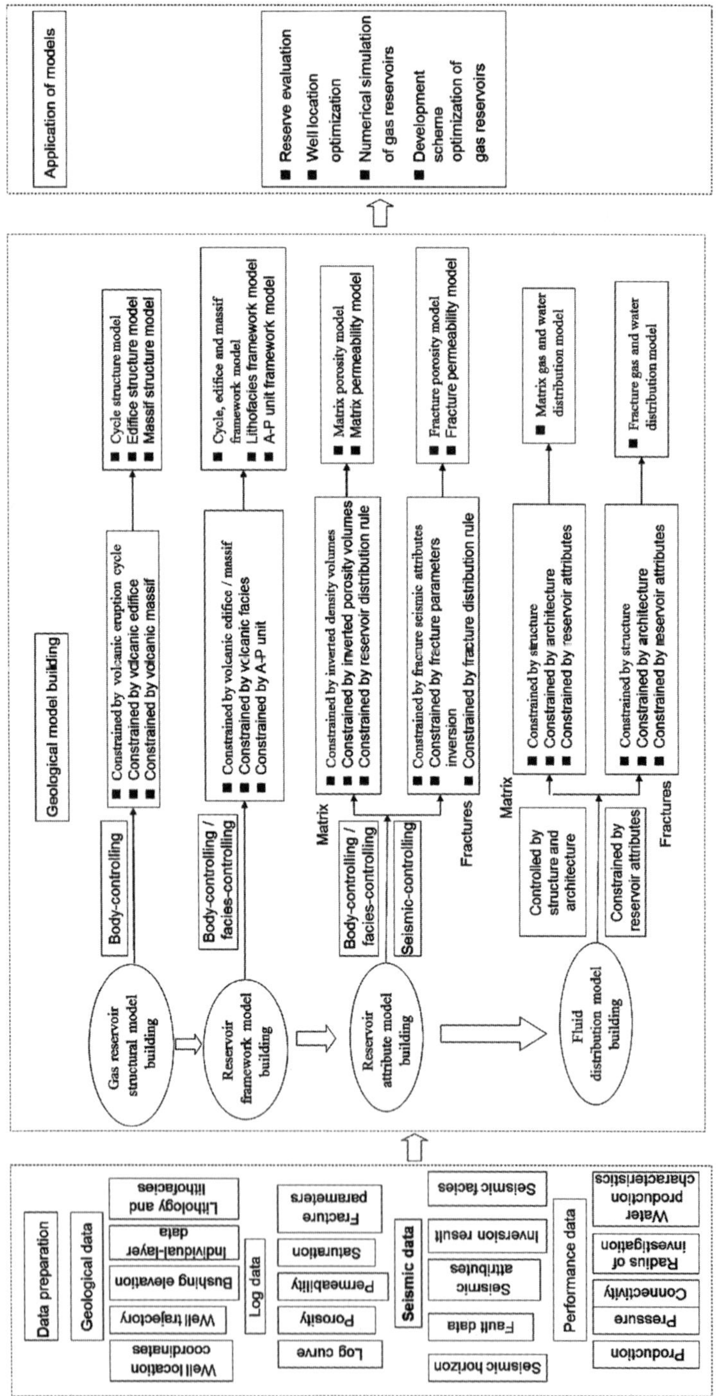

FIGURE 12.1

Technical solutions and process of geological modeling for volcanic gas reservoirs.

2. Reservoir framework model building

Multilevel reservoir framework models are constructed on the basis of structural models, using the characterization results for volcanic eruption cycles, volcanic edifices, volcanic massifs, volcanic facies, and A-P units. These models help reveal the geometric characteristics and superposition relationships at various levels of frameworks as well as their scales, such as thickness, area, and volume.

3. Reservoir attributes model building

On the basis of the development characteristics of pores, cavities, and fractures in volcanic reservoirs, the reservoir parameter models of porosity and permeability for matrixes and fractures are built, respectively, under the constraints of seismic data and volcanic reservoir frameworks [3–5].

4. Fluid distribution model building

Based on the structural model, framework model, and reservoir attribute model already available, the gas and water distribution model is developed first, followed the fluid distribution models for matrixes and fractures under the constraints of structure, architecture, and reservoir attributes.

12.2 GEOLOGICAL MODEL BUILDING FOR VOLCANIC GAS RESERVOIRS

Based on the specific geological characteristics of volcanic gas reservoirs [6,7], multilevel structural models of volcanic rocks, multilevel reservoir framework models, dual-medium attribute models constrained by reservoir frameworks, and fluid distribution models controlled by architectures are constructed using methods that are different from those applied in conventional geological modeling. These models provide a quantitative description of complex architectures; multiple porous media consisting of pores, cavities, and fractures; and complex fluid distribution patterns in volcanic rocks.

12.2.1 Multilevel structural model building techniques

The division of structural models of volcanic gas reservoirs is used as the basis for building structural models at the levels of volcanic eruption cycles, volcanic edifices, and volcanic massifs, respectively.

1. Levels of structural models for volcanic gas reservoirs

Volcanic structural models need to reflect three levels of structural geometry and variation characteristics. Structural models of the first level correspond to volcanic eruption cycles, reflect the regional tectonic characteristics, and depict changes in the distribution and thickness of volcanic rocks. Structural models of the second level correspond to volcanic edifices, reflecting the characteristics of local structures as well as the distribution and scale of gas reservoirs. Structural models of the third level correspond to volcanic massifs, reflect smaller local structures, and constrain the distribution and scale of individual gas-water systems. Therefore, volcanic structural modeling involves the construction of three levels of volcanic gas reservoir structural models by interpreting the structures and dissecting the architectures with the full application of single-well data as well as the horizon data derived from seismic interpretation of eruption cycle, volcanic edifices, and volcanic massifs.

2. Structural models of volcanic eruption cycles

Structure modeling for volcanic eruption cycles is realized mainly through the following six steps:

1. Identifying cycle interfaces or boundaries in every individual well, and tracing and interpreting seismic horizons according to volcanic eruption cycle markers
2. Establishing a velocity field and converting seismically interpreted horizon data of time domain for each eruption cycle into the horizon data of depth domain
3. On the basis of structural fault interpretation, introducing fault data files and building a 3D volcanic fault model
4. On the basis of the cycle divisions in wells, creating grid values of elevation of the surface of every volcanic cycle by Kriging interpolation, using cycle surface data as constraints between wells (Figure 12.2)
5. Calibrating eruption cycle structural planes and combining faults to develop the structural model (Figure 12.3)
6. Characterizing the structural model; defining the structural geometry of each eruption cycle, volcanic edifice, and volcanic massif of the volcanic rocks; revealing the spatial distribution of volcanic faults; and determining the structural highs of volcanic development as well as the locations of favorable gas reservoirs

As shown in Figure 12.2, there is a large-scale anticline and three small-scale anticlines in the upper eruption cycle of Well Block XX9. The structural highs of the large-scale anticline are the favorable development areas for gas reservoirs because volcanic rocks at structural highs are easily weathered and denudated to form weathered crust reservoirs. The three small-scale anticlines are the next favorable areas.

3. Volcanic edifice structural model

The steps used for volcanic edifice structural modeling are essentially the same as those used for volcanic eruption cycles, but there are two differences between them: (1) the volcanic edifice model reflects the local structural geometry characteristics of the major components of the volcanic edifice

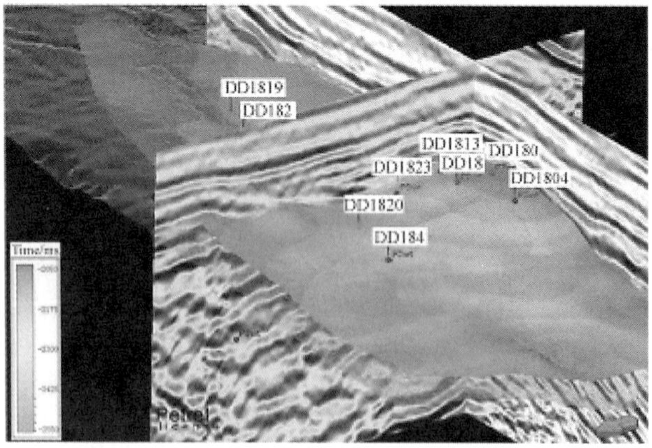

FIGURE 12.2

Correlation diagram of a cycle structural plane versus a seismic profile.

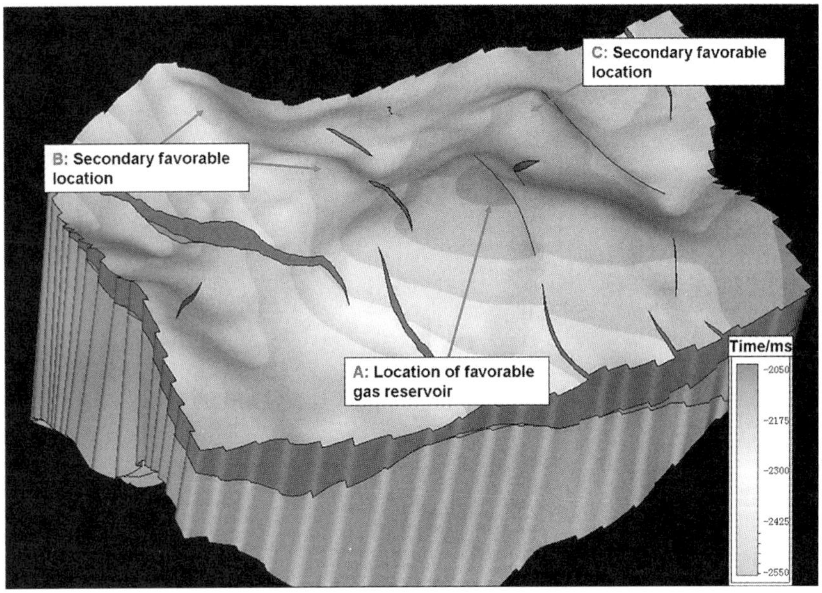

FIGURE 12.3

Top-surface structure model of the volcanic eruption cycle.

such as craters and periclinal structures; and (2) the volcanic edifice model characterizes in detail the overall structural features of the volcanic edifice, the trend of structural change, and the scale of the volcanic edifice (Figures 12.4 and 12.5).

Volcanic edifice structural modeling aims to characterize the scale and geometry of volcanic edifices and to provide the foundation for evaluating the scale of gas reservoirs and the volume of formation water. As shown in Figure 12.5, the top-surface structure of the volcanic edifice in Well Block DD10 clearly displays the configuration of a volcanic edifice. The scale of gas reservoirs is clearly

FIGURE 12.4

Correlation between an edifice structural plane and a seismic profile.

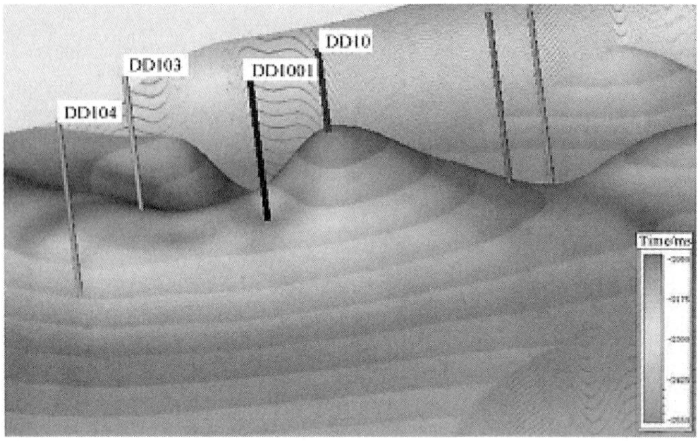

FIGURE 12.5

Volcanic edifice top-surface structural model.

controlled by the volcanic edifice. The structural high of the gas reservoirs is located near the crater penetrated by Well DD10. Well DD1001 encountered the gas-bearing structural high near the crater. Wells DD103 and DD104 encountered structural lows far from the crater with aquifers. This demonstrates that the gas-water system in gas reservoirs is controlled by the volcanic edifice, with a typical "upper gas and lower water" configuration.

4. Volcanic massif structural model

Volcanic massif structural model building involves three aspects: (1) it characterizes the shape and structural change of individual volcanic massifs and the scale of volcanic massifs (Figure 12.6); (2) when multiple volcanic massifs are superposed on one other, it can truthfully reflect the spatial distribution and superposition relationships of the massifs as well as the connectivity between adjacent massifs; and (3) the spatial distribution and scale of volcanic massifs control the spatial distribution and scale of gas-water systems.

The volcanic massif structural model provides a basis for investigating the gas-water distribution and aquifer size, evaluating producible reserves, and classifying the development layer series.

The thee gas-water systems developed in Well Block XX21 are located in three volcanic massifs, respectively, where Wells XX21, XX23, and XX231 are located (see Figure 12.6). These volcanic massifs are disconnected from one other, and the gas-water interfaces and fluid properties of the corresponding gas-water systems are thus significantly different.

12.2.2 Multilevel reservoir framework model building technique

The distribution of volcanic reservoirs is controlled by each level of reservoir frameworks. To accommodate the complex volcanic rock architectures and rapid changes in their thickness and geometry at each level of the reservoir framework, reservoir framework models are built for each level of volcanic gas reservoirs with the constraints of the architectural unit division in individual wells and seismically interpreted architectures.

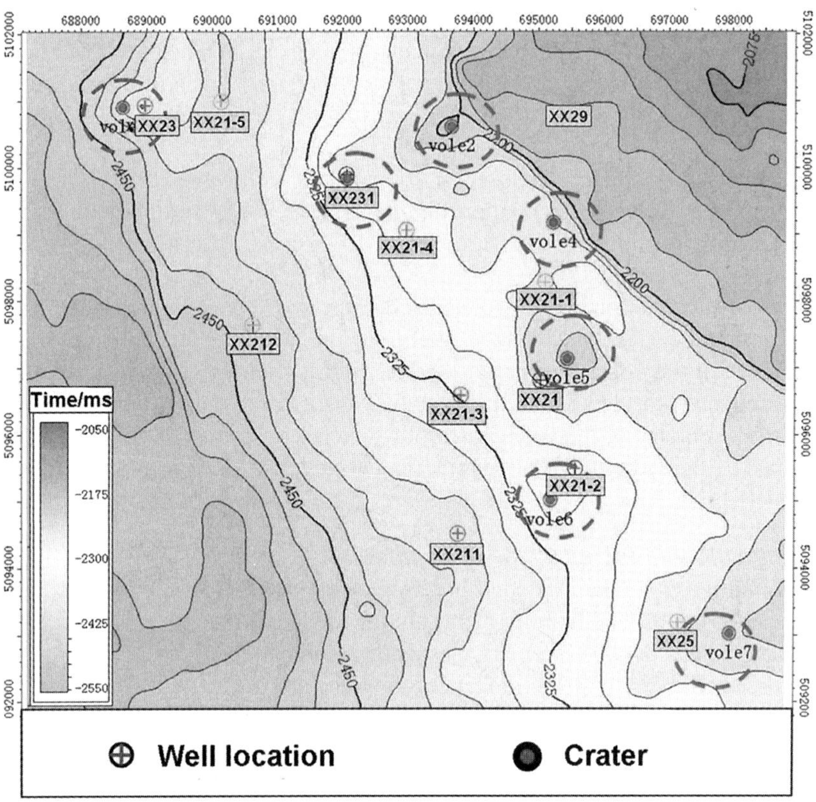

FIGURE 12.6

Microstructural model controlled by volcanic massifs.

1. Levels of the reservoir framework model

In terms of the scale, reservoir framework models can be classified into five levels. The first level is the volcanic eruption cycle framework model, which involves multiple volcanic edifices from multiple eruption stages, vertically covering the entire volcanic succession and laterally extending for tens to hundreds of kilometers. The second level is the volcanic edifice framework model, which corresponds to multiple volcanic massifs formed through eruptions of the same crater, with a rock succession that is from hundreds to thousands of meters thick and laterally extends for several kilometers to tens of kilometers. The third level is the volcanic massif framework model, which deals with multiple, genetically related lithofacies assemblages, ranging from a hundred to several hundreds of meters thick and laterally extending for hundreds to thousands of meters. The fourth level is the volcanic facies framework model, which is up to tens of meters thick and laterally extend for hundreds of meters; this modeling level directly reflects the distribution of volcanic reservoirs and the quality of reservoir attributes. The fifth level corresponds to the A-P unit framework model, with thicknesses ranging from several meters to tens of meters and laterally extending from tens of meters to over a hundred meters.

2. Building volcanic eruption cycle, edifice, and massif framework models

Based on the established multilevel volcanic structural models as well as the constraints of structural model surfaces, the reservoir framework models are constructed for volcanic eruption cycles, volcanic edifices, and volcanic massifs on the basis of the spatial closure surface at each level of reservoir frameworks.

Framework models of volcanic eruption cycles, volcanic edifices, and volcanic massifs closely reflect the geometry and superposition relationships at various levels of stratigraphic frameworks. They can provide characteristic parameters of various levels of stratigraphic frameworks, such as length, width, height, and slope (Figure 12.7).

The area of volcanic eruption cycles, volcanic edifices, and volcanic massifs can be calculated by the addition of grid areas at various levels of reservoir framework models within their corresponding range; the thickness of the volcanic eruption cycles, volcanic edifices, and volcanic massifs can be determined by subtracting the depth of the top boundary from that of the bottom boundary in the framework model; and the volume of the volcanic eruption cycles, volcanic edifices, and volcanic massifs can be determined by the addition of grid volumes between the top and the bottom structural planes in framework models.

3. Establishing volcanic facies framework models

On the basis of multilevel structural models and framework models, volcanic facies 3D space distribution models can be established by fully incorporating the identification and characterization results of volcanic facies, guided by the volcanic facies distribution patterns. Volcanic facies models can be built using two methods: deterministic model building and stochastic model building.

The establishment of volcanic facies models with the deterministic approach mainly involves the following four steps: (1) considering volcanic massifs as the study units and subdividing the eruption rhythm within the massifs; (2) performing a coarsening treatment of lithofacies types of each eruption rhythms according to dominant lithofacies; (3) predicting lithofacies distribution for each eruption rhythm within the massifs, based on the lithofacies division in single wells, the results of profile facies

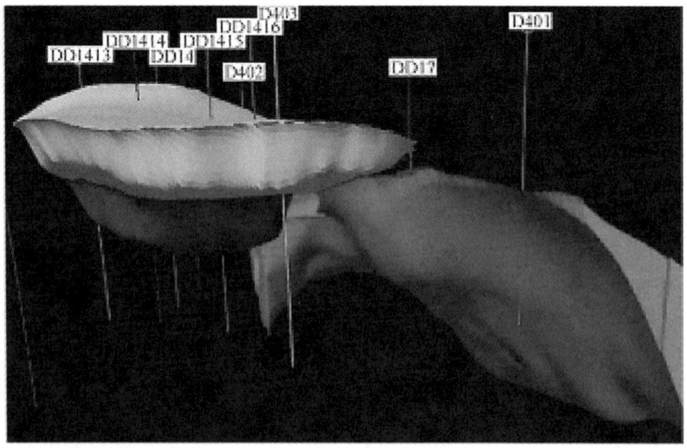

FIGURE 12.7

Volcanic massif framework model.

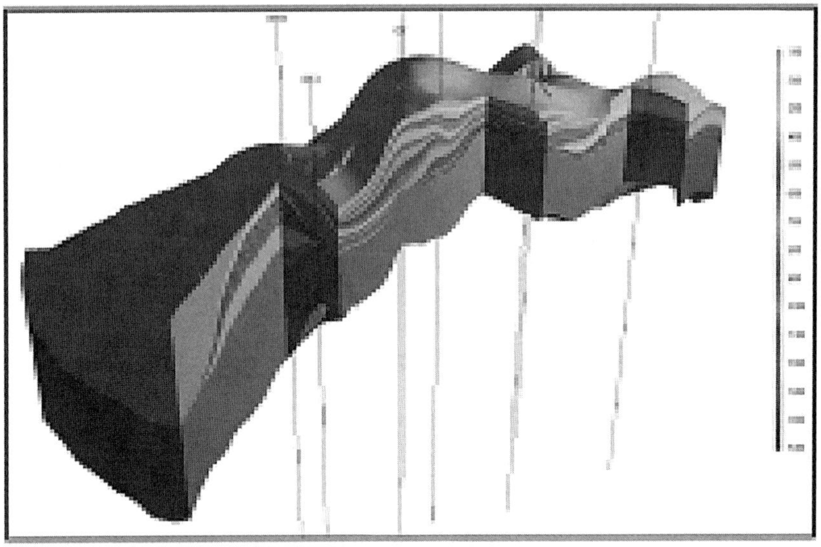

FIGURE 12.8

Volcanic facies framework model.

analysis and the seismic attributes such as waveform classification and root-mean-square (RMS) amplitude; and (4) building body control–based volcanic facies 3D geological models on the basis of lithofacies identification results in wells, using volcanic massif frameworks as constraints between wells, and using the results of volcanic facies prediction for each eruption rhythm as the co-constraints (Figure 12.8).

The specific procedure for establishing volcanic facies models using the stochastic modeling approach includes (1) determining the actual quasi-variation function with modeling data such as single-well lithofacies classification for well points and interwell lithofacies characterization and (2) building the volcanic facies stochastic prediction models using algorithms such as sequential indicator simulation, based on the established discrete models. The volcanic facies stochastic prediction models tend to have strong ambiguity and thus limited utility for guiding production because volcanic facies change abruptly and the superposition relationships are complex.

The establishment of volcanic facies framework models under the constraint of volcanic edifices and volcanic massifs can help reveal the geometry and superposition relationships of various types of volcanic facies; obtain the characteristic parameters of different volcanic facies such as length, width, and thickness; and accordingly reveal the area and volume of different volcanic facies.

Volcanic facies framework models provide fine constraints for establishing A-P unit framework models and also provide the basis for predicting favorable facies belts in reservoirs (see Figure 12.8).

4. Establishing volcanic A-P unit framework models

Volcanic A-P unit framework models form the core of the volcanic multilevel reservoir framework models. The model building process includes the following steps: (1) identifying and classifying A-P units by well and determining the top-bottom interfaces of each A-P unit; (2) making a preliminary determination of the A-P units in volcanic gas reservoirs based on established volcanic massif and volcanic facies framework models and the results of geological interpretation, logging interpretation, and seismic inversion; (3) forming a dynamic correlation, identification, and classification of A-P units based on the analysis of probing radius, fluid properties, and pressure system; (4) determining the spatial distribution of A-P units based on the unit division in wells and through integrated dynamic and static analysis; and (5) building body control–based and facies control–based volcanic gas reservoir A-P unit models according to the A-P unit identification and characterization results, treating well-interpreted A-P units as well points and volcanic massifs and facies as the constraints between wells (Figure 12.9).

The A-P unit models include two key components. First, these models qualitatively describe the geometry, superposition relationships, and spatial distribution patterns of volcanic A-P units. Second, these models quantitatively characterize the distribution area, thickness, and volume scale of volcanic A-P units.

A-P unit models serve as a guide for designing well placement, reservoir connectivity analysis, and well-controlled reserve evaluation in volcanic gas reservoirs. As shown in Figure 12.9, the distribution of volcanic A-P units in Well Block XX21 is characterized by irregular stripe and embayment patterns. In particular, type I A-P units are mostly located at structural highs near the crater, are commonly isolated in a small scale, and are the preferred area for well placement; type II A-P units have good continuity and are mainly located at structural subhighs near the crater; and type III A-P units mostly have an irregular narrow stripe pattern, located at structural lows away from the crater. Due to their limitation by factors such as a lower structural location and poor

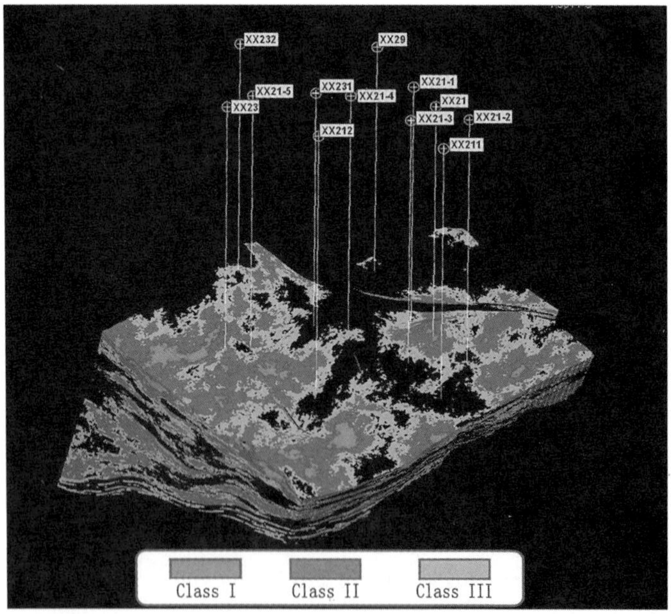

FIGURE 12.9

Diagram of the 3D distribution of volcanic A-P units.

petrophysical properties, type II and type III A-P units should be considered differently from type I A-P units when selecting development well types and establishing technological policies for development.

12.2.3 Attribute model building constrained by framework models

Reservoir attribute models for matrix and fracture systems are built by taking into consideration the characteristics of volcanic reservoirs (e.g., the development of pores, cavities, and fractures), variable types of pore-fracture combinations, and significant differences in accumulation and permeation capacity, and by applying the constraints of five-level reservoir framework models and the "body-controlling inversion" attributes volume, which will provide a reliable basis for reserve calculation and numerical simulation.

1. Building matrix porosity and permeability models

Volcanic reservoirs have extremely strong heterogeneity compared with conventional sedimentary reservoirs. As a result, volcanic reservoir attribute models are built under multiple constraints of geological conditions with the combination of well data and seismic data. The constraint conditions include (1) architectural constraints (attribute models are built under the constraints of volcanic eruption cycles, volcanic edifices, and volcanic massif framework models, which will ensure the objectivity and validity of volcanic reservoir distribution), (2) volcanic facies constraints (volcanic facies are important controlling factors for the petrophysical properties of reservoirs, and attribute models built under the constraints of volcanic facies can effectively improve their prediction accuracy), (3) A-P unit constraints (A-P units control the petrophysical properties, connectivity, and scale of reservoirs, and attribute models constrained by the A-P unit framework can improve the accuracy of attribute prediction in reservoirs with different accumulation and permeation capacities), and (4) body-controlling inversion constraints (considering the strong heterogeneity, abrupt change in interwell petrophysical properties, and low well controlling levels, volcanic attribute models will be more accurate if they are built with interwell constraints of density and wave impedance data volumes obtained from body-controlling seismic inversion).

Matrix porosity modeling is constrained by the five-level reservoir framework models, using the following three model-building methods: (1) using body-controlling inversion to obtain density and wave impedance inversion data volumes, which can be converted into porosity volumes based on the relationship among density, wave impedance, and porosity, and building deterministic matrix porosity models with porosity data volumes; (2) establishing a deterministic matrix porosity model with porosity volumes obtained from direct porosity inversion under body-controlling conditions; and (3) using individual-well-interpreted matrix porosity data as the basis for well points, using the density and wave impedance volumes obtained from seismic inversion as covariables between wells, and building matrix porosity stochastic models with stochastic simulation algorithms (Figure 12.10).

Matrix permeability modeling is constrained by the five-level reservoir framework models, using the following two methods: (1) using body-controlling inversion to obtain density and wave impedance inversion data volumes that can be converted into porosity volumes based on the relationships among density, wave impedance, and porosity, converting porosity volumes into matrix permeability volumes according to the relationship between matrix porosity and permeability, and building deterministic matrix permeability models with permeability data volumes; and (2) applying body-controlling

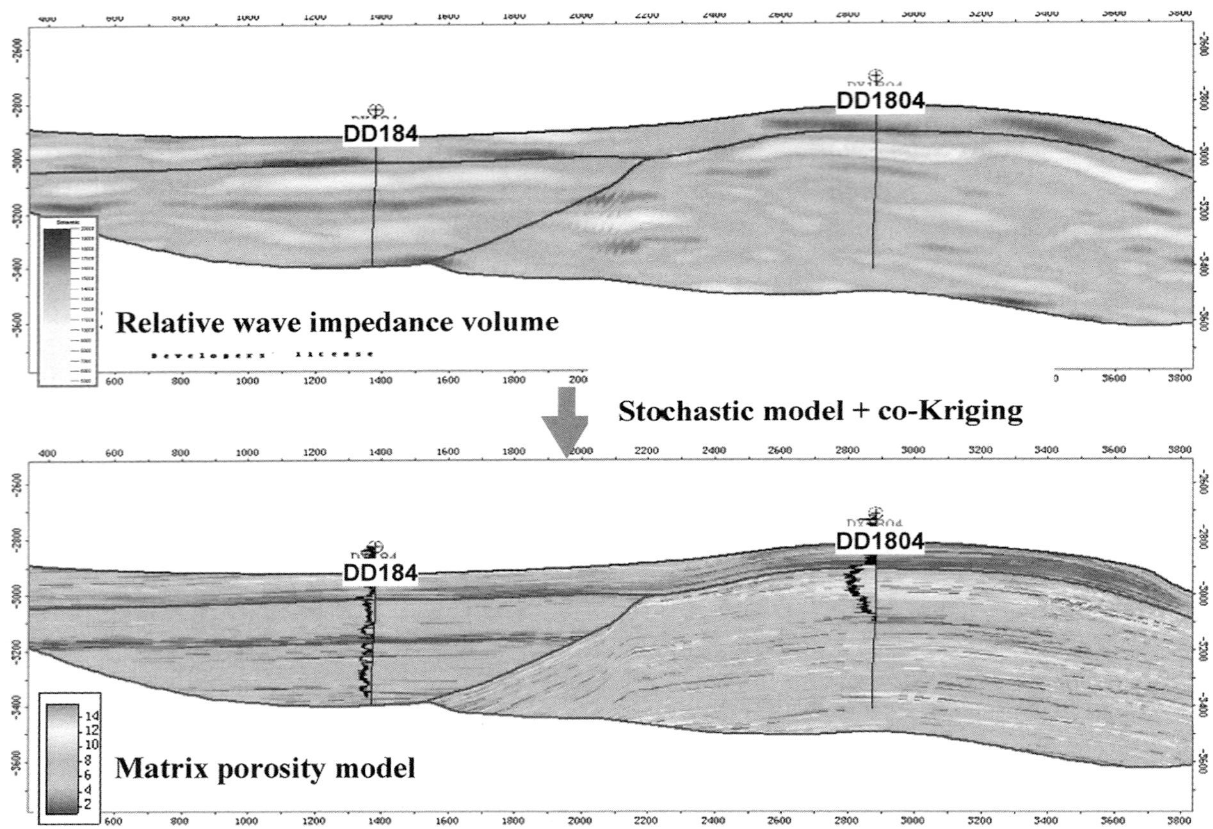

FIGURE 12.10

Prediction of matrix porosity through stochastic simulation using wave impedance as a covariable.

inversion constraints by using individual-well-interpreted matrix permeability data as the basis for well points, using the density and wave impedance volumes obtained from body-controlling seismic inversion as covariables between wells, and building matrix permeability stochastic models with stochastic simulation algorithms.

Matrix attributes models are built by above-mentioned methods and are constrained by the five-level reservoir framework models. These models can be used to demonstrate effectively the petrophysical properties of the volcanic matrix and reveal the spatial change of reservoir petrophysical properties.

Matrix attributes models serve as a foundation for optimizing favorable reservoir development areas and well location designs. As shown in Figure 12.11, the volcanic matrix porosity model constrained by the reservoir framework not only clearly displays the high-porosity areas and the planar distribution of porosity in volcanic reservoirs but also reveals the difference in reservoir properties between various volcanic massifs, lithofacies belts, and A-P units.

2. Building fracture porosity and permeability models

Fractures directly affect the microflow mechanisms and macroproduction performance characteristics of volcanic reservoirs. In volcanic gas reservoir modeling today, it remains a challenging task to find appropriate approaches to build fracture attribute models that objectively reflect the spatial distribution characteristics of fracture porosity and permeability. To improve the accuracy of fracture attribute models, fracture porosity and permeability models must be developed in accordance with the five-level reservoir framework model and using body-controlling inversion attribute volumes as interwell constraints.

Compared with conventional modeling techniques, this model-building method has the following advantages: (1) architectural constraints (volcanic rock architectures are important for controlling the distribution of reservoirs and, therefore, building volcanic reservoir fracture attribute models under the constraints of various levels of framework models will ensure the rationality of volcanic fracture

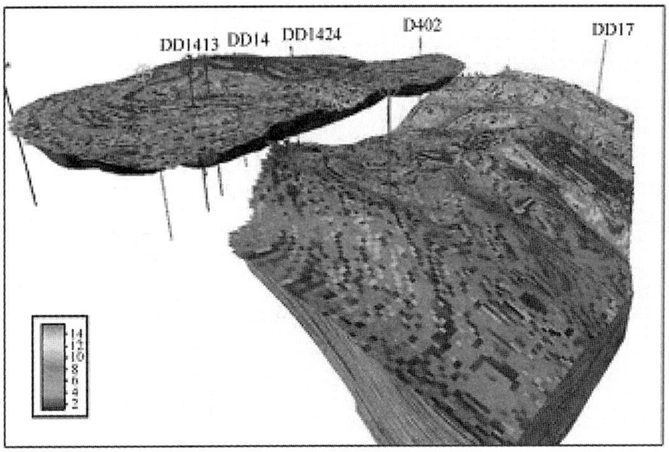

FIGURE 12.11

Diagram of 3D distribution of matrix porosity.

distribution); (2) volcanic facies constraints (a volcanic facies is important for controlling fracture types and the degree of fracture development, thus building fracture attribute models under the constraint of volcanic facies can improve the prediction accuracy of fracture attribute models); (3) A-P unit constraints (the degree of fracture development and the accumulation-permeation capacity of fractures in various A-P units are quite different and, therefore, building fracture attribute models under the constraint of A-P unit frameworks will facilitate the prediction of fracture attributes in different types of A-P units); and (4) body-controlling inversion constraints (considering the rapid interwell changes in fracture properties, the prediction accuracy of interwell fracture attributes can be effectively improved by constraining the interwells conditions with body-controlling fracture sensitivity attribute volumes or fracture parameter inversion results when the volcanic fracture attribute model is developed).

The concepts and approaches for fracture attribute modeling and matrix attribute modeling are essentially identical. However, seismic data volumes that serve as one of the interwell constraint conditions involve two types (i.e., sensitivity attributes and fracture parameter data volumes). Therefore, there are some differences in their specific modeling methods.

1) Types of seismic data volumes with interwell constraints

In constructing volcanic fracture attribute models, the seismic data volumes that are used as interwell constraints involve two types: the first comprises fracture sensitivity attributes that are extracted under body-controlling conditions and have a good correlation with fracture properties, such as coherence attributes and dip angle attributes; and the other comprises fracture parameter data volumes obtained from body-controlling inversion, such as fracture density, the fracture development index, and so on.

2) Building fracture attribute models

Fracture sensitivity attributes are used as interwell constraint conditions. First, correlation analysis is performed between fracture parameters obtained from single-well interpretation and single-attributes/composite attributes obtained from cross-well seismic traces (Figure 12.12).

Second, the seismic attributes that have a good correlation with fracture properties are determined on the basis of an attribute correlation analysis. For the XX gas field, the correlation analysis indicates that the seismic attributes have a good correlation with fracture properties including the coherence attribute, dip angle attribute, formation dip angle deviation attribute, and composite attribute volumes obtained by integrating multiple sensitivity attributes using the neural network method.

Finally, under the constraint of the five-level reservoir framework models, the fracture attribute model is built using single-well-interpreted fracture property data as the basis for well points and optimized fracture sensitivity attributes as constraints between wells (Figures 12.13 and 12.14).

The result of a fracture parameter inversion is used as an interwell constraint. Guided by the five-level reservoir framework models, the fracture attribute model is built with single-well-interpreted fracture property data as the basis for well points and treating inverted data volumes obtained from "body-controlling" inversion (fracture density, fracture development index, etc.) as co-constraints between wells.

The established fracture attribute models can be used to assess the spatial distribution of volcanic fractures, define the favorable areas of fracture development, and reveal the spatial change patterns of fracture porosity and permeability.

Fracture attribute models provide the geological basis for optimizing and enhancing effective reservoirs and horizontal well designs. As shown in Figure 12.14, the volcanic fracture porosity model

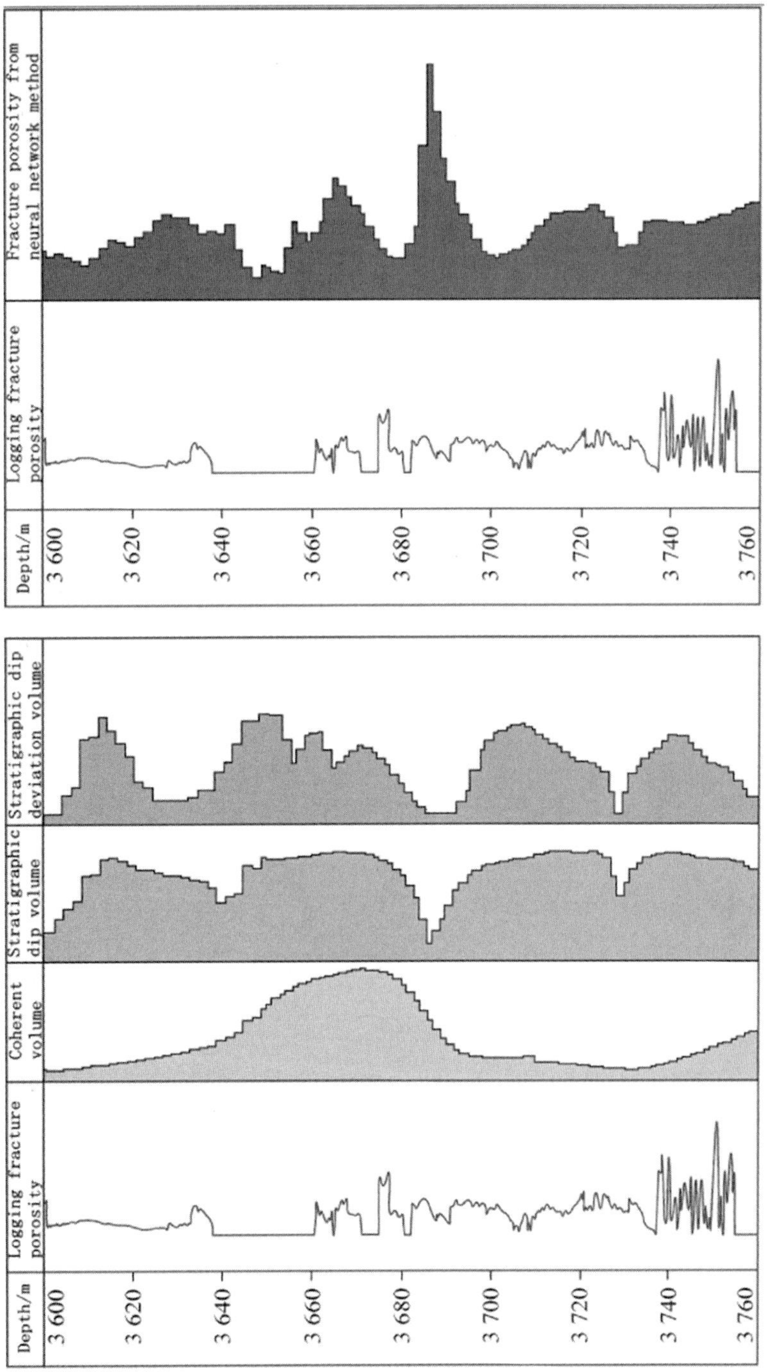

FIGURE 12.12

Comparison of correlations between logging fracture porosity and seismic attributes. (a) Correlation between logging fracture porosity and specific seismic processing data volume. (b) Correlation between logging fracture porosity and fracture porosity by the neural network method.

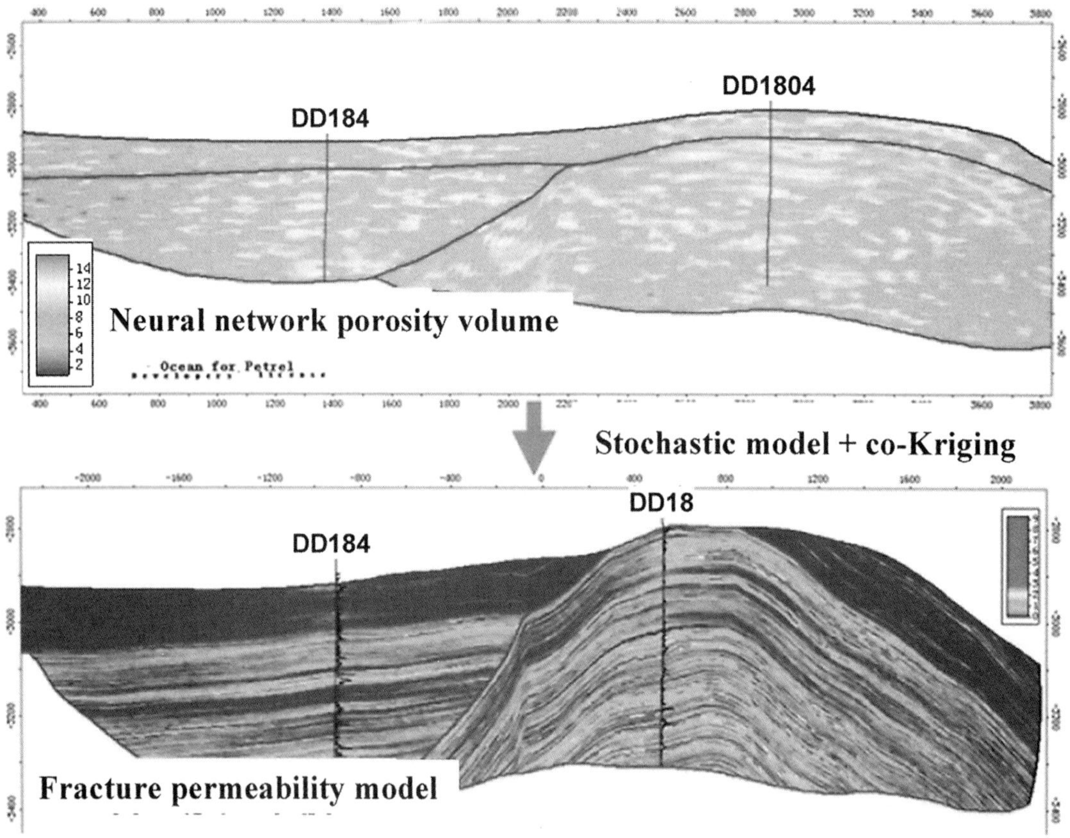

FIGURE 12.13

Prediction of fracture permeability through stochastic simulation with neural network as a covariable.

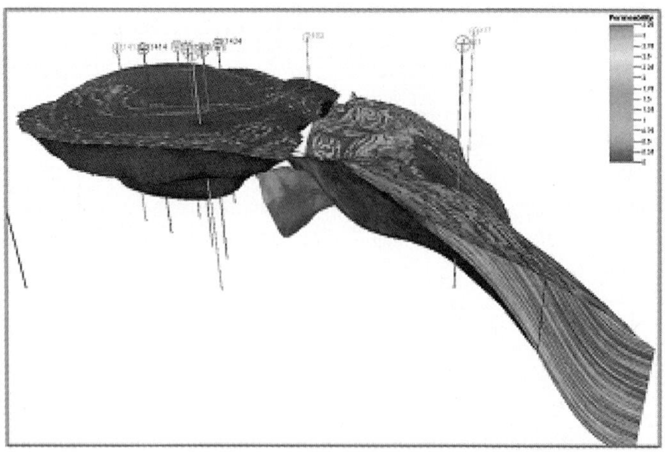

FIGURE 12.14

Diagram of 3D space distribution of fracture permeability.

demonstrates the high-value areas and planar distribution conditions of volcanic fractures and reveals the fracture difference in different volcanic massifs and lithofacies belts, thereby providing the basis for optimal well placement.

12.2.4 Fluid distribution model building under architectural control

Volcanic gas reservoirs are characterized by complex water-gas relationships and significant variability in gas saturation states. The gas-water distribution is controlled by multiple factors, such as structure, architecture, and reservoir media. In building a fluid distribution model, therefore, the following three constraint conditions should be taken into consideration: (1) volcanic gas reservoirs are controlled entirely by structures and show the typical "upper gas and lower water" configuration. To ensure the rationality of the macroscopic distribution of fluid, the constraint of the volcanic eruption cycle framework model needs to be reflected during the model-building process; (2) volcanic edifices control the distribution and scale of gas reservoirs, whereas volcanic massifs control the distribution and scale of gas-water systems; consequently, building a fluid distribution model under the constraints of volcanic edifice and volcanic massif framework models can improve the accuracy of fluid distribution models; and (3) volcanic reservoirs consist of dual media of matrixes and fractures, and the fluid distribution and gas saturation in various media are significantly different. In developing a volcanic gas reservoir fluid distribution model, therefore, the matrix and fracture systems shall be treated differently to build a corresponding fluid distribution model for each system so as to improve the accuracy of the fluid distribution model.

1. Gas and water distribution models

Volcanic gas reservoirs consist of multiple levels of stratigraphic frameworks. Each level poses a different type of control on gas-water distribution. Based on the identification result of individual-well gas layers and water zones, therefore, the gas and water distribution model is built first under the constraints of the volcanic eruption cycle and multilevel volcanic framework models to reveal the upper gas and lower water distribution characteristics in volcanic gas reservoirs; next, the gas and water distribution model of volcanic edifices is built to reveal the distribution and scale of gas reservoirs controlled by volcanic edifices; finally, the gas and water distribution model of volcanic massifs is constructed to reveal the distribution patterns of gas and water systems and the scale of water bodies controlled by volcanic massifs. The multilevel gas and water distribution models thus established using the previously described methods will serve as the foundation for establishing matrix and fracture gas saturation models.

2. Building matrix gas saturation models

Volcanic matrix gas saturation models are built on the basis of the gas and water distribution model, with gas saturation values from log interpretation used as the basis for well points, volcanic internal structures as constraints between wells, and matrix attribute data volumes as co-constraints (Figure 12.15).

3. Building fracture gas saturation models

Geological investigation and log interpretation indicate that the original gas saturation of a fracture system is typically 95%. On the basis of gas and water distribution models, therefore, the volcanic fracture gas saturation model can be built by assigning a 95% gas saturation value to fractures above the gas-water contact and a 0% gas saturation value to fractures below the gas-water contact.

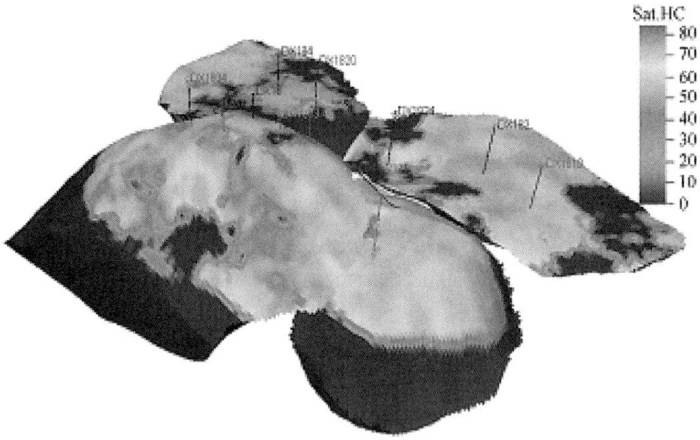

FIGURE 12.15

Volcanic matrix gas saturation model.

The fluid distribution model helps delineate the gas-water distribution patterns controlled by various levels of stratigraphic frameworks in volcanic gas reservoirs; characterizes the thickness, area, and volume of both gas layers and water zones; and effectively determines the volume of both gas and water bodies.

The fluid distribution model provides the basis for the reserve evaluation, water zone size analysis, well placement, and numerical simulation of gas reservoirs. Figure 12.15 shows that the gas saturation in volcanic gas reservoirs is controlled mainly by the structure and petrophysical properties, and the gas saturation is generally high at structural highs and low at structural lows. In addition, the gas-water contact varies in different volcanic massifs.

The concepts, constraint conditions, and specific approaches for modeling volcanic gas reservoirs are clearly different from those for the geological modeling of sedimentary reservoirs, and they pose greater challenges. Therefore, reservoir model building for volcanic gas reservoirs still has some deficiencies and awaits improvements in future research.

12.3 APPLICATIONS AND RESULTS

Three-dimensional geological models that reflect the complex architectures and dual-medium nature of volcanic rocks are the integrated expressions of volcanic gas reservoir characterization and can be used mainly for reserves evaluation, well location design, and the numerical simulation of gas reservoirs.

12.3.1 Evaluating original gas in place (OGIP) of gas reservoirs

OGIP calculation in 3D geological models uses the accumulation method, which directly computes the rock volume, pore volume, and original gas in place (OGIP) of each grid cell, then adds up the values of all grid cells, and finally determines the OGIP of gas reservoirs. This method is more reliable than the grand averaging methods (e.g., the volumetric method). Analysis and correlation between the OGIP calculated from geological models and that calculated by the volumetric method constitute an important technique for OGIP evaluation and scale verification.

12.3.2 Guiding well location optimization and horizontal well trajectory design

Geological models can help identify the "sweat spots" of volcanic reservoirs and optimize the location and type of development wells. Also, 3D geological models can be used to design, track, and adjust horizontal well trajectories, thereby effectively improving the drilling success rate of development wells and the success rate of reservoir hits by horizontal wells.

12.3.3 Providing digital models for numerical simulation

Geological models can be coarsened to provide digitized geological models for numerical simulation, then historical matching and development index prediction can be conducted to guide the appropriate development of volcanic gas reservoirs.

References

[1] Qiu Y. Oil and gas reservoir evaluation technology. Rev ed. Beijing: Petroleum Industry Press; 1997, p. 62.
[2] Qiu Y. Reservoir geological models//China transactions of oil and gas reservoir research papers. Beijing: Petroleum Industry Press; 1993, p. 1–7.
[3] Wu S, Jin Z, Huang C, et al. Reservoir model building. Beijing: Petroleum Industry Press; 1999.
[4] Wu S, Li Y. Reservoir model building: current situation and development prospect. Mar Petrol Geol 2007;12(3):55–60.
[5] Lv X, Wang D, Jiang H. Reservoir geological model and random model building technique. Petroleum Geology and Oilfield Development in Daqing 2000;19(1):10–6.
[6] Wu S, Zhang Y. Geological constraint principles improving the accuracy of reservoir stochastic model building. Journal of the University of Petroleum, China 2001;25(1):55–8.
[7] Damslesh E. A two-stage stochastic model applied to a North Sea reservoir. Journal of Petroleum Technology 1992;44(4):402–86.

Index

Note: Page numbers followed by *f* indicate figures and *t* indicate tables.

F

G

H

I

J

K

L

Q

R

T

W

X

Y

Z